W9-BXQ-613

Thematic Cartography and Geovisualization

This comprehensive and well-established cartography textbook covers the theory and the practical applications of map design and the appropriate use of map elements. It explains the basic methods for visualizing and analyzing spatial data and introduces the latest cutting-edge data visualization techniques. The fourth edition responds to the extensive developments in cartography and GIS in the last decade, including the continued evolution of the Internet and Web 2.0; the need to analyze and visualize large data sets (commonly referred to as Big Data); the changes in computer hardware (e.g., the evolution of hardware for virtual environments and augmented reality); and novel applications of technology.

Key Features of the Fourth Edition:

- Includes more than 400 color illustrations and it is available in both print and eBook formats.
- A new chapter on Geovisual Analytics and individual chapters have now been dedicated to Map Elements, Typography, Proportional Symbol Mapping, Dot Mapping, Cartograms, and Flow Mapping.
- All chapters have been revised and extensive revisions have been made to the chapters on Principles of Color, Dasymetric Mapping, Visualizing Terrain, Map Animation, Visualizing Uncertainty, and Virtual Environments/ Augmented Reality.
- All chapters include Learning Objectives and Study Questions.
- Provides more than 250 web links to online content, over 730 references to scholarly materials, and an additional 540 references available for Further Reading.

There is ample material for either a one- or two-semester course in thematic cartography and geovisualization. This textbook provides undergraduate and graduate students in geoscience, geography, and environmental sciences with the most valuable up-to-date learning resource available in the cartographic field. It is a great resource for professionals and experts using GIS and Cartography and for organizations and policy makers involved in mapping projects.

Thematic Cartography and Geovisualization

Fourth Edition

Terry A. Slocum, Robert B. McMaster, Fritz C. Kessler,
and Hugh H. Howard

CRC Press
Taylor & Francis Group
Boca Raton London New York

CRC Press is an imprint of the
Taylor & Francis Group, an **informa** business

Note about the cover: This dasymetric map is based on raw block group population counts, but has been refined by incorporating limiting ancillary features such as residential zoning, parks, schools, airports, etc. It provides a more accurate picture of where people live by "placing" people in much smaller areas, areas in which they actually live, as opposed to being evenly distributed in the much larger original block group polygons. More detail on the approach for creating this map can be found in Chapter 16.

Fourth edition published 2023
by CRC Press
6000 Broken Sound Parkway NW, Suite 300, Boca Raton, FL 33487-2742

and by CRC Press
4 Park Square, Milton Park, Abingdon, Oxon, OX14 4RN

CRC Press is an imprint of Taylor & Francis Group, LLC

© 2023 Taylor & Francis Group, LLC

Third edition published by Pearson [2008]

Reasonable efforts have been made to publish reliable data and information, but the author and publisher cannot assume responsibility for the validity of all materials or the consequences of their use. The authors and publishers have attempted to trace the copyright holders of all material reproduced in this publication and apologize to copyright holders if permission to publish in this form has not been obtained. If any copyright material has not been acknowledged please write and let us know so we may rectify in any future reprint.

Except as permitted under U.S. Copyright Law, no part of this book may be reprinted, reproduced, transmitted, or utilized in any form by any electronic, mechanical, or other means, now known or hereafter invented, including photocopying, microfilming, and recording, or in any information storage or retrieval system, without written permission from the publishers.

For permission to photocopy or use material electronically from this work, access www.copyright.com or contact the Copyright Clearance Center, Inc. (CCC), 222 Rosewood Drive, Danvers, MA 01923, 978-750-8400. For works that are not available on CCC please contact mpkbookspermissions@tandf.co.uk

Trademark notice: Product or corporate names may be trademarks or registered trademarks and are used only for identification and explanation without intent to infringe.

ISBN: 978-0-367-71270-9 (hbk)
ISBN: 978-0-367-71370-6 (pbk)
ISBN: 978-1-003-15052-7 (ebk)

DOI: 10.1201/9781003150527

Typeset in Times
by KnowledgeWorks Global Ltd.

Access the Support Material: www.routledge.com/9780367712709

Dedication

In memory of George F. Jenks

Contents

Part II   Mapping Techniques

Part III Geovisualization

Preface

JUSTIFICATION FOR A NEW EDITION

We are pleased to offer this fourth edition of *Thematic Cartography and Geovisualization* published by CRC Press. We deem a fourth edition particularly appropriate given the extensive developments that have taken place in cartography and GIScience since the third edition was first published in 2009. One development is the continued evolution of the Internet and the associated World Wide Web, which is commonly referred to as Web 2.0. One characteristic of Web 2.0 is the ability to harness the collective intelligence of Web users. In the case of geography, we often refer to this collective intelligence as "volunteered geographic information" or VGI; an example of VGI is OpenStreetMap (openstreetmap.org/). The second characteristic of Web 2.0 is lightweight programming models that enable the creation of "map mashups" by combining geographic data from one source with a map from another source. For example, HealthMap uses data feeds from sources such as the World Health Organization and others to plot disease outbreaks on an interactive world map (healthmap.org/en/). The third characteristic of Web 2.0 is the ability to work with social media such as Facebook and Twitter. Although social media sites may not seem relevant to cartography, geographers and other social scientists often analyze and map the character of social media content. For instance, Mapbox allows users to map geotagged tweets from Twitter accounts using code listed at mapbox.com/blog/twitter-map-every-tweet/. The fourth edition considers these and other developments in Web 2.0.[1]

Another development in cartography and GIScience is the need to analyze and visualize very large data sets, commonly referred to as Big Data. Traditionally, cartographers might have considered a large data set to be the 3,000+ counties of the United States. Today, however, we may need to map millions or hundreds of millions of data records. *Forbes* reported in 2018 that we produce 2.5 quintillion bytes (or 10^{18} bytes) of data every day.[2] These data might be derived from the Web (e.g., Twitter data associated with a national election), from a series of remotely sensed images showing temporal changes in the growth of a metropolitan area, or GPS records showing the locations of vehicles within a truck fleet on an hourly basis over a week-long period. The need to analyze and visualize Big Data has led to the development of an entirely new discipline—"visual analytics"—or in the case of geography, "geovisual analytics." Another new field that focuses on the analysis and visualization of Big Data is data science, which recognizes that much of Big Data has a spatial component and therefore relies on maps to create effective visualizations.[3] We argue that cartography, with its long-standing tradition in visually communicating data, is well equipped to offer effective strategies for helping the data science community analyze and visualize Big Data.

At the same time that the Web has been evolving, there have been developments in computer hardware; in addition to desktop computers and laptops, we now have tablets, smartphones, and sensors (such as wearable computers). We now also have the ability to store and analyze data in the "cloud" (in cartography, this is often referred to as "mapping in the cloud"). In theory, Web 2.0 enables the Web to appear on any hardware device, but the nature of a legible and usable map will obviously differ among these devices (e.g., a map that is legible in a desktop environment may not be legible on a smartphone). Another important development in hardware is the increased capability to create an immersive environment that is like the real world—this has spawned continued developments in "virtual reality" and "augmented reality."

There have also been developments in the application of computer technology that have excited the mapping community. For example, story maps (or digital storytelling with maps) are a combination of maps, text, images, and multimedia content that narrates a story in a web browser. Story maps allow the user to actively learn about an issue (for example, the homeless problem in Los Angeles[4]) through a compelling and interactive web interface. The resurgence of artificial intelligence (AI) is another development that has the potential to affect cartography. AI attempts to automate human decisions and tasks such as human vision and speech. One of the common AI tools is machine learning, where computers learn semi-autonomously by analyzing data. Loosely speaking, this same idea is how Google is able to "recommend" other products in your browser based on past Internet searches. Although AI is still a developing technology, its applications for understanding patterns in spatial data and learning about those patterns could prove beneficial to map users.

Aside from the impressive mapping capabilities that these developments have enabled, they have also spurred a resurgence of job growth in the cartographic field. For instance, the U.S. Bureau of Labor Statistics reported that between 2018 and 2028, cartography jobs were expected to see a 15 percent growth rate.[5] This expected job growth is also receiving attention in the media. Reputable news outlets including *The Washington Post*, *The New York Times*, and *Parade* magazine have all featured recent stories about the importance of cartography and employment gains in the field.[6] This fourth edition of our textbook equips educators with the most valuable up-to-date learning resource available in the cartographic field.

FEATURES OF THE NEW EDITION

You will find the look and feel of this new edition a dramatic change from the earlier third edition. Both printed and eBook versions are now available, and you will find that color is available throughout the textbook. In fact, the vast majority of our illustrations are now in color (the previous edition was limited to a 48-page color insert). The eBook version is particularly advantageous, as the numerous web links can easily be accessed by clicking on them; for those who purchase the printed version, we provide all web links on the home page of our book at www.routledge.com/9780367712709.

If you examine the table of contents, you will see that the overall structure of this edition consists of an introductory chapter and three major parts: (I) Principles of Cartography, (II) Mapping Techniques, and (III) Geovisualization. This is similar to the structure for the third edition, except that Chapter 1 now stands by itself. We have eliminated the chapters on Web Mapping and Trends in Research and Development. A dedicated chapter on Web Mapping seemed unnecessary given that we now make heavy use of web links throughout the textbook. Some of the material from Trends in Research and Development has been moved into other chapters, and we now cover recent developments within chapters where those developments have occurred. We have added a new chapter on Geovisual Analytics, reflecting the growth in this area that we noted above.

We have broken several of the chapters into separate chapters that are more focused on particular topics. First, the Map Elements and Typography chapter has been split into two chapters: one entitled Map Elements and one entitled Typography. While typography is an important component of map elements, it is a field unto itself with its own history, rules, guidelines, and conventions. Separation of the topics allows us to add important content to each, without bloating a single chapter. Second, the Proportional Symbol and Dot Mapping chapter has been split into chapters entitled Proportional Symbol Mapping and Dot Mapping. Although both of these techniques are based on point symbols, they are utilized for different purposes. Finally, we have split the Cartograms and Flow Maps chapter into separate Cartograms and Flow Mapping chapters. These are distinctly different topics and we now have sufficient material to make each of these a separate chapter. Although all chapters have been updated, you will find that the most extensive changes have been made to the following chapters: Principles of Color, Dasymetric Mapping, Cartograms, Flow Mapping, Visualizing Terrain, Map Animation, Visualizing Uncertainty, and Virtual Environments/Augmented Reality.

The structure of individual chapters has been enhanced by including a set of Learning Objectives at the beginning of each chapter and a set of Study Questions at the end of each chapter. Answers to the Study Questions can be found on the homepage for the book (www.routledge.com/9780367712709) at the password-protected site, which is available to instructors who are using the textbook in courses.

We feel that our textbook continues to provide a range of unique features that distinguish it from other cartographic textbooks. The following is a summary of some key features:

- The chapter on the history of thematic cartography (Chapter 2) now includes sections on innovations in European thematic cartography and the characteristics of the Maps and Society paradigm.
- Provides a thorough foundation to various statistical and graphical approaches that can be used in concert with mapping techniques (Chapter 3). Other textbooks cover some of this material, but not as thoroughly as ours.
- Clearly contrasts differing approaches for symbolizing spatial data. Many textbooks present individual mapping techniques, but they generally fail to contrast the different approaches (see Section 4.6).
- The chapter on data classification (Chapter 5) clearly explains (and illustrates) the differences among various data classification techniques, including Jiang's novel head/tails method; the related technique of cluster analysis is covered in Chapter 22.
- A separate chapter on generalization (Chapter 6) considers a variety of basic generalization operations and now reviews some of the recent developments in generalization.
- Includes an extensive introduction to Earth's coordinate system and map projections, including a chapter on how to select an appropriate projection (Chapters 7–9). No other textbook provides step-by-step guidelines on how to select map projections for specific map purposes.
- The chapter on map elements (Chapter 11) includes the most comprehensive description of map elements (and their appropriate use) of any cartographic textbook.
- The chapter on cartographic design (Chapter 13) provides the student with clear descriptions of various aspects of effective, efficient map design, with an emphasis on the practical application of design theories.
- Discusses approaches for selecting appropriate color schemes for choropleth maps. Other books cover this material, but they do not consider the broad range of factors, nor do they include as many sample maps (Chapter 15).
- An expanded chapter on dasymetric mapping (Chapter 16) now includes several practical examples of the creation of dasymetric maps.
- Discusses various algorithms for interpolating spatial data, including kriging (see Chapter 17). Generally, cartographic or GIS texts do not cover

this material in the depth that we do. An exception is Burrough et al.'s GIS text *Principles of Geographical Information Systems.*

- Includes an extensive discussion of bivariate and multivariate mapping (Chapter 22), which is not covered in other cartographic texts in such depth.
- Chapter 23 summarizes a broad range of approaches for symbolizing Earth's topography and now includes a section on "Issues in Creating Shaded Relief."
- Chapters 24–28 discuss the cutting-edge topics of animation, data exploration, geovisual analytics, visualizing uncertainty, and virtual environments/augmented reality. No other text has individual chapters on so many cutting-edge topics.
- Approximately 740 references are included in the textbook, and the Further Readings online resource provides approximately 540 additional references (www.routledge.com/9780367712709).

There is ample material here for either a one- or two-semester course in thematic cartography and geovisualization. For a one-semester course, we suggest Chapters 1, 3, 4, 5, 6, 7, 8, 11, 12, 13, 15, 17 (excluding kriging), 18, 24, and 25. For a two-semester course, Part I and the initial portion of Part II (through Chapter 18) could be covered in the first semester, and the remainder of Part II and all of Part III could be covered in the second semester. We have tried to write each chapter so that it can be used independently; thus, skipping a chapter generally should not prevent subsequent chapters from being understood. Where preceding material is essential, we have referred the reader back to the appropriate sections.

Note about the cover: This dasymetric map is based on raw block group population counts but has been refined by incorporating limiting ancillary features such as residential zoning, parks, schools, airports, etc. It provides a more accurate picture of where people live by "placing" people in much smaller areas, areas in which they actually live, as opposed to being evenly distributed in the much larger original block group polygons. More detail on the approach for creating this map can be found in Chapter 16.

NOTES

1 Tim O'Reilly, "What is Web 2.0: Design patterns and business models for the next generation of software," 2005, http://oreilly.com/web2/archive/what-is-web-20.html.

2 Bernard Marr. "How Much Data Do We Create Every Day? The Mind-Blowing Stats Everyone Should Read." *Forbes*, Forbes Magazine, September 5, 2019, https://www.forbes.com/sites/bernardmarr/2018/05/21/how-much-data-do-we-create-every-day-the-mind-blowing-stats-everyone-should-read/#5f7b0e5960ba.

3 Cathy O'Neill and Rachel Schutt. 2014. *Doing Data Science.* Sebastopol, CA: O'Reilly Media.

4 https://storymaps.arcgis.com/stories/400d7b75f18747c4ae1ad22d662781a3.

5 See https://www.bls.gov/ooh/architecture-and-engineering/cartographers-and-photogrammetrists.htm.

6 For an example, see https://parade.com/468377/kmccleary/the-8-fastest-growing-careers-of-2016/.

Acknowledgments

The fourth edition of this textbook would not have been possible without the help of numerous people who assisted in creating high-resolution illustrations, utilizing various software applications, editing sections of our manuscript, interpreting research, and making material available via the Web. We especially would like to thank the following individuals: Gennady Andrienko, Natalia Andrienko, Glenn Brauen, Eddie Bright, Daniel Carr, Anna Dmowska, Jerry Dobson, Ron Eyton, Sara Fabrikant, Sven Fuhrmann, Amy Griffin, Mark Harrower, Lorenz Hurni, Bernhard Jenny, Bin Jiang, Patrick Kennelly, Caglar Koylu, Menno-Jan Kraak, Alan MacEachren, Michael Peterson, Robert Roth, Ian Ruginski, Hanan Samet, Guillaume Touya, Christoph Traun, David Wong, Ningchuan Xiao, and Stephen Yoder. We sincerely thank Irma Britton for encouraging us to develop a contract with CRC Press. Irma assisted not only in developing this contract, but in responding to our numerous questions during the early stages of the manuscript creation process. Others who assisted in the book production process included Kalie Koscielak, Shannon Welch, and Iris Fahrer. Kalie was particularly helpful in assisting us in the copyediting and proofing stages of book production.

Terry Slocum thanks Michael Dobson and George Jenks for nurturing his interest in cartography and the undergraduate and graduate students at the University of Kansas who helped him retain a love for maps over a 35-year period. Terry also thanks Arlene Slocum for providing space on the kitchen table in a small cabin in Alaska where he could happily revise drafts of this book.

Robert B. McMaster is very appreciative of his University of Minnesota geography colleagues, in particular John Borchert, Philip Porter, Mark Lindberg, and Dwight Brown, who instilled the idea that maps remain a critical part of the geographers' life. He thanks his mentors, Mark Monmonier, who helped to launch Robert's career in mapping, and George Jenks, for the many many hours of discussion, debate, and counsel during his graduate career. He also gives thanks to the three women at home, Susanna, Keiko, and Katherine, who run his life, which is a good thing.

Fritz Kessler thanks Terry Slocum, Alan MacEachren, and Hugh Blömer for their irreplaceable stewardship in cultivating his excitement for and awareness about cartography. Fritz also extends gratitude to the late John Snyder, who made the realm of map projections accessible and fascinating.

Hugh Howard thanks Hans-J. Meihoefer for acting as a mentor and role model in cartography and geographic education. He also thanks Terry Slocum for providing a pathway to his doctorate in Geography and providing the opportunity to co-author this textbook and oversee production of the illustrations. In addition, Hugh acknowledges his present and former students for constantly reigniting his love of maps. And with gratitude, Hugh thanks his wife, Tracy, for her never-ending love and support.

About the Authors

Terry A. Slocum is an Emeritus Professor with the University of Kansas, where he taught cartography and statistics for 35 years and chaired the Department of Geography for 8 years. His research interests have included data exploration, map animation, visualizing uncertainty, stereoscopic displays, history of thematic mapping, and color usage on maps. He has published in numerous refereed journals, including *Cartography and Geographic Information Science*, *Cartographica*, *The Cartographic Journal*, *Annals of the American Association of Geographers*, *Journal of Geoscience Education*, and *Journal of Geography*. Professor Slocum has been affiliated with six grants from the U.S. National Science Foundation and received two Teacher Appreciation Awards from the Center for Teaching Excellence at the University of Kansas. He has chaired 14 dissertation and thesis committees and served on more than 75 dissertation and thesis committees.

Robert B. McMaster is Vice Provost and Dean of Undergraduate Education and Professor of Geography at the University of Minnesota—Twin Cities. His research interests include automated generalization, environmental risk assessment, Geographic Information Science and society, and the history of U.S. academic cartography. He has authored or edited seven books on cartography and GIS, and his papers have been published in *The American Cartographer*, *Cartographica*, *The International Yearbook of Cartography*, *Geographical Analysis*, *Cartography and GIS*, and the *International Journal of GIS*. He served as editor of the journal *Cartography and Geographic Information Systems* from 1990 to 1996. He has served as President of the United States' Cartography and Geographic Information Society, President of UCGIS, and Vice President of the International Cartographic Association. Robert served a three-year term on the National Research Council's Mapping Science Committee (2005–2008). In 2010, he was named *GIS Educator of the Year* by the University Consortium on Geographic Information Science. In 2013 he was named a Fellow of UCGIS.

Fritz C. Kessler is a Teaching Professor at Penn State. His teaching interests span cartography, statistics, and geography of health. His research focus spans several topics in cartography that include map projections, geometric and geopotential datums, history of thematic mapping, and data exploration. He has published papers in numerous refereed journals, including *Cartography and Geographic Information Science*, *Cartographica*, *Cartographic Perspectives*, *Annals of the American Association of Geographers*, *Journal of Geography*, and *GeoJournal*. He also co-authored a book with Dr. Sarah Battersby (at Tableau) titled *Working with Map Projections: A Guide to their Selection*. He is a former President of the North American Cartographic Information Society (NACIS) and a board member of the Cartography and Geographic Information Society (CaGIS). His cartographic background is not limited to academia but has evolved through several professional positions including Ohio University's Cartographic Center, USGS Water Resource Division, Intergraph Corporation, R. R. Donnelley and Sons, and the University of Kansas' T. R. Smith Map Library.

Hugh H. Howard is a professor of GIS at American River College in Sacramento, California. He is the GIS Coordinator, Geosciences Department Chair, and currently teaches five GIS courses, including Cartographic Design for GIS. Hugh earned his Ph.D. in Geography from the University of Kansas specializing in cartographic design, and developed an expert system software application to aid students in designing better maps. In 2019, he won an Excellence in Education award from the California Geographic Information Association (CGIA), and a Lifetime Achievement in Geospatial Two-Year College Education award from the GeoTech Center (an NSF-funded National Geospatial Technology Center of Excellence). Hugh has worked as a cartographer for the U.S. Forest Service, the City of San Francisco, CB Richard Ellis, and Cartographics. He also taught GIS and managed GIS labs at Stanford University and San Francisco State University.

1 Introduction

1.1 OVERVIEW

This book focuses on thematic mapping and the associated expanding area of geographic visualization (or geovisualization). A **thematic map** is used to display the spatial pattern of themes (or attributes)—a familiar example is the temperature map shown on weather-based web pages (e.g., https://www.weather.gov/forecastmaps), in daily newspapers, or on the nightly TV newscast; the theme, in this case, is the predicted high temperature for the day. Another example would be a **choropleth map** that depicts the death rates due to COVID-19 for countries of the world (https://covid19.who.int/); in this case, the theme is the death rates due to COVID-19. With respect to thematic maps, our overarching goal in this textbook is twofold. First, our intent is to provide guidelines for the effective design of thematic maps. For instance, given that you wish to map death rates due to COVD-19, we could ask, "what is an appropriate symbology to use?" The choropleth map symbology is one possible solution, but we will see that there are numerous other possibilities. A second overarching goal is to enable you to critically evaluate the design of maps that you see and utilize. For instance, you might evaluate whether the symbolization method and colors used on the COVID-19 map properly reflect the underlying data values.

In Section 1.3, we contrast the thematic map with the **general-reference map**, which focuses on geographic location as opposed to the spatial pattern (e.g., a topographic map might show the location of rivers). In Section 1.4, we consider the different uses for thematic maps: to provide **specific information** about particular locations, to provide **general information** about spatial patterns, and to **compare patterns** on two or more maps. We will argue that although a thematic map can provide both specific and general information, providing general information about spatial patterns is the fundamental reason for constructing a thematic map, as precise specific information can be acquired from a table, or if an interactive map is utilized, the precise value can be acquired through interactive methods, such as a mouse-over operation.

As we suggested above, an important function of this book is to assist you in selecting and implementing appropriate techniques for symbolizing spatial data. For example, imagine that you wish to depict the amount of forest cleared for agriculture in each country during the preceding year and have been told that the number of acres of forest cleared by country is available on the Web. You wonder whether additional data (e.g., the total acres of land in each country) should be collected and how the resulting data should be symbolized. In Section 1.5, we present five basic steps that will assist you in tackling such problems and ultimately will enable you to communicate the desired information to map readers. Although some have criticized this **map communication model** approach, it is our experience that carefully considering these steps can assist you in creating a well-designed map that meets your desired purpose.

Like many disciplines, the field of cartography has undergone major technological changes. As recently as the 1970s, most maps were still produced by manual and photomechanical methods, whereas today, nearly all maps are produced using computer technology. Since the 1950s, this computer technology has evolved from the use of mainframe computers to desktop and laptop computers to smartphones and wearable computing devices. Paralleling this growth in computer hardware has been the evolution of the Internet and associated **World Wide Web** (or **Web**). Today's version of the Web is termed **Web 2.0**. In Section 1.6, we discuss some of the major characteristics of Web 2.0 and then consider some of the consequences of this technological change for the discipline of cartography, including (1) the ability of virtually anyone to create maps (the **democratization of cartography**); (2) the ability to create maps that would have been difficult or impossible to create using manual methods, such as animated maps; (3) the ability to explore geographic data in an interactive graphics environment, and the associated ability to visualize very large data sets (**Big Data**), such as millions of Twitter feed records; and (4) the ability to create realistic representations of the environment (i.e., to create a **virtual reality** or **virtual environment**) and to utilize the notion of **augmented reality** (i.e., enhancing our view of the real world through computer-based information).

In Section 1.7, we consider the origin and definition of the term geographic visualization. The term "visualization" has its roots in **scientific visualization**, which was developed outside geography to explore large multivariate data sets, such as those associated with medical imaging, molecular structure, and fluid flows. Borrowing from these ideas, geographers created the notion of **geographic visualization** (or **geovisualization**), which can be defined as a private activity in which unknowns are revealed in a highly interactive environment. Communication on traditional static maps involves the opposite: It is a public activity in which knowns are presented in a noninteractive environment. In this book, we will cover issues related to both geovisualization and communication.

Although our emphasis in this book is on cartography, you should be aware of developments in the broader realm of geographic information science (GIScience), which can be considered to include cartography and the techniques of

DOI: 10.1201/9781003150527-1

geographic information systems (GIS), remote sensing, and quantitative methods. In Section 1.8, we consider the capability provided by GIS and remote sensing, which allow us to create detailed maps more easily than was possible with manual techniques. GIS accomplishes this through its extensive spatial analysis capabilities, and remote sensing allows us to "sense" the environment, particularly outside our normal visual capabilities (e.g., detecting previsual levels of vegetation stress). The major development in quantitative methods relevant to cartography is that of **exploratory spatial data analysis (ESDA)**, which has close ties with the notion of data exploration that cartographers utilize.

While technological advances have had a major impact on cartography, the discipline has also experienced changes in its philosophical outlook. In Section 1.9, we consider the increasing role that cognition now plays in cartography. Traditionally, cartographers approached mapping with a *behaviorist view*, in which the human mind was treated like a black box. Today cartographers take a *cognitive view*, in which they hope to learn why symbols work effectively. Finally, in Section 1.10, we consider the role that social and ethical issues can play in cartography. We will see that maps often have hidden agendas and meanings and that the process of **deconstruction** can help us uncover these hidden agendas and meanings. More generally, the process of critically evaluating maps is known as **critical cartography**; although we do not focus on critical cartography in this textbook, we encourage you to explore this subdiscipline of cartography through the references that we point you to and the Further Readings.

1.2   LEARNING OBJECTIVES

- Define the term thematic map and contrast it with the general reference map.
- Summarize the kinds of information that can be acquired from thematic maps.
- List the five basic steps for communicating map information and indicate why it is important to consider these steps.
- Discuss four consequences of technological change in cartography.
- Define geographic visualization and contrast it with map communication.
- Explain why the term geographic visualization fails to cover the breadth of potential mapping approaches and consider the alternative term cybercartography.
- Cartography is one "technique" often used by geographers. List the other geographic techniques and indicate how these techniques can be useful to the cartographer.
- Contrast the notion of a behaviorist view with a cognitive view as applied to cartography.
- Explain how the notions of data exploration and critical cartography overlap one another.

1.3   WHAT IS A THEMATIC MAP?

Cartographers commonly distinguish between two types of maps: general-reference and thematic. **General-reference maps** are used to emphasize the location of spatial phenomena. For instance, topographic maps, such as those produced by the U.S. Geological Survey (USGS), are general-reference maps. On topographic maps, readers can determine the location of streams, roads, houses, and many other natural and cultural features. In contrast, **thematic maps** (or **statistical maps**) are used to emphasize the spatial pattern of one or more geographic **attributes** (or **variables**), such as population density, family income, and daily temperature maxima. A common thematic map is the **choropleth map**, in which **enumeration units** (or data-collection units such as states) are colored to represent different magnitudes of an attribute, such as the percentage voting for Joe Biden in the 2020 U.S. presidential election (Figure 1.1). A variety of thematic maps are possible, including proportional symbol, isarithmic, dot, and flow; Figure 1.2 illustrates a **flow map** of the top ten international flight routes in February of 2021. A major purpose of this book is to introduce you to these and other types of thematic maps, as well as to methods used in designing and constructing them.

Although cartographers commonly distinguish between general-reference and thematic maps, they do so largely for the convenience of categorizing maps. The general-reference map also can be viewed as a thematic map in which multiple attributes are displayed simultaneously; thus, the general-reference map can be termed a *multivariate thematic map*. Furthermore, although the major emphasis of general-reference maps is on the *location* of spatial phenomena, they can also portray the *spatial pattern* of a particular attribute (e.g., the pattern of drainage shown on a USGS topographic sheet).

1.4   HOW ARE THEMATIC MAPS USED?

Thematic maps can be used in three basic ways: to provide **specific information** about particular locations, to provide **general information** about spatial patterns, and to **compare patterns** on two or more maps. As an example of specific information, map A of Figure 1.1 indicates that between 44.9 and 50.6 percent of those voting in Georgia voted for Biden in the 2020 U.S. presidential election. As another example, Figure 1.2 indicates that between 109 and 111 thousand people flew from New York to Santiago in February 2021. Obtaining general information requires an overall analysis of the map. For example, map B of Figure 1.1 illustrates that a low percentage of people voted for Biden in the central, northwestern states and the Great Plains states, whereas a higher percentage voted for Biden in the extreme western states and in the northeastern states. We will explore these voting patterns in greater detail in Chapter 20, where we introduce the **cartogram**, a common technique for portraying election results. As another example of general information, Figure 1.2 reveals that the top ten international flight

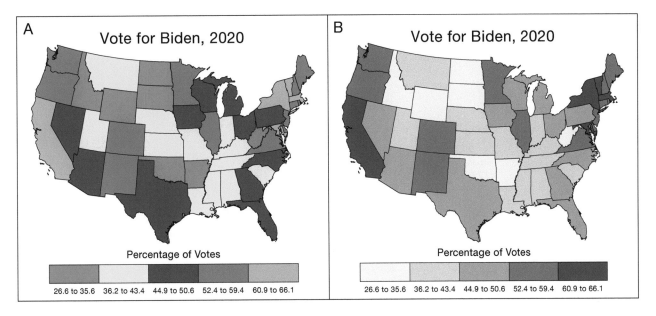

FIGURE 1.1 The choropleth map: an example of a thematic map. Map A uses illogically ordered hues, whereas map B uses logically ordered lightnesses of a single hue. Although map A might allow the reader to discriminate easily between individual states, it does not permit the reader to perceive the overall spatial pattern as readily as map B.
(Data source: https://cookpolitical.com/2020-national-popular-vote-tracker.)

routes appeared to occur between various places in the Caribbean and major cities in the United States and Paris in Europe, and between cities in the Middle East. For a discussion of these travel routes, see https://www.cnn.com/travel/article/world-busiest-air-routes-february-2021/index.html.

A pitfall for naive mapmakers is that they often place inordinate emphasis on specific information. Map A of Figure 1.1 is illustrative of this problem. Here one can discriminate the data classes based on strikingly different colors

and thus determine which class each state belongs in (as we did for Georgia), but it is difficult to acquire information about spatial patterns without carefully examining the legend. In map B, the reverse is true: Determining class membership is more difficult because the classes are all lightnesses of blue, but the spatial pattern of voting is readily apparent because there is a logical progression of legend colors.

As an illustration of pattern comparison, consider the **dot maps** of corn and wheat shown in Figure 1.3. Note that

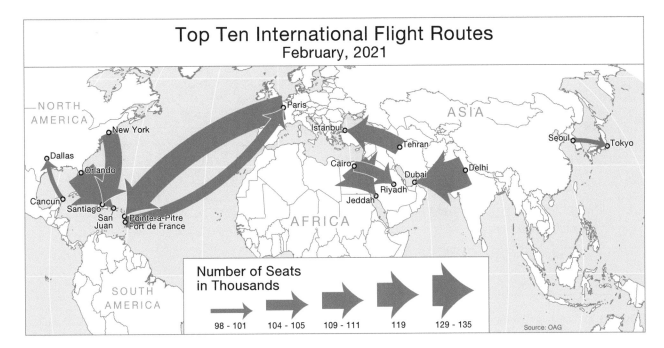

FIGURE 1.2 A flow map: An example of a thematic map. (Data source: https://www.cnn.com/travel/article/world-busiest-air-routes-february-2021/index.html.)

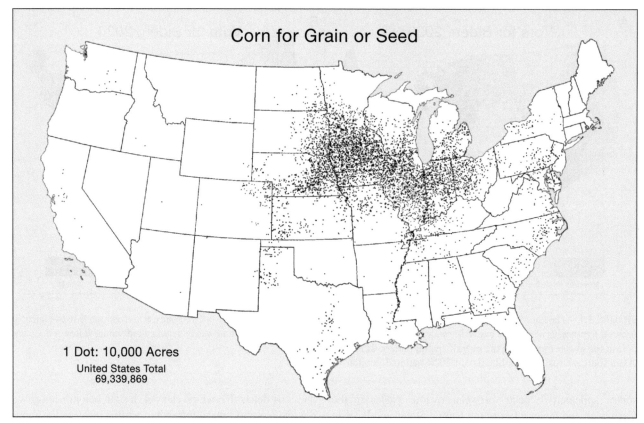

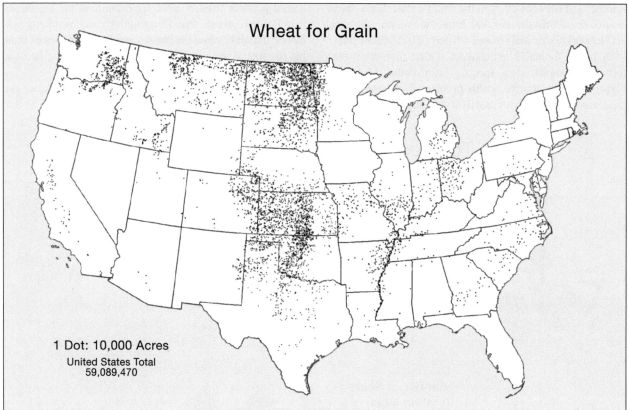

FIGURE 1.3 Dot maps illustrating pattern comparison, one of the fundamental ways in which thematic maps are used. (From U.S. Bureau of the Census 1995.)

the patterns on these two maps are quite different. Corn is concentrated in the traditional Corn Belt region of the Midwest, whereas wheat is concentrated in the Great Plains, with a less notable focus in the Palouse region of eastern Washington. Conventionally, a comparison of patterns such as these was limited by their fixed placement on pages of paper atlases, but interactive graphics (oftentimes via the Web) now allow us to readily compare arbitrarily selected distributions.

Two further issues are important when considering the ways in which thematic maps are used. First, one should distinguish between information acquisition and memory for mapped information.[1] Thus far, we have focused on **information acquisition** or acquiring information while the map is being used. We can also consider **memory for map information** and how that memory is integrated with other spatial information (obtained through either maps or fieldwork). For example, a cultural geographer might note that houses in a particular area are built predominantly of limestone. Recalling a geologic map of bedrock, the geographer might mentally correlate the spatial pattern of limestone in the bedrock with the pattern of limestone houses.

A second important issue is that terms other than specific and general can be found in the cartographic literature. We have used specific and general (developed by Alan MacEachren 1982) because they appear frequently in the literature. Others have developed a more complex set of terms. For example, Philip Robertson (1991, 61) distinguished among three kinds of information: values at a point, local distributions characterized by "gradients and features," and the global distribution characterized by "trends and structure"; Robertson argued that these levels corresponded closely with Jacques Bertin's (1981) elementary, intermediate, and superior levels.[2]

1.5 BASIC STEPS FOR COMMUNICATING MAP INFORMATION

In this section, we consider basic steps involved in communicating map information to others. For instance, imagine that you wish to create a map for a term paper or that you are working for a local newspaper and need to create a map for a major news story. Traditionally, the basic steps for communicating map information were taught within the framework of **map communication models** (e.g., Robinson et al. 1984, 15–16; Dent 1996, 12–14). Although such models have received criticism (e.g., MacEachren 1995, 3–11), their use often leads to better designed maps. The map communication model that we use is shown in Figure 1.4 as a set of five idealized steps. Let's examine these steps by assuming that you wish to map the distribution of population in the United States based on data provided by the U.S. Bureau of the Census.

Step 1. Consider what the real-world distribution of the phenomenon might look like. One way to implement this step is to ask yourself, "What

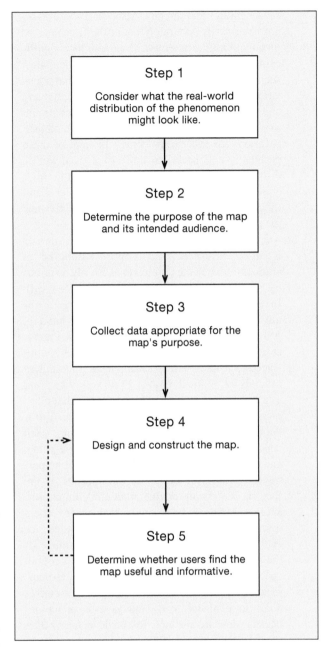

FIGURE 1.4 Basic steps for communicating map information to others.

would the distribution of the phenomenon look like if I were to view it while traveling across the landscape?" In the case of our population example, you might know (based on your travels or knowledge as a geographer) that a large percentage of people are concentrated in major cities and that such cities are much more densely populated than rural areas.

Often, however, it is unrealistic to presume a single "objective" real world. In the case of population, "correct" population values are unknown for several reasons. The U.S. Bureau of the Census is never able to make an exact count of population; after every census, some city officials dispute the census figures for their city. Also, census figures do not necessarily accurately count the homeless or illegal aliens (the latter would account for a significant percentage of the population in some states, such as California).[3] Another problem is that population obviously varies locationally during the day and throughout the year—that is, people commute to work, travel to the beach on weekends, and take vacations far away from home.[4] In spite of these problems, it is useful to think of a "real-world" distribution. Such an approach forces you to think about the distribution at its most detailed level and then to decide what degree of generalization meets the purpose of the map.

Step 2. *Determine the purpose of the map and its intended audience.* One purpose would be to attempt to match the real world as closely as possible (within the constraints of the map scale used). In the case of population, you might want to distinguish clearly between urban and rural population. Another purpose might be to map the distribution at a particular geographic level (say, the county level); such views are often sought by government officials for political reasons. From your viewpoint as a mapmaker, it is important to realize that mapping at a particular geographic level can introduce errors into the resulting map because each enumeration unit is represented by a single value, and thus the variation within units cannot be portrayed. This error might be unimportant if the focus is on how one enumeration unit compares to another, but it can be a serious problem if readers infer more from the map than was intended; for example, readers might erroneously assume that the population density is uniform across a county on a choropleth map. The key point is that mapmakers often display data at the level of a convenient political unit (e.g., county, state, or nation) because data are available for that level rather than consider the purpose of the map.

The nature of your intended audience may also play an important role in designing the map. For instance, if you are utilizing thematic maps in a grade school textbook, you might select **pictographic symbols** that are particularly suggestive of the phenomenon being mapped (e.g., pizzas to represent the amount of pizza consumed in census tracts throughout a city). In many instances, however, it is difficult to anticipate exactly who will read the map (such as when designing maps for a daily newspaper).

Step 3. *Collect data appropriate for the map's purpose.* In general, spatial data can be collected from primary sources (e.g., field studies) or secondary sources (e.g., Census data). For something close to the real-world view of population, you would likely consult the U.S. Census of Population for information on urban and rural population; additionally, you would collect ancillary data that could assist you in locating the population data within rural areas. For a county-level view of population, the Census figures for individual counties would suffice.

Step 4. *Design and construct the map.* Designing and constructing the map involves not only selecting an appropriate symbology (e.g., using a dot map rather than a choropleth map) but also selecting and positioning the various map elements (e.g., title, legend, and source) so that the resulting map is both informative and visually pleasing. This step is a complex one that involves assessing the following sorts of questions:

1. How will the map be used? Will it be used to portray both general and specific information?
2. What is the spatial dimension of the data? For instance, are the data available at points, do they extend along lines, or are they areal in nature?
3. At what level are the data measured—nominal, ordinal, interval, or ratio?
4. Is data standardization necessary? If the data are raw totals, do they need to be adjusted?
5. How many attributes are to be mapped?
6. Is there a temporal component to the data?
7. Are there any technical limitations? For example, a journal might not be willing to reproduce maps in color.
8. What are the characteristics of the intended audience? Is the map intended for the general public or for professional geographers? What limitations might particular members of the audience have (e.g., do any have color vision impairments)?
9. What are the time and monetary constraints? For example, creating a high-quality dot map will cost more than creating a choropleth map, regardless of the technical capabilities available.

A full consideration of these questions will occupy the rest of this book. For now, consider two maps that could result from efforts to create a population map of the United States: a combined proportional symbol–dot map (Figure 1.5) for the real-world view and a choropleth map for the county-level view (Figure 1.6). The proportional symbol–dot map is illustrative of how one can attempt to match the mapped spatial distribution to the real world. Note that the overall population is split into urban and rural categories and that the urban population is further subdivided into "urbanized areas" and "places outside urban areas." Here a great deal of effort has been made to create a map that shows the variation of population over geographic space— we can even see the variation of population density

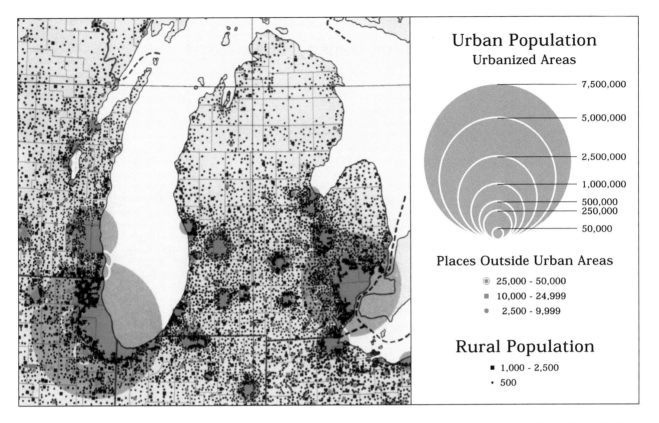

FIGURE 1.5 A combined proportional symbol—dot map that attempts to represent what the population might look like in the "real world." (After U.S. Bureau of the Census 1970.)

within individual counties (these are represented by the brown lines). In contrast, on the county-level map in Figure 1.6, it is impossible to see the variation of population density within individual counties.

In examining Figure 1.6, you will note that we have used the overall U.S. population density of 90 people per square mile as a breakpoint in the data classification.[5] Inspiration for this approach was a population density map that Cynthia Brewer and Trudy Suchan utilized in their atlas of diversity in the United States (see p. 11 of https://usa.usembassy.de/etexts/soc/diversity.pdf). For ease in reading the map, other class breaks were based on convenient rounded numbers. We also used density per square mile (rather than density per square kilometer) as most people in the United States find this easier to interpret. In Chapter 5, we will discuss a wide variety of approaches for classifying data.

Step 5. Determine whether users find the map useful and informative. Possibly the most important point to keep in mind is that you are designing the map for others, not for yourself. For example, you might find a particular color scheme pleasing, but you should ask yourself whether others also will find it attractive. Ideally, you should answer such questions by getting feedback about the map from potential users. Here we are not suggesting that you conduct a formal **user study** but rather that you get some informal input from several potential users.[6] Admittedly, time and monetary constraints make this task difficult, but it is necessary to undertake because you could discover not only whether a particular mapping technique works but also the nature of information that users acquire from the map. Moreover, if you plan to employ the map to illustrate a particular concept (as for a class lecture), then you will want to know whether users acquire this concept when viewing the map.

If your analysis reveals that the map is not useful and informative, then you will need to redesign the map. This possibility is shown as a dashed line in Figure 1.4, where it is presumed that you would return to step 4 and design a new map. It is conceivable that you also might have to return to an earlier step, but it is more likely that you will have to modify some design aspect, such as the color scheme.

Unfortunately, naive mapmakers are unlikely to follow the five steps we have outlined. Instead, their decisions are frequently based on readily available data and mapping software. As an example, imagine that for a term paper, a student wished to map the distribution of population that we have been discussing. Rather than considering steps 1 and 2 of the model, the student might simply use state totals, either because fewer numbers would have to be entered into

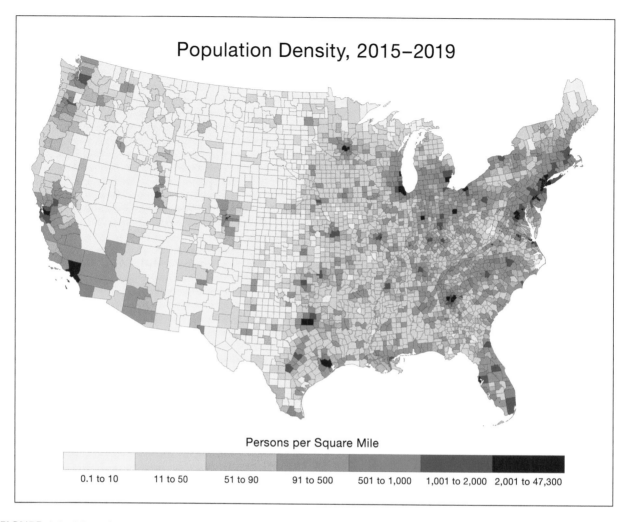

FIGURE 1.6 Map of population density for 2015–2019 for U.S. counties that would be created by following the basic steps for communicating map information to others. (From American Community Survey; https://www.census.gov/programs-surveys/acs.)

the computer or because the data are readily available (e.g., via a Web site). Furthermore, in step 4, the student might choose a choropleth map because the software for creating choropleth maps is readily available. Presuming that the student had collected data in raw-total form, the choropleth map would be a poor choice because it requires standardized data (as we will see in Chapter 4).

1.6 TECHNOLOGICAL CHANGE IN CARTOGRAPHY AND ITS CONSEQUENCES

Ever since the late 1950s, the field of cartography has undergone major technological change, evolving from a discipline based on pen and ink to one based on computer technology. In summarizing historical developments in thematic mapping over the course of the twentieth century, Slocum and Kessler (2015) divided the twentieth century into four periods: Pre-Computer Era (1900–1958), Mainframe Computer Era (1959–1976), Desktop Computer Era (1977–1990), and Internet Era (1991 onward). Although one could argue that we are still in the Internet Era, beginning in the early

2000s, the nature of the associated Web evolved to a new level that is often termed **Web 2.0** (O'Reilly 2005; O'Reilly and Battelle 2009).

One characteristic of Web 2.0 is the ability to harness the collective intelligence of Web users. Outside the realm of geography, this is often referred to as **crowdsourcing**, as the intent is to take advantage of the "wisdom of the crowd." A classic example of crowdsourcing is the free encyclopedia Wikipedia (https://en.wikipedia.org/wiki/Main_Page), which now has more than 57 million articles in more than 300 languages—clearly, such an effort would not be possible without the approximately 120,000 people who regularly contribute to Wikipedia. When content collected via the Web has geographic locations (geotags) associated with it, we refer to it as **volunteered geographic information (VGI)** (Goodchild 2007; Sui et al. 2013); the most widely referenced example of VGI is OpenStreetMap (openstreetmap.org/), an editable map of the world created by volunteers. We will examine the character of contributions to OpenStreetMap in greater depth in Chapter 26. Although the term VGI is sometimes used to refer to any geotagged information collected via the Web, this is arguably inappropriate as such information oftentimes is not volunteered (as in the case of Twitter data). In such cases, a more appropriate term would be **contributed geographic information (CGI)** (Harvey 2013).

The second characteristic of Web 2.0 is lightweight programming models that enable the creation of **map mashups** by combining geographic data from one source with a map from another source. For example, HealthMap uses data feeds from sources like the World Health Organization and others to plot disease outbreaks on an interactive world map (healthmap.org/en/).[7] The third characteristic of Web 2.0 is the ability to work with social media sites such as Facebook and Twitter. Although such social media sites may not seem relevant to cartography, geographers, and other social scientists often analyze and map the character of such social media content. For instance, Mapbox allows users to map geotagged tweets from Twitter accounts (mapbox.com/blog/twitter-map-every-tweet/).

Clearly, these are exciting times for cartography. In addition to the above capabilities, the Web can serve as a source of locational and attribute data, static and animated maps, software for creating thematic maps and for exploring data, tools for developing software, electronic atlases, and teaching material (Table 1.1). All of this can be accessed from either our desktop computer or from a mobile device that we take into the field. Interactive software for visualizing our world such as Google Earth is now available to virtually anyone who has access to computing technology.

What are some of the consequences of this technological change that has occurred and the capability provided by the Web? One consequence is that map production is no longer the sole province of trained cartographers, as anyone with access to the Web can create maps.[8] On the one hand, this is desirable, as it provides the potential for a much larger pool of potential mapmakers (i.e., it enables a **democratization of cartography** (Morrison 1997)). On the other hand, the resulting maps produced by untrained users may not be well designed and accurate. The maps shown in Figure 1.1 illustrate a simple example of design problems that can arise. Map A uses an illogical set of unordered hues, whereas map B uses logically ordered lightnesses from the same hue. Although map A allows users to discriminate easily between individual states, it does not readily permit perception of the overall spatial pattern, which is one of the major reasons for creating a thematic map.

Another error commonly committed by naive mapmakers is illustrated in Figure 1.7. Map A suggests that forested land is more likely to be found in the western part of the United States, whereas map B suggests a pattern with an eastern dominance. Map B portrays the more accurate distribution because it is based on standardized data (i.e., the number of acres of forested land in relation to the area of each state); in contrast, map A is based on raw totals (i.e., the number of acres of forested land). Choropleth maps of raw totals tend to portray large areas as having high values of a mapped attribute; in the case of map A, readers might incorrectly interpret the dark shades as indicating a high proportion of forested land.

One purpose of this book is to explain how to avoid these and other map symbolization problems. For example, Chapter 15 discusses how to select appropriate color schemes for choropleth maps, and Chapter 4 introduces the need for data standardization. An alternative to cartographic instruction is the development of **expert systems** in which a computer automatically makes decisions about symbolization by using a knowledge base provided by experienced cartographers. Angeliki Tsorlini and colleagues (2017) review a number of previous efforts to develop expert systems for thematic mapping and describe a sophisticated rule-based wizard that they developed to assist users in selecting appropriate thematic mapping techniques for a broad range of spatial data types. Their wizard is now included in the Swiss OCAD software (https://www.ocad.com/wiki/ocad/en/index.php?title=Thematic_Map).

TABLE 1.1

Capabilities of the Web Available for Cartography

Function	Example	Address (URL)
Locational data	Longitude and latitude of fast food restaurants	https://data.world/datafiniti/fast-food-restaurants-across-america
Attribute data	U.S. census data	https://www.census.gov/data.html
Static maps	Far & Wide's 100 Amazing World Maps	https://www.farandwide.com/s/amazing-world-maps-74d6186e6d0e414b
Animated maps	Examples from Infogram	https://infogram.com/blog/map-examples-from-the-web/
Software for creating thematic maps	Mapbox Studio	https://www.mapbox.com/mapbox-studio
Software for exploring data	Tableau Public	https://public.tableau.com/en-us/s/
Tools for developing software	D3	https://d3js.org/
Electronic atlases	The Atlas of Canada	https://www.nrcan.gc.ca/maps-tools-and-publications/maps/atlas-canada/10784
Teaching material	GIS&T Body of Knowledge	https://gistbok.ucgis.org/

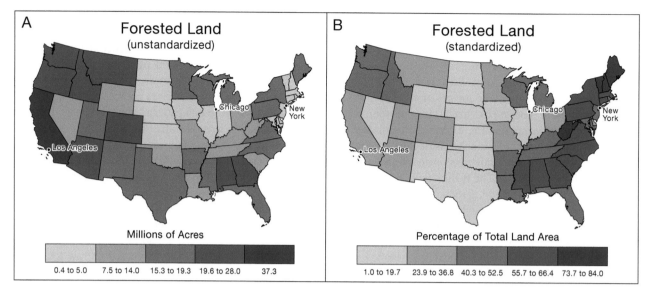

FIGURE 1.7 A comparison of the effect of data standardization. Map A is based on raw totals (i.e., the number of acres of forested land), whereas map B is based on standardized data (i.e., the number of acres of forested land relative to the area of each state). Map A is misleading because states with large areas tend to have more forests. (Powell et al. 1992.)

Although expert systems appear to have great potential for cartography, you will find that most GIS and cartography software vendors do not fully incorporate expert systems. Possibly this is because cartographers do not always agree on the rules for symbolization (Wang and Ormeling 1996).[9]

The second consequence of technological change is the ability to produce maps that would have been difficult or impossible to create using manual methods. An early example was the **unclassed map**, introduced by Waldo Tobler (1973).[10] Figure 1.8 compares a traditional classed map (A) with its unclassed counterpart (B). On the **classed map**, data are grouped into classes of similar value, with a progressively darker blue assigned to each class, whereas on the unclassed map, lightnesses of blue are assigned proportionally to each data value. Some people have promoted unclassed maps, arguing that they more accurately reflect the real-world distribution, whereas others have promoted classed maps on the grounds that they are easier to interpret. We consider this issue in more detail in Chapter 15.

Animated maps (or maps characterized by continuous change while the map is viewed) are particularly representative of the capability of modern computer technology. Although the notion of animated mapping has been around at least since the 1930s, it wasn't until the early 1990s

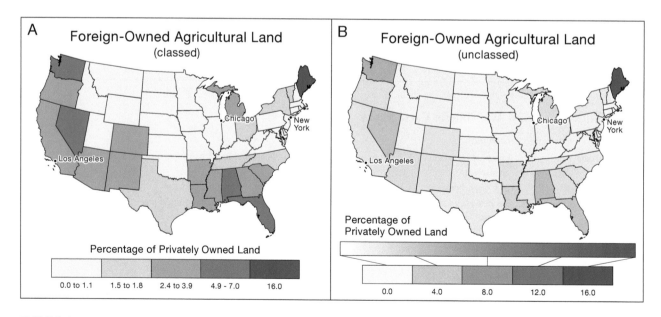

FIGURE 1.8 A comparison of (A) classed and (B) unclassed maps. On the classed map, states are grouped into classes of similar value, with a lightness of blue assigned to each class, whereas on the unclassed map, lightnesses of blue are selected proportional to the data value associated with each state. (Johnson et al. 2010.)

that cartographers began to recognize its full potential (Campbell and Egbert 1990). Probably the most common forms of animation are those representing changing cloud cover, precipitation, and fronts on daily weather reports, but animations of a wide range of spatial phenomena are now available via the Web (numerous examples can be found at https://infogram.com/blog/map-examples-from-the-web/). Although animations are often eye-catching, a key question is whether they actually provide useful spatial information: Do they work? We will consider this issue in Chapter 24 where we examine numerous animations that have been developed.

A third consequence of technological change is that it alters our fundamental way of using maps. With the communication model approach, cartographers generally created one "best" map for users. In contrast, interactive graphics now permit users to examine spatial data dynamically and thus develop several different representations of the data—a process termed **data exploration**. The software package MapTime (described in more detail in Chapter 25) exemplifies the nature of data exploration. MapTime permits users to explore data using three approaches: animation, **small multiples** (in which individual maps are shown for each time period), and **change maps** (in which an individual map depicts the difference between two points in time). The examination of population values for 196 U.S. cities from 1790 to 1990 (a data set distributed with MapTime) illustrates the different perspectives provided by these approaches. For example, an animation reveals major growth in northeastern cities over most of the period, with an apparent drop for some of the largest cities from about 1950 to 1990. In contrast, a change map, showing the percent of population gains or losses between 1950 and 1990, reveals a distinctive pattern of population decrease throughout the Northeast (Figure 1.9). One of the keys to data exploration is that displays such as these can be created in a matter of seconds.

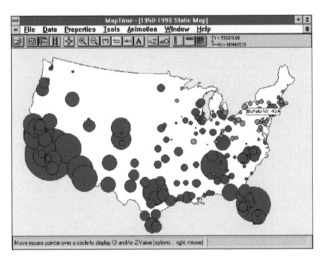

FIGURE 1.9 A change map created using MapTime. Red and blue circles represent population decreases and increases, respectively, between 1950 and 1990. Note the distinctive region of population losses in the Northeast; this pattern was not revealed in an animation of the data over this time period. (Courtesy of Stephen C. Yoder.)

In recent years, the notion of data exploration has evolved into **geovisual analytics**, which combines the visualization capability of data exploration with the computational power of computers to assist us in making sense of large geospatial datasets, or so-called **Big Data**. Examples of Big Data include remote sensing data collected by satellites, location data associated with cell phones and in-vehicle navigation systems, and Twitter feed data. Such data can result in millions (or even billions) of records that cannot be visualized directly—rather, it is often necessary to perform computational processing prior to visualizing the data. We will consider geovisual analytics and associated Big Data in detail in Chapter 26.

The fourth consequence of technological change is the capability to immerse oneself in a 3-D environment; for example, you might don an Oculus (https://www.oculus.com/) **head-mounted display (HMD)** and take a flight of your choosing through the Grand Canyon.[11] The images that you see might appear highly realistic and thus far removed from conventional abstract maps, but it is also possible to create abstract representations of data in 3-D space and explore these representations with similar technology. Such capability gets us into the realm of **virtual reality (VR)**, or what we will term **virtual environments**. Closely aligned with virtual environments is the notion of **augmented reality**, in which computer-based information is combined with our normal view of the real world. For instance, imagine that you are a physical geographer studying vegetation changes in a particular region, and you wish to examine the current vegetation in the field and compare it with past vegetation patterns. Traditionally, you would accomplish this by taking maps into the field and comparing them with current vegetation patterns. With augmented reality, however, you can don a *wearable computer* (and associated specialized viewing hardware) and *see* historic vegetation patterns overlaid on the present-day landscape. We will consider the evolving technologies of virtual environments and augmented reality in Chapter 28.

1.7   WHAT IS GEOVISUALIZATION?

Outside geography, the term *visualization* has its origins in a special issue of *Computer Graphics* authored by Bruce McCormick and colleagues (1987). To McCormick et al., the objective of **scientific visualization** was "to leverage existing scientific methods by providing … insight through visual methods" (3). Work in scientific visualization extends far beyond the realm of spatial data, which geographers deal with, to include topics such as medical imaging and visualization of molecular structure and fluid flows. A classic reference is Peter Keller and Mary Keller's (1993) *Visual Cues: Practical Data Visualization*, which provides numerous examples of the use of scientific visualization; Helen Wright (2007) provides a more recent summary of the field. The most recent developments in scientific visualization can be found in the proceedings of the Institute of Electrical and Electronics Engineers (IEEE) Visualization conference, which has been held every year since 1990. Perhaps not surprisingly, the 2021 conference had a heavy focus on visual analytics (http://ieeevis.org/year/2021/welcome).

The notion of **information visualization** is closely related to scientific visualization. Information visualization focuses on the visual representation and analysis of abstract information that may not have inherent spatial parameters. A classic reference on information visualization is Stuart Card and colleagues (1999) *Readings in Information Visualization: Using Vision to Think*. A summary of more recent work can be found in Robert Spence's (2014) *Information Visualization: An Introduction*. Although the abstract information associated with information visualization is often presented with corresponding abstract graphics (e.g., node-link graphs), it is possible to create a "geographic map" of such information. For example, Figure 1.10 shows one view of the relations among books purchased on Amazon that was developed by Emden Gansner and colleagues (2010). Here the names of "countries" are assigned as a function of the themes of books found at those locations. The space being mapped here is not real-world geographic space but the space of the relations between books purchased (i.e., two books are placed near one another if an individual purchasing one book tends to also purchase the other book).

Although those outside geography were the first to popularize visualization, the idea existed in cartography since at least the 1950s (MacEachren and Taylor 1994, 2). As a result, cartographers struggled to define the term

visualization from the standpoint of geography. Two basic definitions emerged. The first definition was a broad one that encompassed both paper and computer-displayed maps. According to Alan MacEachren and colleagues (1992, 101):

> Geographic visualization [can be defined] as the use of concrete visual representations—whether on paper or through computer displays or other media—to make spatial contexts and problems visible, so as to engage the most powerful human information-processing abilities, those associated with vision.

Using this definition, geographic visualization could be applied to the visual analysis of a paper map created by pen-and-ink methods or to the visual analysis of a map created on an interactive graphic display.

The second definition, and the one that we will use, was based on MacEachren's (1994, 6) cartography-cubed representation of how maps are used, which is shown in Figure 1.11.[12] In this graphic, visualization is contrasted with communication along three dimensions: private versus public, revealing unknowns versus presenting knowns, and the degree of human–map interaction. Based on this diagram, MacEachren argued that **geographic visualization** is a private activity in which unknowns are revealed in a highly interactive environment, whereas *communication* is the opposite: a public activity in which knowns are presented in a noninteractive environment.

As an example of geographic visualization, MacEachren employed Joseph Ferreira and Lyna Wiggins' (1990)

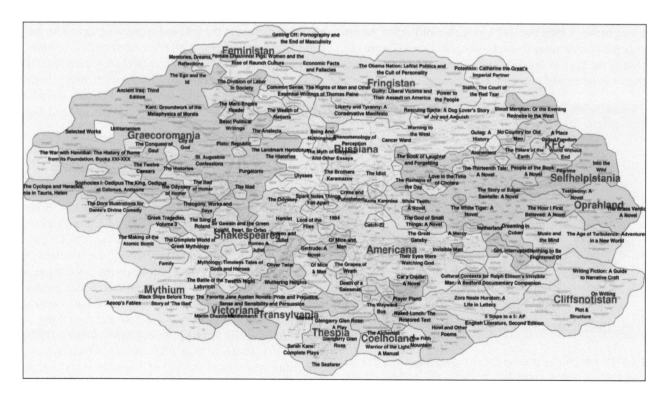

FIGURE 1.10 A "map" of the relations among books purchased on Amazon. The names of "countries" are assigned based on the books that fall within their boundaries. (Republished with permission of The Institute of Electrical and Electronics Engineers, Incorporated (IEEE), from "Visualizing graphs and clusters as map," by E. R. Gansner, Y. Hu, and S. G. Kobourov, Vol. 30, no. 6, 2010; permission conveyed through Copyright Clearance Center, Inc.)

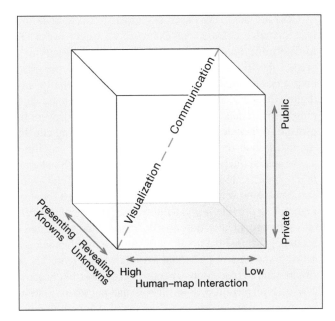

FIGURE 1.11 Alan MacEachren's cartography-cubed representation of how maps are used. Note that visualization is contrasted with communication along three dimensions: private versus public, revealing unknowns versus presenting knowns, and the degree of human–map interaction. (After MacEachren 1994, 7.)

"density dial," in which class break points on choropleth maps were manipulated to identify and enhance spatial patterns. The idea is that users of the density dial could discover previously unknown spatial patterns. In contrast, MacEachren argued that communication is exemplified by the "you are here" maps used to locate oneself in a shopping mall. MacEachren stressed that certain map uses do not fit neatly into either category (thus the need for the cartography-cubed representation). For example, Gail Thelin and Richard Pike's (1991) shaded relief map shown in Figure 1.12 (which we will discuss in Chapter 23) fits into the communication realm because it is available to a wide readership (making it "public") and because the user cannot interact with the paper version of it.[13] The Thelin and Pike map, however, fits visualization in the sense that it can reveal unknowns; for instance, many readers would be unfamiliar with the flat-iron-shaped plateau in the north central part of the map—the Coteau des Prairies. The net result is that the Thelin and Pike map falls in the upper right front corner of MacEachren's diagram.

The notions of map communication models and data exploration introduced previously can be associated with communication and visualization, respectively, in MacEachren's cartography-cubed representation. Thus, when using the five steps of the communication model presented in Section 1.5, the intention generally is that the map is being made for the general public, there will be low human–map interaction, and the focus is on presenting knowns. In data exploration, the emphasis is on revealing unknowns via high human–map interaction in a private setting; in this sense, the word "exploration" could easily be substituted for "visualization" in MacEachren's graphic.

FIGURE 1.12 A small-scale version (approximately 1:21,400,000) of the USGS map *Landforms of the Conterminous United States: A Digital Shaded-Relief Portrayal.* (Developed by Thelin and Pike (1991); for a pdf of this map, see https://pubs.usgs.gov/imap/i2206/.)

On an informal basis, visualization has been used to describe any recently developed, novel method for displaying data. Thus, cartographers have placed animation and virtual environments under the rubric of "visualization." Additionally, novel methods that might result in static maps (e.g., Daniel Dorling's cartograms; see Chapter 20) are also placed under the heading of "visualization." MacEachren eventually simplified the term *geographic visualization* to **geovisualization** (e.g., MacEachren et al. 1999).

One limitation of the term geovisualization is its implication that our interpretation of spatial patterns is solely a function of our sense of vision. As we will see in Chapter 4, however, cartographers have begun to explore how our other senses (e.g., sound, touch, and even smell) might be used to interpret spatial patterns. Thus, it would seem desirable to use a term that is more inclusive. One term that has been proposed (by Fraser Taylor) is **cybercartography**, which, Taylor argues, includes the following elements:

- It is multisensory ([it includes] sound, touch, smell, vision, and even taste) and multimodal;
- It uses multimedia forms (e.g., Web 2.0 and mobile devices);
- It is interactive and "map users" can increasingly become "map creators";
- It is an information/analytical package and an organizational framework for the emerging products and processes of the Web 2.0 era of social computing;
- Cybercartographic products are compiled by teams of individuals from a variety of different disciplines, not just cartographers;
- Cybercartography is applied to a wide range of topics, responding to societal demands, including topics not usually "mapped";
- The creation of cybercartographic products involves new research and development partnerships with government and industry (Taylor 2009, 2–3).[14]

Taylor and colleagues have developed numerous products that reflect their views of the notion of cybercartography (see https://gcrc.carleton.ca/index.html). Although the term cybercartography has not been widely adopted by cartographers, it seems clear that ultimately we will need a term that is more inclusive than geovisualization.

1.8   RELATED GISCIENCE TECHNIQUES

The picture of cartography today would not be complete without mentioning the broader realm of geographic information science (GIScience), which can be considered to include cartography and the related techniques of geographic information systems (GIS), remote sensing, and quantitative methods. In this section, we consider not only how cartographers can use these techniques but also how knowledge of cartography can provide those working with these techniques the ability to create more effective cartographic products.

GISs are computer-based systems that are used to analyze spatial problems. For example, we might use a GIS to determine optimal bus routes in a school district or to predict the likely location of a criminal's residence based on a series of apparently related crimes. Traditionally, most digital thematic maps were created using thematic mapping software (e.g., MapViewer).[15] Today, GIS software has evolved to include considerable thematic mapping capability, in addition to its inherent spatial analysis capabilities. These spatial analysis capabilities also enable GIS to handle more sophisticated mapping problems than traditional thematic mapping software. Dot maps (such as those shown in Figure 1.3) provide a good illustration of the capability of GIS software. Ideally, dots on a dot map should be located as accurately as possible to reflect the underlying phenomenon. Thus, in the case of the wheat map shown in Figure 1.3, we would want to place dots where wheat is most likely grown (say, on level terrain with fertile soil) and not where it cannot be grown (in water bodies). Traditional thematic mapping software did not accomplish this, as dots would normally be placed solely on the basis of enumeration units (e.g., counties). In contrast, GIS software enables a large number of factors (or layers in GIS terminology) to be accounted for. We consider this process further in Chapter 19.

The basic purpose of remote sensing is to record information about Earth's surface from a distance (e.g., via satellites and aircraft). For instance, we might use remote sensing to determine temporal changes resulting from forest fires (e.g., acreage burned, effects of erosion, and regrowth) or to determine the health of crops. The importance of remote sensing to cartography can be illustrated by again presuming that we wish to map the distribution of wheat across the United States. Rather than use a GIS approach, in which layers of related information are considered, we could use remote sensing to determine the precise location of wheat fields. For instance, Dana Peterson and colleagues (2009) found that wheat fields could be identified with an accuracy in excess of 90 percent by using remotely sensed imagery for multiple time periods. In Chapters 16 and 19, we'll consider how remotely sensed imagery (and associated land use/land cover information) can be used to enhance both dasymetric and dot mapping, respectively.

Quantitative methods are used in the statistical analysis of spatial data. For instance, we might develop an equation that relates the death rate due to drunk driving to various attributes that we think might explain the spatial variation in the death rate, such as the severity of laws that penalize drunk driving, the extent to which the laws are enforced, the percentage of the population that are members of various religious groups, the traffic density, and many other attributes. The major development in quantitative methods relevant to cartography is that of **exploratory spatial data analysis (ESDA)**, which refers to data exploration techniques that accompany a statistical analysis of the data. As an example, we might explore a map of predicted deaths due to drunk driving to see how the pattern is affected by various attributes that we include in the model. We only touch on ESDA in this book (in Chapter 25) because a thorough understanding requires a more complete background in statistics than is typical for the introductory cartography student.

Thus far, we have considered how cartographers can benefit from a knowledge of other geographic techniques. Let's now consider how those working with these other techniques can also benefit from a knowledge of cartography. For example, imagine that you are working in the GIS department of a city and wish to examine the distribution of auto thefts and that you create a choropleth map depicting the number of auto thefts for each census tract in the city. Based on the high incidence of auto thefts in a contiguous set of census tracts, you recommend to the police department that they focus their patrols in those areas. Unfortunately, your solution might be inappropriate because you failed to consider the population (and possibly the number of cars owned) in each census tract. Instead of mapping the raw number of

auto thefts, you probably would want to adjust for the population (or the number of cars owned) in each tract. This is another example of the data standardization problem that we mentioned earlier. Effective use of GIS requires an understanding of basic cartographic principles such as these.

As another example of how those working in other areas can benefit from cartography, imagine that you are a remote sensor working on the GreenReport (https://greenreport-kars.ku.edu/), which uses remotely sensed images to depict the health of crops and natural vegetation throughout the United States. Figure 1.13A illustrates a basic greenness map that you might use to represent current vegetation conditions. One can argue that this is a logical color scheme because dark green represents "High Biomass," whereas dark brown indicates "Low Biomass." Figure 1.13B depicts a change map that might be used to compare current conditions to those two weeks

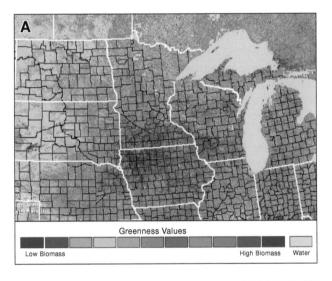

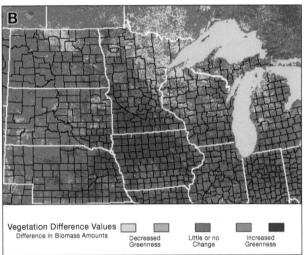

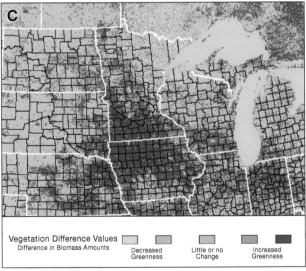

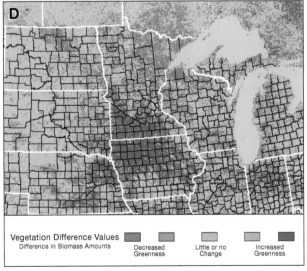

FIGURE 1.13   Potential images to be used with the GreenReport: (A) a basic greenness map used to represent current vegetation conditions; (B) a change map in which the "Little or no Change" category might be confusing because of its relation to the colors for Decreased Greenness; (C) a change map in which a gray shade is used for the "Little or no Change" category to avoid confusion; (D) a change map in which gray is used for the "Little or no Change" category, and colors not used on the greenness map are used to depict changes in greenness. (Courtesy of Kansas Applied Remote Sensing, University of Kansas.)

earlier. Note that in this case, the "Little or no Change" category appears related to the two Decreased Greenness categories to its left, which could confuse a map reader. An improved symbology might use a gray tone for the "Little or no Change" category, as shown in Figure 1.13C. An alternative symbology would be to use completely different colors to represent the changes in greenness; for example, Figure 1.13D uses shades of blue to represent increased greenness and orange and red shades to represent decreased greenness. The latter symbology could be desirable because users could associate certain colors with the raw image and a different set of colors with the changes.

1.9   COGNITIVE ISSUES IN CARTOGRAPHY

Understanding the role that cognition plays in cartography requires contrasting cognition with perception. *Perception* deals with our initial reaction to map symbols (e.g., that a symbol is there, that one symbol is larger or smaller than another, or that symbols have different colors). In contrast, *cognition* deals not only with perception but also with our thought processes, prior experiences, and memory.[16] For example, contour lines on a topographic map can be interpreted without looking at a legend because of one's past experience with such maps. Alternatively, one might correlate the pattern of soils on a particular map with the distribution of vegetation seen on a previous map, making use of our memory of mapped information.

The principles of cognition are important to cartographers because they can explain why certain map symbols work (i.e., communicate information effectively). Traditionally, cartographers were not so concerned with why symbols worked but rather with determining which symbols worked best. This was known as a *behaviorist view*, in which the human mind was treated like a black box. Today, cartographers are more likely to utilize a *cognitive view*, in which they hope to determine *why* symbols work effectively. A cognitive view provides a theoretical basis for map symbol processing that not only assists in explaining why particular symbols work but also provides a basis for evaluating other map symbols, even those not yet developed.

To illustrate the difference between the behaviorist and cognitive views, consider an experiment in which a cartographer wishes to compare the effectiveness of two color schemes for numerical data: five hues from a yellow-to-red progression and five hues from the electromagnetic spectrum (red, orange, yellow, green, and blue). Let us presume that the results of such an experiment reveal that the yellow–red progression works best. The traditional behaviorist would report these results but probably would not provide any indication as to why one sequence worked better than another. In contrast, the cognitivist would consider how color is processed by the eye–brain system, possibly theorizing that the yellow–red progression works best because of the **opponent-process theory** (see Chapter 10). The effectiveness of the spectral hues might be hypothesized on

the grounds that different hues will appear to be at slightly different distances from the eye and thus will form a logical progression (e.g., red will appear nearer than, say, blue, as discussed in Chapter 17). Spectral hues might also be considered effective because of their common use on maps and the likelihood that readers have experience in using them.

An important concept of cognition is the three types of memory: iconic memory, short-term visual store, and long-term visual memory.[17] **Iconic memory** deals with the initial perception of an object (in our case, a map or portion thereof) by the retina of the eye (see Chapter 10 for a detailed discussion of the retina). Calling this "memory" is somewhat of a misnomer because it exists for only about one-half second and because we have no control over it. Visual information initially recorded in iconic memory is passed on to the **short-term visual store**. Only selected information is passed on at this stage; for example, the boundary of Texas shown in Figure 1.14 will likely be simplified to some extent in moving from iconic to a short-term visual store. Keeping information in short-term visual store requires constant attention (or activation). This is accomplished by a rehearsal of the items being memorized (e.g., staring at the map of Texas and telling yourself to remember its shape).

After the information has been rehearsed in the short-term visual store, it is ultimately passed on to **long-term visual memory** (Figure 1.14). Note that arrows are shown going in both directions between short-term visual store and long-term visual memory. When something is initially memorized, information must be moved from the short-term store to long-term memory; when something

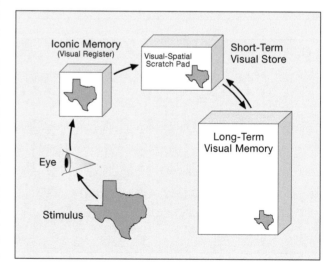

FIGURE 1.14   Three forms of memory used in cartography. A perceived map is initially stored in iconic memory (the retina of the eye). The image is then passed to the short-term visual store in the brain (where it is rehearsed). Finally, information is stored for later use in long-term visual memory in the brain. (From Peterson, Michael P., *Interactive And Animated Cartography*, 1st Edition, © 1995, 27. Reprinted by permission of Pearson Education, Inc., Upper Saddle River, NJ.)

is retrieved from memory, the opposite is the case. As an example of the latter, imagine that you are shown a map of Texas and asked to indicate which state it is. To make your decision, you must retrieve the image of Texas from long-term memory and compare it with the image in the short-term store to make your decision. We have covered here only some of the basic concepts of cognitive psychology that are necessary for understanding this text. For a broader overview from the perspective of cartography, see Peterson (1995, Chapter 2) and MacEachren (1995).

1.10 SOCIAL AND ETHICAL ISSUES IN CARTOGRAPHY

Although maps, and especially interactive digital maps, have tremendous potential for visualizing spatial data, we should also consider various social and ethical issues associated with their use. The notion of social and ethical issues in cartography was first developed in the context of *postmodernism*. Those who subscribe to a postmodernist view believe that problems can best be approached from multiple perspectives or viewpoints; for example, in a study of an urban neighborhood, a postmodernist would want to acquire not only the perspective of those in positions of political and economic power, but also the perspectives of as wide a sample of inhabitants as possible: men and women, children, the elderly, each of the social classes, ethnic and racial groups, or any others who might shed light on the dynamics of the neighborhood (Cloke et al. 1991, 170–201).

An important notion of postmodernism is that a text (or map) is not an objective form of knowledge but rather has numerous hidden agendas or meanings. **Deconstruction** enables us to uncover these hidden agendas and meanings. Brian Harley (1989), widely recognized as one of the early proponents of deconstruction in cartography, stated:

> Deconstruction urges us to read between the lines of the map … to discover the silences and contradictions that challenge the apparent honesty of the image. We begin to learn that cartographic facts are only facts within a specific cultural perspective. (3)

In addition to discussing the importance of deconstruction, Harley argued that maps are instruments of power and that this power was largely controlled by Western-centric political structures that used maps to influence knowledge and space.

As an example of deconstruction, consider Mark Bockenhauer's (1994) examination of the official state highway maps of Wisconsin over a seven-decade period. Although the major purpose of highway maps is presumably to assist a motorist in getting from one place to another, Bockenhauer (17) argued that "three dominant 'cultures' [could] be identified as factors influencing the map product: a transportation/modernization culture, a culture of promotion, and a subtle, beneath-the-surface culture of dominion." As an example of the transportation/

modernization culture, Bockenhauer recounted the removal of surfaced county roads from recent editions of the map, and thus the greater emphasis on "getting us from here to there by encouraging travelers onto the freeways" (21). An illustration of highway maps as promotional devices was the encouragement of highway travel by government officials; for example, in the 1989–1990 edition, Governor Tommy Thompson showcased himself next to a replica of a Duesenberg automobile, a high-end automobile from the early 1900s. Finally, an example of the culture of dominion was the portrayal of women on the maps: "Among the most common and prominent images appearing on the … maps … are those of women in swim suits and fishermen. Nearly all of the photos of people enjoying Wisconsin fishing … are of white men …. [The] women seem to be part of the package of 'pleasure' offered to white men in Wisconsin" (24). Although some would disagree with Bockenhauer's interpretations of these maps, it is clear that maps can convey information other than their supposed primary purpose.

We do not focus on postmodernism and map deconstruction in this book, but both mapmakers and map users need to recognize their importance. Mapmakers must realize that maps can communicate unintended messages and that the data they have chosen to include on a map or the method of symbolizing that data might be a function of the culture of which they are a part. Conversely, map users must recognize that a single map might depict only one representation of a spatial phenomenon (e.g., a map of percent forest cover is only one representation of vegetation), and that other representations may provide alternative and perhaps useful perspectives on the mapped data.

Note that there is some overlap between the notions of data exploration and postmodernism because both promote the notion of multiple representations of data. In the context of data exploration, "multiple representations" refers to the various methods of symbolizing the data (e.g., using MapTime to display population data as both an animation and a change map). The postmodernist would likely support this approach because it concurs with the notion that there is no single "correct" way of visualizing data. Additionally, however, the postmodernist would be interested in purposefully seeking out the multiple meanings and potentially hidden agendas found in a particular thematic map.

In recent years, the term **critical cartography** has been used to describe the cartography practiced by those sharing views similar to Brian Harley.[18] An overview paper by Chris Perkins (2018) makes clear, however, that critical cartography is much more comprehensive than Harley's original vision. Not only does critical cartography consider that maps can have multiple meanings and hidden agendas and that maps can be instruments of power, but it includes approaches that mapmakers can utilize to create maps that appear to counter traditional mapping practices (e.g., they might be *radical*, *disruptive*, *protesting*, or *subversive*) (81).

The notions of postmodernism and map deconstruction also led to an examination of the social and ethical implications of the broader field of GIScience (Sui 2004); an examination of such issues became known as **critical GIS** (Harvey et al. 2005). Early on, those taking a critical view were concerned that, because of its quantitative and empirical nature, GIS failed to properly consider the everyday world that individuals live in, that the cost and complexity of GIS software prevented access by the full range of the population, and, consequently, that GIS was largely under the control of those in positions of political and economic power. To understand the nature of such concerns, consider the role that GIS might play in tracking the whereabouts of individuals. A variety of companies now market wristbands that can be affixed to an individual and locked or unlocked remotely, enabling another party to track the movement of that individual in real time via the global positioning system (GPS) and GIS. Although such technology is arguably useful for tracking children and the elderly, its availability raises interesting ethical questions. For instance, what if a child does not wish to be tracked (not surprising for a teenager)? And should spouses be able to track one another? Jerry Dobson and Peter Fisher (2003) warned us about the perils of such technology, noting that if a transponder were added to the wristband, it would be possible to administer a form of punishment to the individual. They termed the net result **geoslavery**, suggesting that the results would be far worse than George Orwell's 1984, as one "master" could potentially monitor and enslave thousands of people. We will further discuss how ethics and power play into GIScience in Chapter 2.

Although those involved with critical GIS are concerned about the potential misuse of GIS, they also note that effective visualizations can be produced if GIS is used with care. For instance, in a paper dealing with feminist visualization, Mei-Po Kwan (2002) described how she mapped the spacetime paths of African-American women in Portland, Oregon. Kwan argued that "not only do the homes and workplaces of these women concentrate in a small area … but their activities' locations are much more spatially restricted when compared to those of all other gender/ethnic groups" (654). We will examine Kwan's work in more depth in Chapter 24. As another example, Jeremy Crampton (2004) criticized the traditional choropleth map because "it produces a view of human life as crammed into pre-given political units." Utilizing the work of Holloway and colleagues (1999), Crampton illustrated how a consideration of ancillary information such as land use/land cover can be used to create a much more detailed dasymetric map. We will consider more recent dasymetric mapping efforts in Chapter 16.

1.11 SUMMARY

This textbook focuses on thematic maps and the burgeoning area of geographic visualization (or geovisualization). Thematic maps display the spatial pattern of themes (or attributes), such as the murder rates for U.S. cities or yearly precipitation totals for major cities around the world. Thematic maps can be used in three ways: to provide

specific information about particular locations, to provide general information about spatial patterns, and to compare patterns on two or more maps. Imagine that you created a choropleth map showing the suicide rate in each census tract of a city. An example of specific information would be comparing a symbol on the map with those shown in the legend to determine the suicide rate associated with a particular census tract. An example of general information would be examining the overall distribution of colors representing the suicide rates to see where high rates seem to cluster. You might then wish to compare the pattern of suicides on this map with the pattern of some other attribute that you think explains suicides, such as the use of a particular drug.

We discussed five basic steps that you can utilize when designing a map to communicate spatial information to others: (1) Consider what the real-world distribution of the phenomenon might look like; (2) Determine the purpose of the map and its intended audience; (3) Collect data appropriate for the map's purpose; (4) Design and construct the map; and (5) Determine whether users find the map useful and informative. Unfortunately, naïve mappers often ignore these steps and map data that is conveniently available; furthermore, they often utilize software (and associated default options) that also are conveniently available. The end result is likely a map that is inappropriate given the desired map purpose and intended audience.

We noted that since the late 1950s, the field of cartography has been characterized by continuous technological change, moving from a discipline based on pen and ink to one in which today, virtually all maps are designed and produced using computer-based techniques. Particularly important in this process has been the development of the Internet and associated Web 2.0 technology. Three characteristics of Web 2.0 have been of particular importance to cartography: (1) the ability to harness the collective intelligence of Web users (for geographers, this is commonly termed volunteered geographic information (VGI)); (2) lightweight programming models that enable the creation of map mashups by combining geographic data from one source with a map from another source; and (3) the availability of geotagged social media sites (such as Twitter) that enable geoscientists to map the character of associated social media content.

These technological advancements have several important consequences. One consequence is that virtually anyone can now make maps—this is known as the democratization of cartography. This is exciting because those interested in making maps no longer have to rely on professional cartographers, but it means that the resulting maps may not be well designed or accurate. A second consequence is the ability to produce maps that would have been difficult or impossible to create using manual methods. An example of this is the unclassed map, in which each data value can theoretically be represented by a different symbol (e.g., a data set with 100 different data values might be depicted using 100 different lightnesses of blue on a choropleth map).

A third consequence is that interactive graphics now permit users to examine spatial data dynamically and thus develop several different representations of the data—a process termed data exploration. Some potential forms of representation include animation, small multiples (in which individual maps are shown for each time period), and change maps (in which an individual map depicts the difference between two points in time). One of the keys to data exploration is that these different forms of representation be created in a matter of seconds. The notion of data exploration has evolved into geovisual analytics, which combines the visualization capability of data exploration with the computational power of computers to assist us in making sense of large geospatial datasets, or so-called Big Data.

The fourth consequence of technological change is the capability to immerse oneself in a 3-D environment (e.g., by donning a head-mounted display (HMD))—these are termed virtual environments or virtual reality. Such environments can appear either realistic (e.g., we might travel through a representation of Yosemite National Park) or abstract (e.g., we might travel through a representation of changing ozone levels in the atmosphere). Closely aligned with virtual environments is augmented reality, in which computer-based information is used to enhance our view of the real world. For example, we might don a wearable computer (and associated specialized viewing hardware) and actually see the historic urban landscape overlaid on the present-day urban environment.

Using the cartography-cubed representation of how maps are used (Figure 1.11), geovisualization can be contrasted with map communication along three dimensions: private versus public, revealing unknowns versus presenting knowns, and the degree of human–map interaction. Geovisualization is a private activity in which unknowns are revealed in a highly interactive environment, whereas communication is the opposite: a public activity in which knowns are presented in a noninteractive environment.

We examined several geographic techniques that are important to cartography. GIS is important because of its inherent ability to analyze and display layers of spatial data. This layering and related analytic abilities will be especially important in dot and dasymetric mapping. Remote sensing provides information about Earth's surface that may not be available in more traditional published data sources derived through surveys of people. For example, we might use satellite remote sensing to determine the success of the corn crop in the state of Iowa and compare that with the amounts of corn harvested, as reported by farmers in an agricultural survey. The major development in quantitative methods relevant to cartography is that of exploratory spatial data analysis (ESDA), which refers to exploratory techniques that accompany a statistical analysis of the data. For example, we might explore a map of deaths due to AIDS to see how the pattern is affected by various attributes that we include in a model predicting deaths due to AIDS.

We stressed the importance of taking a cognitive view of cartography, as opposed to a behaviorist view. In the behaviorist view, the human mind was treated like a black box, in which cartographers were not so concerned with why symbols worked but rather with determining which symbols worked best. In contrast, in a cognitive view, cartographers want to determine why symbols work effectively. A cognitive view provides a theoretical basis for map symbol processing that not only assists in explaining why particular symbols work but also provides a basis for evaluating other map symbols, even those not yet developed.

Finally, we examined the notion of deconstruction and the broader area of critical cartography. The process of deconstruction allows us to uncover hidden agendas and meanings in maps and possibly recognize any underlying power structures that were used to create the map. Critical cartography takes a broader view of cartography than this, potentially providing avenues for mapmakers to be radical, disruptive, protesting, or subversive.

1.12   STUDY QUESTIONS

1. Which of the following would you consider to be thematic maps? Explain your choices. (A) a digital road map available through a mobile app like Google Maps; (B) a map using pictographic gun symbols to represent the total number of guns owned by households in each county of Minnesota; (C) a map that utilizes lightnesses of blue to represent different depths to the ocean floor; and (D) a map showing the locations of major cities and major river systems around the world.

2. Discuss why the colors shown in Figure 1.1A are inappropriate for mapping the percentage voting for Biden in 2020.

3. Imagine that a GIS instructor asks students in her class to create thematic maps of a topic that they are interested in. A student interested in COVID-19 death rates in his home state of Missouri uses the Internet and finds that the death rates are conveniently available at the county level. The student then uses Mapbox Studio (https://www.mapbox. com/mapbox-studio) to create a choropleth map of the death rates. Finally, the student shows the resulting map to several classmates to determine whether they are able to get useful information from the map. Discuss to what extent the student has fulfilled the five steps given in Figure 1.4.

4. A geographer decides to evaluate the spatial character of tweets (from Twitter) associated with statements about "masking" during the COVID-19 pandemic. For this purpose, the geographer collects all tweets using terms related to masking and creates maps using the geotags associated with each tweet. Explain whether you would consider this a form of "volunteered geographic information."

5. Define the term "democratization of cartography" and make arguments for and against its usefulness.

6. Imagine that you used the unclassed map shown in Figure 1.8B for a term paper dealing with foreign-owned agricultural land in the United States. (A) Why might the instructor of the class in which you submit the term paper have difficulty interpreting this map in its present static form? (B) How could data exploration software be utilized to assist in interpreting the data depicted on this map?

7. Imagine that you have been examining the issue of death rates due to drunk driving and that you write an article about this issue for your local newspaper. Along with your article, you include a static choropleth map of death rates for the United States at the county level that you feel illustrates key "hot spots" that you point out in your article. In addition, you also provide an interactive version of the map that allows readers to examine the data on their own and create alternative choropleth maps of the data. Discuss how the static and interactive maps would be positioned in MacEachren's cartography-cubed illustration shown in Figure 1.11.

8. Go to the GreenReport (https://greenreport-kars. ku.edu/) and note two kinds of maps that it produces: the Greenness Map and Difference Map 1. Select the year 2021 and 4/20 – 5/3 within that year. Discuss the logic of the color schemes used on each of the two maps and summarize what each map tells you (use the "larger images" for your analysis).

9. Three-dimensional prism maps are constructed by raising enumeration units to a height proportional to the data. Imagine that a cartographer compares the effectiveness of such maps with more conventional choropleth maps and finds that users are able to remember spatial patterns more effectively on the three-dimensional prism maps. Given what has been stated thus far in this question, does it appear that the cartographer has utilized a behaviorist view or a cognitive view? Explain your choice.

10. Choose an official state highway map from your home state. Many states have Department of Transportation websites from which their official state highway map can be downloaded as a PDF. Use the map to deconstruct its content looking for any hidden meanings or messages within the map. Explain those hidden meanings or messages that you find.

NOTES

1 Technically, memory for mapped information would be equivalent to what psychologists term *long-term memory*, but for simplicity, cartographers frequently use just the term *memory*.

2 For a more recent sophisticated approach to the kinds of user tasks associated with thematic maps, see Chapter 3 of Andrienko and Andrienko (2006).

3 Note that the US Constitution requires that all people living in the United States be counted every ten years (both citizens and non-citizens); see https://www.prb.org/resources/u-s-2020-census-faq/.

4 For a statistical approach for handling mobile populations, see Li (1998).

5 For a discussion of population density in the United States, see https://tinyurl.com/UnderstandingUSPopDensity.

6 For an introduction to user studies, see van Elzakker and Ooms (2018).

7 For early examples of map mashups, see Liu and Palen (2010).

8 The process of creating maps on the Web is sometimes referred to as either **neogeography** or **neocartography**, reflecting the fact that those involved often have either little or no geographic or cartographic training.

9 For additional efforts related to expert systems, see Armstrong (2019) and Degbelo et al. (2020).

10 The earliest choropleth maps were actually unclassed, but they rapidly gave way to classed maps (see Robinson 1982, 199).

11 Another example of an HMD can be found at https://birdly.com/.

12 MacEachren used the term (Cartography)³; we use the word *cubed* to avoid confusion with the superscript 3.

13 Note that a digital version of this map with limited interactive capability is now available at https://pubs.usgs.gov/imap/i2206/usa_shade.pdf.

14 After pp. 2–3 of Taylor, D. R. F. (2009) "Some new applications in the theory and practice of cybercartography: Mapping with indigenous people in Canada's North." *Proceedings of the 24th International* Cartography *Conference*, Santiago, Chile, 1–11, Creative Commons Attribution 4.0 License, https://www.ica-conference-publications.net/licence_and_copyright.html.

15 As of 2021, Golden Software indicated it no longer would sell MapViewer, but that some of its features would be added to another of its products, Surfer (https://www.goldensoftware.com/products/surfer).

16 For a more in-depth discussion of perception and cognition, see Goldstein and Brockmole (2017) and Farmer and Matlin (2019), respectively. For a cartographic view of perception and cognition, see Griffin (2018).

17 For a cartographic perspective on these terms, see Peterson (1995).

18 For critical cartographers' views of Harley's work, see the special issue of *Cartographica* (issue no. 1, 2015) focusing on Harley's article "Deconstructing the map."

REFERENCES

Armstrong, M. P. (2019) "Active symbolism: toward a new theoretical paradigm for statistical cartography." *Cartography and Geographic Information Science 46*, no. 1:72–81.

Andrienko, N., and Andrienko, G. (2006) *Exploratory Analysis of Spatial and Temporal Data: A Systematic Approach.* Berlin: Springer-Verlag.

Bertin, J. (1981) *Graphics and Graphic Information-Processing.* Berlin: Walter de Gruyter.

Bockenhauer, M. H. (1994) "Culture of the Wisconsin official state highway map." *Cartographic Perspectives* no. 18:17–27.

Campbell, C. S., and Egbert, S. L. (1990) "Animated cartography: Thirty years of scratching the surface." *Cartographica 27*, no. 2:24–46.

Card, S. K., Mackinlay, J. D., and Shneiderman, B. (eds.). (1999) *Readings in Information Visualization: Using Vision to Think*. San Francisco, CA: Morgan Kaufmann.

Cloke, P., Philo, C., and Sadler, D. (1991) *Approaching Human Geography: An Introduction to Contemporary Theoretical Debates*. London: Paul Chapman.

Crampton, J. W. (2004) "GIS and geographic governance: Reconstructing the choropleth map." *Cartographica 39*, no. 1:41–53.

Degbelo, A., Sarfraz, S. and Kray, C. (2020) "Data scale as cartography: A semi-automatic approach for thematic web map creation." *Cartography and Geographic Information Science 47*, no. 2:153–170.

Dent, B. D. (1996) *Cartography: Thematic Map Design* (4th ed.). Dubuque, IA: William C. Brown.

Dobson, J. E., and Fisher, P. F. (2003) "Geoslavery." *IEEE Technology and Society Magazine 22*, no. 1:47–52.

Farmer, T. A., and Matlin, M. W. (2019) *Cognition* (10th ed.). New York: Wiley.

Ferreira, J. J., and Wiggins, L. L. (1990) "The density dial: A visualization tool for thematic mapping." *Geo Info Systems 1*, no. 0:69–71.

Gansner, E. R., Hu, Y. and Kobourov, S. G. (2010) "Visualizing graphs and clusters as maps." *IEEE Computer Graphics and Applications 30*, no. 6:54–66.

Goldstein, E. B. and Brockmole, J. (2017) *Sensation and Perception* (10th ed.) Boston, MA: Cengage Learning.

Goodchild, M. F. (2007) "Citizens as sensors: The world of volunteered geography." *GeoJournal 69*, no. 4:211–221.

Griffin, A. L. (2018) "Cartography, visual perception and cognitive psychology." In *The Routledge Handbook of Mapping and Cartography*, ed. by A. J. Kent and P. Vujakovic, pp. 44–54. New York: Routledge.

Harley, J. B. (1989) "Deconstructing the map." *Cartographica 26*, no. 2:1–20.

Harvey, F. (2013) "To volunteer or to contribute locational information? Towards truth in labeling for crowdsourced geographic information." In *Crowdsourcing Geographic Knowledge: Volunteered Geographic Information (VGI) in Theory and Practice*, ed. by D. Sui, S. Elwood and M. Goodchild, pp. 31–42. Dordrecht: Springer.

Harvey, F., Kwan, M.-P., and Pavlovskaya, M. (2005) "Introduction: Critical GIS." *Cartographica 40*, no. 4:1–4.

Holloway, S., Schumacher, J., and Redmond, R. L. (1999) "People and place: Dasymetric mapping using ARC/INFO." In *GIS Solutions in Natural Resource Management: Balancing the Technical-Political Equation*, ed. by S. Morain, pp. 283–291. Sante Fe, NM: OnWord Press.

Johnson, L. A., Feather, C. A., and Schultz, L. (2010) Foreign Holdings of U.S. Agricultural Land Through December 31, 2010. Washington, DC: U.S. Department of Agriculture.

Keller, P. R., and Keller, M. M. (1993) *Visual Cues: Practical Data Visualization*. Los Alamitos, CA: IEEE Computer Society Press.

Kwan, M.-P. (2002) "Feminist visualization: Re-envisioning GIS as a method in feminist geographic research." *Annals of the Association of American Geographers 92*, no. 4:645–661.

Li, M. Y. (1998) "Research in statistical methods for mobile population (SMMP)." *The Cartographic Journal 35*, no. 2:155–164.

Liu, S. B. and Palen, L. (2010) "The new cartographers: Crisis map mashups and the emergence of neogeographic practice." *Cartography and Geographic Information Science 37*, no. 1:69–90.

MacEachren, A. M. (1982) "The role of complexity and symbolization method in thematic map effectiveness." *Annals of the Association of American Geographers 72*, no. 4:495–513.

MacEachren, A. M. (1994) "Visualization in modern cartography: Setting the agenda." In *Visualization in Modern Cartography*, ed. by A. M. MacEachren and D. R. F. Taylor, pp. 1–12. Oxford, England: Pergamon.

MacEachren, A. M. (1995) *How Maps Work: Representation, Visualization, and Design*. New York: Guilford.

MacEachren, A. M., and Taylor, D. R. F. (1994) *Visualization in Modern Cartography*. Oxford, England: Pergamon.

MacEachren, A. M., Buttenfield, B. P., Campbell, J. B., et al. (1992) "Visualization." In *Geography's Inner Worlds: Pervasive Themes in Contemporary American Geography*, ed. by R. F. Abler, M. G. Marcus and J. M. Olson, pp. 99–137. New Brunswick, NJ: Rutgers University Press.

MacEachren, A. M., Wachowicz, M., Edsall, R., et al. (1999) "Constructing knowledge from multivariate spatiotemporal data: Integrating geographical visualization with knowledge discovery in database methods." *International Journal of Geographical Information Science 13*, no. 4:311–334.

McCormick, B. H., DeFanti, T. A., and Brown, M. D. (1987) "Visualization in scientific computing." *Computer Graphics 21*, no. 6.

Morrison, J. L. (1997) "Topographic mapping for the twenty-first century." In *Framework for the World*, ed. by D. Rhind, pp. 14–27. Cambridge: Geoinformation International.

O'Reilly, T. (2005) "What is Web 2.0: Design patterns and business models for the next generation of software." http://oreilly.com/web2/archive/what-is-web-20.html.

O'Reilly, T. and Battelle, J. (2009) "Web squared: Web 2.0 five years on." *Web 2.0 Summit*, San Francisco, CA. https://www.kimchristen.com/wp-content/uploads/2015/07/web2009_websquared-whitepaper.pdf.

Perkins, C. (2018) "Critical cartography." In *The Routledge Handbook of Mapping and Cartography*, ed. by A. J. Kent and P. Vujakovic, pp. 80–89. London: Routledge.

Peterson, M. P. (1995) *Interactive and Animated Cartography*. Englewood Cliffs, NJ: Prentice Hall.

Peterson, D., Whistler, J., Bishop, C., et al. (2009) "The Kansas next-generation land use/land cover mapping initiative." *ASPRS 2009 Annual Conference*, pp. 1–12, Baltimore, MD: American Society for Photogrammetry and Remote Sensing.

Powell, D. S., Faulkner, J. L., Darr, D. R., et al. (1992) "Forest resources of the United States, 1992." United States Department of Agriculture, Forest Service, Rocky Mountain Forest and Range Experiment Station, *General Technical Report RM-234 (Revised)*.

Robertson, P. K. (1991) "A methodology for choosing data representations." *IEEE Computer Graphics & Applications 11*, no. 3:56–67.

Robinson, A. H. (1982) *Early Thematic Mapping in the History of Cartography*. Chicago: University of Chicago Press.

Robinson, A. H., Sale, R. D., Morrison, J. L., et al. (1984) *Elements of Cartography* (5th ed.). New York: Wiley.

Slocum, T. A., and Kessler, F. C. (2015) "Thematic mapping." In *Cartography in the Twentieth Century*, ed. by M. Monmonier, pp. 1500–1524. Chicago: University of Chicago Press.

Spence, R. (2014) *Information Visualization: An Introduction* (3rd ed.). Cham, Switzerland: Springer.

Sui, D., Goodchild, M., and Elwood, S. (2013) "Volunteered geographic information, the exaflood, and the growing digital divide." In *Crowdsourcing Geographic Knowledge: Volunteered Geographic Information (VGI) in Theory and Practice*, ed. by D. Sui, S. Elwood and M. Goodchild, pp. 1–12. Dordrecht: Springer.

Sui, D. Z. (2004) "GIS, cartography, and the 'Third Culture': Geographic imaginations in the computer age." *The Professional Geographer 56*, no. 1: 62–72.

Taylor, D. R. F. (2009) "Some new applications in the theory and practice of cybercartography: Mapping with indigenous people in Canada's North." *Proceedings of the 24th International Cartography Conference*, Santiago, Chile, 1–11.

Thelin, G. P., and Pike, R. J. (1991) Landforms of the Conterminous United States—A Digital Shaded-Relief Portrayal, Map I-2206. Washington, DC: U.S. Geological Survey. (Scale 1:3,500,000.)

Tobler, W. R. (1973) "Choropleth maps without class intervals?" *Geographical Analysis 5*, no. 3:262–265.

Tsorlini, A., Sieber, R., Hurni, L., et al. (2017) "Designing a rule-based wizard for visualizing statistical data on thematic maps." *Cartographic Perspectives*, no. 86:5–23.

U.S. Bureau of the Census (1970) "Population distribution, urban and rural, in the United States: 1970." *United States Maps, GE-50, No. 45.*

U.S. Bureau of the Census (1995) *Agricultural Atlas of the United States*. Washington, DC: U.S. Government Printing Office.

van Elzakker, C. P. J. M., and Ooms, K. (2018) "Understanding map uses and users." In *The Routledge Handbook of Mapping and Cartography*, ed. by A. J. Kent and P. Vujakovic, pp. 55–67. London: Routledge.

Wang, Z., and Ormeling, F. (1996) "The representation of quantitative and ordinal information." *The Cartographic Journal 33*, no. 2:87–91.

Wright, H. (2007) *Introduction to Scientific Visualization*. London: Springer.

Part I

Principles of Cartography

2 A Historical Perspective on Thematic Cartography[1]

2.1 INTRODUCTION

This chapter covers several aspects related to the history of cartography and, more specifically, thematic cartography. Section 2.3 provides a brief historical overview of the broad field of cartography. We will see that the earliest known map dates back 4,500 years and was produced on a clay tablet, obviously a different medium from the computer screens we use today! Although the Egyptians and the Chinese produced some of our earliest maps, many of our modern cartographic ideas date to the time of the Greeks. For example, the Greek Claudius Ptolemy discussed basic principles of cartography including the construction of projections. As in many disciplines, cartography flourished during the Renaissance (e.g., the first globe was created by Martin Behaim during that period). In the eighteenth and nineteenth centuries, the first national surveys of countries were completed and methods for base mapping were developed. There is a rich history around the development of topographical mapping, including the survey by Harvey (1980).

Section 2.4 provides an overview of the historical development of thematic cartography. Although much of the development of thematic cartography came about in the nineteenth and twentieth centuries, we'll see that significant advances took place years earlier—for example, Edmund Halley, of comet fame, developed statistical maps of trade winds and monsoons and developed a predictive map of the path of an eclipse over England. With the compilation of national-level statistics, the field of social cartography, which involves mapping human phenomena, particularly various aspects of population, came into being. As a result, the world saw the development of many thematic mapping techniques, including the dot, choropleth, isarithmic, and even the dasymetric method in the 1800s. Halley and the Frenchmen Charles Joseph Minard and D'Angeville were prominent early thematic cartographers.

In Section 2.5, we discuss developments in the history of U.S. academic cartography by considering four major time periods: early beginnings, the post-war era, the growth of second-generation programs, and the integration of GIScience (Table 2.1). The early beginnings period, spanning from the early part of the twentieth century to the 1940s, represents what might be called nodal activity, in which academic cartography was centered at only two or three institutions and involved individuals who were not necessarily educated in cartography. An example was John Paul Goode at the University of Chicago. The post-war era, from the 1940s to the 1960s, saw the building of first-generation core programs that incorporated multiple faculty, strong graduate programs, and PhD students who ventured off to create their own programs. Three core programs stand out—those of the University of Wisconsin, University of Kansas, and University of Washington. The third period, from the 1960s to the 1980s, involved the growth of second-generation programs, such as those at UCLA, Michigan, and Syracuse, whose core faculty had mostly studied at the first-generation programs. It was in this period that academic cartography emerged as a true subdiscipline of geography. Since the 1980s, cartography has become more integrated with GIScience, though recently, there has been a resurgence in focus on cartography, geovisualization, and representation.

Section 2.6 provides a brief review of some of the major developments in thematic cartography emerging from the long and distinguished European research activity. The field of thematic cartography dates back many centuries in Western Europe, and the early innovations stemmed from the development of national censuses that provided demographic data and surveys of biological information such as vegetation. European cartographers led the way in the development of map symbolization, terrain representation, label placement, generalization, and many other areas. Interestingly, Europeans also developed the first topographic map series, such as the Carte de Cassini in France and Ordnance Survey in Great Britain.

In Section 2.7, we consider the notion of paradigms in cartography, focusing on the paradigm of **analytical cartography**, which considers the mathematical concepts and methods underlying cartography and includes topics such as cartographic data models, coordinate transformations, map projections, interpolation, and analytical generalization. Waldo Tobler is widely recognized as the founder of analytical cartography. A second paradigm reviewed is the newer work on **maps and society**, which represents the emerging ideas on power and maps, maps and privacy, and the access to maps by the public.

2.2 LEARNING OBJECTIVES

- Understanding the growth of thematic cartography in the twentieth century
- Identifying the major paradigms of American thematic cartography
- Comprehending the contributions of European cartography

DOI: 10.1201/9781003150527-3

- Understanding the major academic programs where thematic cartography developed
- Recognizing the basic developments in nineteenth-century European thematic cartography
- Understanding the relationship between maps and society, in particular, privacy and access

TABLE 2.1

The Four Major Periods of U.S. Academic Cartography

Early Beginnings (Early 1900s to Mid-1940s)

J. Paul Goode

Erwin Raisz

Guy-Harold Smith

Richard Edes Harrison

Post-War Era of Building Core Programs (Mid-1940s to Mid-1960s)

University of Wisconsin

University of Kansas

University of Washington

Growth of Second-Generation Programs (Late 1960s to Late 1980s)

UCLA

University of Michigan

University of South Carolina

Syracuse University

Curricula Integrated with Geographic Information Systems/Science (1990s)

University of California, Santa Barbara

SUNY Buffalo

University of South Carolina

2.3 A BRIEF HISTORY OF CARTOGRAPHY

Cartography has a long and distinguished history that dates back thousands of years. Although this chapter will focus on some of the major developments in thematic cartography, you should be aware of the rich literature on the broader subject of cartography, including R. V. Tooley's *Maps and Map-Makers* (1987), Lloyd Brown's *The Story of Maps* (1979), Norman Thrower's *Maps and Civilization* (2008), Leo Bagrow and R. A. Skelton's *History of Cartography* (1964), and P. D. A. Harvey's *The History of Topographical Maps* (1980). You can also refer to the remarkable series on the history of cartography produced by the History of Cartography Project at the University of Wisconsin (http://www.geography.wisc.edu/histcart/).

The earliest known map, dating back 4,500 years, was created in ancient Babylon. This map, produced on a clay tablet, depicts a river valley (most probably the Euphrates), mountains, and a larger body of water (Raisz 1948, 5). But many cultures began mapping in ancient times, including the Egyptians, who focused on land surveys for taxation purposes, and the Chinese, who had already established basic principles of cartography in the first few centuries CE. Some of these principles included rectilinear divisions, orientation, accurate indications of distances, indications of lower and higher altitudes, and attention to the right and left angles of bends in roads (Raisz 1948, 6).

Many of our modern cartographic ideas date to the Greeks, who in the fourth century introduced the ideas of the spherical shape of the Earth; the concept of the Equator and the divisions of the Earth into hot, temperate, and frigid zones; and an accurate measurement of the size of the Earth. It was Eratosthenes of Cyrene (276–195 BCE), the head of the Library of Alexandria, who, using some basic knowledge of Earth–Sun geometry, calculated the circumference of the Earth to be 250,000 stadia, or 40,500 kilometers (25,170 miles), just 425 kilometers (264 miles) from the true value of 40,075 kilometers (24,906 miles). As it turned out, it was a series of compensating errors that had led to this incorrect estimation (see Section 7.4.1). It was also during this period that Claudius Ptolemy of Alexandria (90–168 CE) produced his seminal *Geographia* volumes; the eighth volume discussed basic principles of cartography, including the construction of projections. Accompanying these volumes were numerous maps that located many of the known places on Earth (Raisz 1948, 10).

Important cartographic works from the Roman period included The Peutinger Table—really, a scroll 21-feet long and 1-foot high—that depicted the worldwide road network of that time. It was, for all practical purposes, a type of premodern road map, or more specifically a TripTik® (https://triptik.aaa.com/home/index.html). Although cartography slipped backward during the Middle Ages, with the period's significant focus on ecclesiastical knowledge in its maps, such as the common T-in-O map (Obis Terrarum), this period did produce the portolan chart, which appears to have been the basis for precise water navigation. It was also during the Middle Ages that Edrisi, an Arabic cartographer, produced a world map that was based on Ptolemy's representations. However, it wasn't until the Renaissance that cartography began to flourish once again, with the Age of Discoveries (https://www.worldatlas.com/articles/what-was-the-age-of-exploration-or-the-age-of-discovery.html); developments in printing and engraving that led to the widespread dissemination of cartographic knowledge; the creation of the first globe (which still exists today); and the prolific creation of atlases by the Dutch. Specific examples of cartographic accomplishments from this period are the Martin Behaim globe of 1492, the printing press, and the Dutch atlases of Mercator, Ortelius, Hondius, and Blaeu.

The eighteenth and nineteenth centuries also saw the completion of the first national surveys of countries and the development of methods for base mapping. The key to accurate base mapping is the method of triangulation, whereby a series of measured baselines and angles are calculated. In 1744, César Francois Cassini completed one of the very first baseline projects in France—the first triangulation, which became the foundation for the complete survey of that country. The survey itself, also designed by Cassini, was completed in the late 1700s at a scale of 1:86,400 and was composed of 182 sheets depicting the cultural and natural features of France. Another of the early national surveys was the Ordnance Survey of Great Britain, published in 1801 at a scale of 1:63,360.

2.4 HISTORY OF THEMATIC CARTOGRAPHY

The history of thematic mapping—the mapping of spatial distributions—is more recent, dating largely to the nineteenth and twentieth centuries, although we will see that there were developments before the nineteenth century. Many of the original developments occurred in Europe, driven by the acquisition of new forms of statistical data. As described by Robinson (1982, 32):

> The development of a concern for science and for a universal system of measure from the mid-seventeenth to the mid-nineteenth century in Europe was accompanied by a parallel growth of interest and competence in statistics and statistical method, especially in physical science. In the realm of social affairs some of this was based on theoretical aspects, for example probability theory, some on problems resulting from growing populations and industrial growth, and some simply on the burgeoning interest in descriptive geography. For a considerable period in the two centuries the concern focused on such topics as population, fertility, mortality, and the comparative characteristics of nations. (Republished with permission of University of Chicago Press—Books, from *Early Thematic Mapping in the History of Cartography* by A. H. Robinson, 1982, permission conveyed through Copyright Clearance Center, Inc.)

One can find numerous examples of creative thematic cartography during the seventeenth century, but perhaps the most interesting work is that of Edmund Halley, a seminal member of the British scientific community (Thrower 1969; Robinson 1982). Halley's most significant discovery, of course, was the comet named after him, but he also created a series of ingenious statistical maps depicting the trade winds and monsoons (1686), the path of the eclipse of the Sun (1715), and, perhaps the most famous, his 1701 chart of compass variations in the Atlantic Ocean that uses isarithmic lines (isogones) (Figure 2.1). The 1715 eclipse

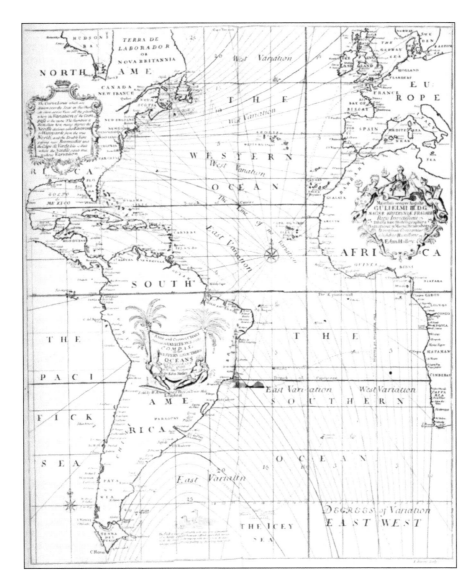

FIGURE 2.1 Halley's 1701 chart of compass variations in the Atlantic Ocean using the isarithmic technique. (From Robinson, Arthur H. 1982. *Early Thematic Mapping in the History of Cartography*. University of Chicago Press; originally acquired from the U.S. Library of Congress.)

map, which depicts the duration and timing of the eclipse as it passes across England, was widely distributed throughout England to help the population understand the path of this phenomenon.

Much of the early work in thematic cartography focused on the mapping of natural phenomena and was published in major atlases. Some of this early work concentrated on the portrayal of relief such as those in Heinrich Berghaus's *Physikalischer Atlas* (1845), J. L. Dupain-Triel's 1791 map of France, and Alexander Johnston's 1852 map of "The Mountains, Table Lands, Plains, & Valleys of Europe" (Robinson 1982). In this same period, maps of the biological world began to appear. Some depicted agricultural distributions, such as Schouw's 1823 map of cereals (rye, wheat, and maize) and the 1839 world map of cultivated plants in Berghaus's *Physikalischer Atlas*. One of the most interesting maps is that of the world distribution of plants that depicts Schouw's 25 phytogeographic regions. Robinson (1982, 103–105) pointed out that many of the cartographic techniques used for the mapping of natural vegetation and cultivated plants—including using colors, patterns, boundaries of extents, and isolines—were straightforward, while the acquisition of data was more difficult.

2.4.1 THE RISE OF SOCIAL CARTOGRAPHY

With the acquisition of national-level statistics, the field of *social cartography* rapidly evolved. Although social cartography involves mapping all types of human phenomena (e.g., language, religion, ethnicity, economic activity, and transportation), various aspects of the population were the focus in nineteenth-century Europe. Some of the seminal works included Frère de Montizon's map of the population of France, dated 1830 (using the dot-mapping technique; see Figure 2.2); the 1848 world map of population density and food habits in Berghaus's *Physikalischer Atlas* (using the choropleth technique); D'Angeville's 1836 map of the number of persons per square myriamètre (using the choropleth technique; see Figure 2.3); Harness's 1837 map of the number of persons per square mile in Ireland (using the dasymetric technique); and Petermann's 1848 map of cholera in the British Isles (a map based on continuous tones; see Figure 2.4) (Robinson 1982). Some of the more interesting works of this era were produced by Charles Joseph Minard, who was trained as an engineer but then became interested in geography. As documented in Robinson, Minard produced maps of the amount of butcher's meat sent to Paris from each department (1858), the extent of the markets for foreign oil and coke (1961), the volume of French wines exported by sea (1865; see Figure 21.2), and the movement of travelers on the principal railways of Europe (1865). Minard was a prolific and creative cartographer who used a variety of innovative cartographic methods including segmented graduated circles, colored regions, and graduated flow maps. During the rise of social cartography, cartographers and statisticians also began to map health data such as diseases and sanitation, the most famous being Dr John Snow's map depicting the spread of cholera. It is interesting to study these (mostly European) early examples of thematic maps, as they utilized symbology that we still use today and that are discussed in this book (e.g., choropleth, dot, proportional circle, isoline, and flow maps).

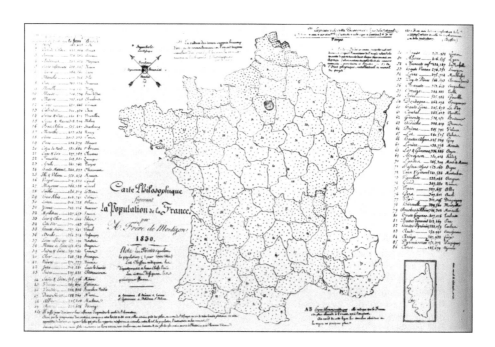

FIGURE 2.2   Frère de Montizon's 1830 dot map of the population of France; each dot represents 10,000 people. (From Robinson, Arthur H. 1982. *Early Thematic Mapping in the History of Cartography*. University of Chicago Press; originally acquired from Bibliothèque Nationale, Paris.)

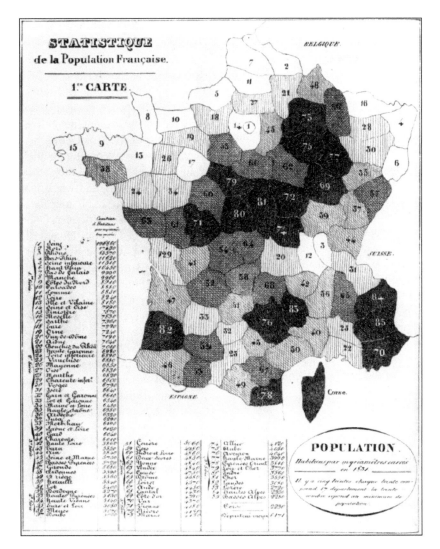

FIGURE 2.3   D'Angeville's 1836 choropleth map of the number of persons per square myriamètre (100 square kilometers); in contrast to the convention followed today, the lightest tone represented the highest population density, and the darkest tone represented the lowest population density. (From Robinson, Arthur H. 1982. *Early Thematic Mapping in the History of Cartography.* University of Chicago Press; originally acquired from Bibliothèque Nationale, Paris.)

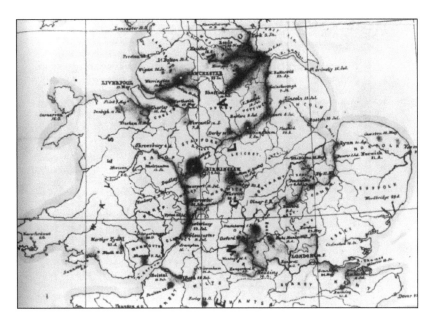

FIGURE 2.4   Petermann's 1848 continuous-tone map of cholera in the British Isles; the original map was tinted in red, with darker shades representing a higher mortality ratio. (From Robinson, Arthur H. 1982. *Early Thematic Mapping in the History of Cartography.* University of Chicago Press; originally acquired from the British Library.)

2.5 HISTORY OF U.S. ACADEMIC CARTOGRAPHY

Thematic cartography morphed into a true academic discipline during the twentieth century. A rich history of ideas and a complex relationship with other disciplines, most often geography, provide a strong interdisciplinary base for cartography. The remainder of this chapter will focus on some of the major activities in the United States, where academic thematic cartography grew out of geography departments. But academic thematic cartography also existed in other places, particularly Europe, where, for instance, many key ideas in terrain representation and label placement (Imhof 1975), map symbolization (Bertin 1981, 1983), cartographic design (Keates 1973), and generalization (Eckert 1921) were formulated. It was also the European cartographers who initiated the development of the International Cartographic Association (ICA) in the late 1950s, with the first International Cartographic Conference held in 1961 in Paris. Today, European cartographers continue to provide leadership in many areas of the field, and thematic cartography flourishes in Europe.

2.5.1 Period 1: Early Beginnings

2.5.1.1 John Paul Goode

Although basic training in cartography started as early as 1900 in the United States, "It could be argued that the first genuine American academic cartographer was John Paul Goode at the University of Chicago" (McMaster and Thrower 1987, 346). Goode (Figure 2.5) was one of the most professionally active geographers of his time. During the early part of the twentieth century, Goode became a charter member of the Association of American Geographers (AAG), served as coeditor of the *Journal of Geography* from 1901 through 1904, helped organize the Geographic Society of Chicago, and was appointed by President Taft to assist in leading a U.S. tour for a distinguished group of Japanese financiers. However, Goode is best known for the development of Goode's Homolosine Projection, first presented at the AAG's meetings in 1923, and the development of *Goode's World Atlas*, also published in 1923 and now in its 21st edition. As an example of his teaching, in 1924, Goode taught a course at the University of Chicago entitled "A Course in Graphics," which included four major themes: The Graph, The Picture, Preparation of Illustrative Material for Reproduction, and Processes of Reproduction. It is clear that Goode used this course as a prerequisite to the more traditional cartography class that he taught.

Goode's influence was extended through his students at Chicago. Most of these students did not devote themselves to cartography specifically, but some were able to influence the course of the field through their positions in the private sector, government, and academia. Two of Goode's students at the University of Chicago, Henry Leppard (Universities of Chicago, Washington, and UCLA) and

FIGURE 2.5 J. Paul Goode, who was arguably the first genuine American academic cartographer. (After McMaster and Thrower 1991. First published in *Cartography and Geographic Information Systems* 18(3), p. 151. Reprinted with permission from the Cartography and Geographic Information Society.)

Edward Espenshade (Northwestern University), devoted their careers to cartographic education and continued Goode's work with both base map development and the many generations of the *Goode's World Atlas*, published by Rand McNally.

2.5.1.2 Erwin Raisz

Erwin Raisz (Figure 2.6) was the leading American academic cartographer between the time of Goode's death and the emergence of major graduate programs in Wisconsin, Kansas, and Washington. Born in Hungary in 1893, Raisz immigrated to the United States after World War I. While working on his PhD in geology at Columbia, Raisz studied with Douglas Johnson, a William Morris Davis—trained geomorphologist at Harvard who had strong interests in the construction of block diagrams and the representation of landscapes.[2] As an instructor at Columbia, Raisz offered the first cartography course there.

In 1931, Raisz published "The Physiographic Method of Representing Scenery on Maps" in the *Geographical Review*, where he expressed his own individualized approach to landscape representation (Raisz 1931). Based on this work and on the recommendation of his mentor, Douglas Johnson, Davis himself offered Raisz the position

FIGURE 2.6 Erwin Raisz, who was a leading American academic cartographer between the time of Goode's death and the emergence of major graduate programs in cartography. (After McMaster and Thrower 1991. First published in *Cartography and Geographic Information Systems* 18(3), p. 153. Reprinted with permission from the Cartography and Geographic Information Society.)

of lecturer in cartography at the Institute of Geographical Exploration at Harvard. During the 1930s, Raisz continued to publish and work on his physiographic techniques. In 1938, he published the first edition of the influential book *General Cartography*, which was to remain the only general English textbook on cartography for 15 years (Raisz 1938). The book was part of the geography series published by McGraw-Hill, which at the same time was publishing such classics as Finch and Trewartha's *Physical Elements of Geography*, Platt's *Latin America*, and Whitbeck and Finch's *Economic Geography*. In addition to the predictable emphasis on "representation," the contents of *General Cartography* were what one would expect to find at this time period: projections, lettering, composition, and drafting. A significant part of Raisz's book also covered the history of cartography, with sections on manuscript maps, the renaissance of maps, the reformation of cartography, and American cartography. What is most interesting is the relatively small amount of text that was devoted to "statistical mapping" as we know it today.

Although the geography department at Harvard folded in 1947 (Smith 1987), Raisz remained in the Boston area, teaching at Clark University until 1961, and finished his career at the University of Florida, where he published his second textbook, *Principles of Cartography*, in 1962. What is curious about Raisz is that he never held a regular academic appointment. Thus, he was unable to produce a generation of students who would perpetuate his brand of cartography. Raisz is best known for his production of "landform" maps of various parts of the world. His "Landform Outline Map of the United States" (1954), perhaps one of the best examples of cartography in the twentieth century, has become a standard reference in U.S. geography classes (see Section 23.4.3 for a discussion of this map). Robinson and Sale (1969) asserted that landform maps or physiographic diagrams, such as those created by William Morris Davis, A. K. Lobeck, Guy-Harold Smith, and Raisz, were "one of the more distinctive contributions of American cartography" (187).

2.5.1.3 Guy-Harold Smith

Cartography was taught as early as 1925 at Ohio State University—a class in Map Construction and Interpretation was offered by Fred Carlson—making this one of the oldest cartography courses in the country. In 1927, Guy-Harold Smith (Figure 2.7), a recent PhD recipient from the University of Wisconsin (A. K. Lobeck was his advisor), took over the cartography program and taught for nearly 40 years until his retirement in 1965. Although chair of the department for 29 years, Smith was also a prolific thematic cartographer, producing his famous "Relative Relief Map of Ohio" and "Population Map of Ohio" using graduated spheres (see Figure 18.3). A talented teacher, his best-known student was Arthur Robinson, who created the influential program in cartography at the University of Wisconsin, which we describe later in this chapter.

FIGURE 2.7   Guy-Harold Smith, who was an early thematic cartographer who produced a famous map of graduated spheres. Arthur Robinson was one of Smith's students. (After McMaster and Thrower 1991. First published in *Cartography and Geographic Information Systems* 18(3), p. 153. Reprinted with permission from the Cartography and Geographic Information Society.)

2.5.1.4   Richard Edes Harrison

Richard Edes Harrison (Figure 2.8), who was born in Baltimore in 1901, was the son of Ross Granville Harrison, one of the most distinguished biologists of his time. Although Harrison graduated with a degree in architecture from Yale University in 1930, his interests soon turned to scientific illustration, and in the years after completing his degree, he eventually was attracted to cartography. He drew his first map for *Time* magazine in 1932. This initial exposure to mapping piqued his curiosity, and he soon became a freelance cartographer for *Time* and *Fortune* magazines. In the late 1940s, Harrison flew to Syracuse University once a week to teach a course in cartography (George Jenks was one of his students). Although not formally an educator, Harrison nonetheless influenced the discipline of cartography through his specific technique and intrinsic cartographic abilities. He can also be considered one of the first "popular" cartographers because of his work in media mapping.

2.5.2   Period 2: The Post-War Era and the Building of Core Academic Programs

A seminal event in the evolution of American academic cartography—the first meeting of the Committee on Cartography—was organized by Erwin Raisz on April 6,

FIGURE 2.8   Richard Edes Harrison, who was one of the first "popular" cartographers because of his work in media mapping; George Jenks was one of his students.

1950, at Clark University during the national meetings of the AAG. Based on the results of this important conference, it was clear at the time that cartography was positioning itself in relation to geography and other disciplines even though it was still rather descriptive and atheoretical. Fortunately, a series of academic cartographers with strong interests in more conceptual and theoretical issues emerged during the 1950s and led the development of basic research programs. Although only a sporadic set of institutions had offered courses in cartography before World War II, the demands of the war accelerated the development of cartographic curricula. Over the following decade (1950–1960), three major programs would emerge at the Universities of Wisconsin, Kansas, and Washington.

2.5.2.1   University of Wisconsin

Although Arthur Robinson (Figure 2.9) traced the teaching of cartography at the University of Wisconsin, Madison, back to 1937 (1991, 156), the key beginning occurred when Robinson was hired in 1945 after completing a stint in the Geography Division of the Office of Strategic Services (OSS) in World War II. In his faculty position at Wisconsin, Robinson was responsible for establishing a cartography and map use instructional program that, at the outset, included two basic cartography courses (i.e., introductory and intermediate cartography) as well as an aerial photo interpretation course. Later, other courses were added including Seminar in Cartography, Cartographic Production, and Use and Evaluation of Maps. These were followed by another series of courses in Map Projections and Coordinate

FIGURE 2.9 Arthur Robinson, who was responsible for developing the cartography program at the University of Wisconsin.

Systems, Problems in Cartography, Computer Cartography, History of Map Making, and Cartographic Design. In the late 1960s, the cartography staff increased when Randall Sale became associate director of the University of Wisconsin Cartographic Laboratory.

In a 1979 paper on World War II's influence on cartography, Robinson wrote about these early years:

> In the development of cartography in American academic geography, probably the most notable event prior to World War II was the publication of Erwin Raisz's *General Cartography*, in 1938. By the mid-1930s the majority of graduate students in geography (probably few if any undergraduates) took one course in cartography. Mine, at Wisconsin, came before Raisz's book appeared and our "textbooks" were Deetz and Adams' *Elements of Map Projections* and Loebeck's *Block Diagrams*. (From p. 97 of "Geography and cartography then and now," by A. H. Robinson, *Annals of the Association of American Geographers*, 1979, publisher Taylor & Francis Ltd, http://www.tandfonline.com, reprinted by permission of the publisher.)

Arthur Robinson established himself as the unofficial "Dean" of American academic cartographers and built the cartography program at the University of Wisconsin into the very best in the United States during the 1970s and early 1980s. His seminal volume *The Look of Maps*, based on his doctoral dissertation at Ohio State University, was arguably the seed for three decades of cartographic research.

Robinson established the first American journal in cartography, *The American Cartographer*, in 1974; wrote six volumes of *Elements of Cartography*; and was president of the International Cartographic Association. Robinson also influenced several generations of students who themselves ventured off and established graduate programs in cartography. Robinson and Sale guided the cartography program at Madison until 1968, when Joel Morrison, a Robinson PhD, and later Philip Muehrcke, joined them. Both brought strong mathematical expertise to the program. Thus, in the mid-1970s, when many geography departments were struggling to maintain a cartography program with a single faculty member, Wisconsin had four individuals, separate BS and MS degrees in cartography, and the very best cartography laboratory (within a geography department). It was a cartographic tour de force. The cartography program at Wisconsin has produced many—perhaps more than any other program—master's and doctoral students who have gone on into the private, government, and academic sectors.

2.5.2.2 University of Kansas

The cartography program at the University of Kansas was started and nurtured for more than 35 years by George Jenks (Figure 2.10). Jenks received his PhD in agricultural geography from Syracuse University and also studied with Richard Edes Harrison, the cartographer for *Time* and *Fortune* magazines, at Syracuse. A significant event in

FIGURE 2.10 George Jenks, who was responsible for developing the cartography program at the University of Kansas.

Jenks's career, and for the program itself, was his receiving a grant from the Fund for the Advancement of Science that allowed him to visit all major mapmaking establishments of the federal government as well as a number of quasi-public laboratories in 1951–1952. He incorporated the information he had collected during this grant year into an *Annals of the Association of American Geographers* paper entitled "An Improved Curriculum for Cartographic Training at the College and University Level" and adopted it for the cartography program at Kansas.

The grant enabled Jenks (1953) to identify a series of key problems for cartographers, including: (1) mass-production techniques had to be improved; (2) new inks, papers, and other materials were needed; and (3) additional personnel had to be trained. Jenks pointed out that, at the time, several factors impeded cartographic training: inexperienced instructors, poorly equipped cartographic facilities and map libraries, limited research and limited access to research, and too much emphasis on theory. It is interesting that Jenks, who spent much of his research career building cartographic theory in design, symbolization, and classification, would make such an argument. But he wrote, "too little time and effort has been spent on the practical application of theory. Theorizing on art does not make an artist, knowledge of medical theory does not make a qualified doctor, and talking about maps (and listening to lectures on cartography) does not mean that the student can execute a map" (Jenks 1953, 319).

The grant also led Jenks to propose the following five-course core sequence in cartography:

Course 1. Elementary training in projections, grids, scales, lettering, symbolization, and simple map drafting.

Course 2. The use, availability, and evaluation of maps.

Course 3. Planning, compiling, and constructing small-scale maps, primarily subject-matter maps.

Course 4. Planning, compiling, and constructing large-scale maps, primarily topographic maps.

Course 5. Nontechnical training in the preparation of simple manuscript maps for persons wishing the minimum in the manipulative aspects of cartography.

The importance of Jenks's landmark study cannot be overemphasized. Cartography had emerged from World War II as a true discipline, in part due to the great demand for war-effort maps and mapping. Both those who had been practicing before the war, such as Arthur Robinson and Erwin Raisz, and those who emerged after, such as George Jenks and John Sherman, realized that comprehensive cartographic curricula could be maintained within geography departments. Jenks's study, in parallel with the previously described efforts by Erwin Raisz and the AAG, provided the intellectual infrastructure for those attempting to build cartography programs at universities.

Other significant influences on Jenks's early career were his relationship with John Sherman of the University of Washington and their establishment in the 1960s of the National Science Foundation-funded Summer Institutes for College Teachers. The institutes, nine weeks each in length, were designed to educate college professors in the modern techniques of cartography. Jenks wrote, "We were surprised to find that a number of professors had been assigned arbitrarily by their deans or chairman to teach mapmaking that fall. Moreover, several were going to have to teach without a laboratory, equipment, or supplies. These activities greatly enhanced my teaching and were the basis for numerous changes in our curriculum" (Jenks 1963, 163).

Despite the presence of faculty at Kansas who had interests closely related to cartography—in particular, statistics and remote sensing—at the end of the 1960s, Jenks was still the only cartographer on the staff. Robert Aangeenbrug, who had strong interests in computer cartography and urban cartography and was the director of two of the International Symposium on Computer-Assisted Cartography (Auto-Carto) conferences, joined the Kansas faculty in the 1960s. Thomas Smith, who had arrived in the department as its second hire in 1947, established coursework during the 1970s and 1980s in the history of cartography.

The Kansas program grew rapidly in the 1970s. As explained by Jenks (1991), "George McCleary joined the staff, and with his help we renovated and broadened the offerings in cartography. More emphasis was placed on map design and map production, and new courses at the freshman and sophomore level were added" (164). During this period, Jenks initiated research projects on 3-D maps, eye-movement research, thematic map communication, and geostatistics. By the end of the 1970s, Jenks had turned his attention to cartographic line generalization, and he supervised 10 PhDs, 15 MA candidates, and four postdoctoral cartographers, most of whom pursued research related to thematic cartography and symbolization. Many of these individuals accepted academic appointments, continued the "Jenks school," and made significant contributions to the field of cartography through their teaching and research. Jenks continued to teach and be engaged in research until his retirement at Kansas in 1986.

2.5.2.3 University of Washington

Although the first course at the University of Washington that was formally identified as a "cartography" course had been offered by William Pierson in the Geography Department during the 1937–1938 academic year, John Sherman (Figure 2.11) is primarily considered to have developed the cartography program there. Sherman received his MA from Clark University in 1944 and his PhD from Washington in 1947. Interestingly, unlike Jenks and Robinson, who had both received formal training in cartography, Sherman never had any coursework in cartography.

When Donald Hudson joined the staff at the University of Washington in 1951, he implemented a new program for the Geography Department that concentrated on

FIGURE 2.11 John Sherman discussing lunar modeling with a group of students. Sherman was responsible for developing the cartography program at the University of Washington.

Anglo-America, the Far East, economic geography, and cartography. He asked Sherman to lead the cartography concentration and also invited Henry Leppard, who had studied under J. Paul Goode at the University of Chicago, to assist. As explained by Sherman:

> By 1953 six cartography courses were in place, including Maps and Map Reading, Introductory Cartography, Intermediate Cartography, Techniques in the Social Sciences, Map Reproduction, and Map Intelligence. In 1954, Leppard left for UCLA and in 1958 Willis Heath, having completed his Ph.D. in the department, joined Sherman in carrying on the cartography program. (1991, 169)

One noteworthy event in the early history of the program was Heath and Sherman's participation in the Second International Cartographic Conference at Northwestern University sponsored by Rand McNally, held in June 1958. According to Sherman (1991, 169), a group of some 50 international cartographers was able to discuss "the graphic philosophy, functional analysis, and technological developments that were then influencing the field." Based on discussions at the conference, changes and additions were made to the cartography program at Washington. Another event influencing Sherman, and thus the cartography program at Washington, was the Summer Institute for College Teachers in Cartography that Sherman and Jenks led (see our earlier discussion). Later, in 1968, Sherman developed a proposal to establish a National Institute of Cartography, as requested by the National Academy of Sciences/National Research Council (NAS/NRC) Committee on Geography. A panel of prestigious cartographers, including Arch Gerlach, Norman Thrower, Richard Dahlberg, Waldo Tobler, George McCleary, George Jenks, and Arthur Robinson, assisted Sherman. Unfortunately, the proposed institute was never created.

The program Sherman and Heath put together strongly emphasized both design and production. Personal correspondence with Carlos Hagen (a graduate student at Washington during this period) supports this. Hagen wrote, "One thing that particularly impressed me at that time was the importance that John Sherman and Bill Heath gave to a sort of sacred trilogy, 'Drafting-Printing-Reproduction.' In the Latin American and European traditions, these production techniques are certainly not considered part of the academic environment. They are very respected and much appreciated, but generally you will find them not in academia, but in the realm of a very professional and dedicated tradition of craftsmanship" (personal communication 1987). Whereas Sherman's main research interests were in communication, map design, and tactile mapping, many of his doctoral students pursued dissertation topics related to analytical and computer cartography.

2.5.3 PERIOD 3: GROWTH OF SECONDARY PROGRAMS

During the 1970s and 1980s, a series of what might be called secondary programs, many established by PhDs from the Universities of Wisconsin, Kansas, and Washington, were created in the United States. Although not exhaustive, one can point to programs at UCLA with Norman J. W. Thrower (a Wisconsin PhD), Michigan with Waldo Tobler (a Washington PhD), South Carolina with Ted Steinke and Patricia Gilmartin (Kansas PhDs), SUNY Buffalo with Kurt Brassel (a Zurich PhD) and Duane Marble (a Washington PhD), Michigan State with Richard Groop (a Kansas PhD) and Judy Olson (a Wisconsin PhD), Northern Illinois University with Richard Dahlberg (a Wisconsin PhD), Oregon State University with A. Jon Kimerling (a Wisconsin PhD), Syracuse with Mark Monmonier (a Pennsylvania State PhD), Penn State University with Alan MacEachren (a Kansas PhD), and Ohio State with Harold Moellering (a Michigan PhD). Key activities in these departments included Tobler's development of analytical cartography, Thrower's expertise in the history of cartography and remote sensing, Moellering's animated cartography and emphasis on a numerical cartography, Monmonier's statistical mapping, and Olson's work in cognitive research. Each of the institutions developed its own area of expertise in which, unlike the earlier days when students would pursue a general graduate program in cartography, individual graduate programs were tailored to a student's particular research specialty, such as cognitive or analytical cartography.

This period also witnessed rapid growth in academic cartography in terms of faculty hired, students trained, journals started, and professional societies strengthened. It was in this period that cartography emerged as a true academic subdiscipline, nurtured within academic geography departments that had strong research programs and well-established graduate education programs. The pinnacle of academic cartography in the United States was in the mid-1980s when cartography had reached its maximum growth, but the effect of the emerging discipline of GIScience had not yet been felt.

2.5.4   PERIOD 4: INTEGRATION WITH GIScience

The intellectual landscape of cartography has changed significantly over the past quarter-century, in large part due to the rapid growth of GIS and GIScience. Fifteen years ago, the prognosis for a PhD in cartography being able to acquire an academic position was excellent, whereas today's job market seeks out the geographic information scientist, often with an emphasis in analytics and/or data science. One can certainly still study cartography at most major institutions, but the number of cartographic courses has decreased as the number of GIS-related courses has increased. Additionally, the field of geographic visualization, increasingly used by many departments instead of cartography, has caused a further erosion of the professional base of cartography. However, one hope for the discipline is that as GISs become almost ubiquitous in our society, some researchers are realizing that a deeper knowledge of maps, cartography, and map symbolization and design is still crucial and necessary. Kraak and Ormeling (1996), in their textbook *Cartography: Visualization of Spatial Data*, made the following point with respect to the relationship between GIS and cartography:

> There are conflicting views regarding the relations between cartography and GIS, viz. whether GIS is a technical-analytical subset of cartography, or whether cartography is just a data visualization subset of GIS. (16)

There have been major changes in the way cartography is taught in American universities. Some of the most significant changes include (1) closer integration with GIS education; (2) the nearly complete transition to digital methods; (3) less emphasis on procedural programming (e.g., Fortran and Pascal) and greater emphasis on object-oriented, user interface, and Windows programming; (4) greater emphasis on the dynamic aspects of cartography, including animation and multimedia; and (5) greater use of the Internet for acquiring spatial data (e.g., census data) and for online mapping applications.

2.6   EUROPEAN THEMATIC CARTOGRAPHY

In a fashion similar to the development of U.S. academic cartography, there was much thematic cartography innovation in European countries over the past two centuries. In fact, many of the basic methods for visualizing thematic data emerged from Europe, as was seen earlier. Although we cannot comprehensively review all the programs, we will focus on three developments—the Swiss School that was focused in Zurich, the Experimental Cartographic Unit in Great Britain, and the creation of

the visual variables by the pioneering French cartographer Jacques Bertin.

2.6.1   THE SWISS SCHOOL

The major figure that emerged in the development of Swiss thematic cartography was Eduard Imhof. Imhof's cartographic contributions were wide ranging, but his major contributions were in the area of relief representation. Imhof's passion for terrain was the result of his many experiences in his youth, where he took part in mountain expeditions and refined his expertise in drawing. He received his degree in engineering and was appointed as an assistant at the Federal Institute of Technology (ETH) in Zurich upon receiving his degree. In 1925, he was further appointed as an associate professor in topography at ETH as well as in surveying and map drawing. As Ormeling (1986) writes, "… he founded the cartographic Institute on the top floor of the Geodesy building [at ETH] without much fuss or ceremony, the first academic training and research institute in the world."

Imhof's techniques for relief representation are legendary, and he perhaps made more contributions to this field than any other cartographer in modern history. His primary method, known as the "Swiss Manner," applied hill shading using oblique illumination for the upper left-hand side, along with hypsometric tints. Ormeling writes, "Imhof's aerial perspective hypsometric tints are based upon the experience that in normal vision nearby landscape colours are brighter than those further away" (1986, 4).

Imhof also made major contributions to the areas of cartographic design, where he strived for simplicity and established formalized rules for type placement, which became the standard in the field. Imhof published the textbook *Thematische Kartographie* in 1972, was Editor in Chief of the *Atlas der Schweiz*, and was instrumental in establishing the International Cartographic Association (ICA) in 1959, serving as its first president.

2.6.2   THE BRITISH EXPERIMENTAL CARTOGRAPHIC UNIT

The British Experimental Cartography Unit (ECU), a research unit of the Natural Environment Research Council (NERC), was a unique "think tank" that developed innovative digital cartography in the 1960s and 1970s. The unit was led by the forward-looking David Bickmore, who created the ECU while head of the cartography unit at Clarendon Press in Oxford. These were the very early years in the transition from manual to digital methods in cartography, and the software/hardware issues were challenging. The ECU contracted with the scientific, governmental, and commercial sectors, including the Institute of Oceanographic Sciences, Soil Survey, Royal Mail, and Geological Survey (Rhind, 2008). As Rhind (2008) documents, some of the activities listed in the 1969 ECU report include programs for converting digitizer to global coordinates and changing map projections, automatic line following, analysis of the accuracy of manual digitizing, and pedagogical activities (136). Eventually, as with many of these "experimental" research units, they morphed into other entities or withered due to lack of continued funding—the British ECU transitioned into the Thematic Information Services of NERC.

2.6.3 BERTIN AND FRENCH THEMATIC CARTOGRAPHY

Although there are many individuals who have contributed to the development of French thematic cartography, none has been more influential than Jacques Bertin. Bertin, who studied at the Sorbonne, became founder and director of the Cartographic Laboratory of the École pratique des hautes études (EPHE) in 1954 and later director of studies in 1957. In the 1970s, he became head of research at the Centre national de la recherche scientifique (CNRS).

Bertin is best known for his internationally renowned 1967 book, *The Semiology of Graphics: Diagrams, Networks, Maps*, which laid out a theoretical framework for understanding cartographic data and methods for visualization. In 1983, Bertin's work was translated into English by W. Berg. The basic building blocks of the framework were a set of six qualitative and quantitative visual variables (what Bertin called the retinal variables), including size, value, texture, color, orientation, and shape. Bertin recognized that certain visual variables were intrinsically qualitative in nature, such as shape, while others were quantitative, such as value (lightness or darkness of a symbol). Bertin thus identified the level of organization on the plane where variables are *selective, associative, ordered*, or *quantitative* (1983, 48). Bertin's work is filled with examples on the effective and ineffective use of data and variables in both the mapping and graphing domains. The work by Bertin on visual variables has become a standard for explaining the theory of cartographic visualization; for instance, we use his work as a foundation for our discussion of visual variables in Section 4.5. For his influential work, Bertin received the highest honor from the ICA in 1999, the Carl Mannerfelt Gold Medal.[3]

2.7 THE PARADIGMS OF AMERICAN CARTOGRAPHY

As academically oriented graduate programs emerged in the post–World War II period, basic research in cartography accelerated. Although we could document many research paradigms, some of the more substantial efforts were made in communication models, a theory of symbolization and design, experimental cartography, and analytical cartography.[4] A newer paradigm that has emerged over the past few decades focuses on maps and society, or how the democratization of mapping has enabled a much larger segment of society to create and utilize spatial information and maps.

Parallel to their expanded use has been a fuller critique of how these maps may "empower" certain components of society and thus "disempower" others, an area called critical cartography. The remainder of this chapter focuses on two of these paradigms, analytical cartography and maps and society.

2.7.1 ANALYTICAL CARTOGRAPHY

If any one twentieth-century paradigm within cartography had an "intellectual leader," it is that of Analytical Cartography. Waldo Tobler (Figure 2.12) originated (in the 1960s) and nurtured (in the 1970s and 1980s) the idea of mathematical, transformational, and analytical approaches to cartography. Tobler laid out the agenda for Analytical Cartography in his seminal 1976 paper "Analytical Cartography," published in *The American Cartographer*. This paper, and Tobler's ideas, have had a profound effect on American academic cartography.

What exactly is "analytical cartography"? Kimerling, in his 1989 *Geography in America* review of cartography, described it as "the mathematical concepts and methods underlying cartography, and their application in map production and the solution of geographic problems" (697). Analytical Cartography includes the topics of cartographic data models, digital cartographic data-collection methods and standards, coordinate transformations and map projections, geographic data interpolation, analytical

FIGURE 2.12 Waldo Tobler, who was the developer of the notion of analytical cartography.

generalization, and numerical map analysis and interpretation. Tobler's original syllabus described a series of topics steeped in theory and mathematics, and his goal for the course was futuristic (as well as prescient):

> The spirit of Analytical Cartography is to try to capture this theory, in anticipation of the many technological innovations which can be expected in the future; wrist watch latitude/longitude indicators, for example, and pocket calculators with maps displayed by colored light emitting diodes, do not seem impossible. (1976, 29)

What seemed like futuristic ideas at that time are now part of our daily spatial reality.

Tobler finished his PhD in 1961 at the University of Washington under John Sherman, completing a doctoral dissertation entitled "Map Transformations of Geographic Space." While at Washington, Tobler was influenced not only by some of the faculty's (William Garrison, for instance) strong emphasis on quantification but also by the large number of graduate students who were interested in mathematical geography, including Duane Marble, Arthur Getis, Brian Berry, and John Nystuen. In the early 1960s, the Department of Geography at Washington was a center point for the quantitative revolution in geography. Many of its students had enrolled in J. Ross MacKay's Statistical Cartography course, which MacKay taught in the late 1950s. In a personal interview, Tobler (2001) also discussed the influence of Carlos Hagen, a graduate student at Washington who had arrived from Chile in the late 1950s hoping to pursue graduate work in mathematical cartography. Tobler himself actually had little training in formal mathematics but was self-taught and was intrigued by Hagen's work in projections.

After completing his dissertation at Washington, Tobler joined the faculty at the University of Michigan, where his graduate student colleague from Washington, John Nystuen, had also moved. It was at Michigan that Tobler honed his ideas on analytical cartography, in part assisted by a relatively obscure event in American geography: the meetings of the Michigan Interuniversity Community of Mathematical Geographers (MICMOG). Many of the topics presented at these Detroit-based meetings were strongly cartographic in nature, including Gould's "Mental Maps," Perkal's "Epsilon Filtering," and Tobler's own work on generalization. Tobler's work, which significantly influenced the disciplines of both cartography and geography, led to his election into the prestigious National Academy of Sciences, making him the only cartographer to hold that honor.

What emerged from the concept of "Analytical Cartography" was a cadre of individuals working on problems that can be identified as analytical, computational-digital, and mathematical in nature. Some were Tobler's own PhD students or students who enrolled in his classes, such as Stephen Guptill (USGS), Harold Moellering (Ohio State University), and Phil Muehrcke (Universities of Washington and Wisconsin). Others, immersed in the paradigm without

necessarily having obtained a formal education in it, were Mark Monmonier, the author of the first textbook on computer cartography; Carl Youngmann (a Jenks-educated cartographer at Kansas who joined Sherman at Washington); and Jean-Claude Muller (another Jenks student who worked at the University of Georgia, the University of Alberta, the International Training Center in the Netherlands, and the University of Bochum in Germany). Additionally, a large group of individuals who had been educated in the late 1970s through the early 1980s considered themselves computer or analytical cartographers, including Terry Slocum (PhD, University of Kansas), Keith Clarke (PhD, University of Michigan), Nicholas Chrisman (PhD, Bristol), Timothy Nyerges (PhD, Ohio State University), Marc Armstrong (PhD, University of Illinois), Barbara Buttenfield (PhD, University of Washington), and Robert McMaster (PhD, University of Kansas).

A strong argument can be made that the principles of numerical, analytical, and digital cartography became the "core" of modern GISs. For instance, many of the basic ideas in analytical and computer cartography that had been developed at the Harvard Laboratory for Spatial Analysis and Computer Graphics, including the concept of topological data structures, were directly transferred to modern GISs, such as the Environmental System Research Institute's ARC/Info product (Chrisman 1998).

2.7.2 Maps and Society

In the late 1980s and early 1990s, a series of debates started over the "ethics" and "power" of maps. These debates stemmed, in part, from applying the critical theory of deconstruction to the field of cartography. An inspiring work by Harley (1989) deconstructed the field of cartography and the mapping process, critiquing, for example, what information is purposefully omitted from/included on the map given that the "expert" cartographer often has complete agency in these decisions. In fact, there is a long history of how maps can reflect the cartographer's or government's own biases. For example, Mark Monmonier's *How to Lie with Maps* (2018) documents how maps can be deliberately designed and used to distort reality for specific purposes, and his *Rhumb Lines and Map Wars* (2004) focuses on the societal aspects of the use and misuse of the historically important Mercator map projection (e.g., the use of the Mercator projection by the then Soviet government to misrepresent the size of that country emphasizing its dominance on the world's stage—the Mercator projection exaggerates areas in a poleward direction). Maps can also be used as tools of propaganda. Tyner (1982) discusses the Nazi's use of maps as tools of persuasion to help forward their political and military agendas. This debate quickly expanded into the field of geographic information science/systems and become a subfield called GIS and Society, later critical GIS. One of the core arguments was that what *is not* represented on maps might be just as important as what *is* represented. One of the seminal works on this topic was *Ground Truth*

(1995), edited by John Pickles. Writing on the emergence of the geospatial technologies, Pickles stated, "Together these technologies are now transforming our ways of worldmaking and the ways in which geographers and others think about and visualize the places, regions, environments, and peoples of the earth" (vii).

2.7.2.1 Privacy

A core part of the Maps and Society debate has dealt with the privacy of cartographic information. With the amount, and detail, of geospatial information that is now publicly available and that can be mapped, are we violating the privacy of the citizen? For example, Google continually tracts the location of people's whereabouts through their smartphone apps, and the use of Strava (a popular fitness tracking app) revealed the location of "secret" and "sensitive" U.S. Special Forces military camps in Afghanistan. These and other examples of data availability and privacy continue to be at the forefront of today's technology.[5] This is a similar argument that is applied to the vast quantity of data acquired by credit card companies on each of us. The use of these data allows for targeted geomarketing and profiling of individuals. Of course, the government sector has had access to detailed cartographic information for decades. One example is the census data that provide population information down to the block level in some instances. One can also obtain from government agencies detailed information on individual house prices, tax rates, and other characteristics of the property.

2.7.2.2 Power and Access

With the proliferation of cartographic data, some argue that differential power structures are created. Powerful governmental organizations and private sector companies have access to remarkably sophisticated cartographic data, methods of analysis, and visualization techniques. Alternatively, smaller neighborhood groups and the individual have little access to these technologies to provide alternative visualizations. Examples of where such power differentials are accentuated include insurance redlining (where insurance companies delimit "red zones" often in low income and minority neighborhoods) that create higher insurance rates, or gerrymandering, which involves the purposeful alteration of electoral boundaries to favor certain electoral geographies. Other examples include zoning decisions, eminent domain authority, and transportation planning, where government can utilize differential power in making spatial decisions. At the same time, these power differentials have been somewhat lessoned through the democratization of geospatial technologies. Now smaller grassroots organizations have access to the computer hardware and software, and web-based technologies, that enable them to generate their own analyses and maps.

2.7.2.3 Ethics

One of the major critiques of modern digital mapping has been the ethical use of geospatial information. The example of cartographic ethics most commonly used is around military mapping and the use of cartography in warfare. For hundreds of years, maps have been used in military activity, but recent advances in the accuracy of cartographic databases and mapping technology have enabled remarkable accuracy and precision in weapons usage. Despite this claimed accuracy improvement, wrong coordinates or poor map interpretation can still lead to disastrous effects (e.g., incorrect coordinates on a map resulted in the accidental bombing of the China Embassy[6]). But many other ethical questions arise over the mapping of endangered species, hazardous materials sites, and other sensitive types of data.

2.7.2.4 Public Participation GIS/Mapping

An emerging area of cartography has been the creation of the possibility for "public participation mapping/GIS." As mentioned above, there have been remarkable developments in providing access to both cartographic data and the necessary technologies to enable neighborhood groups, non-profit organizations, and individuals to engage in spatial analysis and mapping. One example of the use of public participation mapping is in the creation of neighborhood inventories. Figure 2.13 depicts one such example of a community-based map documenting the relationship among hazardous materials (Toxic Release Inventory (TRI) sites, Land Recycling sites, Petrofund sites), block-level African American population, and institutional population (schools, day care centers, community centers, and retirement homes) for the Phillips Neighborhood in Minneapolis, Minnesota. The objective was to enable the neighborhood residents to visualize the proximity of these toxic sites to vulnerable populations and institutions. An interactive website enabled Phillips residents to select and map many different types of environmental and demographic variables. Such technologies help to empower smaller organizations to create counter scenarios to those created by larger and better-resourced agencies.[7]

Future developments will likely continue to enhance the ability for the public to engage in the creation of cartographic representations to allow for a more equitable community involvement in decision-making. This represents a significant change from the past, where only those expert cartographers had access to the data and mapping technologies.

2.8 SUMMARY

We have seen that many of our modern cartographic ideas date to the time of the Greeks, who introduced the ideas of the spherical shape of the Earth; the concept of the Equator and the divisions of the Earth into hot, temperate, and frigid zones; and an accurate measurement of the size of the Earth. Although cartography slipped backward during the Middle Ages, it flourished again during the Renaissance with the developments of printing and engraving, which led to the widespread dissemination of cartographic knowledge, the creation of the first globe, and the prolific creation of atlases by the Dutch. The eighteenth and nineteenth centuries saw the completion of the first national surveys

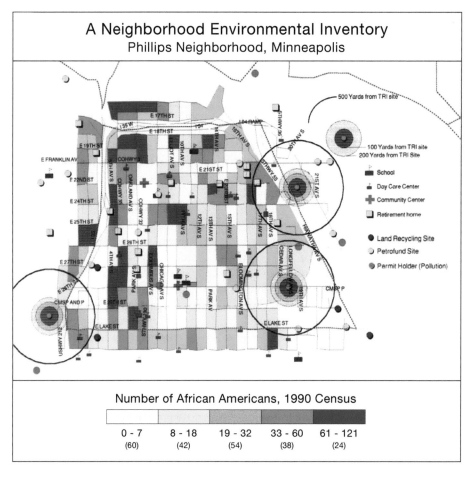

FIGURE 2.13 A community-based map illustrating the relationship among hazardous materials, block-level African American population, and institutional populations. (After McMaster et al. 1997. First published in *Cartography and Geographic Information Systems* 24(3), p. 187. Reprinted with permission from the Cartography and Geographic Information Society.)

of countries and the development of scientific methods for base mapping.

Compared to the history of the broader discipline of cartography, the history of thematic cartography is relatively recent, with major developments not taking place until the nineteenth and twentieth centuries. In the early 1800s, with the compilation of national-level statistics and the evolution of the field of social cartography, the world saw the development of several basic thematic mapping techniques, including the dot, choropleth, and dasymetric methods. Charles Joseph Minard was one of the pioneers of thematic mapping—we'll consider his work in greater depth when we discuss flow mapping in Chapter 21.

The history of U.S. academic cartography can be divided into four periods: the Early Beginnings, the Post-War Era of Building Core Programs, the Growth of Second-Generation Programs, and the Integration of GIScience. In the Early Beginnings period (from the early part of the twentieth century to the 1940s), academic cartography was centered at only two or three institutions and involved individuals who were not necessarily educated in cartography; key figures of this era included John Paul Goode, Erwin Raisz, Guy-Harold Smith, and Richard Edes Harrison. The Post War Era of Building Programs (1940s to 1960s) saw the creation of

three core programs at the University of Wisconsin (Arthur Robinson), the University of Kansas (George Jenks), and the University of Washington (John Sherman). Secondary Growth Programs developed during the third period (1960s to 1980s) included those at UCLA, Michigan, and Syracuse. It was in this period that cartography emerged as a true subdiscipline, nurtured within academic geography departments that had strong research programs and well-established graduate education programs. The integration of cartography with GIScience has been occurring since the 1980s and has raised questions about the direction that cartography will take in the future. While it can be argued that cartographic principles are critical to GIS (Kraak and Ormeling 1996), the number of courses in cartography has decreased as the field of geographic information science/systems—which, many assume, "contains" cartography/representation—has begun to dominate most university curricula.

We can document several research paradigms that came about as graduate programs evolved after World War II, including communication models, a theory of symbolization and design, experimental cartography, analytical cartography, and the recent series of debates on social and ethical issues in cartography. In this chapter, we considered the analytical cartography and maps, and society paradigms. Analytical Cartography involves the mathematical concepts and methods underlying cartography, including topics such as cartographic data models, coordinate transformations and map projections, interpolation, and analytical generalization. As we consider the relationship between Maps and Society, topics such as the "ethics" and "power" of maps, the privacy of cartographic information, and access by the private citizen have emerged. As we move toward a democratization of maps and mapping, public participation mapping has become a reality with the development of easy-to-use mapping software and access to cartographic databases. Such access will allow community and neighborhood groups to create their own maps and enable these groups to produce counter-cartographies based on their own knowledge of their space and environment.

2.9 STUDY QUESTIONS

1. List three types of cartographic symbolization developed in early European thematic cartography?
2. Explain the basic idea of "analytical cartography"?
3. What are the benefits of enabling public access to cartographic databases and mapping technologies?
4. Describe the major cartographic programs that evolved after World War II?
5. Who were two early innovators in American thematic cartography, and what were their major contributions?

NOTES

1 Portions of this chapter were taken from McMaster, R., and McMaster, S. (2002) "A history of twentieth-century American academic cartography." *Cartography and Geographic Information Science 29*, no. 3:305–321. Reprinted with permission from the Cartography and Geographic Information Society and Mark Monmonier.

2 One of Davis's students, A. K. Lobeck, carried on the tradition of using block diagrams to represent landscapes. For an example, see Figure 23.14.

3 For a special journal issue honoring the work of Bertin, see *Cartography and Geographic Information Science*, 46, no. 2, 2019.

4 For a discussion of academic paradigms in cartography, see pp. 1–13 of *Cartography in the Twentieth Century,* ed. by Mark Monmonier, https://press.uchicago.edu/books/HOC/HOC_V6/HOC_VOLUME6_A.pdf.

5 See details of the story here (https://www.theguardian.com/world/2018/jan/28/fitness-tracking-app-gives-away-location-of-secret-us-army-bases). Strava, by default, makes all uploaded data publicly available. The Strava user must manually make the uploaded information "private" to disable public access to the data.

6 The Chinese Embassy in Belgrade was mistakenly bombed after coordinates from an outdated map were used to carry out the mission (https://www.washingtonpost.com/wp-srv/inatl/longterm/balkans/stories/military051099.htm).

7 For a more detailed discussion of this example, see McMaster et al. (1997).

REFERENCES

Bagrow, L., Revised and enlarged by R. A. Skelton (1964) *History of Cartography.* Cambridge, MA: Harvard University Press.

Berghaus, H. (1845, 1848) *Physikalischer Atlas oder Sammlung von Karten.* Gotha: Justus Perthes.

Bertin, J. (1981) *Graphics and Graphic Information-Processing.* Berlin: Walter de Gruyter.

Bertin, J. (1983) *Semiology of Graphics: Diagrams, Networks, Maps.* Madison, WI: University of Wisconsin Press.

Brown, L. A. (1979) *The Story of Maps.* New York: Dover Publications.

Chrisman, N. (1998) "Academic origins of GIS." In *The History of Geographic Information Systems: Perspectives from the Pioneers,* ed. by T. W. Foresman, pp. 33–43. Upper Saddle River, NJ: Prentice Hall.

Eckert, M. (1921) *Die Kartenwissenschaft. 2* volumes. Berlin: Walter de Gruyter & Co.

Harley, J. B. (1989) "Deconstructing the map." *Cartographica 26,* no. 2:1–20.

Harvey, P. D. A. (1980) *The History of Topographical Maps: Symbols, Pictures, and Surveys.* London: Thames and Hudson.

Imhof, E. (1975) "Positioning names on maps." *The American Cartographer 2,* no. 2:128–144.

Jenks, G. F. (1953) "An improved curriculum for cartographic training at the college and university level." *Annals of the Association of American Geographers 43,* no. 4:317–331.

Jenks, G. F. (1963) *Discussion from 1963 Summer Institutes for College Teachers.* Seattle, WA.

Jenks, G. F. (1991) "The history and development of academic cartography at Kansas: 1920–80." *Cartography and Geographic Information Systems 18,* no. 3:161–166.

Keates, J. (1973) *Cartographic Design and Production.* London: Longman.

Kimerling, A. J. (1989) "Cartography." In *Geography in America,* ed. by G. S. Gaile and C. J. Willmott, pp. 686–717. Columbus, OH: Merrill.

Kraak, M.-J., and Ormeling, F. J. (1996) *Cartography: Visualization of Spatial Data.* Essex, England: Addison Wesley Longman.

McMaster, R. B., and Thrower, N. J. W. (1987) "The training of academic cartographers in the United States: Tracing the routes." *Proceedings of the International Cartographic Association,* Morelia, Mexico, pp. 345–359.

McMaster, R. B., Leitner, H. and Sheppard, E. (1997) "GIS-based environmental equity and risk assessment: Methodological problems and prospects." *Cartography and Geographic Information Systems 24,* no. 3:172–189.

Monmonier, M. (2004) *Rhumb Lines and Map Wars: A Social History of the Mercator Projection.* Chicago: University of Chicago Press.

Monmonier, M. (2018) *How to Lie with Maps.* (3rd ed.). Chicago: University of Chicago Press.

Ormeling (1986) Translation of a biography of Imhof that appeared in *Kartografisch Tijdschrift,* https://icaci.org/eduard-imhof-1895-1986/.

Pickles, J. (ed.) (1995) *Ground Truth: The Social Implications of Geographic Information Systems.* New York: The Guilford Press.

Raisz, E. (1931) "The physiographic method of representing scenery on maps." *The Geographical Review 21,* no. 2:297–304.

Raisz, E. (1938) *General Cartography.* New York: McGraw-Hill.

Raisz, E. (1948) *General Cartography* (2nd ed.). New York: McGraw Hill.

Raisz, E. (1962) *Principles of Cartography*. New York: McGraw-Hill.

Rhind, D. W. (2008) "Experimental cartography unit." In *Encyclopedia of Geographic Information Science*, ed. by K. Kemp. Thousand Oaks, CA: SAGE.

Robinson, A. H. (1979) "Geography and cartography then and now." *Annals of the Association of American Geographers 69*, no. 1:97–102.

Robinson, A. H. (1982) *Early Thematic Mapping in the History of Cartography*. Chicago: University of Chicago Press.

Robinson, A. H., and Sale, R. D. (1969) *Elements of Cartography* (3rd ed.). New York: Wiley.

Sherman, J. C. (1991) "The development of cartography at the University of Washington." *Cartography and Geographic Information Systems 18*, no. 3:169–170.

Smith, N. (1987) "Academic war over the field of geography: The elimination of geography at Harvard, 1947–1951." *Annals of the Association of American Geographers 77*, no. 2:155–172.

Thrower, N. J. W. (1969) "Edmund Halley as a thematic geocartographer." *Annals of the Association of American Geographers 59*:652–676.

Thrower, N. J. W. (2008) *Maps and Civilization: Cartography in Culture and Society* (3rd ed.). Chicago: University of Chicago Press.

Tobler, W. R. (1976) "Analytical cartography." *The American Cartographer 3*, no. 1:21–31.

Tooley, R. V. (1987) *Maps and Map-Makers*. London: B.T. Batsford Ltd.

Tyner, J. A. (1982). "Persuasive cartography." *Journal of Geography 81*, no. 4:140–144.

3 Statistical and Graphical Foundation

3.1 INTRODUCTION

The purpose of this chapter is to provide a statistical and graphical foundation for the remainder of the textbook. Although the bulk of the material in this chapter is covered in statistics courses in geography, it is our experience that a review of this material is helpful in understanding material presented later in the textbook.

To illustrate many concepts in this chapter, we will analyze the relationship between the murder rate (number of murders per 100,000 people) and the following attributes for 50 U.S. cities with a population of 100,000 or more in 1990: (1) percentage of families whose income was below the poverty level; (2) percentage of those 25 years and older who were at least high school graduates; (3) the drug arrest rate (number of arrests per 100,000 people); (4) population density (number of people per square mile); and (5) total population (in thousands).[1] The raw data are shown in Table 3.1 (ordered on the basis of murder rate). One problem with these data is that an analysis at the city level might be inappropriate because it fails to account for the variation within a city; it might instead be desirable to look at finer geographic units, such as census tracts, or even at individual murder cases. We chose the city level for analysis, however, because it is easier to relate to individual cities than to individual census tracts. Later in the chapter, we will consider the effect of aggregation of enumeration units on measures of numerical correlation.

3.2 LEARNING OBJECTIVES

- Compute measures of central tendency and dispersion and relate these to graphical displays such as the dispersion graph and histogram.
- Compute a grouped-frequency table and relate it to the construction of the dispersion graph and histogram.
- Explain the concepts of normal and skewed distributions and compute transformations that can convert a skewed distribution into a normal one.
- Contrast two graphical methods for analyzing the relationship between two or more attributes: the scatterplot and the parallel coordinate plot.
- Explain the concepts of correlation and regression and how values of the correlation coefficient can be interpreted.
- Explain the notion of exploratory data analysis and construct and interpret associated stem-and-leaf and box plots.
- Define a geographic center and explain why its computation depends on the scale of the map.

- Explain the concept of spatial autocorrelation and how values of the Moran coefficient are interpreted.
- Compute a complexity face measure for a choropleth map.

3.3 POPULATION AND SAMPLE

In statistics, a **population** is defined as the total set of elements or things that one could study, and a **sample** is the portion of the population that is actually examined (in this book, the number of elements in each will be represented by N and n, respectively). Generally, scientists collect samples because they don't have the time or money to examine the entire population. For example, a geomorphologist studying the effect of wave behavior on beach development would collect data at a series of points along a shoreline rather than examine the entire shoreline.

The data for the 50 U.S. cities shown in Table 3.1 have characteristics of either a population or a sample, depending on one's perspective of the data. Consider first a perspective from the standpoint of the six attributes. Murder rate, drug arrest rate, population, and population density are all based on the entire population of each city (as defined by the census); for example, in the case of murder rate, all murders occurring within a city are considered relative to the entire population of that city. The other attributes, percentage of families below the poverty level and percentage of high school graduates, are based on sampling approximately one of every six housing units (U.S. Bureau of the Census 1994, A-2).

From the perspective of observations, we sampled the 50 cities shown in Table 3.1 from a larger set of 200 cities. Sampling was done in two stages. The first stage involved eliminating cities with political boundaries that extended beyond the limits of where most people live within those cities.[2] This was done because one of the attributes being analyzed, population density, was a function of city area. In the second stage, the remaining cities were ordered on the basis of total population and split into ten classes using Jenks's optimal method (see Chapter 5). A proportional number of cities were randomly sampled from each of the ten classes to obtain a broad range of city sizes that would be representative of those found in the United States.

3.4 DESCRIPTIVE VERSUS INFERENTIAL STATISTICS

Statistical methods can be split into two types: descriptive and inferential. **Descriptive statistics** describe the numerical character of a sample or population. For example, to

DOI: 10.1201/9781003150527-4

TABLE 3.1
Sample Data for 50 U.S. Cities (Sorted on Murder Rate)

City	Murder Rate*	Families below Poverty Level (%)	High School Graduates (%)	Drug Arrest Rate[†]	Population Density[‡]	Total Population (in Thousands)
Irvine, CA	0.0	2.6	95.1	780	2,607	110
Cedar Rapids, IA	0.9	6.6	84.5	110	2,034	109
Overland Park, KS	0.9	1.9	94.1	255	2,007	112
Livonia, MI	1.0	1.7	84.7	665	2,823	101
Lincoln, NE	1.6	6.5	88.3	294	3,033	192
Madison, WI	1.6	6.6	90.6	57	3,311	191
Glendale, CA	1.7	12.3	77.2	452	5,882	180
Allentown, PA	1.9	9.3	69.4	1,078	5,934	105
Tempe, AZ	2.1	7.0	89.9	295	3,590	142
Boise, ID	2.4	6.3	88.6	512	2,726	126
Lakewood, CO	2.4	5.2	88.2	216	3,100	126
Mesa, AZ	3.1	6.9	84.8	223	2,653	288
Pasadena, TX	3.4	11.1	69.8	370	2,727	119
San Jose, CA	4.5	6.5	77.2	1,289	4,568	782
Waterbury, CT	4.6	9.9	66.8	1,326	3,815	109
Springfield, MO	5.0	11.6	77.0	446	2,068	140
Chula Vista, CA	5.2	8.6	75.7	808	4,661	135
St. Paul, MN	6.6	12.4	81.1	260	5,157	272
Arlington, VA	7.0	4.3	87.5	758	6,605	171
Alexandria, VA	7.2	4.7	86.9	834	7,281	111
Portland, OR	7.6	9.7	82.9	1,001	3,508	437
Des Moines, IA	8.3	9.5	81.0	118	2,567	193
Lansing, MI	8.7	16.5	78.3	780	3,755	127
Pittsburgh, PA	9.5	16.6	72.4	723	6,649	370
Yonkers, NY	9.6	9.0	73.6	917	10,403	188
Riverside, CA	9.7	8.4	77.8	1,703	2,916	227
Elizabeth, NJ	10.0	13.7	58.5	929	8,929	110
Berkeley, CA	10.7	9.4	90.3	1,569	9,783	103
Buffalo, NY	11.3	21.7	67.3	580	8,080	328
Raleigh, NC	11.5	7.7	86.6	634	2,360	208
Sacramento, CA	11.7	13.8	76.9	1,555	3,836	369
Tacoma, WA	14.1	12.5	79.3	673	3,677	177
Knoxville, TN	15.2	15.3	70.8	328	2,137	165
Beaumont, TX	16.7	16.6	75.2	693	1,427	114
Winston-Salem, NC	16.8	11.6	77.0	1,343	2,018	143
Montgomery, AL	18.2	14.4	75.7	131	1,386	187
Waco, TX	21.2	19.7	68.4	400	1,367	104
Jackson, MS	22.3	18.0	75.0	693	1,804	197
Savannah, GA	23.9	18.5	70.1	707	2,198	138
Norfolk, VA	24.1	15.1	72.7	624	4,859	261
Los Angeles, CA	28.2	14.9	67.0	1,391	7,426	3,485
Chicago, IL	30.6	18.3	66.0	1,157	12,251	2,784
New York, NY	30.7	16.3	68.3	1,255	23,701	7,323
Houston, TX	34.8	17.2	70.5	555	3,020	1,631
Newark, NJ	40.7	22.8	51.2	1,751	11,554	275
Baltimore, MD	41.4	17.8	60.7	2,063	9,108	736
Gary, IN	55.6	26.4	64.8	261	2,322	117
Detroit, MI	56.6	29.0	62.1	1,052	7,410	1,028
Atlanta, GA	58.6	24.6	69.9	2,330	2,990	394
Washington, DC	77.8	13.3	73.1	1,738	9,883	607

* Murders per 100,000 people.
[†] Arrests per 100,000 people.
[‡] Number of people per square mile.

assess the current president's job performance, you might ask a sample of 500 people, "Is the President doing an acceptable job?" The percentage responding yes, say 52 percent, would be an example of a descriptive statistic. **Inferential statistics** are used to make an inference about a population from a sample. For example, based on the 52 percent figure just given, you might infer that 52 percent of the entire population thinks the President is doing an acceptable job. You would be surprised, of course, if the 52 percent figure truly applied to the population because the figure is based on a sample. To correct this problem, in inferential statistics, it is necessary to compute a *margin of error* (e.g., plus or minus 3 percent) around the sampled value; we often find such errors reported in media coverage of polling. This book primarily focuses on descriptive statistics, although we touch on inferential statistics when we discuss visualizing uncertainty in Chapter 27.

3.5 ANALYZING THE DISTRIBUTION OF INDIVIDUAL ATTRIBUTES

This section focuses on three basic approaches that have been used to analyze the distribution of individual attributes along the number line: tables, graphs, and numerical summaries. Tables and graphs provide a visualization of the data, whereas numerical summaries provide a numerical expression that we hope correlates with our visualization of the data.

3.5.1 TABLES

3.5.1.1 Raw Table

The simplest form of tabular display is the **raw table** in which the data for an attribute of interest are sorted from lowest to the highest value, as for the murder rate data in Table 3.1. Tabular displays are useful for providing specific information about observations (e.g., Buffalo, NY, had a murder rate of 11.3 per 100,000 people in 1990), but they can provide additional information if they are examined carefully. Note, for example, that the sorted values provide the minimum and maximum of the murder rate data (0.0 for Irvine, California, and 77.8 for Washington, DC). With some mental arithmetic, the **range** can be calculated by simply subtracting the minimum from the maximum ($77.8 - 0.0 = 77.8$ in this case).

Raw tables can also reveal any duplicate values and outliers in the data that might have special significance. For murder rate, duplicate values (e.g., 0.9 for Cedar Rapids, Iowa, and Overland Park, Kansas) are unimportant, as they are a function of the number of significant digits reported. For some attributes, however, duplicates can be quite meaningful. For example, an examination of the distribution of Major League Baseball player salaries would reveal numerous duplicates because several players on each team typically earn exactly the same amount, the minimum required salary for all of Major League Baseball. **Outliers**—values

that are quite unusual or atypical—can also be seen.[3] For murder rate, no value is considerably different from the rest (although Washington, DC, is clearly larger), but for the total population attribute, note that several cities are quite large, with New York more than twice the value of any other city in the sample.

Although raw tables are useful for providing specific information, they are not very good at providing an overview of how data are distributed along the number line. For example, in studying Table 3.1, note that roughly half the cities have murder rates below 10.0, but it is difficult to develop a feel for what the overall murder rate distribution really looks like. Grouped-frequency tables are more useful for this purpose.

3.5.1.2 Grouped-Frequency Table

To construct **grouped-frequency tables**, we divide the data range into equal intervals and then tally the number of observations that fall in each interval. Although grouped-frequency tables generally are constructed using software, it is useful to consider the actual steps involved so that you have a clear understanding of what the resulting table reveals. (We will use a similar approach in Chapter 5 to create equal interval classes for choropleth maps.)

Step 1. Decide how many groups (or classes) you want to use. Although you might want to experiment with various numbers of classes, we will use 15 classes.

Step 2. Determine the width of each class (the class interval). The class interval is computed by dividing the range of the data by the number of classes. The following would be the computation for the murder rate data using 15 classes:

$$\frac{Range}{Number\ of\ classes} = \frac{High - Low}{Number\ of\ classes}$$

$$= \frac{77.8 - 0.0}{15} = 5.187.$$

Note that we use more decimal places for the class interval than appear in the actual data to avoid rounding errors that could cause the last class not to match the highest value in the data.

Step 3. Determine the upper limit of each class. The upper limit for each class is computed by repeatedly adding the class interval to the lowest value in the data. The result is on the right-hand side of the Limits column in Table 3.2. Note that the highest calculated limit does not match the highest data value (77.8) exactly because the class interval is not a simple fraction; a more precise class interval would be 5.186666667, but even with this interval, the match would not be exact (the highest value would be 77.80000001).

Step 4. Determine the lower limit of each class. Lower limits for each class are specified so that

TABLE 3.2
Grouped-Frequency Table for the Murder Rate Data

Class	Limits	Number of Observations in Class	Percent in Class	Cumulative Percent
1	0.000–5.187	16	32	32
2	5.188–10.374	11	22	54
3	10.375–15.561	6	12	66
4	15.562–20.748	3	6	72
5	20.749–25.935	4	8	80
6	25.936–31.122	3	6	86
7	31.123–36.309	1	2	88
8	36.310–41.496	2	4	92
9	41.497–46.683	0	0	92
10	46.684–51.870	0	0	92
11	51.871–57.057	2	4	96
12	57.058–62.244	1	2	98
13	62.245–67.431	0	0	98
14	67.432–72.618	0	0	98
15	72.619–77.805	1	2	100

they are just above the highest value in a lower-valued class. This is done so that any observation will fall in only one class. For example, the lowest value in the second class is 5.188, which is 0.001 larger than 5.187, the highest value in class 1.

Step 5. Tally the number of observations falling in each class. These numbers are shown as Number of Observations in Class in Table 3.2. Also shown in the table are the percent and cumulative percent in each class.

In comparison to the raw table, the grouped-frequency table provides a somewhat better overview of the data. For example, Table 3.2 notes that 80 percent of the cities fall in the five lowest murder rate classes. Both raw and grouped-frequency tables, however, do not take full advantage of our visual processing powers; for this, we turn to graphs.

3.5.2 GRAPHS

3.5.2.1 Point and Dispersion Graphs

In a **point graph** or **one-dimensional scatterplot**, each data value is represented by a small point symbol plotted along the number line (Cleveland 1994, 133; in Figure 3.1A, an open circle is used). For the murder rate attribute, the point graph shows that the data are concentrated at the lower end of the distribution (based on the graph, under a rate of about 12). One obvious problem with a point graph is that individual symbols might overlap, thus making the distribution

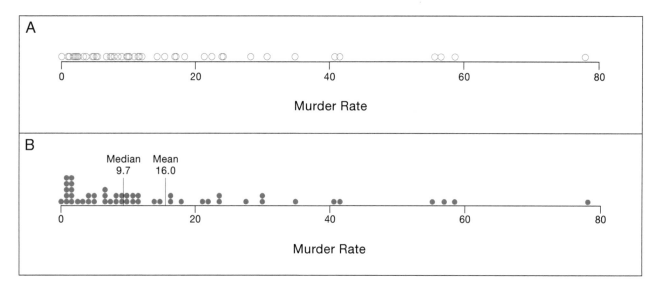

FIGURE 3.1 Point (A) and dispersion (B) graphs for the murder rate data presented in Table 3.1. Note the locations of the median and mean in (B).

difficult to interpret (note the "smearing" in the left-hand portion of Figure 3.1A); duplicate values, in particular, cannot be detected at all by this method. The problem of smearing can be ameliorated to some extent by varying the opacity of each symbol (e.g., we might make each dot 50 percent transparent), but the distribution will still be difficult to interpret.

An alternative to the point graph is the **dispersion graph** (Hammond and McCullagh 1978), in which data are grouped into classes, the number of values falling in each class is tallied, and dots are stacked at each class position (Figure 3.1B). Intervals for classes are defined in a fashion identical to the grouped-frequency table, except that a large number of classes are used. For example, 99 classes were used for Figure 3.1B. The dispersion graph for the murder rate data portrays a distribution similar to the point graph, except that potential confusion in the overlapping areas is eliminated.

Closely allied with the dispersion graph is the **dot plot**, in which dots are also stacked, but the placement of a stack is determined by the raw data values rather than by classes resulting from the grouped frequency table. The basic idea in creating a dot plot is to start at the lowest value in the data set and stack dots at that location as long as a data value is within one dot's width of the lowest data value. When this step is completed, a new stack is started at the location of the data value not within one dot's width of the lowest data value. This process is repeated until all data values have been considered. Leland Wilkinson (1999) developed a widely used algorithm that accomplishes this in an elegant fashion. The advantage of the resulting dot plot (compared to the dispersion graph) is that the dots will be positioned closer to where data values actually occur. As a result, the spacing of the dot stacks may not appear as uniform as they do in a dispersion graph. Since the resulting dot plots look similar to the dispersion graph, we use the term "dispersion graph" to describe any stacked set of dots in this book.

3.5.2.2   Histogram

A **histogram** is constructed in a manner analogous to a dispersion graph, except that fewer classes are generally used, and bars of varying height are used to represent the number of values in each class. One popular approach for determining an appropriate number of bars, k, is to use Sturges rule:

$$k = 1 + (3.3 * \log n) = 1 + (3.3 * \log 50) = 6.6.$$

Here we have shown the result for the murder rate data. Rounding 6.6 to 7 would indicate that 7 bars would be appropriate, which is appreciably less than the 15 we have used. Michael Correll and colleagues (2019), however, show that the Sturges rule can hide important characteristics in the data and thus recommend using roughly twice the number of bars, which supports our use of 15 in this case.

Because the histogram is more commonly used than either the point or dispersion graph, all of the attributes shown in Table 3.1 are graphed using the histogram in Figure 3.2. Looking first at the murder rate histogram, note that the up-and-down nature of the dispersion graph has been smoothed out; there is clearly a peak in the graph on the left, with a decreasing height in bars as one moves to the right.

Histograms are often compared with a hypothetical **normal** (or bell-shaped) **distribution**. For a normal distribution, most of the observations fall near the mean (in the middle of the distribution), with fewer observations in the tails of the distribution (Figure 3.3). Curves representing a normal distribution have been overlaid on the histograms shown in Figure 3.2. Distributions lacking the symmetry of the normal distribution are termed **skewed**. *Positively skewed* distributions have the tallest bars concentrated on the left-hand side (as for total population, murder rate, and population density), whereas *negatively skewed* distributions have the tallest bars on the right-hand side. (There is no distinctive example of a negatively skewed distribution in Figure 3.2, although the percent with high school education has a slight negative skew.)

Because many inferential tests require a normal distribution, raw data are often transformed to make them approximately normal. Transformation involves applying the same mathematical operation to each data value of an attribute; for example, we might compute the $\log_{10}$ of each murder rate value, or alternatively, we could compute the square root of each murder rate value. $\log_{10}$ and square root transformations are commonly used to convert a positively skewed distribution to a normal distribution, with the $\log_{10}$ used for a more severe skew; to illustrate, Figure 3.4 portrays such transformations for the murder and drug arrest data, respectively.

3.5.3   Numerical Summaries

Although tables and graphs are useful for analyzing data, they are prone to differing subjective interpretations, and the limited space of formal printed publications often limits their use. As an alternative, statisticians frequently use numerical summaries, which typically are split into two broad categories: measures of central tendency and measures of dispersion.

3.5.3.1   Measures of Central Tendency

Measures of central tendency are used to indicate a value around which the data are most likely concentrated. Three measures of central tendency are commonly recognized: mode, median, and mean. The **mode** is the most frequently occurring value and is thus generally useful for only nominal data, such as on a land use/land cover map. The **median** is the middle value in an ordered set of data or, alternatively, the 50th percentile because 50 percent of the data are below it. For the murder rate data, the median is 9.7. Note its location in Figure 3.1B. The **mean** is often referred to

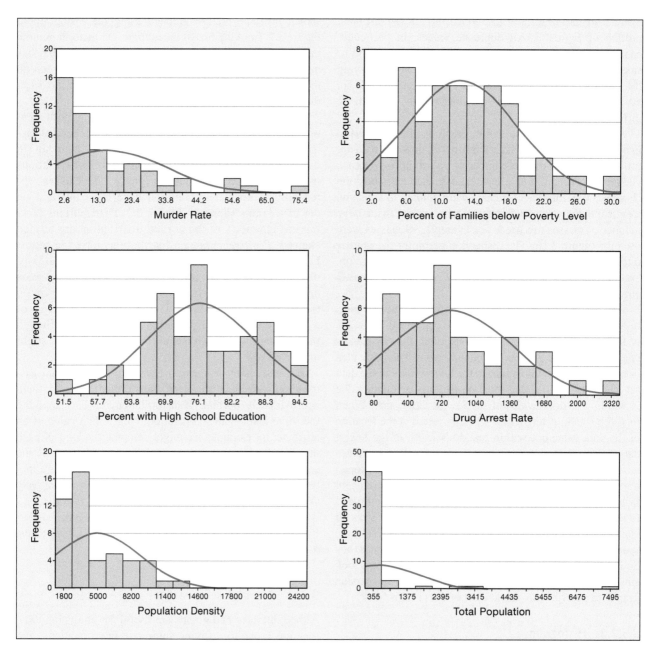

FIGURE 3.2   Histograms for the data presented in Table 3.1.

as the "average" of the data and is calculated by summing all values and dividing by the number of values. Separate formulas are used to distinguish mean values for the sample and population, as follows:[4]

$$\text{Sample}: \bar{X} = \frac{\sum_{i=1}^{n} X_i}{n}$$

$$\text{Population}: \mu = \frac{\sum_{i=1}^{N} X_i}{N},$$

where the X_i are individual data values. Because the data in Table 3.1 are a sample from 200 cities, the sample formula is appropriate in this case. The mean for the murder rate data is 16.0. Note its location in Figure 3.1B. One problem with the mean is that either skew or outliers in the data affect it, whereas the median is resistant to these characteristics. We can see this in Figure 3.1B, in which the median falls where most of the data are concentrated, whereas the mean is pulled to the right by the positive skew.

3.5.3.2   Measures of Dispersion

Measures of dispersion provide an indication of how data are spread along the number line. The simplest measure is

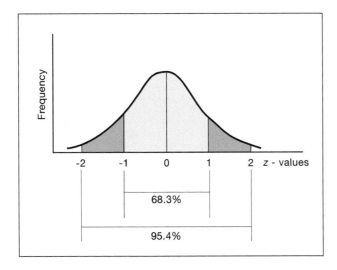

FIGURE 3.3 An example of a normal curve. Histograms will approximate this shape if the data are normal. For a perfectly normal data set, approximately 68 percent and 95 percent of the observations will fall within 1 and 2 standard deviations, respectively, of the mean.

the range, which was defined earlier as the maximum minus the minimum. Obviously, the range is of limited usefulness because it is based on only two values, the maximum and the minimum of the data.

More useful measures of dispersion are the interquartile range and standard deviation, which should be used with the median and mean, respectively. The **interquartile range** is the absolute difference between the 75th and 25th percentiles, or where the middle 50 percent of the data lie. For the murder rate data, the result is $|22.700 - 3.325| = 19.4$. An important characteristic of the interquartile range is that it, like the median, is unaffected by outliers in the data. For example, if the highest murder rate were replaced by a value of 150, the interquartile range would still be 19.4.

In a fashion similar to the mean, separate formulas normally are provided for the sample and population **standard deviation**:

$$\text{Sample}: s = \sqrt{\frac{\sum_{i=1}^{n}\left(X_i - \bar{X}\right)^2}{n-1}}$$

$$\text{Population}: \sigma = \sqrt{\frac{\sum_{i=1}^{N}\left(X_i - \mu\right)^2}{N}}$$

In comparing these formulas, note that they differ principally in the denominator: a value of 1 is subtracted in the sample case but not in the population case. Subtracting a value of 1 is necessary because a sample estimate using just n tends to underestimate the population value (Burt et al. 2009, 113). Using the sample formula, the standard deviation for the murder rate data is 17.5. In contrast to the interquartile range, outliers in the data do affect the standard deviation. To illustrate, replacing the highest murder rate by a value of 150 results in a standard deviation of 24.4, an increase of nearly 40 percent.

3.6 ANALYZING THE RELATIONSHIP BETWEEN TWO OR MORE ATTRIBUTES

As with individual attributes, there are three basic approaches that have been used to analyze the relationship of two more attributes: tables, graphs, and numerical summaries.

3.6.1 TABLES

In the previous section, we saw that considerable information could be derived from raw tables when examining an individual attribute. When trying to relate two attributes,

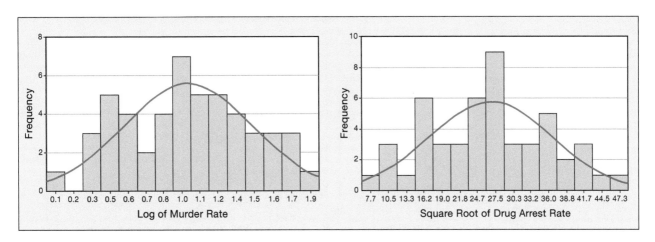

FIGURE 3.4 Histograms of two transformed attributes: log of murder rate and square root of drug arrest rate (compare with corresponding histograms in Figure 3.2).

TABLE 3.3

Tabular Relationship between Murder Rate and Percent of Families Below Poverty Level for Cities in Table 3.1

	Percent Families Below Poverty Level				
Murder Rate	1.70–7.16	7.17–12.62	12.63–18.08	18.09–23.54	23.55–29.00
62.25–77.80			1		
46.69–62.24					3
31.13–46.68			2	1	
15.57–31.12		1	6	3	
0.00–15.56	13	14	5	1	

however, this task becomes difficult. To convince yourself of this, try using Table 3.1 to relate murder rate to percent of families below the poverty level. In general, the attributes appear related (low poverty values are associated with low murder rates, and high poverty values are associated with high murder rates), but it is difficult to summarize the relationship. To simplify the process, we can class both attributes using the grouped-frequency method and create a matrix of the result (Table 3.3). The same general relation between the attributes is still apparent, but now it is more easily seen; also, the matrix reveals that Washington, DC (in the highest murder rate class), does not fit the general trend of the data, which extends from the lower left to the upper right of the table.

3.6.2 GRAPHS

A **scatterplot** is commonly used to examine the relationship of attributes against one another in two-dimensional space. To illustrate, Figure 3.5 portrays scatterplots of murder rate and log of murder rate against percent of families below poverty level (also shown are best-fit regression lines, which we will consider shortly). On scatterplots, *dependent* and *independent* attributes normally are plotted on the *y* and *x* axes, respectively. Since we might hypothesize that murder rate depends, in part, on

poverty, murder rate has been plotted on the *y* axis as the dependent variable.

In examining Figure 3.5A, note the similarity of the distribution of points to the pattern of cells in Table 3.3. Also, note that after transforming murder rate (Figure 3.5B), Washington, DC is not quite so different from the rest of the data; thus, data transformations affect not only the values for individual attributes but also the relationship between attributes.

When the number of observations is large, scatterplots can become difficult to interpret, just as the smearing of dots made the point graph of murder rate hard to interpret. One solution to this problem is the **hexagon bin plot**, which is shown in the right-hand portion of Figure 3.6. Such a plot is created by laying a grid of hexagons onto a scatterplot (see the left-hand portion of Figure 3.6) and then filling each hexagon grid cell with a solid hexagon that is proportional to the number of dots falling in each cell.[5]

To examine the relationship among three attributes simultaneously, the basic scatterplot can be extended to a **three-dimensional scatterplot** by specifying *x, y,* and *z* axes. Many statistical packages have options for creating such plots and even permit rotating them interactively. Our experience is that these plots are difficult to interpret; moreover, they cannot be extended to handle more than three attributes.

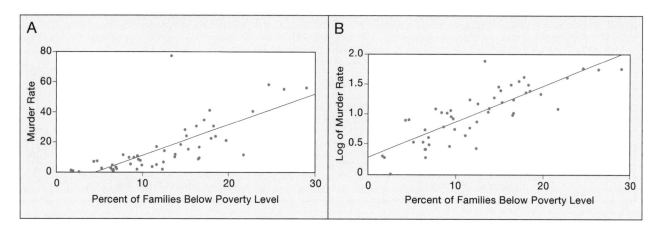

FIGURE 3.5 Scatterplots of (A) murder rate against percent of families below poverty level and (B) log of murder rate against the same attribute. Also shown are best-fit regression lines.

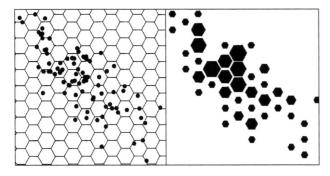

FIGURE 3.6 Hexagon bin plot: an alternative to the scatterplot. The size of the hexagons shown on the right is proportional to the number of dots falling within the hexagons on the left. The x and y axes represent sulfate and nitrate deposition values for sites in the eastern United States. (After Carr et al. 1992. First published in *Cartography and Geographic Information Systems* 19(4), p. 229. Reprinted with permission from the Cartography and Geographic Information Society.)

One method that can handle three or more attributes is the **scatterplot matrix** (Figure 3.7). This graph might at first appear rather complex, but the principles underlying it are quite simple. Note first that the relation between any two attributes is displayed twice, once above the diagonal of attribute names and once below it. This approach enables each attribute to be shown as either an independent or dependent attribute in relation to other attributes. By scanning a row, you can see what happens when an attribute is considered dependent, and by scanning a column, you can see what happens when it is considered independent. If our focus were on log of murder rate as a potential dependent attribute, we would want to examine the first row, where we

see that the strongest relationship appears to be between log of murder rate and percent below poverty level.

A second approach for displaying three or more attributes simultaneously is the **parallel coordinate plot**. In this graph, attributes are depicted by a set of parallel axes, and observations are depicted as a series of connected line segments passing through the axes. To illustrate, Figure 3.8 portrays a parallel coordinate plot for a subset of 10 of the 50 cities shown in Table 3.1. The six attributes in Table 3.1 are represented by six parallel vertical lines, whereas the ten observations are represented by horizontal line segments. Figure 3.8A displays all ten observations, and Figure 3.8B highlights Washington, DC. At the top of the figures are correlation coefficients, or r values, which are described in the following section.

3.6.3 Numerical Summaries

Bivariate correlation-regression is the most widely used approach for summarizing the relationship between two numeric attributes. **Bivariate correlation** is used to summarize the character of the relationship, and **bivariate regression** provides an equation for a best-fit line passing through the data when shown in a scatterplot.

3.6.3.1 Bivariate Correlation

One value commonly computed in bivariate correlation is the **correlation coefficient** or r:

$$r = \frac{\sum_{i=1}^{n}(X_i - \bar{X})(Y - \bar{Y})}{\sqrt{\sum_{i=1}^{n}(X_i - \bar{X})^2}\sqrt{\sum_{i=1}^{n}(Y - \bar{Y})^2}}$$

Table 3.4 shows r values for the transformed sample data. (Note that the table is symmetric about a diagonal extending from upper left to lower right; this occurs because the correlation between attributes A and B is identical to that between attributes B and A.)

Extreme values for r range from -1 to $+1$. A positive r value indicates a *positive relationship* in which increasing values on one attribute are associated with increasing values on another attribute; for example, the r value for log of murder rate and percent below poverty level is 0.82. Conversely, a negative r value indicates a *negative relationship* in which increasing values on one attribute are associated with decreasing values on another attribute; for example, log of murder rate and percent with high school education have a correlation of -0.70. Note that as r values approach -1 or $+1$, the points cluster more tightly about the best-fit line. (Compare Table 3.4 and Figure 3.7, and note that the graph of log of murder rate and percent below poverty level is more tightly clustered than the one for log of murder rate and square root of drug arrest.)

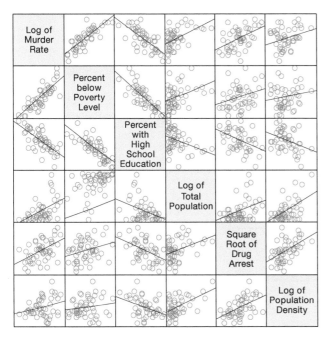

FIGURE 3.7 A scatterplot matrix for the data presented in Table 3.1. Also shown are best-fit regression lines.

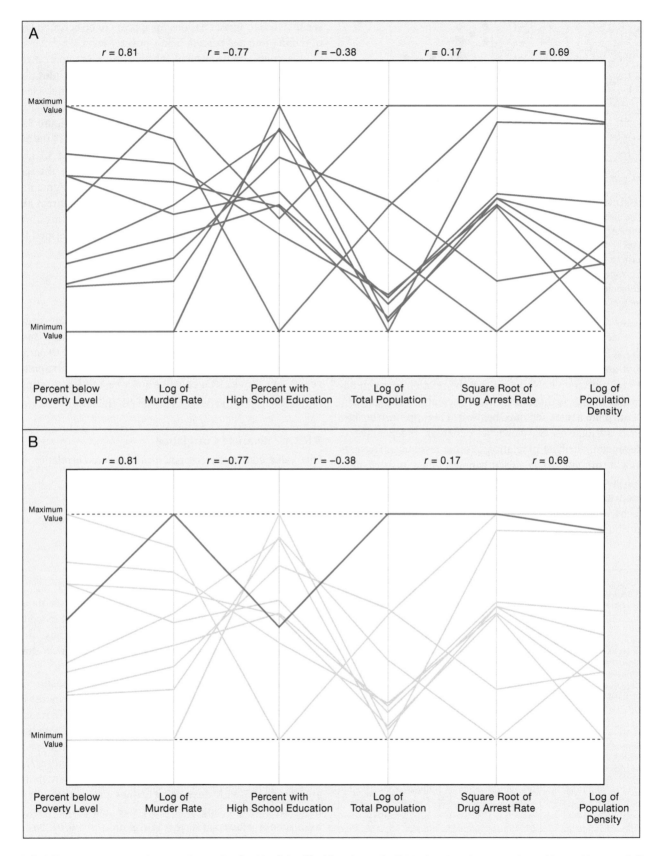

FIGURE 3.8 (A) A parallel coordinate plot for 10 of the 50 cities shown in Table 3.1; (B) the same plot, with Washington, DC, highlighted.

TABLE 3.4

Matrix of Correlation Coefficients for the Transformed City Data

	Log of Murder Rate	Percent Below Poverty Level	Percent with High School Education	Log of Total Population	Square Root of Drug Arrest Rate	Log of Population Density
Log of murder rate	1.00	0.82	−0.70	0.52	0.49	0.27
Percent below poverty level	0.82	1.00	−0.79	0.38	0.29	0.17
Percent with high school education	−0.70	−0.79	1.00	−0.39	−0.44	−0.36
Log of total population	0.52	0.38	−0.39	1.00	0.39	0.54
Square root of drug arrest rate	0.49	0.29	−0.44	0.39	1.00	0.52
Log of population density	0.27	0.17	−0.36	0.54	0.52	1.00

Now that we have introduced the correlation coefficient, we can discuss the meaning of the r values commonly shown in the parallel coordinate plot. To illustrate, Figure 3.9 portrays characteristic shapes of the parallel coordinate plot for particular values of r: parallel line segments indicate an r value of 1; line segments intersecting one another in the middle of the graph indicate an r value of −1; and line segments crossing one another at different angles suggest an r value close to 0. With this information as a basis, we can see whether the r values reported in Figure 3.8 make sense. The highly positively correlated percent below poverty level and log of murder rate ($r = 0.81$ for the 10 cities) have relatively parallel line segments (Washington, DC, is a notable exception). In contrast, line segments for the highly

negatively correlated log of murder rate and percent with high school education ($r = −0.77$ for the 10 cities) tend to cross one another near the middle of the graph. Obviously, the appearance of the parallel coordinate plot will change depending on the order in which the axes are plotted. Ideally, a successful interpretation of the plot requires an interactive program that can easily change the order of the attributes and highlight selected observations. In Chapter 25, we will consider data exploration software that has this sort of capability.

3.6.3.2 Bivariate Regression

In general, the equation for any straight line is $Y_i = a + bX_i$, where X_i, Y_i is a point on the line, a is the y intercept, and b

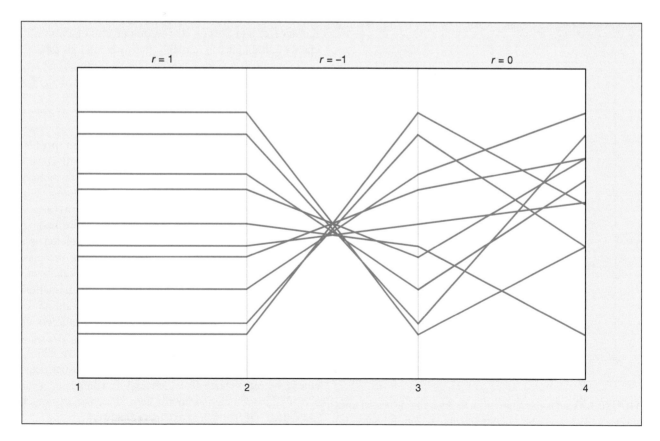

FIGURE 3.9 Characteristic parallel coordinate plots for particular values of r.

is the slope. In regression analysis, X_i and Y_i are raw values for the independent and dependent attributes, respectively, and the line of best fit is defined as $\hat{Y}_i = a + bX_i$, where $\hat{Y}_i$ is a predicted value for the dependent attribute. The best-fit line is found by minimizing the differences between the actual and predicted values, the vertical distances shown in Figure 3.10. The values of a and b that minimize the differences are:

$$a = \bar{Y} - b\bar{X}$$

$$b = \frac{n\sum_{i=1}^{n} X_i Y_i - \sum_{i=1}^{n} X_i \sum_{i=1}^{n} Y_i}{n\sum_{i=1}^{n} X_i^2 - \left(\sum_{i=1}^{n} X_i\right)^2}$$

where $\bar{X}$ and $\bar{Y}$ are the mean of the X and Y values, respectively. The best-fit lines shown in Figures 3.5 and 3.7 were derived using this approach.

3.6.3.3 Reduced Major-Axis Approach

When one does not wish to specify a dependent attribute (as when relating murder rate and race), it makes sense to minimize the vertical and horizontal distances shown in Figure 3.10 simultaneously, effectively minimizing the area of the triangles; this is known as the **reduced major-axis approach** (Davis 2002, 214–218). The equations for this approach turn out to be simpler than those for standard regression. The slope is just the ratio of the standard

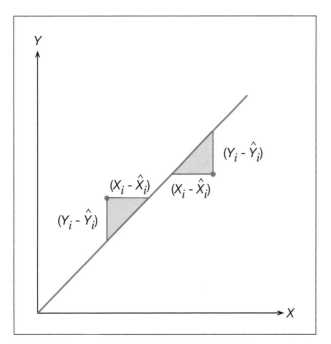

FIGURE 3.10 Possible approaches for determining best-fit lines for a set of data include (1) minimizing vertical distances, (2) minimizing horizontal distances, and (3) minimizing the area of the resulting triangles.

deviations of the attributes, or s_Y/s_X, and the intercept is still $\bar{Y} = b\bar{X}$. In Chapter 22, we will consider how the reduced major-axis approach can be used to create a logical bivariate map.

3.6.3.4 Multiple Regression and Other Multivariate Techniques

When the concept of bivariate correlation-regression is extended into the multivariate realm, it is termed **multiple regression**. In multiple regression, there is still a single dependent attribute, but multiple independent attributes are possible. In a fashion similar to the preceding, it is possible to perform correlation and regression analyses that summarize the relationship between the dependent and independent attributes. For example, we might attempt to explain the murder rate as a function of percent below poverty level, percent with high school education, and the drug arrest rate. A discussion of multiple regression and other multivariate techniques, such as **principal component analysis**, is beyond the scope of this book. You should consult statistical textbooks such as Davis (2002) and Rogerson (2020) for related information.

3.6.3.5 Considerations in Using Correlation-Regression

There are several things that you should consider carefully when using bivariate correlation-regression and multiple regression. One is that high correlations do not necessarily imply a causal relationship. The high correlation between murder rate and poverty that we found could be a result of chance (although it is unlikely that two random data sets could result in a correlation of this magnitude); or some other attribute or set of attributes could be influencing both of these attributes.

A second consideration is that the magnitude of r can be affected by the level at which data have been aggregated, which is known as the **modifiable areal unit problem** (Clark and Hosking 1986; Barrett 1994). Generally, coarser levels of aggregation (e.g., analyzing at the city level rather than at the census tract level) will lead to higher r values because "aggregation reduces the between unit variation in an attribute, making the attribute seem more homogeneous" (Clark and Hosking, 1986, 405). A related issue is that the magnitude of r might be a function of the size and arrangement of enumeration units when the number of enumeration units remains constant (Fotheringham 1998). A solution to the aggregation problem is to examine data at the individual level (looking at individual murders and collecting data regarding the people involved). An example was The Project on Human Development in Chicago Neighborhoods (https://scholar.harvard.edu/sampson/content/chicago-project-phdcn-0), which analyzed data from a variety of Chicago neighborhoods; here, the emphasis was on individual interviews and extensive fieldwork.

A third consideration is that specialized regression techniques have been developed to handle the fact that geographical data tend to be spatially autocorrelated, meaning that like values tend to be located near one another (e.g., a high-income census tract will tend to occur near another high-income census tract). For a discussion of these specialized techniques, see Griffith (1993) and Rogerson (2020). GeoDa (https://spatial.uchicago.edu/geoda) and SpaceStat (https://www.biomedware.com/software/spacestat/) are software packages that have been developed explicitly for handling such techniques. In Chapter 25, we'll discuss GeoDa in the context of data exploration.

Finally, we need to recognize that regression techniques can be applied both globally and locally. By globally, we mean that a regression equation is applied uniformly throughout a geographic region. In contrast, local regression techniques consider the notion that a single model might not be appropriate for an entire region, as local variation might necessitate different models within subregions. As with modifications due to spatial autocorrelation, a full discussion of this issue is beyond the scope of this book. For an overview of the local–global issue applied to regression and other statistical methods, see Fotheringham (1997); a detailed discussion of the related technique of *geographically weighted regression (GWR)* can be found in Fotheringham et al. (2002).

3.7 EXPLORATORY DATA ANALYSIS

One of the most important advances in statistical analysis was John Tukey's (1977) development of **exploratory data analysis (EDA)**. In Chapter 1, we suggested that rather than trying to make one "best" map, interactive graphics systems should provide multiple representations of a spatial data set. In much the same way, Tukey proposed that rather than trying to fit statistical data to standard forms (normal, Poisson, binomial), data should be explored, much as a detective investigates a crime. In the process of exploring data, the purpose should not be to confirm what one already suspects but rather to develop new questions or hypotheses.

One technique representative of Tukey's approach is the **stem-and-leaf plot**, which is depicted in Figure 3.11 using the murder rate data. To construct a stem-and-leaf plot, one first separates the digits of the data values into three classes: sorting digits, display digits, and digits not displayed because of rounding.[6] For the murder rate data, we chose the 10s place as a sorting digit, the 1s place as the display digit, and did not display the 10ths place because we rounded to the nearest whole percent. Sorting digits are placed to the left of the vertical line shown in Figure 3.11 and are known as *stems,* whereas display digits are placed to the right and are known as *leaves.* For Figure 3.11, Tukey's conventional system of asterisks and dots was used to split the 10s place into two parts (leaf values of 0 to 4 are plotted on one row and leaf values of 5 to 9 on the next row). For example, Norfolk's murder rate of 24.1 appears as the fourth leaf ("4") in the fifth row ("2*"), and Los Angeles' rate of 28.2 appears as the only leaf ("8") in the sixth row ("2·").

0	*	0111222222233
0	·	5555777889
1	*	000011224
1	·	5778
2	*	1244
2	·	8
3	*	11
3	·	5
4	*	11
4	·	
5	*	
5	·	679
6	*	
6	·	
7	*	
7	·	8

FIGURE 3.11 Stem-and-leaf plot of the murder rate data presented in Table 3.1.

If you mentally rotate the stem-and-leaf plot counterclockwise so that the stems are on the bottom and then compare it to those graphical methods discussed previously for individual attributes (Figures 3.1 and 3.2), you will note a great deal of similarity to the histogram. Both methods portray a peak on the left side of the graph with a distinct positive skew. Although the graphs are similar, the major advantage of the stem-and-leaf plot is that it is possible to determine approximate values for each observation. For the murder rate data, this might not be particularly useful, but for some data sets, it can be. For example, Burt and Barber (1996, 542–544) described a stem-and-leaf plot that portrays when houses were constructed in a neighborhood. The stem-and-leaf plot revealed that seven houses were built during the post–World War II period (1945–1947) but that no houses were built from 1948 to 1950. This sort of detailed information is unlikely to be clearly portrayed by either a dispersion graph or histogram.

Another technique representative of Tukey's work is the **box plot** (Figure 3.12). Here, a rectangular box represents the interquartile range (the 75th minus the 25th percentile), and the middle line within the box represents the median, or 50th percentile. The position of the median, relative to the 75th and 25th percentiles, is an indicator of whether the distribution is symmetric or skewed; for the murder rate data, the position indicates a positive skew (a tail toward higher values on the number line). The horizontal lines outside the rectangular box represent the maximum and minimum values in the data. The relative positions of these lines also permit an examination of the symmetry of the distribution, and their positions relative to the box indicate how extreme

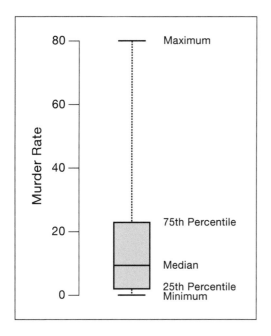

FIGURE 3.12 Box plot of the murder rate data presented in Table 3.1.

the maximum and minimum values are. In this case, their relative positions again suggest a positive skew.[7]

Box plots are most frequently used to compare two or more distributions having the same units of measurement. For example, Figure 3.13 compares murder rate data for cities falling into two distinct regions (the "Deep South" and "California," as defined by Birdsall and Florin 1992). Note that California cities have a distinctly smaller range, a smaller maximum, and an interquartile range that is smaller and shifted toward lower murder rates.

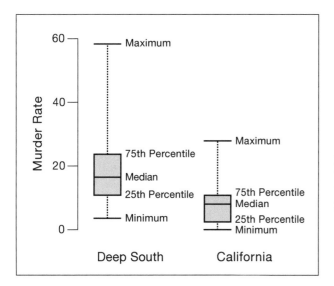

FIGURE 3.13 Box plots comparing murder rate data for cities in the Deep South and California. California cities appear to have a distinctly smaller range, a smaller maximum, and an interquartile range that is smaller and associated with lower murder rates.

It is important to realize that many of the more conventional graphical methods can also be used in an exploratory manner. For example, in a fashion similar to the stem-and-leaf plot, a dispersion graph can uncover nuances in the data not revealed by a histogram. The scatterplot matrix can also be used in an exploratory fashion, especially if the number of attributes is large. Remember that the key to exploratory analysis is to reveal hidden characteristics; the broadest possible range of approaches should be considered for achieving this.

3.8 NUMERICAL SUMMARIES FOR GEOGRAPHIC DATA

This section considers numerical summaries in which spatial location is an integral component.

3.8.1 GEOGRAPHIC CENTER

The **geographic center** (or **centroid**) can be considered the balancing point for a geographic region: Imagine that a map of a region consists of a uniform thin material and we try to find the point at which the region balances on a pencil point. Another way of thinking about the geographic center is that it is the point that minimizes the sum of squared distances from all points within the region to its center. Practically speaking, cartographers use the geographic center when placing a map symbol in the center of an enumeration unit.

Peter Rogerson (2015) provides an extensive discussion of various approaches that have been used to compute a geographic center. For relatively small geographic areas (when the curvature of the Earth is not important), the following formulas are appropriate for determining the geographic center x_c, y_c of a region:

$$x_c = \frac{1}{6A}\left\{\sum_{i=0}^{n-1}(x_i + x_{i+1})(x_i y_{i+1} - x_{i+1}y_i)\right\}$$

$$y_c = \frac{1}{6A}\left\{\sum_{i=0}^{n-1}(y_i + y_{i+1})(x_i y_{i+1} - x_{i+1}y_i)\right\}$$

where A is the area of the region, and the n coordinate pairs are numbered sequentially counterclockwise along the perimeter of the region beginning with 0 and ending with $n-1$.[8] The area of the region can be computed as follows:

$$A = 0.5\left\{\sum_{i=0}^{n-1}x_i y_{i+1} - x_{i+1}y_i\right\}$$

If the region covers a substantial portion of the Earth's surface, then the curvature of the Earth must be taken into account. Rogerson indicates that this can be accomplished by making use of the azimuthal equidistant projection, which is ideal because it preserves distances from the location it is centered on (see Section 8.8.3 for a discussion of this projection). The following is a five-step procedure for determining the geographic center:[9]

Step 1. Make an educated guess for the estimated location of the geographic center.

Step 2. Convert the longitude, latitude point locations defining the boundary of the region to x and y coordinates using the azimuthal equidistant projection centered on the estimated location of the center.

Step 3. Compute the geographic center for the x and y coordinates using the formulas presented above (for areas that do not consider the curvature of the Earth).

Step 4. Convert the geographic center from Step 3 back to a latitude-longitude pair.

Step 5. Using the result from Step 4, return to Step 2 with this as a new estimate for the center.

The procedure terminates when the difference between an estimate for the center and a previous estimate of the center is very small. For our purposes in this book, we will often refer to the geographic center as the "center of an enumeration unit," and we will presume that procedures similar to those we have described above have been used to compute a center.

3.8.2 SPATIAL AUTOCORRELATION AND MEASURING SPATIAL PATTERN

Although maps allow us to visually assess spatial pattern, they have two important limitations: their interpretation varies from person to person, and there is the possibility that a perceived pattern is actually the result of chance factors and thus not meaningful. For these reasons, it makes sense to compute a numerical measure of spatial pattern, which can be accomplished using spatial autocorrelation. In this section, we will consider how spatial autocorrelation can be applied to choropleth maps. For a discussion of measures appropriate for examining patterns on other types of maps, see Unwin (1981) and Davis (2002).

Spatial autocorrelation is the tendency for like things to occur near one another in geographic space. For example, expensive homes likely will be located near other expensive homes, and soil cores with high clay content likely will be found near other soil cores with high clay content. A common measure of spatial autocorrelation is the Moran coefficient (MC), which is defined as

$$MC = \frac{\sum_{i=1}^{n}\sum_{j=1}^{n} w_{ij}\left(X_i - \bar{X}\right)\left(X_j - \bar{X}\right) \Big/ \sum_{i=1}^{n}\sum_{j=1}^{n} w_{ij}}{\sum_{i=1}^{n}\left(X_i - \bar{X}\right)^2 \Big/ n},$$

where $w_{ij} = 1$ if enumeration units i and j are adjacent (or contiguous) and 0 otherwise.[10] Computations for the MC for a hypothetical region consisting of nine enumeration units are shown in Figure 3.14. Within each enumeration

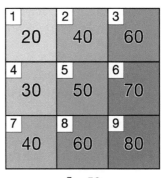

$\bar{x} = 50$

Enumeration Units

i	j	w_{ij}	$(x_i - \bar{x})(x_j - \bar{x})$	=	
1	2	1	(20–50)(40–50)	=	300
1	4	1	(20–50)(30–50)	=	600
2	1	1	(40–50)(20–50)	=	300
2	3	1	(40–50)(60–50)	=	–100
2	5	1	(40–50)(50–50)	=	0
3	2	1	(60–50)(40–50)	=	–100
3	6	1	(60–50)(70–50)	=	200
4	1	1	(30–50)(20–50)	=	600
4	5	1	(30–50)(50–50)	=	0
4	7	1	(30–50)(40–50)	=	200
5	2	1	(50–50)(40–50)	=	0
5	4	1	(50–50)(30–50)	=	0
5	6	1	(50–50)(70–50)	=	0
5	8	1	(50–50)(60–50)	=	0
6	3	1	(70–50)(60–50)	=	200
6	5	1	(70–50)(50–50)	=	0
6	9	1	(70–50)(80–50)	=	600
7	4	1	(40–50)(30–50)	=	200
7	8	1	(40–50)(60–50)	=	–100
8	5	1	(60–50)(50–50)	=	0
8	7	1	(60–50)(40–50)	=	–100
8	9	1	(60–50)(80–50)	=	300
9	6	1	(80–50)(70–50)	=	600
9	8	1	(80–50)(60–50)	=	300

4000

$$\sum_{i=1}^{n}\sum_{j=1}^{n} w_{ij}(x_i - \bar{x})(x_j - \bar{x}) = 4000$$

$$\sum_{i=1}^{n}\sum_{j=1}^{n} w_{ij} = 24$$

$$\sum_{i=1}^{n}(x_i - \bar{x})^2/n = \sigma^2 = 333.333$$

$$MC = (4000/24)/333.333 = 0.50$$

FIGURE 3.14 Computation of the Moran coefficient (MC) for spatial autocorrelation.

unit, the upper left number is an identifier for that unit, and the middle number is the value for a hypothetical attribute. Note that the formula involves multiplying a weight for two enumeration units $\left(w_{ij}\right)$ times the product of the difference between the attribute values for the enumeration units and the mean of the data $\left(X_i - \bar{X}\right)\left(X_j - \bar{X}\right)$. Computations are shown only for adjacent enumeration units because w_{ij} will be 0 for nonadjacent units, and thus the product $w_{ij}\left(X_i - \bar{X}\right)\left(X_j - \bar{X}\right)$ also will be 0.

The formula for MC resembles the formula for the correlation coefficient, r, discussed previously (compare the equation shown here with that for r in Section 3.6.3.1). For both equations, the denominator contains a measure of the variation in the attribute about the mean, and the numerator contains a measure of how adjacent enumeration units covary. Thus, it is not surprising that MC also ranges from -1 to $+1$. A value close to $+1$ indicates that similar values are likely to occur near one another, whereas a value close to -1 indicates that unlike values are apt to occur near one another. Finally, a value near 0 is indicative of no spatial autocorrelation or a situation in which values of the attribute are randomly distributed. "Moderate" values of positive spatial autocorrelation are most frequently observed in the real world (Griffith 1993, 2).

To illustrate how the MC might be used, consider mapping the rates for respiratory cancer for white males in Louisiana counties (Figure 3.15). A visual assessment of this map suggests that there is a strong positive spatial autocorrelation, with high cancer rates occurring both in the southern part of the state and along an east–west strip in the northern part of the state. The MC for this pattern is 0.14, with a probability of less than 3 percent that this value would occur by chance (Odland 1988). The value of 0.14 indicates that the pattern is not quite as strongly positively autocorrelated as a visual examination of the map suggests, but the associated probability indicates that the pattern is significant and worthy of further exploration.

An important recent development in statistical geography is a consideration of spatial autocorrelation in the context of inferential statistics. We touched on this idea earlier in the chapter with respect to regression, but it should be recognized that virtually all traditional inferential statistics (including regression analysis) can be modified to account for spatial autocorrelation. For further discussion of this issue, see Griffith (1993) and Rogerson (2020). You should also recognize that we have only computed a measure of the spatial autocorrelation for the entire map pattern. It is also possible to compute a localized measure for individual enumeration units; for a discussion of this, see Rogerson and Yamada (2009).

3.8.3 Measuring Map Complexity

The notions of spatial autocorrelation and associated measures such as the MC were developed by statistical geographers. Cartographers have also been interested in measures for describing the spatial pattern. As an example, we'll consider some map complexity measures developed by Alan MacEachren (1982). MacEachren defined **map complexity** as "the degree to which the combination of map elements results in a pattern that appears to be intricate or involved" (32). He argued that when a map is used for presentation, very complex maps might hinder the communication of information.

MacEachren's measures are based on Muller's (1974) application of graph theory to choropleth maps. To compute measures of complexity, a base map is treated as a set of faces (enumeration units), edges (the boundaries of the units), and vertices (where the edges intersect; see Figure 3.16A). When mapping a distribution, edges and vertices between like values are omitted (Figure 3.16B and C). Complexity measures are computed by dividing the number of faces, edges, and vertices on the mapped distribution by the corresponding number on the base map:

$$C_F = \frac{\text{Observed number of faces}}{\text{Number of faces on base map}}$$

$$C_V = \frac{\text{Observed number of vertices}}{\text{Number of vertices on base map}}$$

$$C_E = \frac{\text{Observed number of edges}}{\text{Number of edges on base map}}$$

In the example illustrated in Figure 3.16, map C has the higher complexity measures and so would presumably be more difficult for a user to interpret.

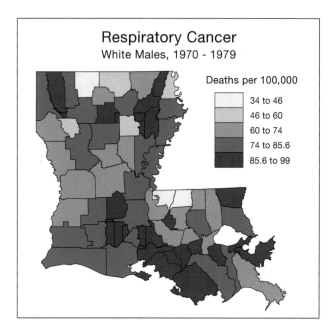

Respiratory Cancer
White Males, 1970 - 1979

Deaths per 100,000

	34 to 46
	46 to 60
	60 to 74
	74 to 85.6
	85.6 to 99

FIGURE 3.15 A map of rates of respiratory cancer for white males in Louisiana counties. The Moran coefficient (a measure of spatial autocorrelation) for this map is 0.14, with a probability of less than 3 percent that this value occurs by chance. Thus, the pattern is significant and worthy of further exploration. (After Odland 1988, 45.)

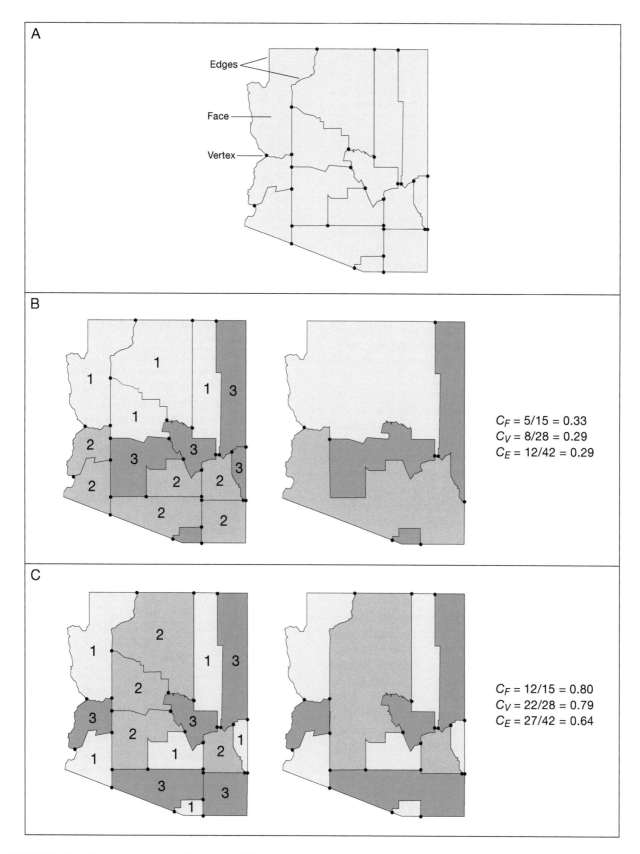

FIGURE 3.16 Computation of MacEachren's (1982) face, vertex, and edge complexity measures. (A) Faces, edges, and vertices for a county-level map of Arizona; (B) three-class map for which low complexity values result; (C) three-class map for which high complexity values result.

We will use the complexity measure for faces in Chapter 5 to examine the complexity of choropleth maps. For complexity measures appropriate for mapping situations that involve a variety of base information (such as those found in OpenStreetmap, https://openstreetmap.org), we encourage you to explore a recent study by Susan Schnur and colleagues (2018).

3.9 SUMMARY

In this chapter, we have examined a variety of techniques (tables, graphs, and numerical summaries) that can be used along with maps to analyze spatial data. These techniques each have their advantages and disadvantages. In the univariate realm, raw tables are useful for providing specific information (e.g., the murder rate for Atlanta, GA, in 1990 was 58.6 per 100,000 people), but they fail to provide an overview of a data set (e.g., the murder rate for U.S. cities in 1990 had a distinct positive skew). Grouped-frequency tables do provide an overview but are not as effective as graphical methods (e.g., the dispersion graph and histogram).

Potential weaknesses of graphical methods, however, are the subjectivity of interpretation and the physical space they require. Numerical summaries (e.g., the mean and standard deviation) are a solution to these problems. A weakness of numerical summaries, though, is that they hide the detailed character of the data; for example, an outlier might be missed if only a numerical summary is used. Clearly, the advantages and disadvantages of these various techniques indicate that they must be combined if spatial data are to be successfully analyzed.

When examining two or more attributes at once, tabular displays are of limited use. Much more suitable are graphical displays, such as the scatterplot and parallel coordinate plot. The numerical method of bivariate correlation-regression can also be useful if one wishes to summarize the relationship between two attributes using a single number or equation. As with univariate data, care should be taken in using bivariate numerical summaries, as they hide the detailed character of the relationship between attributes. An important problem with bivariate correlation is that it is a function of the level of aggregation (i.e., a different correlation coefficient might arise at the census tract level as opposed to the block-group level), an issue known as the modifiable areal unit problem.

Given the emphasis on data exploration in this book, special emphasis in this chapter was placed on exploratory data analysis—that data should be explored, much as a detective investigates a crime. We looked at two common methods for exploratory data analysis: the stem-and-leaf plot and the box plot. Finally, we considered several numerical summaries for which spatial location is an integral part: the geographic center, Moran's coefficient for spatial autocorrelation, and map complexity measures.

3.10 STUDY QUESTIONS

1. Assume that the following are some scientists' predictions of the year when average temperatures will have increased by 3 degrees Celsius in central Alaska as a result of increased CO_2 concentrations: 2050, 2075, 2090, 2100, 2125, 2145, 2155, 2185, 2200, 2400. (A) Using either pencil-and-paper or a software application, construct a point graph of these data. (B) Using terms presented in Chapter 3, discuss the resulting visualization. (C) Treating the data as a sample, compute the mean, standard deviation, median, and interquartile range. (Presume that the 25th and 75th percentiles are 2,092 and 2,178, respectively.) (D) Is the mean or median an appropriate measure of central tendency for these data? Why?

2. Compute a grouped-frequency table of the murder rate data shown in Table 3.1, assuming seven classes of data. Contrast the resulting table with the one shown in Table 3.2.

3. The following are population density values for ten cities selected to span the range of population density values shown in Table 3.1: 1,367, 2,007, 2,198, 2,653, 2,990, 3,508, 3,836, 5,882, 7,410, 8,080, 23,701. (A) Create a point graph of these data. (B) Transform each data value using a $\log_{10}$ transform. Create a point graph of these data and explain whether this seems to be a logical transform.

4. Discuss the r values of 0.81 and 0.17 shown in Figure 3.8 and relate them to lines shown on the parallel coordinate plot.

5. A. Construct a stem-and-leaf plot for the following data indicating the dates that houses were built within a particular neighborhood. Use the tens and hundreds place as sorting digits and the ones place as a display digit. 1,822, 1,881, 1,891, 1,892, 1,893, 1,895, 1,897, 1,900, 1,902, 1,903, 1,903, 1,904, 1,905, 1,911, 1,912, 1,931, 1,932, 1,933, 1,934, 1,946, 1,946, 1,946, 1,946, 1,946, 1,946, 1,946, 1,951, 1,952, 1,953, 1,955, 1,980.

   B. Explain what this plot tells you about the dates of housing construction within this neighborhood.

6. What general approach would you use to determine the geographic center of each of the following countries: Monaco, Switzerland, and Brazil.

7. Imagine that you are analyzing the degree to which neighborhoods in a U.S. city have been racially integrated. Using census blocks as your observations, you compute the Moran coefficients for the percentage Black located within three neighborhoods composed of census blocks and find the following values: 0.07, 0.62, and −0.42. Discuss what these three values tell you about racial integration in these three neighborhoods.

8. Compute a face complexity measure for Figure 3.15. Note that there are 64 parishes (counties) in Louisiana. How would this result compare with a face complexity value of 0.16?

NOTES

1 The murder and drug arrest data were obtained from the *Sourcebook of Criminal Justice Statistics 1991* (Flanagan and Maguire 1992). The remaining data were taken from the *1994 City and County Data Book* (U.S. Bureau of the Census 1994). All data were for either 1989 or 1990.

2 We utilized the "extended city" definition provided by the U.S. Census Bureau; see https://www2.census.gov/geo/pdfs/reference/GARM/Ch12GARM.pdf.

3 In Section 25.5.6, we discuss a formal approach for detecting outliers.

4 $\sum$ is the symbol for summation, indicating that all data values should be summed. Readers unfamiliar with summation notation should consult an introductory statistics book such as Burt et al. (2009, 101–102).

5 For alternative methods of displaying dense data in scatterplots, see Hao et al. (2010), Keim et al. (2010), and Raidou et al. (2019).

6 Tukey (1977) indicated that digits beyond the display digit could either be used for rounding or simply ignored.

7 Numerous variations of the box plot have been developed; for an overview and suggestions for an alternative using color, see Carr (1994). It has also been suggested that the size of areas associated with observations should be taken into account when constructing box plots (Willmott et al. 2007).

8 Also, note that $(x_n, y_n) = (x_0, y_0)$.

9 We have modeled these five steps after Plane and Rogerson (2015, 971). They used them to compute the mean center of an attribute, but the same concept applies for a geographic center, as the x and y coordinates of the attribute are conceptually replaced by the x and y coordinates of the boundary of the region.

10 The weights can be modified to account for differing sizes of enumeration units (Odland 1988, 29–31). We have used weights of 0 and 1 for computational simplicity.

REFERENCES

Barrett, R. E. (1994) *Using the 1990 U.S. Census for Research*. Thousand Oaks, CA: Sage.

Birdsall, S. S., and Florin, J. W. (1992) *Regional Landscapes of the United States and Canada*. New York: Wiley.

Burt, J. E., and Barber, G. M. (1996) *Elementary Statistics for Geographers* (2nd ed.). New York: Guilford.

Burt, J. E., Barber, G. M., and Rigby, D. L. (2009) *Elementary Statistics for Geographers* (3rd ed.). New York: The Guilford Press.

Carr, D. B. (1994) "A colorful variation on box plots." *Statistical Computing & Statistical Graphics Newsletter 5*, no. 3:19–23.

Carr, D. B., Olsen, A. R., and White, D. (1992) "Hexagon mosaic maps for display of univariate and bivariate geographical data." *Cartography and Geographic Information Systems 19*, no. 4:228–236, 271.

Clark, W. A. V., and Hosking, P. L. (1986) *Statistical Methods for Geographers*. New York: Wiley.

Cleveland, W. S. (1994) *The Elements of Graphing Data* (rev. ed.). Summit, NJ: Hobart Press.

Correll, M., Li, M., Kindlmann, G. et al. (2019) "Looks good to me: Visualizations as sanity checks." *IEEE Transactions on Visualization and Computer Graphics 25*, no. 1:830–839.

Davis, J. C. (2002) *Statistics and Data Analysis in Geology* (3rd ed.). New York: Wiley.

Flanagan, T. J., and Maguire, K. (1992) *Sourcebook of Criminal Justice Statistics 1991*. Washington, DC: U.S. Department of Justice, Bureau of Justice Statistics.

Fotheringham, A. S. (1997) "Trends in quantitative methods I: Stressing the local." *Progress in Human Geography 21*, no. 1:88–96.

Fotheringham, A. S. (1998) "Trends in quantitative methods II: Stressing the computational." *Progress in Human Geography 22*, no. 2:283–292.

Fotheringham, A. S., Brunsdon, C., and Charlton, M. (2002) *Geographically Weighted Regression: The Analysis of Spatially Varying Relationships*. West Sussex, England: John Wiley.

Griffith, D. A. (1993) *Spatial Regression Analysis on the PC: Spatial Statistics Using SAS*. Washington, DC: Association of American Geographers.

Hammond, R., and McCullagh, P. (1978) *Quantitative Techniques in Geography: An Introduction* (2nd ed.). Oxford, England: Clarendon.

Hao, M. C., Dayal, U., Sharma, R. K. et al. (2010) "Variable binned scatter plots." *Information Visualization 9*, no. 3:194–203.

Keim, D. A., Hao, M. C., Dayal, U. et al. (2010) "Generalized scatter plots." *Information Visualization 9*, no. 4:301–311.

MacEachren, A. M. (1982) "Map complexity: Comparison and measurement." *The American Cartographer 9*, no. 1:31–46.

Muller, J. C. (1974) "Mathematical and statistical comparisons in choropleth mapping." Unpublished Ph.D. dissertation, University of Kansas, Lawrence, KS.

Odland, J. (1988) *Spatial Autocorrelation*. Newbury Park, CA: Sage.

Plane, D. A. and Rogerson, P. A. (2015) "On tracking and disaggregating center points of population." *Annals of the Association of American Geographers 105*, no. 5:968–986.

Raidou, R. G., Gröller, M. E., and Eisemann, M. (2019) "Relaxing dense scatter plots with pixel-based mappings." *IEEE Transactions on Visualization and Computer Graphics 25*, no. 6:2205–2216.

Rogerson, P. and Yamada, I. (2009) *Statistical Detection and Surveillance of Geographic Clusters*. London and New York: Taylor & Francis Group.

Rogerson, P. A. (2015) "A new method for finding geographic centers, with application to U.S. states." *Professional Geographer 67*, no. 4:686–694.

Rogerson, P. A. (2020) *Statistical Methods for Geography: A Student's Guide* (5th ed.). London: SAGE Publications.

Schnur, S., Bektaş, K., and Çöltekin, A. (2018) "Measured and perceived visual complexity: A comparative study among three online map providers." *Cartography and Geographic Information Science 45*, no. 3:238–254.

Tukey, J. W. (1977) *Exploratory Data Analysis*. Reading, MA: Addison-Wesley.

Unwin, D. (1981) *Introductory Spatial Analysis*. London: Methuen.

U.S. Bureau of the Census (1994) *County and City Data Book: 1994*. Washington, DC: U.S. Government Printing Office.

Wilkinson, L. (1999) "Dot plots." *The American Statistician 53*, no. 3:276–281.

Willmott, C. J., Robeson, S. M., and Matsuura, K. (2007) "Geographic box plots." *Physical Geography 28*, no. 4:331–344.

4 Principles of Symbolization

4.1 INTRODUCTION

The purpose of this chapter is to introduce basic principles involved in symbolizing thematic maps. We will see that there is a broad range of possible symbols (or **visual variables**) that can be used to represent geographic phenomena. For illustrative purposes, we will contrast four common thematic mapping techniques (and their associated visual variables): choropleth (lightness), proportional symbol (size), isopleth (lightness), and dot (location). We will see that selecting one of these four techniques requires that you consider: (1) the nature of the underlying phenomenon (e.g., deciding whether the phenomenon can be conceived as points, lines, areas, or volumes); (2) the level at which the phenomenon is measured (nominal, ordinal, interval, or ratio); and (3) whether you have standardized a dataset representing a phenomenon to account for the area over which the phenomenon is distributed.

Although our emphasis in this book is on approaches for *visualizing* spatial data, you should be aware that our other senses (e.g., sound, touch, and smell) may also be used to represent spatial data. In the last section of the chapter, we touch on some approaches that map designers have used to depict spatial data via these other senses. Here we will consider **abstract sound variables** and a **haptic variable syntax** that are analogous to the visual variables but apply to sound and touch. In Chapter 28, we will consider these others senses in greater depth when we introduce the notion of virtual environments.

4.2 LEARNING OBJECTIVES

- Determine the spatial dimension (point, linear, areal, $2\frac{1}{2}$-D or true 3-D) associated with geographic phenomena.
- Determine the level of measurement (nominal, ordinal, interval, or ratio) for a set of spatial data.
- Select an appropriate visual variable given the spatial dimension of a phenomenon and the level of measurement.
- Distinguish between the actual phenomenon that you wish to map and the data collected to represent that phenomenon.
- Determine whether a data set constitutes raw totals or has been standardized.
- Decide when it is appropriate to use choropleth, proportional symbol, isopleth, and dot maps.
- Select an appropriate visual variable for a choropleth map.
- Suggest approaches for representing spatial data that do not involve our sense of vision.

4.3 NATURE OF GEOGRAPHIC PHENOMENA

4.3.1 SPATIAL DIMENSION

One way to think about geographic phenomena is to consider their extent or **spatial dimension**. For our purposes, we will consider five types of phenomena with respect to spatial dimension: point, linear, areal, $2\frac{1}{2}$-D, and true 3-D.

Point phenomena are assumed to have no spatial extent and are thus termed "zero-dimensional." Examples include weather station recording devices, oil wells, and locations of nesting sites for eagles. Locations for point phenomena can be specified in either two- or three-dimensional space; for example, oil well locations are commonly defined by x and y coordinate pairs (longitude and latitude), whereas nesting sites for eagles might be defined by x, y, and z coordinates (the z coordinate would be the height of the nest above the Earth's surface).

Linear phenomena are one-dimensional in spatial extent, having length but essentially no width. Examples include a boundary between countries and the migration routes of individual whales. Locations of linear phenomena are defined as an unclosed series of x and y coordinates (in two-dimensional space), or an unclosed series of x, y, and z coordinates (in three-dimensional space).

Areal phenomena are two-dimensional in spatial extent, having both length and width. An example would be a lake (assuming that we focus on its two-dimensional surface extent). Data associated with political units (e.g., counties) can also fit this framework because the location of each county can be specified as an enclosed region. In two-dimensional space, areal phenomena are defined by a series of x and y coordinates that completely enclose a region (computer systems generally require that the first coordinate pair equal the last).

When we move into the realm of volumetric phenomena, it is convenient to consider two types: $2\frac{1}{2}$-D and true 3-D. The first of these, $2\frac{1}{2}$-D **phenomena** can be thought of as a surface in which geographic location is defined by x and y coordinate pairs, and the value of the phenomenon is the height above a zero point (or depth below a zero point). Probably the easiest example to understand is the elevation above sea level because we can actually see the surface in the real world; here, height above a zero point is the elevation of the land surface above sea level. A more abstract example would be precipitation falling over a region during the course of a year; in this case, the height of the surface would be the total amount of precipitation for the year.

Another way of thinking about $2\frac{1}{2}$-D surfaces is that they are single-valued in the sense that each x and y coordinate location has a single value associated with it

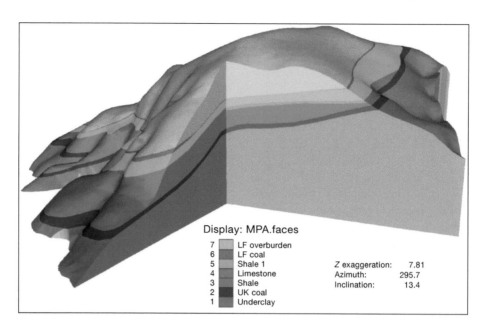

Display: MPA.faces

7	▨	LF overburden
6	▨	LF coal
5	▨	Shale 1
4	▨	Limestone
3	▨	Shale
2	▨	UK coal
1	▨	Underclay

Z exaggeration: 7.81
Azimuth: 295.7
Inclination: 13.4

FIGURE 4.1   Three-dimensional model of an open-pit coal mining site. (From Pennsylvania Department of Environment Protection; courtesy of Dynamic Graphics, Inc., Alameda, CA.)

(an exception would be an overhanging cliff). In contrast, **true 3-D phenomena** are multivalued because each x and y location can have multiple values associated with it. With true 3-D phenomena, any point is specified by four values: an x coordinate, a y coordinate, a z coordinate (which is the height above or depth below sea level), and the value of the phenomenon. Consider mapping the concentration of carbon dioxide (CO_2) in the atmosphere. At any point in the atmosphere, it is possible to define longitude, latitude, height above sea level, and an associated level of CO_2. Figure 4.1 illustrates another true 3-D phenomenon: geologic material underneath the Earth's surface. Although ideally, we should distinguish true 3-D phenomena from $2\frac{1}{2}$-D phenomena; in this book, we sometimes refer to a map of either as a *3-D map*.

It is important to realize that map scale plays a major role in determining how we handle the spatial dimension of a phenomenon. For example, on a **small-scale map** (e.g., a page-sized map of France), places of religious worship occur at points, but on a **large-scale map** (e.g., a map of a local neighborhood), individual buildings would be apparent, and thus the focus might be on the area covered by the place of worship. Similarly, a river could be considered a linear phenomenon on a small-scale map, but on a large-scale map, the emphasis could be on the area covered by the river.

4.3.2 Models of Geographic Phenomena

The notion of spatial dimension is just one way of thinking about how geographic phenomena are arranged in the real world. Another approach is to consider the arrangement of geographic phenomena along discrete–continuous and abrupt–smooth scales. In this section, we define the terms

associated with these continua and show how they provide a useful set of *models of geographic phenomena*, a notion developed by Alan MacEachren and David DiBiase (1991).

The terms "discrete" and "continuous" are often used in statistics courses to describe different types of data along a number line; here, we consider their use by cartographers in a spatial context. **Discrete phenomena** are presumed to occur at distinct locations (with space in between). Individual people living in a city would be an example of a discrete phenomenon; for an instant in time, a location can be specified for each person, with space between individuals. **Continuous phenomena** occur throughout a geographic region of interest. The examples presented previously for $2\frac{1}{2}$-D phenomena would also be considered continuous phenomena. For instance, when considering elevation, every longitude and latitude position has a value above or below sea level.

Discrete and continuous phenomena can also be described as either abrupt or smooth. **Abrupt phenomena** change suddenly, whereas **smooth phenomena** change in a gradual fashion. This concept is most easily understood for continuous phenomena. The number of electoral votes for each state in the United States would be considered an abrupt continuous phenomenon because although each enumeration unit (a state) has a value, there are abrupt changes at the boundaries between states. In contrast, the distribution of total precipitation over the course of a year for a humid region would be a smooth continuous phenomenon because we would not expect such a distribution to exhibit abrupt discontinuities.

Figure 4.2A provides a graphic portrayal of a variety of models of geographic phenomena that result when we combine the discrete–continuous and abrupt–smooth scales. We will discuss the models in detail because considering

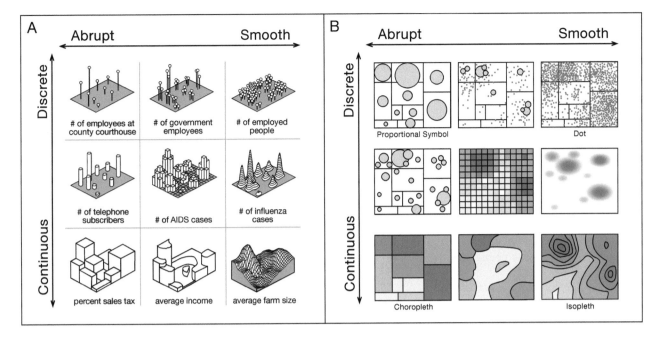

FIGURE 4.2 (A) Models of geographic phenomena arranged along discrete–continuous and abrupt–smooth scales. (B) A set of symbolization methods appropriate for these models. (After MacEachren 1992, 16; courtesy of *Cartographic Perspectives* (North American Cartographic Information Society, publisher) and Alan MacEachren.)

the nature of geographic phenomena is extremely important in selecting an appropriate method of symbolization. First, consider the continuous phenomena shown in the bottom row of Figure 4.2A. Percent sales tax is an obvious *abrupt* continuous phenomenon, as it changes suddenly at the boundary between enumeration units (e.g., one state's sales tax is different from another state's sales tax). In contrast, average farm size is an example of a *smooth*, continuous phenomenon because we would expect it to vary in a relatively gradual fashion (as the climate becomes drier, we would expect the average farm size to increase). Average income falls somewhere between percent sales tax and average farm size on the abruptness–smoothness scale. In some cases, average income would exhibit the abrupt changes of percent sales tax (as at the boundary between urban neighborhoods), while in others, it would exhibit a more gradual change (as one moves up a hill toward a region of more attractive views, average income should increase).

In contrast to the bottom row, the top row of Figure 4.2A represents a range of discrete phenomena. The number of employees located at a county courthouse is clearly an abrupt discrete phenomenon, as there can be only one value for a county and it occurs at an isolated location. In contrast, the number of employed people (based on where they live, as opposed to where they work) is a *smooth* discrete phenomenon because it gradually changes over geographic space. The number of government employees (again, based on where they live) falls somewhere between these phenomena; it might exhibit an abrupt characteristic in the sense that government employees might live near government offices, but it will not exhibit the extreme abruptness of the courthouse example.

The middle row of Figure 4.2A represents phenomena that can be classified as not clearly continuous or discrete and that also span the abruptness–smoothness scale. This row is probably most easily understood by considering the influenza case first. Because influenza is an infectious disease, it should exhibit a smooth character. Although individual influenza cases could be represented at discrete locations, it makes sense to suggest some degree of continuity if we wish to stress the potential of infection. At the other end of the row is the number of subscribers to a particular telephone company. Competition between telephone companies could lead to a distribution that changes abruptly, but that exhibits continuity between the lines of abrupt change. Finally, the number of people with AIDS is in the middle of the diagram. AIDS occupies a more abrupt position than influenza because of its mode of transmission (sexual intercourse, sharing of needles, and blood transfusions).

Figure 4.2B is a set of symbolization methods that MacEachren and DiBiase argued would be appropriate for the models shown in Figure 4.2A. Note that we have labeled the four corner maps (proportional symbol, dot, choropleth, and isopleth), which will be discussed in depth in Section 4.6.

4.3.3 Phenomena versus Data

When mapping geographic phenomena, it is important to distinguish between the actual *phenomenon* and the *data* collected to represent that phenomenon. For example, imagine that we wish to map the percentage of forest cover in South Carolina. If we try to visualize the phenomenon, we can conceive of it as smooth and continuous in some portions of the state where the percentage gradually increases or decreases. In other areas, we can conceive of relatively abrupt changes where the percentage shifts very rapidly (when, say, an urban area is bounded by a hilly, forested region).

One form of data representing the percentage of forest cover is individual percentage values for counties, which can be found in the state statistical abstract for South Carolina (https://dc.statelibrary.sc.gov/handle/10827/19485). We can map these data directly by creating the **prism map** shown in Figure 4.3B. Note that in this case, there are abrupt changes at the boundaries of each county. Such a map might be appropriate if we wished to provide a typical value for each county, but it obviously hides the variation within counties and misleads the reader into thinking that changes take place only at county boundaries.

A potentially better approach is the smooth, continuous map (**fishnet map**) shown in Figure 4.3A; this map indicates that the percentage of forest cover does not coincide with county boundaries but rather changes in a gradual fashion. A still better map would be one that shows some of the abrupt changes that are likely to occur. Creating such a map would require detailed information about the location of forest within the state, as might be available from a remotely sensed image. Our purpose at this point in the text is not to create the most representative map of the phenomenon but to stress that the mapmaker must carefully distinguish between data that have been collected and the phenomenon that is being mapped. Which type of map is used will be a function of both the nature of the underlying phenomenon and the purpose of the map. We consider this issue in greater depth in Section 4.6.

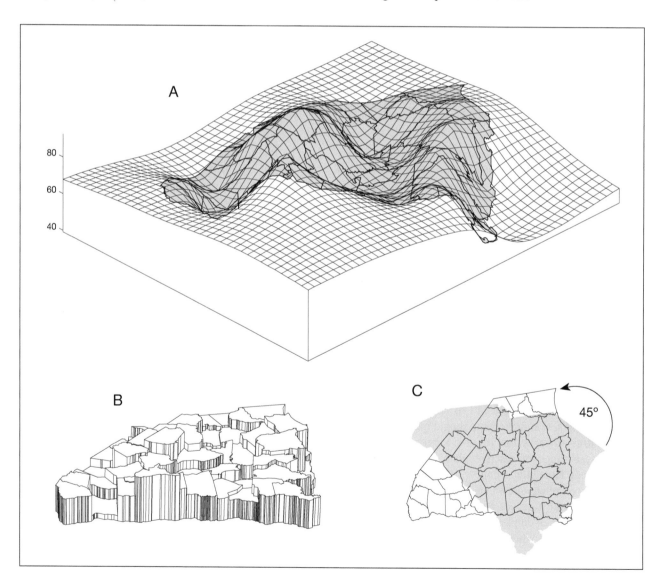

FIGURE 4.3 Approaches for mapping a data set of percentage of forest cover by county for the state of South Carolina: (A) The data are treated as coming from a smooth, continuous phenomenon; (B) the data are treated as an areal phenomenon. Map (C) illustrates that maps (A) and (B) have been rotated 45° from a traditional north-oriented map. For A, values outside the state are extrapolated and thus must be treated with caution.
(Data Source: *South Carolina Statistical Abstract 1994*, https://dc.statelibrary.sc.gov/handle/10827/19485.)

4.4 LEVELS OF MEASUREMENT

When a geographic phenomenon is measured to create a data set, we commonly speak of the **level of measurement** associated with the resulting data. Conventionally, four levels of measurement are recognized—nominal, ordinal, interval, and ratio—with each subsequent level including all characteristics of the preceding levels. The **nominal** level of measurement involves grouping (or categorization) but no ordering. The classic example is religion, in which individuals might be identified as Catholic, Protestant, Jewish, or Other; here, each religious group is different, but one is not more or less religious in value than another. Another example would be classes on a land use/land cover map; for example, grassland, forest, urban, water, and cropland differ from one another, but one class is not more or less in value than another.

The second level of measurement, **ordinal**, involves categorization plus an ordering (or ranking) of the data. For example, a geologist asked to specify the likelihood of finding oil at each of 50 well sites might be unwilling to provide numerical data but would feel comfortable specifying a low, moderate, or high potential at each site. Here three categories (low, moderate, and high) are provided, with a distinct ordering among them. Another example of ordinal data would be rankings resulting from a map comparison experiment. Imagine that you constructed dot maps for ten different phenomena and asked people to compare these maps with another dot map (say, of population) and to rank the maps from "most like" to "least like" the population map. The ten maps ranked by each person would constitute a distinct ordering and thus represent ordinal data.

An **interval** level of measurement involves an ordering of the data plus an explicit indication of the numerical difference between two categories. Classic examples are the Fahrenheit and Celsius temperature scales. Consider temperatures of 20°F and 40°F recorded in Fairbanks, Alaska, and Chattanooga, Tennessee, respectively. These two values are ordered, and they reveal the precise numerical difference between the two cities. One characteristic of interval scales is the arbitrary nature of the zero point. In the case of the Celsius scale, 0 is the freezing point for pure water, whereas, on the Fahrenheit scale, 0 is the lowest temperature obtained by mixing salt and ice. A result of an *arbitrary zero point* is that ratios of two interval values cannot be interpreted correctly; for example, 40°F is numerically twice the value of 20°F, but it is not twice as warm (in terms of the kinetic energy of the molecules). An example of an interval scale familiar to academics is SAT scores, which range from a minimum of 200 to a maximum of 800. Note that it is not possible to say that an individual scoring 800 on an SAT exam did four times better than an individual scoring 200; all that can be said is that the individual scored 600 points better. A geographical example of interval-level data is elevation, where the establishment of mean sea level represents an arbitrary zero point.

A **ratio** level of measurement has all the characteristics of the interval level plus a *nonarbitrary zero point*. Continuing with the temperature example, the Kelvin scale is ratio in nature because at 0°K molecular motion is minimized; thus, a temperature of 40°K is twice as warm as 20°K (in terms of the kinetic energy of the molecules). Ratio data sets are more common than interval data sets. For example, a perusal of the maps shown in this text will reveal that most are based on ratio-level data. Because many symbol forms can be used with both interval and ratio scales, these two levels of measurement are often grouped together and referred to as **numerical data**. The basic scales that we have discussed can also be divided into *qualitative* (nominal data) and *quantitative* (ordinal, interval, and ratio data) scales.[1]

An important extension to the basic levels of measurement is the three kinds of numerical data proposed by J. Ronald Eastman (1986): bipolar, balanced, and unipolar. **Bipolar data** are characterized by either natural or meaningful dividing points. A n*atural* dividing point is inherent to the data and can be used intuitively to divide the data into two parts. An example would be a value of 0 for a percentage of population change, which would divide the data into positive and negative percent changes. A *meaningful* dividing point does not occur inherently in the data but can logically divide the data into two parts. An example would be the mean of the data, which enables differentiating values above and below the mean. **Balanced data** are characterized by two phenomena that coexist in a complementary fashion. An example is the percentages of English and French spoken in Canadian provinces— a high percentage of English-speaking people implies a low percentage of French-speaking people (the two are in "balance" with one another). **Unipolar data** have no natural dividing points and do not involve two complementary phenomena. Per capita income associated with countries of Africa or states of the United States would be an example of unipolar data. Throughout the remainder of this textbook, we will concentrate primarily on unipolar and bipolar data because these are much more common than balanced data.

4.5 VISUAL VARIABLES

The term **visual variables** is commonly used to describe the various perceived differences in map symbols that are used to represent geographic phenomena. The notion of visual variables was developed by the French cartographer Jacques Bertin (1983) and subsequently modified by others. Our approach is similar to that of Alan MacEachren (1994), but differs primarily in our inclusion of 2 $\frac{1}{2}$-D and true 3-D phenomena and the use of the perspective-height visual variable.

In this chapter, we consider only visual variables for static maps. Additional visual variables for animated maps and for depicting uncertainty are covered in Chapters 24 and 27, respectively. The visual variables for static maps

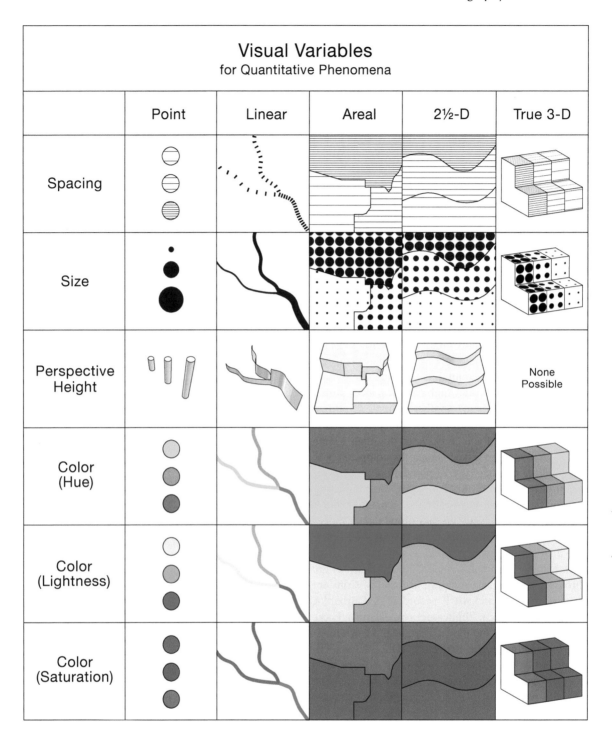

FIGURE 4.4 Visual variables for quantitative phenomena.

are illustrated in Figures 4.4 (for quantitative phenomena) and 4.5 (for qualitative phenomena). Note that the visual variables appear in the rows and that the columns represent the dimensions of spatial phenomena discussed in the preceding section. In discussing the visual variables, we sometimes need to distinguish between the overall *symbol* and the *marks* making up the symbol. For example, note that the spacing visual variable shown for point phenomena in Figure 4.4 consists of circular symbols and that each circle is composed of parallel horizontal marks.

4.5.1 Visual Variables for Quantitative Phenomena

Let's first consider the visual variables for quantitative phenomena—here, the visual variables should reflect either an ordinal, interval, or ratio level of measurement.

4.5.1.1 Spacing

The **spacing** visual variable involves changes in the distance between the marks making up the symbol—a smaller distance between marks suggests a higher data value

(Figure 4.4). Cartographers traditionally used the term *texture* to describe these changes, but we use the term *spacing* because texture has varied usages in the literature.

4.5.1.2 Size

Cartographers have used **size** as a visual variable for quantitative phenomena in two different ways. One has been to change the size of the entire symbol, as is shown for the point and linear phenomena (Figure 4.4). Another is to change the size of individual marks making up the symbol, as for the areal, $2\frac{1}{2}$-D and true 3-D phenomena. This inconsistency might be a bit confusing, but the term *size* seems to reflect quantitatively the visual differences that arise in each case. Note that for areal phenomena, the size of the entire areal unit could also be changed, as is done on cartograms, to be discussed in Chapter 20.

4.5.1.3 Perspective Height

Perspective height refers to a perspective 3-D view of a phenomenon (Figure 4.4). It is interesting to consider some of the potential applications of this visual variable. In the case of point phenomena, oil production at well locations might be represented by raised cylinders above each well, with the cylinder height proportional to well production. For linear phenomena, total traffic flow between two cities over some time period could be represented by a fencelike structure above each roadway, with the height of the "fence" proportional to traffic flow. For areal phenomena, each enumeration unit is raised to a height proportional to the data, just as we did for the forest cover data in South Carolina in Figure 4.3; note that a similar approach is used for $2\frac{1}{2}$-D phenomena. Perspective height cannot be used for true 3-D phenomena because three dimensions are needed to locate the phenomenon being mapped.

4.5.1.4 Hue, Lightness, and Saturation

The visual variables hue, lightness, and saturation are commonly recognized as basic components of color (Figure 4.4). **Hue** is the dominant wavelength of light making up a color (the notion of wavelengths of light and the associated electromagnetic spectrum will be considered in detail in Chapter 10). In everyday life, hue is the parameter of color most often used; for example, you might note that one person has on a red shirt and another a blue shirt. For quantitative phenomena, we should select hues that suggest a quantitative difference; for instance, yellow, orange, and red would be appropriate hues because orange is seen as a mixture of yellow and red (based on opponent-process theory, a topic to be covered in Chapter 10).

Lightness (or **value**) refers to how dark or light a particular color is while holding hue constant; for instance, differing lightnesses of a green hue might be used (Figure 4.4). Differing lightnesses of gray can also be used in the absence of what we commonly would call color. **Saturation** (or **chroma**) can be thought of as a mixture of gray and a pure hue (Figure 4.4). It is the intensity of color; for instance, we might speak of different intensities of colorful shirts.

4.5.2 Visual Variables for Qualitative Phenomena

Now let's consider the visual variables for qualitative phenomena, where visual variables should reflect only a nominal level of measurement.

4.5.2.1 Orientation and Shape

As with the size visual variable, the character of the **orientation** visual variable is a function of the kind of spatial phenomena. For linear, areal, and true 3-D phenomena, orientation refers to the direction of individual marks making up the symbol. In contrast, for point phenomena, orientation refers to the direction of the entire point symbol (Figure 4.5). (Marks of differing direction could be applied to point symbols, but the small size of point symbols often makes it difficult to see the marks.) Because orientation is most appropriate for representing nominal data, we do not recommend using it for $2\frac{1}{2}$-D phenomena, which are inherently numerical. Note that in Figure 4.5, the **shape** visual variable is handled in a fashion similar to orientation.

4.5.2.2 Arrangement

Understanding the **arrangement** visual variable requires a careful examination of Figure 4.5. For areal and true 3-D phenomena, note that arrangement refers to how marks making up the symbol are distributed; marks for some areas are part of a regular arrangement, whereas marks for other areas appear to be randomly placed. For linear phenomena, arrangement refers to splitting lines into a series of dots and dashes, as might be found on a map of political boundaries. Finally, for point phenomena, arrangement refers to changing the position of the white markers within the black symbol. Regardless of how these differing arrangements are created, they seem to suggest qualitative (nominal) differences rather than quantitative ones.

4.5.2.3 Hue

We noted above that color is comprised of hue, lightness, and saturation. Only hue is appropriate for qualitative phenomena, as lightness and saturation differences suggest quantitative differences. Furthermore, we must select color hues that imply qualitative differences. For instance, red, green, and blue would be appropriate choices for qualitative phenomena (as shown in Figure 4.5) because these colors are not easily associated with a sequence of low, medium, and high data values.

4.5.3 Some Considerations in Working with Visual Variables

You should bear in mind that we have utilized only a fraction of the many symbols that could be used to illustrate the visual variables; for example, either circles or squares might be used to depict point phenomena for the size visual variable. A major group of symbols not shown in the figures

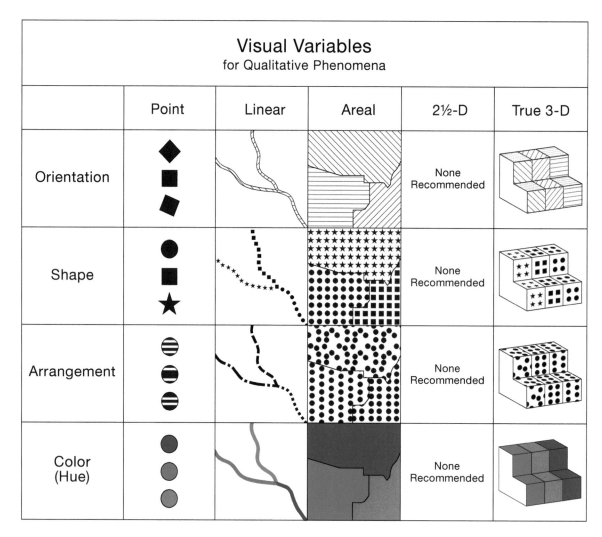

FIGURE 4.5   Visual variables for qualitative phenomena.

are **pictographic symbols**, which are intended to look like the phenomenon being mapped (as opposed to **geometric symbols** such as circles). For instance, Figure 4.6 illustrates the use of different-sized beer mugs to represent the number of microbreweries and brewpubs in each U.S. state. Pictographic symbols are often used in maps intended for children, although, as this example illustrates, they can be meaningful for adults too.

In the figures we have used, the term for each visual variable appears to be a clear expression of the visual differences that we see; for example, in the case of the orientation visual variable for point phenomena (Figure 4.5), we see that one square is at a different orientation than another. Moreover, if we wanted, we could compute a mathematical expression of this difference (that one square is rotated 40° from a vertical, whereas another is rotated 50°). Sometimes, describing the visual difference between symbols is not so easy. For example, try describing the differences between the symbols shown in Figure 4.7. Such symbols are often referred to as differing in "pattern" or "texture" and are frequently used to symbolize nominal data.

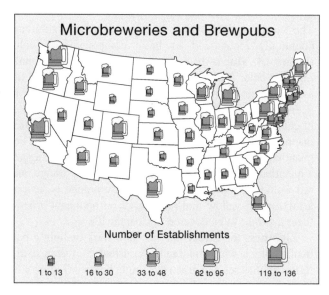

FIGURE 4.6   Using a pictographic visual variable (beer mugs) to represent the number of microbreweries and brewpubs in each U.S. state. (Data are for 2012; for similar data, see https://www.brewersassociation.org/statistics-and-data/state-craft-beer-stats/.)

FIGURE 4.7 Different "patterns" or "textures" that can be created to portray nominal information. Note that these are not readily described in terms of the visual variables shown in Figures 4.4 and 4.5.

It also should be noted that the visual variable location was not explicitly depicted in the illustrations. **Location** refers to the position of individual symbols. We chose not to illustrate this visual variable because it is an inherent part of mapping (e.g., each symbol shown for point phenomena can be defined by the *x* and *y* coordinate values of its center). If location were illustrated, it would be represented by constant symbols (identical dots for point phenomena) that varied only in position.

4.6 COMPARISON OF FOUR COMMON THEMATIC MAPPING TECHNIQUES

In this section, we define and contrast four common thematic mapping techniques: choropleth, proportional symbol, isopleth, and dot. For illustrative purposes, we examine these techniques by mapping data for acres of wheat harvested in counties of Kansas (Table 4.1). Here we consider a basic introduction to these mapping techniques; more advanced concepts are covered in subsequent chapters (15 for choropleth, 18 for proportional symbol, 17 for isopleth, and 19 for dot).

TABLE 4.1

Wheat Harvested in Kansas Counties 2011

County	Acres Harvested	Area of County (in Acres)	% of Land in Wheat
Wyandotte	215	97,024	0.2
Doniphan	932	251,782	0.4
Jefferson	1,500	340,845	0.4
Greenwood	3,600	731,712	0.5
.	.	.	.
.	.	.	.
.	.	.	.
Pratt	158,500	470,432	33.7
Mitchell	152,000	449,146	33.8
McPherson	202,000	574,893	35.1
Sumner	329,000	756,442	43.5
Harper	225,000	512,813	43.9

Source: National Agricultural Statistics Service, United States Department of Agriculture, https://quickstats.nass.usda.gov/; a complete table for all 105 counties can be found in the file Kansas_Wheat_2011_Data.xlsx at www.routledge.com/9780367712709.

4.6.1 CHOROPLETH MAP

A **choropleth map** is commonly used to portray data collected for enumeration units, such as counties or states. To construct a choropleth map, data for enumeration units are typically grouped into classes and a color is assigned to each class. The choropleth map is clearly appropriate when values of a phenomenon change abruptly at enumeration unit boundaries, such as for state sales tax rates. Choropleth maps might also be appropriate when you want the map reader to focus on "typical" values for individual enumeration units, even though the underlying phenomenon does not change abruptly at enumeration unit boundaries. For example, politicians and government officials might use this approach when stressing how one county or state compares with another. Although choropleth maps are commonly used in this fashion, it is important to recognize two major limitations: (1) Such maps do not portray the variation that might actually occur within enumeration units, and (2) the boundaries of enumeration units are arbitrary, and thus unlikely to be associated with major discontinuities in a phenomenon.[2]

An important consideration in constructing choropleth maps is **data standardization**, in which **raw totals** are adjusted for differing sizes of enumeration units. To understand the need to standardize, consider map A in Figure 4.8, which portrays a hypothetical distribution consisting of three distinct regions: S, T, and U. Assume that regions S and T have equal-sized enumeration units, each 16 acres in size. In contrast, assume that region U has enumeration units four times the size of those in S and T, or 64 acres in size.

Let's further assume that the number of acres of wheat harvested from enumeration units in each region is as follows: 0 in S, 16 in T, and 64 in U (these numbers are shown within each enumeration unit in Figure 4.8A). The acres of wheat harvested from each enumeration unit represent raw totals. Mapping these raw totals with the choropleth method produces the result shown in Figure 4.8B (note that higher data values are depicted by a darker color). A user examining this map would likely conclude that because region U is the darkest, it must have more wheat grown in it. Unfortunately, this conclusion would be inappropriate because we have not accounted for the size of enumeration units. One approach to adjust (or standardize) the size of enumeration units is to divide each raw total by the area of the associated enumeration unit; the resulting values are 0/16, or 0 for region S; 16/16, or 1 for region T; and 64/64, or 1 for region U. Mapping these values with the choropleth method results in Figure 4.8C; note that regions T and U now have the same color and thus have the same density of wheat harvested.

In a similar fashion, the Kansas wheat data shown in Table 4.1 can be standardized by dividing the acres of wheat harvested within each county by the area of the corresponding county (in acres). For example, for Wyandotte

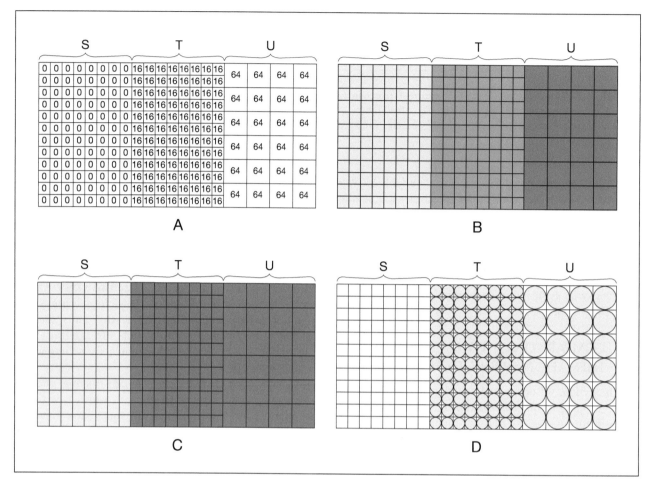

FIGURE 4.8 A hypothetical illustration of the effect of data standardization: (A) raw totals—number of acres of wheat harvested in each enumeration unit; (B) a choropleth map of the raw totals; (C) a choropleth map of standardized data achieved by dividing the raw totals by the area of the corresponding enumeration unit; and (D) a proportional circle map of the raw totals.

County, we divide 215 by 97,024 to yield a proportion of 0.002. If we then multiply by 100, we get the percentage value of 0.2 shown in the last column of the table. Note that the table has been sorted based on this last column. Figure 4.9 portrays maps of both the unstandardized and the standardized Kansas wheat data. The difference between the maps is not as dramatic as the hypothetical example because the areas of counties in Kansas are similar in size, but we can still note how certain counties are affected (for example, in southwestern Kansas, note how the relatively large L-shaped county (Finney) moves from the next to the highest class to the lowest class).

Standardization not only adjusts for the differing sizes of counties but also provides a very useful attribute, namely, the proportion (or percentage) of each county from which wheat was harvested. Such an attribute provides an indication of the probability that one might see wheat being cut at harvest time while driving through the county. The standardized map is shown in Figure 4.10A for comparison with the other methods of mapping discussed in this section. (Note that the *lightness* visual variable has been used for the choropleth map.)

4.6.2 PROPORTIONAL SYMBOL MAP

A **proportional symbol map** is constructed by scaling symbols in proportion to the magnitude of data occurring at point locations. These locations can be *true points*, such as an oil well, or *conceptual points*, such as the center of an enumeration unit for which data have been collected; the latter is the case with the wheat harvested data. In contrast to the standardized data depicted on choropleth maps, proportional symbol maps are normally used to display raw totals. Thus, the magnitudes for acres of wheat harvested are depicted as proportional circles in Figure 4.10C. (Note that the visual variable used here is *size*.)

The raw totals depicted on proportional symbol maps provide a useful complement to the standardized data shown on choropleth maps. Raw totals are important because a high proportion or rate might not be meaningful if there is not also a high raw total. As an example, consider counties of the same size having populations of 100 and 100,000, in which 1 and 1,000 people, respectively, have some rare form of cancer. Dividing the number of cancer cases by the population yields the same proportion of people suffering

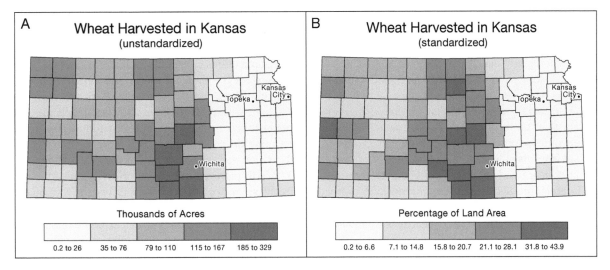

FIGURE 4.9 Standardizing the data for wheat harvested in Kansas counties in 2011: (A) a map of the unstandardized data (the number of acres harvested); and (B) a standardized map resulting from dividing the number of acres harvested by the area of each county. (From National Agricultural Statistics Service, United States Department of Agriculture, https://quickstats.nass.usda.gov/.)

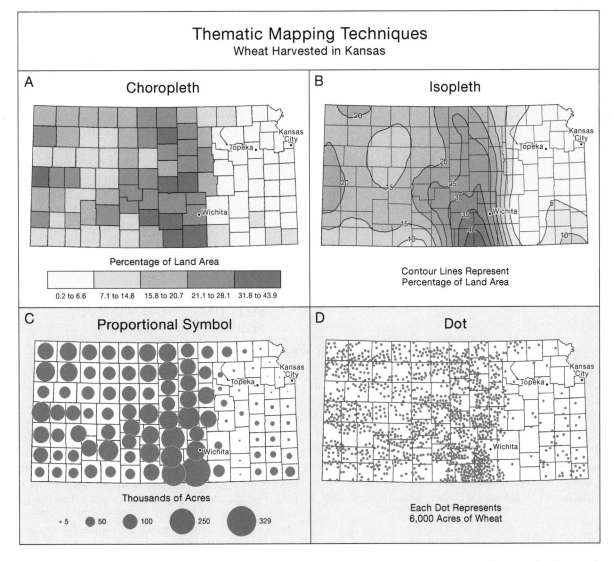

FIGURE 4.10 A comparison of basic thematic mapping techniques: (A) choropleth, (B) isopleth, (C) proportional symbol, and (D) dot maps. The choropleth and isopleth maps are based on the percentage of land area from which wheat was harvested, whereas the proportional symbol and dot maps are based on the total acres of wheat harvested.

from cancer (0.01), but the rate for the less populous county would be of lesser interest to an epidemiologist.

Although the proportional symbol map is a better choice than the choropleth map for depicting raw totals, care should be taken in using it. To illustrate, consider map D in Figure 4.8, which displays the hypothetical wheat data using proportional circles. Note that all circles in region U are larger than those in region T. This could lead to the mistaken impression that counties in region U are more important in terms of wheat production than those in region T. Counties in region U might be more important to a politician in assigning tax dollars (i.e., more wheat harvested indicates that a greater tax is appropriate), but in terms of the density of wheat harvested, regions T and U are identical.

4.6.3 ISOPLETH MAP

An **isarithmic map** (or **contour map**) is created by interpolating a set of isolines between sample points of known values; for example, we might draw isolines between temperatures recorded for individual weather stations. The **isopleth map** is a specialized type of isarithmic map in which the sample points are associated with enumeration units. It is an appropriate alternative to the choropleth map when one can assume that the data collected for enumeration units are part of a smooth, continuous $2\frac{1}{2}$-D phenomenon. For example, in the case of the wheat data, it might be argued that the proportion of land in wheat changes in a relatively gradual (smooth) fashion, as opposed to changing just at county boundaries (as on the choropleth map).

In a fashion similar to a choropleth map, an isopleth map also requires standardized data. Referring again to the hypothetical raw totals shown in Figure 4.8A, imagine drawing contours through such data. High-valued contour lines would tend to occur in region U, where there are high values in the data; but as has already been shown for the choropleth case, region U is really no different from region T. Dividing the raw totals by the area of each enumeration unit would result in standardized data that could be appropriately contoured.

The isopleth map resulting from contouring the standardized Kansas wheat data is shown in Figure 4.10B. (Again, note that the visual variable lightness has been used.) Although this map is likely more representative of the general distribution of wheat harvested than the choropleth map, we would be surprised if the distribution of wheat harvested followed these smooth trends, given the range of attributes that might affect where wheat is grown. Our final technique considered in this section, the dot map, may provide a more realistic map as it enables the consideration of multiple factors in determining where wheat is harvested.

4.6.4 DOT MAP

To create a **dot map**, one dot is set equal to a certain amount of a phenomenon, and dots are placed where that phenomenon is most likely to occur. The phenomenon might actually cover an area or areas (e.g., a field or fields of wheat), but for the sake of mapping, the phenomenon is represented as located at points. Constructing an accurate dot map requires collecting ancillary information that indicates where the phenomenon of interest (wheat, in our case) is likely found. For the wheat data, this was accomplished using the wheat category of a 2005 land use/land cover map (the detailed procedures are described in Chapter 19). The resulting dot map is shown in Figure 4.10D. (In this case, the visual variable *location* is used.) Clearly, the dot map is able to represent the spatial distribution of the underlying phenomenon (wheat that is harvested) with much more accuracy than any of the other methods we have discussed. Also, note that parts of the distribution exhibit sharp discontinuities that would be difficult to show with the isopleth method (which presumes smooth changes).

4.6.5 DISCUSSION

An examination of Figure 4.10 reveals that each of the four maps provides a quite different picture of the spatial variation of wheat harvested in Kansas. Which method is used should depend on the purpose of the map. If the purpose is to focus on "typical" county-level information, then the choropleth and proportional symbol maps are appropriate. The choropleth map provides standardized information, whereas the proportional symbol map provides raw total information. In contrast, the dot and isopleth methods should be considered as two possible solutions for representing an underlying phenomenon that is not coincident with enumeration unit boundaries. In the case of the wheat data, the dot method is probably the more appropriate approach because it captures some of the discontinuities in the phenomenon. The isopleth method, however, could be improved with a finer grid of enumeration units (e.g., townships);[3] of course, this would also be true of the choropleth and proportional symbol maps.

4.7 SELECTING VISUAL VARIABLES FOR CHOROPLETH MAPS

In the preceding section, the visual variable lightness was utilized for the choropleth map. From Section 4.5, we know that there are a number of other visual variables that might be used to represent a quantitative phenomenon that is treated as areal in nature. This section considers how to choose among these visual variables and how to avoid selecting improper qualitative visual variables. The basic solution is to choose a visual variable that appears to conceptually "match" the level of measurement of the data. For illustrative purposes, we again use the Kansas wheat data. Several visual variables that we discuss are illustrated in Figure 4.11. We also consider Figure 4.12, which summarizes the use of visual variables for various levels of measurement. Note that the body of this figure is shaded and labeled to indicate various levels of acceptability: Poor (P), Marginally effective (M), and Good (G).

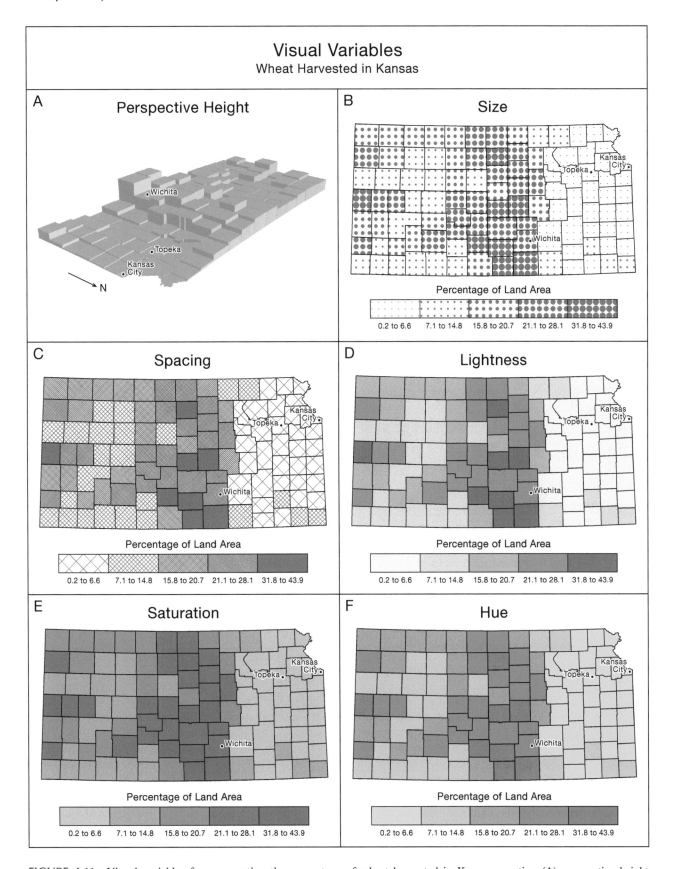

FIGURE 4.11 Visual variables for representing the percentage of wheat harvested in Kansas counties: (A) perspective height, (B) size, (C) spacing, (D) lightness, (E) saturation, and (F) hue.

	Nominal	Ordinal	Numerical
Spacing	P	M[c]	M[c]
Size	P	M	M
Perspective Height	P	M[a]	G[b]
Orientation	G	P	P
Shape	G	P	P
Arrangement	G	P	P
Lightness	P	G	G
Hue	G	G[d]	M[d]
Saturation	P	M	M

P = Poor   M = Marginally Effective   G = Good

[a] Since height differences are suggestive of numerical differences, use with caution for ordinal data.

[b] Hidden enumeration units and lack of a north orientation are problems.

[c] Not aesthetically pleasing.

[d] The particular hues selected must be carefully ordered, such as yellow, orange, red.

FIGURE 4.12   Effectiveness of visual variables for each level of measurement for areal phenomena. (After MacEachren 1994, 33.)

Alan MacEachren (1994, 33) developed a similar figure, which he appeared to apply to all kinds of spatial phenomena. We use Figure 4.12 only for areal phenomena; as an exercise, you might consider developing such a figure for other kinds of phenomena.

Let's consider the perspective height and size visual variables first because they have the greatest potential for logically representing the numerical data depicted on choropleth maps. The use of perspective height produces what is commonly termed a **prism map** (Figure 4.11A). In Figure 4.12, note that perspective height is one of only two visual variables receiving a "good" rating for numerical data. The justification for a "good" rating is that an unclassed map (as is shown here) based on perspective height can portray ratios correctly (a data value twice as

large as another will be represented by a prism twice as high) and that readers will perceive the height of resulting prisms as ratios (Cuff and Bieri 1979).

There are two problems, however, that complicate the extraction of numerical information from prism maps. One is that tall prisms sometimes block smaller prisms. A solution to this problem is to rotate the map so that blockage is minimized; for example, the map in Figure 4.11A has been rotated so that the view is from the lower-valued northeast. A second solution to the blockage problem is to manipulate the map in an interactive graphics environment. If a flexible program is available, it might even be possible to suppress selected portions of the distribution so that other portions can be seen. A third solution is to use the perspective height variable but also symbolize the distribution with another visual variable; for example, Figure 4.11D might be displayed in addition to Figure 4.11A. In Section 15.8, we'll show how these two techniques can be combined into a single map using the illuminated choropleth technique. Another problem with prism maps is that rotation might produce a view that is unfamiliar to readers who normally see maps with north at the top. This problem can be handled to some extent by using an overlay of the base to show the amount of rotation (as in Figure 4.3C).

The size visual variable is illustrated in Figure 4.11B; note that here the size of individual marks making up the areal symbol has been varied. Size might be considered appropriate for representing numerical relations because circles can be constructed in direct proportion to the data (a data value twice another can be represented by a circle twice as large in area). However, we will see in Chapter 18 that a correction factor might have to be implemented to account for underestimation of larger circles.

Although some cartographers (most notably Bertin) have promoted the use of the visual variable size on choropleth maps, two problems are apparent. First, it is questionable whether map users actually consider the sizes of circles when used as part of an areal symbol. Users might analyze circle size when trying to acquire specific information, but it seems unlikely that they would do so when analyzing the overall map pattern. Rather, it is more likely that they perceive areas of light and dark in a fashion similar to the lightness visual variable. Second, many cartographers (and presumably map users) find the coarseness of the resulting symbols unacceptable—they prefer the fine tones shown in Figure 4.11D. The latter problem, in particular, caused us to give the size variable only a moderate rating for portraying numerical data (Figure 4.12).

Note also that we have given both perspective height and size only moderate ratings for portraying ordinal data. The logic is that if these visual variables are intended to illustrate numerical relations, users might perceive such relations when only ordinal relations are intended. Obviously, both visual variables are inappropriate for nominal data because different heights and sizes suggest quantitative rather than qualitative information.

Although other visual variables can be manipulated mathematically to create proportional (ratio) relationships, the resulting symbols cannot be interpreted easily in a ratio fashion. For example, consider the lightness variable shown in Figure 4.11D. It is easy to see that one shade is darker or lighter than another, but it is difficult to establish proportional relations (e.g., that one shade is twice as dark as another). We have given the lightness variable a rating of "good," however, because of its common use by cartographers for representing numerical data. Similar comments can be made for the visual variables spacing, saturation, and hue, with the following caveats: First, note that we have given spacing only a moderate rating for ordinal information because, in our opinion, the symbols are not aesthetically pleasing (Figure 4.11C), and there is the implication that low data

values are qualitatively different from high data values. Second, we have given saturation (Figure 4.11E) only a moderate rating for ordinal information because it is our experience that people have a difficult time understanding what a "greater" saturation means.

We have rated hue as "good" for both nominal and ordinal data because some hues work well for nominal data, and other hues work better for ordinal data. For example, to display different soil types (alfisols, entisols, mollisols), red, green, and blue hues might be deemed appropriate (one of these hues does not inherently represent a higher data value than another). For ordinal and higher-level data, logically ordered hues are necessary; for example, a yellow, orange, and red scheme (Figure 4.11F) is one possibility because orange is seen as a mixture of yellow and red.

The remaining visual variables (orientation, shape, and arrangement) are appropriate only for creating nominal differences. Thus, these three visual variables receive a rating of "good" for nominal data and a rating of "poor" for ordinal and numerical data in Figure 4.12. It should be noted that cartographers are not in complete agreement on the ratings displayed in Figure 4.12, and so you might develop slightly different ratings. For example, one of our students rated the orientation variable "poor" (even for nominal data) because he felt it lacked aesthetic quality and that it was difficult to discriminate among different orientations.

4.8 USING SENSES OTHER THAN VISION TO INTERPRET SPATIAL PATTERNS

As the term "geovisualization" in this book's title implies, the bulk of this book focuses on how people use vision to interpret spatial patterns. Increasingly, however, map designers have begun considering how we might use our other senses, either as an alternative or as a supplement to vision. Using senses other than vision is obviously critical for blind (or visually impaired) individuals, but there are numerous reasons why senses other than vision can be useful even to normally sighted individuals. First, other senses can be useful when the work environment requires that vision be used for tasks besides map interpretation (as in an aircraft cockpit). Second, other senses can assist in map interpretation when visual interpretation is not clear. For instance, Suresh Lodha and colleagues (2000) found that the comparison of regions on choropleth maps was enhanced when both vision and sound were used. Third, multiple senses can be useful for interpreting multivariate data; for instance, Chris Harding and colleagues (2002) used vision, sound, and touch to represent three respective attributes (gravity, age of the oceanic crust, and change of slope) associated with geological structures along the Mid-Atlantic Oceanic Ridge. Below we consider additional examples of how sound, touch, and smell might be utilized to interpret spatial patterns.

4.8.1 SOUND

From a cartographic standpoint, John Krygier (1994) conducted some of the earliest work on using sound as an alternative to vision. Krygier noted two basic types of sound: realistic and abstract. He divided **realistic sounds**, those sounds whose meanings are based on our past experiences, into speech (or narration) and mimetic sound icons (or "earcons"). A simple example of speech is when a computer speaks the value of an enumeration unit rather than displays it as text. Another example of speech is the audio component sometimes included in multimedia encyclopedias (e.g., you might hear, "You are now looking at major Allied troop movements during the Battle of the Bulge"). An example of a mimetic sound icon would be the sounds of fire and wind in an animation illustrating changes in a forested landscape. If we think about realistic sounds in a broad sense, we can conceive of a **soundscape**, which considers the perceived acoustic environment at a landscape scale. For instance, Kimon Papadimitriou and colleagues (2009) describe how we might visualize a soundscape consisting of sounds from human activity, biological organisms, and geophysical processes.

Abstract sounds have no obvious meaning and thus require a legend to explain their use. For example, imagine a map of census tracts with the title "Median Income, 1997" in which different magnitudes of loudness represent different incomes (e.g., a mouse click on a high-income tract would produce a louder sound than a mouse click on a low-income tract). To understand the magnitudes of loudness, a legend indicating that a higher magnitude represents a higher income would have to be provided (if the reader were blind, someone would have to either tell the reader what each magnitude of loudness indicates or give the reader a tactile legend). The process of creating abstract sounds is sometimes referred to as **sonification**.

Analogous to the visual variables that we discussed in Section 4.5, Krygier (1994) identified the **abstract sound variables** shown in Figure 4.13. As Figure 4.13 shows, Krygier argued that the bulk of these variables would be effective for ordinal-level data. Krygier also created a video that illustrated several applications of how abstract sound variables might be used (the video can be downloaded from https://krygier.owu.edu/krygier_html/ohio_university_sound_maps.html). In one application, he combined a choropleth animation of AIDS in the United States from 1980 to 1995 with a loudness variable representing the total number of AIDS cases in each year. When viewing the resulting animation, one gets the feeling of an impending disaster, which is exactly what Krygier intended.

In another application, Krygier created a multivariate display by combining traditional visual variables (lightness and size) with sound variables. For the traditional visual variables, Krygier superimposed proportional circles onto a choropleth map. The proportional circles depicted "median income," whereas the choropleth map displayed the "percentage of population not in the labor force." For

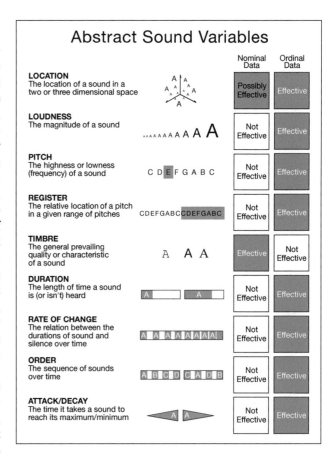

FIGURE 4.13 Abstract sound variables. (After Krygier, J. B., *Sound and Geographic Visualization,* Copyright 1994, p. 153, with permission from Elsevier Science.)

the sound variables, Krygier used a pitch within three different octaves (a register) to display a "drive to work index" and three levels of pitch within each of the three octaves to represent the "percentage poor." He obtained values for the sound variables by pointing at enumeration units. For example, for an enumeration unit with a long drive to work and a low rate of poverty, a high-octave pitch followed by a low pitch within that octave would be heard. Krygier (1994) argued that "After a short period of [use] … it becomes relatively easy to extract the four data variables" (158).

Several other researchers performed early experiments with sound. Peter Fisher (1994) developed software that utilized sound to portray the uncertainty in remotely sensed images. With Fisher's software, a cursor could be moved across a remotely sensed image under either user or automatic control. As the cursor moved, the user heard a sound representing the uncertainty associated with the current pixel location. For example, if duration were used as the sound variable, a long duration would indicate a pixel with a low uncertainty (i.e., high reliability). In using Fisher's software, we found it relatively easy to determine the uncertainty associated with specific pixels but difficult to determine the uncertainty of groups of pixels. This problem might be obviated by also using visual methods to depict uncertainty. For example, we might utilize both sound and a set of grays to depict uncertainty (say, with dark grays representing more uncertain information).

A more recent example of the use of sound is the iSonic software developed by Haixia Zhao and colleagues (2008). They noted that traditional approaches for presenting spatial data to the visually impaired often utilized text readers that speak geographic names and values in alphabetical order, thus preventing a full understanding of the spatial pattern of a data set. As an alternative, iSonic provides a variety of novel approaches to assist in interpreting spatial data. For example, a choropleth map can be interpreted by associating map classes with violin pitches and then playing these pitches in a fixed spatial sequence, from left to right across the map and from the top to the bottom of the map. Alternatively, it is possible to divide a map into regions and play the values for enumeration units in each region. User studies revealed that blind, visually impaired and normally sighted (blindfolded) users could utilize iSonic to acquire both specific and general map information but that they struggled with learning the details of spatial patterns (Zhao et al. 2008; Delogu et al. 2010). For an overview of iSonic, see https://www.youtube.com/watch?v=8hUIAnXtlc4.

As you can see in these examples, most of the research on sound has focused on how *abstract* sounds can be used to represent spatial data. Researchers also have explored how *realistic* sounds might be utilized. An example is Glenn Brauen's (2006) use of sound to demonstrate the variability in Canadian federal election results. In traditional election maps, each electoral district is colored according to the affiliation of the winning candidate. As an alternative, Brauen created a sound map in which the volume of a speech given by each candidate was proportional to the percentage of votes received by that candidate. When a cursor is placed over a voting district, the viewer can hear each of the candidate's voices, with a winning candidate being dominant. In a closely contested election, "several leaders' speeches blend together, simultaneously vying for attention, [which] … reintroduces some of the complexity of the election results into the map" (64).[4]

Recent use of sonification is Joram Schito and Sara Fabrikant's (2018) application to digital elevation models (DEMs). In theory, this should be challenging because DEMs are continuous in a numerical sense (changing elevation values) and in spatial sense (changing latitude and longitude values). However, Schito and Fabrikant found that sighted (blindfolded) users could successfully interpret continuous DEMs and, in fact, performed better than with discrete DEMs. This result suggests that at least some of Krygier's abstract sound variables might be extended into the numerical realm. Shito and Fabrikant also found that the accuracy of DEM interpretation was positively correlated with one's musical ability, technical ability with digital input devices, and experience with topographic mapping. This may indicate that sonified maps for DEMs are not for everyone or that some training may be necessary to achieve maximum benefit from them.

Another recent intriguing application is Carlos Duarte and colleagues' (2018) sonification of the movement of 321 tagged elephant seals over a decade. The key data parameters for the seals were the position (as expressed primarily by longitude) and the spread (the degree of displacement for the group), which were mapped to pitch and volume, respectively, in a 54-minute video. Duarte and colleagues developed a number of interesting observations based on the sonification; for instance, they stated:

> The harmony in the sonification suggests a remarkable degree of coherence in the movement of the animals, both in terms of the zonal displacement along the North Pacific … as well as the expansion and contraction of the spread between the animals involved in each trip …. (5)

It would be interesting to see whether those less familiar with the data would develop these sorts of observations.

4.8.2 Touch (or Haptics)

Strictly speaking, touch is the sensation we experience when our skin comes in contact with an object, such as when we touch a three-dimensional physical model. **Haptics** is a broader term that includes not only the sensation of contact but also the kinesthetic receptors activated in our muscles, tendons, and joints, as in the action of "sweeping the hand across a three-dimensional surface" (Golledge et al. 2005, 340). In practice, however, these terms are often used interchangeably.

There has been a considerable amount of work on haptic maps for the blind and visually impaired (for overviews of recent work, see Koch (2012) and Touya et al. (2019)), but the potential of haptics for normally sighted users has not been fully explored. In response to this limitation, Amy Griffin (2001) proposed the **haptic variable syntax** shown in Figure 4.14. In examining this figure, you can see that Griffin split the haptic variables into three types: *tactile*, *visual analog derived*, and *kinesthetic*. The tactile haptic variables involve only touch; for instance, by placing a finger on a portion of a map, given the appropriate technology, one should be able to feel whether the map is vibrating and how cool or warm it is. The visual analog derived variables are named after the traditional visual variables and again involve only touch; for instance, circular disks of different sizes could be placed on a map, and a user could be asked to judge the differences in the disks' sizes based on touch. The kinesthetic variables involve "a movement of, or a change in the user's position, or in the position of a stimulus relative to the user" (17); for instance, Griffin suggested that increased friction could correspond to increased terrain roughness. Griffin specified that haptic variables, similar to sound variables, should be used only for nominal or ordinal levels of measurement

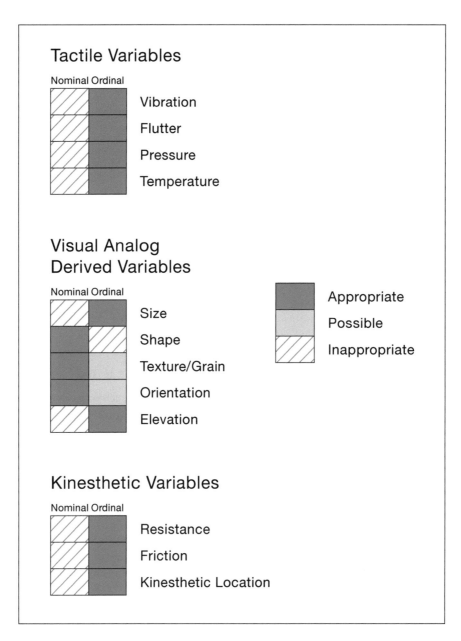

FIGURE 4.14 A haptic variable syntax for touch. (After Griffin 2001, 18; courtesy of *Cartographic Perspectives* (North American Cartographic Information Society, publisher) and Amy Griffin.)

(Figure 4.14). Although Griffin did not actually test the proposed haptic variables with people, numerous applications were suggested such as the following:

> the vibration variable might be used to visualize the output of a seismic model that predicts the intensity of shaking due to an earthquake .... Temperature might be used to represent water temperature in a river, so that the user could track cold currents through space and time. Resistance could be used to represent measures of pollution; the more polluted an area, the more difficult it might be to ... move your hand through the area. (After Griffin 2001, 19; courtesy of *Cartographic Perspectives* (North American Cartographic Information Society, publisher) and Amy Griffin.)

It is important to note that sound and haptics do not need to be used alone but can be combined in what is termed a **multimodal interface**. For instance, Reginald Golledge and colleagues experimented with both sound and haptics in the Haptic Soundscapes project (e.g., Golledge et al. 2005; Rice et al. 2005). As another example, Nikolaos Kaklanis and colleagues (2013) utilized touch and sound to allow the blind and visually impaired to examine OpenStreetMap data.

4.8.3 SMELL

Although it may seem far-fetched to acquire information from maps via a sense of smell, Tracy Lauriault and Gitte

Lindgaard (2006) made a strong case for the potential of using smell to interpret spatial data. They noted that in Western cultures, smell is often placed at the bottom of the sensory hierarchy, but clearly, there are instances in which smell is extremely important, such as when evaluating the effectiveness of winegrowing regions in France. Lauriault and Lindgaard also note that untrained people can recognize only about 16 scents but that specialists (e.g., those in the perfume industry) can recognize hundreds of scents. As an example of the use of smell in cartography, they reference the map *Twin Cities Odorama: A Smell Map of Minneapolis and Saint Paul*. Scents are not actually embedded in this map, but quotes such as the following provide an idea of the importance of smell to inhabitants of the Minneapolis–St. Paul region:

> Minneapolis' Ichiban Japanese Restaurant always smells wonderful when you walk by. It's that distinctive Japanese food smell of Tepanyaki, Sushi and Tempura. It has a very comforting emotional feel to me.

Actually, generating this particular smell could be challenging, but Lauriault and Lindgaard note that hardware and software have been developed for generating smells (for example, see https://en.wikipedia.org/wiki/Digital_scent_technology). As technology evolves and as our willingness to consider the importance of the sense of smell also evolves, it will be interesting to see what role smell plays in cartography.

4.9 SUMMARY

In this chapter, we have introduced basic principles of symbolization that enable us to select appropriate visual variables to represent geographic phenomena. In Figures 4.4 and 4.5, we have seen that the choice of a visual variable is a function of the spatial dimension of the phenomenon (point, linear, areal, $2\frac{1}{2}$-D or true 3-D) and whether the phenomenon is quantitative or qualitative. For the Kansas wheat example illustrated in Figure 4.10, we saw that if we consider the phenomenon areal (associated with counties) and collect quantitative data on acres of wheat harvested, two maps are possible: a proportional symbol map (using the visual variable size) depicting the raw count of acres harvested, and a choropleth map (using the visual variable lightness) depicting the standardized percent of land area in wheat. Alternatively, if we consider the phenomenon $2\frac{1}{2}$-D (as part of a surface), two other maps are possible: an isopleth map (visual variable lightness) for the standardized data and a dot map (visual variable location) for the raw unstandardized data. With the dot map, we are assuming discontinuities in the surface, which is probably more representative of what we find in the real world. In Figure 4.11, we saw a variety of other visual variables that could be used for the Kansas wheat data. In Chapter 15, we will explore some of these visual variables in greater depth when we focus on choropleth mapping.

The focus of this chapter, and ultimately this book, is on approaches for *visualizing* spatial data, i.e., these approaches make use of our sense of vision. We have seen that our other senses (sound, touch, and smell) can also be used to interpret spatial patterns. For instance, if we wished to illustrate the spatial distribution of lumbering in a geographic region, we might use the following to represent increased lumbering as we pointed to various areas on the map: a louder chain saw, a rougher texture (if you wish to show that greater lumbering in a particular area is undesirable), and a stronger odor of cut lumber. Although the use of these other senses is not common, ultimately, they may provide the map designer with a broader and more effective range of tools for depicting spatial patterns for not only the visually impaired but also normally sighted individuals.

4.10 STUDY QUESTIONS

1. For each of the following, indicate the appropriate spatial dimension (point, linear, areal, $2\frac{1}{2}$-D and true 3-D) and justify your choice: (A) the level of sulfur dioxide in the air above an urban area; (B) interstate highways to be shown on a page-size map of California; (C) location of fast-food restaurants in Houston, Texas; (D) the depth to the bottom of Lake Baikal from each point throughout the horizontal extent of the lake; (E) polygons defining regions burned in Colorado during the summer of 2020.

2. Using Figure 4.2, explain where you would place the COVID-19 pandemic on the continuous-discrete and abrupt-smooth scales.

3. For each of the following; specify the level of measurement (nominal, ordinal, interval, and ratio) and justify your choice: (A) population density (in people per square kilometer); (B) occupation (e.g., miner, barber, teacher, etc.) of people in a survey; (C) soils specified as either marginally, moderately, or highly suitable for agriculture; (D) Stream depth with a rod that has equally spaced markings of 10 units between 0 and 100, but 0 does not start at the bottom of the rod.

4. Using Figures 4.4 and 4.5, explain how you would select an appropriate visual variable for each of the following attributes: A) the predominate religion in each country of Europe; B) the precipitation (in centimeters) received throughout India over the course of the year.

5. (A) Assuming that you wish to create a choropleth map, discuss whether it would make sense to map the number of wild horses found in each county of Nevada. (B) Which of the other maps in Figure 4.10 would you recommend as an alternative to the choropleth map? Explain your response.

6. Presume that you are working for a small GIS firm in Louisville, KY, and your supervisor asks you to make a map showing how median income varies across

the city and provides you data on median income for each census tract in the city. Assuming that you can make only one map, explain how you would choose between a choropleth and an isopleth map.

7. Contrast the spatial patterns that you see on the perspective height and lightness maps shown in Figure 4.11. Discuss the advantages and disadvantages that you see for each of these mapping techniques.

8. Explain how the visual variable hue can receive a "Good" rating for both nominal and ordinal data in Figure 4.12.

9. View the video describing the iSonic system (https://www.youtube.com/watch?v=8hUIAnXtlc4) and discuss whether you feel a system similar to iSonic could be used to interpret spatial patterns on a county-level map of the entire U.S. (Try to ignore the quality of the computer speech in the demo as a present-day system would have a higher speech quality.)

NOTES

1 See Chrisman (1998) for several extensions to the four basic levels of measurement that we have covered.

2 See Langford and Unwin (1994) for a more detailed discussion of these limitations.

3 Data at the township level are not released to the general public to protect the confidentiality of individual farm production.

4 An illustration of Brauen's sound map for election results can be found at https://play.library.utoronto.ca/play/05ba9826f9d0c163bd3fcf345aa87ca8. For a more complete set of audio-visual work by Brauen, see https://avcartography.gbrauen.ca/.

REFERENCES

Bertin, J. (1983) *Semiology of Graphics: Diagrams, Networks, Maps*. Translated by W. J. Berg. Madison, WI: University of Wisconsin Press.

Brauen, G. (2006) "Designing interactive sounds maps: Using scalable vector graphics." *Cartographica 41*, no. 1:59–71.

Chrisman, N. R. (1998) "Rethinking levels of measurement for cartography." *Cartography and Geographic Information Systems 25*, no. 4:231–242.

Cuff, D. J., and Bieri, K. R. (1979) "Ratios and absolute amounts conveyed by a stepped statistical surface." *The American Cartographer 6*, no. 2:157–168.

Delogu, F., Palmiero, M., Federici, S. et al. (2010) "Non-visual exploration of geographic maps: Does sonification help?" *Disability and Rehabilitation: Assistive Technology 5*, no. 3:164–174.

Duarte, C. M., Riker, P., Srinivasan, M., et al. (2018) "Sonification of animal tracks as an alternative representation of multi-dimensional data: A northern elephant seal example." *Frontiers in Marine Science 5*:1–9.

Eastman, J. R. (1986) "Opponent process theory and syntax for qualitative relationships in quantitative series." *The American Cartographer 13*, no. 4:324–333.

Fisher, P. F. (1994) "Hearing the reliability in classified remotely sensed images." *Cartography and Geographic Information Systems 21*, no. 1:31–36.

Golledge, R. G., Rice, M., and Jacobson, R. D. (2005) "A commentary on the use of touch for accessing on-screen spatial representations: The process of experiencing haptic maps and graphics." *The Professional Geographer 57*, no. 3:339–349.

Griffin, A. L. (2001) "Feeling it out: The use of haptic visualization for exploratory geographic analysis." *Cartographic Perspectives* no. 39:12–29.

Harding, C., Kakadiaris, I. A., Casey, J. F. et al. (2002) "A multisensory system for the investigation of geoscientific data." *Computers & Graphics 26*:259–269.

Kaklanis, N., Votis, K. and Tzovaras, D. (2013) "Open Touch/Sound Maps: A system to convey street data through haptic and auditory feedback." *Computers & Geosciences 57*:59–67.

Koch, W. G. (2012) "State of the art of tactile maps for visually impaired people." In *True-3D in Cartography: Autostereoscopic and Solid Visualization of Geodata*, ed. by M. Buchroithner, pp. 137–151. Berlin: Springer.

Krygier, J. B. (1994) "Sound and geographic visualization." In *Visualization in Modern Cartography*, ed. by A. M. MacEachren and D. R. F. Taylor, pp. 149–166. Oxford, England: Pergamon.

Langford, M., and Unwin, D. J. (1994) "Generating and mapping population density surfaces within a geographical information system." *The Cartographic Journal 31*, no. 1:21–26.

Lauriault, T. P., and Lindgaard, G. (2006) "Scented cybercartography: Exploring possibilities." *Cartographica 41*, no. 1:73–91.

Lodha, S. K., Joseph, A. J., and Renteria, J. C. (2000) "Audio-visual data mapping for GIS-based data: An experimental evaluation." In *Workshop on New Paradigms in Information Visualization and Manipulation (NPIVM '99)*, ed. by D. S. Ebert and C. D. Shaw, pp. 41–48. New York: The Association for Computing Machinery.

MacEachren, A. M. (1992) "Visualizing uncertain information." *Cartographic Perspectives* no. 13:10–19.

MacEachren, A. M. (1994) *Some Truth with Maps: A Primer on Symbolization & Design*. Washington, D.C.: Association of American Geographers.

MacEachren, A. M., and DiBiase, D. (1991) "Animated maps of aggregate data: Conceptual and practical problems." *Cartography and Geographic Information Systems 18*, no. 4:221–229.

Papadimitriou, K. D., Mazaris, A. D., Kallimanis, A. S. et al. (2009) "Cartographic representation of the sonic environment." *The Cartographic Journal 46*, no. 2:126–135.

Rice, M., Jacobson, R. D., Gollege, R. G. et al. (2005) "Design considerations for haptic and auditory map interfaces." *Cartography and Geographic Information Science 32*, no. 4:381–391.

Schito, J. and Fabrikant, S. I. (2018) "Exploring maps by sounds: Using parameter mapping sonification to make digital elevation models audible." *International Journal of Geographical Information Science 32*, no. 5:874–906.

Touya, G., Christophe, S., Favreau, J.-M. et al. (2019) "Automatic derivation of on-demand tactile maps for visually impaired people: First experiments and research agenda." *International Journal of Cartography 5*, no. 1:67–91.

Zhao, H., Plaisant, C., Shneiderman, B. et al. (2008) "Data sonification for users with visual impairment: A case study with geo-referenced data." *ACM Transactions on Computer-Human Interaction 15*, no. 1:Article no. 4.

5 Data Classification

5.1 INTRODUCTION

Data for choropleth maps are typically grouped into classes, and a particular color is used for each class (a **classed map** results); for instance, in Figure 1.8A, percentage values for foreign-owned agricultural land have been grouped into five classes, and a lightness of blue has been used to depict each class. Although it is possible to not class the data (and let each data value be represented by a different lightness of a color), as in Figure 1.8B (an **unclassed map** results), classed maps are much more common. Why is the classed map more common? Cartographers generally have argued that the classed map is easier to interpret because of the limited number of colors that need to be discriminated and the relative ease with which these can be matched with appropriate legend colors. We consider the question of classed versus unclassed maps in more depth in Chapter 15 when we focus on choropleth mapping. In the present chapter, we focus on five common methods of data classification (equal interval, quantiles, mean-standard deviation, natural breaks, and optimal). In addition, we consider a relatively new method called head-tail breaks. For each method, we discuss the underlying computations and key advantages and disadvantages. We conclude the chapter by considering the potential role of the spatial distribution of the data in classification, a concept that is often ignored when discussing classification.

5.2 LEARNING OBJECTIVES

- Explain why cartographers classify data for choropleth maps.
- Compute class limits for data classification methods.
- Compare the objectives of data classification methods.
- Contrast the advantages and disadvantages of data classification methods.
- Describe how the optimal method of data classification can be used to determine an appropriate number of classes.
- Explain how the spatial distribution of data can be used in classification.

5.3 DATA TO BE CLASSIFIED

For illustrative purposes, we map the percentage of foreign-born residents for counties in Florida in the United States from 2006 to 2010. These data were collected from the American Community Survey (ACS), which is based on a sample of households rather than all households throughout the United States. We map a composite of five years of data to reduce the uncertainty in the data associated with using a sample. Even with this composite of five years of data, there is still uncertainty in the data; in Chapter 27, we will consider methods for depicting this uncertainty. For now, use the following URL for the American Community Survey Sample Size Definitions to read more about the nature of this data and its associated uncertainty: https://tinyurl.com/ACSUncertainty.

The foreign-born data (sorted from low to high data values) and dispersion graphs for this data set are given in Table 5.1 and Figure 5.1, respectively (for now, just consider the top dispersion graph in Figure 5.1). Note that the dispersion graph has a positive skew (the bulk of the data are concentrated to the left), with a distinctive outlier (i.e., Miami-Dade County, which includes the city of Miami).

Before attempting to classify data, it is useful to consider whether the data are bipolar (whether there is either a natural or a meaningful dividing point that can be used to partition the data; see Section 4.4). For example, a data set of "percent population change" has a natural dividing point of zero, which can be used to create two classes: values at or above zero and values below zero. Once data have been split in such a fashion, it might be appropriate to apply one of the methods discussed in this chapter to each subset of data. Even if there is no natural dividing point in the data, it might be desirable to create a meaningful dividing point prior to classifying. For example, we might compute the mean for foreign born in all U.S. counties and use that as a dividing point for the Florida foreign-born data. For this chapter, we treat the data as unipolar and assume that there is no natural or meaningful dividing point that we wish to use.

An important consideration in any method of classification is selecting an appropriate *number of classes*. In Section 5.8.3, we will discuss how the results of optimal classification can be used to select an appropriate number of classes. Ultimately, we chose to use six classes because we knew it would be difficult to distinguish between lightnesses of blue using more than the six classes shown in Figure 5.2.

Another essential consideration is determining the level of data precision that you wish to portray on the map. The level of precision you display should be a function of the initial data you have available, your impression of the quality of the data, and how easily you think readers can interpret the numeric values you provide. Since the American Community Survey provided the data to the nearest tenth of a percent and we felt that readers of this book would be

DOI: 10.1201/9781003150527-6

TABLE 5.1

Percentage of Foreign-Born Population in Florida Counties in 2006–2010 (Sorted from Low to High Data Values)

Observation	County	Foreign-Born (%)
1	Baker	1.0
2	Liberty	1.0
3	Bradford	1.7
4	Wakulla	2.0
5	Taylor	2.1
6	Holmes	2.3
7	Dixie	2.6
8	Gilchrist	2.6
9	Nassau	2.7
10	Franklin	3.1
11	Jackson	3.1
12	Gulf	3.2
13	Columbia	3.3
14	Washington	3.3
15	Calhoun	3.6
16	Madison	3.8
17	Jefferson	3.9
18	Putnam	4.2
19	Lafayette	4.3
20	Levy	4.4
21	Santa Rosa	4.7
22	Walton	4.9
23	Hamilton	5.0
24	Citrus	5.3
25	Bay	5.4
26	Gadsden	5.6
27	Suwannee	5.8
28	Union	5.8
29	Clay	5.9
30	Escambia	5.9
31	Sumter	5.9
32	St. Johns	6.3
33	Leon	6.5
34	Hernando	6.6
35	Okaloosa	6.8
36	Volusia	7.7
37	Marion	7.8
38	Brevard	8.6
39	Lake	8.7
40	Duval	9.0
41	Charlotte	9.2
42	Pasco	9.2
43	Alachua	10.2
44	Martin	10.4
45	Indian River	10.5
46	Polk	10.7
47	Pinellas	11.2
48	Highlands	11.3
49	Seminole	11.5
50	Sarasota	11.6
51	Flagler	12.1
52	Okeechobee	12.1
53	Manatee	12.3
54	Hillsborough	15.1
55	Lee	15.3
56	St. Lucie	15.6
57	Monroe	16.2
58	Glades	16.3
59	Orange	19.1
60	DeSoto	19.6
61	Osceola	19.7
62	Hardee	21.3
63	Palm Beach	22.3
64	Collier	23.6
65	Hendry	27.8
66	Broward	30.9
67	Miami-Dade	51.1

Source: American Community Survey, https://www.census.gov/programs-surveys/acs/data.html.

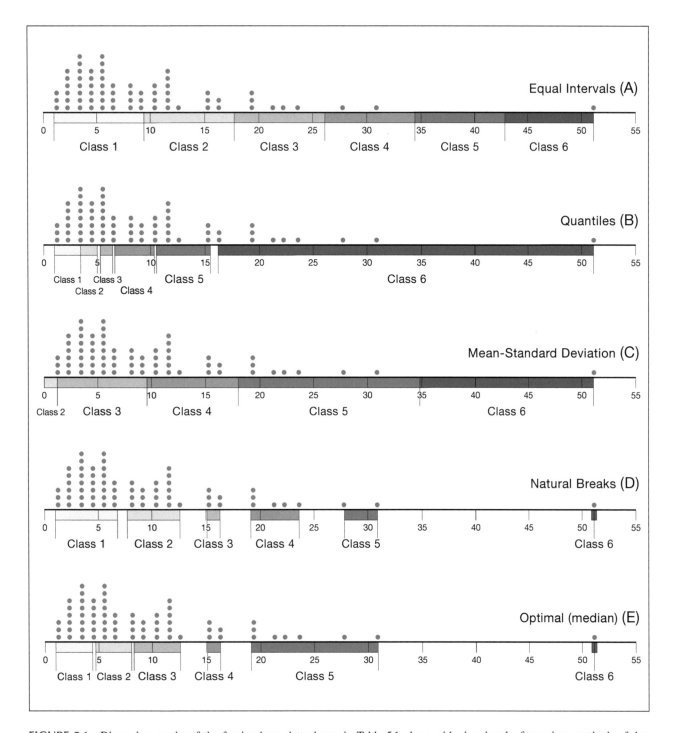

FIGURE 5.1 Dispersion graphs of the foreign-born data shown in Table 5.1 along with class breaks for various methods of data classification.

comfortable with this level of precision, we chose to display data to the nearest tenth of a percent.

5.4 EQUAL INTERVALS METHOD

In the **equal intervals** (or **equal steps**) method of classification, each class occupies an equal interval along the number line. The steps for computation for our six-class map are as follows:

*Step 1. Determine the **class interval** or width that each class occupies along the number line. This*

is computed by dividing the range of the data by the number of classes. The result for the foreign-born data is:

$$\frac{\text{Range}}{\text{Number of classes}} = \frac{\text{High} - \text{Low}}{\text{Number of classes}} = \frac{51.1 - 1.0}{6} = 8.35.$$

Step 2. Determine the upper limit of each class. The upper limit for each class is computed by repeatedly adding the class interval to the lowest value in the data. (For the first class, adding the

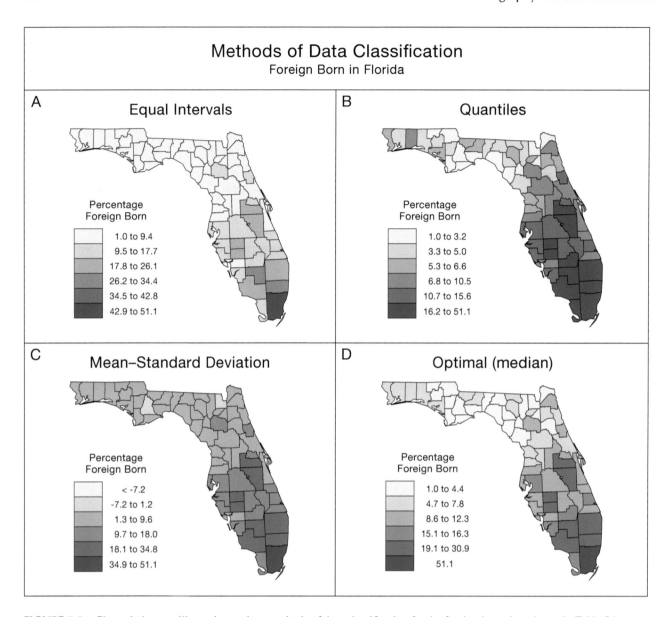

FIGURE 5.2   Choropleth maps illustrating various methods of data classification for the foreign-born data shown in Table 5.1.

class interval 8.35 to 1.0 yields a value of 9.35.)
The result is the right-hand set of numbers in the
Calculated Limits column in Table 5.2.

Step 3. Determine the lower limit of each class.
Lower limits for each class are specified so that
they are numerically just above the highest value

in a lower-valued class (as shown in Table 5.2, the
lower limit of class 2 is 9.36, which is 0.01 more
than the upper limit of class 1).

*Step 4. Specify the class limits actually shown in
the legend.* The class limits actually shown in the
legend should reflect the precision of the raw data
on which the classification is based. Because our
raw data values were rounded to the nearest tenth
of a percent, we also should report class limits to
the nearest tenth of a percent. Thus, we round the
Calculated Limits to create the Legend Limits
shown in Table 5.2.

Step 5. Determine which observations fall in each class.
This involves simply comparing the raw data values
with the Legend Limits from step 4. Figure 5.1A
provides a graphical expression of how the resulting
classes relate to a dispersion graph of the data, and
a map of the classified data appears in Figure 5.2A.

TABLE 5.2
Class Limits for Equal-Intervals Classification

Class	Calculated Limits	Legend Limits
1	1.00 to 9.35	1.0 to 9.4
2	9.36 to 17.70	9.5 to 17.7
3	17.71 to 26.05	17.8 to 26.1
4	26.06 to 34.40	26.2 to 34.4
5	34.41 to 42.75	34.5 to 42.8
6	42.76 to 51.10	42.9 to 51.1

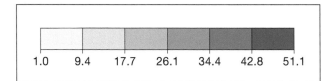

FIGURE 5.3   A legend for the equal-intervals method that takes advantage of the continuous nature of class limits.

An advantage of equal intervals is that the five steps can be completed using a calculator or even pencil and paper. As a result, this method was often favored before mapping software became available. A second advantage is that the resulting equal intervals will, in some cases, be easy for map users to interpret. For example, if you were making a five-class map of "percent urban population," and the data ranged from 0 to 100, the resulting classes would be convenient rounded multiples (0–20, 21–40, 41–60, 61–80, and 81–100). For the foreign-born data, however, note that the legend limits shown in Figure 5.2A do not readily reveal what the class interval might be.

The third advantage of equal intervals is that the legend limits contain no missing values (or gaps): For the foreign-born data, the difference between the upper value in a class and the lower value in the next class is 0.1, which is the precision of the data. Gaps, which we will see can occur in other classification methods (e.g., quantiles), might cause a reader to wonder why some of the data are missing from the legend. A related advantage for equal intervals is that the legend limits can be simplified so that only the lowest and highest values in the data and the upper limit of each class are shown (Figure 5.3). This approach should permit faster map interpretation, but it might also create confusion concerning the bounds of each class (e.g., the reader might wonder whether 9.4 falls in the first or second class).

The major disadvantage of equal intervals is that the class limits fail to consider how data are distributed along the number line. For example, if you inspect the dispersion graph for the foreign-born data (Figure 5.1A), you will note that class 5 seems to be wasted, as this class has no observations within it. On the plus side, however, note that unusual data values (such as the Miami-Dade County outlier) will appear in their own unique class. Thus, although one class is not used for the foreign-born data, the mapped distribution in Figure 5.2A does correctly show Miami-Dade County as being distinctly different from the rest of the data.

5.5   QUANTILES METHOD

In the **quantiles** method of classification, data are rank-ordered and the same number of observations is placed in each class. Different names for this method are used, depending on the number of classes; for example, four-, five- and six-class quantiles maps are referred to as *quartiles*, *quintiles*, and *sextiles*, respectively. To compute the number of observations in a class, the total number of

observations is divided by the number of classes. For the foreign-born data, we have:

$$\text{Number in class} = \frac{\text{Total observations}}{\text{Number of classes}}$$

$$= \frac{67}{6} = 11, \text{ remainder } 1.$$

Thus, we should place 11 observations in each class, and we have 1 observation left over. We arbitrarily chose to place this remaining observation in the first class. Again, keep in mind that this is done using the rank-ordered data.

Because identical data values should not be placed in different classes, ties can complicate the quantiles method. To illustrate, consider the results of using the above formula for an eight-class map.

$$\text{Number in class} = \frac{\text{Total observations}}{\text{Number of classes}}$$

$$= \frac{67}{8} = 8, \text{ remainder } 3.$$

If we assume that the three observations leftover are split among the first three classes, those classes would have nine observations apiece. Thus, observations 1 through 9 would fall in the first class, 10 through 18 would fall in the second class, and 19 through 27 would fall in the third class. Examining Table 5.1, we can see that this is problematic because the value 5.8 occurs at both the 27th and 28th ranked positions, and we don't want 5.8 to occur in more than one class. One solution to this problem would be to include the 28th position in the third class. This is somewhat undesirable, however, because the third class would now contain ten observations, two more than the initial ideal number of eight observations. An alternative solution would be to include both 5.8 values in the fourth class rather than the third class; using this approach, the third class would contain eight observations, and the fourth class would contain nine observations.

Two approaches are possible for defining legend limits for the quantiles method. One is to specify the lowest and highest values of members in a class: The results for our six-class map (assuming that we place 12 observations in the first class and 11 observations in the remaining classes) are shown in Table 5.3 in the column Class Limits Based on

TABLE 5.3

Class Limits for Quantiles Classification

Class	Class Limits Based on the Data	Class Limits with No Gaps
1	1.0 to 3.2	1.0 to 3.2
2	3.3 to 5.0	3.3 to 5.2
3	5.3 to 6.6	5.3 to 6.7
4	6.8 to 10.5	6.8 to 10.6
5	10.7 to 15.6	10.7 to 15.9
6	16.2 to 51.1	16.0 to 51.1

the Data. Although the resulting class limits do reflect the range of actual data falling in each class, there are gaps in the legend limits (e.g., 5.0 to 5.3) that may be disconcerting to the reader. The other approach is to avoid these gaps by computing a class boundary as an average of the highest value in a class and the lowest value of the next class; using this approach for the foreign-born data, in Table 5.3, we can see that the upper limit for the second class would be (5.0 + 5.3)/2, or 5.15, which rounds to 5.2. We prefer the former approach because it more accurately reflects the range of data values falling in a class.

As with equal intervals, an advantage of quantiles is that class limits can be computed manually. A second advantage is that because an equal number of observations fall in each class, the percentage of observations in each class will also be the same. Depending on the number of classes, this might simplify our discussion of the mapped data. For instance, on a five-class map, we can refer to the upper or lower 20 percent of the data, whereas on our six-class map, we have to refer to the upper or lower 16.7 percent of the data, which is not a convenient round number. A related advantage is that the 50th percentile or median (a measure of central tendency in the data; see Section 3.5.3.1) will be logically associated with the classes. For an odd number of classes, the median will fall in the center of the middle class, whereas for an even number of classes, the median will fall between the two middle classes (or very near this point—in our case, the median is the upper limit of the third class, a value of 6.6).

The third advantage of quantiles is that because the class assignment is based on rank order, quantiles are useful for ordinal-level data. For example, if the 50 states of the United States were ranked on "quality of life," the resulting ranks could be split into five equal groups: no numeric information would be necessary to create the classification. A fourth advantage is that if enumeration units are approximately the same size, each class will have approximately the same map area; in Chapter 22, we will see that this trait is useful when comparing maps.

The quantiles method shares the major disadvantage of equal intervals: It fails to consider how the data are distributed along the number line. For example, note that for the foreign-born data, the outlier for Miami-Dade County

is included in the same class with values of considerably lower magnitude (Figure 5.1B). Thus, we have eliminated the problem of empty classes, but distinctly unlike data values have been placed in the same class.

5.6   MEAN–STANDARD DEVIATION METHOD

The **mean–standard deviation** method is one of several classification techniques that do consider how data are distributed along the number line. In this method, classes are formed by repeatedly adding or subtracting the standard deviation from the mean of the data, as shown in Table 5.4.[1] As with the equal-intervals method, both calculated and legend limits can be computed. For our six-class map, Calculated Limits are computed using the mean and standard deviation values listed under Normal Distribution Limits in Table 5.4. For instance, for the first class $\bar{x} - 2s = 9.59 - (2*8.41) = -7.23$. To create the Legend Limits shown in the table, Calculated Limits are adjusted so that identical values cannot fall into two different classes, and the limits of the lowest and highest classes are adjusted to reflect the lowest and highest values in the data.

A major disadvantage of the mean–standard deviation method is that it works well only with data that are close to normally distributed. This is particularly evident with the foreign-born data, in which the lowest class contains solely negative values and therefore has no members (Figure 5.2). One solution to this problem is to transform the data (as described in Section 3.5.2.2), but this is inappropriate if the intention is to examine the *raw* data. Another disadvantage is that the mean–standard deviation method requires an understanding of some basic statistical concepts; a message on the map or in the text indicating that "classes were developed based on the mean and standard deviation" would not be meaningful if one had no statistical training.

A distinct advantage of the mean–standard deviation method, however, is that if the data are normally distributed (or near normal), the mean serves as a useful dividing point, enabling a contrast of values above and below it. This is most effectively accomplished if an even number of classes is used, as with our six-class map (in our case, the rounded mean value is the upper limit of the third class). For a five-class map, the two middle classes could be combined, and

TABLE 5.4

Class Limit Computations for Mean–Standard Deviation Classification for the Florida Foreign-Born Data ($\bar{x} = 9.59$, $s = 8.41$)

Class	Normal Distribution Limits	Calculated Limits	Legend Limits
1	$< \bar{x} - 2s$	< -7.23	< -7.2
2	$\bar{x} - 2s$ to $\bar{x} - 1s$	-7.23 to 1.18	-7.2 to 1.2
3	$\bar{x} - 1s$ to $\bar{x}$	1.18 to 9.59	1.3 to 9.6
4	$\bar{x}$ to $\bar{x} + 1s$	9.59 to 18.00	9.7 to 18.0
5	$\bar{x} + 1s$ to $\bar{x} + 2s$	18.00 to 26.41	18.1 to 26.4
6	$> \bar{x} + 2s$	> 26.41	51.1

the mean would fall in the middle of the resulting class. Another advantage of the mean–standard deviation method is that the legend limits contain no gaps that might confuse the reader.

5.7 NATURAL BREAKS

In **natural breaks** classification, graphs (e.g., a dispersion graph or histogram) are examined visually to determine logical breaks (or, alternatively, clusters) in the data. Stated another way, the purpose of natural breaks is to minimize differences between data values in the same class and to maximize differences between classes. Later we will see that this is also the objective of the optimal method, but with the optimal method, the classification is done using a numerical measure of classification error, whereas, with natural breaks, the classification is accomplished subjectively by the mapmaker.

To illustrate the computation of natural breaks, consider how we might divide the foreign-born data into six classes (you should examine the dispersion graph associated with the natural breaks method in Figure 5.1). The highest value in the data (51.1) appears to be quite different from the rest of the data, so we will place it in a class by itself. The two values near 30 appear to belong together, and so we will group those to create a second class. Next, we have six values concentrated near 20, three just below 20 and three above 20, but below 25; these arguably form a third class. Then we have a cluster of five values just above 15, which appear to form a fourth class. The remaining data seem to form two clusters, one extending from 1 to about 7 and one from about 8 to 13. We can see in this example that an obvious problem with natural breaks is that decisions on class limits are subjective and, therefore, can vary among mapmakers.

5.8 OPTIMAL

The **optimal** classification method is a solution to the subjectivity of natural breaks. The optimal method places similar data values in the same class by minimizing an objective measure of classification error. To illustrate, consider how a small hypothetical data set of nine values would be classified by the quantiles and optimal methods using three classes (Table 5.5). The quantiles method assigns the same number of observations to each class (three, in this case) and thus places similar values in different classes (e.g., 14 appears with 31 and 32 in class 2, even though 14 is more similar to 11, 12, and 13, the members of class 1). In contrast, the optimal method places similar values in the same class (the first class consists of 11, 12, 13, and 14, whereas the second class consists of 31, 32, and 33).

One measure of classification error commonly used in the optimal method is the sum of absolute deviations about class medians (ADCM). Computing this measure involves calculating the median of each class, the sum of absolute deviations of class members about each class median, and then adding the resulting sums of absolute deviations. For example, for quantiles, the median in the first class is 12 (as discussed in Section 3.5.3.1, the median is simply the middle value in an ordered set of data), and the sum of absolute deviations for the class is $|11 - 12| + |12 - 12| + |13 - 12| = 2$. The sum of absolute deviations for all classes is $2 + 18 + 67 = 87$. In contrast, ADCM for the optimal method is $4 + 2 + 1 = 7$, which is obviously a smaller value and thus indicative of a better classification.

The data for this hypothetical example were selected so that the results would be clear-cut. In the real world, the desired minimum-error classification is normally not obvious, so researchers have developed computer-based algorithms for determining solutions. Here we consider two algorithms: the Jenks–Caspall and the Fisher–Jenks.

5.8.1 THE JENKS–CASPALL ALGORITHM

The **Jenks–Caspall algorithm**, developed by George Jenks and Fred Caspall (1971), is an empirical solution to the problem of determining optimal classes. As with the example shown in Table 5.5, we assume that we wish to minimize the total map error (ADCM).[2] The algorithm begins with an arbitrary set of classes (say, the quantiles classes shown in Table 5.5), calculates a total map error, and attempts to reduce this error by moving observations between adjacent classes. Observations are moved using reiterative and forced cycling. In *reiterative cycling*, movements are accomplished by determining how close an observation is to the median of another class; for example, for the quantiles data in Table 5.5, the value 14 is closer to the median of class 1 (12) than 13 is

---

TABLE 5.5

Computing the Sum of Absolute Deviations about Class Medians (ADCM)

Raw Data: 11, 12, 13, 14, 31, 32, 33, 99, 100

Quantiles Classification			Optimal Classification		
Class	Values	Error	Class	Values	Error
1	11, 12, 13	2	1	11, 12, 13, 14	4
2	14, 31, 32	18	2	31, 32, 33	2
3	33, 99, 100	67	3	99, 100	1
ADCM = 87			ADCM = 7		

to the median of class 2 (31), so 14 would be moved to the first class. Movements based on the relation of observations to class medians are repeated until no further reductions in total map error can be made.

In *forced cycling*, individual observations are moved into adjacent classes, regardless of the relation between the median value of the class and the moved observation. After a movement, a test is made to determine whether any reduction in total map error has occurred. If error has been reduced, the new classification is considered an improvement, and the movement process continues in the same direction. Forcing is done in both directions (from low to high classes and from high to low classes). At the conclusion of forcing, the reiterative procedure described earlier is repeated to see whether any further reductions in error are possible. Although this approach does not guarantee an optimal solution (a minimum total sum of absolute deviations for all classes), Jenks and Caspall indicated that they were "unable to generate, either purposefully or by accident, a better … representation in any set of data."[3]

5.8.2 THE FISHER–JENKS ALGORITHM

In contrast to the empirical approach used by Jenks and Caspall, the Fisher–Jenks algorithm has a mathematical foundation that guarantees an optimal solution. Walter Fisher (1958) was responsible for developing the mathematical foundation, and George Jenks (1977) introduced the idea to cartographers—hence the term Fisher-Jenks algorithm. Cartographers generally have chosen to recognize only Jenks for this contribution, so you might find the algorithm referred to as "Jenks's optimal method."

To understand the Fisher–Jenks algorithm, it is worthwhile to consider how an optimal solution might be computed using brute force. Imagine that you wanted to develop an optimal two-class map of the data 1, 3, 7, 11, and 22. With such a small data set, it is easy to list all possible two-class solutions and compute associated error measures (Table 5.6). If the process is so simple for a small data set,

it would seem that for large data sets, a computer could be used to determine an optimal solution by simply considering all possibilities. Unfortunately, for large data sets, the number of possible solutions becomes prohibitively large; for example, Jenks and Caspall calculated that for the 102 counties of Illinois, there would be over 1 billion possible seven-class maps.

Rather than consider all solutions, the Fisher–Jenks algorithm takes advantage of the mathematical foundation provided by Fisher, which states that any optimal partition is simply the sum of optimal partitions of subsets of the data. We illustrate this concept by considering some initial steps for handling the data 1, 3, 7, 11, and 22 (Table 5.7). For computational simplicity, we use the median (and associated sum of absolute deviations); another version of the algorithm uses the mean (and associated sum of squared deviations about the mean).

Step 1 involves computing the sum of absolute deviations about the class median for any ordered subset of the raw data, ignoring how these subsets might fit into a particular classification. Let $D(i, j)$ represent the sum of absolute deviations for the ith through the jth observation; for example, the sum of absolute deviations for the first through third observations (the subset 1, 3, and 7) is $D(1,3) = |1 - 3| + |3 - 3| + |7 - 3| = 6$. This result appears in row 1, column 3, of the matrix shown in step 1 of Table 5.7.[4]

In step 2, the optimal solution for a two-class map of the complete data set is computed, along with optimal two-class solutions for subsets of the data. Together, these are termed the *optimal two partitions*. Calculations for the optimal two-class map for the complete data set are shown in part (a) of step 2 (Table 5.7), and some of the results for subsets of the data are shown in part (b) of step 2. Note how these results make use of the results from step 1; for example, the possible solution 1 3 | 7 11 22 in step 2(a) uses the computations $D(1,2) + D(3,5) = 2 + 15 = 17$. Looking at the results shown for step 2(a) in Table 5.7, you can see that the best solution is to place 22 in a class by itself, or 1 3 7 11 | 22, as this choice produces the smallest ADCM value of 14. Although the

---

TABLE 5.6
Computing ADCM for All Potential Two-Class Maps

Raw Data: 1, 3, 7, 11, 22

	Solution 1			Solution 2	
Class	Values	Error	Class	Values	Error
1	1	0	1	1, 3	2
2	3, 7, 11, 22	23	2	7, 11, 22	15
ADCM = 23			ADCM = 17		
	Solution 3			Solution 4	
Class	Values	Error	Class	Values	Error
1	1, 3, 7	6	1	1, 3, 7, 11	14
2	11, 22	11	2	22	0
ADCM = 17			ADCM = 14		

Solution 4 is optimal because it has the smallest total error (ADCM = 14).

TABLE 5.7

Initial Steps in the Fisher–Jenks Algorithm for Optimal Data Classification

Step 1. Compute the sum of absolute deviations about the class median for all ordered subsets of the data. The following matrix shows the sum of absolute deviations about the median for the ith through the jth observation ($D(i,j)$); for example, if $i = 1$ and $j = 3$, then $D(1,3) = |1 - 3| + |3 - 3| + |7 - 3| = 6$. (Note that this result appears in the first row and the third column of the matrix.)

	*j*th Observation				
*i*th Observation	1	2	3	4	5
1	0	2	6	14	29
2		0	4	8	23
3			0	4	15
4				0	11
5					0

Step 2. Compute all optimal two partitions.

(a) The results for the optimal two-class map of the complete data set are as follows:

1 | 3 7 11 22
$D(1,1) + D(2,5) = 0 + 23 = 23$
1 3 | 7 11 22
$D(1,2) + D(3,5) = 2 + 15 = 17$
1 3 7 | 11 22
$D(1,3) + D(4,5) = 6 + 11 = 17$
1 3 7 11 | 22
$D(1,4) + D(5,5) = 14 + 0 = 14$ (Optimal result)

(b) The following are results for the first four values for optimal two-class partitions of subsets of the data:

1 | 3 7 11
$D(1,1) + D(2,4) = 0 + 8 = 8$
1 3 | 7 11
$D(1,2) + D(3,4) = 2 + 4 = 6$ (Tied for optimal result)
1 3 7 | 11
$D(1,3) + D(4,4) = 6 + 0 = 6$ (Tied for optimal result)

Optimal two partitions also would be computed for 1 3 7, 3 7 11, 3 7 11 22, and 7 11 22

---

subset calculations shown in step 2(b) are not used in determining the optimal two-class map, they are used to determine optimal classifications for maps with a greater number of classes. The process shown here would be repeated for subsequent steps with a greater number of classes, thus taking advantage of computations that have already been done.

The results of applying the Fisher-Jenks algorithm to the foreign-born data are shown in the Optimal dispersion graph in Figure 5.1E. Note how the results compare with the natural breaks approach: Classes 1 and 2 for natural breaks have been split into three classes, whereas classes 4 and 5 for natural breaks have been combined into a single class for the optimal method. Overall, however, there is more similarity between these two methods than there is between the other methods and the optimal method. This is why you will find that the optimal method is sometimes referred to as "Jenks's natural breaks" method.

5.8.3 ADVANTAGES AND DISADVANTAGES OF OPTIMAL CLASSIFICATION

The obvious advantage of the optimal method is that it considers, in detail, how data are distributed along the number line. It is the "best" choice for classification when the intention is to place like values in the same class (and unlike values in different classes) based on the position of values along the number line.

Another advantage is that the optimal method can assist in determining an appropriate number of classes. When the median is used as the measure of central tendency, this is accomplished by computing the **goodness of absolute deviation fit (GADF)**, which is defined as:

$$GADF = \frac{ADCM}{ADAM}$$

where ADCM is the sum of absolute deviations about class medians for a particular number of classes, and ADAM is the sum of absolute deviations about the median for the entire data set. An analogous measure can be computed when the mean is used as the measure of central tendency (and the error in a class is the sum of squared deviations about the mean) and is known as the **goodness of variance fit (GVF)**.

GADF ranges from 0 to 1, with 0 representing the lowest accuracy (a one-class map) and 1 representing the highest accuracy. If there are no ties in the data, then a GADF value of 1 will result only when each observation is a separate class (an *n*-class map will result, where *n* is the number of classes). Ties can, however, reduce the number of classes needed to achieve a GADF of 1. For example, in the case of the foreign-born data, only 58 classes are needed (although there are 67 data values, there are 9 ties).

It is important to note that an *n*-class map is equivalent to an "unclassed map." This might be confusing because the term *unclassed* suggests no classes, and there are actually *n* classes on an *n*-class map. "Unclassed" is commonly used to indicate that no classing or grouping has been applied to the data; for instance, it is not necessary to run the optimal program to create an unclassed map.

We can use the GADF values to assist in determining an appropriate number of classes by plotting the GADF values against the number of classes, as shown in Figure 5.4. This graph can be interpreted in two ways. One approach is to look for a point at which a curve fit to the data begins to flatten out. In the case of the foreign-born data, there is some subjectivity of interpretation, but the curve seems to start to flatten at six classes and more definitely at eight classes. A flattening indicates that a larger number of classes would not substantially reduce the classification error.

Another approach for interpreting the graph is to determine the number of classes for which the GADF first exceeds a certain value, say, 0.80 (i.e., the accuracy is 80 percent). For the foreign-born data, this approach yields a six-class map (Figure 5.4). Note, however, that if a more stringent value were used, say, 0.90, a ten-class map would be required. Admittedly, both of these approaches are subjective, but they are an improvement over choosing an arbitrary number of classes.

Another factor to consider in interpreting the graph is that a map with more classes will be more difficult to

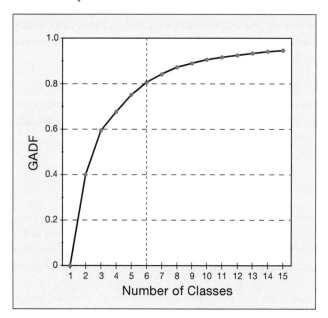

FIGURE 5.4 Graph of the number of classes plotted against GADF values. The curve appears to flatten out at about either six or eight classes, depending on how one interprets the graph.

interpret, as a greater number of areal symbols will need to be differentiated. Thus, we chose six classes for comparing classification methods in this section because we knew it would be very difficult to distinguish a greater number of classes using the lightnesses of blue shown in Figure 5.2.

In addition to helping you determine an appropriate number of classes for the optimal method, the GADF technique could also be used to compute the accuracy of other classification methods and thus determine whether those methods might be appropriate. For example, if you were to compute similar GADF values for optimal and quantiles, you might choose the quantiles method because it provided a similar total error measure, and it would be easier for the user to understand how the class limits were created.

Disadvantages of the optimal method include the difficulty of understanding the concept and the appearance of gaps in the legend. Another disadvantage is that mapping software sometimes does not include the optimal method as an option, although this is less problematic than it was historically.

5.9 HEAD/TAIL BREAKS: A NOVEL CLASSIFICATION METHOD

Recently, Bin Jiang (2013) has developed a new data classification method known as **head/tail breaks**, which Jiang argues is appropriate for data that are *heavy-tailed* (they have a strong positive skew).[5] Although the foreign-born data that we use is only moderately positively skewed, we feel that it is sufficiently skewed to illustrate the head/tail method.

The term heavy-tailed is applied to a plot of the rank of data values along the *x*-axis against the attribute being mapped along the *y*-axis, as shown in Figure 5.5 for the foreign-born data. Note that the highest data value is ranked first, and the lowest data value is ranked last. The prominent peak on the left of the graph is termed the *head* and the curve of values to the right is termed the *tail*. Since high data values tend to occur infrequently and low data values tend to occur frequently, the curve typically has a heavy tail to the right. Another way of saying this is that there are far more small data values than large data values. Biang argues that the focus should be on the low-frequency high data values because they tend to have the highest impact.

Computations for creating classes using the head/tail method are relatively simple. One partitions data values into two parts using the mean as a breakpoint; this process continues iteratively for values above the mean as long as the data above the mean continue to be heavy-tailed (or the head contains only one observation). Although various rules-of-thumb could be developed for determining whether a distribution is heavy-tailed, Jiang considers that a distribution is heavy-tailed as long as less than 40 percent of the data fall above the mean.[6]

Table 5.8 shows the result of applying the head/tail method to the foreign-born data. In step 1, the 67 counties are split into two classes using the mean of 9.6 as a

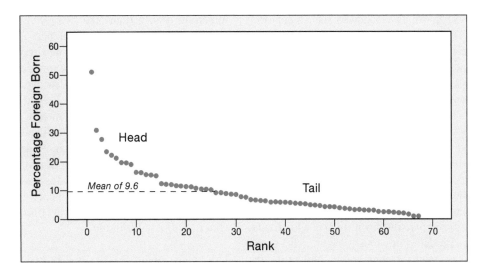

FIGURE 5.5   Plot of the rank of data values against the magnitude of data values for the foreign-born data (the highest data value is ranked 1). Such plots are often used to illustrate heavy-tailed distributions. The mean of the data (9.6) is shown by the dashed line—note that it divides the data into two groups, 42 counties below the mean and 25 counties above the mean.

TABLE 5.8
Steps in Classification for the Head/Tail Method Using the Foreign-Born Data

Step	Number of Counties	Number in Head	Percent in Head	Number in Tail	Mean
1	67	25	37	42	9.6
2	25	9	36	16	17.5
3	9	3	33	6	26.2
4	3	1	33	2	36.6

breakpoint. Note that 25 counties are above 9.6 (in the *head*) and 42 counties are below 9.6 (in the *tail*) (this result is also shown in Figure 5.5). In step 2, the 25 counties above the mean in step 1 are split into two classes using the mean of these 25 counties as a breakpoint. This process continues in steps 3 and 4 and stops at step 4 because the head consists

of only one county and thus can be split no further. Note that at each stage of the division, the percentage in the head was less than 40 percent.

Figure 5.6A shows a map resulting from applying the head/tail method to the foreign-born data. Note that the upper limit of the first four classes corresponds to the mean values

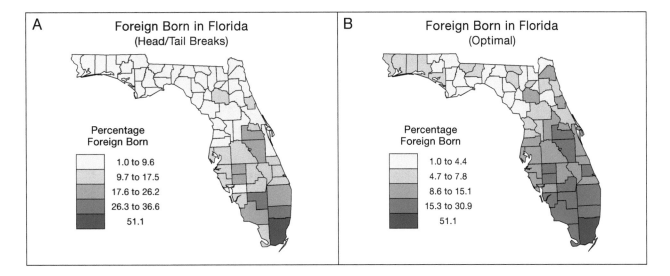

FIGURE 5.6   A comparison of head/tail breaks and optimal classification methods for the foreign-born data for a five-class map. Note how the high-frequency lower data values are shown as two classes on the optimal map but as only one class on the head/tail breaks map.

computed in Table 5.8. We have followed Jiang's approach for specifying the lower limit of classes 2, 3, and 4, as they are simply 0.1 greater than the mean for the preceding class. For class 5, we chose to enter a single data value to emphasize that the class consists of just that value. In viewing the resulting map, Jiang recommends that you consider lower-valued classes as a background for the higher-valued classes; for instance, the lowest class ranging from 1.0 to 9.6, serves as the background for the 9.7 to 17.5 class. Jiang terms the resulting relationship between classes a *scaling hierarchy.*

An interesting characteristic of the head/tail method is that it automatically suggests an appropriate number of classes—in this case, a five-class map is produced. Since it makes little sense to compare the resulting map with the six-class maps that we produced previously, Figure 5.6B shows an optimal five-class map for comparison. Obviously, the make-up of the classes on the two maps is quite different. Jiang argues that the head/tail method is more appropriate for heavy-tailed distributions because it portrays the underlying scaling hierarchy, as each class is defined by attempting to visually emphasize the infrequent number of observations in the head over the more frequent observations in the tail. Note that the lowest class on the head-tail map contains

numerous observations and completely encompasses the two lowest classes on the optimal map.

We have emphasized that the head/tail method is advantageous when one wishes to examine the scaling hierarchy associated with heavy-tailed distributions (heavily positively skewed distributions). A second advantage is that the number of classes is automatically determined by the process used to terminate the classification procedure (the head should contain no more than 40 percent of the observations). A third advantage is the relative ease of computing class limits. A potential disadvantage is that the class limits specified in the legend will not match the actual limits of data falling in each class; rather, the limits include the mean values used in partitioning.

5.10 CRITERIA FOR SELECTING A CLASSIFICATION METHOD

In discussing various methods of classification, we have pointed out numerous criteria that might be used to judge the methods' usefulness. Figure 5.7 summarizes these criteria and rates each classification method as very good, good, or poor on each measure (and acceptable or unacceptable

	Equal Intervals	Quantiles	Mean SD	Optimal	Head/Tail Breaks
Considers distribution of data along a number line	P	P	G[a]	VG	VG
Ease of understanding concept	VG	VG	VG	G[b]	VG
Ease of computation	VG	VG	VG	VG[c]	VG
Ease of understanding legend	VG[d]	P[e]	G	P[e]	G
Legend values match range of data in a class	P	VG	P	VG	P
Acceptable for ordinal data	U	A	U	U	U
Assists in selecting number of classes	P	P	P	VG	VG

P = Poor   G = Good   VG = Very Good

A = Acceptable   U = Unacceptable

[a] Rating would be poor if data are not normal.

[b] Only a good rating is assigned because of the fairly complex nature of the algorithm.

[c] The optimal method does require the use of a computer.

[d] Only a good rating would be appropriate if round numbers are not used.

[e] Using rounded values may produce a good rating; some data distributions may mimic an equal interval map, thus producing a good or very good rating.

FIGURE 5.7   Criteria for selecting a method of classification.

in the case of "acceptable for ordinal data"). One problem with any rating system is that it is a function of the computer environment the mapmaker has available and the knowledge of the map user. For Figure 5.7, we assume that computer software is available for creating all of the classification methods and that the map user is a college-level student with a basic foundation in introductory statistics.

Note that "Ease of understanding legend" is a function of whether or not there are gaps in the legend: Remember that gaps between class limits can make the legend difficult to understand. The equal-intervals method receives a very good rating on this criterion because not only are there no gaps, but the rounded intervals can be easy to understand (e.g., 0–25, 26–50, etc.). You might avoid the problem of gaps by creating continuous legends for all classification methods (as in Figure 5.3). Remember, though, that this approach will not indicate the actual range of values falling in a class (the latter is dealt with in the criterion "Legend values match range of data in a class").

An analysis of Figure 5.7 reveals that there is no single best method of classification. Although the optimal approach is often touted as the best method, it is best only in terms of grouping like values together (as a function of their position along a number line). Clearly, there are several other criteria for which it is not the best. It is noteworthy that the recently developed head/tail method scored quite well on many of the criteria. Remember, though, that this technique is intended for dealing with heavy-tailed distributions.

Ultimately, you must consider the purpose of the map and the knowledge of the intended audience before selecting a classification method. A good illustration of the role of map purpose and the intended audience is the approach that Cynthia Brewer and Trudy Suchan utilized in the atlas *Mapping Census 2000: The Geography of U.S. Diversity* (https://usa.usembassy.de/etexts/soc/diversity.pdf).[7] Rather than use one of the classification methods presented in this chapter, Brewer and Suchan used meaningful breaks (e.g., a percentage figure for the entire United States), rounded breaks, and breaks that were identical across a set of maps. Brewer argued that the resulting map set was much more useful than a "map-by-map optimization approach."[8]

5.11 CONSIDERING THE SPATIAL DISTRIBUTION OF THE DATA

One limitation of the classification methods that we have introduced is that they do not consider the spatial distribution of the data (where the data values are located geographically). Rather, the methods only consider the values of an attribute along the number line. The argument can be made, however, that a choropleth map can be simplified (and thus more easily interpreted) if we do consider the spatial distribution of the data in the data classification process.

To illustrate how the spatial distribution of the data might be included in a data classification, consider Figure 5.8. The top portion of the figure lists raw data for 16 hypothetical

enumeration units, and the bottom portion portrays two maps of these data. The map on the left is an optimal classification; note that breaks for this map occur between 6 and 9 and between 13 and 16. How might we simplify the appearance of this map? One potential solution is shown in the right-hand map, where we have shifted the values 9 and 16 from the second and third classes to the first and second classes, respectively. Note that the map has a slightly lower GADF (0.70 versus 0.76), but the map has a simpler appearance. We can express the simpler appearance by computing the complexity face measure C_F (see Section 3.8.3). C_F is computed by dividing the number of faces on the mapped distribution by the number of faces on the base map (the number of enumeration units). For the optimal map, we have 10/16 = 0.63, while for the optimal map with a spatial constraint, we have 6/16 = 0.38. We have thus lowered the complexity of the map considerably.

The simple example that we have described in Figure 5.8 can be criticized because it is a small dataset and we have considered only one possible manipulation of that dataset. How can we expand this notion to larger datasets and consider multiple criteria for modifying the classification? In response to this sort of question, Marc Armstrong and colleagues (2003) developed a data classification approach that considers four criteria for classifying a dataset: (1) minimizing the *error variance* in each class (this is essentially the optimal method, except that the mean is the measure of central tendency and squared deviations are computed about the mean), (2) minimizing the difference between classes for contiguous enumeration units (i.e., minimizing what is termed the *boundary error*), (3) maximizing the overall spatial autocorrelation in the resulting classed map (see Section 3.8.2 for a discussion of spatial autocorrelation), and (4) equalizing the area in each class (which is often useful when we wish to

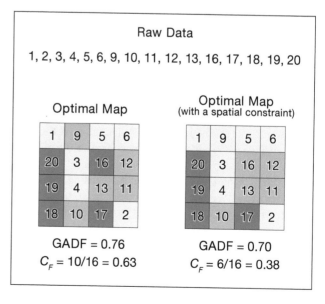

FIGURE 5.8 Applying a spatial constraint to the optimal classification approach.

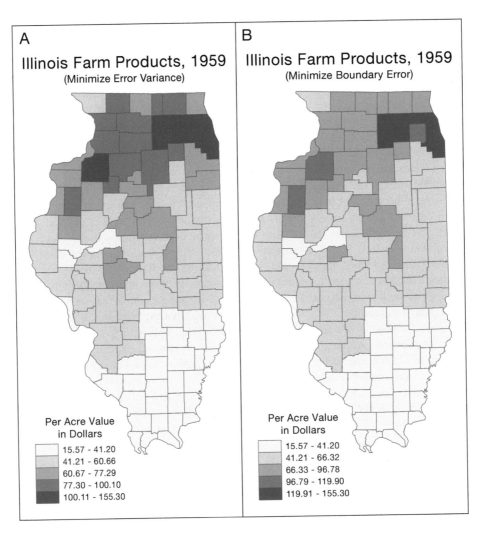

FIGURE 5.9 Comparison of maps minimizing the error variance (an optimal classification) and minimizing the boundary error. The boundary error map arguably produces a simpler spatial pattern and the GVF is only slightly lower (0.921 as opposed to 0.931). (After Armstrong et al. 2003.)

compare choropleth maps—see Chapter 22). Note that the latter three of these are all spatial constraints.[9]

Figure 5.9 compares the results of two of the above methods using a classic farm products data set that both Jenks and Caspall (1971) and Armstrong et al. (2003) utilized. The map on the left minimizes the error variance (it provides an optimal solution), while the map on the right minimizes the boundary error. The map on the right appears to have a simpler appearance (as a result of minimizing the boundary error). We can ask, though, how much has the error variance been affected by focusing instead on boundary error. We can determine this by computing the GVF for both maps. As discussed in Section 5.8.3, GVF provides a measure of fit when the mean is used as the measure of central tendency, and squared deviations are computed about the mean. GVF for the optimal map is 0.931 and GVF for the minimum boundary error map is 0.921. Thus, we have lost very little in the GVF by minimizing the boundary error, and the result is arguably a more easily interpreted map.

In early work, Ningchuan Xiao and Marc Armstrong (2006) developed the software ChoroWare to assist users in evaluating the tradeoffs in classifying data using the above four criteria. More recently, Xiao has developed an approach based on Jupyter Notebooks (https://jupyter.org/); Xiao's application can be found at https://github.com/compgeog/choropleth-tradeoffs. One notebook produces all possible classifications, and another notebook allows you to explore the tradeoffs of using the four classification approaches. Although programming expertise is not required to utilize these notebooks, those with a programming background will be able to modify the code and thus develop their own analyses and visualizations of the data.

5.12 SUMMARY

Data classification involves grouping raw data into classes, with each resulting class depicted by a different symbol. Data classification is particularly appropriate for choropleth maps because of the difficulty of differentiating areal symbols (e.g., lightnesses of a single hue) on an unclassed map.

In this chapter, we focused on five data classification methods: equal intervals, quantiles, mean-standard deviation, optimal, and head/tail breaks. Equal intervals (in which each class occupies an equal portion of the number line) is potentially appropriate when the intervals produce rounded values that are easy to interpret (e.g., 0–20, 21–40, 41–60, etc.). Quantiles (in which the same number of observations fall in each class) are desirable because we can easily refer to a fixed percentage of observations in each class. Quantiles are the only method that can be used for ordinal data. A limitation of equal intervals and quantiles is that they do not consider the distribution of data along the number line.

Mean-standard deviation and optimal are two classification methods that do consider the distribution of the data along the number line. Mean-standard deviation accomplishes this by utilizing the mean and standard deviation, two parameters commonly used in association with the normal distribution. As such, the mean-standard deviation approach should only be used for data that are close to normally distributed. The optimal method considers the distribution of the data by placing similar data values in the same class. The optimal method is *optimal* in the sense that it minimizes the sum of ADCM. Another advantage of the optimal method is that it can be used to determine an appropriate number of classes by examining a plot of the GADF against the number of classes.

Head/tail breaks is a novel classification method that uses the mean of the data to recursively divide the data as long as the data above the mean are heavy-tailed (strongly positively skewed). In addition to automatically determining an appropriate number of classes, head/tail breaks are unique in the sense that each lower-valued class should be viewed as a background for higher-valued classes.

One limitation of most traditional data classification methods is that they fail to consider the *spatial distribution* of the data. Fortunately, three techniques have been developed that do consider the spatial distribution of the data: (1) minimizing the difference between classes for contiguous enumeration units (i.e., minimizing the *boundary error*), (2) maximizing the overall spatial autocorrelation in the classed map, and (3) equalizing the area in each class. You can experiment with these techniques and compare them with an optimal approach that minimizes the error variance by utilizing Xiao's Jupyter Notebooks at https://github.com/compgeog/choropleth-tradeoffs.

5.13 STUDY QUESTIONS

1. Use the five steps discussed in Section 5.4 to create the class limits for a five-class equal intervals map for the Florida foreign-born data.
2. A. Recreate Table 5.3 for a five-class quantiles map of the Florida foreign-born data. Add a column to the table indicating the number of observations falling in each class.
   B. Compare the class limits from the table that you have developed with the dispersion graph of the data shown in Figure 5.1B. Explain whether your limits appear to pay attention to the distribution of the data.
3. If the histograms in Figure 3.2 of Chapter 3 represented a graphical depiction of data for a choropleth map, which datasets would you deem appropriate for mean-standard deviation classification?
4. Discuss how you would develop classes for a five-class natural breaks map using the Florida foreign-born data.
5. Compute ADCM for the following equal intervals classification: Class 1 with 11, 12, 13, 14, 31, 32, and 33; Class 2 with no members; and Class 3 with 99 and 100. Contrast the magnitude of your result with those for quantiles and optimal classification methods shown in Table 5.5.
6. Use Figure 5.4 to present arguments for and against creating an eleven-class map for the foreign-born data.
7. If the histograms in Figure 3.2 of Chapter 3 represented a graphical depiction of data for a choropleth map, indicate which data set you would deem most appropriate for head/tail breaks and explain why.
8. Use the criteria shown in Figure 5.7 to discuss the effectiveness of the optimal and head/tail breaks data classification methods.
9. Compute complexity face measures, C_F, for the two maps shown in Figure 5.9. (Note that there are 102 enumeration units in the base map.) Explain whether the resulting measures make sense in terms of the ease of interpreting the two maps.
10. Use a search engine such as Google images (https://www.google.com/imghp?hl=en) to find a choropleth map that you find interesting. Answer the following questions: (A) Indicate the number of classes used and whether you feel readers could easily extract information from the map. (B) Examine the class limits shown in the legend and indicate whether they seem to follow the guidelines suggested in this chapter. (C) Attempt to determine the source of the data and the method of classification. (D) Overall, do you feel that the map is effective? Explain your response.

NOTES

1 We use mean and standard deviation formulas appropriate for a sample; because of the relatively large number of observations, similar results would be obtained with population formulas.
2 Jenks and Caspall used the mean as a measure of central tendency and computed the sum of absolute deviations about the mean.

3 Jenks and Caspall (1971, 236).
4 The algorithm as described here was taken from Hartigan (1975).
5 Jiang indicates that traditional approaches for handling heavy-tailed distributions include arithmetic and geometric series (see Kraak and Ormeling 2021 for a discussion of these methods).
6 Jiang has developed a 2.0 version of head/tail breaks that continues the classification process as long as the *average* percentage in the head is less than 40 percent (personal communication).
7 See Brewer (2001) for a discussion of the approaches used to create the atlas.
8 Brewer (2001, 225).
9 Traun and Loidl (2012) describe an alternative approach that combines a measure of spatial autocorrelation (Moran's I scatterplot) with optimal classification.

REFERENCES

Armstrong, M. P., Xiao, N., and Bennett, D. A. (2003) "Using genetic algorithms to create multicriteria class intervals for choropleth maps." *Annals of the Association of American Geographers 93*, no. 3:595–623.

Brewer, C. A. (2001) "Reflections on mapping Census 2000." *Cartography and Geographic Information Science 28*, no. 4:213–235.

Fisher, W. D. (1958) "On grouping for maximum homogeneity." *Journal of the American Statistical Association 53*, December:789–798.

Hartigan, J. A. (1975) *Clustering Algorithms*. New York: Wiley.

Jenks, G. F. (1977) *Optimal Data Classification for Choropleth Maps. Occasional Paper No. 2*. Lawrence, KS: Department of Geography, University of Kansas.

Jenks, G. F., and Caspall, F. C. (1971) "Error on choroplethic maps: Definition, measurement, reduction." *Annals of the Association of American Geographers 61*, no. 2:217–244.

Jiang, B. (2013) "Head/tail breaks: A new classification scheme for data with a heavy-tailed distribution." *The Professional Geographer 65*, no. 3:482–494.

Kraak, M.-J., and Ormeling, F. (2021) *Cartography: Visualization of Geospatial Data*. (4th ed.). Boca Raton, FL: CRC Press.

Traun, C., and Loidl, M. (2012) "Autocorrelation-Based Regioclassification: A self-calibrating classification approach for choropleth maps explicitly considering spatial autocorrelation." *International Journal of Geographical Information Science 26*, no. 5:923–939.

Xiao, N., and Armstrong, M. P. (2006) "ChoroWare: A software toolkit for choropleth map classification." *Geographical Analysis 38*, no. 1:102–120.

6 Scale and Generalization

6.1 INTRODUCTION

This chapter presents basic concepts in scale and generalization. Scale is a fundamental geographic principle, but there is often confusion about the exact meaning of **geographic scale**, **cartographic scale**, and **data resolution**. Geographers think about scale as the area covered, where large-scale studies cover a large region such as a state. Cartographers think about scale mathematically and use the **representative fraction (RF)** to express the relationship between map and Earth distance. For instance, most national map series are established at a specific scale, such as the French 1:25,000 BD Topo and the U.S. Geological Survey (USGS's) 1:24,000 series. Data resolution refers to the granularity of the data, such as the pixel size of a remote sensing image. Directly related to the concept of scale are the ideas of generalization and modifying the information content so that it is appropriate at a given scale. It would not be possible, for instance, to depict the street network for the entire United States with the country mapped at a scale that would fit on one page—only major highways could be depicted.

Section 6.3 introduces the concepts of geographic and cartographic scale and covers how scale controls the amount of map space and thus the appropriate information content. Section 6.4 provides some basic definitions of **generalization**, including a discussion of some fundamental generalization operations. Section 6.5 discusses several conceptual models of the generalization process. One of the more complete models divides the process into "why," "when," and "how" components of generalization. Section 6.6 describes the many operations that have been designed for the generalization process, including **simplification**, **smoothing**, **aggregation**, **amalgamation**, **collapse**, **merging**, **refinement**, **exaggeration**, **enhancement**, and **displacement**. We also discuss in detail simplification, which involves "weeding" unnecessary data from a linear or areal feature, thus retaining the critical points defining that feature. Section 6.7 illustrates a practical application of generalization using the National Historical Geographic Information System (NHGIS; http://www.nhgis.org). To conclude, Section 6.8 provides a review of some of the newest developments in the field of scale and generalization.

6.2 LEARNING OBJECTIVES

- Understand the concepts of geographic scale, cartographic scale, and resolution
- Comprehend how the representative fraction (RF) works

- Explain the basic models of generalization
- Describe the fundamental operators of generalization
- Understand the details of line simplification algorithms (e.g., a local vs. a global algorithm)
- Realize that spatial data structures can support automated generalization

6.3 GEOGRAPHIC AND CARTOGRAPHIC SCALE

Scale is a fundamental concept in all of science and is of particular concern to geographers, cartographers, and others who are interested in geospatial data. Astronomers work at a spatial scale of light-years, physicists work at the atomic spatial scale in mapping the Brownian motion of atoms, and geographers work at spatial scales from the human to the global. Within the fields of geography and cartography, the terms **geographic scale** and **cartographic scale** are often confused. Geographers and other social scientists use the term *scale* to mean the extent of the study area, such as a neighborhood, city, region, or state. Here, "large scale" indicates a large area—such as a state—whereas "small scale" represents a smaller entity—such as a neighborhood. Climatologists, for instance, talk about large-scale global circulation in relation to the entire Earth; in contrast, urban geographers talk about small-scale gentrification of part of a city. Alternatively, the cartographic scale is based on a strict mathematical principle: the **representative fraction (RF)**. The RF, which expresses the relationship between map and Earth distances, has become the standard measure for map scale in cartography. The basic format of the RF is quite simple: RF is expressed as a ratio of map units to Earth units (with the map units standardized to 1). For example, an RF of 1:25,000 indicates that 1 unit on a map is equivalent to 25,000 units on the surface of Earth. The elegance of the RF is that the measure is unitless—in our example, the 1:25,000 could represent inches, feet, or meters. Of course, in the same way, that ½ is a larger fraction than ¼, 1:25,000 is a larger scale than 1:50,000. Similarly, a scale of 1:25,000 depicts relatively little area but does so in much greater detail, whereas a scale of 1:250,000 shows a larger area in less detail. Thus, it is the cartographic scale that determines the mapped space and the level of geographic detail possible. At the extreme, architects work at very large scales, perhaps 1:100 (where individual rooms and furniture can be depicted), whereas a standard globe might be constructed at a scale of 1:30,000,000, allowing for only the most basic of geographic detail to be provided. As noted in Chapter 11, certain design issues have to be considered

DOI: 10.1201/9781003150527-7

when representing scales on maps, as do a variety of methods for representing scale, including the RF, the verbal statement, and the graphical bar scale.

Let's assume that one wishes to produce a map at 1 inch equals 10 miles. What would be the representative fraction, or RF? Since 1 mile is equivalent to 63,360 inches, 10 miles equals $10 \times 63,360$, or 633,600 inches. So the exact RF would be 1:633,600. Of course, the standard mapping scales (representative fractions) for many countries differ, with the USGS having established the base map at 1:24,000, or 1 inch equals 2,000 feet. European mapping agencies have used larger scales—1:10,000 and 1:20,000, since the size of the countries allows for many fewer map sheets even at the larger scale.

The term **data resolution**, which is related to scale, indicates the granularity of the data that are used in mapping. When mapping population characteristics of a city (an urban scale), the data can be acquired at a variety of resolutions, including census blocks, block groups, tracts, and even minor civil divisions (MCDs). Each level of resolution represents a different "grain" of the data. Likewise, when mapping biophysical data using remote sensing imagery, a variety of spatial resolutions are possible based on the sensor. Common grains are 79 meters (Landsat Multi-Spectral Scanner), 30 meters (Landsat Thematic Mapper), 20 meters (SPOT HRV multispectral), and 1 meter (Ikonos panchromatic). "Low resolution" refers to coarser grains (e.g., counties), and "high resolution" refers to finer grains (e.g., blocks). Cartographers must be sure that they understand the relationships among geographic scale, cartographic scale, and data resolution and how these relationships influence the information content of the map.

6.3.1 MULTIPLE-SCALE DATABASES

Increasingly, cartographers and other geographic information scientists require the creation of multiscale/multiresolution databases from the same digital data set. This assumes that one can generate, from a master database, additional versions at a variety of scales. The need for such multiple-scale databases results from the requirements of users. For instance, when mapping census data at the county level, a user might wish to see significant detail along the boundaries. Alternatively, when using the same boundary files at the state level, less detail is needed. Because the generation of digital spatial data is extremely expensive and time-consuming, one master version of the database is often created, with smaller scale versions being generated from this master database. A good example of this is the standard mapping of the National Historical Geographic Information System (NHGIS). The master database for the NHGIS is at a scale of 1:150,000, which is the original scale of the TIGER files and the digital line graphs (DLGs) of the USGS. From this base scale, additional databases may be generated through the application of generalization operations. All versions of this multiple-scale database are accessible via the NHGIS Web site (http://www.nhgis.org).

6.4 DEFINITIONS OF GENERALIZATION

6.4.1 DEFINITIONS OF GENERALIZATION IN THE MANUAL DOMAIN

Generalization is the process of reducing the information content of maps because of scale change, map purpose, intended audience, and/or technical constraints. For instance, when reducing a 1:24,000 topographic map (large scale) to 1:250,000 (small scale), some of the geographical features must be either eliminated or modified because the amount of map space is significantly reduced. Of course, all maps, to some degree, are generalizations, as it is impossible to represent on a map all features from the real world, no matter what the scale. A quote from Lewis Carroll's (1893) *Sylvie and Bruno Concluded* nicely illustrates this concept:

> "That's another thing we've learned from your Nation," said Mein Herr, "map making. But we've carried it much further than you. What do you consider the largest map that would be really useful?"
>
> "About six inches to the mile."
>
> "Only about six inches!" exclaimed Mein Herr. "We very soon got to six yards to the mile. Then we tried a hundred yards to the mile. And then came the grandest idea of all! We actually made a map of the country on a scale of a mile to the mile!"
>
> "Have you used it much?" I enquired.
>
> "It has never been spread out yet," said Mein Herr. "The farmers objected: they said it would cover the whole country, and shut out the sunlight! So now we use the country itself, as its own map, and I assure you it does nearly as well."

Cartographers have written on the topic of cartographic generalization since the early part of the twentieth century. Max Eckert, the seminal German cartographer and author of *Die Kartenwissenschaft*, wrote in 1908, "In generalizing lies the difficulty of scientific map making, for it no longer allows the cartographer to rely merely on objective facts but requires him to interpret them subjectively" (347). Other cartographers also have struggled with the intrinsic subjectivity of the generalization process as they have attempted to understand and define cartographic generalization. For instance, in 1942, John K. Wright argued that "Not all cartographers are above attempting to make their maps seem more accurate than they actually are by drawing rivers, coasts, form lines, and so on with an intricacy of detail derived largely from the imagination" (528). Wright identified two major components of the generalization process: simplification, which is the reduction of raw information that is too intricate, and amplification, which is the enhancement of information that is too sparse. The idea that generalization could be broken down into a logical set of processes such as simplification and amplification has become a common theme in generalization research.

Erwin Raisz (1962), for example, identified three major components of generalization: combination, omission, and simplification. Arthur Robinson and colleagues (1978) identified four components: selection, simplification, classification, and symbolization. In Robinson et al.'s model, the selection was considered a preprocessing step to generalization itself. Selection allowed for the identification of certain features and feature classes, whereas generalization applied various operations such as simplification. We discuss the Robinson model in more detail in Section 6.5.1.

6.4.2 Definitions of Generalization in the Digital Domain

Robert McMaster and Stuart Shea (1992) noted that the generalization process itself accomplishes the following: reduces the amount of data that needs to be stored; helps to modify features with changing map scale; and assists in simplifying (for instance, classifying) and symbolizing data. One definition of generalization is the process of deriving, from an original data source, a symbolically or digitally encoded cartographic data set through the application of spatial and attribute transformations. Of course, in applying these transformations (through generalization operations), we try to be mindful of the map's purpose and intended audience. For instance, is the map designed for a transportation analysis or for a bike route, and what is the impact on the generalization operations?

Students of modern cartography should realize that much of what we know today about the generalization process is the result of what cartographers over the centuries learned through manual cartography, where maps were made with pen and ink. When changing scale, cartographers would carefully study a feature at the original scale and then decide how the feature needed to be modified for the new scale. Did it need to be moved (displaced) to avoid colliding with another feature? Did it need to be simplified and/or smoothed? A series of techniques were developed over hundreds of years to help cartographers generalize maps and map features. Today, we attempt to replicate manual generalization with computers; thus, the term *digital generalization* is now appropriate.

6.5 MODELS OF GENERALIZATION

To better understand the complexity of generalization, researchers have attempted to design conceptual models of the process. Some researchers have focused on fundamental operations and the relationships among them, whereas other researchers have created complex models.

6.5.1 Robinson et al.'s Model

Arthur Robinson and colleagues (1978) developed one of the first formal models or frameworks to better understand the generalization process. They separated the process into two major steps: selection (a preprocessing step) and the actual process of generalization, which involves the geometric and statistical manipulation of objects. *Selection* involves the identification of objects to retain in (or eliminate from) the map or database. For instance, in developing content for a thematic map, often a minimal amount of base material is selected, such as major roads, political boundaries, or urbanized areas. Detailed base information, such as place names and hydrologic networks, is often left off a thematic map because this base information is not deemed critical. On the other hand, considerable base information is often selected for detailed topographic maps, as this information is deemed critical. *Generalization* involves the processes of simplification, classification, and symbolization. Simplification is the elimination of unnecessary detail from a feature; classification involves categorizing objects; and symbolization is the graphic encoding. Simplification will shortly be discussed further, and both classification and symbolization are covered in other chapters of this book.

6.5.2 McMaster and Shea's Model

Attempting to create a comprehensive conceptual model of the generalization process, McMaster and Shea (1992) identified three significant components: the conceptual objectives, or *why to generalize*; the cartometric evaluation, or *when to generalize*; and the fundamental operations, or *how to generalize* (Figure 6.1). We will now discuss why and when we generalize; in Section 6.6, we will consider a number of issues related to how we generalize.

6.5.2.1 Why Generalization Is Needed: The Conceptual Objectives of Generalization

The conceptual objectives of generalization include reducing complexity, maintaining spatial accuracy, maintaining attribute accuracy, maintaining aesthetic quality, maintaining a logical hierarchy, and consistently applying the rules

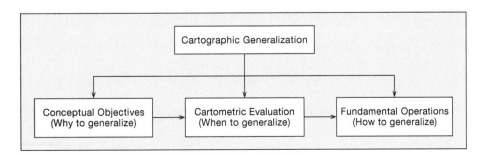

FIGURE 6.1 An overview of McMaster and Shea's model of generalization.

of generalization. Reducing complexity is perhaps the most significant goal of generalization. The difficulty for the cartographer is relatively straightforward: How does one take a map at a scale of, perhaps, 1:24,000 and reduce it to a scale of 1:100,000? More important, the cartographer faces the difficulty of reducing the information content so that it is appropriate for the scale. Obviously, the complexity of detail that is provided at a scale of 1:24,000 cannot be represented at 1:100,000; some features must be eliminated and some details must be modified. For centuries, through considerable experience, cartographers developed a sense of what constituted appropriate information content. Figure 6.2 nicely illustrates this point. The top portion of this figure depicts the same feature—a portion of Ponta

Creek in Kemper and Lauderdale Counties, Mississippi— at four different scales: 1:24,000, 1:50,000, 1:100,000, and 1:250,000. These features, digitized by Philippe Thibault (2002) in his doctoral dissertation, effectively show the significantly different information content that is visible as one reduces the scale from 1:24,000 to 1:250,000. The bottom part of the illustration depicts an enlargement of the smaller scale features to match the features at 1:24,000. Note, for instance, the enlargement of the 1:250,000 scale line by 1,041.67 percent to match Ponta Creek at 1:24,000. In this case, the cartographer has manually generalized Ponta Creek through a series of transformations—including simplification, smoothing, and enhancement (as described later)—as a holistic process, unlike current computer

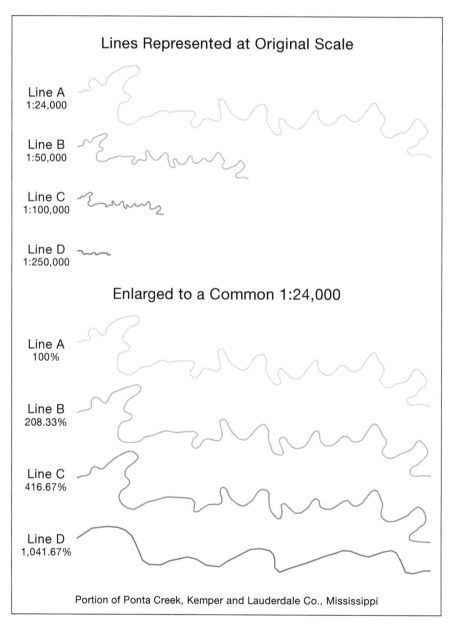

FIGURE 6.2   Depiction of Ponta Creek (in Mississippi) at four different scales. (Courtesy of Philippe Thibault.)

approaches that require a separation of these often linked processes. The set of decisions required to generalize cartographic features based on their inherent complexity is difficult if not impossible to quantify, although, as described next, several attempts have been made over the past two decades.

Clearly, there is a direct and strong relationship among scale, information content, and generalization. John Hudson (1992) explained the effect of scale by indicating what might be depicted on a 5 by 7-inch map:

- A house at a scale of 1:100
- A city block at a scale of 1:1,000
- An urban neighborhood at a scale of 1:10,000
- A small city at a scale of 1:100,000
- A large metropolitan area at a scale of 1:1,000,000
- Several states, at a scale of 1:10,000,000
- Most of a hemisphere, at a scale of 1:100,000,000
- The entire world with plenty of room to spare at a scale of 1:1,000,000,000

He explained that these examples, which range from largest ($1:10^2$) to smallest ($1:10^9$) span eight orders of magnitude and a logical geographical spectrum of scales. Geographers work at a variety of scales, from the very large—the neighborhood—to the very small—the world. Generalization is a key activity in changing the information content so that it is appropriate for these different scales. However, a rough guideline that cartographers follow is that scale change should not exceed $10\times$ existing scale. Thus, if you have a scale of 1:25,000, you should use the scale only for generalizations up to 1:250,000. Beyond 1:250,000, the original data are "stretched" beyond their original fitness for use, although the nature of digital data (in some cases considered scaleless) is changing our thinking on the nature of the spatial scale.

Two additional theoretical objectives that are important in generalization are maintaining the spatial accuracy of features and maintaining the attribute accuracy of features. Spatial accuracy deals primarily with the geometric shifts that necessarily take place in generalization. In simplification, for instance, coordinate pairs are deleted from the data set. By necessity, this shifts the geometric location of the features, creating "error." The same problem occurs with feature displacement, in which two features are pulled apart to prevent a graphical collision. The goals of the generalization process are to minimize this shifting and to maintain as much spatial accuracy as possible. Attribute accuracy deals with the subject being mapped, such as population density or land use. For instance, classification, a key component of generalization, often degrades the original "accuracy" of the data through data aggregation.

6.5.2.2   When Generalization Is Required

In a digital cartographic environment, it is necessary to identify those specific conditions when generalization will be required. Although many such conditions can be identified, four fundamental conditions include:

1. Congestion
2. Coalescence
3. Conflict
4. Complication

As explained by McMaster and Shea (1992), *congestion* refers to the problem, under scale reduction, of too many objects being compressed into too small a space, which results in overcrowding due to high feature density (Figure 6.3). Significant congestion results in decreased communication of the mapped message—for instance when too many buildings are in close proximity. *Coalescence* refers to the condition in which features graphically collide due to scale change. In these situations, features actually touch. This condition thus requires the implementation of the displacement operation, as discussed shortly. *Conflict* results when, due to generalization, an inconsistency between or among features occurs. For instance, if the generalization of a coastline eliminates a bay with a city located on it, either the city or the coastline must be moved to ensure that the urban area remains on the coast. Such spatial conflicts are difficult both to detect and to correct. The condition of *complication* is dependent on the specific spatial configurations that exist in a defined space. An example is a digital line that changes in complexity from one part to the next, such as a coastline (like Maine's) that progresses from very smooth to very crenulated. In this context, Barbara Buttenfield (1991) demonstrated the use of line-geometry-based structure signatures as a means of controlling the tolerance values, based on complexity, in the generalization process. Later, we provide details on other techniques for detecting changes in linear complexity.

One issue important to determining when to generalize is measuring the amount of generalization that either has been done or is planned. Two basic types of measures can be identified: procedural and quality assessment.

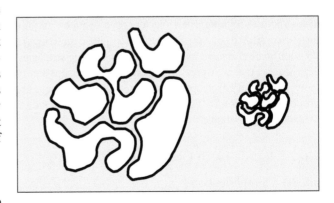

FIGURE 6.3   Example of spatial congestion with scale change.

Procedural measures are those needed to invoke and control the process of generalization. Such measures might include those that: (1) select a simplification algorithm, given a certain feature class; (2) modify a tolerance value along a feature as the complexity changes; (3) assess the density of a set of polygons being considered for agglomeration; (4) determine whether a feature should undergo a type change (e.g., area to point) due to scale modification; and (5) compute the curvature of a line segment to invoke a smoothing operation. *Quality assessment measures* evaluate both individual operations, such as the effect of simplification and the overall quality of the generalization (e.g., poor, average, and excellent). Several studies have discussed mathematical and geometric measures, including those by Buttenfield (1991), McMaster (1986; 1987), and Plazanet (1995).

6.6 THE FUNDAMENTAL OPERATIONS OF GENERALIZATION

6.6.1 A FRAMEWORK FOR THE FUNDAMENTAL OPERATIONS

In the McMaster and Shea model discussed earlier, the third major component involves the fundamental operations, or how to generalize. Most of the research in generalization assumes that the process can be broken down into a series of logical operations that can be classified according to the type of geometry of the feature. For instance, a simplification operation is designed for linear features, whereas an amalgamation operator works on areal features. Table 6.1 provides a framework for the operations of generalization, dividing the process into those activities needed for raster- and vector-mode processing. Geographical features are normally represented in either a "vector" or a "raster" format inside of a computer. The vector representation uses *x-y* coordinate pairs to represent point features such as a house (a single *x-y* coordinate pair), a line feature such as a river (a string of connected *x-y* coordinate pairs), or an areal feature such as a park boundary (a string of *x-y* coordinate pairs in which the first pair matches the last pair). The raster approach uses a matrix of cells of a given resolution (e.g., 30 meters) to represent features. Many standard GIS books (e.g., Lo and Yeung 2007) describe these two data structures in more detail. Vector-based operators require more complicated strategies because they operate on strings of *x-y* coordinate pairs and require complex searching strategies. The next section will provide a more detailed discussion of individual vector-based operations (raster-based operations are not covered in this text). Figure 6.4 provides graphic depictions of some key operations.

6.6.2 VECTOR-BASED OPERATIONS

6.6.2.1 Simplification
Simplification is the most commonly used generalization operator. The concept is relatively straightforward because,

TABLE 6.1
A Framework for Generalization Operations

Raster-mode generalization	Vector-mode generalization
Structural generalization	Point feature generalization
Simple structural reduction	Aggregation
Resampling	Displacement
Numerical generalization	Line feature generalization
Low-pass filters	Simplification
High-pass filters	Smoothing
Compass gradient masks	Displacement
Vegetation indices	Merging
Numerical categorization	Enhancement
Minimum-distance to means	Areal feature generalization
Parallelopiped	Amalgamation
Maximum-likelihood	Collapse
classification	Displacement
Categorical generalization	Volume feature generalization
Merging (of categories)	Smoothing
Aggregation (of cells)	Enhancement
Non-weighted	Simplification
Category-weighted	Holistic generalization
Neighborhood-weighted	Refinement
Attribute change	

Source: After McMaster and Monmonier (1989) and McMaster (1989b).

at its most basic level, it involves a "weeding" of unnecessary coordinate data. The goal is to retain as much of the geometry of the feature as possible while eliminating the maximum number of coordinates. Below, we provide more detail on the simplification process.

6.6.2.2 Smoothing
Although often assumed to be identical to simplification, **smoothing** is actually a much different process. The smoothing operation shifts the position of points to improve the appearance of the feature (Figure 6.4). Smoothing algorithms relocate points in an attempt to plane away small perturbations and capture only the most significant trends of the line (McMaster and Shea 1992). As with simplification, there are many approaches for the process—a simple classification is provided in Table 6.2. Research has shown that a careful integration of simplification and smoothing routines can produce a simplified yet aesthetically acceptable result (McMaster 1989a).

6.6.2.3 Aggregation
As depicted in Figure 6.4, **aggregation** involves merging multiple point features, such as a cluster of buildings. This process involves grouping point locations and representing them as areal units. The critical problems in this operation are determining the density of points needed to identify a cluster to be aggregated and

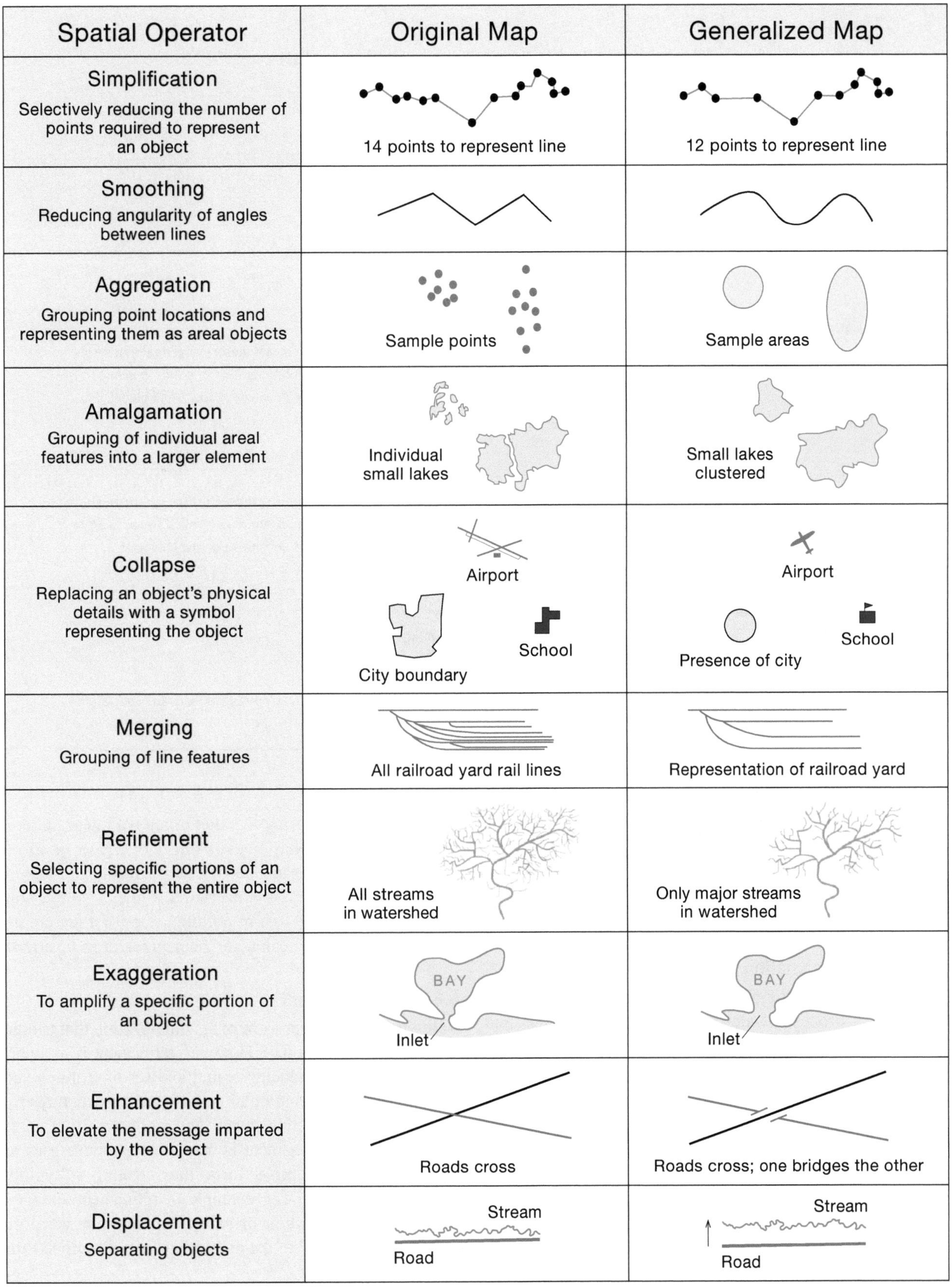

Spatial Operator	Original Map	Generalized Map
Simplification Selectively reducing the number of points required to represent an object	14 points to represent line	12 points to represent line
Smoothing Reducing angularity of angles between lines		
Aggregation Grouping point locations and representing them as areal objects	Sample points	Sample areas
Amalgamation Grouping of individual areal features into a larger element	Individual small lakes	Small lakes clustered
Collapse Replacing an object's physical details with a symbol representing the object	Airport City boundary · School	Airport Presence of city · School
Merging Grouping of line features	All railroad yard rail lines	Representation of railroad yard
Refinement Selecting specific portions of an object to represent the entire object	All streams in watershed	Only major streams in watershed
Exaggeration To amplify a specific portion of an object	BAY Inlet	BAY Inlet
Enhancement To elevate the message imparted by the object	Roads cross	Roads cross; one bridges the other
Displacement Separating objects	Stream Road	Stream Road

FIGURE 6.4 The fundamental operations of generalization. (After McMaster and Shea 1992; courtesy of Philippe Thibault.)

TABLE 6.2

A Classification of Algorithms Used to Smooth Cartographic Features

Category 1:	Weighted averaging
	Calculates an average value based on the positions of existing points and neighbors, with only the endpoints remaining the same; maintains the same number of points as the original line; algorithms can be easily adapted for different smoothing conditions by adjusting tolerance values (e.g., number of points used in smoothing); all algorithms use local or extended processors.
	Examples: Three-point moving average
	Five-point moving average
	Other moving average methods
	Distance-weighted averaging
	Slide averaging
Category 2:	Epsilon filtering
	Algorithm uses certain geometrical relationships between the points and a user-defined tolerance to smooth the cartographic line; endpoints are retained, but the absolute number of points generated for the smoothed line is algorithm dependent; approaches are local, extended local, and global.
	Examples: Epsilon filtering
	Brophy algorithm
Category 3:	Mathematical approximation
	Develop a mathematical function or series of mathematical functions that describe the geometrical nature of the line; number of points on the smoothed line is variable and is controlled by the user; retention of the endpoints and of the points on the original line is dependent on the choice of algorithms and tolerances; function parameters can be stored and used to later regenerate the line at the required point density; approaches are local, extended local, and global.
	Examples: Local processing: cubic spline
	Extended local processing: b-spline
	Global processing: Bezier curve

Source: After McMaster and Shea 1992, *Generalization in Digital Cartography*, pp. 86–87. Copyright American Association of Geographers (http://www.aag.org).

specifying the boundary around the resulting cluster. The most common approach is to triangulate the points (i.e., create triangles among neighboring points) and determine the density of the triangles (e.g., a grouping of smaller triangles might represent a cluster for aggregation) (Jones et al. 1995).

6.6.2.4 Amalgamation

Amalgamation is the process of fusing together nearby polygons, a process needed for both noncontinuous and continuous areal data (Figure 6.4). A noncontinuous example is a series of small islands in close proximity that have size and detail that cannot be depicted at a smaller scale. A continuous example is census tract data, where several tracts with similar statistical attributes can be joined together.

6.6.2.5 Collapse

The **collapse** operation involves the conversion of geometry. For instance, it might be that a complex urban area is collapsed to a point due to scale change and is resymbolized with a geometric form such as a circle. A complex set of buildings might be replaced by a simple rectangle—which might also involve amalgamation.

6.6.2.6 Merging

Merging involves fusing together groups of linear features, such as parallel railway lines or edges of a river or stream (Figure 6.4). Merging can be viewed as a form of collapse in which an areal feature is converted to a line. A simple solution is to average the two or multiple sides of a feature and to use this average to calculate the new feature's position.

6.6.2.7 Refinement

Refinement is another form of resymbolization that is much like collapse (Figure 6.4). However, refinement is an operation that involves reducing a multiple set of features such as roads, buildings, and other types of urban structures to a simplified representation rather than a conversion of geometry. The key to refinement is that complex geometries are resymbolized to a simpler form, thus creating a "typification" of the objects. The example of refinement shown in Figure 6.4 is a selection of part of a stream network that depicts the "essence" of the distribution in a simplified form.

6.6.2.8 Exaggeration

Exaggeration is one of the more commonly applied generalization operations. Often it is necessary to amplify a specific part of an object to maintain clarity in scale reduction.

The example in Figure 6.4 depicts the exaggeration of the mouth of a bay, which would close under scale reduction.

6.6.2.9   Enhancement

Enhancement is a symbolization change that emphasizes the importance of a particular object. For instance, the delineation of a bridge under an existing road is often portrayed as a series of cased lines, which assists in emphasizing that feature over another.

6.6.2.10   Displacement

Displacement is perhaps the most difficult of the generalization operations because it requires complex measurement (Figure 6.4). The problem can be illustrated with a series of cultural features in close proximity to a complex coastline. Assume, for example, that a highway and a railroad follow a coastline in close proximity, with a series of smaller islands offshore. In the process of scale reduction, all features would tend to coalesce. The operation of displacement would pull the features apart to prevent this coalescence. What is critical in the displacement operation is the calculation of a displacement hierarchy because one feature will likely have to be shifted away from another (Nickerson and Freeman 1986; Monmonier and McMaster 1990). A description of the mathematics involved in displacement can be found in McMaster and Shea (1992).

6.6.3   THE SIMPLIFICATION PROCESS

Most simplification routines utilize complex geometrical criteria (distance and angular measurements) to select **critical points**, those points that are significant in defining the structure of a linear or areal feature. A general classification of simplification methods consists of five approaches: independent point routines, local processing routines, constrained extended local processing routines, unconstrained extended local processing routines, and global methods (Table 6.3). **Independent point routines** select coordinates based on their positions along a line, nothing more. For instance, a typical *n*th point routine might select every third point to quickly weed out unnecessary coordinate data. Although computationally efficient, these algorithms

---

TABLE 6.3

A Classification of Algorithms Used to Simplify Cartographic Features

Category 1:   Independent point algorithms
               Do not account for the mathematical relationships with the neighboring pairs, operate
                 independent of topology.
               Examples: *n*th point routine
                       Random selection of points

Category 2:   Local processing routines
               Utilize the characteristics of the immediate neighboring points in determining significance.
               Examples: Distance between points
                       Angular change between points
                       Jenks's algorithm (distance and angular change)

Category 3:   Constrained extended local processing routines
               Search continues beyond immediate coordinate neighbors and evaluates sections of a line.
               Extent of search depends on distance, angular, or number of points criterion.
               Examples: Lang algorithm
                       Opheim algorithm
                       Johannsen algorithm
                       Deveau algorithm
                       Roberge algorithm
                       Visvalingam algorithm

Category 4:   Unconstrained extended local processing routines
               Search continues beyond immediate coordinate neighbors and evaluates sections of a line.
               Extent of the search is constrained by geomorphological complexity of the line, not by
                 algorithmic criterion.
               Example: Reumann–Witkam algorithm

Category 5:   Global routines
               Considers the entire line, or specified line segment; iteratively selects critical points.
               Example: Douglas–Peucker algorithm

Source: After McMaster and Shea 1992, *Generalization in Digital Cartography*, p. 73. Copyright American
            Association of Geographers (http://www.aag.org).

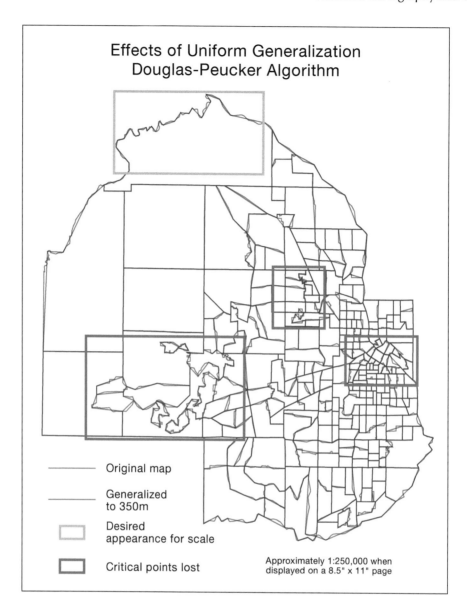

FIGURE 6.5   The Douglas–Peucker algorithm applied to Hennepin County, Minnesota. (Republished with permission of John Wiley & Sons—Books, from McMaster, R. B., and Sheppard, E. (2004) "Introduction: Scale and Geographic Inquiry," p. 9. In *Scale and Geographic Inquiry: Nature, Society and Method*, ed. by E. Sheppard and R. B. McMaster, permission conveyed through Copyright Clearance Center, Inc.)

are crude in that they do not account for the true geomorphological significance of a feature. **Local processing routines** utilize immediate neighboring points to assess the significance of the point. Assuming a point to be simplified as x_n, y_n, these routines evaluate the point's significance based on its relationship to immediate neighboring points x_{n-1}, y_{n-1} and x_{n+1}, y_{n+1}. This relationship is normally determined by either a distance or an angular criterion, or both. **Constrained extended local processing routines** search beyond the immediate neighbors and evaluate larger sections of lines, again, normally determined by distance and angular criteria. Certain algorithms search around a larger number of points, perhaps two, three, or four, in either direction, whereas other algorithms use more complex criteria. **Unconstrained extended local processing routines**

also search around larger sections of a line, but the search is terminated by the geomorphological complexity of the line, not by algorithmic criteria. Finally, **global algorithms** process the entire line feature at once and do not constrain the search to subsections. The most commonly used simplification algorithm—the Douglas–Peucker—takes a global approach and processes a line "holistically." Details of the Douglas–Peucker algorithm can be found in McMaster (1987) and McMaster and Shea (1992), and comparisons of the various algorithms can be found in McMaster (1987). Table 6.3 provides details on algorithms that can be found in each of the five categories.

The effect of the Douglas–Peucker algorithm can be seen in Figures 6.5 and 6.6, where the algorithm is applied to Hennepin County, Minnesota, using a 350-meter tolerance value.

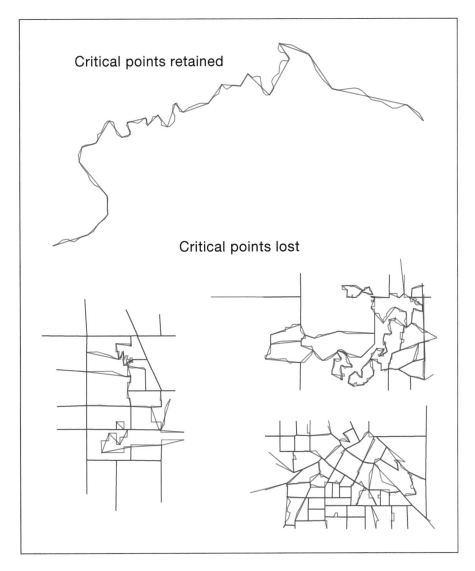

FIGURE 6.6 An enlargement of the Douglas–Peucker algorithm along parts of Hennepin County boundaries showing critical points retained and lost. (Courtesy of National Historical Geographic Information System.)

A tolerance value allows the user to decide how much detail the algorithm eliminates. The original spatial data, taken from the U.S. Bureau of the Census TIGER files, is in light gray, whereas the generalized feature is in red. Note that many of the original points have been eliminated, thus simplifying the features. Unfortunately, the effects of this approach—as with most generalization processes—are not consistent, given that the algorithm behaves differently depending on the geometric character or geomorphological significance of the feature. In areas that are more "natural," such as streams and rivers, the simplification produces a relatively good approximation. For instance, note that a reasonable set of critical points seems to have been retained for the northern boundary of Hennepin County, which follows the Mississippi River. However, in urban areas in the central portion of the map, where the census geography follows the rectangular street network, the results are less satisfactory. In many cases, logical right angles are simplified to diagonals.

A significant problem of the generalization process involves the identification of appropriate tolerance values for simplification. Unfortunately, this is mostly an experimental process in which the user must test and retest values until an acceptable solution is empirically found. As explained previously, cartographers often turn to measurements to ascertain the complexity of a feature and to assist in establishing appropriate tolerance values.

Figure 6.7 depicts the calculation of one specific measurement, the *trendlines*, for the Hennepin County data set. The trendlines for a digitized curve are based on a calculation of angularity, or where the lines change direction. Where a curve changes direction—for example, from left to right—a mathematical inflection point is defined (theoretically, the point of no curvature). The connection of these inflection points, which indicates the general "trend" of the line, is called the trendline. The complexity of a feature can be approximated by looking at the trendlines for the entire feature or for the entire data set. A simple measure of

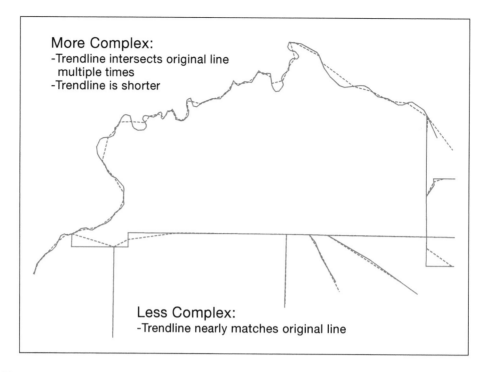

FIGURE 6.7 The calculation of a trendline. (Courtesy of National Historical Geographic Information System.)

complexity derived from the trendline is the trendline/total length of a line or the *sinuosity* of a feature. Along relatively straight line segments, with little curvilinearity, the trendline will be very close to the curve, and the trendline/total length ratio will be nearly equal to 1.0 (e.g., the relatively straight line near the middle of Figure 6.7). However, a highly complex curve, such as the northern border of Hennepin County, will deviate significantly from the trendline, and the ratio will be distinctly less than 1.0. Thus, the greater the difference between the actual digitized curve and the trendline, the more complex the feature.

6.7 AN EXAMPLE OF GENERALIZATION

To illustrate a practical application of generalization, we will consider how information is generalized when using the NHGIS database. Much of this database was acquired from USGS digital line graphs that were digitized at a scale of 1:150,000. Consider Figure 6.8, which depicts three different representations of coastlines and county boundaries along the Florida Gulf Coast from Tampa Bay northward. Figure 6.8A shows the NHGIS raw data at a scale of 1:2,000,000. Here we can see excessive detail along the

FIGURE 6.8 An NHGIS series of generalizations for the Florida coastline, assuming a target scale of 1:2,000,000: (A) the NHGIS raw data (based on USGS digital data at a scale of 1:150,000; (B) an NHGIS-produced generalization; (C) a generalization produced by the U.S. Bureau of the Census. (From Schroeder and McMaster (2007), Creative Commons Attribution 4.0 License, https://www.ica-conference-publications.net/licence_and_copyright.html.)

coast, where lines coalesce and converge, a problem that is created by having far too much information (too many coordinate pairs) for the space. The NHGIS-produced generalization is depicted in Figure 6.8B. In particular, note the reduction of detail in the three inset boxes (i, ii, and iii), where the coastline is simplified, islands are eliminated, and the overall complexity is reduced. The third illustration (Figure 6.8C) shows a generalization produced by the U.S.

Bureau of the Census. Although also a simplified version, this map might also be considered an overgeneralization of the features in which too much of the critical geomorphological character has been eliminated. Note in particular inset ii, where the important peninsula has been eliminated.

A second set of illustrations (Figure 6.9) depicts the NHGIS's capability of generating generalized boundaries for a variety of target scales. The scales selected for

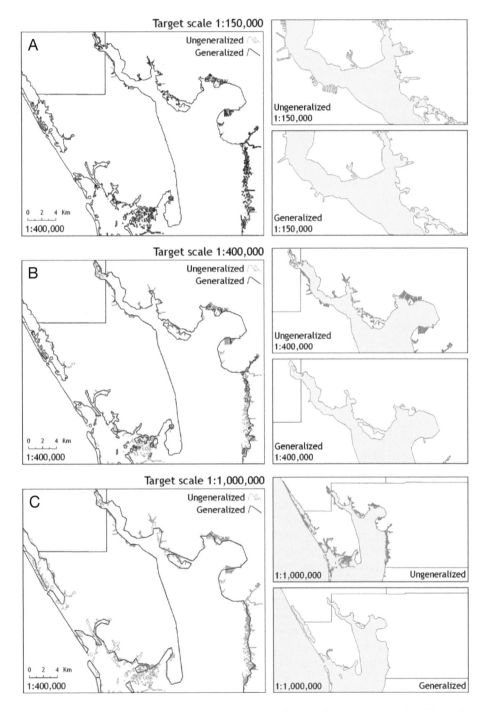

FIGURE 6.9 NHGIS's capability of generating generalized boundaries for a variety of target scales. The scales selected for these illustrations are 1:150,000 (A), 1:400,000 (B), and 1:1,000,000 (C), each depicting the coastline around Charlotte Harbor, Florida. (From Schroeder and McMaster (2007), Creative Commons Attribution 4.0 License, https://www.ica-conference-publications.net/licence_and_copyright.html.)

these illustrations are 1:150,000 (Figure 6.9A), 1:400,000 (Figure 6.9B), and 1:1,000,000 (Figure 6.9C), each depicting the coastline around Charlotte Harbor, Florida. Note that the main map on the left of each figure superimposes the generalized data over the ungeneralized data at a scale of 1:400,000 for each of the target scales. The related maps on the right of each figure illustrate how the ungeneralized and generalized data would appear at the target scales of 1:150,000, 1:400,000, and 1:1,000,000. Note how the detail along the coastline has been generalized for the given scale.

The NHGIS generalization system (Schroeder and McMaster 2007) works as follows: The main processing is divided into two parts. The first part eliminates small areas (e.g., islands, small parts of multipart features, and slivers caused by historical boundary changes) according to measures of area and area/perimeter (McMaster 1987). The second part generalizes boundaries in four steps:

1. Join feature parts that touch each other at only one node in order to simplify topological relationships. Conceptually, this is similar to transforming a figure-8 into an hourglass figure.
2. Apply two simple line simplification algorithms to remove insignificant points or those vertices that contribute little to the geographical character of a boundary. The first algorithm removes vertices connecting nearly parallel segments, effectively "straightening" lines that are nearly straight anyway. The second algorithm is the Douglas–Peucker (1973) algorithm, implemented using a low tolerance. This is primarily a preprocessing step that has little visual impact at the target scale but reduces line complexity in order to speed up the following, more complex operations.
3. Complete line generalization using an altered version of the Visvalingam–Whyatt (1993) algorithm with many modifications designed to maintain boundary smoothness, avoid topological conflicts, and prevent overreduction of small features.
4. Eliminate node wedges, which are narrow spaces found where multiple feature edges intersect. This requires a separate step because the above line generalization operations are applied only to individual edges, which are boundaries that have either no neighbors or one neighbor on each side in each feature class (e.g., a boundary between two census tracts or a coastline between a census tract and the ocean). Step 4 will therefore remove long, narrow "extensions" that occur along a single edge but will not remove such features if they lie between two connected edges.

An important component within each of these steps is the maintenance of correct topology, which requires additional operations to prevent generalized boundaries from intersecting one another. Additional details on the NHGIS

generalization process can be found in Schroeder and McMaster (2007).

6.8 NEW DEVELOPMENTS IN CARTOGRAPHIC GENERALIZATION

Research and development in the area of Scale and Generalization continues, with the majority of new work in generalization coming from European researchers and agencies, such as the Institute Geographique National (IGN) in France, the University of Zurich, and multiple universities in Germany and the United Kingdom. In the following section, some of the concepts and terms might require students to pursue the further readings section. For example, in discussing how quadtrees and R-trees might support scale change and generalization, we will not explain these concepts here, but the further readings will provide a full description of a quadtree.

6.8.1 MEASUREMENT OF SCALE CHANGE

One research area that continues is around measurement. Hanna Stigmar and Lars Harrie (2011) develop and utilize a series of measures for describing the legibility of maps. The three types of measures in their study included: (1) the amount of information, which is determined by the amount and size of map objects; (2) the spatial distribution, which is determined by the density and distribution of map objects; and (3) the object complexity. Measures of the amount of information were typified by a number of objects, number of vertices, object line length, and object area. Spatial distribution measures included spatial distribution of objects, spatial distribution of vertices, degree of overlap, number of neighbors, and local density. Complexity measures were represented by object size, line segment size, angularity, line connectivity, and polygon shape. After applying these measures to several maps (at scales of 1:50,000 and 1:10,000), Stigmar and Harrie concluded that the measures that showed the best correspondence between perceived legibility and calculated legibility were the number of vertices, object line length, local density, proximity indicator, and degree of overlap. They also discovered that a combination of measures can provide a better reflection of the legibility than a single measure.

6.8.2 FULLY AUTOMATED GENERALIZATION

Research into more fully automated generalization continues. Guillaume Touya and Jean-François Girres (2013) developed an extension to the original concept of ScaleMaster (which enabled users to formalize how to choose cartographic features from different data sets; see Brewer and Buttenfield 2007) to a ScaleMaster 2.0 that enables the integration of datasets representing different levels of detail (or different points of view) for the same geographical features—thus enabling *automatic multiscale generalization*. The authors explain that "In the

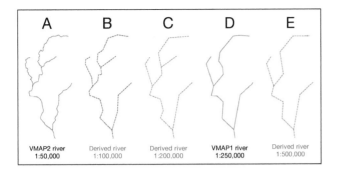

FIGURE 6.10 Illustration of the automatic generation of continuous representations of a river theme in ScaleMaster 2.0. (B) and (C) were derived from (A) (a 1:50,000 scale database), and (E) was derived from (D) (a 1:250,000 scale database). Note the resulting smooth progression in generalization over the various scales. (After Touya and Girres 2013. First published in *Cartography and Geographic Information Systems* 40(3), p. 197. Reprinted with permission from the Cartography and Geographic Information Society.)

original ScaleMaster, a ScaleLine described the generalization rules for one data theme across the scales. In the proposed model, the ScaleLine is composed of scale intervals (e.g., 1:25,000 to 1:50,000)" (Touya and Girres 2013, 194). In explaining the model, the authors use the example of the "river line" theme, which can be derived from river lines from one data set. Alternatively, river areas can be derived from another dataset. Each of these cases is monitored by a different ScaleMasterElement. In their model, the ScaleMasterElement is an essential component, as it contains the generalization rules and required enrichments (such as an attribute-based query that selects the appropriate data for the scale interval). The ScaleMaster 2.0 model is applied to a database with many themes at three levels: ~1:1,000,000, ~1:250,000, and ~1:50,000. The basic generalization functions of Douglas–Peucker filtering, Gaussian smoothing, polygon merging, polygon skeletonization, strokes-based road selection, and contour selection were applied to geographical features including roads, rivers, lakes, contours, and built-in areas with positive results (Figure 6.10 illustrates an example for a river theme).

6.8.3 Data Models for Generalization

Fundamental to much of the work in generalization is the development of an appropriate spatial data structure. Several data structures to support effective scale change and generalization have been developed and applied for creating multiple scale versions of maps. Although a detailed discussion of spatial data models is outside the scope of this book, Peter van Oosterom and colleagues (2014) provide an excellent review of multi-scale data models that support generalization. Some of these multi-scale/variable-scale vector data structures include the Strip-tree, Multi-Scale Line tree, Arc-tree, and the Binary Line Generalisation tree (BLG-tree). Several other data structures better suited to

support generalization should be mentioned. van Oosterom et al. (2014) identify the multi-scale vector reactive BSP-tree structure as an improvement since it supports spatial indexing and multiple level of detail. A second superior model is a Reactive Tree, a dynamic data structure that supports the spatial organization of all map objects and multiple scales. However, it is the topological GAP tree (van Oosterom 1995), in combination with the BLG-tree and the Reactive tree, that is recommended since it avoids the problems of redundant storage and slivers near the boundary of two neighboring areas. With the proliferation of web-based mapping, a major quality of this structure is that it enhances the continuous or smooth zooming of map data.

Pia Bereuter and Robert Weibel (2013) propose the application of the quadtree data structure to support generalization. Although again outside the scope of this book, a quadtree structure allows for space to be divided (tessellated) into binary shrinking quadrants. Basically, a square is divided into four squares, and each of the four squares is divided again (and again) until, at the smallest level (a leaf of the quadtree), it is homogeneous with respect to the attribute being represented (e.g., a soil type). In supporting the quadtree model, the authors explain that this data structure supports algorithms for generalization operations, including selection, simplification, aggregation, and displacement of point data. Additionally, the quadtree spatial index (the referencing index for the quadtree) can further be used to generate local and global measures. These measures can assist in determining the density and proximity of points across map scales and thus enable control of the operation of generalization algorithms, which is consistently one of the challenging problems (272). Our primary objective with this summary of these models is to emphasize the importance of effective and flexible data structures in supporting scale change and generalization.

6.8.4 New Forms of Cartographic Data

Cartographic generalization traditionally has focused on institutional types of data from established agencies— USGS, U.S. Bureau of the Census, the French IGN, and the British Ordnance Survey, to name a few—but the field has witnessed the proliferation of data gathered by individuals, and from "non official sources." This type of data, referred to as volunteered geographic information (VGI), poses unique challenges for scale change and generalization. Examples of VGI include temperature observations from individuals, observations of traffic conditions, and restaurant/hotel reviews. Social media has greatly accelerated the generation of VGI. As pointed out by Monika Sester and colleagues (2014), there are differences with VGI and institutional data, including the thematic content and use; availability, coverage, and homogeneity; timeliness of data; scale range; quality, reliability, trust, and liability; redundancy; user-centric information; and spatial reference and geometric representation. Of course, the VGI must be mapped onto a standard base map and it is unlikely the two data sets have the same level of detail—either the base map data or VGI will have lesser or greater detail (Sester et al. 2014). In order to harmonize these two very different types of data, generalization is necessary to create spatial consistency among the different data and to visualize the data at a smaller scale. Sester et al. apply generalization in several case studies with different types of VGI. One of these case studies provides a comparison between the application of generalization approaches within the VGI structure of the OpenStreetMap project to that of an official National Mapping Agency (NMA) database. OpenStreetMap is the quintessential example of VGI—a global spatial dataset populated with various forms of localized knowledge (https://www.openstreetmap.org/about). As would be expected, the spatial accuracy and data quality vary significantly with these data. Whereas generalization with an NMA database uses an organized and systematic approach to create distinct and predetermined scales, the OSM generalization will be idiosyncratic based on the "creator" of the knowledge. These are two very different approaches, and the VGI-based generalization results in new challenges and thinking about scale change and generalization.

6.9 SUMMARY

Scale is a fundamental process in geography and cartography, as it requires the cartographer to select the appropriate information content, given the map's purpose and intended audience. The processes used to manipulate (i.e., change the scale of) the spatial information are collectively known as **generalization**. For centuries cartographers have tackled the problem of generalization in the manual domain, but conversion to the digital world has created many new challenges. This chapter has provided a discussion of the various forms of scale including **cartographic scale**, which is depicted with a **representative fraction (RF)** such as 1:24,000. It should be noted, however, that geographers and other spatial scientists often conceptualize scale very differently, such as climatologists and ecologists use of the term large scale to describe global processes.

We reviewed major definitions and models of the generalization process. Generalization models include those by Robinson and colleagues and those by McMaster and Shea. The McMaster and Shea model, which was designed for the generalization process in a digital environment, includes three significant components: the conceptual objectives, or *why to generalize*; the cartometric evaluation, or *when to generalize*; and the fundamental operations, or *how to generalize*.

In the context of how to generalize, we considered a broad range of fundamental operations, including **simplification** ("weeding" out unnecessary coordinate data, thus retaining the **critical points** that define a linear or areal feature); **smoothing** (shifting points to improve the appearance of a feature); **aggregation** (merging multiple point features to create an areal feature); **amalgamation** (combining individual areal features to create a larger areal

feature); **collapse** (replacing an object's physical details with a symbol representing the object); **merging** (grouping linear features); **refinement** (selecting portions of a feature to represent the entire feature); **exaggeration** (amplifying a specific portion of a feature, as in the mouth of a bay); **enhancement** (changing the symbolization to emphasize the importance of a feature, as in using cased lines for a bridge); and **displacement** (moving features apart to prevent coalescence). We considered a variety of simplification algorithms that have been developed, ranging from **local processing routines** that utilize only immediate neighboring points to **global algorithms** that consider an entire linear feature. The most commonly used algorithm has been the Douglas–Peucker routine, a global algorithm.

We provided an example of generalization using ongoing work at the NHGIS, housed at the University of Minnesota. We saw that creating an acceptable generalization at a particular target scale requires the application of not just one but a variety of algorithms and a consideration of the linkages between lines that constitute areal features (i.e., one must carefully consider the topology of spatial features).

Ongoing work in the area of scale and generalization continues. Much of the core research has been completed at European institutions (both university and government), although other projects are underway in other parts of the world. Research continues in the areas of fully automated generalization, measurement, spatial data structures designed to support generalization, and new types of spatial data. Interestingly, the area of generalization has brought together researchers from several different disciplines, including geography, cartography, geographic information science, and computer science.

6.10 STUDY QUESTIONS

1. Define geographic scale, cartographic scale, and data resolution.
2. If on a map 1 inch equals 5 miles on Earth, what is the representative fraction (RF)? What are the advantages of the RF?
3. Describe the model of generalization developed by Arthur Robinson and colleagues.
4. List and describe five fundamental operations of generalization.
5. List one example for each of the following simplification algorithms: local processing routine, global processing routine. What are the advantages of each?
6. Provide one example of a spatial data structure that supports generalization.

REFERENCES

Bereuter, P., and Weibel, R. (2013) "Real-time generalization of point data in mobile and web mapping using quadtrees." *Cartography and Geographic Information Science 40*, no. 4:271–281.

Brewer, C. A., and Buttenfield, B. P. (2007) "Framing guidelines for multi-scale map design using databases at multiple resolutions." *Cartography and Geographic Information Science 34*, no 1:3–15.

Buttenfield, B. P. (1991) "A rule for describing line feature geometry." In *Map Generalization: Making Rules for Knowledge Representation*, ed. by B. P. Buttenfield and R. B. McMaster, pp. 150–171. London: Longman.

Carroll, L. (1893) *Sylvie and Bruno Concluded*. New York: Dover Publications (1988).

Douglas, D. H., and Peucker T. K. (1973) "Algorithms for the reduction of the number of points required to represent a digitized line or its character." *The Canadian Cartographer 10*(2):112–123.

Eckert, M. (1908) "On the nature of maps and map logic." *Bulletin of the American Geographical Society 40*, no. 6:344–351. Translated by W. Joerg.

Hudson, J. C. (1992) "Scale in space and time." In *Geography's Inner Worlds: Pervasive Themes in Contemporary American Geography*, ed. by R. F. Abler, M. G. Marcus, and J. M. Olson, pp. 280–300. New Brunswick, NJ: Rutgers University Press.

Jones, C. B., Bundy, G. L., and Ware, J. M. (1995) "Map generalization with a triangulated data structure." *Cartography and Geographic Information Systems 22*, no. 4:317–331.

Lo, C. P., and Yeung, A. K. W. (2007) *Concepts and Techniques of Geographic Information Systems* (2nd ed.). Upper Saddle River, NJ: Pearson Prentice Hall.

McMaster, R. B. (1986) "A statistical analysis of mathematical measures for linear simplification." *The American Cartographer 13*, no. 2:330–346.

McMaster, R. B. (1987) "The geometric properties of numerical generalization." *Geographical Analysis 19*, no. 4:103–116.

McMaster, R. B. (1989a) "The integration of simplification and smoothing routines in line generalization." *Cartographica 26*, no. 1:101–121.

McMaster, R. B. (1989b) "Introduction to 'numerical generalization in cartography.'" *Cartographica 26*, no. 1:1–6.

McMaster, R. B., and Monmonier, M. (1989) "A conceptual framework for quantitative and qualitative raster-mode generalization." *GIS/LIS'89 Proceedings*, Volume 2, Orlando, FL, pp. 390–403.

McMaster, R. B., and Shea, K. S. (1992) *Generalization in Digital Cartography*. Resource Publications in Geography. Washington, DC: Association of American Geographers.

Monmonier, M. S., and McMaster, R. B. (1990) "The sequential effects of geometric operators in cartographic line generalization." *International Yearbook of Cartography 30*:93–108.

Nickerson, B. G., and Freeman, H. R. (1986) "Development of a rule-based system for automated map generalization." *Proceedings, Second International Symposium on Spatial Data Handling*, Williamsville, NY: International Geographical Union Commission on Geographical Data Sensing and Processing, Seattle, WA, pp. 537–556.

Plazanet, C. (1995) "Measurement, characterization and classification for automated line feature generalization." *AUTO-CARTO 12 (Volume 4 of ACSM/ASPRS '95 Annual Convention & Exposition Technical Papers)*, Bethesda, MD, pp. 59–68.

Raisz, E. (1962) *Principles of Cartography*. New York: McGraw-Hill.

Robinson, A. H., Sale, R. D., and Morrison, J. L. (1978) *Elements of Cartography* (4th ed.). New York: Wiley.

Schroeder, J. P., and McMaster, R. B. (2007) "The creation of a Multiscale National Historical Geographic Information System for the United States Census." *Proceedings, 23rd International Cartographic Conference*, Moscow, Russia, CD-ROM.

Sester, M., Arsanjani, J. J., Klammer, R. et al. (2014) "Integrating and generalising volunteered geographic information." In *Abstracting Geographic Information in a Data Rich World, Lecture Notes in Geoinformation and Cartography*, ed. by D. Burghardt, C. Duchêne, and W. Mackaness, pp. 119–155. Switzerland: Springer International Publishing.

Stigmar, H., and Harrie, L. (2011) "Evaluation of analytical measures of map legibility." *The Cartographic Journal 48* no. 1:41–53.

Thibault, P. (2002) "Cartographic generalization of fluvial features." Unpublished Ph.D. dissertation, University of Minnesota, Minneapolis, MN.

Touya, G. and Girres J. F. (2013). "ScaleMaster 2.0: A ScaleMaster extension to monitor automatic multi-scales generalizations." *Cartography and Geographic Information Science 40*, no. 3:192–200.

van Oosterom, P. (1995) "The GAP-tree, an approach to 'on-the-fly' map generalization of an area partitioning." In *GIS and Generalization: Methodology and Practice*, ed. by J. Müller, J. Lagrange, and R. Weibel, pp. 120–132. London: Taylor & Francis.

van Oosterom, P., Meijers, M., Stoter, J. et al. (2014) "Data structures for continuous generalisation: tGAP and SSC." In *Abstracting Geographic Information in a Data Rich World, Lecture Notes in Geoinformation and Cartography*, ed. by D. Burghardt, C. Duchêne, and W. Mackaness, pp. 83–117. Switzerland: Springer International Publishing.

Visvalingam, M., and Whyatt, D. (1993) "Line generalization by repeated elimination of points." *The Cartographic Journal 30*(1):46–51.

Wright, J. K. (1942) "Map makers are human." *The Geographical Review 32*, no. 4:527–544.

7 The Earth and Its Coordinate System

7.1 INTRODUCTION

This chapter provides fundamental material related to the shape and size of Earth and the nature of its coordinate system. Knowledge of these concepts is essential for understanding the material on map projections covered in Chapters 8 and 9.

We begin the chapter by introducing the basic characteristics of Earth's coordinate system, which is comprised of an imaginary network of lines called the **graticule**. The graticule is further composed of lines of **latitude** and **longitude**, which crisscross Earth's surface, allowing point locations to be uniquely described in terms of a **sexagesimal system** (degrees, minutes, and seconds). The arrangement of this coordinate system is based on Earth's position relative to the sun as well as its alignment with celestial objects such as the North Star. This alignment establishes the **Equator** as the reference line for latitude values, which start at the Equator and run 90° to the North Pole and 90° to the South Pole. On the other hand, the **Prime Meridian**, passing through Greenwich, England, serves as the reference line for longitude values that run 180° east and 180° west of this line.

Finally, we introduce the field of **geodesy**, the science of understanding and measuring Earth's size and shape. Approximately 2,000 years ago, the Greek Eratosthenes was able to calculate the circumference of Earth as 40,500 kilometers (25,170 miles), a figure not far from our present-day value of 40,075 kilometers (24,906 miles). Determining the correct shape of Earth has proved more problematic, as it wasn't until about 300 years ago that Sir Isaac Newton suggested that Earth is an **oblate spheroid** or **ellipsoid**, bulging at the Equator due to **centrifugal force**. This ellipsoid concept extends to the reference ellipsoid as a solid body that more accurately defines Earth's shape than a simple spherical assumption. Advances in satellite technology have led to the concept that Earth is not a smooth, mathematically definable surface (as described by the **reference ellipsoid**) but due to differences in gravitational forces, can be modeled as a **geoid**—a shape that Earth would take on if the world's oceans were allowed to flow over the land, adjusting to the gravitational differences across its surface and creating a single undisturbed water body. A reference ellipsoid and a geoid are components that comprise a **geodetic datum**. A geodetic datum establishes the origin for latitude and longitude values as well as elevation and can be divided into a horizontal and vertical datum. More specifically, a **horizontal datum** is based on a reference ellipsoid that establishes the origin for latitude and longitude values for Earth's address system. On the other hand, a **vertical datum** is defined by the geoid, which establishes mean sea level and thus elevations across Earth's surface. One example of a geodetic datum that is used in the United States is the **North American Datum 1983** (**NAD83**).

7.2 LEARNING OBJECTIVES

- Explain that the graticule is an imaginary network of lines composed of latitude and longitude and is based on a sexagesimal system of angular measurement.
- Recognize that the equator serves as the origin for latitude and that the Prime Meridian serves as the origin for longitude.
- Define the nature of distance and direction on Earth's spherical surface as represented by great and small circles and azimuths and loxodromes, respectively.
- Explain the significant historical events and personalities that have led to the present understanding of Earth's size and shape.
- Describe the component parts of a reference ellipsoid.
- Explain the differences between the various reference ellipsoids that have been used to model Earth's size and shape.
- Explain how the geoid better reflects Earth's shape than a spherical model.
- Discuss how a specific reference ellipsoid and a geoid are combined together to create a geodetic datum.

7.3 BASIC CHARACTERISTICS OF EARTH'S GRATICULE

Addresses, zip codes, and telephone area codes are just some of the many ways that places on Earth's surface can be located. Although useful, such approaches are limited because they have no uniformity (e.g., each country has its own system of postal codes), they locate areas rather than specific points, and they do not completely cover Earth (no area codes exist over the world's oceans). To overcome these limitations, a system composed of latitude and longitude known as the **graticule** was developed and is a critical foundation for understanding the concepts on map projections that appear in Chapters 8 and 9.

To understand the graticule, we start by recalling that Earth rotates about its axis, called the **axis of rotation**, which is aligned with the **North Star** (Polaris) and passes through the **North** and **South Poles** (Figure 7.1). Perpendicular to

DOI: 10.1201/9781003150527-8

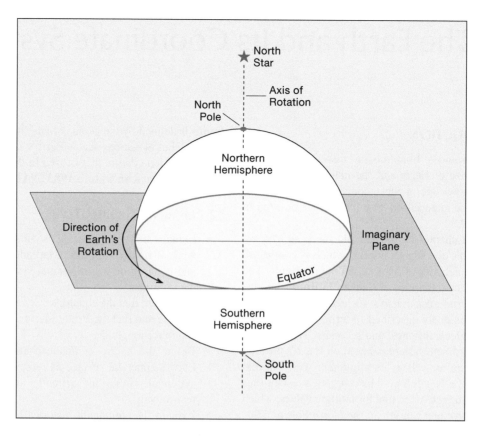

FIGURE 7.1   The Equator is formed by the intersection of an imaginary plane perpendicular to and bisecting Earth's axis of rotation, which is aligned with the North Star. The Equator divides Earth into Northern and Southern Hemispheres.

this axis and positioned halfway between the poles, we can envision an imaginary plane passing through Earth's center. The trace of this plane on Earth's circumference creates a circle called the **Equator**, which divides Earth into the **Northern** and **Southern Hemispheres**.

It is interesting to note that the alignment of Earth's axis of rotation with the North Star is not exact and is purely coincidental. In fact, the North Star is actually a small distance away from Earth's axis of rotation. Moreover, the distance between the axis of rotation and the North Star is moving from year to year due in part to Earth's change in orbital path and axial tilt. Because of this constant movement, the North Star will someday no longer provide utility as a pole star.

7.3.1   LATITUDE

Describing a point on Earth's surface requires that a location's latitude and longitude be known with respect to an origin. In the case of **latitude**, the Equator serves as a convenient origin because it divides Earth into two equal halves. Figure 7.2A illustrates lines of latitude, shown as thin red lines, with the Equator represented by the thicker red line. Because lines of latitude are parallel to each other, they are often called **parallels**. Latitude values are reported in angular measurements of **degrees**, **minutes**, and **seconds**. Similar in concept to units of time, this **sexagesimal system** has a base unit of 60 (the decimal system's base unit is 10), where each degree is divided into 60 minutes and

each minute is divided again into 60 seconds. In this system, the ° symbol denotes the number of degrees, a single quote (′) indicates minutes, and double quotes (″) specify the number of seconds. There are 90° of latitude north and south of the Equator (designated as 0°) for a total of 180° from pole to pole. It is customary to apply the terms North and South to designate latitude locations above or below the Equator. In some cases, plus (+) and minus (−) signs are attached as a prefix to the degree values, indicating latitude locations above and below the Equator, respectively. Thus, the latitude of the Washington Monument can be specified as 38° 53′ 22″ N or +38° 53′ 22″.

7.3.2   LONGITUDE

Longitude, using the same sexagesimal system as latitude, measures location east or west of an origin called the **Prime Meridian**, which has an associated value of 0° longitude coinciding with the Royal Observatory in Greenwich, England. In Figure 7.2B, lines of longitude (also called **meridians**) are shown as thin red lines, with the Prime Meridian represented by the thicker red line. Lines of longitude run north–south from pole to pole but are measured east or west of the Prime Meridian. There are 180° of longitude east and west of the Prime Meridian for a total of 360° of longitude. As with latitude, plus and minus signs are attached as a prefix to indicate location east or west of the Prime Meridian, respectively.[1] Thus, the longitude

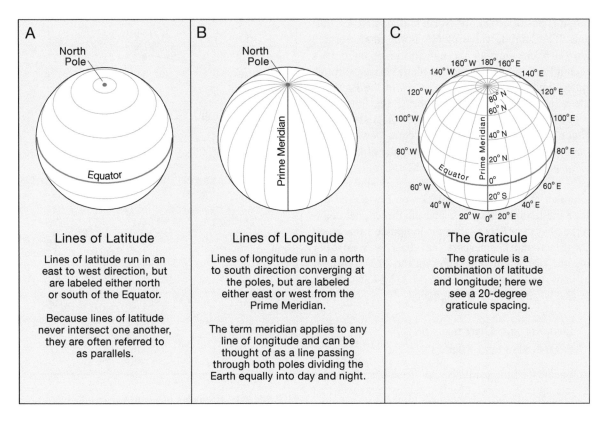

FIGURE 7.2   Latitude and longitude on Earth's surface.

of the Washington Monument can be specified as 77° 02′ 7″ W or –77′ 02′ 7″. The 180th meridian is directly opposite the Prime Meridian and partially coincides with the **International Dateline** (the line dividing days of Earth's rotation). Figure 7.2C shows the graticule—lines of latitude and longitude in combination with a spacing of 20°.

To utilize latitude and longitude when locating positions on Earth's surface, two quantities must be known: the angular distance from the Equator to a given location (the latitude) and the angular distance from the Prime Meridian to a given location (the longitude). Figure 7.3 illustrates how such angles are determined. To illustrate, assume the point

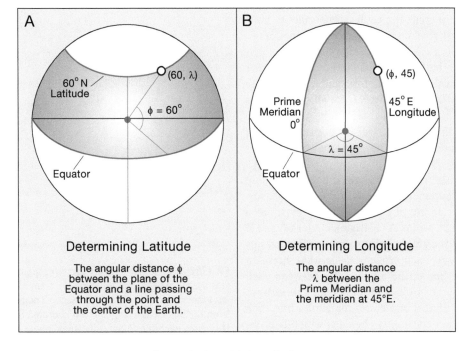

FIGURE 7.3   Determining the angular values of (A) latitude and (B) longitude.

in question is positioned at 60° North latitude and 45° East longitude. The latitude value of 60° is the angle formed between (1) a line passing through the point on Earth's surface and Earth's center, and (2) a plane passing through the Equator and Earth's center (the plane of the Equator). Latitude is usually symbolized by the Greek letter ϕ as shown in Figure 7.3A.[2] Similarly, Figure 7.3B shows that a longitude value of 45° is the angle formed between (1) a plane passing through the point and the center of Earth and (2) a plane passing through the Prime Meridian. Longitude is usually symbolized by the Greek letter λ. The conventional manner of referencing a location on Earth's surface is to state the latitude value first and the directional counterpart (i.e., North or South), then the longitude value along with its directional counterpart (i.e., East or West). For example, the U.S. population center for the 2020 Census is located in Wright County, Missouri, at 37° 24′ 56″ North latitude and 92° 20′ 47″ West longitude.[3]

7.3.3 DISTANCE AND DIRECTIONS ON EARTH'S SPHERICAL SURFACE

The arrangement of the parallels and meridians on Earth's surface sets up two important relationships: distances and directions. These relationships are important to cartographers because various types of maps and the projections that serve as their frameworks are used for accurately representing distances and directions.

The shortest distance between any two points is a straight line. On Earth's curved surface, however, the shortest distance between two points is an arc of a **great circle**. A great circle results from the trace of the intersection of any plane and Earth's surface as long as the plane passes through Earth's center. Because lines of longitude are traces of planes that intersect Earth's surface and its center, all meridians are great circles. It is important to note, however, that not all great circles coincide with meridians. The Equator is also a great circle because its plane intersects Earth's center, but no other parallels intersect Earth's center. Rather, all other parallels form circles on Earth's surface called **small circles**. However, not all small circles coincide with a parallel. A small circle results when a plane passes through Earth's curved surface but does not intersect Earth's center. Figure 7.4 illustrates great and small circles on Earth's surface.

Great circles also establish an important relationship regarding the measurement of directions on Earth's surface. In Figure 7.5, an arc of a great circle is shown as a thin red line between points A (Fairbanks, Alaska) and B (Tokyo, Japan). As the great circle arc crosses Earth's surface, it also intersects each sequential meridian between A and B. At a, b, and c, the angle made between each meridian and the great circle arc is called an **azimuth** (or direction).[4] Along a great circle arc, the azimuth at each meridian intersection constantly changes, which is problematic for navigators. For instance, airplanes ideally fly along great circles, the more direct route, but their flight

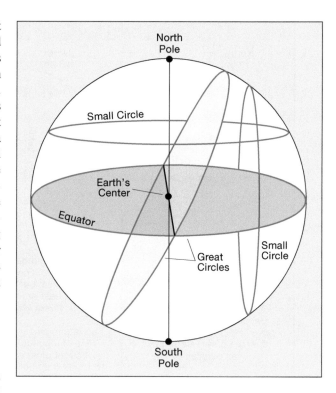

FIGURE 7.4 Examples of great and small circles on Earth's surface.

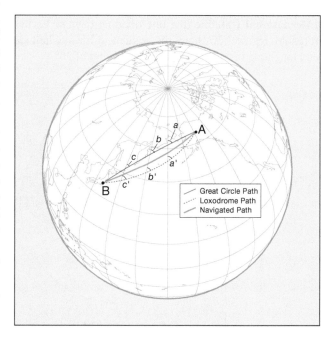

FIGURE 7.5 A great circle arc from point A to point B creates a constantly changing angle (or azimuth) along each meridian that the great circle arc intersects. The angle at a is different from the angles at b and c. The loxodrome from point A to point B crosses each meridian at the same angle at a', b', and c'. The navigated course (thick orange line) results from a series of short loxodromes that approximate the great circle path.

paths must be constantly adjusted because the azimuths along the great circle arc are also constantly changing. From a navigation standpoint, following an azimuth means having to constantly alter a course and is not very practical.

Lines on Earth's surface that have constant direction can also be described. **Loxodromes** (or **rhumb lines**) are special lines that intersect all meridians at the same angle. A simple example will illustrate this concept. The Equator and all small circles that coincide with parallels that intersect meridians at right angles are loxodromes. A loxodrome can be constructed, however, that is not aligned with a parallel or meridian; for instance, Figure 7.6 shows that a line constructed on Earth's surface from A to B so that it intersects all meridians at the same constant angle will, if continued beyond point B at the same angle, spiral toward the North Pole. This constant angle relationship is due to the convergence of the meridians at the poles. In a more practical example, Figure 7.5 shows the loxodrome path between points A and B as a dashed red line. As the loxodrome path intersects each meridian, notice that the angle at a′ is the same as the angle at b′ and c′. Loxodromes are a preferred course for navigation due to the minimum number of alterations to the course. Although navigating along a great circle route across Earth's surface is the most direct route, it is not practical for navigators to constantly change their course. Following the loxodrome reduces the number of course corrections the navigator must make. By following the loxodrome, however, time and distance are added to the journey. In most instances, the great circle route will be approximated by a series of short loxodromes that necessarily lengthen the travel time but also facilitate the navigation process.

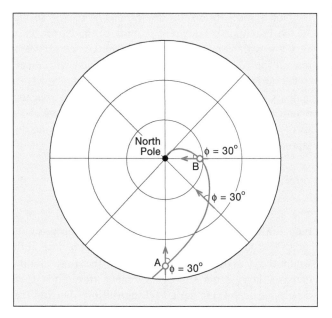

FIGURE 7.6 A line drawn from point A to point B crossing each meridian at a constant angle is called a loxodrome. If extended, this line will continue to spiral toward the North Pole.

7.4 DETERMINING EARTH'S SIZE AND SHAPE

One of the more important historical scientific pursuits has been determining Earth's size and shape—a pursuit that is still ongoing in the field of **geodesy**. You might think that after 2,000 years of scientific inquiry, scientists would have a firm grasp on Earth's size and shape. To some extent, they do, but due to uneven gravitational forces, tides, plate motion, glacial ice melting, and a host of other factors, Earth's interior and exterior constantly change, making it difficult to define its exact size and shape.

7.4.1 EARTH'S SIZE

There is a rich history on the efforts involved in deriving estimates of Earth's size. One early investigator was Eratosthenes (276–195 BCE), who worked at the Alexandria library in Egypt. To determine Earth's circumference, Eratosthenes applied the following formula for calculating the circumference of a circle (C)

$$C = 360° \cdot d/\phi,$$

where d is the distance between two locations on Earth's surface, and ϕ is the angular (or latitude) difference separating the two locations. Eratosthenes selected Syene (present-day Aswan, Egypt) and a city to the north, Alexandria, Egypt, as the two locations because these cities were assumed to be aligned along a meridian, allowing the distance to be computed more accurately.

Modern scholars are unsure of how Eratosthenes arrived at his estimate of distance; some assume that the distance was an educated guess, whereas others argue that he based it on the number of days it took for a camel to walk between the two cities! In any case, he arrived at a value of approximately 810 kilometers (500 miles). To obtain the angular or latitude difference between these cities, Eratosthenes knew that at noon on a day that has the greatest amount of daylight in the Northern Hemisphere (the Summer Solstice), the sun was directly overhead a local well in Syene. On that same day and time, he measured an angle cast by a shadow in Alexandria and calculated it to be 7° 12′. Through a simple geometric relationship, the angle he measured on Earth's surface at Alexandria was the same as the one that would be measured at Earth's center (Figure 7.7). Substituting these values into the equation, he arrived at an estimated circumference of 40,500 kilometers (25,170 miles):

$$C = 360° \cdot 810 \text{ kilometers}/7°12′$$

$$C = 40,500 \text{ kilometers}.$$

This result is very close to contemporary measurements of Earth's size—40,075 kilometers (24,906 miles).

Although a remarkable feat for his time, there is considerable doubt placed on the accuracy of Eratosthenes's measurement, as his method introduced many unaccounted errors.

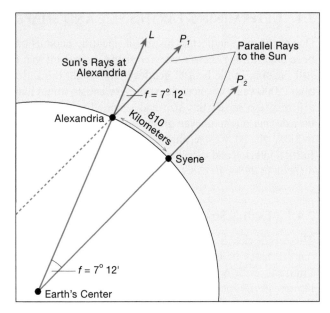

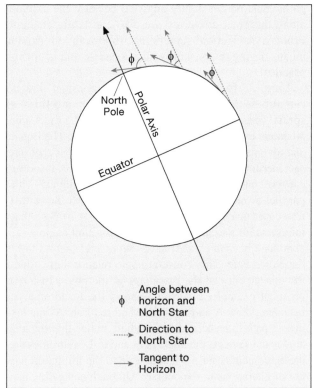

FIGURE 7.7 Method employed by Eratosthenes to estimate Earth's size. Through geometry, a line L that intersects another line P_1 creates an angle ϕ. If that line continues to Earth's center and intersects another line P_2 parallel to P_1, the angle formed at the second intersection is the same as the angle at the first intersection.

FIGURE 7.8 Positions on Earth's surface suggesting a spherical surface. Note that the position of the North Star is relative and not to scale.

We have already mentioned the problem of determining distance, but there were other problems. For instance, Syene and Alexandria are *not* along the same meridian—Aswan is located at 32° 53′ 56″ E, and Alexandria is located at 29° 55′ 09″ E—an east–west angular difference of 2° 58′ 47″. We also know that Syene is located at 24° 5′ 15″ North latitude and not along the Tropic of Cancer, which is at 23° 30′ 0″ North latitude, and thus the sun's rays were not perfectly vertical. Moreover, there is no agreement on the exact conversion between modern-day miles or kilometers and the units Eratosthenes used.[5] Regardless of his results, or modern-day interpretations thereof, his measurement should be regarded as one of the first to scientifically investigate Earth's size.

7.4.2 EARTH'S SHAPE

Similar to investigations of Earth's size, there has been a constant quest to arrive at a definitive description of Earth's shape. Recorded evidence of these investigations comes from various Greek scholars, such as Pythagoras (sixth century BCE), who relied on philosophy to deduce the spherical nature of Earth's surface. Pythagoras argued that Earth must be a sphere because the sphere was considered the perfect shape and the Greeks inhabited a perfect world. Other scholars, such as Aristotle (fourth century BCE) and Archimedes (third century BCE), made assumptions about Earth's shape through direct observation. For instance, Aristotle noticed that as a ship sailed toward the horizon, the ship's hull disappeared first rather than becoming smaller and smaller.

From our everyday experience, we, just as individuals have for over 2,000 years, see evidence suggesting Earth is a sphere. One simple way to see this shape is to note that during a lunar eclipse, the edge of Earth's shadow is a circular arc. A more involved proof is the relative position of the North Star with respect to locations on Earth's surface. At the Equator, the apparent position of the North Star is very low with respect to the horizon,[6] but as one progresses toward the North Pole, the apparent position with respect to the horizon rises, until at the North Pole, the North Star appears directly overhead (see Figure 7.8). This phenomenon can be explained only if Earth has a spherical surface—a fact that early navigators utilized for locating their latitude in the Northern Hemisphere.

7.4.2.1 The Prolate versus Oblate Spheroid Controversy

The notion of Earth as a perfect sphere existed for more than 1,000 years after the early Greeks first proposed the idea. The idea was not challenged until the late 1600s when the Cassini family undertook the first comprehensive, large-scale survey of France through a newly developed technique called **triangulation**.[7] During their survey, Jacques Cassini encountered the fact that one degree of latitude in the northern part of France was not the same length as one degree of latitude in the southern part of France. He concluded that Earth is not a perfect

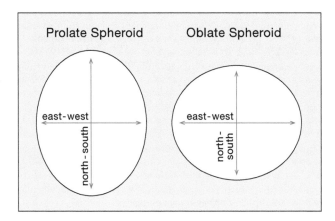

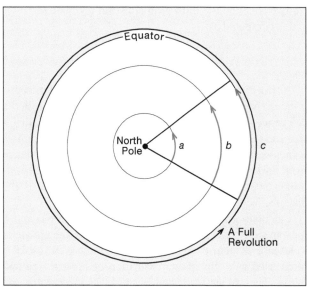

FIGURE 7.9 The prolate spheroid, shown on the left, has an east–west axis that is compressed or flattened relative to the north–south axis. The oblate spheroid, on the right, shows the opposite relationship. Both figures are greatly exaggerated to show the differences.

FIGURE 7.10 The notion of centrifugal force is supported by differing speeds at different latitudes—positions near the Equator (c) must travel farther and, therefore, faster than those near the pole (a).

sphere but rather a **prolate spheroid**—an ellipse that bulges in the north–south direction and is compressed in the east–west direction (Figure 7.9).

About the same time period, other pieces of evidence regarding Earth's gravitational forces were being investigated that would later be paramount in providing a more exact description of its shape. For instance, Jean Richer (1630–1696) examined the periods, or motions, of pendulums at different locations on Earth. While on expedition to Cayenne, French Guyana, he found that a pendulum swung more slowly there than it did in Paris. From this, he theorized that gravity must be weaker at Cayenne and, therefore, that Cayenne is further from Earth's center than Paris, which contradicted Cassinis' prolate spheroid assumption.

In concert with Richer and other gravity investigations, Sir Isaac Newton (1642–1727) was promoting his laws of gravitational motion. His 1687 treatise stated that every object in the universe attracts every other object with a force that is equal to the product of their masses divided by the square of the distance between them. He applied his gravitational concept to Earth's shape and suggested that Earth's rotating mass creates a **centrifugal force** that essentially pushes mass away from its center. He substantiated his assumption by noting that different latitudes on Earth's surface travel at different speeds. For example, in Figure 7.10, we can see that each latitudinal position completes one full rotation per day, but that locations at the Equator have much farther to travel and therefore must travel faster. Thus, according to Newton's law, the centrifugal force must be stronger at the Equatorial regions, which, he proposed, creates an **oblate spheroid** where the Equator bulges but the poles are flattened (Figure 7.9).

Whether Earth is a prolate or an oblate spheroid became the focus of a heated debate among the scientific community in the early 1700s. To resolve the debate, l'Académie Royale des Sciences in Paris organized two expeditions—one to

the equatorial region of Ecuador and another to Lapland (the border between Sweden and Finland) to measure the length of one degree of latitude on Earth's surface. The expeditions took more than nine years to complete but finally reported that one degree of latitude at the Equator is 111.321 kilometers, whereas at Lapland, the same distance is 111.900 kilometers. Thus, these findings proved Newton correct—Earth is an oblate spheroid.

The fact that Earth was described as an oblate spheroid provided scientists a new figure to describe Earth's shape in simple mathematical terms. Basically, an oblate spheroid can be described as an ellipse, which has a semimajor axis (symbolized by a) and a semiminor axis (symbolized by b) (Figure 7.11). The degree to which the ellipse deviates from

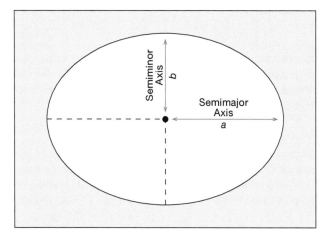

FIGURE 7.11 The notion of a reference ellipsoid and associated axis parameters.

a circle is called the **flattening constant** (symbolized by *f*) and is computed as:

$$f = (a - b)/a.$$

A two-dimensional ellipse can be rotated about its semiminor axis, arriving at a three-dimensional figure called a **reference ellipsoid**, which can be used to describe Earth's size and shape. On a reference ellipsoid, the semimajor axis coincides with the *equatorial radius*, which is the distance from Earth's center to a point on the Equator, and the semiminor axis coincides with the *polar radius*, which is the distance from Earth's center to the North Pole.

Calculating *f* requires knowing the values for *a* and *b* for a specific reference ellipsoid. For example, the Geodetic Reference System of 1980 (GRS80) reference ellipsoid is often used as a global approximation of Earth's size and shape. The values for *a* and *b* are 6,378,137 meters and 6,356,752.3141 meters, respectively. Note that, for the GRS80 reference ellipsoid, the value of *a* is exact and the value of *b* has been rounded to four decimal places. Substituting the values of *a* and *b* into the equation, we have the following expression:

$$f = (6,378,137 - 6,356,752.3141)/6,378,137$$

$$f = 0.003352810...$$

Rather than express *f* as a small decimal value, it is more common to take the reciprocal, so we would perform the following operation:

$$1/f = 1/0.003352810...$$

$$1/f = 298.257222...$$

Armed with the idea of a reference ellipsoid, various countries undertook national surveys to define the semimajor and semiminor radii values of ellipsoids that best fit local areas. Figure 7.12 illustrates this concept, where a reference ellipsoid's semimajor and semiminor radii have been defined so as to "fit" Australia, and Table 7.1 lists the semimajor and semiminor values of several prominent historical reference ellipsoids. In total, dozens of reference ellipsoids have been defined, and currently, there are more than 30 in use by various countries around the world.

7.4.2.2 Reference Ellipsoid and the Graticule

Although latitude was defined in general terms at the beginning of this chapter, it should be recognized that the same location can take on different latitude values, depending on the chosen reference ellipsoid. Figure 7.13A shows geodetic and geocentric latitudes on a reference ellipsoid. **Geodetic latitude** computed on a reference ellipsoid is measured by an angle that results when a perpendicular line at the ellipsoid's surface is drawn toward Earth's center. In every

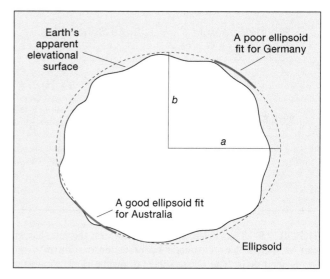

FIGURE 7.12 A comparison between Earth's irregular surface and a smooth reference ellipsoid surface. (This illustration is greatly exaggerated to show the effects.)

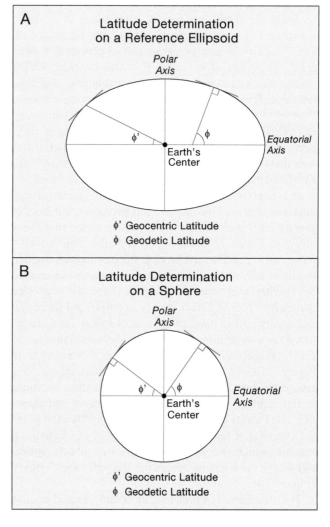

FIGURE 7.13 A comparison between geocentric and geodetic latitudes on (A) a reference ellipsoid and (B) a sphere.

TABLE 7.1

Several Common Ellipsoids Developed for National Mapping Programs. Values Taken from the Geodetic Parameter Dataset Hosted by the European Petroleum Survey Group

Ellipsoid Name	Year of Inception	Semimajor Axis (meters)	Semiminor Axis (meters)	Flattening	Where Used
Airy	1830	6,377,563.44	6,356,256.91	1/299.324	Great Britain
Everest	1830	6,377,276.35	6,356,075.42	1/300.802	India
Clarke	1866	6,378,206.4	6,356,583.8	1/294.978	North America
International	1924	6,378,388.0	6,356,911.95	1/297.0	Select areas
Krassovsky	1940	6,378,245.0	6,356,863.03	1/298.3	Soviet Union
World Geodetic System (WGS)	1972	6,378,135.0	6,356,750.5	1/298.26	U.S. Defense Dept.
Geodetic Reference System (GRS)	1980	6,378,137.0	6,356,752.3141	1/298.25720	World
World Geodetic System (WGS)	1984	6,378,137.0	6,356,752.3142	1/298.25722	World

Source: epsg.org/home.html.

instance, except when the location in question is directly over the Equator or one of the poles, the perpendicular line will never intersect Earth's center. Rather, the line will pass through the equatorial plane at some other location. This result occurs because an ellipsoid does not have a constant radius. The angle measured at the intersection of the line and the equatorial plane is called the geodetic latitude. On the other hand, **geocentric latitude** computed on a reference ellipsoid is measured by an angle that results when a line at Earth's surface is drawn intersecting the plane of the Equator at Earth's center. Given the same point on any given reference ellipsoid, the geocentric and geodetic latitudes will be off by a small amount.[8] This amount is significant for creating accurate maps (e.g., large-scale topographic maps) of local areas. For instance, surveyors commonly use geodetic latitude when referencing their survey positions because their surveys are tied to a specific reference ellipsoid for the country in which they are surveying. Geocentric latitudes are commonly used for global phenomena, such as satellite ground tracking maps.

On a spherical surface, geodetic and geocentric latitudes are computed using the same methods previously outlined. Note, however, in Figure 7.13B that geodetic and geocentric latitude computations produce the same result on a spherical surface—both latitude computations measure the angle that results when a line at Earth's surface is drawn intersecting the plane of the Equator at Earth's center.

Because the ellipsoid is not perfectly spherical, the length of one degree of geodetic latitude is not constant—the flattening at the poles means that one degree of latitude here is longer than one degree at the Equator. It is customary to measure one degree of latitude beginning at the Equator and moving toward a pole, with the measurement centered on a multiple of, for example, 10° (Figure 7.14A). Table 7.2 lists lengths of one degree of latitude for the GRS80 reference ellipsoid (one of the few reference ellipsoids that model the entirety of Earth's shape and size) at 10° intervals.

The National Geodetic Survey (NGS) has developed the interactive online Geodetic Toolkit called Inverse (https://geodesy.noaa.gov/cgi-bin/Inv_Fwd/inverse2.prl) that can compute the lengths of one degree of latitude for the GRS80 reference ellipsoid (or other ellipsoids of your choosing) in one-degree increments.

On the sphere, the distance between successive degrees of latitude is constant and is derived by first calculating Earth's circumference and then dividing the result by 360°. So, assuming Earth to have a circumference of 40,075 kilometers (24,906 miles), the length of one degree of latitude is approximately 111.2 kilometers (69.1 miles).

Due to the convergence of meridians at the poles, the length of one degree of longitude becomes progressively

TABLE 7.2

The Length of One Degree of Latitude as Measured along a Meridian Based on the GRS80 Reference Ellipsoid

Latitude	Kilometers	Statute Miles
0°	110.574	68.708
10°	110.608	68.729
20°	110.704	68.789
30°	110.852	68.881
40°	111.035	68.994
50°	111.229	69.115
60°	111.420	69.228
70°	111.562	69.321
80°	111.660	69.382
90°	111.694	69.403

Note: Values were computed using INVERSE, a free program available from the National Geodetic Survey at https://geodesy.noaa.gov/cgi-bin/Inv_Fwd/inverse2.prl. Values are rounded to the nearest thousandths place.

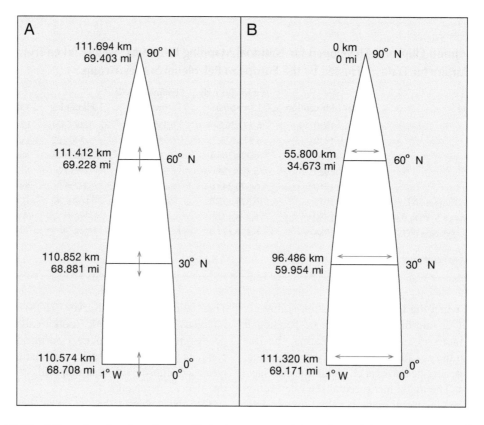

FIGURE 7.14 (A) The different lengths of one degree of latitude as measured along the meridian centered at 0° 30′ West longitude on the GRS80 reference ellipsoid. (B) The different lengths of one degree of longitude as one travels from the Equator to the North Pole as computed on the GRS80 reference ellipsoid.

smaller as one travels from the Equator to the poles, regardless of whether we assume a spherical Earth. This notion is illustrated in Figure 7.14B. Table 7.3 lists lengths of one degree of longitude for the GRS80 reference ellipsoid. As with computing one degree of latitude, the online program Inverse can compute the lengths of one degree of longitude for the GRS80 reference ellipsoid in one-degree increments.

7.4.2.3 The Geoid

Up to the late 1950s, suitable reference ellipsoids were developed on Earth's surface utilizing surveying instruments that required visual contact between the instruments. As such, these surveys tended to be very localized in nature, for example, focusing on an individual country, and thus no single reference ellipsoid could describe the entire Earth in very precise terms. In addition, each new reference ellipsoid had its own center and *a*, *b*, and *f* parameter values, which made a given reference ellipsoid unique for a specific country. Thus, it was difficult for other countries to adopt existing reference ellipsoids that were previously defined for a specific country. Satellite measurements have changed this, as models describing Earth's size and shape are now computed based on Earth's center of mass for the whole planet. This new idea allows a single reference ellipsoid to be used that gives a better fit to the entire Earth than previously possible.

The first U.S. satellite measurements of Earth's shape were from the National Aeronautical and Space Administration's (NASA) Vanguard rocket program in the late 1950s. If Earth was perfectly spherical, satellites would

TABLE 7.3

Lengths of One Degree of Longitude as Measured along a Specific Parallel

Longitude	Kilometers	Statute Miles
0°	111.320	69.171
10°	109.639	68.127
20°	104.647	65.025
30°	96.486	59.061
40°	85.394	53.061
50°	71.696	44.550
60°	55.800	34.673
70°	38.187	23.728
80°	19.394	12.051
90°	0	0

Note: Values were computed using INVERSE, a free program available from the National Geodetic Survey at https://geodesy.noaa.gov/cgi-bin/Inv_Fwd/inverse2.prl. Values are rounded to the nearest thousandth place.

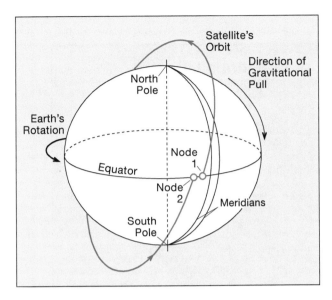

FIGURE 7.15 The precession of the nodes. Due to Earth's elliptical nature, the orbital path of a satellite is pulled toward the equator due to differences in Earth's gravity, causing the satellite's path to predictably cross a different meridian with each orbit. (After Maling 1992.)

operate in an orbital path around Earth, and that path would coincide with a plane. Deviations from this plane due to variations in Earth's gravitational forces resulted in a predictable motion called *precession of the nodes*. Figure 7.15 illustrates this concept. The amount of change that occurred with each orbital path was measured very precisely, providing an indication of the gravitational forces acting on Earth's surface and thus a measure of Earth's shape. In fact, the Vanguard rocket program showed a slight bulge in the Northern Hemisphere's polar areas and a slight depression around the Southern Polar areas, giving rise to the concept of a *pear-shaped Earth*. The North Pole was reported to be about 10 meters (32.8 feet) farther from the Equator than previously thought, and the South Pole was found to be resting approximately 30 meters (98.4 feet) closer to the Equator than an elliptical Earth model suggested (Maling 1992, 14).

As satellite measurements have continued, a more descriptive and informative term has been used to describe Earth's shape—the **geoid** (meaning Earth-like). Understanding the geoid concept is somewhat difficult because it is not directly observable. Generally speaking, the geoid is defined as a smooth, undulating surface Earth would take on if the oceans were allowed to flow freely over the continents—without currents, tides, waves, or other forces—which would create a single, undisturbed water body. This water body would then be free to adjust to differences in the gravitational forces (i.e., centrifugal force caused by Earth's rotation) as well as the uneven distribution of Earth's mass that exists at its surface. Thus, this new surface would have *undulations* (i.e., peaks and valleys) that reflect the influence of gravity. One can think

of this "new" surface as coinciding with mean sea level, which varies in height by some 100 meters (328 feet) worldwide.

To compute a geoidal surface requires knowing, among other parameters, the specific nature of the gravitational forces that exist across Earth's surface. These forces, called gravitational anomalies, are not evenly distributed, nor are they equal in their intensity, because Earth's crust contains different rock densities (e.g., metamorphic and igneous rocks are more likely to contain the iron-rich mineral magnetite and thus influence gravity to a greater extent than sedimentary rocks), but the anomalies can be measured and mapped. At every location on Earth's surface, a line can be placed so that it is perpendicular to the local gravitational anomaly. As shown in Figure 7.16, the line perpendicular to the geoid will not coincide with a line that is perpendicular to the chosen reference ellipsoid's surface. A geoidal surface (or equipotential[9] surface) can be created from these perpendicular gravitational measurements. In general terms, Earth's crust is thicker over continents, where the geoid typically rises but falls over the oceans, where the crust is thinner. However, even the smooth undulation in the geoid surface has minor highs and lows that can rise as much as 60 meters (196.8 feet) in some areas. Figure 7.17 shows one geoid model (GEOID18) for the conterminous United States developed by the NGS. The "18" part of GEOID18 refers to the year that the datum was realized or published by the National Geodetic Survey. In this case, GEOID18 is the geoid model that was published in 2018.

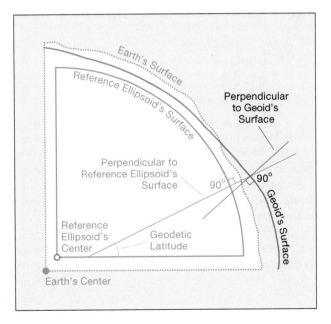

FIGURE 7.16 The relationship among surfaces representing Earth, a reference ellipsoid, and a geoid. Note that when determining geodetic latitude, a line perpendicular to the reference ellipsoid is not perpendicular to the geoid.

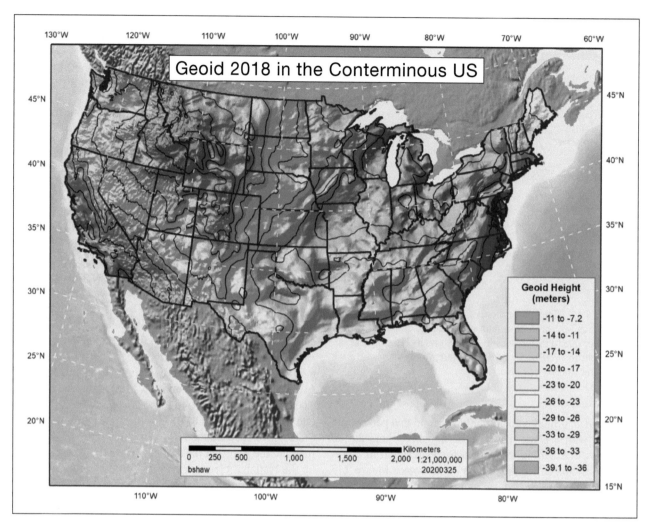

FIGURE 7.17 GEOID18 is a geoid model of the conterminous United States. In GEOID18, geoid heights range from a low of approximately 39 meters (127.95 feet), represented by blue in the Atlantic Ocean, to a high of 7 meters (22.96 feet), shown as red in the Rocky Mountains. Note that, for the lower 48 United States, all geoid height values are negative since the geoid surface is below the reference ellipsoid surface. In Alaska, that relationship is reversed. (From the National Geodetic Survey; https://geodesy.noaa.gov/GEOID/GEOID18/.)

7.4.2.4 Geodetic Datum

The relationship between the geoid and a chosen reference ellipsoid is important for a variety of mapping applications, especially large-scale mapping projects such as aeronautical charting, where coordinate or elevation accuracy is important. A **geodetic datum**[10] is a model that describes the location, elevation, and scale relationships with respect to an origin on Earth's surface. There are two components to a geodetic datum: horizontal and vertical. A **horizontal datum** specifies locations in terms of geodetic latitude and longitude based on a chosen reference ellipsoid relative to an origin, which coincides with the Equator for latitude values and the Prime Meridian for longitude values. The horizontal datum specifies a precise description of the parameters for the reference ellipsoid so that the semimajor axis (equatorial radius) and flattening constant for Earth are known, thus providing an accurate means to establish latitude and longitude. A **vertical datum** provides a zero surface that enables elevations to be

determined and, as mentioned earlier, is tied to the concept of mean sea level, which relates to the geoid. Although these two datum concepts have been historically separate entities because the surfaces to which they refer are different, they are commonly tied together today because they are conceptually and operationally related. For example, when surveying a new highway, surveyors need to accurately provide latitude and longitude based on a specific horizontal datum and a derived elevation value determined through the geoid.

Historically, the United States used the Clarke Ellipsoid of 1866 reference ellipsoid and the geoid whose origin is located at Meades Ranch in Kansas (located near the geographic center of the United States) for its large-scale topographic mapping program. Taken together, the Clarke 1866 reference ellipsoid and the geoid origin of Meades Ranch comprise the **North American Datum of 1927** (**NAD27**). Mapping activities at a world scale call for a a global-based rather than local-based reference ellipsoid. To meet this

TABLE 7.4

A Listing of Parameter Values Comparing the Reference Ellipsoids of NAD27 and NAD83

Datum Parameters	NAD27	NAD83
Reference ellipsoid name	Clarke 1866	Geodetic Reference System 1980 (GRS80)
Semimajor axis (a)	6,378,206.4 meters	6,378,137.0 meters
Semiminor axis (b)	6,356,583.8 meters	6,356,752.3 meters
Flattening	1/294.978698	1/297.257222
Intended use	North America	Worldwide

Note: Values take from Snyder (1987).

need, a new reference ellipsoid, the Geodetic Reference System (GRS), was recommended by the International Association of Geodesy and subsequently adopted by the NGS,[11] creating the North American Datum of 1983 (NAD83). NAD83 utilizes the Geodetic Reference System of 1980 as the reference ellipsoid whose center is near Earth's center of mass. Table 7.4 contrasts NAD27 and NAD83 on several key parameters. As a result of changing from NAD27 to NAD83, most locations within North America have updated latitude and longitude values. This readjustment resulted in the publication of updated coordinate data for approximately 250,000 geodetic control survey points or benchmarks throughout the United States and yielded shifts in coordinate values that exceed 400 meters (1,312.3 feet) in some parts of the country.[12]

In a similar manner, vertical datums used in the United States have also experienced updates as new techniques to measure gravity across Earth's surface, for example, have been developed. An early vertical datum used in the United States was the **National Geodetic Vertical Datum of 1929** (**NGVD29**). This datum was developed from observing mean sea level at 26 tidal gauges (21 in the United States and 5 in Canada) located in harbors along the coastal waters. Observations of water heights at these gauge stations were recorded over an approximate 19-year period (235 lunar months) which corresponds to a **Metonic cycle**. It is known that tides along coastal waters rise and fall in a temporal manner due to gravitational interactions between the sun, moon, and Earth as each orbits the other. The Metonic cycle allows these interactions to reach a maximum and minimum from which an adjusted tidal height observations is computed. As a result, observations at all tidal stations throughout the United States had a common reference or zero elevation. However, the assertion that local mean sea level was constant at any gauge stations was fundamentally flawed as mean sea level is not constant between any two stations. Later analysis confirmed that fixing a zero elevation at these tidal gauge stations introduced considerable error in elevation values across the United States.[13]

To address this situation, a new vertical datum was developed: The **North American Vertical Datum of 1988** (**NAVD88**). Recall that NGVD29 relied on local sea level from a dispersed network of tidal gauge stations. As a result, NGVD29 did not coincide with a single gravitational surface. In other words, NGVD29 did not correspond to the geoid. To correct this, NAVD88 relied upon thousands of new geodetic control stations whose positions and elevations were accurately surveyed. These newly surveyed control stations were connected to a single tidal station located at Father Point, Québec, Canada, along the Gulf of Saint Lawrence. As a result, NAVD88 models a gravitational surface that presents an approximation of a geoid and removes the inconsistency of forcing mean sea level to several tidal stations. This new approach provides greater elevation accuracy than NGVD29 could. Figure 7.18 illustrates the vertical differences between the elevations represented by NGVD29 and NAVD88. Note that the map shows an east-to-west trend in the elevation differences with the northwest Pacific coast exhibiting greater changes in height differences than along the Atlantic coast.

While NAVD88 was an improvement, the development of the elevation surface was not directly tied to Earth's gravitational surface across the United States. Added to this was increased awareness that satellite geodesy, especially Global Navigation Satellite Systems (GNSS), such as GPS, demonstrated that both NAD83 and NAVD88 contained significant positional and elevational errors. This latter condition stemmed from the fact that NAD83 and NAVD88 were derived using manual surveying methods based on a passive or static geodetic control station network (benchmarks that once placed in the ground were not regularly re-surveyed, and therefore the coordinate and elevation information was static and outdated).

Taking advantage of GNSS and other observational and measurement technology, the NGS plans to modernize NAD83 and NAVD88 as part of the **National Spatial Reference System** (**NSRS**). The NSRS, the fundamental spatial reference system for the United States, ties together two ideas: geometric datum and geopotential datum. A **geometric datum** (or terrestrial reference frame) provides information about latitude, longitude, ellipsoid heights, and crustal velocities. Ellipsoid height is the perpendicular distance from the reference ellipsoid surface to Earth's surface. An ellipsoid height can be determined via GNSS signal information. **Crustal velocities** report the speed at which Earth's crustal plates are moving and are important when determining accurate coordinate values. The new geometric datum is called the **North American Terrestrial Reference Frame 2022 (NATRF2022)**. A **geopotential datum** provides information about the geoid, gravity, and elevations. NAVD88 will be replaced by the **North American-Pacific Geopotential Datum (NAPGD2022)**. Coupled with NAPGD2022 will be a new geoid model (GEOID2022) that provides geoid height (the distance measured along a line perpendicular from the geoid to the reference ellipsoid). In addition, NAPGD2022 will provide orthometric heights (elevations commonly reported on topographic maps).

Two limitations of NAD83 are addressed with the adoption of NATRF2022. First is the lack of alignment

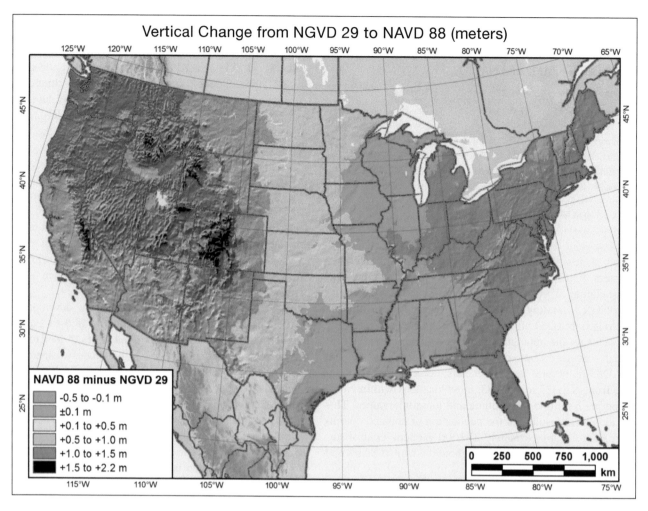

FIGURE 7.18 An illustration showing the range of elevation differences between NGVD29 and NAVD88. The legend represents the differences for the Continental United States. (Source: ngs.noaa.gov.)

of the GRS80 with Earth's true center of mass. Satellite measurements show that the GRS80 reference ellipsoid is approximately 2 meters away from Earth's true center of mass. Second, Earth's crust is comprised of continental plates that are in constant motion. Combined together, the misalignment of GRS80 axes and crustal plate motion causes problems when using a static datum like NAD83. NATRF2022 will correct for this static nature by reporting how much a given plate has moved since a specific time stamp or epoch. Depending on the location in the United States, NATRF2022 will change latitude, longitude, and ellipsoid height by between 1 and 2 meters. Similarly, NAPGD2022 will address two important errors associated with NAVD88. First, a network of passive benchmarks for survey control takes an enormous amount of time and resources to maintain and is quite cost prohibitive. Second, there are ongoing vertical changes in elevation (such as ground subsidence or glacial rebound) that need to be addressed in any vertical datum. NAPGD2022 will monitor the time-varying nature of gravity across North America and is anticipated to deliver up to 1cm of elevation accuracy.

7.4.2.5 Geodetic Datums and Thematic Cartography

Although reference ellipsoids, geoids, geodetic datums, and geodesy are certainly worthwhile topics of investigation, they are limited in significance to large- and medium-scale mapping applications that involve accurate measurements of distances, directions, or areas. For example, a ship's navigator uses a nautical chart to accurately plot a ship's intended course so that its precise location can be known at any instant. In the realm of thematic mapping, the focus is on examining spatial patterns and distributions, and so knowledge of precise direction, distance, and area is not critical.

To illustrate the limited applicability of datums to thematic mapping, consider a small-scale thematic mapping exercise that requires symbolizing data for each country of the world. Assume for this problem that the model of Earth is the GRS80 reference ellipsoid as opposed to a spherical model of Earth. In this case, the actual difference in length between the two reference ellipsoids' semiaxes is approximately 21.3 kilometers (13.2 miles), or the circumference associated with the polar axis is about 135 kilometers (83.3 miles) smaller than that for the equatorial axis (40,075 − 39,940 = 135 km), and as we calculated earlier, the flattening is very close to 1/298. If you were to draw a page-sized map of the world with the GRS80 parameters, the Equator would be represented by a line 12.75 centimeters (7.9 inches) long. At this size, the difference between the equatorial and polar radii of the GRS80 reference ellipsoid parameters would be about 0.21 mm. Because 0.21 mm is also the approximate width of a line that can be used to represent fine detail on maps, the flattening would be hidden by the width of the pen used to draw world outlines. So, the selection of a more accurate model of Earth using the GRS80 reference ellipsoid would not produce a visual difference in the final appearance of the map when compared to the spherical model. This is an important conclusion because it permits the assumption that Earth can be regarded as spherical for thematic applications. This assumption drastically simplifies the mathematics involved when creating a map projection.

While the choice of an appropriate datum is likely to be inconsequential to thematic mapping, the reliance on software for creating thematic maps usually does require that the spatial data used have a datum definition. If your mapping project involves more than one data set, the data also may be cast in different datums. In order for your data to display correctly on screen, a single common datum must be defined. Therefore, even cartographers who typically produce thematic maps need to have a working knowledge of datums.

7.5 SUMMARY

In this chapter, we have examined the basic concepts behind Earth's size and shape and its geographic coordinate system. We learned that the graticule serves as the framework for the imaginary network of latitude and longitude lines, allowing any point location on Earth's surface to be uniquely identified. The Equator serves as the dividing line for the Northern and Southern Hemispheres and references all latitude locations north or south of that line. On the other hand, the Prime Meridian divides Earth into Eastern and Western Hemispheres and is the reference line for all longitude locations east or west of that line. Although the graticule is important for referencing spatial activity that takes place on Earth's surface, a more important concept, especially with regard to mapmaking, is the changes to the graticule's appearance that occur as a result of map projection, a topic addressed in Chapters 9 and 10.

Earth's size and shape are important considerations in determining locational accuracy. Historically, beginning with the work of Eratosthenes, Earth's shape evolved from a simple spherical assumption to a more complex shape called an oblate spheroid. This figure of Earth was developed after Newton proposed his gravitational laws in which a rotating body produces centrifugal force, causing a bulging of Earth at the Equator.

When creating a model of Earth, geodesists frequently utilize a reference ellipsoid, which is a smooth, mathematically definable figure that flattens at the poles and bulges at the Equator. Numerous reference ellipsoids have been developed, each defined by different parameters for the semimajor and semiminor axes, with the objective of trying to accurately map specific geographic extents (usually countries). In recent years, the reference ellipsoid concept has been updated to reflect accurate measurements of gravitational variation across Earth's surface. This variation has been modeled as a geoid, the two-dimensional curved surface Earth would take on if the oceans were allowed to flow freely over the continents. A datum combines the geoid and a specific reference ellipsoid to produce a reference for horizontal locations defined by latitude and longitude (established by the reference ellipsoid) and vertical elevations (defined by the geoid). The North American Datum of 1927 (NAD27) historically served as the basis on which the U.S. topographic mapping program was established but has been updated to the North American Datum of 1983 (NAD83). Ongoing satellite-based technologies (such as GNSS, GRACE-FO, and GRAV-D) have prompted new accurate modeling of Earth's horizontal and vertical datums, which will be updated to the National Spatial Reference System.

7.6 STUDY QUESTIONS

1. Think about address systems other than latitude and longitude that are used to locate places on Earth's surface. Explain how they are organized to provide location information. What are their advantages and limitations?

2. The Prime Meridian runs through the Royal Observatory in Greenwich, England. The Royal Observatory has an interesting website that discusses how the Prime Meridian has "moved" throughout history and what explanations are at the root of the movement (https://www.rmg.co.uk/royal-observatory/history). Discuss two of the key movements and why they occurred.

3. Go to https://www.kavas.com/blog/great-circle-and-rhumbline.html and explore the differences between a rhumb line and a great circle path on a map. On the map, two points appear symbolized by large red markers. The red line connecting the two points represents the great circle path, while the green line represents the rhumb line path. Using your mouse, you can move each marker to a new location and visualize the great circle and rhumb line paths that connect the two markers. Experiment moving the two markers and answer the following questions: At what marker locations do the great circle and rhumb line paths overlap each another? What causes this overlap to occur? When is there the greatest difference between the

overlap in the lines? What causes this difference to occur?

4. This chapter discussed several individuals such as Eratosthenes, Pythagoras, Aristotle, Cassini, Richter, and Newton and their contributions to developing and understanding Earth's shape and size. Identify two other prominent individuals who were instrumental in developing an understanding of Earth's size and shape. For each individual, explain their contribution to the development of understanding Earth's shape and size.

5. Go to the European Petroleum Survey Group's Geodetic Parameter Dataset at (epsg.org). Use the Search window and enter "ellipsoid." Along the tabs that are returned, click on Datums and a list of reference ellipsoids will return. From the list of reference ellipsoids produced, find two reference ellipsoids that were not covered in this chapter. For each reference ellipsoid, what are the values for the semimajor axis, semiminor axis, and flattening, and how do each compare to the Clarke 1866 and GRS80 reference ellipsoid parameter values? For each of the two new reference ellipsoids that you chose, explain who developed the ellipsoid and indicate what purpose for which the ellipsoid was developed?

6. Aside from GNSS, GRAV-D (an NGS initiative) and GRACE-FO (a NASA program) are two remote sensing technologies that are useful in measuring gravity across Earth's surface and are playing a role in the development of a new vertical datum to replace NAVD88. Learn more about GRAV-D at https://geodesy.noaa.gov/GRAV-D and GRACE-FO at https://grace.jpl.nasa.gov/mission/grace/. Using information contained on these websites, explore both of these technologies and indicate the advantages each offers in measuring gravity and defining the geoid.

7. This chapter discussed the NGS and the role it played in developing and updating horizontal and vertical datums that are used in the United States. Other countries have similar kinds of organizations that help direct and maintain their own datum development. For example, the following organizations publish information about the development and maintenance of their datums:
   - Australian Geodetic Survey (http://www.ga.gov.au/scientific-topics/positioning-navigation/geodesy),
   - Survey of India (https://surveyofindia.gov.in/),
   - Bundesamt für Kartographie und Geodäsie (https://www.bkg.bund.de/DE/Home/home.html).

Choose one organization and explain how it is responsible for developing and maintaining its geodetic datums.

8. Use the NGS's Coordinate Conversion and Transformation Tool—NCAT (https://www.ngs.noaa.gov/NCAT/) to determine how much of a horizontal shift there has been in latitude and longitude for your location between the NAD27 and NAD83(1986) datums. To use the tool, you will need to enter the latitude and longitude values where you are located. For example, you could try this point location first: 39.3191513889 north latitude and −82.1037805556 west longitude. Note, these values are formatted in decimal degrees and represent a geodetic control station in Athens, OH. Experiment with other points in different parts of the United States. Where do you think the greatest/least change in coordinate values are between these two datums in the United States? Use the NCAT tool to confirm your assumptions.

NOTES

1 The "+ = east" and "− = west" convention is not universal. Different countries and mapping organizations sometimes assign "+ = west" and use a complete 360° system of longitude rather than dividing the world into two 180° halves.

2 The adoption of ϕ is not universal as a symbol for latitude. The Greek letters φ and ψ are also used to reference latitude.

3 The National Geospatial Agency GeoNames server allows you to look up the latitude and longitude coordinates of places throughout the world (https://geonames.nga.mil/namesviewer/). There is a very rich and complex history in the development of latitude and longitude. For two excellent books on this topic, see Ariel and Berger (2006) and Monmonier (2004).

4 Azimuths are customarily measured in a clockwise fashion starting with geographic north as the origin and passing through 360°. This definition of azimuth is not universally followed. For instance, azimuths can be measured from magnetic north rather than from geographic north.

5 The units in which Eratosthenes reported distance were stadia, which originated in ancient Greece. NASA provides an estimate of converting stadia to kilometers as 1 stadia = 0.15 km. (https://imagine.gsfc.nasa.gov/features/cosmic/earth_info.html).

6 Although there are several definitions of horizons (Defense Mapping Agency 1981), the meaning here is the general term, which means the apparent or visible junction of Earth and sky as seen from any position on Earth's surface.

7 Triangulation is a method of surveying where two points having known locations are established and the distance between them is measured. This line is called the baseline. Next, using one end of the baseline, an angle is measured to a distant point. At the other end of the baseline, an angle is measured to the same distant point closing the triangle. Using the angles of this triangle, the length of any unknown side can be computed. Other triangles are established from this initial triangle, creating a triangulated network.

8 The difference between geodetic and geocentric latitude as measured on the same reference ellipsoid reaches a maximum of approximately 11′ 45″ at 45° North and South latitude (Van Sickle 2017).

9 The term equipotential surface is defined as a surface having the same potential of gravity everywhere.

10 The concept of "geodetic datum" is being replaced in the literature by "terrestrial reference frame."

11 See the NGS home page (http://www.ngs.noaa.gov/) for further information on this agency and its activities. For more detailed publications concerning horizontal and vertical datums, geodesy, and geoid models, see the listing of NGS publications at https://geodesy.noaa.gov/library/.

12 The USGS and NGS published the *North American Datum of 1983: Map Data Conversion Tables* that reported the change in latitude and longitude coordinate values expressed in NAD27 to NAD83 as they relate to the use of mapping and charting products at scales of 1:10,000 and smaller (United States Geological Survey 1989).

13 Differences in heights for the conterminous United States between NAVD88 and NGVD29 range from 40 cm to 150 cm. In Alaska, the differences range from approximately 94 cm to 240 cm. In more geologically "stable" areas of the United States, height differences are generally less than 1 cm.

REFERENCES

Ariel, A., and Berger, N. A. (2006) *Plotting the Globe: Stories of Meridians, Parallels, and the International Date Line.* Westport, CT: Praeger Publishers.

Defense Mapping Agency (1981) *Glossary of Mapping, Charting, and Geodetic Terms* (4th ed.). Washington, DC: U.S. Department of Defense.

Maling, D. H. (1992) *Coordinate Systems and Map Projections.* New York: Pergamon.

Monmonier, M. S. (2004) *Rhumb Lines and Map Wars: A Social History of the Mercator Projection.* Chicago: University of Chicago Press.

Snyder, J. (1987) *Map Projections: A Working Manual.* Washington, DC: United States Geological Survey.

United States Geological Survey (1989) *North American Datum of 1983, Map Data Conversion Tables.* Denver, CO: United States Geological Survey.

Van Sickle, J. (2017) *Basic GIS Coordinates.* (3rd ed.). Boca Raton, FL: CRC Press.

8 Elements of Map Projections

8.1 INTRODUCTION

The process of projecting the spherical Earth onto a flat surface is accomplished through a **map projection**. One of the more complex processes in cartography, the map projection plays an important role in the map's appearance (e.g., it determines shapes of landmasses and the arrangement of the graticule) as well as the kinds of map uses that are possible (e.g., accurately measuring distances as opposed to visualizing the spatial variation of a data set symbolized by dots).

The projection process begins conceptually with the **reference globe** (a model of Earth at some chosen scale) and a **developable surface** (a mathematically definable surface onto which the landmasses and graticule are projected from the reference globe). Traditionally, the cartographer set the scale of the reference globe and then chose one of the developable surfaces (**cone**, **plane**, or **cylinder**) onto which the landmass and graticule would be projected. Once projected, the developable surface was "unrolled," producing a map. Today, computers and mathematical equations have automated this laborious task with software capable of producing hundreds of projections. Although the reference globe and developable surface are no longer relied on when creating projections, they help conceptually to understand the projection process and are referred to throughout this chapter.

Computers make creating projections a rather trivial process compared to just a few years ago, but understanding the mathematical equations involved in creating a projection remains important if projection software is to be utilized effectively. In addition to understanding the mathematical equations associated with projections, you need to comprehend the various characteristics that a given projection possesses. These characteristics include the **class**, **case**, and **aspect** of a projection. The projection's class refers to the developable surface used in creating the projection. Thus, the cone produces the **conic class**, the plane produces the **planar class**, and the cylinder produces the **cylindrical class** of map projections. The developable surface can be positioned so that it touches the reference globe along one point or line creating the **tangent case** or two lines creating the **secant case**. The developable surface can also be centered over any location on the reference globe. If centered over one of the poles, a **polar aspect** results; if the center is somewhere along the Equator, an **equatorial aspect** results; or, if the center is not at either pole or along the Equator, an **oblique aspect** results.

Regardless of how carefully a projection is constructed, distortion is an inevitable consequence. One way to understand distortion is to realize that, unlike on Earth, the scale varies considerably across a projection and thus a map's surface. This variation in scale can be analyzed across a projection through the **scale factor**. This numeric assessment explains how much departure there is at a given location from the scale that is found at that same location on Earth's surface. The scale factor is also a useful underlying concept for **Tissot's indicatrix**, which provides a visual means of showing how distortion varies at point locations across a projection. In addition, the indicatrix provides a quantitative analysis of distortion that describes the amount and type of distortion that occurs at points across the map. When the indicatrix is mapped across a projection, certain distortion patterns become apparent and can be classified according to the projection's class. Examining these distortion patterns provides insight on how suitable a particular projection's class, case, and aspect are for minimizing distortion over a specific geographic area.

Although no projection can avoid some form of distortion, most projections minimize distortion in either areas, angles, distances, or direction. These are the projection properties. **Equivalent projections** preserve areal relationships, whereas **conformal projections** preserve angular relations. Preserving distance and directional relationships are accomplished on **equidistant** and **azimuthal projections**, respectively. **Compromise projections** preserve no specific property but rather strike a balance among various projection properties.

8.2 LEARNING OBJECTIVES

- Describe the different developable surfaces and how they contribute to forming the projection classes.
- Explain the difference between a tangent or a secant case of a projection.
- Recognize the importance of choosing the location of an appropriate standard point, line, or lines for a map projection.
- Explain the role that choosing an appropriate aspect has on the appearance of a map projection's graticule and landmass arrangement.
- Describe how the scale factor controls the type and distribution of distortion across a projection's surface.
- Explain how Earth's spatial relationships of areas, angles, distances, and directions are preserved through the projection properties.
- Explain how computers have changed the way in which map projections are created and viewed for mapping tasks.

DOI: 10.1201/9781003150527-9

8.3   THE MAP PROJECTION CONCEPT

When trying to create a model of Earth's spherical surface, a globe is an obvious possibility because globes preserve areal, angular, distance, and directional relationships. Unfortunately, globes of any size are rather inconvenient to carry around and store, and detailed representations of even large countries would require rather sizable globes. Using portions of a globe to reproduce a specific country would reduce some of the bulkiness, but measuring, for example, distances across a globe's curved surface is difficult.

Maps have four distinct advantages over globes. First, the two-dimensional nature of maps makes many cartometric activities (e.g., measuring areas, plotting a course, and calculating direction) much easier than on the curved surface of a globe. Second, maps can show considerable detail for a given landmass. Consider the United States Geological Survey's (USGS) 1:24,000 topographic map series. This series maps portions of the United States in greater detail than could be found on even the largest globe. Third, most maps are of a dimension that makes them easy to work with and very portable — for example, 1:24,000 sheets are roughly 56 cm × 66 cm (22 in. × 26 in.). Fourth, maps are less costly than globes to produce and purchase. A single USGS 1:24,000 map sheet costs about $8.00, whereas a 30 cm (12 in.) diameter globe is priced around $50.00.

Although maps have numerous advantages over globes, they suffer from distortion, which is a natural consequence that results when Earth's two-dimensional curved surface is projected to a map—that is, when a map projection is created. The **map projection** is one of the most important concepts in cartography because the way in which Earth's two-dimensional curved surface is projected directly impacts the appearance of the graticule (lines of latitude and longitude in combination) and landmasses and, ultimately, the kinds of uses for which a map can be applied.

8.4   THE REFERENCE GLOBE
AND DEVELOPABLE SURFACES

The techniques employed to create map projections have changed considerably over the past 30 years. Prior to the advent of computers, map projections were laboriously created through manual drafting techniques. Two key concepts from the pre-computer era are the **reference globe** and the **developable surface**. These conceptual aids were used to assist the mapmaker in envisioning how Earth's two-dimensional curved surface would be projected to create a map. Although computers have replaced manual drafting techniques, the reference globe and developable surface are still useful in explaining how a map projection is created.

When the cartographer begins creating a map projection, an initial step involves conceptually reducing Earth's size to a smaller, imaginary reference globe. The final size of this globe is set according to the final map scale, which is usually dictated by the map purpose. Figure 8.1 shows the relationship between Earth and the reference globe, which

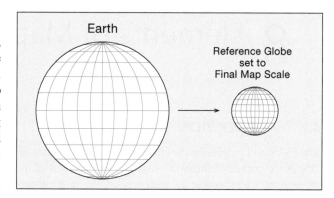

FIGURE 8.1   The relationship between Earth and the reference globe. Note that the sizes of Earth and reference globe are not to scale.

is a model of Earth at a reduced scale that is used to project the graticule and landmasses onto the map.

A developable surface is a simple mathematical surface that can be flattened to form a plane without compressing or stretching any part of the original surface. There are three developable surfaces: **cylinders**, **cones**, and **planes**. To project the graticule from the reference globe to a map, first, a developable surface is conceptually placed over the reference globe, touching it along either a meridian or parallel or at one point (Figure 8.2); second, the graticule and geographic landmasses are projected onto the developable surface. Afterward, the developable surface is "unrolled," revealing the graticule and landmasses.

Figure 8.3 provides a detailed view of how the reference globe and developable surface can be used to create a map projection. Conceptually speaking, a projection is created by first shining an imaginary light on the reference globe or some portion of it. The light source, called the **point of projection**, can be located in one of several positions with respect to the reference globe. For example, in Figure 8.3, the point of projection changes from the center of the reference globe (Figure 8.3A) to a position on the side of the reference globe opposite where the developable surface touches the reference globe (Figure 8.3B), to a point at infinity (Figure 8.3C). The light casts an image of the graticule and landmasses onto the developable surface (in this case, a plane), which touches the reference globe at one point, creating a projection. Note how changing the location for the point of projection influences the spacing of the graticule.

8.5   THE MATHEMATICS OF MAP
PROJECTIONS

The reference globe and developable surfaces are conceptual aids that help illustrate the projection process, but they are not used to create projections today. Rather, the field of mathematics is utilized to create projections, and so it is important to understand some of the basic mathematical manipulations used to project Earth onto a map.

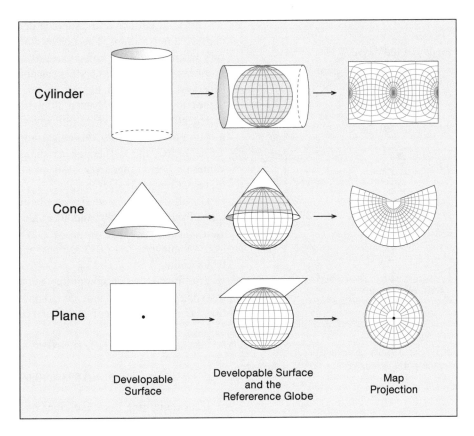

FIGURE 8.2   How developable surfaces and the reference globe are used to create map projections.

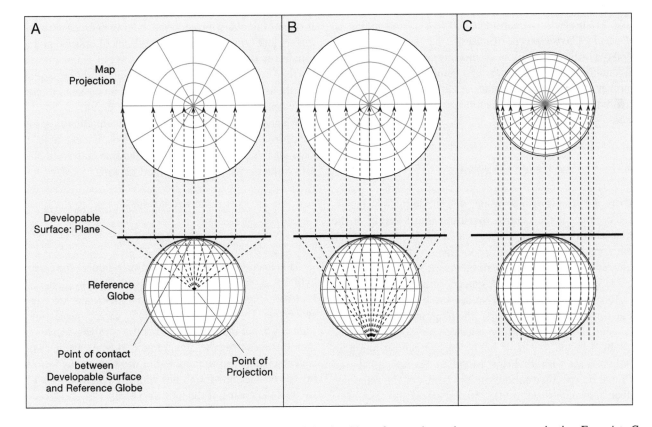

FIGURE 8.3   A detailed view of how the reference globe and developable surface can be used to create a map projection. From A to C, the point of projection changes from the center of the reference globe to a point on the reference globe opposite where the developable surface touches the reference globe to a point at infinity.

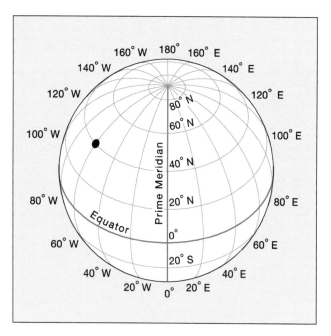

FIGURE 8.4   A point on Earth's curved surface described by the graticule. In this case, the point is located at −60° W longitude and 40° N latitude.

For illustrative purposes, assume that you want to transform a point on Earth's curved surface to a flat map and that the point in question is defined by the latitude and longitude values of 40° N and 60° W, as shown in Figure 8.4. Our goal is to project this coordinate pair to a corresponding set of x and y **Cartesian coordinates**.

At a minimum, all map projections involve at least two mathematical equations: one defining the x value and another defining the y value. A simple map projection involving two equations is as follows:

$$x = R * (\lambda - \lambda_0)$$

$$y = R * (\phi),$$

where λ is the longitude value, ϕ is the latitude value, λ_0 is the value of the **central meridian** (the location of the projection's east–west center), and R is the radius of the reference globe. The x and y coordinate values commonly referred to as **plotting coordinates**, specify the longitude and latitude values in a Cartesian coordinate system.

To compute the x and y values using these equations, four steps are necessary. First, the longitude value (λ_0) for the central meridian must be selected. Any longitude value can be selected, but if 0° is chosen, the central meridian will coincide with the Prime Meridian. Second, a value of R must be specified to indicate the radius of the reference globe (ultimately setting the final scale of the map). For example, if a world map was to be created with a principal scale of 1:30,000,000, then a reference globe with a radius (R) of 21.24 cm (8.36 in.) would be used. To simplify the computations that follow, we assume an R value of 1.0.

Third, all latitude and longitude degree values must be converted into radians. This conversion is especially necessary when using map projection software or computer programming languages, as neither can compute trigonometric functions specified in degree values. Converting degree values to radians involves multiplying the degree measurement by the constant $\pi/180$, where π is approximately 3.1415. For instance, 90° in radians equals $90° \times (\pi/180) = 1.5707$ or $\pi/2$, and 180° equals π radians. Because longitude values on Earth range from −180° to +180° and all x values represent longitude degree values, x values range from $-\pi$ to $+\pi$. On the other hand, because latitude values on Earth range from −90° to +90° and all y values represent latitude values, y values range from $-\pi/2$ to $+\pi/2$. For instance, −60° W converts to −1.047 radians, and 40° N converts to 0.698 radians.

The fourth step in computing x and y values involves inserting longitude and latitude radian values into the equations. Using our values of −1.047 and 0.698, we have:

$$x = 1.0 * (-1.047 - 0) = -1.047$$

$$y = 1.0 * 0.698 = +0.698.$$

If all remaining point locations on Earth's curved surface defined by latitude and longitude were computed (e.g., every 15°) using the above equations, the plate carrée[1] cylindrical projection would be generated (as shown on the right side of Figure 8.5). The plate carrée projection is one of the oldest projections, having been developed by the ancient Greeks. Note that the lines of latitude and longitude for the plate carrée are spaced at equal intervals. Furthermore, although all lines of longitude are equal in length (as they should be), they do not converge to points representing the poles as they ideally should.

We can expand on the simple pair of equations just introduced by adding trigonometric functions. For example, consider what happens if we compute the sine of the latitude for all y values—we would have the following:

$$x = R * (\lambda - \lambda_0)$$

$$y = R * \sin \phi.$$

This mathematical transformation produces the Lambert cylindrical projection developed in 1772 by Johann H. Lambert (Figure 8.6). Note that the meridians are equally spaced as they are on the plate carrée projection in Figure 8.5, but that the spacing of the parallels decreases as their distance from the Equator increases. As an alternative, we can change just the spacing of the x values by introducing the cosine function to our simple equations. This time, we take the cosine of the latitude value:

$$x = R * (\lambda - \lambda_0) * \cos \phi$$

$$y = R * \phi.$$

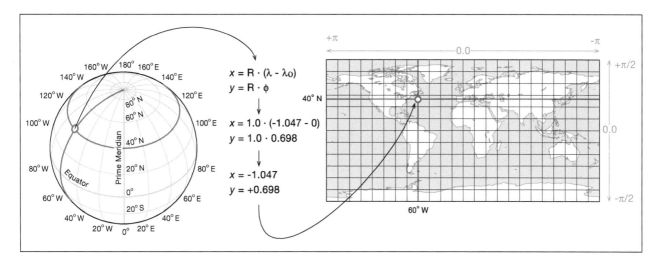

FIGURE 8.5 The mathematics of a map projection transforms a point on Earth's curved surface represented by a latitude and longitude pair to *x* and *y* coordinate pairs on a map. In this case, the plate carrée cylindrical projection results. The ranges of *x* and *y* coordinate values in radians are shown in gray along the sides of the map projection.

The sinusoidal pseudocylindrical projection results as shown in Figure 8.7. The term pseudocylindrical is applied here because this projection shares only some of the visual characteristics of the graticule with the plate carrée cylindrical projection, namely that the lines of latitude are parallel and, in this case, equally spaced. However, the meridians of this projection are curved lines converging at the North and South Poles.

Now, combining the cosine and sine functions together yields the following equations:

$$x = R*\left(\lambda - \lambda_0\right)*\cos\phi$$

$$y = R*\sin\phi.$$

This set of mathematical equations produces the polycylindrical projection shown in Figure 8.8. In this case, some of the graticule characteristics of the previous projections should be apparent. For instance, the meridians are curved

as in the sinusoidal projection and the parallels are parallel as in the other two cylindrical projections.

8.6 MAP PROJECTION CHARACTERISTICS

Through manipulation of the mathematical equations, numerous projections are possible, where each projection has specific characteristics, making each one useful for a particular mapping purpose. This section discusses the various characteristics of map projections based on class, case, and aspect.

8.6.1 CLASS

Earlier, we saw that the developable surface concept was useful in illustrating the manner in which the graticule and landmasses on the reference globe are projected to the map. This concept is also a constructive way to describe the overall shape and appearance of the graticule after the

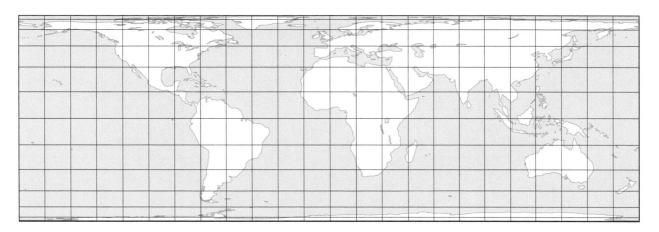

FIGURE 8.6 The Lambert cylindrical projection with a 15° graticule spacing.

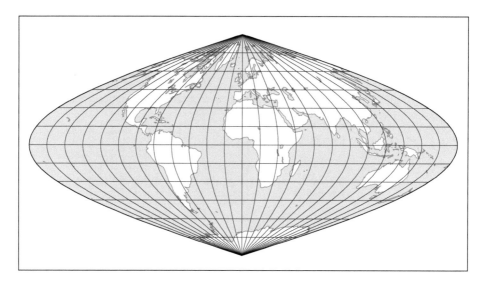

FIGURE 8.7   The sinusoidal pseudocylindrical projection with a 15° graticule spacing.

projection process is complete, referred to as the projection's **class**. The three common map projection classes are **cylindrical**, **conic**, and **planar**.

The cylindrical class of projections results from wrapping the developable surface of a cylinder around the reference globe, projecting the landmasses and graticule onto the cylinder, and then unrolling the cylinder. On cylindrical projections, lines of longitude typically appear as straight, equally spaced, parallel lines, whereas lines of latitude appear as straight parallel lines that intersect the lines of longitude at right angles (Figure 8.9A and B). The spacing of the parallels distinguishes one cylindrical projection from another.

The conic class of projections results from wrapping the developable surface of a cone around the reference globe, projecting the landmasses and graticule onto the cone, and then unrolling the cone. On conic projections, lines of longitude typically appear as straight lines of equal length radiating from a central point (usually one of the poles), whereas lines of latitude appear as concentric circular arcs centered about one of the poles (Figure 8.9C and D). The overall shape of most conic projections can be described as a pie wedge, where a pie would be a full circle. Note that in the right-hand portion of Figure 8.9C, the pie wedge is slightly more than one-half, whereas, in Figure 8.9D, the pie wedge is slightly less than one-half. The angular extent of the pie wedge and the spacing of the parallels distinguish one conic projection from another.

The planar class of projections results from positioning the developable surface of a plane next to the reference globe and projecting the landmasses and graticule onto the plane.[2] On planar projections, lines of longitude typically appear as straight lines that radiate from the center (when the center of the projection is one of the poles), and lines of latitude appear as concentric circles centered about a point, for example, one of the poles (Figure 8.9E and F). The spacing of the parallels distinguishes one planar projection from another.

There are many variations of the three developable surfaces that create other interesting projections. Like the cylindrical

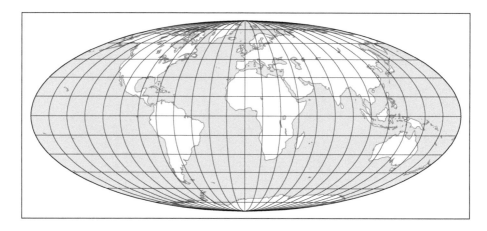

FIGURE 8.8   A polycylindrical projection with a 15° graticule spacing.

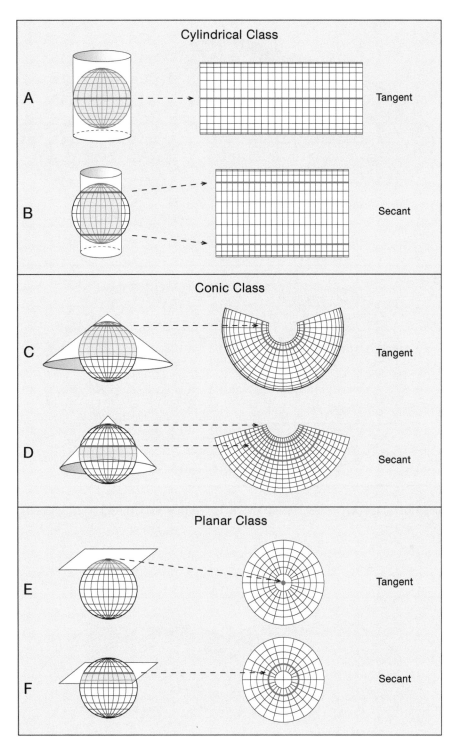

FIGURE 8.9 A comparison of the different classes of projections: (A–B) cylindrical, (C–D) conic, and (E–F) planar. Also shown are the tangent and secant cases for each class.

projections, the pseudocylindrical class of projections shows parallels as straight, nonintersecting lines. However, pseudocylindrical projections are visually distinguished from cylindrical projections by curved meridians that are equally spaced along every parallel and converge at the poles, which are represented by points or lines. Figure 8.10A illustrates the Craster parabolic pseudocylindrical projection with the meridians converging to points, and Figure 8.10B shows the

Eckert III pseudocylindrical projection with meridians converging to lines representing the poles.

The polyconic class of projections conceptually employs multiple cones rather than one cone. Polyconic projections display curved lines (nonconcentric circular arcs) representing the parallels in a fashion similar to the conic projections, but the meridians curve toward the central meridian. Figure 8.11 illustrates the rectangular polyconic projection.

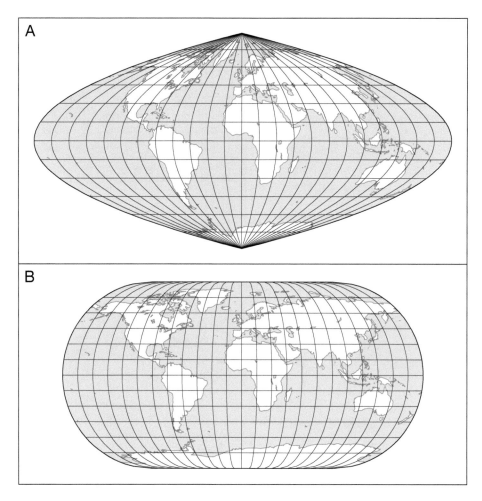

FIGURE 8.10 A comparison of the (A) Craster parabolic and (B) Eckert III pseudocylindrical projections. Each projection has a 15° graticule spacing.

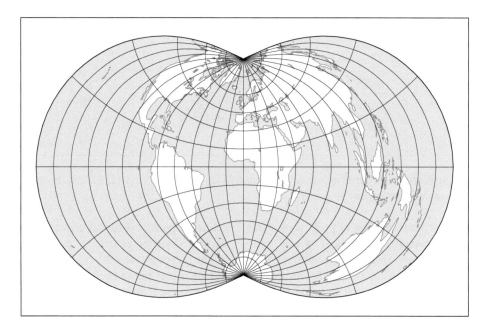

FIGURE 8.11 The rectangular polyconic projection with a 15° graticule spacing.

There are many other variations of the basic developable surface concept. However, unlike the cylinder, cone, and plane, each variation requires a mathematical rather than a visual approach to explain how each projection is developed.

8.6.2 CASE

The case of a projection relates to how the developable surface is positioned with respect to the reference globe and is either tangent or secant. A brief example illustrates the tangent idea. Conceptually speaking, imagine a ball rolling across the floor. At any given time, there is exactly one point in common between the ball and the floor. This point of contact is called the point of tangency. If all points of contact were connected together (either on the ball or on the floor), a line of tangency would result. In the tangent case of a map projection, the reference globe touches the developable surface along only one line or at one point. Parts A, C, and E of Figure 8.9 illustrate the tangent case for the cylindrical, conic, and planar map projections, respectively. The secant case of a projection occurs when the developable surface passes through the reference globe. In the secant case of the cylindrical and conic map projections (Figure 8.9B and D), there are two secant lines, whereas, in the case of the planar projection, there is one secant line (Figure 8.9F).[3]

For a given projection class, the mapmaker usually has a variety of choices for the desired specific line(s) or point of tangency. For instance, with the conic class, any line of latitude can be selected as the tangent line. In the secant case of the conic class, any two lines of latitude can be selected. If these lines are nonequally spaced on opposite sides of the Equator, equally spaced on the same side of the Equator, or nonequally spaced on the same side of the Equator, the parallels and meridians take on the familiar cone shape. If, however, the two lines are equidistant from and on opposite sides of the Equator, the parallels and meridians appear as straight parallel lines. The scale along secant and tangent lines and points of tangency is equal to the principal scale of the reference globe. Thus, secant and tangent lines are called **standard lines**,[4] and points of tangency are called **standard points**. All other lines and points will have either a larger or a smaller scale than the principal scale of the reference globe. This scale variation across the reference globe also determines how scale changes across the map's surface. Figure 8.12 illustrates the concept of a standard line and its impact on scale variation across a map. In the figure, a portion of the reference globe is represented by the dashed line, and the developable surface is represented by the solid orange and blue lines. Note that the developable surface cuts the reference globe, creating two standard lines that have the same principal scale as the reference globe. The area between the standard lines projected to the developable surface (shown by a solid blue line) has a compressed scale, whereas the areas beyond the standard lines (shown by the orange lines) have an exaggerated scale. Figure 8.13 shows the effect of selecting different tangent and secant cases for the Euler conic projection, developed by Leonhard Euler

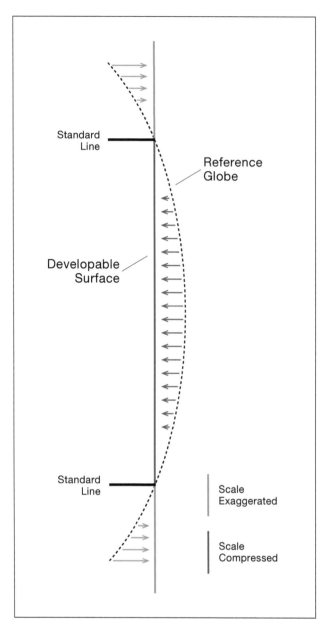

FIGURE 8.12 The effect of standard lines on scale variation on a map.

in 1777. Figure 8.13A shows the tangent case with only one standard line at 25° N, and Figure 8.13B shows the secant case with two standard lines located at 25° N and 50° N. Note that for both figures, the meridians converge toward the poles and the parallels form curves concave toward the North Pole. Figure 8.13C shows the secant case with standard lines at −25° S and −50° S. Note that when placing the standard lines in the southern latitudes, the parallels are shown as curved lines that concave toward the South Pole, and the meridians converge to the South Pole.

The choice of the tangent or secant case, as well as their placement with respect to the reference globe, not only controls the distribution of scale across the map's surface but also impacts the shape of the landmasses and the arrangement of the graticule. For instance, in Figure 8.13A, the

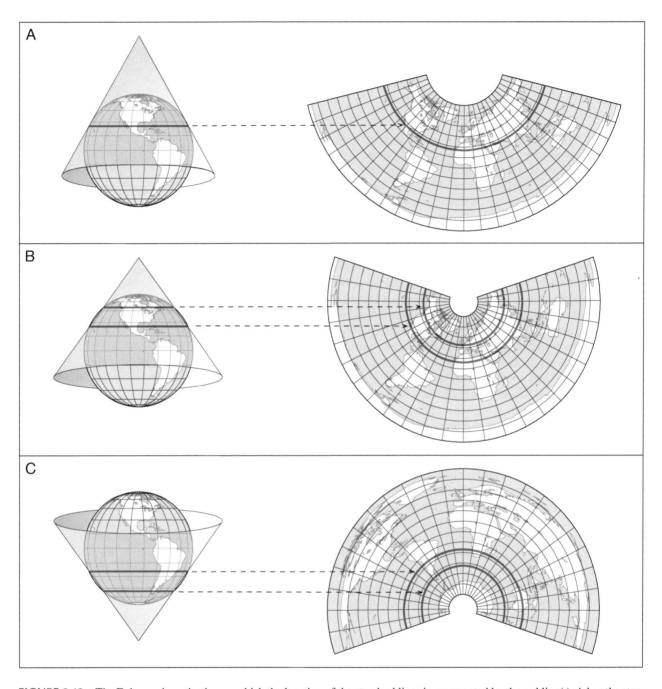

FIGURE 8.13 The Euler conic projection on which the location of the standard lines is represented by the red line(s). A has the standard line tangent to 25° N. B has standard lines secant at 25° N and 50° N. C has standard lines secant at –25° S and –50° S.

convergence of the meridians at the North Pole is not as great as the meridian convergence in Figure 8.13B. As a result, landmasses in the northern latitudes are not as stretched in an east–west direction in Figure 8.13B. Figures 8.13A and B also illustrate that the placement of the standard lines in the northern latitudes greatly distorts the landmasses in the extreme Southern Hemisphere (e.g., Antarctica is stretched out the length of the map). In contrast, when both standard lines are placed in the Southern Hemisphere (as in Figure 8.13C), the landmasses there are not as distorted, but the landmasses in the Northern Hemisphere are greatly distorted.

8.6.3 Aspect

The aspect of a projection concerns the placement of the projection's center with respect to Earth's surface. The aspect is important as it allows the map maker to center the geographic area to be mapped in the projection. In elementary terms, a projection can have one of three aspects: **equatorial**, **oblique**, and **polar**. An equatorial aspect is centered somewhere along the Equator, a polar aspect is centered about one of the poles, and an oblique aspect is centered somewhere between a pole and the Equator. The aspect of the projection can be defined more precisely in

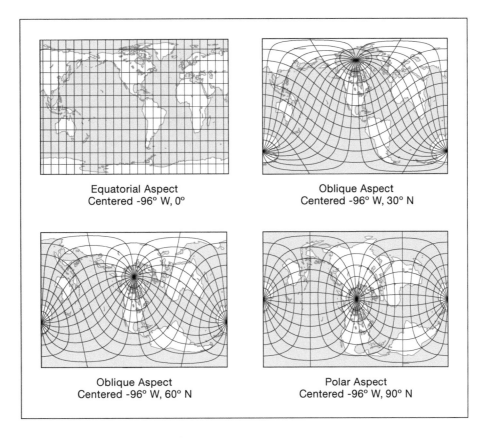

FIGURE 8.14 Four aspects of a cylindrical projection.

terms of the placement of the central meridian and the **latitude of origin**. Figure 8.14 shows four different aspects for the equidistant cylindrical projection. For these maps, the central meridian is located at –96° W, and the latitude of origin begins at 0° (an equatorial aspect) and moves northward every 30° of latitude (oblique aspects) until the projection is centered over the North Pole (a polar aspect).

8.7 DISTORTION ON MAP PROJECTIONS

An important consequence of map projections is the distortion created in the resulting map. More specifically, we can define **distortion** as the alteration of the size of Earth's landmasses and the arrangement of Earth's graticule when they are projected to the two-dimensional flat map. There are numerous ways in which distortion can be examined and analyzed on a projection, many of which are covered in a thorough review by Karen Mulcahy and Keith Clarke (2001).

8.7.1 A VISUAL LOOK AT DISTORTION

One way to examine distortion is to visually compare the sizes of landmasses and the arrangement of the graticule on the spherical Earth to how the landmasses and graticule arrangement appear on the map projection. To illustrate, Figure 8.15A shows an orthographic projection presenting a view over the North Atlantic Ocean similar to what

one would see if looking at Earth from space. Note the size of Greenland relative to the United States in this figure. Now compare the size of Greenland to the United States on the Mercator projection in Figure 8.15B; obviously, Greenland is now relatively much larger than the United States. As a point of reference, Greenland has a land area of 2,166,086 km² which is smaller than the United States at 9,158,960 km², but on the Mercator projection, Greenland is represented as larger than the United States. In general, you can see that landmasses in the upper latitudes on the Mercator projection are considerably exaggerated—a limitation when using this projection for maps of the world.

Depending on the way in which a projection is developed, distortion can also cause Earth's graticule to take on a number of different appearances. For instance, on Earth's surface, all meridians converge to the poles. This relationship is preserved on the orthographic projection, but on the Mercator projection in Figure 8.15B, all meridians are equally spaced straight lines. As a result of this nonconvergence, the graticule on the Mercator projection is stretched compared to its representation on the orthographic projection.

A visual inspection of a map projection is useful in providing a general overview of the distortion present on the map, but a more sophisticated approach is needed to quantitatively analyze the amount and kind of distortion throughout a projection. Moreover, once the quantitative analysis is complete, there remains the need to visualize the distribution

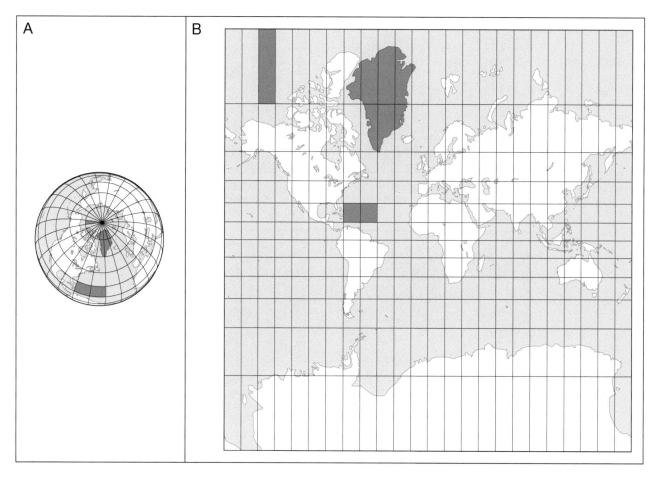

FIGURE 8.15 (A) The orthographic projection gives the appearance of Earth as if viewed from space. (B) On the Mercator projection, note the exaggerated size of the landmasses, especially in the upper latitudes. Also, the meridians do not converge at the poles, and the spacing of the lines of latitude increases from the Equator to the poles. Although it might not appear so, both projections are shown at the same scale.

of distortion. There have been several approaches to quantitative distortion analysis that have been developed, but our discussion focuses on one of the more common—Tissot's indicatrix.

8.7.2 SCALE FACTOR

To understand the details of Tissot's indicatrix, we begin with an introduction to the **scale factor** (**SF**), which is a numerical assessment of how the map scale at a specific location on a map compares with the map scale at a standard point or along standard line(s). Figure 8.9 showed that the scale on the map is the same as the reference globe only where the developable surface comes in contact with the reference globe at a standard point or along standard line(s). In all other locations, depending on the way the projection is mathematically created, the scale is either exaggerated or compressed.

The SF at any given location is computed using the following formula:

$$\text{Scale factor} = \frac{\text{Local scale}}{\text{Principal scale}},$$

where *local scale* is the scale computed at a specific location and *principal scale* is the scale computed at a standard point or along standard line(s). Using this formula, we can compute how much deviation exists between the local scale and the principal scale at any location on a given projection. To illustrate, assume that we are working with the quartic authalic projection (suitable for world maps) shown in Figure 8.16 and that this projection has a principal scale of 1:235,043,988 along the Equator, which is the standard line. Now assume you wish to compute the SF at some point along the parallel at 30° N, say, between 0° and −15° W (shown in Figure 8.16 as *a*). You would first calculate the local map scale between 0° and 15° W, which can be found using the following steps: Measuring along the parallel at 30° N between 0° and 15° W, assume you arrive at a map distance of 0.65 cm (0.256 in.). To compute the corresponding Earth distance for that same distance, visit Inverse[5] to compute the distance for one degree of longitude for 30°, which is 96,486.28 meters. Because Earth's distance between 0° and 15° is 15°, we need to multiply 96,486.28 by 15°. Doing so, we arrive at an Earth distance of 1,447,294.2 m, or 1447.2942 km.

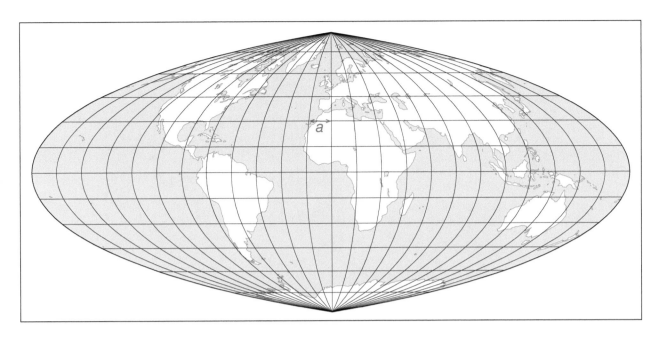

FIGURE 8.16 The quartic authalic projection with a 15° graticule spacing. The letter *a* references the location at which the scale factor is computed.

Next, substituting the 0.65 cm and 1447.2942 km into the following basic scale equation, we obtain the local scale:

$$\text{Local scale} = \frac{\text{Earth Distance}}{\text{Map Distance}}$$

$$\text{Local scale} = \frac{1447.2942 \text{ km}}{0.65 \text{ cm}}$$

Before solving, the units in the equation must be the same, so we convert km to cm:

$$1447.2942 \text{ km} * \frac{100,000 \text{ cm}}{1 \text{ km}} = 144,729.420 \text{ cm}$$

Therefore,

$$\text{Local scale} = \frac{144,729,420 \text{ cm}}{0.65 \text{ cm}} = 1 : 222,660,646.$$

The local scale along the 30° parallel is thus computed as 1:222,660,646, which is larger than the principal scale along the Equator of 1:235,043,988. Knowing the principal scale of the map projection (1:235,043,988) and the local scale along the 30° N latitude (1:222,660,646), we can compute the SF at this location. Substituting the different scale values into the SF equation, we arrive at the following:

$$SF = \frac{1 : 222,660,646}{1 : 235,043,988} = 1.056$$

$$SF = \frac{1/222,660,646}{1/235,043,988} = \frac{235,043,998}{222,660,646} = 1.056.$$

The computed *SF* value suggests that the local scale has been exaggerated, meaning that the distances on Earth's curved surface along 30° N latitude have been enlarged with respect to its true distance. On the other hand, if a local scale was computed to be 1:240,000,000 and the principal scale remained at 1:235,043,988, then the SF would be 0.979, and a compressed scale would result. Obviously, if we had a location on the same projection with a local scale of exactly 1:235,043,988, the *SF* would be 1.0, indicating that there has been no change in scale at that location.

8.7.3 Tissot's Indicatrix

Nicolas Tissot, a French mathematician, was one of the pioneers in analyzing distortion found on map projections. Tissot (1881) developed the **indicatrix**, which provides a graphical representation of distortion at various points across a projection. The indicatrix, as found on the reference globe's surface where all spatial relationships are preserved, results in no distortion and is called the unit circle.[6] In other words, at each and every point on the reference globe, an infinitely small circle exists that can be described as a unit circle having a radius (or SF) of 1.0, as shown in Figure 8.17. Even though the indicatrix is displayed as a unit circle of some areal extent, it is important to remember that the appearance of the indicatrix on the reference globe is restricted to measurements made around infinitely small points and is not considered valid when discussing landmasses or water bodies of any great extent.

After projection, the indicatrix has two characteristics that allow the amount and kind of distortion to be analyzed. First, the indicatrix is defined by two specific radii called the semimajor (*a*) and semiminor (*b*) axes. By convention and through mathematical manipulation, the semimajor axis is aligned in the direction of the maximum SF, and the semiminor axis is aligned in the direction of the minimum SF. These two axes are always perpendicular to one another.

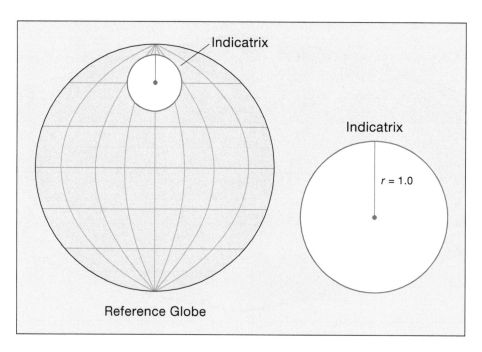

FIGURE 8.17   The indicatrix and its characteristics as shown on the reference globe. The radius of the indicatrix on the reference globe is 1.0.

Second, on the indicatrix, there exist two lines called *l* and *m* that intersect, forming an angle called 2ω (Figure 8.18). The specific changes in the SFs along *a* and *b*, as well as the alteration of 2ω after projection provide a useful means to analyze distortion.

There are four general instances of the indicatrix that result after a projection is created. In the first instance, there is no change in the SFs along *a* and *b*, so the indicatrix remains a unit circle (Figure 8.18A), indicating no distortion. The second instance is where the SFs along *a* and *b* change unequally, creating an ellipse, but the area of this ellipse is equal to the area of the unit circle. In Figure 8.18B, there is an increase in the SF along *a*, but the SF along *b* decreases proportionately. This leads to a change in the angle 2ω and so the projection is said to possess *angular distortion*. Note, however, that the area of the unit circle is preserved, as found on the reference globe, so there is no areal distortion. In the third instance, the SFs along *a* and *b* change in the same manner (e.g., *a* = 2.0 and *b* = 2.0). Here, the indicatrix is still a circle, but it is larger (or smaller) in size than the unit circle found on the reference globe (Figure 8.18C), indicating *areal distortion*. Because the angle 2ω has been preserved, there is no angular distortion. The fourth instance is where *a* and *b* change unequally, producing an ellipse, and the area of the ellipse does not equal the unit circle (Figure 8.18D). In this case, there is both areal and angular distortion.

When the indicatrix is plotted on a projection (e.g., every 15°), the type and amount of distortion can be visualized across the projection. For example, Figure 8.19 shows the indicatrix for the Winkel tripel projection. In this case, note that along the central meridian at approximately 44° 28′ 25″ N/S is the only location where the indicatrix is a circle and the SFs along both *a* and *b* are 1.0. At all other locations, the indicatrix takes on an elliptical shape, suggesting that there is angular distortion throughout the projection. In a similar light, the ellipses do not appear to have the same areal extent, indicating that areal relations are not preserved in the Winkel tripel projection.

8.7.4   Distortion Patterns

Figure 8.9 showed that the developable surface contacts the reference globe at either a single standard point or along one or two standard lines. Where the standard point or standard line(s) contact the reference globe, no distortion in scale is present—the SF at these locations is 1.0. However, we learned that at other locations throughout the map projection, scale varies. This scale variation can be mapped, producing characteristic distortion patterns that are unique to the specific class and case of projections, as shown in Figure 8.20, where darker tints of red indicate greater distortion.

8.7.5   Using Geocart to Visualize Distortion Patterns

Recall from Section 8.7.3 that Tissot's indicatrix was presented as a graphical way to represent changes in scale factors (i.e., distortion) that occurs at infinitesimal points across a projection's surface. There are two concerns that can be raised regarding the usefulness of the indicatrix as a symbolization method to represent distortion: visual appearance and shape recognition. The visual appearance of the indicatrices on a projection greatly influences how the reader interprets the distortion patterns across the

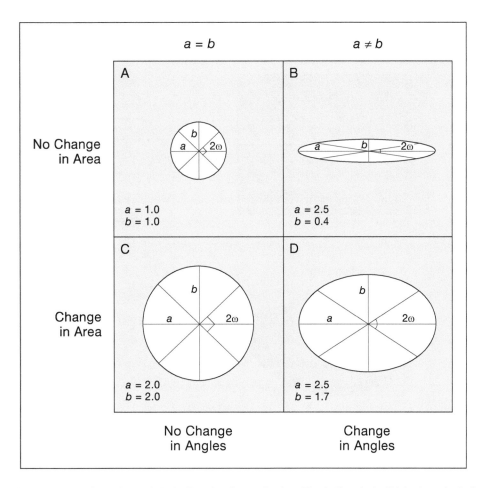

FIGURE 8.18 The possible configurations of the indicatrix after projection. The indicatrix in (A) is the unit circle as found on the reference globe, and thus there is no distortion in scale factors, areal, or angular relations. In (B), the scale factors are not equal; the area of the original indicatrix has been preserved, and note the change in 2ω suggesting angular distortion. The scale factors in (C) are greater than 1.0 but equal to one another, and thus the area is exaggerated, but there is no angular distortion. In (D), there is both angular and areal distortion.

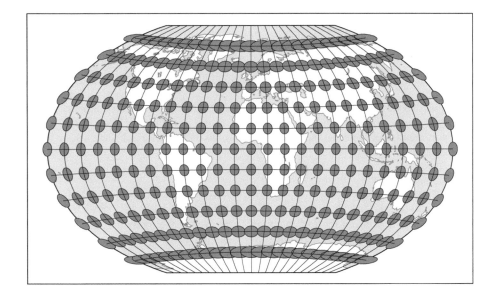

FIGURE 8.19 The indicatrix on the Winkel tripel projection. Note that the indicatrix is a circle only along the central meridian at approximately 44° 28′ 25″ N/S. The indicatrix takes on a variety of elliptical shapes and sizes throughout the remainder of the projection, suggesting that this projection preserves neither areas nor angles. The indicatrix is plotted every 15°.

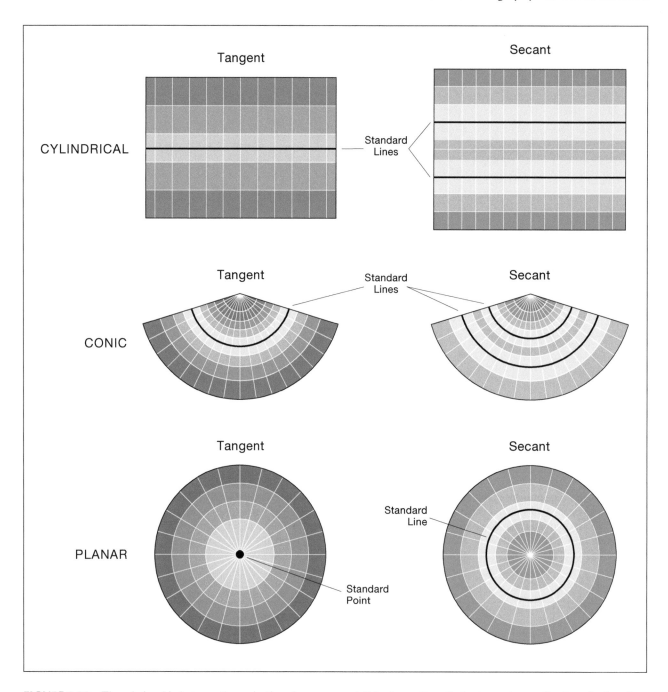

FIGURE 8.20 The relationship between the projection class, case, and distortion pattern. In the tangent case of each projection class, the standard line or point is the location where the scale factor is 1.0, shown by the thicker black line or point. In the secant case of each projection class, the standard lines (note that the planar class only has one standard line) have a scale factor equal to 1.0. Darker red shading indicates increasing distortion.

projection's surface. For example, the indicatrix is usually plotted along the graticule in equal intervals (e.g., every 15° of latitude and longitude). While plotting the indicatrix at equal intervals gives a reasonable overview of the distortion pattern across the projection's surface, a reader cannot be certain about the distortion that occurs in between that intervals. In contrast, it is rather easy for someone to look at the indicatrices in Figure 8.24, see that they are all circular, and conclude that this is a conformal projection (angular relations are preserved). However, looking at the

indicatrices in Figure 8.19 to determine the projection property is not as easy. It is challenging for the reader to visually inspect the indicatrices in Figure 8.19 to determine, for example, if the area across all indicatrices is equal and thus an equal area projection. This identification is made difficult as the indicatrices in the upper latitudes of Figure 8.19 overlap. A similar situation appears in Figure 8.23, where the indicatrices in the upper latitudes are stretched in the east-to-west extent to such a degree that their shapes are no longer readable. In both Figures 8.19 and 8.23, reliance on the varied shapes and sizes of the indicatrices as visual cues to the distortion occurring on these projections is challenging and because of this challenge, the reader would have difficulty identifying whether the preserved projection property is equal area, equidistant, or compromise.

Addressing the concerns in using Tissot's indicatrix to symbolize distortion, one could argue for a symbolization method that uses a smooth and continuous approach to representing distortion changes across the projection's surface. Using such a symbolization method would remove the issue of symbol overlap to better recognize subtle changes in the overall indicatrix shape and show distortion as a continuous surface that exists everywhere across the projection. Moreover, unique colors could be assigned to different types of distortion so that a conformal projection, for example, could be easily distinguishable from an equal area projection.

Geocart is map projection software that represents distortion across a projection's surface using a smooth continuous symbolization. The software permits the user to display specific kinds of distortion (e.g., angular, area, and scale) using a magenta-green-grey continuous color scheme. Figure 8.21A shows shades of magenta, with darker shades representing greater amounts of angular distortion on the Eckert III pseudocylindrical compromise projection. Using this continuous color scheme as a visual guide, the areas around the Equator are relatively free of angular distortion, but the increasing amounts of magenta in the upper latitudes suggest greater levels of angular distortion. Figure 8.21B shows the areal distortion pattern on the Eckert III projection. Here, continuous tones of green are used to represent areal distortion across the projection's surface. Unlike the angular distortion pattern shown in Figure 8.21A, two bands of low areal distortion appear in the mid-latitudes. These low distortion bands coincide with the location of standard lines on this projection at 35° 58′ N and S latitude. Areal distortion increases away from these standard lines reaching a maximum at the poles. Figure 8.21C shows grey tones to represent scale distortion across the projection's surface. The color pattern shown in this figure illustrates that the maximum scale distortion is located near the upper latitudes. That the upper latitudes show the greatest distortion on this projection for angle, area, and scale should make sense as the poles are represented as lines rather than points. Figure 8.21D shows a mixture of magenta, green, and grey, symbolizing the combination of angular, areal, and scale distortion on the Eckert III. Based on the appearance of the color patterns shown in the previous figures, the equatorial area shows more areal distortion (green tones are prominent), while the mid-latitudes contain more angular distortion (magenta tones are more prominent). Between 30° and 45° N/S, hues of reddish-brown appear, while beyond 60° N and S, the reddish-brown hues become darker and desaturated until almost black at the poles suggesting the upper latitudes contain high levels of overall distortion.

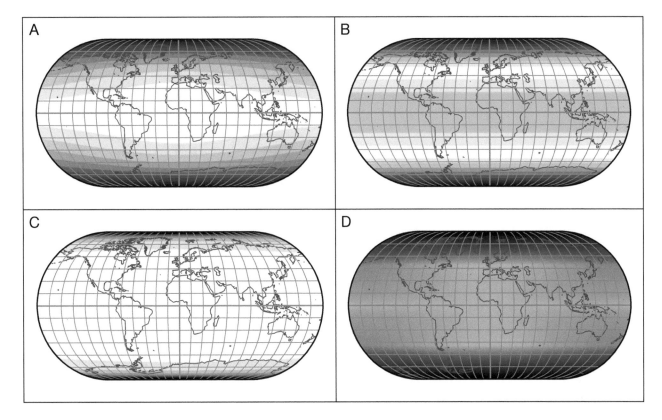

FIGURE 8.21 Smooth continuous shades used to symbolize angular (A), areal (B), scale (C), and overall (D) distortion on the Eckert III pseudocylindrical compromise projection. The distortion maps were produced by Geocart (http://www.mapthematics.com).

8.8 PROJECTION PROPERTIES

A map projection is said to possess a specific property when it preserves one of the spatial relationships (i.e., areas, angles, distances, and directions) found on Earth's surface. The preservation of a particular property is achieved by controlling the SFs throughout the projection. For instance, projections that preserve either areas or angles throughout the projection are called **equivalent** (equal area) and **conformal**, respectively. Projections also are capable of preserving several special properties. When all distances from a particular location are correct, then the projection is said to be **equidistant**. **Azimuthal projections**, which were introduced earlier as planar projections, preserve directions or azimuths from one central point to all others. **Minimum error projections** possess no specific property, but by mathematically optimizing SFs, they achieve lower overall distortion across a projection than can be achieved when one property is preserved. We now consider each of the projection properties in detail.

8.8.1 Preserving Areas

Equivalent projections preserve landmasses in their true proportions, as found on Earth's surface. To illustrate equivalency, consider the landmasses in Figure 8.22, in which we see four samples of Greenland that have been taken from different projections (each projection has the same approximate principal scale). Parts A–C of Figure 8.22 are from equivalent projections, whereas Figure 8.22D is from a nonequivalent projection. To preserve areal relations in Figure 8.22A–C, the shapes of landmasses have been distorted (e.g., in Figure 8.22A and C, the northern portion of

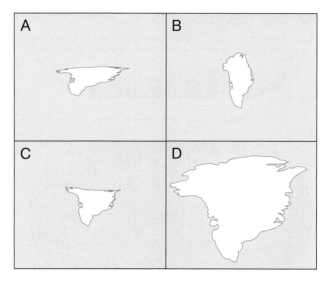

FIGURE 8.22 Greenland depicted on three different equivalent projections—(A) the Eckert IV pseudocylindrical, (B) the Lambert azimuthal equivalent, and (C) the Albers equivalent conic—and one nonequivalent projection, (D) the Miller cylindrical projection.

Greenland is expanded in an east-to-west direction compared to its representation on Earth).

To ensure that areas are preserved on equivalent projections, the SFs must be controlled so that each indicatrix contains the same area. For instance, Figure 8.23 shows the indicatrix on the Albers equivalent conic projection, which possesses two standard lines at 30° N and 45° N latitude. On this projection, the indicatrix appears as a circle along each standard line. At all other locations, the indicatrix takes on an elliptical shape, suggesting that the scale factors a and b change across this projection. In the upper latitudes, the ellipses display a contorted appearance, elongated in an east–west direction and compressed in a north–south direction. Recall that for equivalence, b must decrease proportionally as a increases. The difference in lengths of a and b indicates angular distortion at these locations. Regardless of the appearance of the ellipses, all are of the same size, ensuring that areas are preserved across the projection.

8.8.2 Preserving Angles

Conformal projections preserve angular relationships around a point by uniformly preserving scale relations about that point in all directions. This concept requires some explanation because the term conformal has been misunderstood and misused by many. It often has been interchanged with orthomorphic, meaning "correct shape," and this has conveyed the assumption that conformal projections preserve shapes of entire landmasses (both large and small). Unfortunately, conformal maps do not preserve the shapes of landmasses per se. Rather, the preservation of shapes is found only at infinitely small points.

Conformal projections preserve angular relations by ensuring that SFs change along a and b at the same rate. As Figure 8.24 shows, the indicatrix throughout the Lambert conformal conic projection is displayed as a circle, with a and b necessarily changing uniformly in size. At all locations, however, a and b remain equal to one another, permitting 2ω to maintain the same angle as found on the reference globe.

8.8.3 Preserving Distances

Scale cannot be preserved throughout a map projection. However, scale can be preserved on a projection in limited ways by varying the SFs during the projection process. Recall that equivalent and conformal projections manipulate SFs so that areas and angular relations are preserved throughout equivalent and conformal projections, respectively. On the other hand, equidistant projections preserve scale in the following manners: Along a single line from one point (usually placed at the center of the projection) to any other point, between two points,[7] or along standard line(s). As a consequence of preserving SFs in limited ways, scale can vary considerably on equidistance projections, depending on the size of the mapped area. For instance, the

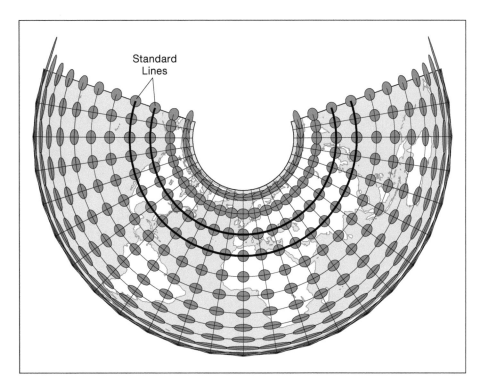

FIGURE 8.23   Indicatrices for the Albers equivalent conic projection, which has standard lines at 30° N and 45° N latitude (shown by the thicker black lines). Indicatrices are circles along the standard lines; for all other locations, indicatrices are elliptical in shape but retain the same area as indicatrices along the standard lines.

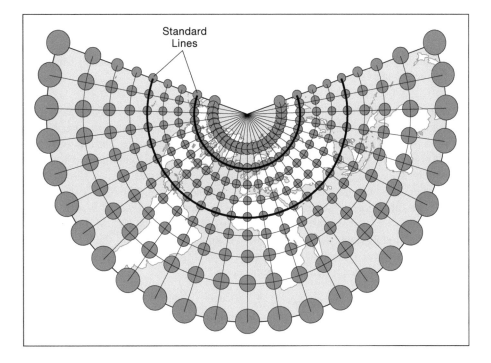

FIGURE 8.24   Indicatrices for the Lambert conformal conic projection, which has standard lines at 15° N and 60° N latitude. All indicatrices are circles (indicating no angular distortion), but the indicatrices change in size when moving away from the standard lines (indicating areal distortion).

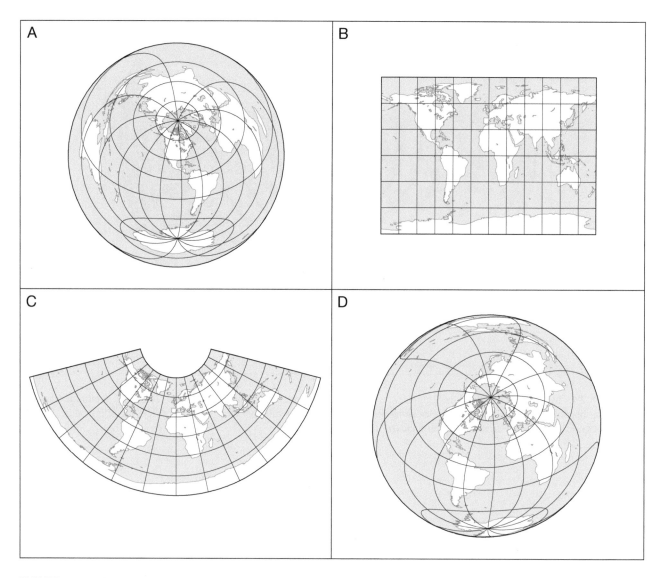

FIGURE 8.25 Four equidistant projections: (A) the azimuthal equidistant, (B) the equirectangular cylindrical, (C) the Euler equidistant conic, and (D) the two-point (or doubly) equidistant azimuthal.

azimuthal equidistant projection (Figure 8.25A) portrays the correct scale from the projection's center point along a straight line to any other point on the map. Locations further from the azimuthal equidistant projection's center contain increasing amounts of scale exaggeration. Examples of other equidistant projections include the equirectangular cylindrical and Euler (pronounced "oiler") equidistant conic (Figure 8.25B and C), where scale is correct along all meridians; and the doubly equidistant azimuthal (Figure 8.25D), which preserves scale along a straight line from either of two central points—in this case between Washington, DC, and London, England.

As a departure from preserving scale from one of the ways described in the previous paragraph, **variable scale map projections** present an interesting view of Earth. As the name suggests, the SFs change considerably across a variable scale map projection but can be controlled to yield interesting and useful results. For example, assume a map-maker wishes to focus on a certain area of the map and

show the relationship of that area to the larger geographical context. To accomplish this, a variable scale map projection would be appropriate. On this type of map projection, the central area to be emphasized is shown at a larger scale, and the peripheral area is shown at a smaller scale. For example, Figure 8.26 shows the Hägerstrand logarithmic projection centered at 96° W, 40° N (the U.S. Midwest). On this projection, the U.S. Midwest is shown at a larger scale compared to the rest of the landmass, whose scale becomes increasingly smaller toward the map's periphery.

8.8.4 Preserving Directions

Directions or azimuths can be preserved on azimuthal projections. In this kind of projection, directions are preserved from the center of the map to any other point on the map. When measuring azimuths from the center of an azimuthal projection, all straight lines drawn or measured to distant points also represent great circle routes. As such,

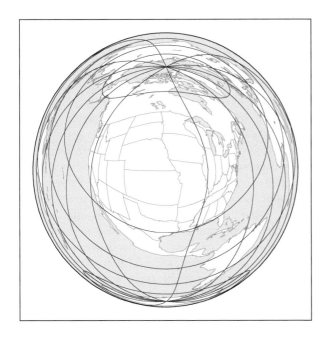

FIGURE 8.26 The Hägerstrand logarithmic projection with a 30° graticule spacing. Here, the appearance of the map is determined by taking the logarithmic distance from the center to all other map locations.

azimuthal projections have been used extensively for navigation and pinpointing locations.

Azimuthal projections are among the oldest known projections, some having been developed by the ancient Greeks. Figure 8.27 shows five common azimuthal projections: (A) Lambert equivalent, (B) stereographic conformal, (C) equidistant, (D) orthographic, and (E) gnomonic. Each azimuthal projection is conceptually developed in a similar fashion by placing a plane tangent to one point on the reference globe. As such, the SF on these azimuthal projections is only true at the center point and is greater than 1.0 at all other locations. It is useful to note that the azimuthal projections, in addition to preserving direction, each preserve another property: The stereographic is conformal, the Lambert is equal area, the azimuthal equidistant is equidistant only along the meridians (here, the SF is set to 1.0 so that one can measure the distance from the center of the projection to any other point), and the gnomonic shows all great circle routes as straight lines. The orthographic displays a spherical appearance suggesting the roundness of Earth, which is a correct representation of Earth if viewed from a distant point in space.

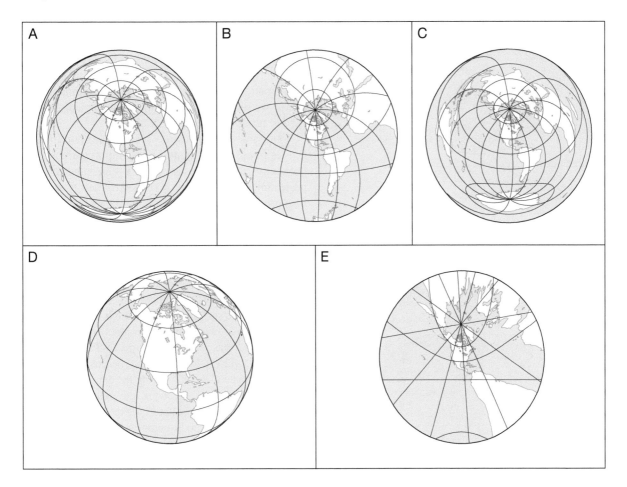

FIGURE 8.27 Five common azimuthal projections each centered at −96° W and 40° N: (A) Lambert equivalent, (B) stereographic conformal, (C) equidistant (scale is preserved from the center to any point on the map), (D) orthographic, and (E) gnomonic (allows all great circles to be drawn as straight lines).

8.8.5 Compromise Projections

In both equivalent and conformal mapping, the sizes and shapes of landmasses are often visually distorted, sometimes to the point of being unrecognizable. A solution to this problem is **compromise projections**, which manipulate the SFs so that the extreme angular and areal distortion found on equivalent and conformal projections are not present. Thus, a compromise projection strikes a balance between the distortion in area present on conformal projections and the angular distortion that is common on a purely equivalent projection. With compromise projections, even though no projection property is completely preserved, the combined areal and angular distortion is usually less than if a single property was preserved, and the resulting map generally gives a better visual representation of landmasses.

Probably the most notable compromise projection is the Robinson. Its popularity increased dramatically when the National Geographic Society in 1988 made the decision to replace the Van der Grinten I projection (Figure 8.28A)

with the Robinson (Figure 8.28B) as the projection for their world maps. Even though both the Van der Grinten I and the Robinson are compromise projections, the National Geographic Society believed the Robinson presented a better visual impression of the shape of the landmasses, and so it was employed up to 1998 when it was replaced with the Winkle tripel (Figure 8.28C), another compromise projection.

A variation on the compromise projection concept is the **minimum error projection**. As previously discussed, each projection property has the objective of preserving one of the spatial relationships (area, angles, directions, and distances) on the final map. Although this is often advantageous for specific mapping applications, in some cases, the preservation of one property often leaves considerable areas, angles, or distances exaggerated in other parts of the map. In some mapping applications, such as general reference maps, there is no need to preserve a specific property. Rather, the desire might be to lessen the consequences of having landmasses modified from how they appear on the reference globe, and so a minimum error projection is used.

8.9 SUMMARY

In this chapter, we examined the fundamental elements of map projections. We began with a discussion of the reference globe and the developable surface as conceptual aids in understanding how Earth's curved two-dimensional surface is projected onto a map. Today, computer software and mathematics are heavily involved in developing projections, and so we illustrated how mathematical equations could be used to produce a map projection.

Using software to create a projection is not difficult, but it does require understanding the importance of the basic projection characteristics of class, case, and aspect. The projection's class is directly related to the developable surface used in creating a projection—cones, planes, and cylinders produce conic, planar, and cylindrical projections. The case of a projection relates to the way in which the developable surface is positioned with respect to the reference globe. When the developable surface touches the reference globe, the tangent case results, where only one line (or point) of contact is shared between the two surfaces. When the developable surface passes through the reference globe, the secant case results, where contact is along either one or two lines. The projection's aspect relates to the way in which the developable surface can be positioned anywhere on the reference globe. Changing the aspect allows the projection to be centered at one of the poles (a polar aspect), along the Equator (an equatorial aspect), or somewhere in between a pole and the Equator (an oblique aspect).

Distortion is an inevitable consequence of the projection process and involves the alteration of scale about a point on a map compared to that same point on Earth. An important consideration in determining the distortion at each point on the map is the scale factor, which is the ratio of the local scale to the principal scale. Closely associated with

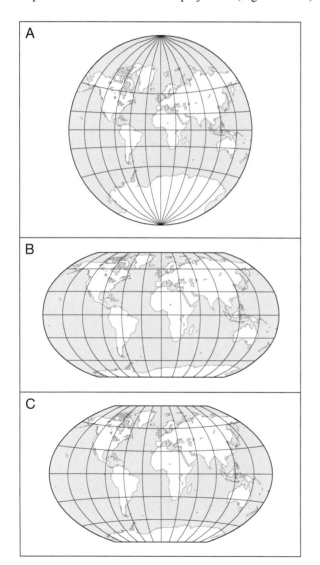

FIGURE 8.28 Three compromise projections: (A) Van der Grinten I, (B) Robinson, and (C) Winkel tripel.

the scale factor is Tissot's indicatrix, which is one of the more common graphical devices used to visualize distortion across the map. If the indicatrix is mapped across the projection, definite distortion patterns emerge that in part reflect the nature of the developable surface that was conceptually used to create the projection.

Important spatial relationships on Earth include the ability to measure distances, directions, areas, and angles. Each projection has certain properties that allow these spatial relationships to be preserved on a map. Preserving areal and angular relations is made possible through equivalent and conformal projections. Equidistant and azimuthal projections permit distances and directions, respectively, to be measured in specific ways across a projection. One consequence of preserving a specific property is that the other properties are often distorted (e.g., on conformal projections, areal relations are distorted). Therefore, compromise projections have been developed that do not preserve any specific property.

8.10   STUDY QUESTIONS

1. Select one of the map projections discussed in Chapter 8. For the map projection of your choosing, use a search engine such as Google to find out who developed it and for what purpose.

2. For the projection, you selected in question 1, find an example of a map that uses that projection. Indicate the nature of the data shown on the map that you've chosen, and discuss whether you think the given projection is appropriate for the data.

3. Map projections have been designed specifically to fulfill some purpose. An interesting purpose is to determine the qibla from anywhere around the world. The qibla is the direction one should face toward the Kaaba in the Sacred Mosque in Mecca, which is used by Muslims for various religious ceremonies. For a quick history of the qibla is and its importance, check out this article (https://amerisurv.com/2010/09/25/not-what-but-where-is-qibla/) before continuing. Anywhere outside of Mecca, a Muslim must face the appropriate direction and projections have been developed to meet this purpose. For example, a retro-azimuthal projection can be used to help a user determine the correct direction to face at any location toward Mecca. The Craig retro-azimuthal projection is discussed at https://www.mapthematics.com/ProjectionsList.php?Projection=241. Using this website, explain how the retro-azimuthal projection differs from an azimuthal projection that was discussed in Chapter 8.

4. The QiblaFinder is an online interactive website that takes advantage of Google Maps to visualize the qibla direction. Using this website (https://www.qiblafinder.org/), enter your location to see the angle that you would face toward Mecca. Then, notice the red path shown on Google Maps.

Why are there two directions provided? Determine what the tracing of this red path (not the reported angle) represents.

5. Map projections have occasionally been the subject of heated debates. One such map projection that has seen considerable debate has been the Peters projection. Here are two interesting articles that discuss the Peters projection, its development, and controversy:

https://www.jstor.org/stable/40571673

https://www.theguardian.com/education/2017/mar/19/boston-public-schools-world-map-mercator-peters-projection

Use information from these websites, present arguments for and against the use of the Peters projection. What are the advantages and disadvantages of this projection? Do you feel that the criticisms stated in the articles are justified?

6. The website Compare Map Projections (available here: https://map-projections.net/index.php) allows users to visually compare the graticule arrangement of two map projections. At the same time, the changes to landmasses that result from each map projection are also directly comparable. You will compare two pairings of map projections. For the first pairing, look at the options under Compare and select the via Selection Form option. On the page that appears, choose the Lambert cylindrical projection under Projection No. 1: header and then select the Robinson projection under Projection No. 2: header.

   a. For the first comparison, select the Lambert cylindrical and Trystan Edwards projections.

   b. For the second comparison, select the Putnins P1 and Robinson projections.

For each projection that you selected,

   • use the Silhouette Map feature and explain how each projection changes the appearance of the landmass. For example, which projection stretches or compresses the landmass the greatest in an east-west or north-south direction, and which locations experience this stretching or compression?

   • use the Tissot Indicatrix, 30° feature, and the indicatrices to explain the distortion pattern that is shown across each projection. In your response, explain where the greatest and least amount of distortion is located. Discuss the evidence you used from the indicatrices to support your description of the pattern distortion. Use information from the indicatrices and explain what projection property is being preserved on each projection and how you arrived at your conclusion.

7. Use the site Worldmapgenerator (https://www.worldmapgenerator.com/en/) to customize your own map projection. Once the site has loaded,

click on the Continue button. On the next screen, click on the Your Viewpoint option. Then, click on the green arrow in the lower right-hand corner of the next screen to appear. Once the new page loads, notice that six icons appear along the left-hand side of the window. Click on the second icon from the top (PROJECTION) which allows you to choose a projection. Use the following list of options. Choose Cylindrical as the type of projection and Miller as the projection. You should see the Miller projection appear in the right-hand portion of the window. Click on the top icon (CENTERING) which will allow you to change the projection's center. To recenter the map, use your mouse to either click on the map and move it or use the slider bars to specify a value for the lambda (longitude), phi (latitude), or gamma symbol (rotation). As a suggestion, you could recenter the map according to your present geographic location. Once finished with these steps, answer these questions:

- what is the projection's *class*?
- what is the *aspect* of this re-centered projection?
- what is the *property* of this projection?
- where is the least distortion located on this projection? What evidence from the map did you use to determine this?
- where is the greatest distortion located on this projection? What evidence from the map did you use to determine this?

NOTES

1 The naming convention of map projections adopted for this book follows the recommended guidelines set forth by the International Cartographic Association's Commission on Map Projections. The guidelines recommend that only proper names associated with a map projection be capitalized (http://www.kartografija.hr/proj-wiki/index.php/Projection_Names).

2 Some texts refer to the planar class of projections as azimuthal. We have chosen to use planar because it is easier for the student to relate to the developable surface concept (you can see that a plane is used).

3 In the strictest sense, a secant case for the planar class is not possible as the graticule would take on strange shapes and landmasses at the projection's center would not be displayed, but it is included here for conceptual completeness. See Richardson (1987) for a complete discussion.

4 In many cases, standard lines coincide with lines of latitude and are referred to as standard parallels. Note that standard line is a general term—a standard line can also coincide with small circles or meridians.

5 Inverse is part of the National Geodetic Survey's Geodetic Toolkit, which is available at https://geodesy.noaa.gov/TOOLS/Inv_Fwd/Inv_Fwd.html.

6 In mathematics, a unit circle radius equals 1 unit, which takes on special meaning when solving trigonometric functions.

7 In technical terms, any two points that are exactly opposite each other on Earth's surface are called antipodes. For example, a location at 39° N and −83° W would have its antipode located at −39° S and 97° E (180° − 83°).

REFERENCES

Mulcahy, K. A., and Clarke, K. C. (2001) Symbolization of map projection distortion: A review. *Cartography and Geographic Information Science 28*, no. 3:167–182.

Richardson, R. (1987) "Secant cases of azimuthal projections: Myth and reality." *Yearbook of the Association of Pacific Coast Geographers 49*:65–78.

Tissot, N. A. (1881) *Memoire sur la Representation des Surfaces et les Projections des Cartes Geographiques.* Paris: Gautier-Villars.

9 Selecting an Appropriate Map Projection

9.1 INTRODUCTION

To assist you in applying the concepts discussed in Chapters 7 and 8, this chapter presents an overview of methods for selecting an appropriate projection for a variety of thematic maps. We begin by reviewing several common approaches to selecting map projections. We then focus on a **projection selection guideline** developed by John Snyder because it offers the greatest utility to beginning map designers. Specifically, Snyder's guideline utilizes three separate tables that initially recommend projections based on the extent of the geographic area to be mapped: (1) world, (2) hemisphere, and (3) continent, ocean, or smaller region. Each table is further subdivided according to additional considerations made during the projection selection process.

To explore the utility of Snyder's selection guideline, we discuss how an appropriate projection can be selected for five sample data sets. For our first data set, we imagine creating a world map of literacy rates by country. In this case, we choose to create a choropleth map and thus need to select a projection that preserves areal relations. Here we introduce the advantages of using an **interrupted projection** for world maps, on which the graticule has been "cut," creating lobes that encompass specific geographic areas (e.g., South America).

The second data set focuses on creating a map showing the population distribution of Russia at the oblast and kray levels. In this case, we combine proportional symbols and dot maps so that we can depict both urban and rural populations, respectively. Here we need to select a projection that preserves areal relations so that the map user can properly compare dot distributions in different geographic areas.

The third data set examines migration from Europe and Asia to the United States. This map's objective is twofold. First, we want to show the general migration route over which immigrants traveled; second, we want the map user to see the spherical nature of Earth over which the immigrants traveled. To meet the first requirement, we choose graduated flow lines as the symbology, which enables the map user to visualize the number of immigrants from the different Asian and European countries, as well as the general routes taken. To meet the second requirement, it is important for the map to communicate the spatial proximity between the United States, Europe, and Asia; thus, we focus on planar map projections that present a hemisphere.

The fourth data set focuses on a small geographic area: Kansas. Specifically, the data set encompasses the paths of F4 and F5 tornadoes across Kansas during the past 50 years. We symbolize tornado paths as arrows—the length of the arrow indicates the distance the tornado traveled across the ground, the direction of the arrow indicates the general direction the tornado took, and the color of the arrow represents the severity of the tornado. Selecting projections for this small geographic area highlights the fact that regardless of the projection property chosen (conformal vs. equidistant), small geographic areas are equally well represented by most projections.

The fifth data set discusses the selection of an appropriate projection to map the direction, distance, and flight path from Fairbanks, AK to Seoul, South Korea. This data set calls for projections that preserve direction and distance, and a projection that preserves all great circles as straight lines (thus establishing an appropriate flight path). This data set shows how Snyder's guideline can be used to select a projection that is focused on the map projection property instead of the geographic area of focus.

After mapping the five data sets, we discuss five key criteria that should be considered when selecting a map projection for a thematic data set. First, in most cases, the mapmaker should select a projection with the lowest distortion. Second, amounts of distortion can be kept small by aligning the geographic area (or data set) under consideration with the standard line(s) or by positioning the map's center with the standard point. Third, as the amount of geographic area under consideration increases (e.g., mapping Kansas compared to mapping the United States), distortion becomes a more important consideration. Fourth, just because a projection has been frequently used (e.g., it has appeared in prominent atlases) does not mean that the projection is suitable for your specific application. Fifth, an often overlooked feature of map projections is their influence on the overall map design (e.g., choosing curved rather than straight meridians for a world map to make the map more visually appealing).

This chapter concludes with an example of how map projections have been integrated into the Web. The Projection Wizard is an interactive web-based tool that considers the geographic area to be mapped and the projection property and then recommends one or more candidate projections that a user can consider adopting when making a map.

9.2 LEARNING OBJECTIVES

- Utilize Snyder's map projection selection guidelines to select an appropriate map projection for a world data set.

DOI: 10.1201/9781003150527-10

- Utilize Snyder's map projection selection guidelines to select an appropriate map projection for a data set covering a hemisphere.
- Utilize Snyder's map projection selection guidelines to select an appropriate map projection for a data set covering a continent, ocean, or smaller region.
- Utilize Snyder's map projection selection guidelines to select an appropriate map projection for a data set requiring a special map projection property.
- Explain how selecting a projection for a specific mapping task may involve going beyond Snyder's guidelines and include a numerical analysis of distortion to assist in choosing a map projection.
- Explain one general consideration that is not addressed in Snyder's guidelines for selecting map projections and how it would impact the choice of a map projection for a world map.
- Explain the impacts of using the Web Mercator for map purposes involving measurements of distance and direction on a map.
- Discuss the advantages that Projection Wizard has compared to using Snyder's analog map projection selection scheme.

9.3 POTENTIAL SELECTION GUIDELINES

Selecting an appropriate map projection is one of the more complex tasks in cartography. In essence, when selecting a projection, a match must be made between the map's purpose and the various projection properties and characteristics. This task is difficult because of the many variables involved when creating a map, such as a map's scale, the amount of Earth to be mapped, the level of generalization, and the thematic symbology used. Similarly, any given projection has numerous characteristics (i.e., class, case, and aspect) and associated properties (i.e., equivalent, conformal, equidistant, and azimuthal). Rarely will a single projection have all the characteristics and properties necessary to satisfy all variables involved in the mapmaking process.

In an attempt to recommend projections for specific map purposes, various **projection selection guidelines** have been developed. Although the format of each guideline differs, the general purpose of each is to provide the mapmaker with one or more projections that can reasonably be applied to a particular purpose. For example, a simple selection guideline discussed by Frederick Pearson (1984) related the choice of projection to the range of latitude depicted on the map. He recommended cylindrical projections for equatorial regions lying 30° on either side of the Equator; conic projections for midlatitude regions between 30° and 65°, and planar projections for polar areas above 65°. The logic of Pearson's guideline rests on the fact that the location of the standard point or line(s) of a projection class lies within the geographic area recommended for that class, hence ensuring that distortion is low throughout the

region of interest. For instance, those areas lying along the Equator are matched with cylindrical projections that have one standard line coinciding with the Equator or two standard lines equally spaced on either side of the Equator. Although Pearson's guideline provides a starting point for selecting projections, it generally is not very useful, as it does not consider the map's purpose.

Arthur Robinson and colleagues (1995) described another simple guideline for selecting projections based on the relationship between projection properties and the intended map purpose. They recommended conformal projections for analyzing, measuring, or recording angular relationships, as in navigation, piloting, and surveying; suitable projections include the Mercator, transverse Mercator, Lambert conformal conic, and stereographic. They recommended equivalent projections when comparing areas across a map, a common purpose of thematic maps. For example, a dot map's primary goal is to visually compare different dot distributions across geographic areas, and this comparison is greatly facilitated if the geographic areas are represented in their correct proportions. Equivalent projections that Robinson et al. recommended include the cylindrical equivalent, Lambert's azimuthal, Albers and Lambert equivalent conic, and most equivalent pseudocylindrical projections. They also noted that when recording and tracking the direction of movement, planar projections are useful, including the orthographic (for showing Earth as if viewed from space), azimuthal equidistant (for showing correct directions and distances from the center of the projection to any other point), and the gnomonic (for showing that all great circles are represented by straight lines).

9.3.1 SNYDER'S HIERARCHICAL SELECTION GUIDELINE

Although the preceding approaches are conceptually useful, we feel that neither aforementioned guideline provides the level of detail nor the flexibility needed to adequately select an appropriate projection. Thus, we focus on a guideline developed by John Snyder (1994) that permits the mapmaker to select a map projection for many mapping purposes. Snyder presented a hierarchical list of suggested projections that are organized according to the region of the world to be mapped, the projection property (e.g., equivalent, conformal, equidistance, azimuthal), and the projection characteristic (e.g., class, case, and aspect). The advantage of a hierarchical approach is that the mapmaker can begin with a broad question (e.g., extent of geographic region to be mapped) and logically proceed to more detailed questions about the mapping situation (e.g., an oblique, polar, or equatorial aspect). In so doing, the mapmaker is led down a particular path through the hierarchy until an appropriate projection is recommended.

9.3.1.1 World Map Projections

Table 9.1 presents Snyder's map projection guideline for creating a map of the entire world. The mapmaker begins by choosing from among the following projection properties:

TABLE 9.1
Snyder's Map Projection Guideline Showing Projections for Mapping the World

Region Mapped	Property	Characteristic	Named Projection
World	Conformal	Constant scale along Equator	Mercator
		Constant scale along a meridian	Transverse Mercator
		Constant scale along an oblique great circle	Oblique Mercator
		No constant scale anywhere on the map	Lagrange
			August
			Eisenlohr
	Equivalent	Noninterrupted	Mollweide
			Eckert IV & VI
			McBryde or McBryde-Thomas
			Sinusoidal
			Other miscellaneous pseudocylindricals
		Interrupted	Hammer (a modified azimuthal)
			Any of the above except Hammer
			Goode's Homolosine
		Oblique aspect	Briesemeister
			Oblique Mollweide
	Equidistant	Centered on a pole	Polar azimuthal equidistant
		Centered on a city	Oblique azimuthal equidistant
	Straight rhumb lines		Mercator
	Compromise distortion		Miller cylindrical
			Robinson pseudocylindrical

conformal, equivalent, equidistant, straight rhumb lines, and compromise distortion. If we imagine that a *conformal projection* is desired, then we see that the mapmaker has several projections from which to choose according to the scale variation across the projection. For instance, when the map purpose requires that a constant scale be shown along the Equator, a meridian, or an oblique great circle, then the Mercator, transverse Mercator, and oblique Mercator projections should be chosen, respectively. If constant scale is not a requirement, then the following conformal projections are recommended: Lagrange, August, and Eisenlohr (Figure 9.1). The Lagrange projection shows the world within a circle—an advantage over the stereographic projection, which commonly shows only one hemisphere. The August and Eisenlohr projections, both epicycloidal[1]

projections, are useful because they are conformal throughout—on many conformal projections, the property of conformality is not maintained at the poles. A shortfall of the projections in Figure 9.1 is the rapidly increasing scale distortion from the projection's center. For instance, note that Antarctica's appearance in Figure 9.1 is greatly exaggerated (in reality, South America's area extent is roughly 19 million km^2, whereas Antarctica's is only 14 million km^2).

Looking at Table 9.1 again, we can see that if an *equivalent projection* is desired, there are three possible characteristics: noninterrupted, interrupted, and oblique aspects. An **interrupted projection** shows the graticule as "cut" along specific meridians, creating one or more lobes, as shown by the Eckert V pseudocylindrical projection in Figure 9.2. Each **lobe** has its own central meridian that creates an area

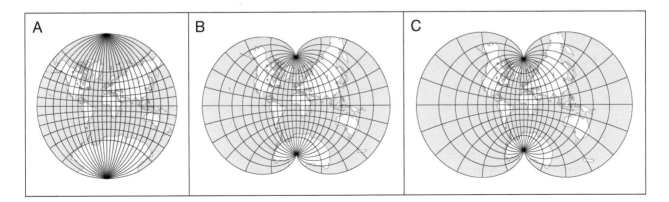

FIGURE 9.1 Different conformal projections of the world. The Lagrange (A) presents the world in a circle, whereas the August (B), and Eisenlohr (C) are epicycloids.

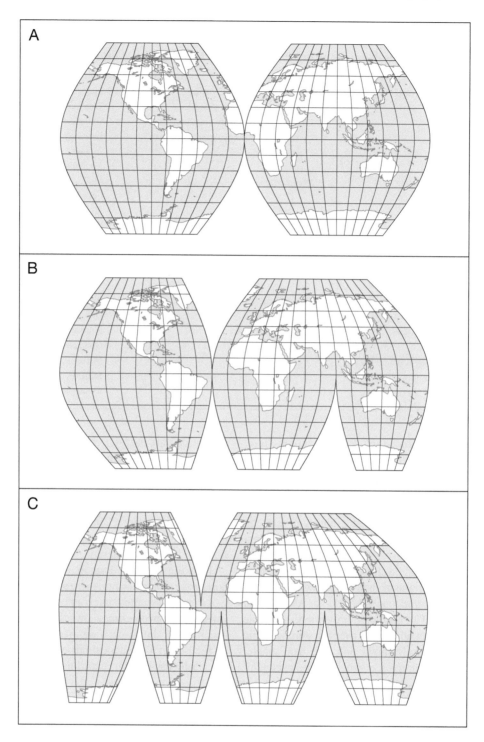

FIGURE 9.2   The Eckert V pseudocylindrical projection showing different placement of cuts.

of lower distortion than if the projection was not interrupted. The number of lobes and placement of each cut are determined by the mapmaker. In Figure 9.2A, two hemispheres are created (the cut is placed along the Prime Meridian and 180°). In contrast, in Figure 9.2B, the cuts are placed along the Atlantic and Indian Oceans so that the continents have lower distortion. In Figure 9.2C, the cuts are again placed along the oceans, but an additional cut is made in the southern Pacific Ocean. Although the cuts in parts B and C of

Figure 9.2 are positioned over water to provide continuity of the landmasses, the mapmaker, depending on the map's purpose, might choose to place the cuts over land, preserving the continuity of the oceans.

The oblique aspect (a projection with a center that is neither the pole nor the Equator) is warranted when geographic areas not aligned along the Equator or centered at the pole must be brought to the center of the projection. For example, Figure 9.3 shows the Briesemeister modified

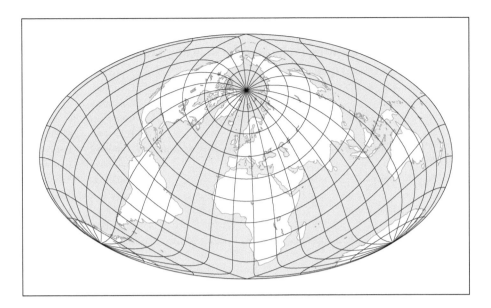

FIGURE 9.3   The Briesemeister modified azimuthal projection centered at 10° E and 45° N.

azimuthal projection,[2] which is equivalent and is centered over northwestern Europe at 10° E and 45° N, thus bringing northern Europe, the Norwegian Sea, and Scandinavia to the map's center.

Finally, in examining Table 9.1, we can see that certain projections are recommended for the projection properties *equidistant*, *straight rhumb lines*, and *compromise distortion*. Straight rhumb lines are a special property that only a few projections possess; for instance, on the Mercator projection, all rhumb lines (i.e., lines of constant compass bearing) appear as straight lines; this is obviously needed in certain mapping applications such as navigation. Some projections do not have a specific property and were described as compromise projections in Chapter 8. These projections typically have lower overall distortion than projections that preserve a single property. In many cases, compromise projections are developed specifically to match the shape of landmasses to their appearance on a globe, as in the Robinson projection (Figure 8.28B).

9.3.1.2   Map Projections for a Hemisphere

Table 9.2 presents Snyder's projection guidelines for maps of a hemisphere. In general, planar projections are suited for mapping hemispheres because a standard point is placed at the center of the projection, and the projection can be

easily recentered over any area of importance. Planar projections are interesting in that many of them combine the azimuthal property (showing directions correctly from the projection's center to all other points) with an additional property (e.g., the stereographic projection is both conformal and azimuthal), whereas most projections possess only one useful property (e.g., the Albers conic projection is exclusively equivalent). When showing a hemisphere, the following properties are options: *conformal*, *equivalent*, *equidistant*, and *global look*. For each property, note that a specific planar projection is recommended.

9.3.1.3   Map Projections for a Continent, Ocean, or Smaller Region

Table 9.3 presents Snyder's guidelines for maps of a continent, ocean, or smaller region. Note that this hierarchy includes an initial subdivision that distinguishes between the predominant directional extent of the landmass: east–west, north–south, oblique (not aligned in an east–west or north–south direction), and equal extent in all directions. This subdivision matches the position of the standard line(s) or points to the directional extent of the landmass, ultimately leading to a reduction of distortion for the geographic area under consideration. For instance, a map of Canada, which has a considerable east–west extent, will have lower distortion when the standard lines can be positioned to coincide with the east–west extent. In contrast, Antarctica, a landmass that is approximately of equal extent in all directions, will be more appropriately mapped with a single standard point as distortion increases radially with distance from this point.

After examining the directional extent of the landmass, the mapmaker should consider the location of the landmass to be mapped. For instance, given a landmass of predominant east–west extent on Earth's surface, we can see in Table 9.3 that the region can be located along the Equator

TABLE 9.2

Snyder's Map Projection Guideline Showing Projections for Mapping a Hemisphere

Region Mapped	Property	Named Projection
Hemisphere	Conformal	Stereographic conformal
	Equivalent	Lambert azimuthal equivalent
	Equidistant	Azimuthal equidistant
	Global look	Orthographic

TABLE 9.3

Snyder's Map Projection Guideline Showing Projections for Mapping a Continent, Ocean, or Smaller Region

Region Mapped	Directional Extent	Location	Property	Named Projection
Continent, ocean, or smaller region	East-West	Along the Equator	Conformal	Mercator
			Equivalent	Cylindrical equivalent
		Away from the Equator	Conformal	Lambert conformal conic
			Equivalent	Albers equivalent conic
	North-South	Aligned anywhere along a meridian	Conformal	Transverse Mercator
			Equivalent	Transverse cylindrical equivalent
	Oblique	Anywhere	Conformal	Oblique Mercator
			Equivalent	Oblique cylindrical equivalent
	Equal extent	Polar, equatorial, or oblique	Conformal	Stereographic
			Equivalent	Lambert azimuthal equivalent

or away from the Equator. If the landmass is located along the Equator, one of the cylindrical projections that have the Equator as a standard line is suitable (e.g., Mercator or cylindrical equivalent). The choice here is a function of whether conformality or equivalence is more important. If the landmass's east–west extent is located away from the Equator (e.g., the midlatitudes), then a conic projection with one or two standard lines is a suitable choice (e.g., Lambert conformal conic or Albers equivalent conic). Again, the choice is a function of whether conformality or equivalence is more important.

9.3.1.4 Map Projections for Special Properties

Tables 9.1 through 9.3 were organized so that the primary consideration in selecting a map projection was to first determine the geographic area of focus. The guideline presented in Table 9.4 departs from this organization in that the selection criterion is primarily based on a desired special property. The first special property is that all rhumb lines drawn on the projection appear as straight lines. From Section 7.3.3, a rhumb line is a path of constant compass bearing and is usually desired by ship or aircraft navigators who wish to chart their course. For this reason, the Location/Geographic Extent is specified as Ocean. It is

important to note, however, that, especially for aircraft travel, landmasses could also be considered as a valid location. Regardless, the named projection is the Mercator, which allows a rhumb line to be drawn as a straight line anywhere on the projection's surface. The second special property is that all great circle routes appear as straight lines on the projection. This special property is again useful to ship and aircraft navigators who wish to chart the path of the shortest distance between two points. In most instances, the charted course will fall within the geographic extent of a hemisphere. However, it is possible that a charted course, particularly flight paths, could extend beyond a hemisphere. To meet this special property, the gnomonic projection is recommended. The third special property deals with mapping tasks that require scale to be preserved. The location of the scale preservation is based on one of three centered locations: at a pole (or any other point), along the Equator, or somewhere away from a pole or the Equator. An equidistant projection is recommended in all three scale considerations. For correct scale centered on a pole or any other point, the azimuthal equidistant projection is appropriate; for mapping tasks that require correct scale along the Equator, the plate carrée projection is suggested; and for instances where correct scale is needed in the midlatitudes, the equidistant conic projection is suitable.

TABLE 9.4

A Portion of Snyder's Map Projection Guideline, Showing Projections for Special Properties

Special Property	Location/Geographic Extent	Named Projection
Straight rhumb lines	Ocean	Mercator
Straight great circle routes	Less than a hemisphere	Gnomonic
Correct scale	Centered at a pole or any other point	Azimuthal equidistant
	Centered along the Equator	Plate carrée
	Centered away from a pole or the Equator	Equidistant conic

9.4   EXAMPLES OF SELECTING PROJECTIONS

In this section, we utilize Snyder's guideline as a foundation for selecting projections for five fictitious mapping situations. For each mapping situation, we discuss the logic of Snyder's guideline and why the projections he recommended are appropriate. Additional insights into selecting projections not covered by Snyder's guidelines are also explored.

9.4.1   Mapping World Literacy Rates

For our first example, imagine that we wish to create a world map illustrating the distribution of literacy rates by country. We begin our selection process by considering the data, symbolization method, intended audience, and overall purpose for the map. The literacy rate is normally computed as the percentage of all people who can read. Although the underlying phenomenon of literacy rates is arguably spatially continuous in nature, we presume that we wish to examine the phenomenon at the level of countries for which data are available.[3] Given our desire to portray data at the country level, choropleth symbolization is suitable. We further presume that the purpose is to create a map that allows for a comparison of literacy rates by country around the world—noting the concentrations of high and low literacy rates and permitting us to examine the spatial pattern. We also presume that we wish to highlight those areas of the world with low literacy rates.

With our purpose in mind, we now consider Snyder's projection guideline. Because our focus is on the world, we start with Table 9.1. Our first step is to select the desired projection property. Given our choice of choropleth symbolization, a critical goal is to preserve the relative areas of each enumeration unit as each would appear on a globe. This sort of thinking leads us to choose an equivalent projection such as the Mollweide shown in Figure 9.4A.

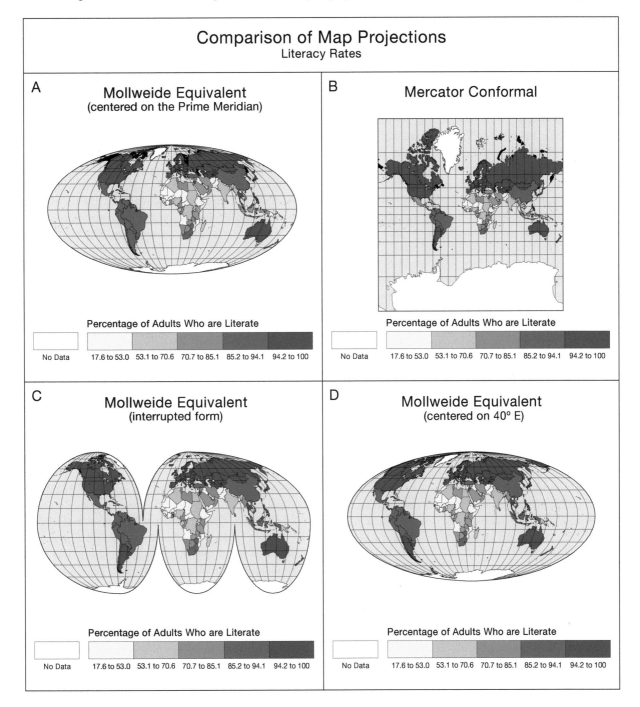

FIGURE 9.4 World literacy rates mapped on the (A) Mollweide equivalent projection, (B) Mercator conformal projection, (C) an interrupted Mollweide projection, and (D) a Mollweide projection that has been recentered along the Equator at 40° E.

For comparison, Figure 9.4B shows a Mercator conformal projection, which is inappropriate for this situation. Note how the Mercator projection emphasizes the high literacy rate category in the upper latitudes, suggesting that it dominates the world.

Once the property of equivalence is decided, the mapmaker has to choose the characteristic of the projection: noninterrupted, interrupted, or oblique aspect. As mentioned earlier, an interrupted projection reduces distortion across a geographic area by creating one or more lobes. Two factors to consider when deciding on the appropriateness of an interrupted projection are the intended audience and where the data exist on the map. Recall that interrupted projections remove the continuity of the oceans or landmasses, which could make the map difficult to comprehend. If the audience is assumed to have limited geographic knowledge (e.g., elementary school children), an interrupted projection would be inappropriate. On the other hand, if the audience is assumed to have good geographic knowledge (e.g., policymakers), then an interrupted projection would be acceptable. In deciding whether an interrupted projection is appropriate, a second factor is the distribution of the data across the map. In the literacy rate example, the data are concentrated over the land and not over the oceans. Because the data are concentrated over land, the continuity of the oceans can be sacrificed to create an interrupted projection. Therefore, the cuts can be placed over the oceans, with each region of interest being placed in a different lobe, as in Figure 9.4C. Comparing Figure 9.4C with the noninterrupted Mollweide projection in Figure 9.4A, we see that the cuts along the oceans create a discontinuity that might make the map difficult for those with limited geographic knowledge to interpret. However, it is also important to note that in Figure 9.4C, the landmasses toward the projection edges (northwest Canada, Chile, Argentina, and Australia) appear less distorted than they do in Figure 9.4A.

Another consideration is whether to specify an oblique aspect. Most world maps employ an equatorial aspect, where the map is centered at the Equator along a chosen central meridian—usually selected so that either the Atlantic or Pacific Ocean is central to the map. In cases where the geographic area of interest is not along the Equator, the projection's center must be moved to allow the geographic area of interest to be brought to the center of the map—an advantage of the oblique aspect. Recall that on most projections, lower amounts of distortion are usually found at a projection's center. This is especially true on world maps where, for example, polar areas suffer the greatest distortion when a point somewhere along the Equator is the projection's center. To determine whether an oblique projection is necessary, we need to again examine the geographic distribution of the data and determine if a particular geographic area of Earth has high or low literacy rates that are important and, if so, where they are. Based on our original statement, we

wish to focus on those areas with low literacy rates in an attempt to highlight their situation. In reviewing the literacy data, Central Africa and portions of Central Asia have the lowest literacy rates and should be brought to the center of the map. These geographic areas are situated along the Equator, and thus an oblique aspect is not called for. However, we should consider recentering the projection over the area of interest by selecting a new central meridian. In this case, the area of low literacy rates primarily falls near 40° E, and so a new central meridian is chosen to coincide with 40° E, as shown on the Mollweide projection in Figure 9.4D.

An additional thought on selecting an appropriate projection not directly addressed by Snyder is the impact that the projection has on the overall map design. One of those characteristics was whether the poles were represented by a point or a line. In Figure 9.5, the quartic authalic and Eckert IV equivalent projections are shown. Notice that the quartic authalic projection (Figure 9.5A) represents the poles as points, whereas the Eckert IV projection (Figure 9.5B) represents the poles as lines. On the quartic authalic projection, landmasses in the upper latitudes are compressed and difficult to see (especially at smaller scales) due to the convergence of the meridians to a point. On the other hand, on the Eckert IV projection, the landmasses in the upper latitudes are stretched in an east–west direction due to the nonconvergence of the meridians, making recognition of the landmasses more apparent, but distorting their overall shapes.

To help determine whether a point or a line should represent the poles, we can examine the map's geographic focus. Because our area of focus is Africa and Central Asia and not the upper latitude areas, it might make sense to use a projection where the poles are represented by a point. According to Snyder's guideline, the Mollweide, Boggs, and sinusoidal projections all represent the poles as a point. On the Mollweide projection, the meridians curve to the poles more gently than on the Boggs or sinusoidal, and the landmasses are not as compressed in the upper latitudes, thus making it a suitable choice for the literacy rate data.

Research has investigated the influence that map projections have on map design. For example, Bojan Šavrič and colleagues (2015) compared readers' preferences for world map projections based on several visual characteristics of the projections and their graticules between professional map makers/users and general map readers. It is interesting to note that results from their study reported that general map readers preferred projections whose poles are represented by points, but other factors (e.g., curves used to represent meridians) seemed to confound this preference. On the other hand, professional map makers/users favored poles represented as lines. This ambiguity suggests that, for example, the level of experience with making and using maps may infer a preference of how poles should be represented on world map projections.

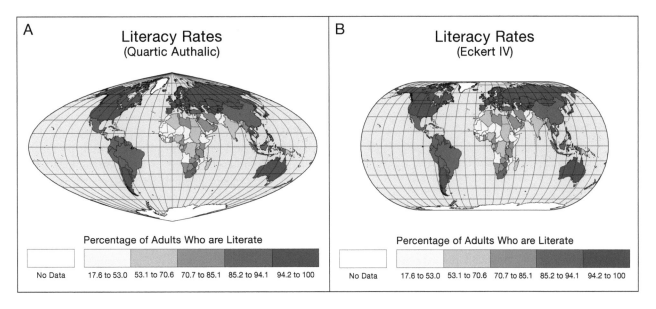

FIGURE 9.5   The literacy rate data shown on the (A) quartic authalic and (B) Eckert IV equivalent pseudocylindrical projections. The quartic authalic represents the poles as points, and the Eckert IV represents the poles as lines.

9.4.2   MAPPING RUSSIAN POPULATION DISTRIBUTION

For our second example, we consider mapping the population distribution in Russia for 1987. As in the previous example, we begin the process of selecting a projection by considering the data, symbolization method, intended audience, and overall purpose of the map. We assume that this map is to be included in a textbook for high school students who are taking an introductory course in geography. In this context, the map will serve a supporting role in the discussion of Russia and its cultural geography. The population can be thought of as discrete entities (individual people) but, in some cases, can be assumed to be more continuous in nature (e.g., a visit to New York City can give the impression that people are everywhere). Thus, in conceptual terms, population distribution is highly variable. In Russia, there are very dense urban centers and vast expanses of emptiness. To capture this variation, a combined proportional symbol–dot method seems appropriate, as it facilitates the observance of high population concentrations (urban areas) and expanses of low population (e.g., in central Siberia).

Starting with Snyder's guideline, we turn toward the continent, ocean, or smaller region selection guideline (Table 9.3). The first selection criterion to be determined is the directional extent of the landmass. Russia clearly has a greater east–west than a north–south extent (Figure 9.6): The east–west extent is approximately 5,163 miles (ranging from 27° E along the Baltic border to 170° W in the Bering Sea), whereas its north–south extent is 2,764 miles (ranging from 40° N along the Azerbaijan border to 83° N in the Arctic Ocean). Moreover, Russia is positioned away from the Equator—60° N is the approximate central latitude of the country's north–south extent. According to Table 9.3, this calls for a conic projection.

To select a specific conic projection, we need to consider the appropriate projection property. Snyder's guideline (Table 9.3) offers conformal or equivalent properties for the conic projections. Because we are creating a combined graduated symbol–dot map, an equivalent projection is necessary to ensure that map users will correctly interpret the relationship between the phenomenon represented by the dots and the geographic area in which it is contained. To demonstrate the importance of an equivalent projection, examine Figure 9.7, which shows a fictitious dot map for some enumeration unit. In Figure 9.7A, the enumeration

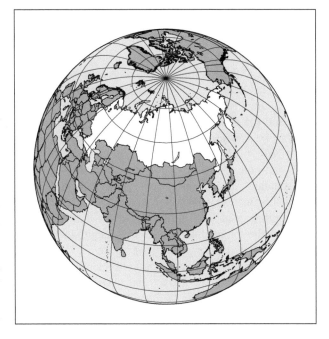

FIGURE 9.6   An orthographic projection showing the considerable east–west extent of Russia.

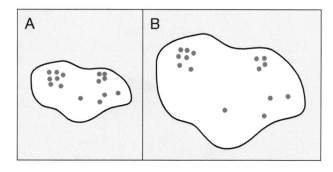

FIGURE 9.7 A fictitious data set represented through dot symbolization on (A) an equivalent projection and (B) a conformal projection.

unit is shown on an equivalent projection, whereas in Figure 9.7B, the same enumeration unit is shown on a conformal projection. Although the number of dots does not change (each map has 15 dots), there obviously is a difference in the density of dots over the given area.

Snyder recommended the Albers equivalent conic projection for thematic maps of landmasses that have a predominant east–west extent and are located away from the Equator because most conic projections have one or two standard lines that coincide with lines of latitude that can be placed to minimize the distortion of the mapped area. Although there are no universally accepted rules on the placement of standard parallels on conic projections, the overall goal is to select latitudes that will result in the lowest distortion over the mapped area. One approach is to place one standard parallel at one-sixth the latitudinal distance from the southern latitude limit of the map and the other standard parallel at one-sixth the latitudinal distance from the northern latitude limit of the map.[4] For example, Russia's southern latitude limit is approximately 41°43′ N (found between the border of Russia and Azerbaijan), and the northern latitude limit is approximately 81°53′ N (along the northern border of Ostrov Komsomolets—the northern most island of the Severnaya Zemlya islands in the Arctic Ocean). This latitude range is close to 40°10′. One-sixth of this latitudinal range is 6.7° or 6°42′. Given the latitudinal range and the general placement guideline presented earlier, the two standard parallels would be placed at 48°25′ N (i.e., 41°43′ +6°42′) and 75°11′ N (i.e., 81°53′ −6°42′).[5]

In addition to the selection and placement of suitable standard parallels, most conic projections have other parameters that can be modified. One parameter is the central meridian, which should be chosen so that it is at the center of the longitudinal range. The longitude limits of Russia are approximately 163° (ranging from 27° E, Ostrov Gogland—a small island in the Gulf of Finland—to 170° W, the boundary between Russia and Alaska). Given this range, the central meridian for Russia would fall at approximately 108°30′ E. Another parameter for most conic projections is the central latitude, which is the latitude that falls halfway between the northern and southern latitudinal limits.[6] Assuming the northern and southern latitude values

of 41°43′ N and 81°53′ N, respectively, the central latitude would be placed at 61°48′ N. Using these values for the Albers conic projection parameters, a combined graduated symbol–dot map of the population distribution of Russia is shown in Figure 9.8.

It is interesting to note that the Albers equivalent projection (Figure 9.9A) was introduced by Heinrich Albers in 1805 but was not frequently employed until the 1900s when it became the primary projection used by the U.S. Geological Survey for maps of the conterminous United States.[7] The standard parallels for these maps were placed at 29°30′ N and 45°30′ N and resulted in a maximum scale error throughout the map of no more than 1.25 percent (Deetz and Adams 1945). Other countries, such as Russia, also have utilized a single conic projection for the atlas mapping of their entire country. For example, the Bol'shoy Sovetskiy Atlas Mira relied heavily on the Kavrayskiy IV equidistant conic projection (Figure 9.9B) for atlases of the former Soviet Union (Snyder 1993).

9.4.3 Mapping Migration to the United States

For our third example, we focus on an appropriate projection for a series of maps (a small multiple) showing migration to the United States. We assume the small multiple will be included as part of an atlas describing a history of U.S. ethnicity and will be published by the U.S. Census Bureau for the general public. Although the atlas will have maps showing migration from all populated continents, this specific small multiple is designed to show the number of individuals who migrated to the United States from European and Asian countries during each 10-year period between 1960 and 2000. A goal is not to simply show the number of immigrants per country but to provide a general sense of the route by which the migration took place. Thus, we argue that the flow map is an appropriate symbolization method.

Turning to Snyder's selection guideline, we come across the problem that the mapped area does not neatly fit any of his predefined geographic categories. The closest geographic category that our mapped area fits into is a hemisphere, which is covered in Table 9.2. In this case, given the source areas for the immigrants (European and Asian countries) and their ultimate destination (the United States), the Northern Hemisphere is the logical geographic area to be mapped. Because the United States should figure prominently in the center of the map and because it is located neither along the Equator nor at a pole, the projection calls for an oblique aspect.

Examining Table 9.2, we must determine which projection property is desirable for our data. Conformality preserves angles at infinitesimally small points throughout the mapped area, but this property is not required for our data. Equivalent projections preserve areal relations throughout the mapped area. In the previous two examples (world literacy rates and Russian population distribution), there was a justifiable need for equivalent projections. However, the enumeration units of the migration data are not directly

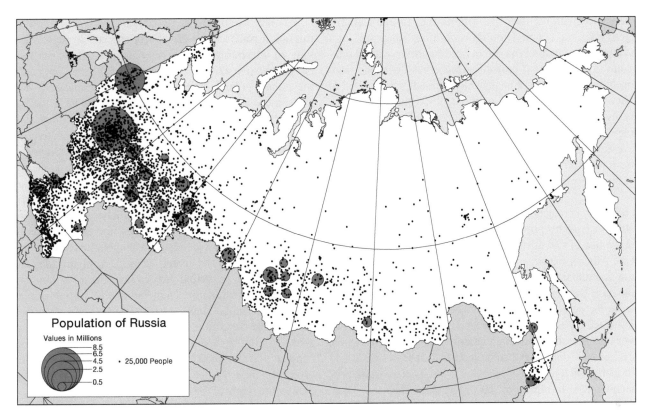

FIGURE 9.8   A combined proportional symbol–dot map showing the population distribution across Russia on an Albers equivalent projection. (Data Source: United Nations Educational Scientific and Cultural Organization (UNESCO), 1987, through United Nations Environment Program (UNEP)/Global Resource Information Database (GRID)-Geneva at https://unepgrid.ch/en/our-resources.)

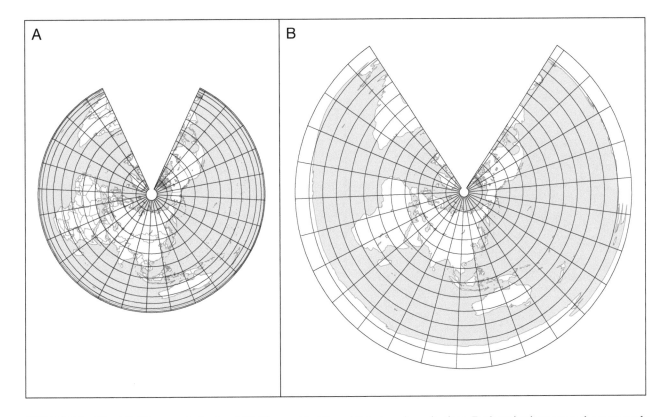

FIGURE 9.9   The (A) Albers equivalent and (B) Kavrayskiy IV equidistant conic projections. Both projections are at the same scale and have standard parallels at 47° N and 62° N.

associated with the flow line symbol (i.e., unlike a choropleth map, the enumeration units on a flow map are not a symbol); thus, there is no direct need to preserve areal relations. An equidistant projection preserves distances from one point to all other points but also is not directly related to our map purpose. Global look projections (the orthographic being among them) give the appearance of Earth as if looking down on it from outer space, which often produces an eye-catching map. However, global look projections are typically limited to showing a hemisphere and have considerable distortion near the map edges.

Although no specific property appears to be an obvious choice for a projection for the migration data, a mapping situation such as this provides an opportunity to be creative and make use of the flexibility that planar projections provide. One planar projection appropriate for the migration data is the azimuthal equidistant. Although discounted earlier on the basis that our data do not require directions or distances to be measured correctly, the projection does have interesting qualities that make it suitable for the migration data. For instance, the azimuthal equidistant projection has lower overall scale distortion than projections that preserve a single property, such as equivalence. As a result, equidistant projections tend to minimize the distortion of shapes of landmasses. For this reason, Maling (1992, 109) indicated that "equidistant map projections are often used in atlas maps, strategic planning maps and similar representations of large parts of Earth's surface." Another interesting quality of the azimuthal equidistant projection is that it displays the entire world (in its default configuration), whereas most planar projections are capable only of showing a single hemisphere. Although the azimuthal equidistant projection can show the entire world (Figure 9.10A), it can easily be cropped to focus on specific geographic areas, as shown in Figure 9.10B.

Another useful planar projection that would allow the migration routes to be effectively shown, as well as provide an interesting perspective view of Earth, is the vertical perspective azimuthal projection. With this projection, the amount of land that is shown can be manipulated—note that Figure 9.11A shows much more land within the border of the projection than Figure 9.11B, which focuses on North and South America. The vertical perspective azimuthal projection also can be used to zoom in on a portion of Earth's surface, which is similar to using a camera to zoom in on an object. For instance, Figure 9.12 illustrates a zoom for North and Central America; notice the dramatic effect this projection has in emphasizing the United States as the surrounding area fades into the background. Obviously, with a projection such as this, the scale changes dramatically from the center outward. Applying the vertical perspective azimuthal projection to the migration data, we produced the map shown in Figure 9.13. Here, the map's focus is the United States, which is clearly the convergence point of the flow lines. The shapes of the landmasses are shown similar to their appearance on the globe so as to not confuse potential map users about the geographic area. The projection also captures the user's attention by providing a compelling view of the data.

9.4.4 Mapping Tornado Paths across Kansas

In the fourth projection selection example, we focus on the paths that category F4 and F5[8] tornadoes took across Kansas from 1950 to 2000. As before, we begin by considering the data, symbolization method, intended audience, and overall map purpose. We assume that this map is to be created for weather experts who are interested in learning about counties in Kansas that historically have

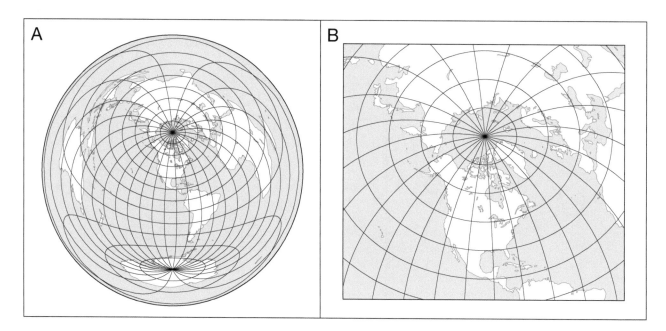

FIGURE 9.10   (A) The azimuthal equidistant planar projection showing the entire world and (B) the same projection cropped to focus on the migration data.

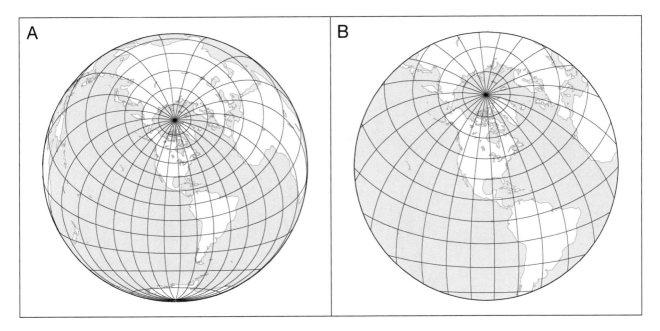

FIGURE 9.11   Two perspective views of Earth using the vertical perspective azimuthal projection.

been hit by tornadoes. Specifically, they are interested in using this map to visualize the precise path and distance along the ground each tornado took. This map requirement focuses on a geographic area to be mapped that is comparatively smaller than the other map examples we have considered.

Next, we consider which of Snyder's categories to use for our mapping situation. Because we are dealing with a smaller region for this data set, we use Table 9.3, which handles selecting a projection for a continent, ocean, or

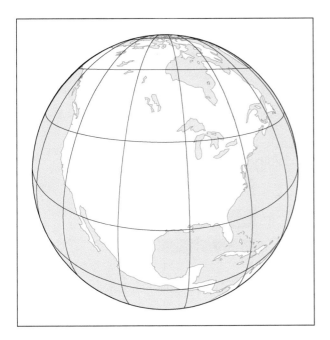

FIGURE 9.12   The vertical perspective azimuthal projection zoomed into North and Central America.

smaller region as the geographic area to be mapped. The first step in applying Table 9.3 is to determine the directional extent of the region to be mapped. In this case, Kansas has a considerably greater east–west than north–south extent. In fact, Kansas is about twice as long in an east–west extent (about 400 miles/644 kilometers vs. about 200 miles/321 kilometers). In the second step, we determine the location of Kansas. Lying between 37° N and 40° N, Kansas clearly is positioned away from the Equator. For the third step, we decide on the appropriate projection property. Notice that for an east–west extending landmass that is located away from the Equator, Snyder's table lists conic projections that are either conformal or equivalent. Because the data focus on the tracings of tornado paths across Kansas and their angles and distances, we can consider conformal and equidistant projections, respectively. We can eliminate equivalent projections because this property is not appropriate—we are not interested in preserving areal relations. Recall that conformal projections preserve accurate angles about individual point locations (e.g., where a tornado touches down) but do not preserve scale. Correct scale preservation is available through equidistant projections. However, equidistant projections restrict the way in which distances are preserved (e.g., all distances are correct from the center of the map to any other point). Thus, as in the previous example, there is no single suitable map projection for this particular mapping situation. Either a conformal or equidistant conic projection might be sufficient, but trade-offs are associated with each projection property and resulting distortion.

To help solve this dilemma between the conformal and equidistant properties, using various distortion measures, we will compare the differences in the amount of distortion between the Lambert conformal conic and the equidistant

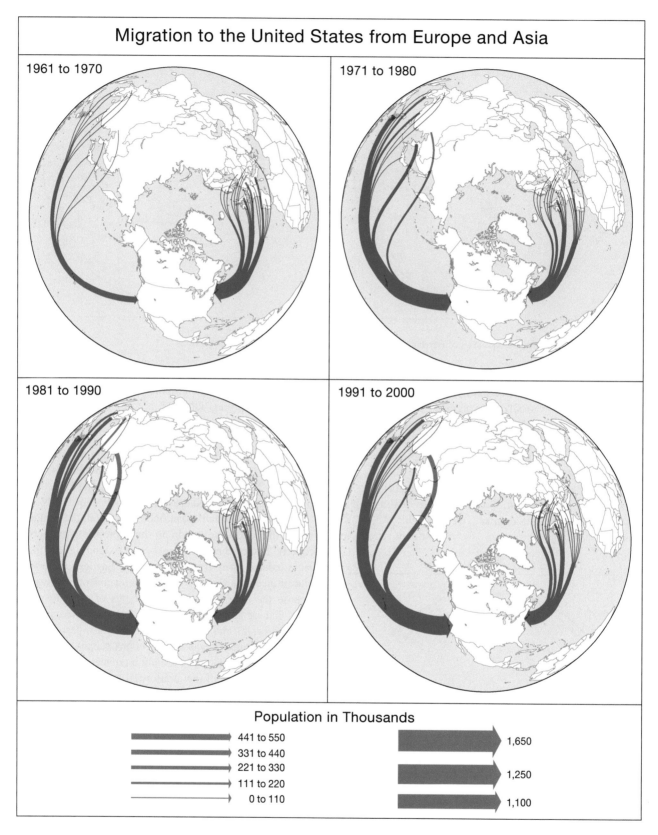

FIGURE 9.13 The vertical perspective azimuthal projection showing migration to the United States from Europe and Asia during different time periods.

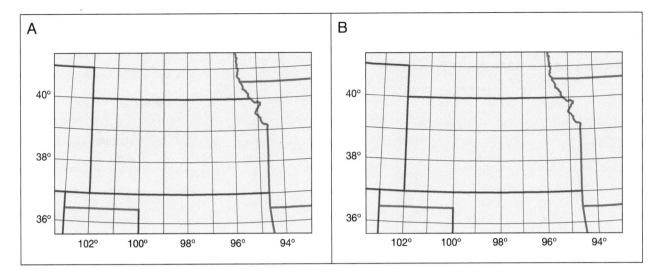

FIGURE 9.14   The (A) Lambert conformal conic and (B) equidistant conic projections of Kansas.

conic projections. To begin our distortion analysis, examine Figure 9.14, where we see the Lambert conformal conic (Figure 9.14A) and the equidistant conic (Figure 9.14B) projections for Kansas. The north–south latitudinal limits range from approximately 35° N to 42° N, and the east–west longitudinal limits range from about 90° W to 104° W. Each projection shares the same conic parameters of a central meridian placed at 98° W, a central latitude at 38°30′ N, and two standard parallels—one at 40°50′ N and the other at 36°10′ N.[9] Although the projections differ in their respective properties, there appears to be nothing that visually distinguishes the mapped appearance of Kansas on the two projections.

Now examine Figure 9.15, which shows Tissot's indicatrix for the same mapped area. Recall that the indicatrix appears as a circle of different sizes on conformal projections (i.e., circles are increasingly larger the greater their distance from the standard point or lines), but it

generally appears as ellipses on an equidistant projection. Figure 9.15A does indeed show the indicatrix as a series of circles on the Lambert conformal conic projection, but Figure 9.15B also portrays the indicatrix as a series of circles on the equidistant projection, suggesting that for this data set, the choice of projection property does not have a significant impact on the visual display of the data. We can determine whether the geometric shapes in Figure 9.15 are actually circles or ellipses by examining Table 9.5, which shows various distortion values for the Lambert conformal and equidistant conic projections. Beginning on the left side, the first two columns in the table list every degree of latitude from 33° N to 45° N along 98° W longitude. Also listed for each projection are a, the scale factor for the semimajor axis; b, the scale factor for the semiminor axis; area distortion; and angular distortion values.

In examining the table, note that the scale factors for a and b on the Lambert conformal conic projection at any

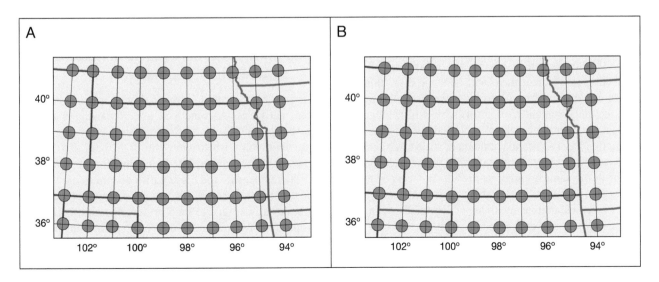

FIGURE 9.15   (A) The Lambert conformal conic and (B) equidistant conic projections of Kansas, showing Tissot's indicatrix at a 1° spacing.

TABLE 9.5

Values for *a*, *b*, Area Distortion, and Angular Distortion for the Lambert Conformal and Equidistant Conic Projections between Latitudes 33°N and 45°N

| Latitude | Lambert Conformal Conic | | | | Equidistant Conic | | | |
	a	*b*	Area Distortion	Angular Distortion	*a*	*b*	Area Distortion	Angular Distortion
33°	1.004	1.004	1.007	0.0°	1.004	1.000	1.004	0.21°
34°	1.002	1.002	1.004	0.0°	1.002	1.000	1.002	0.12°
35°	1.000	1.000	1.000	0.0°	1.001	1.000	1.001	0.06°
36°	1.000	1.000	1.000	0.0°	1.000	1.000	1.000	0.0°
37°	1.000	1.000	1.000	0.0°	1.000	0.999	0.999	0.03°
38°	0.999	0.999	0.998	0.0°	1.000	0.999	0.999	0.05°
39°	0.999	0.999	0.998	0.0°	1.000	0.999	0.999	0.05°
40°	1.000	1.000	0.999	0.0°	1.000	1.000	1.000	0.02°
41°	1.000	1.000	1.000	0.0°	1.001	1.000	1.001	0.0°
42°	1.001	1.001	1.002	0.0°	1.002	1.000	1.001	0.06°
43°	1.002	1.002	1.005	0.0°	1.003	1.000	1.002	0.14°
44°	1.004	1.004	1.008	0.0°	1.005	1.000	1.004	0.23°
45°	1.006	1.006	1.012	0.0°	1.007	1.000	1.016	0.34°

given latitude are equal, a characteristic of conformal projections. The scale factors equal 1.0 at the location of the standard parallels. On this projection, the areal distortion is not severe—ranging from 1.012 (a very slight exaggeration of areas) to 0.998 (a very slight compression of areas). There is no angular distortion on this conformal projection—only values of 0.0° are reported in the angular distortion column. For the equidistant conic projection, the values for *a* and *b* range from 0.999 around the standard parallels to 1.007 toward the top of the map. The amount of areal distortion is slightly more (ranging from 1.016 to 0.999) than on the Lambert conformal conic projection (1.012–0.998). There is also no more than 0.34° (approximately one-third) of a degree of angular distortion on the equidistant conic projection over this geographic area, which is rather low.

After reviewing the distortion values, we can see that neither projection preserves scale throughout the geographic area of interest, which means that regardless of which projection is selected, there will be some distortion present in any distance measurement taken from the map, but that the measurement error at this scale will be negligible. Because neither the Lambert conformal nor the equidistant projection has a substantial amount of scale error across the mapped area, computing distances on either projection will result in the same approximate result. In terms of angular distortion, however, the Lambert conformal conic projection preserves angular relations, which facilitates measuring the direction of each tornado's path.

Based on the preceding discussion, we selected the Lambert conformal conic projection to map the tornado data (Figure 9.16). We should stress again that given the small geographic area of interest, the intended map scale, and the tornado data set, the choice of projection is rather

inconsequential. However, we chose conformality over equidistance because on conformal projections, angular relations are preserved at every point, and the errors from any distance measurement on this projection are negligible. Note in Figure 9.16 that the data are represented as flow lines, where the origin of the flow line indicates the location where each tornado initially touched down. The flow line extent is the distance over which the tornado traveled, and the arrowhead indicates the direction of the tornado's path. The point of the arrowhead indicates where the tornado dissipated. Examining the pattern of tornado paths, we see that the general trend is from the southwest to the northeast and that the southcentral portion of the state appears to have a concentration of the most intense tornado activity for the time period.

9.4.5 Mapping a Flight Path from Fairbanks, AK to Seoul, South Korea

For our fifth example, we deal with several measurement activities. Assume you are preparing to fly from Fairbanks, AK to Seoul, South Korea. In preparation for your trip, you wish to view the flight path as if looking down on it from space, determine the direction your flight will take, measure the flight path's distance, and visualize the flight path's great circle route. To carry out these mapping tasks, you turn to map projections and their special properties. We will map the flight path using mapping software ArcGIS Pro. To compare the results obtained using mapping software, we will also map the flight path and determine the distance using Google Maps.

Reviewing Snyder's guideline, you might consider using either the "continent, ocean, or smaller region" or

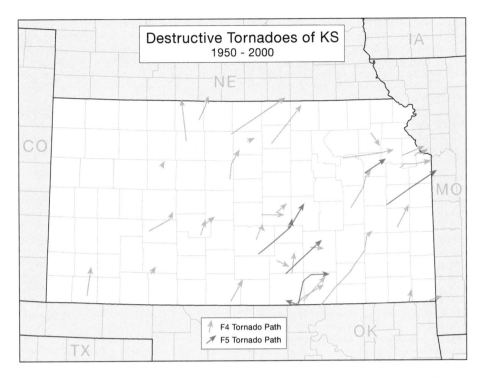

FIGURE 9.16 The Lambert conformal conic projection showing the paths of F4 and F5 tornadoes across Kansas from 1950 to 2000. (Data Source: National Weather Service, Storm Prediction Center, https://www.spc.noaa.gov/gis/svrgis/.)

the hemisphere category. This mapping example, however, does not conveniently fit into either of these categories causing difficulty in deciding which table to use. Furthermore, this example demonstrates the difficulty in categorizing a geographic extent to help select a map projection when the map purpose involves a specific kind of measuring task. Using this mapping task, we will explore how Snyder's selection guidelines can be used to select a map projection based primarily on the required property for a measurement task rather than the geographic area of focus.

9.4.5.1 Mapping the Flight Path from Space

We begin with choosing a projection that allows you to map the flight path between Fairbanks and Seoul as if looking down on it from space. Armed with automatic flight control and navigation systems that control altitude and azimuth, aircraft fly from their origin to their destination along a great circle, which is the shortest distance between two points on a curved surface. For this mapping task, we need a projection to help visualize the spatial location of the two cities and the flight path that connects them. Looking down on Earth from space, one could expect that these two cities are located in a hemisphere (both cities are located in the Northern Hemisphere), and thus we can refer to Table 9.2 for assistance in our selection. According to the available projection properties listed in Table 9.2, the desirable property is a "global look," as this will show a hemisphere and allow the spatial location of the flight path to be viewed (as if looking down on it from space). For this property, Snyder

recommends the orthographic projection, which is often used to show Earth's features as if looking down on them from space. Figure 9.17A shows the orthographic projection centered on Fairbanks, creating an oblique aspect. Note that the visual center of the map in Figure 9.17A is different than the projection's center. As explained in Chapter 8, centering a planar projection on a point allows the various properties associated with a given planar projection to be applied to each measuring task. The solid red line connects the two cities showing the flight path.

9.4.5.2 Mapping the Flight Path's Direction

The orthographic projection, like all planar projections, is azimuthal. Thus, we can consider how this projection can assist in determining the direction or azimuth of our flight. From Section 8.8.4, you learned that azimuthal projections preserve directions from their center to any other point. Recall also from Chapter 7 that azimuth is a direction measured as an angle clockwise from the North Pole to a point. In Figure 9.17A, we have added a thin black line connecting Fairbanks and the North Pole; we can utilize this line to determine the azimuth by computing the angle between it and our original line connecting Fairbanks and Seoul. We arrive at an angle of 284°. This is the true azimuth or direction the plane will fly between the two cities. It is important to note that while the orthographic projection is useful for preserving directions, it is not equidistant. To demonstrate this, you measure the distance between these two cities and arrive at 5,200.1 kilometers. The true distance between the two cities, as we will see shortly, is 6,087.8 kilometers.

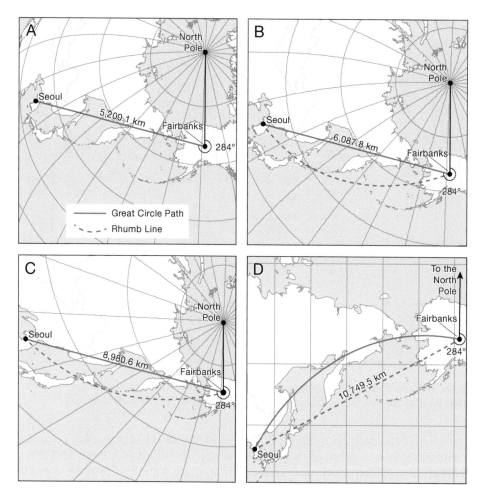

FIGURE 9.17 Four projections depicting the path between Fairbanks, AK and Seoul, South Korea: orthographic (A), azimuthal equidistant (B), gnomonic (C), and Mercator (D). The reported distance values shown in each figure were measured using ArcGIS Pro v 2.8 cast on the WGS84 datum. The maps are not set to the same scale.

9.4.5.3 Mapping the Flight Path Distance

For the next mapping task, you want to measure the distance between Fairbanks and Seoul along the path that the flight will take. Here, we can explore the utility of Table 9.4, which focuses on selecting a projection according to the preservation of a special property. We see that Snyder lists correct scale as one of the special properties, and this property is needed for your task. As you learned in Chapter 8, scale is highly variable across a projection's surface but can be controlled in specific ways. According to Section 8.8.3, you learned that equidistant projections preserve distances along all meridians or from a single point to all other points. In this case, the two cities of your flight are not aligned along a meridian, but the geographic area of focus can be centered over a point (in this case, the origin of your flight – Fairbanks), and the distance can then be measured to any other point (Seoul). From Table 9.4, we see that the location/geographic extent for our mapping task matches the "Centered at a Pole or other point" option. For this option, Snyder recommends the azimuthal equidistant projection. In Figure 9.17B, the azimuthal equidistant projection is shown with a thicker solid line connecting

Fairbanks and Seoul. Carefully measuring the distance between these cities directly on the map, you arrive at 6,087.8 kilometers. This corresponds to the true distance between these two cities. Comparing this value to the distance shown on the orthographic projection confirms that the orthographic projection does not preserve distances. Since the azimuthal equidistant projection is azimuthal, you measure the azimuth between the two cities. You end up with the same 284° value as you found with the orthographic projection.

9.4.5.4 Mapping the Great Circle Flight Path

Continuing with our examination of the flight path, you want to show the actual great circle path the aircraft will fly between Fairbanks and Seoul. When examining the solid red lines representing the great circle paths in Figure 9.17, it is important to note that the projection class and its property influence the overall appearance of the "length" of each path. For example, by design, the Mercator projection shows all rhumb lines as straight. As a consequence of this design, great circle paths appear as curved lines longer than the corresponding rhumb line. In comparison, the gnomonic and azimuthal equidistant projections show all great circle paths from the map's center as straight lines. On these two projections, rhumb lines, however, appear as curved lines.

Returning again to Table 9.4, there is a special property that shows all great circles on a projection as straight lines: Snyder suggests that the gnomonic azimuthal projection is suitable for mapping great circle routes as straight lines (which was also discussed in Section 8.8.4). In Figure 9.17C, the gnomonic projection is displayed with a solid red line drawn between the two cities and a black line connecting Fairbanks to the North Pole. From the line drawn on the projection between the two cities, you note that your plane will fly over the Bering Strait, Eastern Siberia, Sea of Okhotsk, and Sikhote Alin Mountains, before reaching Seoul. In order to achieve this representation, however, the gnomonic projection greatly distorts distances. If you measure the distance between the two cities on this projection, you find that the distance is 8,980.6 kilometers, which is considerably greater than the true distance of 6,087.1 kilometers.

9.4.5.5 Mapping the Rhumb Line

Although not part of your original mapping task, mapping the rhumb line or path of constant compass bearing is a common navigational practice. We will explore this additional task to further highlight the utility of using Snyder's selection guideline that focuses on a special property. Table 9.4 shows that representing straight rhumb lines on a map is possible using a special property that is preserved by the Mercator projection. You learned in Section 8.8.2 that conformal projections preserve angular relationships throughout the map and that the Mercator is a conformal projection. In addition to being conformal, the Mercator's special property is to show all rhumb lines as straight on the projection's surface.

To demonstrate the utility of the Mercator as a navigational tool, if you measure the angle between the two cities on the Mercator using the North Pole as the reference, you will find that the angle between them is 284° which happens to be the same value as found on the azimuthal projections. Figure 9.17D shows this angular measurement on the Mercator projection. In order to preserve angular relations on the Mercator projection, the ability to measure distances on the map is poor. If you measure the rhumb line distance between the two points on the Mercator (represented as the dashed red line), you arrive at a value of 10,749.5 kilometers which is considerably larger than the true distance found on the azimuthal equidistant projection.

The additional property that the Mercator possesses is that any rhumb line shown on the projection is represented as a straight line, as shown in Figure 9.17D. If you look closely, the rhumb line path (shown on the projection as a dashed red line) crosses each meridian at a constant angle. However, the great circle path (shown on the projection as a solid curved red line) constantly changes angles as it intersects each meridian. By comparison, the gnomonic projection shows the reverse relationship where great circles are shown as straight lines and rhumb lines are curved (see Figure 9.17C).

In the era before onboard electronic navigation systems (e.g., auto-pilot) and Global Navigation Satellite System (GNSS), charting a flight path was a more involved process. The process began with the navigator plotting the intended flight path on the gnomonic projection that correctly showed the shortest distance path. While navigating a great circle has always been the preferred route since it is the shortest distance, following the great circle was not practical as navigators would need to constantly alter the aircraft's course. Therefore, once the great circle path was drawn on the gnomonic projection, the navigator would approximate that same path on the Mercator by a series of short straight rhumb lines. This process would result in a lengthening of the travel distance but would simplify the navigational task by minimizing the number of alterations to the navigated path. On-board aircraft navigation systems and GNSS have relegated this manual process to a historical practice.

9.4.5.6 Mapping the Flight Path Using Google Maps

We have seen in this fifth mapping example that each projection offered specific properties that assisted in the tasks associated with mapping your flight. To carry out these various mapping tasks, you would likely need GIS software to create each of the different projections and then carefully use the provided tools to carry out the different measurement activities. While this is certainly possible, many individuals do not have access to GIS software and would likely turn to the Web for assistance. One such popular web-based mapping tool is Google Maps. Let's use Google Maps to measure the distance of your flight path, compare it to the true distance value obtained previously, and point out a caution when using online Web mapping services.

We saw that the Mercator projection offers advantages of preserving angles and rhumb lines but is not well suited for measuring distances. Yet, starting with Google Maps, the Mercator projection has been incorporated into many online Web mapping services. Imagine that you decide to measure the distance of your flight path using Google Maps. You carry out the measurement task, and the reported value is 6,073.27 kilometers (see Figure 9.18). It is interesting to note that Google Maps and other online mapping services do not report the planimetric distance (the distance you would obtain if you measured the straight-line distance between two points directly on a map). Rather, as shown in Figure 9.18, the computed distance is determined along the great circle path, which initially seems appropriate. However, when you compare the 6,073.27-kilometer value with the correct 6,087.8-kilometer value computed previously, you see a numeric difference.

The difference in the distance values arises in part from the way in which the coordinate values are matched to the projection equations used by Google Maps. Remember from Chapter 7 that Earth's shape can be described as either a sphere or an ellipsoid. A spherical model assumes that

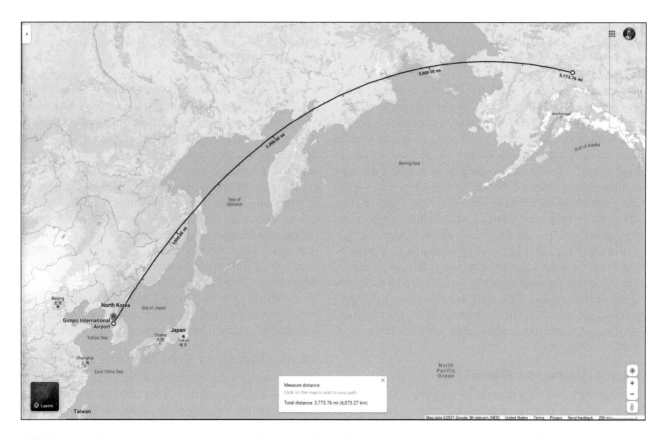

FIGURE 9.18   The great circle distance between Fairbanks, AK and Seoul, South Korea, as computed using Google Maps.

any point on Earth's surface is a constant radius, while an ellipsoidal model has a constantly changing radius. From Section 7.4.2, you also learned that, depending on the chosen Earth model, the same latitude location on that model's surface can be described using geodetic and geocentric latitude. Thus, when coordinate values are collected, the user can choose between Earth modeled as a sphere or ellipsoid, where ellipsoidal coordinates are more accurate reflections of latitude locations.

Map projection formulas are also written to reflect coordinates captured under a spherical or ellipsoid model. For example, if coordinate values are based on an ellipsoid model, then the corresponding ellipsoidal form of the projection equations should be used to calculate the x and y plotting coordinates (see Section 8.5). The chosen Earth model and form of projection equations used can create a difference in distance measurement. This difference can appear in Web mapping services such as Google Maps and result in a mismatch between the coordinate values based on the selected Earth model and the form of the projection equation implemented to calculate the x and y plotting coordinates. Google Maps and other mapping services use coordinate values taken from an ellipsoidal Earth model but compute the x and y plotting coordinates using the spherical form of the Mercator projection equations – producing the Web Mercator. One reason this is done is that the spherical form of projection equations is computationally simpler resulting in faster calculations and computer processing time.

While mixing ellipsoidal coordinates and spherical equations is not a suggested practice for all mapping tasks, the consequences of doing so are miniscule for the rather low-accuracy measurement requirements of a typical user of online Web mapping services. In most cases, individuals use online mapping services for large-scale mapping tasks (e.g., zooming in to neighborhoods or getting driving directions over very small distances) and do not need to worry about differences between the Mercator and the Web Mercator. Of course, for mapping tasks requiring a high degree of measurement accuracy (e.g., for navigational tasks), the ellipsoidal form of a given projection should be implemented.

9.4.6   Discussion

The five projection examples each provided interesting opportunities to see how map projections can be selected. In summary, there are three key objectives that we would like to emphasize. First, in most cases, it is the objective of the mapmaker to select a projection for which distortion is the lowest. Of course, the kinds of distortion that should be considered depend on the purpose of the map (e.g., for many thematic activities, equivalent projections are suitable). Second, amounts of distortion can be kept small by aligning the geographic area (or data set) under consideration with the standard line(s) or by positioning the map's center with the standard point. Third, as the extent of the geographic area under consideration increases (e.g., from

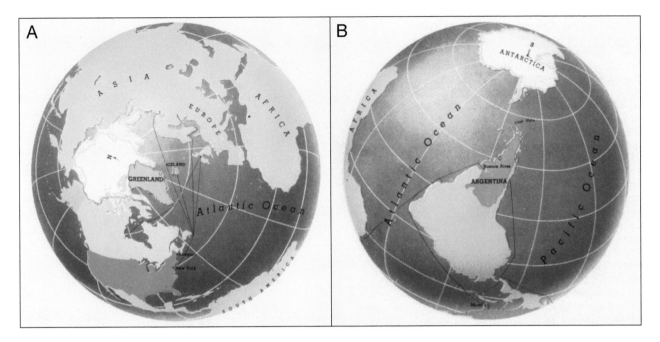

FIGURE 9.19   The orthographic projection showing (A) North America and its proximity to Europe and (B) a view of South America and Antarctica. (From pp. 52–53 of Harrison, R. E. (1944) *Look at the World: The Fortune Atlas for World Strategy.* New York, NY: Alfred A. Knopf.)

country to hemisphere), distortion becomes a more important consideration. For example, we noted in the map of Kansas tornado paths that the variation in the scale factors was very small on either projection. Had we, however, used the Lambert conformal or equidistant conic projection to map the entire world, we would have noted extreme distortion of the scale factors toward the edge of each projection, especially on the Lambert conformal conic. Thus, when mapping smaller geographic areas such as a state, variation in distortion values is less of a concern than for larger geographic areas.

In addition to these three key objectives, there are two additional important considerations to keep in mind when selecting map projections. First, the past popularity of a projection's use (e.g., in prominent atlases) does not mean that the projection is suitable for your specific application. In many cases, a projection's historical usage was dictated by its availability and not its suitability. One distinctive example is the overuse of the Mercator projection as the basis for thematic maps (e.g., showing world population distribution). Second, an often-overlooked element of map projections is their influence on overall map design. Although this idea is more difficult to grasp, there are clear enhancements that projections can make with regard to map design. For example, Richard Edes Harrison, in his *Look at the World* atlas, used the orthographic projection to present some powerful images of World War II illustrating spatial relationships between the United States and the rest of the world. For instance, Figure 9.19A shows the spatial proximity of North America to Europe, and Figure 9.19B illustrates the proximity of South America and Antarctica. Harrison's use of the orthographic and other azimuthal

projections throughout *Look at the World* was a significant departure from atlases that relied on the Mercator projection. By using azimuthal projections, especially the orthographic, Harrison attempted to show how spatially connected the continents are.

9.5   WEB-BASED INTERACTIVE MAP PROJECTION SELECTION

Bojan Šavrič and colleagues (2016) developed Projection Wizard (https://projectionwizard.org/) as a freely available web-based interactive tool to help users select map projections. Projection Wizard is based on the organizational structure of Snyder's hierarchical map projection selection guidelines that were presented earlier in this chapter. Figure 9.20 shows the interface of the Projection Wizard's main window. Along the left-hand side of the window is where the user specifies the desired projection property of equal-area, conformal, equidistant, or compromise. The user can enter the latitude and longitude of the geographic extent that corresponds to their mapped area, or they can use the box shown on the map to select the bounds of the desired area. In the lower left-hand corner, one or more named projections are listed that correspond to the chosen projection property and the geographic area that has been selected. In the lower right-hand corner, a map appears that reflects the recommended map projection and parameters.

Figure 9.20 shows the recommended projections for Section 9.4.5.3. In this example, the blue box approximates the latitude and longitude extents of the mapped area (Fairbanks, AK and Seoul, South Korea). The selected projection property is Equidistant. The two recommended

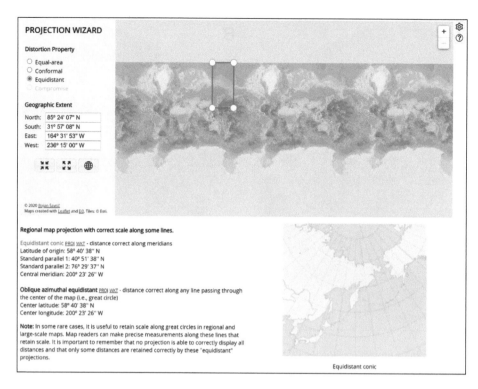

FIGURE 9.20 The recommended map projections from Projection Wizard for the Fairbanks, AK to Seoul, South Korea flight example. (Courtesy of Bojan Šavrič.)

projections include the equidistant conic and oblique azimuthal equidistant. The equidistant conic would not be very useful for calculating the distance between the two cities as distances on this projection are only preserved along the meridians. Since the two cities are not aligned along the same meridian, the correct distance could not be calculated on a map that is cast using this projection. The oblique azimuthal equidistant projection is the better choice as this projection does allow accurate distance measurements from the projection's center. One modification would be needed, however, to measure accurate distances between these two cities. Presently, the reported center latitude and longitude values shown in Figure 9.20 are the latitude and longitude values positioned midway along the great circle path connecting two adjacent corners of the blue box and passing through its center. To accurately measure the distance between the two cities in our example, the user would have to adjust the center latitude and longitude values to coincide with Fairbanks, AK.

9.6 SUMMARY

In this chapter, we examined several commonly referenced map projection selection guidelines, noting that they are limited in helping novice cartographers through the often-confusing process of selecting a projection. In contrast, we feel that Snyder's guideline is well organized and presents a logical hierarchy for selecting an appropriate projection. Snyder's guideline begins with a focus on the geographic area to be mapped: world; hemisphere; and continent, ocean, or smaller region. Once the geographic area to be

mapped has been selected, projection properties and characteristics are closely examined.

We relied on Snyder's guideline to assist in the selection of appropriate map projections for five thematic data sets. The first data set focused on world literacy rates. In selecting a projection for this data set, we focused on the world category, chose to preserve areal relationships (to ensure that areas were preserved in their correct proportions for the choropleth symbolization method and that no area visually dominated the map), and chose an interrupted projection (to reduce distortion by placing cuts over the oceans where data were not present). The result was an interrupted Mollweide projection. In addition, we considered whether the poles should be represented by points or lines, a topic not directly addressed by Snyder's guideline.

In the second data set, we looked at the population distribution of Russia. In this instance, we created a combined proportional symbol–dot map to represent the spatial variation of Russia's population. Because the geographic area of interest was Russia, we utilized Snyder's guidelines for a continent, ocean, or smaller region. Noting that Russia has a considerable east–west extent, we selected a conic projection. Moreover, we chose the Albers equivalent conic projection to enable the reader to properly compare areas of differing densities.

The third data set focused on migration patterns from Europe and Asia to the United States. The objective of this map was to show the routes by which immigration took place, as well as the spatial proximity between the sources of migration and the United States—essentially encompassing a hemisphere. To help us select a projection for this data set, we utilized Snyder's table listing various planar projections suitable for representing a hemisphere. We noted that planar projections offer a considerable range of design options not found in other map projection classes. In the end, we selected the vertical perspective azimuthal projection to highlight the migration paths and the spatial proximity of Europe and Asia to the United States, as well as the spherical nature of Earth.

The fourth data set involved a historical look at the most destructive tornado paths across Kansas. Here, we utilized Snyder's selection guideline for a continent, ocean, or smaller region and selected a conic projection. However, the small size of the geographic area prompted us to investigate the nature of the distortion pattern across the mapped area. In the distortion analysis, it became clear that neither an equidistant nor a conformal conic projection offered any substantial benefit because such a small geographic area was under consideration. Although we did choose the conformal conic as the preservation of angles matched one of the requirements of the map purpose.

The fifth data set required specific visual and measurement tasks to be completed for a flight between Fairbanks, AK and Seoul, South Korea. These tasks included viewing the flight path as if looking down on it from space, determining the direction the flight will take, measuring the flight path's distance, and visualizing the flight path's great circle route. To accomplish these tasks, the orthographic, azimuthal equidistant, and gnomonic planar projections, respectively, were chosen. In each case, the projections were azimuthal (preserved directions) but also allowed an additional projection property such as equidistant to be utilized. The flexibility of the Mercator projection for measuring directions was also shown to be useful for this mapping task. The distance between the two cities was also determined using Google Maps, one of many online mapping services that uses the Web Mercator projection. The differences between the Mercator and the Web Mercator projection were examined.

We summarized five key criteria that you should consider when selecting map projections. First, in most cases, it is the objective of the mapmaker to select a projection for which distortion is the lowest. Second, amounts of distortion can be kept small by aligning the geographic area (or data set) under consideration with the standard line(s) or by positioning the map's center relative to the standard point. Third, as the amount of geographic area under consideration increases, distortion becomes a more important consideration. Fourth, just because a projection has seen considerable exposure (e.g., use in prominent atlases) does not mean that the projection is suitable for your specific application. Fifth, an often-overlooked feature of the map projection is its influence on overall map design—a topic that is not well studied by cartographers.

This chapter concluded with an examination of Projection Wizard, which is an online and interactive tool to help select a map projection. Built on the foundation of Snyder's hierarchical map projection selection guideline, a user specifies a geographic extent of the map and a desired projection property. With that information provided by a user, Projection Wizard recommends one or more suitable projections for the map.

9.7   STUDY QUESTIONS

1. Visit https://tinyurl.com/IronSteelMap to see a map of the world iron and steel trade in 1937. This map comes from the *U.S. Army Service Forces Manual*, 1943. To map this data set at the world scale, the atlas uses the Miller cylindrical projection. Using Snyder's hierarchical map projection selection guidelines, work through the process of selecting a projection for this data set. Based on the results of your efforts, explain the factors you considered in your selection decision and whether you agree with their selection of the Miller cylindrical to map the iron and steel trade.

2. Have you ever considered how the selection of a specific map projection distorts the shapes and sizes of landmasses? Do you think your mental map of world landmasses' shapes and sizes is accurate? On which map projection do you think your mental map is based? Visit this website (http://www.maps.ugent.be/) to test the accuracy of your mental map. Complete the test and see how you perform.

3. Now that you have completed the test in question 2 read through this article (https://biblio.ugent.be/publication/8689207). Based on their findings, explain whether the selection of a map projection influences a person's mental map.

4. Visit https://www.giss.nasa.gov/tools/gprojector/help/projections/Sinusoidal.png to view the sinusoidal projection. Note that the sinusoidal projection shows the poles represented as points and the meridians as sinusoidal curves. Visit https://www.giss.nasa.gov/tools/gprojector/help/projections/Eckert4.png to view the Eckert IV projection. The Eckert IV projection represents the poles using lines and uses elliptical curves to represent meridians. Using these projections, comment on how each projection impacts the visual appearance of the graticule and landmasses. Then, discuss which of these projections you prefer for world maps.

5. Utilize Projection Wizard (https://projectionwizard.org/#) and revisit the map projection selection examples that were discussed in this chapter: world literacy rates, Russian population, migration to the United States, tornado paths across Kansas, and visualizing and measuring a flight path between Fairbanks, AK and Seoul, South Korea. Using one of these examples, answer the following questions:

   a. Explain the projection property that is important for your chosen example.

   b. Describe the geographic extent of the area to be mapped.

c. List the map projection(s) that Projection Wizard recommends for your chosen example.

d. Were any of the projections specified by Projection Wizard the same as those chosen in this chapter? If so, then list which ones.

e. If other projections were listed by Projection Wizard as suitable, what advantages do you think they have over the projections that were chosen in this chapter?

f. Were there any other considerations that you feel relevant and would like to share as a result of working with the Projection Wizard for this example?

NOTES

1 An epicycloidal projection is conceptually developed as a curve traced by a point on a circle's circumference of radius *a* that rolls along the outside of a fixed circle having a radius of 2*a*.

2 The Briesemeister projection, always centered at 10° E and 45° N, is a special oblique case of the Hammer modified azimuthal projection.

3 We utilized the CIA World Factbook, which is available at https://www.cia.gov/the-world-factbook/.

4 Although the "one-sixth" rule will not yield the lowest distortion over the mapped area per se, the procedure will result in low distortion without the complex mathematics associated with finding the standard parallels that result in the lowest distortion over the mapped area.

5 For more sophisticated mathematical approaches to selecting standard parallels, see Snyder (1993) and Maling (1992).

6 Note that although the central latitude definition might seem obvious, it is a parameter that is available when defining or changing projection parameters in many Geographic Information System (GIS) software programs because the central latitude is used to set the origin for the x and y plotting coordinates but has no impact on the visual appearance of the map.

7 *The National Atlas*, published by the U.S. Geological Survey in 1970, contains numerous maps drawn on the Albers equivalent projection.

8 The Fujita scale is used to indicate the intensity of a tornado, where F0 is the least intense and F5 is the most intense according to the damage inflicted by the tornado.

9 The location of these standard parallels was set at one-sixth the distance from the limiting parallels.

REFERENCES

Maling, D. (1992) *Coordinate Systems and Map Projections* (2nd ed.). Oxford, England: Pergamon.

Pearson, F. (1984) *Map Projection Methods*. Blacksburg, VA: Sigma Scientific.

Robinson, A. H., Morrison, J. L., Muehrcke, P. C. et al. (1995) *Elements of Cartography* (6th ed.). New York: Wiley.

Šavrič, B., Bernhard J., White, D. et al. (2015) "User preferences for world map projections." *Cartography and Geographic Information Science 42*, no. 5:398–409.

Šavrič, B., Jenny, B., and Jenny, H. (2016) "Projection Wizard – An online map projection selection tool." *The Cartographic Journal 53*, no. 2:177–185.

Snyder, J. P. (1993) *Flattening the Earth: Two Thousand Years of Map Projections*. Chicago: University of Chicago Press.

Snyder, J. P. (1994) *Map Projections: A Working Manual*. Washington, DC: U.S. Government Printing Office.

10 Principles of Color

10.1 INTRODUCTION

The increasing use of color on maps necessitates that map-makers understand the proper use of color. To assist you in developing this understanding, this chapter covers several issues related to the use of color on maps. In Section 10.3, we consider how either reflected or emitted light from maps passes through the eye and is processed by the eye-brain system. We will see that the processing of color involves two theories. In the **trichromatic theory**, color perception is a function of the relative stimulation of three types of **cones** (red, green, and blue) in the **retina** of the eye. In the **opponent-process theory**, color perception is based on a lightness–darkness channel and two opponent color channels: red–green and blue–yellow. As a result of these theories, we need to be concerned about **simultaneous contrast** (that the perceived color of an area can be affected by surrounding colors) and **color vision impairment** (that approximately 4 percent of the population, primarily males, has some form of color deficiency).

In Section 10.4, we cover five **color models** or systems for specifying colors. The **RGB** and **CMYK** models provide hardware specifications for creating colors: red, green, and blue light on graphic displays (in the case of RGB), and cyan, magenta, yellow, and black ink or toner on printed maps (in the case of CMYK). In contrast, the **HSV** and **Munsell** color models are user-oriented because they are based on how we perceive colors (using attributes such as hue, lightness, and saturation). The **CIE** color model is an international standard for color that allows users to specify color in a form that enables someone else to reproduce that same color in a different environment; this is especially useful when you wish to communicate to others the precise color that you used. The ideal is a **perceptually uniform color model** in which equal steps in the color model correspond to equal visual steps. We will see that only the Munsell and some variants of the CIE model are perceptually uniform.

In Section 10.5, we consider several tools and principles that allow you to utilize color effectively in practical map design situations. One useful tool is the **color wheel**, in which colors are arranged in a circular fashion, typically as a progression of hues from the electromagnetic spectrum. We will see that arranging colors in this fashion can allow us to make appropriate map design decisions. For instance, **harmonious colors**, similar colors that do not create visual tension, appear next to one another on the color wheel. Other useful tools are the various **tints**, **shades**, and **tones** that can be utilized to modify a base color. We will provide explicit definitions for these commonly misused terms

and illustrate how they can be used in map design. Finally, we cover some cartographic conventions for using color to symbolize both qualitative and quantitative data.

10.2 LEARNING OBJECTIVES

- Describe the nature of visible light and its relationship to the electromagnetic spectrum.
- Identify the basic features of the eye relevant to cartography.
- Explain two basic theories of color perception: the trichromatic theory and the opponent-process theory.
- Define simultaneous contrast and explain how it affects the perception of color on maps.
- Recognize that not everyone has normal color vision and identify common color vision problems.
- Describe the characteristics of basic color models used to specify colors.
- Describe complementary and harmonious colors and identify them on color wheels.
- Distinguish between tints, shades, and tones.
- List various qualitative color conventions.
- Describe quantitative color conventions and how they relate to different data types.

10.3 HOW COLOR IS PROCESSED BY THE HUMAN VISUAL SYSTEM

10.3.1 VISIBLE LIGHT AND THE ELECTROMAGNETIC SPECTRUM

We see maps as **visible light**, whether it is emitted from a computer screen or reflected from a paper map. Visible light is a type of **electromagnetic energy**, which is a waveform having both electrical and magnetic components (Figure 10.1).[1] The distance between two wave crests is known as the **wavelength of light**. Because visible wavelengths are small, they are typically expressed in nanometers (nm), which are 1 billionth of a meter. Visible wavelengths range from 380 to 760 nm. Figure 10.2 relates visible light to other forms of electromagnetic energy that humans deal with; the complete continuum of wavelengths is called the **electromagnetic spectrum**.

We have all seen or read about how a prism splits sunlight into the color spectrum (red, orange, yellow, green, blue, indigo, and violet). This phenomenon occurs because the visible portion of sunlight consists of a broad range of wavelengths instead of being concentrated at a particular wavelength. Different colors arise in a prism as a function

DOI: 10.1201/9781003150527-11

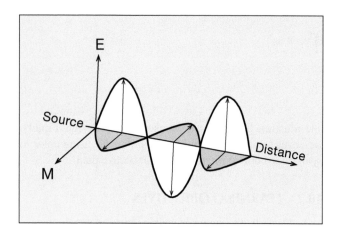

FIGURE 10.1 Electromagnetic energy is a waveform having both electrical (E) and magnetic (M) fields. Wavelength is the distance between two crests.

of how much each wavelength is bent, with shorter wavelengths (e.g., blue) bent more than longer wavelengths (e.g., red). Note that the colors in the visible portion of the spectrum in Figure 10.2 are arranged from short to long wavelength (from violet to red) and thus match the colors we might see using a prism.

10.3.2 STRUCTURE OF THE EYE

The basic features of the eye that concern cartographers are shown in Figure 10.3. After passing through the **cornea** (a

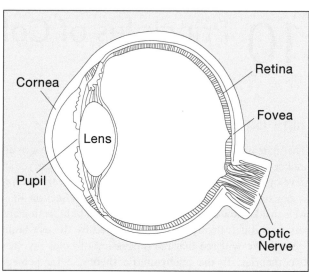

FIGURE 10.3 Basic features of the eye relevant to cartography.

protective outer covering) and the **pupil** (the dark area in the center of our eye), light reaches the **lens**, which focuses the light on the **retina**. Changing the shape of the lens, an automatic process known as **accommodation**, focuses images. As we age, our lenses become more rigid, and our ability to accommodate thus weakens. Generally, around the age of 45, our ability to accommodate becomes so weak that corrective lenses (glasses or contacts) are necessary. The **fovea** is the portion of the retina where our visual acuity is

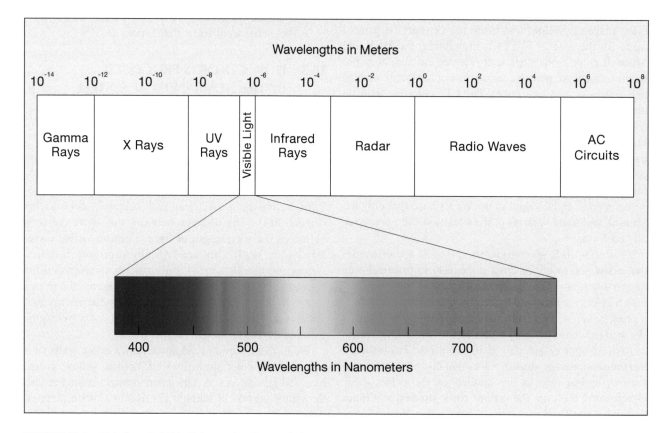

FIGURE 10.2 Relation of visible light to other forms of electromagnetic energy.

the greatest. The **optic nerve** carries information from the retina to the brain and creates what is commonly termed our **blind spot**.[2]

An enlargement of the retina is shown in Figure 10.4. Note that it consists of three major layers of nerve cells (rods and cones, bipolar cells, and ganglion cells), along with two kinds of connecting cells (*horizontal* and *amacrine* cells) that enable cells within the major layers to communicate with one another. **Rods** and **cones** are specialized nerve cells that contain light-sensitive chemicals called **visual pigments**, which generate an electrical response to light.

The concentration of cones is greatest at the fovea, and the highest concentration of rods is about 20° on either side of the fovea. Overall, there are about 120 million rods and 6 million cones in each of our eyes.

Cones function in relatively bright light and enable color vision, whereas rods function in dim light and play no role in color vision. The cones are of primary interest to cartographers because most maps are viewed in relatively bright light (an exception would be maps viewed in the dim light of an aircraft cockpit). Physiological examination of cones taken from the eye of a person with normal color vision

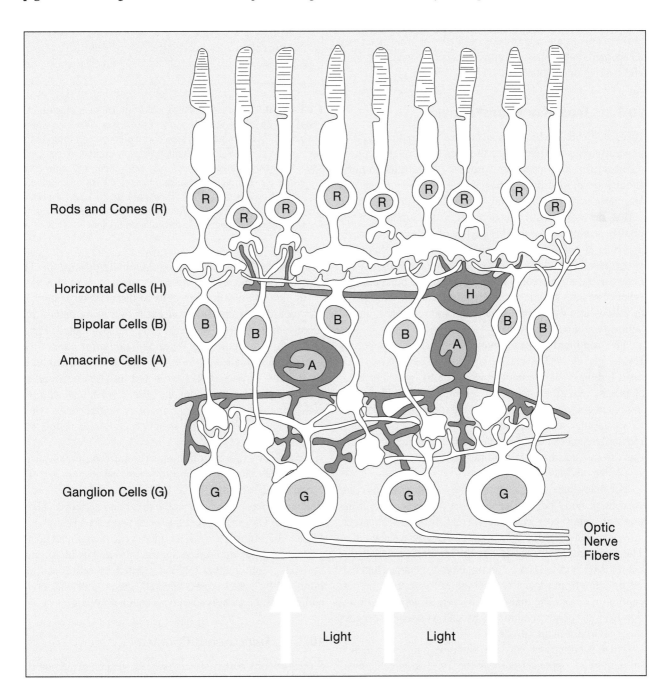

FIGURE 10.4 Major layers of cells found in the retina. (Adapted from J. E. Dowling and B. B. Boycott, 1966, "Organization of the primate retina: electron microscopy," *Proceedings of the Royal Society of London*, 166, Series B, pages 80–111, Figure 23; courtesy of the Royal Society and John E. Dowling.)

reveals three distinct kinds based on the wavelength to which they are most sensitive: short (blue), medium (green), and long (red) (Bowmaker and Dartnall 1980).[3] The distribution and sensitivity of these three kinds of cones vary in the retina: Although blue cones cover the largest area, they are the least sensitive, thus making blue inappropriate for small map features (MacEachren 1995, 56).

The major function of the **bipolar** and **ganglion cells** (Figure 10.4) is to merge the input arriving from the rods and cones. Although there are about 126 million rods and cones, there are only about 1 million ganglion cells. Considerable convergence must take place between the rods and cones and the ganglion cells; each single ganglion cell corresponds to a group of rods or cones, or what is termed a **receptive field**. These receptive fields are circular in form and overlap one another.

10.3.3   Theories of Color Perception

Psychology textbooks (e.g., Goldstein and Brockmole 2017) generally consider two major theories of color perception: trichromatic and opponent process. The **trichromatic theory**, developed by Thomas Young and championed by Hermann von Helmholtz and James Clerk Maxwell in the 1800s, presumes that color perception is a function of the relative stimulation of the three types of cones (blue, green, and red). If only one type of cone is stimulated, that color is perceived (e.g., a red light would stimulate primarily red cones, and thus red would be perceived). The perception of other colors is a function of the relative ratios of stimulation (a yellow light would stimulate green and red cones, and so yellow would be perceived).

The **opponent-process theory**, originally developed by Ewald Hering (1878), states that color perception is based on a lightness–darkness channel and two opponent color channels: red–green and blue–yellow. Colors within each opponent color channel are presumed to work in opposition to one another, meaning that we do not perceive mixtures of red and green or blue and yellow; rather, we see mixtures of pairs from each channel (red–blue, red–yellow, green–blue, and green–yellow).

For many years, proponents of the two theories of color perception hotly debated each theory's merits, presuming that only one theory could be correct. It is now apparent, however, that both can help explain the way we see color. The trichromatic theory is correct in the sense that our color vision system is based on three types of cones and that information from these cones combines to produce the perception of color. The manner, however, in which information from the cones combines is based on opponent-process theory (Goldstein and Brockmole 2017).

There is both psychophysical and physiological evidence in support of opponent-process theory. The psychophysical evidence comes from the seminal work of Leo Hurvich and Dorothea Jameson (1957), which showed that a color of an opposing pair could be eliminated by adding light for the other color in the pair; for example, when yellow light

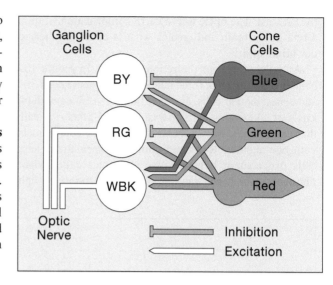

FIGURE 10.5   A model of how color information reaching the cones can be converted to opponent processes. The blue–yellow channel (BY) is excited by green and red cones and inhibited by blue cones; the red–green channel (RG) is excited by red cones and inhibited by green cones; and red, green, and blue cones stimulate the lightness–darkness channel (WBK, for white to black). (After Eastman, 1986. First published in *The American Cartographer*, 13(4), p. 326. Reprinted with permission from the Cartography and Geographic Information Society.)

is added to blue light, the blue eventually disappears. The physiological evidence is based on an analysis of how electrical signals pass through cells in the nervous system. In this regard, an important concept is that nerve cells fire at a constant rate even when they are not stimulated. Firing above this constant rate is termed *excitation*, and firing below it is termed *inhibition*. By studying electrical activity in cells, physiologists have noted linkages between the blue, green, and red cones and the bipolar and ganglion cells; for example, a red light might excite red cones, which in turn excite bipolar and ganglion cells (De Valois and Jacobs 1984).

Although experts in human vision are reasonably sure that certain colors are in opposition to one another and that excited and inhibited nerve cells play a role, the precise linkage between the cones and bipolar and ganglion cells is unknown. One model that has been suggested is shown in Figure 10.5. In this model, the blue–yellow channel (BY) is excited by green and red cones and inhibited by blue cones; the red–green channel (RG) is excited by red cones and inhibited by green cones; and red, green, and blue cones stimulate the lightness–darkness channel (WBK).

10.3.4   Simultaneous Contrast

Simultaneous contrast occurs when the perceived color of an area is affected by the color of the surrounding area, as illustrated for lightness in Figure 10.6A. Here the grays in the central boxes are physically identical, but the gray on the left appears lighter. This occurs because a gray

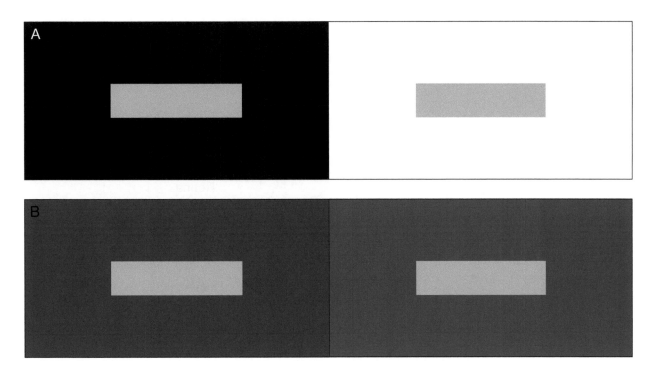

FIGURE 10.6 Illustrations of simultaneous contrast: (A) the central gray strips are physically identical, but the one surrounded by black appears lighter; (B) again, the central strips are the same gray, but they appear to shift toward the opponent color of the surrounding color (e.g., when surrounded by blue, the gray appears yellowish).

surrounded by black shifts toward a lighter value, whereas the same gray surrounded by white shifts toward a darker gray. Note that in this case, the shifts are toward the opposite side of the lightness–darkness channel in the opponent-process model. Also, note that this would hold true if the two lightnesses of the same hue were used (e.g., if a light red were surrounded by a dark red).

When different hues are used, the apparent color of an area will tend to shift toward the opponent color of the surrounding color, as shown in Figure 10.6B. On the left, when the surround is green, the gray appears reddish; in contrast, on the right, when the surround is blue, the gray appears yellowish.

Simultaneous contrast is believed to come about because of the receptive fields mentioned earlier. Receptive fields are not uniform; rather, distinctive centers and surrounds characterize them, with visual information in the surround having an impact on the information found in the center. For a detailed discussion of how simultaneous contrast operates, see Hurvich (1981).

10.3.5 Color Vision Impairment

Up to this point, we have assumed map readers with normal color vision. There is, however, a substantial number of people who have some form of **color vision impairment**. The highest percentages are found in the United States and Europe (approximately 4 percent, primarily males), and the lowest incidence (about 2 percent overall) appears in the Arctic and the equatorial rainforests of Brazil, Africa, and New Guinea (Birren 1983).

The bulk of the color vision–impaired can be split into two broad groups: **anomalous trichromats** and **dichromats**. These groups are distinguished on the basis of the number of cone cells that are operational. Anomalous trichromats have three types of cone cells (red, green, and blue), but one type is deficient. In contrast, dichromats have only two types of working cone cells (i.e., one of the three types is absent). For both groups, the most common problem is distinguishing between red and green; for anomalous trichromats, there is some difficulty, whereas, for dichromats, the colors cannot be distinguished.

The two broad groups are commonly divided into subgroups as a function of the dysfunction of cones or the lack thereof. Anomalous trichromats are divided into three subgroups: protanomaly, deuteranomaly, and tritanomaly depending on whether the red, green, or blue cones are deficient. Deuteranomaly is the most common of these subgroups, accounting for approximately 5 percent of males worldwide.[4] Similarly, dichromats are divided into protanopes, deuteranopes, and tritanopes, depending on whether the red, green, or blue cones are absent. Protanopes and deuteranopes are the most common of these subgroups, each accounting for approximately 1 percent of males worldwide (Jenny and Kelso 2007).[5]

10.3.6 Beyond the Eye

It is important to realize that the eye is part of the larger visual processing system shown in Figure 10.7. Note first that information leaving the eyes via the optic nerves crosses over at the *optic chiasm*; up to that point, information from

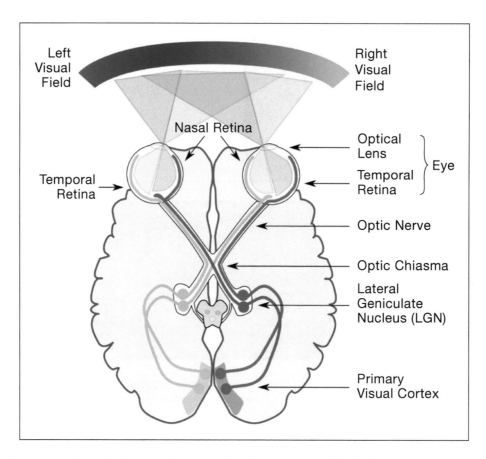

FIGURE 10.7   An overview of the visual processing system viewed from underneath the brain.
(Adapted from https://commons.wikimedia.org/wiki/File:Human_visual_pathway.svg (lettering was modified to match the format of the other figures in the textbook); licensed under Creative Commons License (CC BYAttribution-ShareAlike 4.0 International).)

each eye is separate, but pathways beyond this point contain information from both eyes. After passing through the optic chiasm, each pathway enters the *lateral geniculate nucleus (LGN)*. Physiological experiments with animals reveal that opponent cells similar to those found within the retina are also found in the LGN (De Valois and Jacobs 1984). Interpretation of the visual information begins in the **visual cortex**, the first place where information from both eyes is handled.[6] As with the LGN, our knowledge of processing in this area is largely a function of physiological experiments with animals. Probably the most significant of these is the work of David Hubel and Torsten Wiesel, who received the 1981 Nobel Prize for their efforts. They found three kinds of specialized cells in the visual cortex: *simple cortical cells*, which respond best to lines of particular orientation; *complex cells*, which respond to bars of particular orientation that move in a particular direction; and *end-stopped cells*, which respond to moving lines of a specific length or to moving corners or angles. Not only did Hubel and Wiesel discover these different kinds of cells, but they also mapped out where they occur within the visual cortex (Goldstein and Brockmole 2017).

Although such findings are certainly significant, researchers have not yet been able to explain how the brain handles a complex real-world situation, such as a map. Functional magnetic resonance imaging (fMRI) may ultimately provide some insight into this issue. fMRI utilizes an MRI scanner, which generates strong magnetic fields that pass through the human body; you might be familiar with MRI scanners because they are commonly used to diagnose sports-related injuries. In the case of maps, we are interested in what goes on in the human brain, where changes in the metabolic demands of nerve cells lead to changes in blood oxygenation levels, which can be measured by an MRI scanner. Since changing metabolic demands of nerve cells are an indication of which area of the brain is activated, it is, at least theoretically, possible to associate a particular map use task with a particular area of the brain. Amy Lobben and colleagues (2009, 2014) have begun to explore the use of fMRI to understand how map images are stored in the brain. Lobben et al. (2009, 168) argue that "… while traditional behavioral experimental design has provided the foundation for much of the past, current, and future cartographic … research, we posit that the use of additional data-gathering methods, and of fMRI specifically, will provide cartographers with additional support for existing and developing theories of map design and map use."

10.4 MODELS FOR SPECIFYING COLOR

This section considers five **color models** that have been used to specify colors appearing on maps: RGB, CMYK, HSV, Munsell, and CIE. The RGB and CMYK models are hardware-oriented because they are based on hardware specifications of red, green, and blue light on graphic displays, or cyan, magenta, yellow, and black ink on printed maps. In contrast, the HSV and Munsell models are user-oriented because they are based on how we perceive colors (using attributes such as hue, lightness, and saturation). The CIE model is neither hardware- nor user-oriented; however, it is "optimal" in the sense that if you provide someone with the CIE coordinates of a color you created, that person, in theory, should be able to create exactly the same color.

10.4.1 THE RGB MODEL

In the **RGB** model, colors are specified based on the intensity of red, green, and blue light. Red, green, and blue lights are referred to as the **additive primary colors** (Figure 10.8). Primary colors come in groups of three, and can be mixed together to create a wide range of other colors. Primary colors cannot be created by mixing other colors in the same color model. The term "additive" refers to the fact that when red, green, and blue lights are added together, white light results. Other colors are created by adding red, green, and blue light in different combinations and intensities. The range of intensities can be represented as a cube, as shown in Figure 10.9. Here achromatic grays (or completely desaturated colors) are found along the diagonal extending from "White" to "Black." In general, lighter colors are found around the White point of the cube, and darker colors are found around the Black point of the cube. As you move away from the White–Black line, you move toward more saturated colors; for example, at the "Red" point, you would be at the maximum saturation of red. Finally, note

that hues are arranged in a hexagonal fashion around the White–Black line. The latter can be seen most easily if you look directly down the diagonal from White to Black, as shown in Figure 10.9B.

The RGB model has the advantage of correlating with the method of color production on graphic displays (e.g., a pixel on an LCD monitor has an intensity of red, green, and blue light), but it has two major disadvantages. One is that common notions of hue, saturation, and lightness are not inherent in the model; although we used these terms to describe the model, they are not used to specify colors. Another disadvantage is that equal steps in RGB color space do not correspond to equal visual steps (RGB is not a perceptually uniform color model); for example, the color with RGB values of 125, 0, 0 will not appear to fall midway between the colors with RGB values of 0, 0, 0 and 250, 0, 0. Typically, you will find that an incremental change in low RGB values represents a smaller visual difference than the same incremental change in high RGB values. In spite of these disadvantages, RGB values frequently are used as an option for specifying color in software packages, presumably because of their long history and the consequent familiarity that many users have with them.

10.4.2 THE CMYK MODEL

Because printed maps are based on reflected (as opposed to emitted) light, they create color using a subtractive (as opposed to an additive) process. The three **subtractive primary colors** are cyan, magenta, and yellow (Figure 10.8); black ink, however, is utilized when a true black is desired. Together, cyan, magenta, yellow, and black compose the **CMYK** color model. The term "subtractive" refers to the fact that when white light strikes cyan, magenta, and yellow pigments (ink, toner, etc.), portions of white light are subtracted before being reflected back to the eye. Other colors are created by adding cyan, magenta, and yellow in

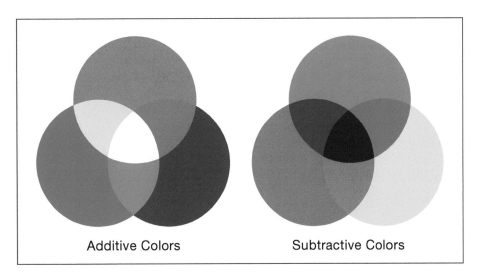

Additive Colors   Subtractive Colors

FIGURE 10.8 Principles of additive and subtractive color. For additive color, overlapping red, green, and blue lights reveal how cyan, magenta, yellow, and white can be created. For subtractive color, the reverse is the case: cyan, magenta, and yellow combine to produce red, green, blue, and black. To obtain a true black with subtractive colors, it is often necessary to add a black layer.

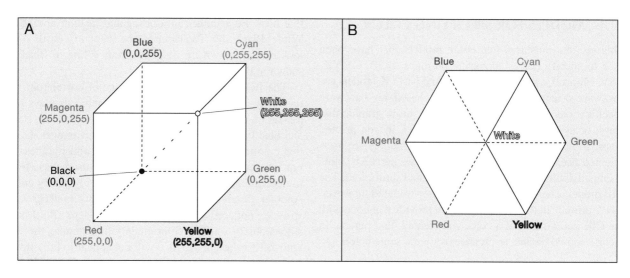

FIGURE 10.9   (A) Schematic form of an RGB color cube for specifying color. (It is assumed that red, green, and blue colors each have a maximum intensity of 255.) (B) The cube viewed looking directly down the diagonal from White to Black.

different combinations and intensities. If we think of cyan, magenta, and yellow in a fashion similar to how we think of red, green, and blue, then it is also possible to conceive of the CMY portion of CMYK as a cube: A certain percentage of cyan, magenta, and yellow would correspond to a particular point in the cube. Black, however, would need to be added to create true lightnesses of gray within the cube. Given the analogy to the RGB cube, it makes sense that CMYK will share the same disadvantages: there is a lack of relation to common color terminology, and equally spaced colors in the model will not correspond to equal visual steps (CMYK is not perceptually uniform). In Chapter 14, we consider in more detail how CMYK colors are used to create colors for printed maps.

10.4.3   THE HSV MODEL

In contrast to RGB and CMYK, the **HSV** model is more intuitive from a map design standpoint because it allows users to work directly with hue, saturation, and value (lightness). Color space in HSV is represented as a hexcone, as shown in Figure 10.10. The logic of the hexcone is apparent if you compare it with the color cube for RGB shown in Figure 10.9B; note that the hexagonal structure of the hues in the cube is retained in the hexagonal structure at the base of the hexcone. Value changes occur as you move from the apex of the cone to its base, whereas saturation changes occur as you move from the center to the edge of the cone.

The intuitive notions of hue, saturation, and value in HSV have led to its common use in software. Although HSV is commonly used, it also has disadvantages. One is that different hues having the same value (V) in HSV will not all have the same perceived value. As an example, consider the base of the cone, where the highest-value green and red are found. If you create such colors on your monitor, green will appear lighter than red. In a similar fashion, different hues having identical saturations (S) will not have the same perceived saturations. HSV also shares a disadvantage noted

for RGB: Selecting a color midway between two colors will not result in a color that is perceived to be midway between those colors (HSV is not perceptually uniform).

10.4.4   THE MUNSELL MODEL

The **Munsell** color model is a user-oriented system that was developed prior to the advent of computers. Munsell colors are specified using the terms *hue, value* (for lightness), and *chroma* (for saturation). The general structure of the model (Figures 10.11 and 10.12) is similar to that of HSV (that is, hues are arranged in a circular fashion around the center, chroma increases as one moves outward from the center, and value increases from bottom to top). Note, however, that in contrast to HSV, the Munsell model is asymmetrical; for example, if

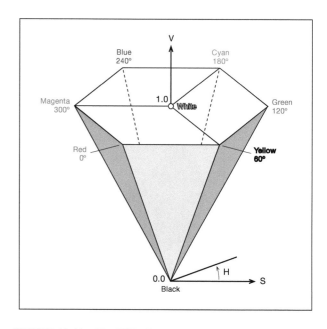

FIGURE 10.10   The HSV (hue, saturation, and value) color system represented as a hexcone.

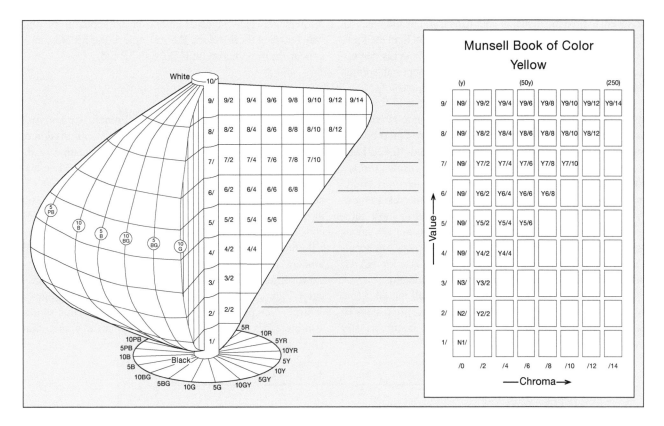

FIGURE 10.11 Three-dimensional representation of the Munsell color solid. A vertical slice through yellow is shown in detail. (Courtesy of Leo M. Hurvich.)

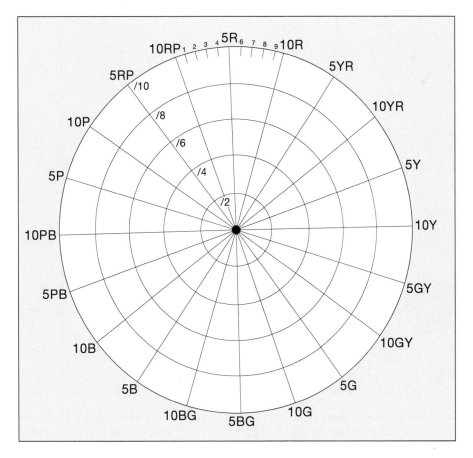

FIGURE 10.12 A horizontal slice through the Munsell color solid shown in Figure 10.11.

you were to hold the model in your hands, you would note that the lightest green would be higher on the model than the lightest red. The asymmetry occurs because the model is perceptually based (i.e., in the real world, the lightest possible green does appear brighter than the lightest possible red).

Ten major Munsell hues are recognized, and these are split into five principal (represented by a single letter, such as Y for yellow) and five intermediate (represented by two letters, such as YR) hues (Figures 10.11 and 10.12). Each major hue is also split into ten subhues (consider the ten subhues shown for R (on the edge of the outermost circle) in Figure 10.12). Values range from 0 to 10 (darkest to lightest), and chromas range from 0 to 16 (least to most saturated). Due to the asymmetry of the model, not all values and chromas occur for each hue. Munsell colors are represented symbolically as H V/C; thus, 5R 5/14 is a distinct red of moderate value and high saturation (a crimson).

An important characteristic of the Munsell model is that equal steps in the model represent equal perceptual steps (it is perceptually uniform). Thus, a color that is numerically

midway between two other colors should appear to be midway between those colors. For example, color 5R 5/5 should appear midway between 5R 2/2 and 5R 8/8.

10.4.5 THE CIE MODEL

CIE is an abbreviation for the French *Commission International de l'Eclairage* (International Commission on Illumination). In theory, careful color specification in the CIE model means that anyone in the world should be able to recognize and reproduce a desired color. CIE colors can be specified in several ways (e.g., *Yxy*, L*u*v*, L*a*b*) but in all cases, a combination of three numbers is used. We consider the *Yxy* model (commonly referred to as the 1931 CIE model) first because it forms the basis of other CIE models. In the *Yxy* model, the x and y coordinates define a two-dimensional space within which hue and saturation vary (Figure 10.13). Note that hues are arranged around a central *white point* (or *equal-energy point*) and that saturation increases as one moves outward from the white point.

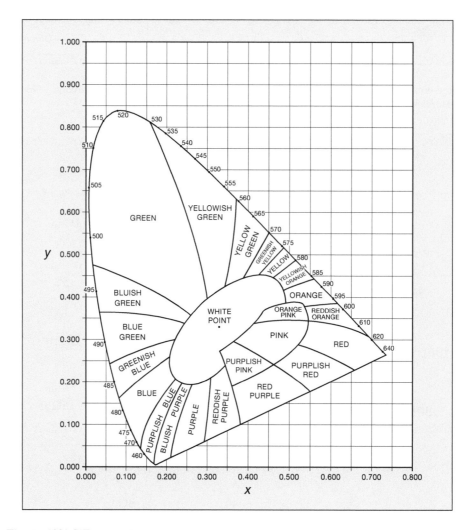

FIGURE 10.13 The *Yxy* 1931 CIE system. Hues are arranged in a continuum around a central white point. Saturation is at a maximum on the edge of the horseshoe and at a minimum at the white point. Numerical values on the edge of the horseshoe represent wavelengths in nanometers. Because differing lightnesses cannot be shown by this diagram, a three-dimensional diagram is required, as in Figure 10.14. (Adapted with permission from Kelly (1943) © The Optical Society.)

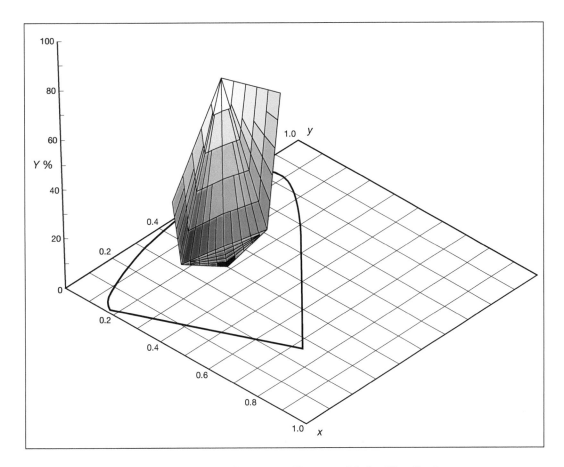

FIGURE 10.14   A three-dimensional view of the *Yxy* CIE system. (Courtesy of A. Jon Kimerling.)

The *Y* portion of the model provides the third dimension—the lightness or darkness component (Figure 10.14).

The structure of the *Yxy* model is similar to both HSV and Munsell (all have hues arranged in a circular fashion, desaturated colors in the middle, and a vertical lightness axis), but note that in CIE, hues and saturations are not related in a simplistic fashion to the *x* and *y* axes. The reason for this can be found in the manner in which CIE was established. CIE was developed using the notion that most colors can be defined by a mixture of three colors (roughly speaking, we can call these red, green, and blue). The appropriate combination of three colors needed to match selected colors was determined using human observers (the average response of the observers was termed the *standard observer*). To understand the matching process, imagine that you are asked to view a screen on which a single circle is projected using a standard light source. In the top portion of the circle, a test color appears, and in the bottom portion, you manipulate the three colors to produce a color identical to the test one. If you repeated this task for many test colors, you would discover that various amounts of the three colors would be required to make appropriate matches.

Results of CIE matching experiments are shown in Figure 10.15. The three curves correspond to the three colors combined in the experiments. The *x*-axis represents the wavelength of the test color, whereas the *y*-axis represents the relative magnitudes of the three colors needed to

match the test color. For example, a test color at 530 nm would require 0.005 of blue, 0.203 of green, and −0.071 of red (Wyszecki and Stiles 1982, 750); these are known as *tristimulus values*. The negative value for red is necessary because, in some cases, you would find that you could

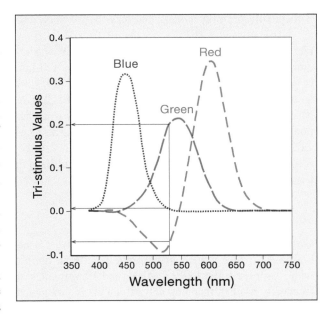

FIGURE 10.15   Curves representing the relative proportion of CIE primaries needed to match test colors at various wavelengths.

not match the test color with any combination of the three colors; to achieve a match, you would have to mix one of the three with the test color, and this is recorded as a minus value in the graph. To avoid having to work with negative values, the developers of the CIE model transformed the results shown in Figure 10.15 to purely positive values, which are commonly referred to as *X*, *Y*, and *Z*. To get coordinates for the *Yxy* system, the *X*, *Y*, and *Z* values were converted to proportional values:

$$x = X / (X + Y + Z)$$

$$y = Y / (X + Y + Z)$$

$$z = Z / (X + Y + Z)$$

Because these proportions add to 1, it is not necessary to plot *z* in the *Yxy* system (*z* would be $1 - (x + y)$). One problem with plotting proportional values is the elimination of information about lightness or darkness. (We get the same proportions when *X*, *Y*, and *Z* are all 10 units as we do when they are all 20 units.) This problem was handled in CIE by arbitrarily assigning the lightness information to *Y* and plotting this as the third dimension, as shown in Figure 10.14 (Hurvich 1981, 284).

It should also be noted that the CIE *Yxy* coordinate values can be adjusted to account for the lighting conditions under which the colors are viewed. For example, you might want to consider the potential effects of natural sunlight vs. fluorescent room light. This is a capability that is not generally included in other color models.

One problem with the 1931 CIE diagram is that colors are not equally spaced in a visual sense, as was the case for the RGB, CMYK, and HSV models. Fortunately, perceptually uniform CIE models have been developed. The two perceptually uniform models that have been widely used are **CIELAB** and **CIELUV**. In the 1970s, CIELAB was recommended for printed material and CIELUV was recommended for graphic displays. However, today CIELAB is much more commonly used. Mark Fairchild (2013, 81), an expert in color appearance models, notes that "CIELAB has become almost universally used for color specification … At this time there appears to be no reason to use CIELUV over CIELAB." In Chapter 15, we will describe a method for unclassed choropleth mapping that utilizes the CIELAB model.[7]

10.4.6 Discussion

Given the variety of color models presented here, it is natural to ask which models mapmakers need to be familiar with. At present, the RGB, CMYK, and HSV models are frequently used in mapping software. This is problematic because the RGB and CMYK models do not relate easily to our common notions of hue, saturation, and lightness, and all three models are not perceptually uniform. The Munsell model is not generally found in mapping software, but it is useful because it relates well to our notion of hue, saturation, and lightness, and it is perceptually uniform. The CIE model, although not commonly used in mapping software, is important to understand because you may see references to CIE colors when reading the results of cartographic research; for example, a recent article by Alžběta Brychtová and Arzu Çöltekin (2017) utilized the uniform color model CIEDE2000 to measure the difference between colors for classes on maps.

10.5 TERMINOLOGY AND PRINCIPLES IN THE PRACTICAL USE OF COLOR

In this section, we cover a number of terms and concepts that can be utilized in the choice and specification of colors on a map.

10.5.1 Color Wheels

Color wheels are graphical devices that allow us to see relationships between colors. The colors in a color wheel can be arranged in just about any order, but they normally follow a general progression of hues as taken from the visible portion of the electromagnetic spectrum (as shown in Figure 10.2). One type of color wheel allows us to view relationships between the additive primary colors (RGB) and the subtractive primary colors (CMY). This is called the **RGB/CMY color wheel** (Figure 10.16), where red, green, and blue are separated from each other by cyan, magenta, and yellow. Each color can be created by mixing the two adjacent colors. For example, cyan is a mixture of green and blue light; red is a mixture of magenta and yellow pigment (ink or toner).

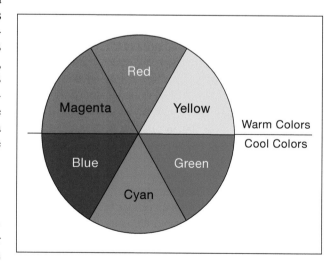

FIGURE 10.16 The RGB/CMY color wheel. Here red, green, and blue are separated from one another by cyan, magenta, and yellow. Note that the warm colors are magenta, red, and yellow, and the cool colors are blue, cyan, and green.

The RGB/CMY color wheel can be split into two halves: the "warm" colors and the "cool" colors. **Warm colors** such as red and yellow are associated with warm things, like fire or the sun, and tend to advance or jump out at the map user.[8] **Cool colors** like blue and green are associated with cool things, like ice and vegetation, and tend to recede or move away from the map user. These characteristics can be used to establish figure-ground contrast, as described in Chapter 13. For instance, a warm color like orange can be used to emphasize thematic symbols, and a cool color like blue can be used to de-emphasize base information, such as roads. The warm/cool difference can also be used to represent features on a map that tend to be either warm or cool in reality, such as cyan for water bodies and tan (yellow plus black) for deserts.

It is interesting to note that ideas about "warm" and "cool" as just described are based purely on human perception and not on the physical properties of color. For instance, blue is perceived as being a cool color, but blue light is actually very high on the **color temperature scale**, which is used to categorize colors as they are emitted from an object (a black body) at different temperatures (Figure 10.17). Red, on the other hand, is very low on the color temperature scale. Color on maps is intended to be perceived by humans and not measured according to wavelength, so color choices should be guided more by our perceptions of warm and cool colors than by the actual color temperatures.

The color wheel in Figure 10.16 can also be used to identify complementary and harmonious colors. **Complementary colors** are opposite each other on the

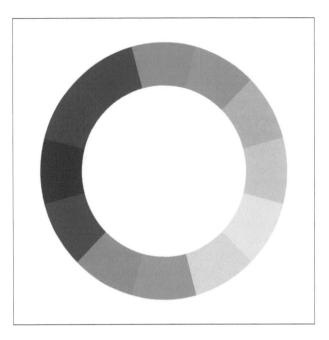

FIGURE 10.18 A red–yellow–blue (RYB) color wheel. This "nonscientific" color wheel provides a more useful series of harmonious colors than the RGB/CMY color wheel.

color wheel, and they produce gray when mixed in equal proportions—they essentially cancel each other out when mixed. When used together on the same map, complementary colors can create visual tension. They can be used to attract attention to one another or to represent features on a map that tend to be opposites, such as red for fire and cyan for ice. **Harmonious colors** are next to each other on the color wheel; they are very similar colors that do not create visual tension. In general terms, colors are harmonious if they are located on the same half of the color wheel.

The RBG/CMY color wheel is valuable for its ability to reveal the relationships between colors, but it is overly simple compared to more "artistic" color wheels. The color wheel in Figure 10.18 is far more detailed, providing a larger collection of colors and more subtle differences between them. It is considered to be an **RYB color wheel**, where RYB refers to red, yellow, and blue—colors that have traditionally been considered primary colors for use in painting. The RYB color model predates modern color theory and modern printing technologies and is considered to be "non-scientific." It produces a relatively small number of other colors when mixed, as opposed to the RGB and CMY color models. In addition, colors on opposite sides of this color wheel will not produce gray when mixed. Despite these drawbacks, the RYB color wheel can be useful for identifying harmonious colors that subtly traverse the colors of the rainbow. More specifically, it is valuable for identifying colors for part-spectral, sequential color schemes that are commonly used on quantitative thematic maps (such as a yellow, orange, red hue-value color scheme). These color schemes will be described in more detail in Section 10.5.4.

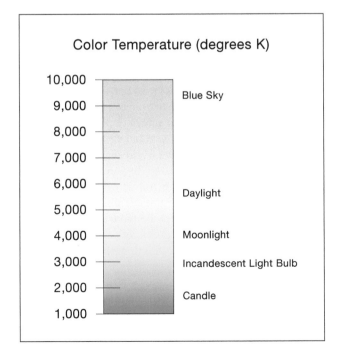

FIGURE 10.17 Color temperatures in relation to common features and phenomena, contradicting intuitive senses of "warm" and "cool."

10.5.2 Tints, Shades, and Tones

The terms tint, shade, and tone are used frequently when discussing color. A **tint** is a lighter version of a particular color, such as blue, as shown in Figure 10.19A. In this example, blue is considered to be the "base color," which is the color that tints will be based on. If the blue is pigment-based, it can be lightened by adding white pigment, thus creating a tint. More commonly, if printing on a white surface, simply printing less of the blue pigment in conjunction with a halftone screen (see Chapter 14) will produce a similar tint. Various other methods can be used to create tints, depending on the color model being used. For example, a base color of cyan that is created by mixing green and blue light can be lightened by simply adding red light. Tints are commonly used when creating monochromatic, sequential color schemes that are commonly used on quantitative thematic maps (such as a light green to dark green color scheme). Tints are also useful when establishing figure-ground contrast, as described in Chapter 13. For instance, thematic symbols such as proportional circles can be emphasized by "screening back" or making lighter, base information such as roads that surround the proportional circles.

A **shade** is a darker version of a base color achieved by adding small amounts of black. For instance, a base color of orange (Figure 10.19B) can be used to produce shades by adding increasing amounts of black, resulting in "burnt" oranges. As with the creation of tints, various approaches can be taken to create shades, depending on the color model

being used. Shades, as with tints, can be used to produce monochromatic, sequential color schemes for use on quantitative thematic maps (such as a yellow to dark yellow or tan, color scheme). Shades are also useful when establishing figure-ground contrast. For example, an area of interest, such as a particular county, can be emphasized by "shading" or making darker the surrounding counties.

A **tone** is a color that has been "muted" through the addition of equal parts of black and white or gray (Figure 10.19C). In other words, it involves the desaturation of a base color. The final tone depends on the amount of gray used, and tones may be lighter or darker than the original hue according to how the lightness of the color is also changed. Tones can add a sense of subtlety and sophistication to a map and are commonly used when symbolizing qualitative categories of data, such as vegetation and land use. Lighter tones are particularly useful when symbolizing large areas of base information, such as parkland, as the addition of gray can reduce the visual impact of pure green and de-emphasize the areas in the establishment of figure-ground contrast.

10.5.3 Qualitative Color Conventions

A **color convention** is a method of using a color that is commonly used and which is typically successful when symbolizing certain types of map features in particular situations. A classic example is the use of cyan for water features (Figure 10.20A). Linear features such as rivers, the boundaries of water bodies such as lakes, and type that identifies water features, should be symbolized with 100 percent cyan. Water bodies should be symbolized with a light tint of cyan, typically in the range of 10 percent to 30 percent (20 percent in the case of Folsom Lake). This convention works very well on printed maps but tends to look overly bright when depicted using the RGB color model on computer screens and projection devices. Shades of cyan work better on these devices. For example, the addition of small amounts of black to cyan can darken pure cyan when displayed on a computer screen or projector.

Parks, forests, or densely-vegetated regions should be symbolized with a tint or tone of green (Figure 10.20B). A light tint of green in the 10 to 30 percent range works well. Alternatively, the green tint can be further muted by creating a tone—adding a small amount of gray by reducing the saturation. Park boundaries can be symbolized with a 100 percent green, or a tint of green, or perhaps with no color at all. Some park boundaries are very complex, and a dark boundary can become overbearing—especially if the park is acting as base information, which should not attract attention. In contrast with parks, forests, and densely-vegetated regions, arid regions are appropriately symbolized with yellows and tans, which connote an absence of lush, green vegetation (Figure 10.20C). Be aware that the distinction between parks and arid regions is not always clear, as many "parks" are located in arid regions.

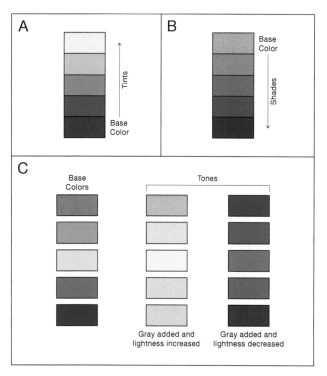

FIGURE 10.19 (A) Lighter tints, (B) Darker shades, and (C) Tones. Tones include neutral gray and can be either lighter or darker than a base color.

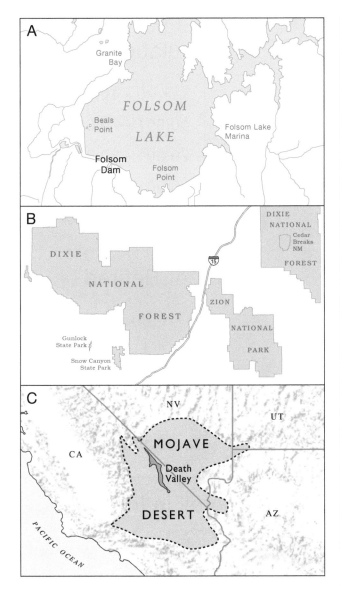

FIGURE 10.20 Qualitative color conventions for (A) water features and descriptive type, (B) parks and forests, and (C) arid regions.

The concept of warm and cool colors, and their association with real-world features or phenomena, was introduced in a previous section. These associations can be exploited when choosing colors for certain map features. For example, reds and yellows work well when representing flame symbols on a map of forest fires (Figure 10.21A) because most people associate fire—or warm things in general—with red and yellow. Similarly, a map that depicts snow cover or glaciers can take advantage of the association between ice and blue or cyan colors (Figure 10.21B).

Qualitative categories of data, such as land cover or geological units, should be symbolized using colors that are based upon distinctly different hues, as illustrated in Figure 10.22. (The colors in the Land Cover legend shown here are taken from the USGS National Land Cover Database, which surprisingly, represents water with a non-cyan color.) Colors for qualitative categories should

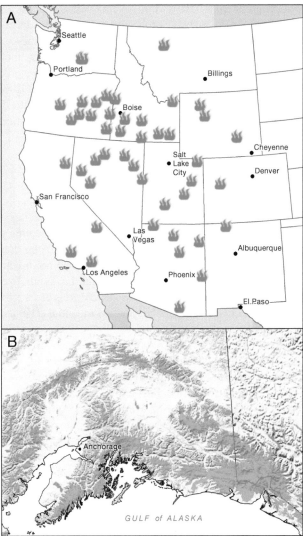

FIGURE 10.21 Qualitative color conventions for (A) "warm" fires and (B) "cool" snow and ice.

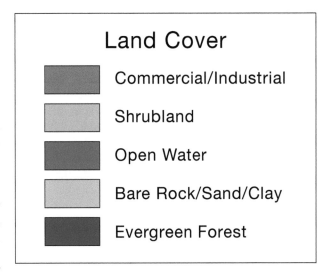

FIGURE 10.22 Qualitative categories with unordered hues represent distinctly different land covers.

not progress from light to dark, as this would suggest that quantitative data are being represented. In addition, the colors should not be arranged in the same order as they are found in the color spectrum. This is especially important in the legend, where sequentially ordered colors such as red, orange, yellow, green, blue, indigo, and violet (ROYGBIV) might be interpreted as representing quantitative data. In short, qualitative categories of data should be symbolized using **unordered hues**.

10.5.4 QUANTITATIVE COLOR CONVENTIONS

Color conventions for representing quantitative data are covered in detail in Chapter 15. Appropriate color schemes are chosen, in part, by first identifying the specific type of quantitative data that are being represented. Unipolar data, or numbers that have no dividing point and do not involve two complementary phenomena, should be symbolized using a **sequential color scheme**, which progresses from light to dark colors. A monochromatic sequential scheme is composed of a single base color and lighter tints (or darker shades), as illustrated for population in Figure 10.23A. A part-spectral sequential color scheme can also be used, which progresses from light to dark, but is composed of a small number of harmonious colors that are adjacent on the color wheel (Figure 10.23B). Use of the full spectrum (ROYGBIV), also known as the **rainbow color scheme**, is not recommended because this approach does not progress from light to dark, and is, therefore, more difficult to interpret as representing numbers that increase from low to high (Figure 10.23C).

Bipolar data and balanced data are appropriately symbolized using a **diverging color scheme**. Bipolar data consists of numbers that have a natural or meaningful dividing point, such as elevations above or below zero (Figure 10.24A). Balanced data consists of numbers that represent two phenomena that exist in a complementary fashion, such as the number of Democratic votes vs. Republican Votes in an election (Figure 10.24B). A diverging color scheme consists of two hues that progress from light to dark with distance from a common light or neutral color. It is essentially two sequential color schemes that sprout from a common value: the natural or meaningful dividing point, which is commonly either zero or one. In Figure 10.24A, the dividing point is the elevation of zero. In Figure 10.24B, the dividing point is one hundred. More specifically, the value found at the break between blue and red is 100. At this point, there are 100 Democratic votes for every 100 Republican votes. In Figure 10.24C, the dividing point is one, at which the number of Asian people is equal to the number of Black people.

10.5.5 THEME-ORIENTED COLOR SCHEMES[9]

In addition to the qualitative and quantitative color conventions described above, thought should be given to the theme of a particular map and whether a particular color

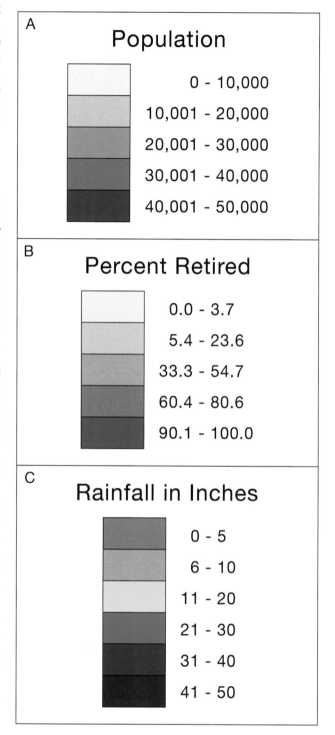

FIGURE 10.23 Quantitative color conventions: (A) monochromatic sequential, (B) part-spectral sequential, and (C) rainbow color schemes.

scheme might help support that theme. Common examples of theme-oriented color schemes include the use of:

- Pink and light blue to represent girls and boys
- Red and blue to represent the Republican and Democratic parties
- Cyan and blues to represent water-oriented themes

- Gray tones to represent historical data
- Greens to represent "green sector" jobs
- Dark reds to represent red wine growing regions

The **traffic light color scheme** is worthy of mention. This color scheme is analogous to the green, yellow, and red of traffic lights, and it can work especially well when representing traffic-related data, such as traffic accidents. A disadvantage of this color scheme is that the progression from green, to yellow, to red, does not progress from light to dark. In other words, it is not a sequential color scheme, and it can be difficult to interpret—particularly for individuals suffering from red-green color blindness.

It is also possible to symbolize ethnic or racial groups using colors that are sometimes associated with those groups. We do not recommend this, as it can reflect insensitivity to those people being represented.

10.6 SUMMARY

Visible light emitted by a computer screen or reflected from a paper map passes through the cornea (protective outer covering of the eye) and the lens, which focuses the light on the retina. Our retina is composed of rods and cones, which are specialized nerve cells. Cones play a key role in color vision, and there are three distinct types based on the wavelength of light to which they are most sensitive: short (blue), medium (green), and long (red). Bipolar and ganglion cells merge the input arriving from the rods and cones and pass this information on to the brain via the optic nerve.

The colors that we perceive are a function of two theories of color perception. The trichromatic theory states that color perception is a function of the relative stimulation of the blue, green, and red cones. The opponent-process theory states that color perception is based on a lightness-darkness channel and two opponent channels: red–green and blue–yellow. The blue–yellow channel is excited by green and red cones and inhibited by blue cones; the red-green channel is excited by red cones and inhibited by green cones; and red, green, and blue cones stimulate the lightness–darkness channel.

Simultaneous contrast occurs when the perceived color of an area is affected by the color of a surrounding area. In general, the color of the internal area will shift toward the opponent color of the surrounding color (e.g., a gray surrounded by yellow will appear blueish. Color vision impairment affects approximately 4 percent of the population (mostly males) worldwide. The most common problem is a confusion of red and green colors. Those with the most severe deficiencies are known as dichromats and are missing one of the three basic types of cones (red, green, or blue).

We considered five types of color models: RGB, CMYK, HSV, Munsell, and CIE. RGB and CMYK are commonly used for specifying colors in digital and printed maps, respectively. Unfortunately, neither of these models is perceptually based, and so equal steps in the color space do not correspond to equal steps in the visual sense (they are not perceptually uniform color models). HSV does use common terminology for perceiving color (hue, saturation, and value), but it also is not perceptually uniform. The Munsell color model is perceptually uniform, but it is not commonly used in digital environments. The CIE system is an

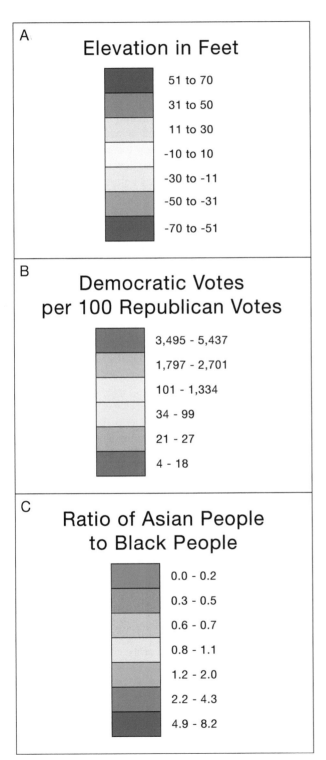

FIGURE 10.24 Diverging color schemes with different dividing points: (A) zero elevation, (B) 100 votes, and (C) ratio equal to one.

international standard for color, and variants of the basic CIE system (e.g., CIELAB and CIEDE2000) are perceptually uniform. We'll discuss the various color models in greater depth in Chapter 15.

Color wheels can be used to see relationships between the additive and subtractive primary colors and between the complimentary and harmonious colors. Some colors are considered "warm" and advance, while others are "cool" and recede, attributes that can be used to characterize certain features and phenomena and to establish figure-ground contrast.

Tints, shades, and tones are aspects of color that can make base colors lighter, darker, or muted (less saturated). They can be used to emphasize or de-emphasize map features and to establish figure-ground contrast.

Qualitative color conventions can be used to symbolize base information (map features used as reference) in an intuitive manner and to identify unique qualitative categories. Quantitative color conventions are based on the unipolar, bipolar, and balanced data types and consist of sequential and diverging color schemes.

10.7 STUDY QUESTIONS

1. Define the term "cone" and discuss the three types of cones found in the human retina.

2. Explain why it is inappropriate to say that "the colors that humans see is simply a function of the relative stimulation of the red, green, and blue cones in the retina."

3. Imagine that the classes for a legend on a choropleth map extend from dark blue to a light gray to a bright yellow. (A) If a light gray enumeration unit on the map is surrounded by bright yellow enumeration units, will the light gray appear to match the light gray shown in the legend? Explain your response. (B) If a light blue enumeration unit on the map is surrounded by dark blue enumeration units, will the light blue appear to match the same light blue shown in the legend? Again, explain your response.

4. Imagine that you are teaching a short course on cartography for a GIS firm. For fun, you decide to test the 20 students in the class to see whether anyone has a color deficiency. You find that only one student, a female, has a color deficiency. (A) Explain whether this result surprises you, given what you have learned about color vision impairment in our textbook. (B) The student returns to class at the end of the short course and tells you that she had further testing done by an ophthalmologist, who says that she has a mild case of deuteranomaly. Explain what this means and what the impact might be for reading maps.

5. Mapping software often provides you the option of entering RGB values for map classes. Presume that you wish to create a five-class isarithmic map and that you have selected RGB values of 0, 250, 250 and 0, 50, 50 for the lowest and highest classes, respectively. Explain whether it would be appropriate to use RGB values of 0, 150, 150 for the middle class.

6. If you are using the Munsell color model to create colors for a choropleth map, and you use Munsell colors 5R 2/6 and 5R 8/6 for the lowest and highest classes, what color should you use for the middle class? Justify your response.

7. If you are given CIE colors in a *Yxy* format, why should you be cautious in using these colors in a mapping context?

8. Assume that you are designing a map that will be printed. Which two subtractive primary colors should you mix to produce the color red? If you are mixing colors for a map that will be displayed, which two additive primary colors should you mix to produce the color cyan?

9. What is the name used to describe colors that are found on opposite sides of the RGB/CMY color wheel? Describe the characteristics of such colors in terms of "warm" and "cool" and give an example of how each color could be used to establish figure-ground contrast.

10. Tints and harmonious colors can both be used to represent quantitative data on thematic maps. Provide the terms used to describe the color schemes that are based on (1) the use of tints and (2) the use of harmonious colors. Give specific examples of each.

11. Describe what a qualitative color convention is and provide specific CMYK values that should be used when symbolizing the following base information: a lake with a river tributary in a forested region.

12. Quantitative color conventions are based upon the type of quantitative data being represented. Provide the names of the color schemes that should be used when representing unipolar, bipolar, and balanced data. Also, list the colors that could be used in each case.

NOTES

1 Light also consists of photons (packets of energy), which behave as particles when light strikes a surface (Birren 1983, 20).

2 For a simple demonstration of the effect of the blind spot, see http://www.colorcube.com/illusions/blndspot.htm.

3 Hubel (1988, 163–164) indicated that technically, the terms *violet, green*, and *yellowish-red* would probably be more appropriate.

4 We focus on the percentage of males because the percentage of females is negligible.

5 For further information on color vision impairment, see https://www.color-blindness.com/.

6 The visual cortex is divided into various areas; technically, the *primary visual cortex* is the first place where information from both eyes is handled.

7   Although CIELAB is widely used, further improvements in the CIELAB uniform model have been made; for instance, CIEDE2000 is an improved technique for computing the difference between two colors in the CIELAB model (Fairchild 2013, 83).

8   We will consider this principle in greater depth when we discuss the color stereoscopic effect in Section 17.7.2.

9   This idea is also covered in Section 15.4.6, dealing with color associations.

REFERENCES

Birren, F. (1983) *Colour.* London: Marshall Editions Limited.

Bowmaker, J. K., and Dartnall, H. J. A. (1980) "Visual pigments of rods and cones in a human retina." *Journal of Physiology* 298:501–511.

Brychtová, A., and Çöltekin, A. (2017) "The effect of spatial distance on the discriminability of colors in maps." *Cartography and Geographic Information Science 44*, no. 3:229–245.

De Valois, R. L., and Jacobs, G. H. (1984) "Neural mechanisms of color vision." In *Handbook of Physiology (Section 1: The Nervous System)*, ed. by J. M. Brookhart and V. B. Mountcastle, pp. 425–456. Bethesda, MD: American Physiological Society.

Eastman, J. R. (1986) "Opponent process theory and syntax for qualitative relationships in quantitative series." *The American Cartographer 13*, no. 4:324–333.

Fairchild, M. D. (2013) *Color Appearance Models* (3rd ed.). Chichester, West Sussex: Wiley.

Goldstein, E. B., and Brockmole, J. (2017) *Sensation and Perception* (10th ed.). Boston, MA: Cengage Learning.

Hering, E. (1878) *Zur Lehre vom Lichtsinne.* Vienna, Austria: Gerold.

Hurvich, L. M. (1981) *Color Vision.* Sunderland, MA: Sinauer Associates.

Hurvich, L. M., and Jameson, D. (1957) "An opponent-process theory of color vision." *Psychological Review 64*, no. 6:384–404.

Hubel, D. H. (1988) *Eye, Brain, and Vision.* New York: Scientific American Library.

Jenny, B., and Kelso, N. V. (2007) "Color design for the color vision impaired." *Cartographic Perspectives*, no. 58:61–67.

Kelly, K. L. (1943) "Color designations for lights." *Journal of the Optical Society of America 33*, no. 11:627–632.

Lobben, A., Lawrence, M., and Olson, J. M. (2009) "fMRI and human subjects research in cartography." *Cartographica 44*, no. 3:159–169.

Lobben, A., Lawrence, M., and Pickett, R. (2014) "The map effect." *Annals of the Association of American Geographers 104*, no. 1:96–113.

MacEachren, A. M. (1995) *How Maps Work: Representation, Visualization, and Design.* New York: Guilford.

Wyszecki, G., and Stiles, W. S. (1982) *Color Science: Concepts and Methods, Quantitative Data and Formulae* (2nd ed.). New York: Wiley.

11 Map Elements

11.1 INTRODUCTION

This chapter introduces common **map elements** that are employed in the creation of thematic maps. The general goal is to provide guidance in the creation of efficient, attractive maps that effectively communicate geographic information. Specific goals include guidance in choosing which map elements to employ and direction on how to effectively implement map elements. Map elements are closely related to the topics of typography and cartographic design, which are covered in the following two chapters. This chapter begins with a clarification of widely misunderstood terms associated with the alignment and centering of map features. A firm understanding of these terms is essential in understanding the remainder of the chapter.

It is the cartographer's job to select appropriate map elements and to implement them appropriately within the context of **available space**. Decisions regarding map elements should be made according to the needs of the **map user**, who represents the map's intended audience. We provide detailed descriptions of map elements, together with rules and guidelines related to their implementation. The elements are introduced in the order we recommend they be placed when constructing a map: the frame line and neat line, mapped area, inset, title and subtitle, legend, data source, scale, and orientation.

This chapter was not written with specific software applications in mind. Graphic design applications such as Adobe Illustrator typically provide the greatest control over graphics and type, but recent advances in the design capabilities of Geographic Information Systems (GIS) have narrowed the gap, allowing you to produce high-quality maps in a user-friendly GIS environment. The examples in this chapter reflect simple thematic maps, although the principles set forth can also be applied to more complex thematic maps and general reference maps.

11.2 LEARNING OBJECTIVES

- Differentiate between horizontal and vertical alignment and centering of map elements.
- List the eight map elements in their recommended order of placement and explain why they are placed in this order.
- Describe each map element in terms of its purpose and general characteristics.
- Decide which map elements should be used and which should be omitted for a particular mapping project.
- Describe each map element in terms of its preferred style, size, and location in relation to other map elements.
- Compare certain map elements in terms of relative type sizes.

11.3 ALIGNMENT AND CENTERING

Before we begin, it is important to clarify terms associated with the alignment and centering of map features. Alignment and centering of map features is achieved either through visual approximation or measurement. Terms such as *visually centered* refer to the placement of map features through visual approximation so that they *look* centered within an area. Alignment and centering is also achieved through measurement, resulting in map features that are precisely aligned or centered. If the term "visually centered" appears, it refers to visual approximation; all other references to alignment and centering in this chapter are based on measurement, as described next.

Terms such as "horizontally aligned to right" and "vertically centered" are used repeatedly in this chapter to describe spatial relationships between map features. These terms are widely misunderstood and warrant clarification. Map features such as symbols and type are often arranged in relation to imaginary lines, as illustrated in Figure 11.1. Horizontal alignment and centering (Figure 11.1A) involves the side-to-side movement of map features in relation to a *vertical* line (the square, circle, star, and "Black Mountain" label are moved in relation to the dashed line). Vertical alignment and centering (Figure 11.1B) involves the top-to-bottom movement of map features in relation to a *horizontal* line. Be aware that the terms *horizontal* and *vertical* are often confused in this context. For example, the square, circle, star, and "Black Mountain" label in Figure 11.1A are *vertically distributed* (arranged from top to bottom), but the alignment and centering are achieved through side-to-side (horizontal) movement. In Figure 11.1C, the alignment and centering controls of popular graphic design and GIS applications are presented. Although the terminology differs slightly from application to application, the concepts just presented hold true. A firm grasp of these concepts and terms is crucial to fully understanding this chapter and will help you when creating maps.

11.4 COMMON MAP ELEMENTS

The various purposes a map can serve, together with the wide variety of mapping techniques available, can lead to the perception that every map is completely unique. Despite the great variety of maps in the world, it is important to

DOI: 10.1201/9781003150527-12

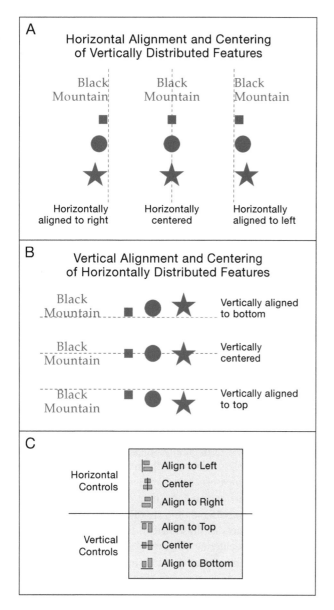

FIGURE 11.1 (A) Options for horizontal alignment and centering. (B) Options for vertical alignment and centering. (C) Alignment and centering controls similar to those found in software applications.

recognize that most are created from a common set of **map elements**. These map elements represent the building blocks of cartographic communication: the transmission of geographic information through the use of maps. The following list presents the most common map elements, each of which is described in detail beginning in Section 11.4.1.

1. Frame line and neat line
2. Mapped area
3. Inset
4. Title and subtitle
5. Legend
6. Data source
7. Scale
8. Orientation

These eight map elements are listed in order of decreasing size—elements toward the top of the list occupy the most space and elements toward the bottom occupy the least. These map elements are also listed in a logical progression and represent the order in which we recommend they be placed when constructing a map. Larger map elements are placed first and smaller map elements are placed last. For example, the frame line occupies the largest area, establishes the size and shape of the initial **available space**, and should be placed first. Available space refers to areas on a page or screen that can be used to place map elements. It is also referred to as "white space." The mapped area and inset are among the largest map elements and should be placed after the initial available space has been defined by the frame line; the title, subtitle, and legend are intermediate in size and should be placed in the space that remains after placing the mapped area and inset; smaller map elements such as the data source, scale, and orientation (typically a north arrow), should be placed last in the space that remains.

Cartographic design, described in detail in Chapter 13, requires that you decide which map elements to include on a particular map. In addition to choosing appropriate map elements, you need to decide on the most appropriate implementation of each element. The content, style, size, and position of each map element must be carefully considered, even before map construction begins.

Proper choice and implementation of map elements are governed in large part by the purpose of the map and its intended audience. The **map user** represents the intended audience; virtually every decision you make should be made in reference to the needs of the map user. The proper choice and implementation of map elements result in a minimization of "map noise," which refers to unnecessary or inappropriate symbolization, design, and typography that interfere with the map user's ability to interpret the map. In Figure 11.2A, map noise interferes with the map's ability to communicate—the noise overpowers the message and hinders the map user's ability to interpret the map. The map in Figure 11.2B is stripped of distracting map noise and is better able to effectively communicate the message: relative life expectancies in African countries.

The rules and guidelines presented here are based on convention, research, common sense, and, to a lesser degree, the opinions of the authors based on many years of experience. We believe it is important to build a foundation of knowledge and experience by following these specific rules and guidelines. After a foundation has been built, we encourage you to consider different points of view and to "break the rules" in creative ways. Always be prepared to explain or defend your reasoning when choosing and implementing map elements. In addition, keep these three points in mind: (1) Think carefully about every action you take when choosing and implementing map elements, (2) Do not implement something simply because you saw it on another map—the world is flooded with examples of poor cartographic choices, and (3) Do not passively accept default

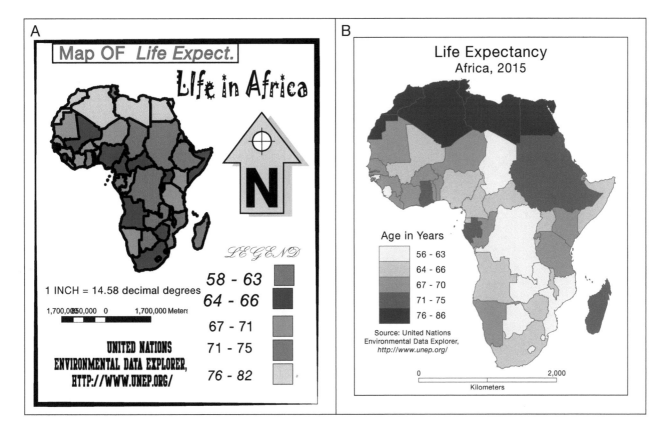

FIGURE 11.2   (A) Excessive map noise hinders communication. (B) Absence of map noise allows for more efficient communication of the map's message.

settings that are built into software applications—defaults are often created by computer programmers with little or no training in cartographic design.

11.4.1   FRAME LINE AND NEAT LINE

The frame line and neat line help to organize the map's contents and to define its extent. A **frame line** encloses all other map elements (Figure 11.3A); it is similar to a picture frame because it focuses the map user's attention on everything within it. The frame line should be the first map element placed because it defines the initial available space that all other map elements will be placed within. A **neat line** might also appear *inside* a frame line. The neat line is used to crop (i.e., clip or limit the extent of) the mapped area, as it does to the San Francisco Bay area in Figure 11.3B. A frame line should be used in most situations; a neat line is used when the mapped area needs to be cropped. In certain cases, a frame line can also act as a neat line, enclosing all map elements *and* cropping the mapped area (Figure 11.3C).

The style of the frame line and neat line should be subtle. A single thin, black line should be used; slightly thicker lines are appropriate when working with larger formats, such as wall maps and posters. Many large-format maps (e.g., wall maps of the world) employ ornate frame lines, which often detract attention from more important map elements, such as the mapped area. Overly thick or ornate lines should be avoided for the same reason that you would avoid choosing a gaudy picture frame for a fine painting or photograph (Figure 11.3D).

The size (width and height) and position of the frame line depend on the desired map size and page dimensions. The size and position of the neat line are normally dictated by the frame line, the mapped area, and the other map elements.

11.4.2   MAPPED AREA

The **mapped area** is the region of Earth being represented, as illustrated by Kenya in Figure 11.4. It consists of visually dominant **thematic symbols** that directly represent the map's theme (we say little about thematic symbols in this chapter, as we focus on them in Chapter 13 and Part II of this text). The mapped area can also include **base information** that provides a geographic frame of reference for the theme. Base information includes, but is not limited to, place names, transportation routes, and landmarks. In Figure 11.4, the blue areas are thematic symbols representing the percentage of all primary schools that are private within each district of Kenya. Nairobi, the nation's capital, and the railroads represent base information that provides a geographic frame of reference for people who might be familiar with them but might not be familiar with the district boundaries. In this case, the base information also suggests a possible relationship between the capital, the railroads, and the districts having higher percentages of private primary schools.

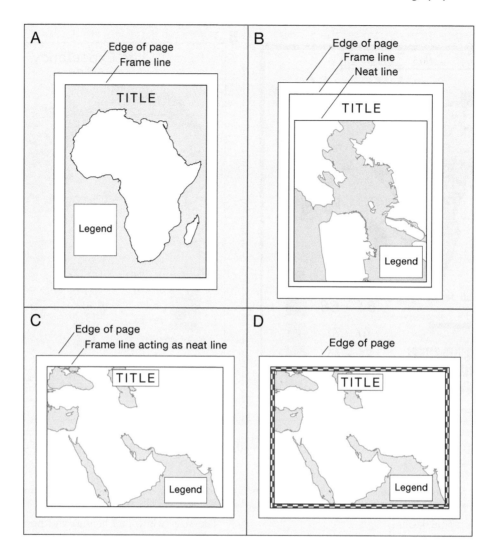

FIGURE 11.3   (A) The frame line encloses all map elements. (B) The neat line crops the mapped area. (C) A frame line can also act as a neat line. (D) Thick or ornate lines should be avoided. (Note that the edge of the page is illustrated to put the frame and neat lines into context. The edge of the page is not illustrated subsequently in this chapter.)

Base information that illuminates the map's theme in this manner is of particular value. As another example, consider Figure 11.9. Four types of base information are included here: interstate highways, city boundaries, city names, and the state line. Without this base information, the thematic symbols (precipitation contours and the associated tints of cyan between contour lines) would be virtually meaningless—the map user would be unable to associate them with a particular geographic region.

In some cases, the mapped area appears as a "floating" geographic region disconnected from neighboring regions, as illustrated by Kenya in Figure 11.4. The use of a floating mapped area, or "closed form," produces available space that often eases the placement of other map elements. However, it also removes the region of interest from its geographic context, making the map more abstract, and possibly confusing the map user. In other cases, a neat line is used to crop the mapped area, as illustrated by the Kansas City region in Figure 11.9. A cropped mapped area represents the region of interest

within its geographic context but can make the placement of other map elements more difficult due to a lack of available space. Consideration of the map's purpose, the geographic region of interest, and other map elements will determine whether the mapped area should be floating or cropped.

The size of the mapped area is dependent on several factors, including the page size, the widths of margins, and the space needed for other map elements. A general guideline is to make the mapped area as large as possible within the available space without being "too close" to the frame line and leaving ample room for the remaining map elements (Figure 11.5A). Maximum size is important because the mapped area—thematic symbols in particular—is instrumental in communicating the map's information. Figure 11.5B represents a mapped area that does not take full advantage of the available space, whereas Figure 11.5C illustrates a mapped area that is too large: It touches the frame line and leaves inadequate space for the title and legend.

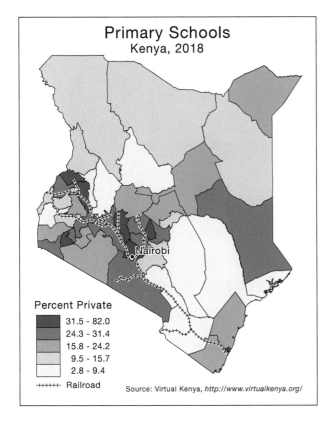

FIGURE 11.4   The mapped area is composed of thematic symbols (blue areas in this example) and base information (Nairobi and railroads in this example).

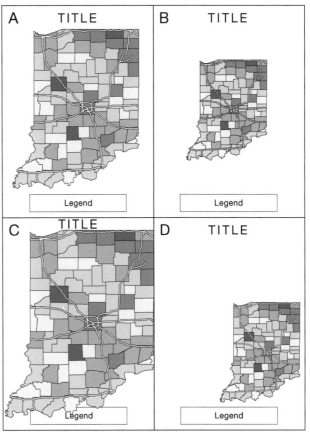

FIGURE 11.5   Sizing and position of the mapped area. (A) Appropriate sizing and position. (B) Insufficient size, but appropriate position. (C) Excessive size, but appropriate position. (D) Insufficient size and inappropriate position. (Note the inclusion of interstate highways, which act as base information.)

The position of the mapped area is dependent on several factors, including the shape of the geographic region, page dimensions, and the other map elements. If possible, the mapped area should be visually centered, both horizontally (side to side) and vertically (top to bottom) within the available space, as defined by the frame line. The mapped areas in Figures 11.5A and 11.5B are appropriately centered; the mapped area in Figure 11.5D is off-center both horizontally and vertically. A properly centered mapped area can lend a sense of balance to a map, but irregularly shaped geographic regions, together with other factors, often make centering impossible. The country of Chile is a classic example of a geographic region that is commonly positioned off-center to accommodate additional map elements.

11.4.3   INSET

An **inset** is a smaller map included within the context of a larger map. Insets can serve several purposes: (1) to show the primary mapped area in relation to a larger, more recognizable area (a locator inset), as illustrated in Figure 11.6A; (2) to enlarge important or congested areas (Figure 11.6B); (3) to show topics that are related to the map's theme, or different dates of a common theme represented by smaller versions of the primary mapped area, as illustrated by the

2000 and 2010 insets in Figure 11.6C; and (4) to show areas related to the primary mapped area that are in a different geographic location, or cannot be represented at the scale of the primary mapped area, as illustrated by Alaska and Hawaii in Figure 11.16.

The style of the inset can vary. In Figure 11.6A, the inset is relatively subtle; its only purpose is to help orient the map user. In parts B and C of Figure 11.6, the insets take on a central focus and attract attention along with the primary mapped area. The size and position of the inset are equally variable, depending on the purpose of the inset, the size of the map, and the other map elements.

11.4.4   TITLE AND SUBTITLE

Most thematic maps require a **title**, although a title is sometimes omitted when a map is used as a figure in a written document, assuming that the theme is clearly expressed in the figure caption. A well-crafted title can draw attention to a map, however, and thus we recommend using a title in virtually all situations—even when the theme is reflected in a figure caption. The title of a

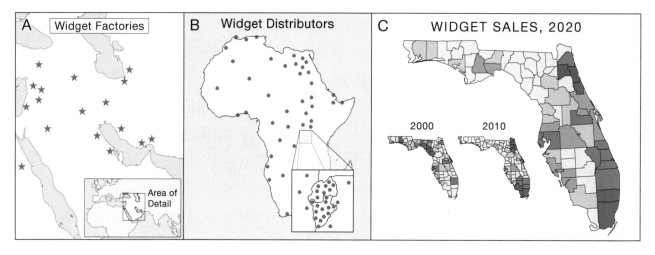

FIGURE 11.6   Insets serve several purposes: (A) locating the primary mapped area; (B) enlarging an important or congested area; and (C) showing related themes or dates. An additional purpose is showing related areas (Figure 11.16).

thematic map should be a succinct description of the map's theme (Figure 11.7A), whereas the title of a general reference map is normally a statement of the geographic region being represented. Unnecessary words should be omitted from the title, but care should be exercised to avoid abbreviations that the map user might not understand. For example, the title "A Map of the Population Density of New Hampshire Counties in the Year 2020" is excessively wordy, and "N.H. Pop. '20" is abbreviated to the point of being cryptic; however, "Population Density" or "Population Density in New Hampshire, 2010" clearly expresses the theme in only a few words (Figure 11.7B). Notice in the last example that the theme, geographic region, and year all appear in the title, and that the theme is aptly stated *before* the region and year. The use of the word *map* in a title is a statement of the obvious and should be avoided.

The **subtitle**, if employed, is used to further explain the title. The name of the region and the data collection date are common components of the subtitle (Figure 11.7C). "New Hampshire, 2010" would be an appropriate subtitle for the "Population Density" title of the previous example. The name of the geographic region is often omitted when the cartographer feels that the map user will easily identify the region. For example, most residents of the United States will recognize the shape of the United States, and most readers of an article focusing on Japan will recognize the shape of Japan.

In the name of legibility, the style of the title and subtitle should be plain. Avoid italics, underlines, ornate type styles, and even bold (Figure 11.7D). Although the use of bold type can emphasize the title and subtitle, it is normally not required if appropriate type sizes are chosen. Use a mask (a subtle bounding box) around the title and subtitle *only* if it is necessary to cover the underlying mapped area to improve legibility, as illustrated in Figure 11.9. The unnecessary use of bounding boxes around map elements is a common source of map noise.

A

Long Term Debt
AVERAGE AGE, 2020
Museums of Modern Art

B

A Map of the Population Density of
New Hampshire Counties
in the Year 2020

N.H. Pop, '20

Population Density in
New Hampshire, 2020

C

Population Density
New Hampsire, 2020

Number of Chickens
Harper County

BIRTH RATE INCREASE
2000 – 2020

Subtitle horizontally centered below title

D

Long Term Debt

AVERAGE AGE, 2021

Museums of Modern Art

FIGURE 11.7   (A) Appropriate titles. The title of a thematic map succinctly describes the map's theme. (B) A wordy title (top), a short but cryptic title (middle), and an appropriate title (bottom). (C) Titles with subtitles. The subtitle is used to further explain the title. (D) Avoid italics, underlines, ornate type styles, and bold.

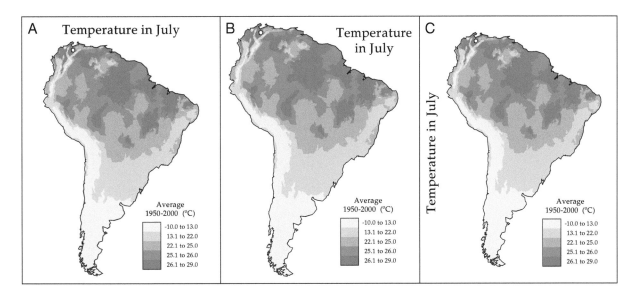

FIGURE 11.8 (A) The title placed at the top center. (B) The title centered in available space after the mapped area is enlarged. (C) Vertical title for the specific case of a rolled map.

The title should generally be the largest type on a map; the subtitle should be visibly smaller than the title. Both the title and subtitle should be limited to one line each in most cases. If possible, the title should be placed toward the top of the map, where the map user is accustomed to seeing titles, and be horizontally centered within the frame line (Figure 11.8A). In Figure 11.8B, the title is placed in the triangular portion of available space to the upper right of South America. This position allows the mapped area to be enlarged, taking full advantage of its available space. Placement of the title in vertical fashion is not recommended unless the map is intended to be rolled lengthwise, in which case, the title can be read when the map is rolled (Figure 11.8C). Large maps are often rolled so that they can easily be transported and stored. A case can also be made for keeping the title directly above the legend to relieve the map user from having to jump back and forth between the title and the legend (Monmonier 1993). Many large-format wall maps make use of this technique because distances between the title and legend can be great. The subtitle, if used, should be located directly below the title and should be horizontally centered with it, as illustrated in Figure 11.7C. The horizontal centering of multiple lines of type creates a "self-balancing" block of type and is used frequently in cartography.

11.4.5 Legend

The **legend** is the map element that defines all of the thematic symbols on a map. Symbols that are self-explanatory or not directly related to the map's theme are normally omitted from simple thematic map legends. In contrast, legends for general reference maps, such as Forest Visitor maps issued by the U.S. Forest Service, often define all symbols found on the map—even if they are self-explanatory. General guidelines for legend design are provided in this section; legend design and content for particular thematic map types are described in Part II of this text.

The style of the legend should be clear and straightforward. Use a subtle mask around the legend only if it is necessary to mask the underlying mapped area, as illustrated in Figure 11.9. Special care should be exercised to ensure that symbols in the legend are *identical* to those found within the mapped area. This includes size, color, and orientation if possible. The legend in Figure 11.10A contains symbols that differ significantly from those in the mapped area; the legend symbols in Figure 11.10B are appropriately identical to those in the mapped area. Additional problems with the legend in Figure 11.10A include an excessively thick line around the mask, inconsistency in type characteristics, poorly aligned symbols and type, and the express bus route symbol—a linear feature that is inappropriately "trapped" inside of an areal symbol (the rectangle). A well-designed legend is self-explanatory and does not need to be identified with "Legend" or "Key" labels like that seen in Figure 11.10A.

Representative symbols should be placed on the left and defined to the right, as illustrated in Figure 11.11A. This arrangement is customary in English dictionaries, which are read from left to right; it allows the map user to view the symbol first and then its definition. In most cases, representative symbols should be vertically distributed (placed in sequence from top to bottom) and horizontally centered with one another, as illustrated in Figure 11.11B. Symbols should be vertically centered with their definitions[1] (Figure 11.11B). Textual definitions such as "Community Garden" and definitions consisting of individual numbers such as "25,000" should be horizontally aligned to the left (Figure 11.11B).

Ranges of numbers are normally separated by a hyphen, which is compact, or the word "to," which can prevent the

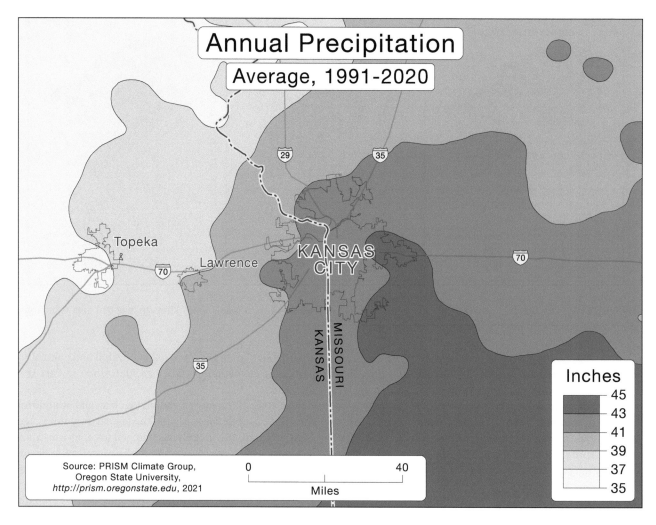

FIGURE 11.9 The title and subtitle placed at the top center. Note how masks (subtle bounding boxes) are used to cover the underlying mapped area. Base information provides a frame of reference for the thematic symbols.

placement of consecutive hyphens when negative numbers are represented (Figure 11.11C). For clarity, spaces should be included to the right and left of each separator.[2] Definitions containing numbers of 1,000 or greater should incorporate commas (thousand separators), and decimal numbers smaller than 1 should incorporate a leading zero, as illustrated, respectively, by "1,021.8" and "0.4" in Figure 11.11C. Integer values are preferred for "whole" things, such as people or farms; decimal values are suitable when representing ratios, percentages, etc. (Figure 11.11D). For readability, use as few decimal places as possible while retaining the meaning of the values.

Space considerations might require the legend to be oriented in a horizontal fashion, with the definitions horizontally centered below the symbols they represent and the symbols vertically aligned to the bottom, as illustrated in Figure 11.11E. This approach is used frequently in this textbook, particularly with mapped areas such as Kansas that are rectangular in shape.

Notice in Figure 11.11 that all the point and line symbols are separated from one another, whereas the areal symbols

(the rectangles, or "boxes") are either separated from one another (Figure 11.11A) or are connected and share a common boundary (Figure 11.11D). Areal symbols within the mapped area are almost always connected and share common boundaries, suggesting that legend rectangles should also be connected. However, the type of data represented by the rectangles also influences whether they should be connected or separated. We suggest that legend rectangles should be connected when representing quantitative magnitudes of a single attribute (e.g., differences in per capita income) because connected rectangles help to emphasize the idea that a gradation of values (associated with a single attribute) is being represented. Legend rectangles should be separated when representing qualitative categories of data (e.g., forest, desert, and tundra). This is because separated rectangles help to reinforce the idea that distinctly different entities are being represented, as opposed to magnitudes of a single attribute.

Areal symbols are sometimes represented in legends by irregular, amorphous polygons, as illustrated in Figure 11.11F. Irregular polygons can look more natural

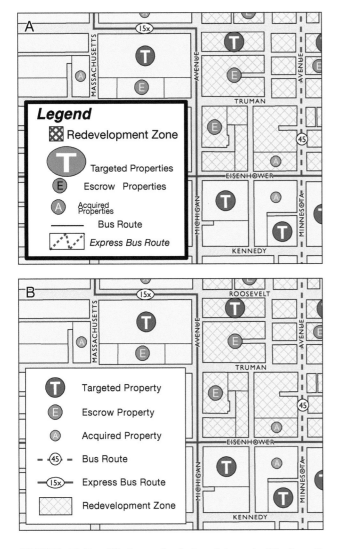

FIGURE 11.10 (A) A poorly designed legend. (B) A well-designed legend that accurately represents symbols found within the mapped area.

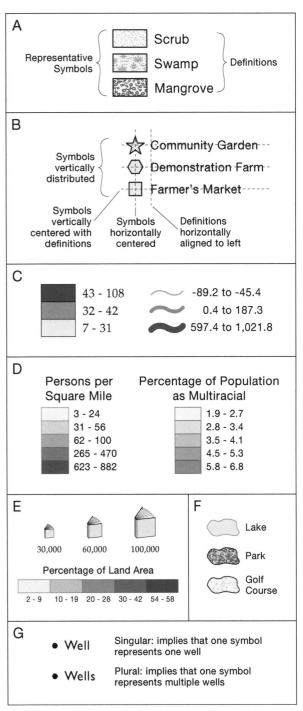

FIGURE 11.11 (A) The legend is composed of representative symbols and definitions. (B) Distribution and alignment of symbols and definitions. (C) Ranges of numbers and numeric formatting. (D) Integer values for whole people and decimal values for percentages of multiracial population. (E) Horizontally oriented legends. (F) Irregular polygons used to represent areal data. (G) Singular vs. plural definitions.

and are most appropriate when representing qualitative categories of areal data such as lakes and golf courses. They should be separated because they represent unique categories of data.

A symbol that represents a singular feature within a mapped area (e.g., a point symbol that represents one well) should be singular in the legend, not plural (Figure 11.11G). A common example of this is a map that includes a legend definition of "Roads." In almost all cases, each road symbol in the mapped area represents a single road, not multiple roads, and should be defined in the legend as "Road" instead of "Roads."

A **legend heading** is often included and is used to further explain the map's theme. The unit of measure (for quantitative data) and the enumeration unit are common components of the legend heading. In Figure 11.11D, the unit of measure is included in each legend heading: square mile and percentage. It is more efficient to place "percentage" just once in the legend heading instead of in each legend category with repeated occurrences of the % symbol. The legend heading should normally be placed above the legend and can be horizontally centered with it; multiple lines of type in the legend heading can be horizontally centered with one another (Figure 11.12A). The horizontal

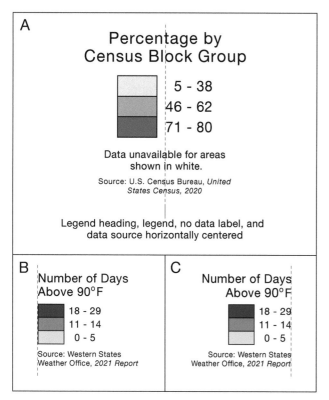

FIGURE 11.12 (A) Typical choropleth legend with legend heading and "no data" label. The data source appears at the bottom. (B) Legend heading, legend, and data source horizontally aligned to the left. (C) The same objects horizontally aligned to the right.

centering of the legend heading, legend content, and related map elements such as the data source creates a group of objects that is self-balancing; it is neither "left-heavy" nor "right-heavy" and can lend to an overall sense of balance in a map. Balance is further described in Section 13.3.5. The same objects are sometimes horizontally aligned to the left (Figure 11.12B) or to the right (Figure 11.12C), but the self-balancing effect is lost.

A category that represents an absence of areal data can be represented on the mapped area in a neutral color such as white or with a subtle texture that is not used in any other category. A simple note below the legend can be used to inform the map user of the "no data" category, as illustrated in Figure 11.12A. A potential problem with this method is that very light gray areas might be confused with the white no-data category, and textures might draw undue attention. Another option is to place a small dot within no-data areas; the dot must then be identified as representing no data in the legend. One problem with this method is that a point symbol is used to represent an areal feature, possibly confusing the map user.

Legend symbols are often organized into groups based on a particular criterion. For example, Figure 11.13 represents three possible methods of grouping symbols that might appear on a map having to do with natural gas

pipeline facilities. In Figure 11.13A, two groups are formed according to whether the symbols represent natural or cultural features. In Figure 11.13B, three groups are formed according to the general geometric form of the symbols: point, line, or area. In Figure 11.13C, two groups are formed according to whether the symbols are thematic in nature and directly related to the map's theme (the upper group), or represent base information (the lower group). This last approach is probably the most appropriate for thematic maps that include base information in the legend.

The legend should be large enough to allow the map user to employ it easily but should not be so large as to occupy vast areas of space or to challenge thematic symbols in the mapped area. The legend heading should be smaller than the subtitle, and the definitions should be smaller than the legend heading, as illustrated in Figure 11.12A.

The position of the legend is dependent on available space, as defined by the other map elements; if possible, the legend should be visually centered within a larger portion of available space. For example, in Figure 11.14, space is available to the left and right of southern South America. Each section of the legend, which has been split in two because of the large number of categories, is visually centered within its available space.

11.4.6 DATA SOURCE

The **data source** allows the map user to determine where the thematic data were obtained. For example, in Figure 11.14, the currency information was acquired from a Web site entitled Nations Online. The particular Web page that contains currency information on the Nations Online Web site (www.nationsonline.org/oneworld/currencies.htm) is considered to be the published data source. Sources of base information such as roads or administrative boundaries are normally omitted from the data source on thematic maps. The data source should be formatted in a manner similar to that of a standard bibliographic reference but is often more concise and less formal. Evaluation of the intended audience's needs will dictate whether to include a fully detailed data source or a simplified version. Many map users have no intention of tracking down a data source but are likely to want a general idea of where the data were obtained. Most data sources in this textbook are simplified versions.

Many data sources are actually compilations of data that have been acquired from a wide variety of original sources. For example, *ArcGIS Data and Maps* is a common source of data that includes data from original sources such as the U.S. Census Bureau, the U.S. Geological Survey, and various international mapping agencies. It is the cartographer's choice whether to specify the original data source or the source of the data compilation.

The style of the data source should be plain and subtle, as illustrated in Figure 11.15. The words "Data Source:" or "Source:" should be included to avoid ambiguity: The data source indicates where the data came from, not map

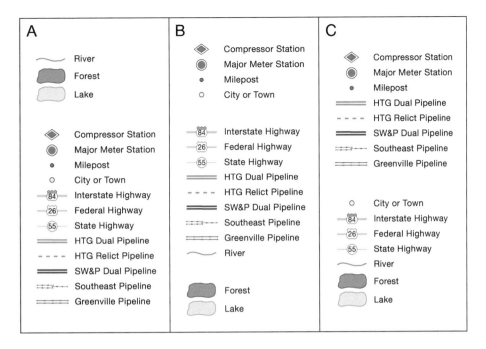

FIGURE 11.13 Grouping of legend symbols according to different criteria: (A) natural and cultural groups; (B) point, line, and polygon groups; and (C) thematic symbol and base information groups.

authorship. If necessary, a separate block of type can be used to indicate map authorship. If the data were obtained from a publication (a book, periodical, data file, website, etc.), the name of the publication should be italicized (as in a bibliographic reference); all other type should be normal.

If the data were obtained from a preexisting map, special methods of citation should be employed based on the type of source map used (Clark et al. 1992). In most cases, multiple lines of type in a data source should be horizontally centered with one another (Figure 11.15A). The data source

FIGURE 11.14 The legend is visually centered within the available space if possible. In this example, the legend has been split into two sections, each centered within its available space.

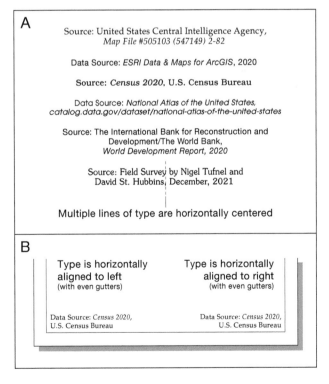

FIGURE 11.15 (A) Data sources. The data source indicates where thematic data were obtained. It is among the smallest type on a map. (B) Alternatives to horizontally centering multiple lines of type in a data source.

should be among the smallest type on a map; its purpose is to inform the curious, not to attract attention. Optimally, the data source is placed directly below the legend to which it refers, and is horizontally centered with the legend heading and legend body, as illustrated in Figure 11.12A. It is also common to place the data source in the lower-right corner (inside the frame line) with the type horizontally aligned to right or in the lower-left corner with the type horizontally aligned to the left (Figure 11.15B). In this case, the spaces between the data source and the frame line are called "gutters" and should be of equal size.

Related to the data source is information regarding the coordinate system and map projection used. The type of projection can become increasingly important to the map user as larger geographic regions are represented and as distortion of the mapped area increases. Include a separate block of type indicating the projection (and information related to the projection) if you feel it will help the map user to interpret the map more accurately. Like the data source, blocks of type related to map authorship and map projections should be among the smallest on the map.

11.4.7 Scale

The **scale** indicates the amount of reduction that has taken place on a map or allows the map user to measure distances. Scale is further described in Chapter 6.

As discussed in Chapter 6, the **representative fraction** (e.g., 1:24,000) is a ratio of map distance to Earth distance, which indicates the extent to which a geographic region has

been reduced from its actual size. For example, on a map with a scale of 1:24,000, one unit of distance on the map (e.g., 1 cm) represents 24,000 of the same units on Earth, regardless of the unit of measure used. The representative fraction becomes invalid if the map on which it appears is enlarged or reduced and cannot easily be used to measure distances. For these reasons, the representative fraction does not appear on most thematic maps.

Although not actually spoken (it is a block of type), the **verbal scale** reads like a spoken description of the relationship between map distance and Earth distance. "One inch to the mile" means that 1 inch of distance on the map represents 1 mile on Earth. Like the representative fraction, the verbal scale becomes invalid if the map on which it appears is enlarged or reduced. Distances on a map can be determined with accuracy using the verbal scale, but only in conjunction with a ruler. Rough measurements can be taken by estimating the length of an inch, centimeter, etc.

The **bar scale**, or scale bar, resembles a ruler that can easily be used to measure distances on a map. Its ability to indicate distance, together with its ability to withstand enlargement and reduction of a map, make it the preferred format to include on a thematic map. You should normally include a bar scale on a thematic map if distance information can enhance the map user's understanding of the theme. The map in Figure 11.14 lacks a bar scale because distance information lends no insight into the type of currency each country uses. However, the map in Figure 11.16 includes bar scales because distance information can potentially

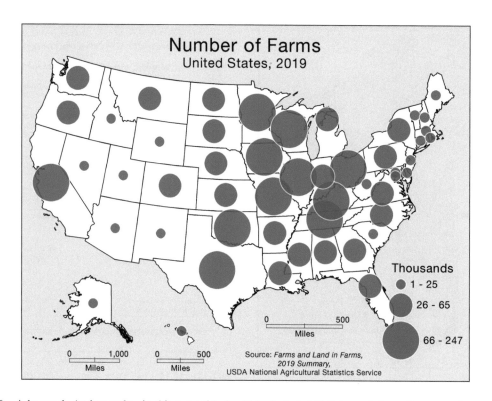

FIGURE 11.16 A bar scale (or bar scales, in this example) should be included if distance information can enhance the map user's understanding of the theme.

enhance the map user's understanding of the theme. For example, general measurements of the state of Colorado taken with the bar scale yield an area of approximately 100,000 square miles. Taking into consideration the minimum number of farms in Colorado's category (26,000), it can be estimated that there is at least one farm per every four square miles, on average. This information is directly related to the map's theme and would be difficult to acquire without a bar scale.

The bar scales in Figure 11.16 also lend insight into the relative amounts of reduction that the primary mapped area and two insets have undergone. For example, compare the bar scale for the contiguous 48 states with the bar scale for Alaska. The bar scale for Alaska is shorter, yet it represents twice the distance in miles, indicating that Alaska is quite large and has been reduced in size much further than the contiguous 48 states.

Bar scales lose much of their utility when used on maps in which the scale changes greatly from one location to another, such as on a map of the entire world. A bar scale should be employed with caution in such cases because it will allow the map user to measure distances accurately only along certain lines (e.g., standard parallels or principal meridians). A variable bar scale is useful with certain map projections (e.g., Miller Cylindrical), as it reflects changes in scale in relation to latitude (Figure 11.17A). A basic understanding of map projections and associated scale changes is essential in providing the map user with a bar scale that is actually useful and accurate.

The maximum distance value represented in a bar scale should always be round and easy to work with, as illustrated in Figure 11.17B. A maximum value of 400 miles is preferred to 437 miles (quick, what's one-quarter of 437?). Decimal values such as 327.75 are difficult to work with and should be avoided in favor of integers. Incorporate a unit of measure that is appropriate for the intended audience. For example, most people raised in the United States will be more familiar with miles than with kilometers, whereas people from outside the United States will be most familiar with kilometers. Incorporate both miles and kilometers if you determine that the map user will benefit from having both (Figure 11.17E). In addition, try to choose a unit of measure that is appropriate for the maximum value on the bar scale. For example, small units of measure such as meters or feet can result in excessively large maximum values such as 10,000,000 meters as opposed to 10,000 kilometers, which is more easily understood. Similarly, a bar scale with a maximum value of 0.18 miles could also be represented in feet, with a maximum value of 1,000.

The bar scale should incorporate a small number of intermediate tick marks to aid in taking measurements. It is important that every tick mark has a round, easy-to-work-with value, even if the value is not indicated. In Figure 11.17C, scale values in parentheses would not actually appear on the bar scale; in the first example, the values are round and easy-to-work-with, while those in the second example are not.

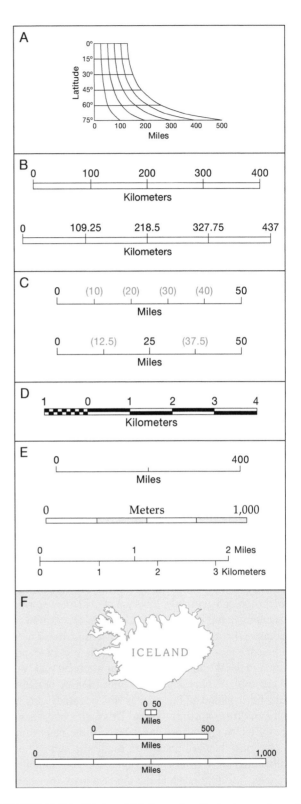

FIGURE 11.17   (A) A variable bar scale reflects changes in scale in relation to latitude. (B) Round, easy-to-work-with bar scale values and awkward decimal values. (C) Round, easy-to-work-with intermediate tick mark values and awkward decimal values (these values are implied (in gray) and thus would not appear on the map). (D) Bar scale incorporating an "extension scale" to the left of zero and the inappropriate "checkerboard" style. (E) Subtle, simple, well-designed bar scales. (F) The 500-mile bar scale represents the most appropriate length in this example.

Some bar scales include an "extension scale," which consists of numbers and tick marks positioned to the left of zero, as illustrated in Figure 11.17D. The extension scale can be useful when employing a method of measuring map distances using a string but is also a source of confusion for many map users, who expect numbers to the left of zero to be negative (as on a number line). Potential confusion, together with the fact that small, intermediate distance ticks can be included to the right of zero, argue against the use of extension scales on thematic maps.

Like the data source, the bar scale should be subtle; its purpose is to inform the curious, not to attract attention. Bulky and complex designs that attract attention, such as the popular "checkerboard" style in Figure 11.17D, should be avoided in favor of subtle and simple designs (Figure 11.17E). Line weights should be fine, and type should be among the smallest on a map. Avoid the use of bold and italic type styles.

The bar scale should be long enough to be useful but not so long that it is cumbersome. Figure 11.17F illustrates three bar scales of differing lengths, each of which is accurate. The 50-mile bar scale is too short to be useful, and the 1,000-mile scale is cumbersome—it is actually wider than the mapped area (Iceland). The 500-mile bar scale is a good solution because it is long enough to be useful but does not take up too much horizontal space. The optimal length is directly tied to the size of the mapped area and to the quest for a maximum distance value that is round and easy to work with. If possible, the bar scale should be placed below the mapped area, where the map user is accustomed to finding it, as illustrated in Figure 11.16.

11.4.8 Orientation

Orientation refers to the indication of north on a map. Orientation can be indicated by a **north arrow** or through the inclusion of a **graticule** (a system of grid lines, normally representing longitude and latitude). It is a common misconception that *every map* needs a north arrow. The orientation of maps with north at the top is a long-standing tradition in the Western world, and it is assumed that "north is at the top" of most maps. An indication of orientation should be included if the map is *not* oriented with geographic or "true" north at the top. This can occur when the shape of the mapped area, together with the page dimensions, require the cartographer to rotate the mapped area, as illustrated by California in Figure 11.18A. Since map users expect to find north at the top of a map, they should be notified if north isn't at the top. Another case in which you would indicate orientation is when the map is intended for use in navigation, surveying, orienteering, or any other function in which the determination of direction is crucial. Finally, an indication of north should be given if geographic features such as roads are oriented in a manner that might confuse the map user. For example, a north arrow should be included when a rectangular street grid is oriented to magnetic north instead of geographic north (Figure 11.18B).

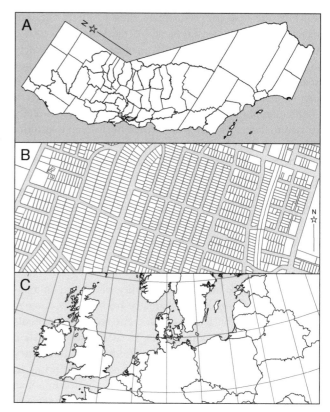

FIGURE 11.18 (A) Use of a north arrow on a map not oriented with north at the top. (B) Indication of geographic north when geographic features are oriented to magnetic north (17° E in this case). (C) Meridians of a graticule indicating the direction of north.

A graticule indicates direction through the orientation of grid lines, typically meridians that run north–south, as illustrated in Figure 11.18C. Notice that the meridians in this example are not parallel to one another—this indicates that the direction of north is variable, depending on which part of the map is focused on. Just as caution should be exercised when employing bar scales on maps with a variable scale, caution should be exercised when placing north arrows on maps in which the direction of north is variable. For example, it would be inappropriate to include a north arrow on the map in Figure 11.18C because it would indicate north for only one particular location, not for the entire mapped area. In addition to orientation, the graticule can also provide positional information. This can be important when a map's theme is in some way related to latitude or longitude. For example, a graticule, or at least marginal tick marks, is often included on thematic maps of natural vegetation because this attribute is strongly influenced by latitude. Labels indicating the values of grid lines should be included when employing a graticule in this fashion. For example, a meridian might be labeled "60° E," and a parallel might be labeled "20° S."

The style of the thematic north arrow and graticule should be simple and subtle. Like the data source and bar scale,

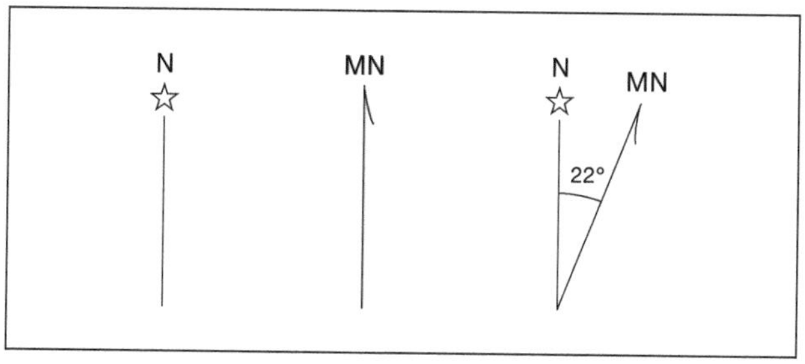

FIGURE 11.19 Subtle and simple north arrows for geographic north (N), magnetic north (MN), and both combined (a compound north arrow).

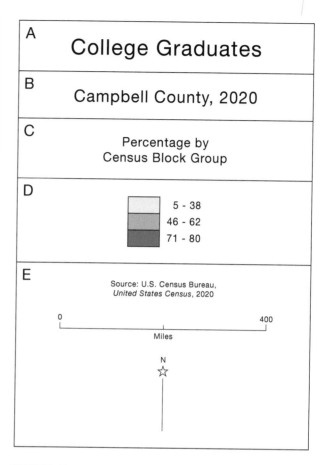

FIGURE 11.20 Relative type sizes for certain map elements. (A) The title is typically the largest type on a thematic map. (B) The subtitle is smaller than the title. (C) The legend heading is smaller than the subtitle. (D) The legend definitions are smaller than the legend heading. (E) Type contained in the data source, scale, and north arrow is smaller than the legend definitions.

these map elements should not attract attention. Line weights should be fine, and type should be among the smallest on a map. Bulky and complex north arrows should be avoided, and only north should be indicated (if necessary, the map user can infer the other cardinal directions). Subtle and simple north arrows are illustrated in Figure 11.19, where a five-pointed star (representing Polaris, the North Star) indicates geographic north, and an arrowhead (representing a compass needle) represents magnetic north. The arrowhead is not the most logical choice for indicating geographic north because compass needles point to magnetic, not geographic, north. Regardless of this logical inconsistency, the arrowhead is commonly used instead of the five-pointed star to indicate geographic north. Compound north arrows indicate both geographic and magnetic north and include the declination (in degrees) between the two. Compound north arrows should be reserved for use on maps intended for navigation with a compass.

The north arrow should be relatively small; it should be large enough to find and use but not so large that it attracts attention. The north arrow should be placed in an out-of-the-way location, preferably near the bar scale.

11.4.9 Relative Type Sizes for Certain Map Elements

Chapter 12 covers typography, or the words that appear on maps, in detail. Type is typically used to identify point, line, and area symbols on a map, but it is also integrated with most of the map elements just described. A normal progression of relative type sizes associated with map elements 4–8 from large to small is provided in Figure 11.20. The largest type on most simple thematic maps should be the title, which describes the map's theme (Figure 11.20A), followed by the subtitle, which commonly provides the geographic region and date (Figure 11.20B). The legend heading is smaller than the subtitle and is used to further explain the map's theme—the unit of measure and enumeration unit in this case (Figure 11.20C). The legend definitions define the representative symbols and are smaller than the legend heading (Figure 11.20D). The data source, bar scale, and north arrow typically contain the smallest type found on a thematic map (Figure 11.20E).

11.5 SUMMARY

In this chapter, we have introduced map elements as the building blocks of cartographic communication and have described their appropriate selection and usage in response to the needs of the map user. The frame line and neat line should be subtle and used to define a map's extent. Thematic symbols directly represent a map's theme and should be visually dominant. They are often used in conjunction with base information in the mapped area. The inset is a smaller map used in the context of a larger map and can serve several purposes. The topic of a thematic map is clearly expressed with a prominent, concise title and is further explained in the subtitle. Map symbols that are not self-explanatory are defined in the legend of a thematic map; representative symbols and their definitions are arranged in various ways according to the type of map and other factors. The map user is told where thematic data were obtained through the use of a subtly designed data source that is similar to a bibliographic reference but often simpler. The bar scale was identified as the most practical tool to allow the map user to make general measurements on a thematic map. A subtle, easy-to-use bar scale should be included if it will help illuminate a map's theme. A subtle north arrow or graticule also can be employed if the map is not oriented with north at the top or if the map user will require directional information.

11.6 STUDY QUESTIONS

1. Differentiate between horizontal and vertical alignment and centering of map elements.
2. List the eight map elements in their recommended order of placement and explain why they are placed in this order.
3. Describe each map element in terms of its purpose and general characteristics.
4. Decide which map elements should be used and which should be omitted for a particular mapping project.
5. Describe each map element in terms of its preferred style, size, and location in relation to other map elements.
6. Compare certain map elements in terms of relative type sizes.

NOTES

1 The legend in Figure 11.9 is an exception. Here, definitions represent the values of the boundaries *between* symbols.
2 An en dash can also be used as a separator. An *en* is a unit of measure equal to half the point size of the type being used. En dashes are half the width of *em* dashes and are used to separate ranges of numbers in written documents without the use of spaces to the right and left of the dash. For clarity in map legends, we recommend the use of the hyphen (which is shorter than the en dash) or the word "to," with spaces on both sides.

REFERENCES

Monmonier, M. S. (1993) *Mapping It Out: Expository Cartography for the Humanities and Social Sciences.* Chicago: University of Chicago Press.

Clark, S. M., Larsgaard, M. L., and Teague, C. M. (1992) *Cartographic Citations, A Style Guide.* Chicago: American Library Association, Map and Geography Round Table, MAGERT Circular No. 1.

12 Typography

12.1 INTRODUCTION

This chapter is dedicated to **typography**, the art or process of specifying, arranging, and designing type. Type is organized according to type family, type style, typeface, and type size. Type can be modified for specific purposes by altering letter spacing, word spacing, kerning, and leading.

General guidelines are provided for the use of type in cartography. These include avoiding the use of decorative type families, the acceptable use of two distinctly different type families, determining a realistic lower limit for type size, the sizing of type to correspond with the relative sizes or importance of real-world features, the need to critically evaluate all aspects of type, and the importance of spell checking. Specific guidelines are provided for point, line, and areal features.

We also provide a description of automated type placement and related **labeling software** that employs aspects of expert systems to place type labels for point, linear, and areal features according to rules and guidelines specified by cartographers.

This chapter was not written with specific software applications in mind. Graphic design applications such as Adobe Illustrator typically provide the greatest control over graphics and type, but recent advances in the design capabilities of Geographic Information Systems (GIS) have narrowed the gap, allowing you to produce high-quality maps in a user-friendly GIS environment. The examples in this chapter reflect simple thematic maps, although the principles set forth can also be applied to more complex thematic maps and general reference maps.

12.2 LEARNING OBJECTIVES

- Define typography and describe the special role it plays in cartography.
- Describe characteristics of type including type family, type style, type face, type size, font, case, and serifs.
- Compare letter spacing, word spacing, kerning, and leading.
- Explain several general typographic guidelines.
- Summarize several specific typographic guidelines for point features, linear features, and areal features.
- Describe automated type placement and provide one advantage and one disadvantage of related labeling software.

12.3 WHAT IS TYPOGRAPHY?

Type, or text, refers to the words that appear on maps. **Typography** is the art or process of specifying, arranging, and designing type. Several of the map elements described in the previous chapter are partly composed of type—the legend, for example—and others, such as the data source, are composed entirely of type. Type can be considered a special sort of symbol or even a map element in its own right. Type that is well designed and smartly applied can make a map easier to understand and more attractive. Type plays an indispensable role in cartography but is often taken for granted or treated as an afterthought. While maps are composed of both graphics and type, we argue that the graphics receive more attention than the type. Graphics, after all, are what differentiate maps from purely written documents. The map in Figure 12.1A is composed of both graphics and type—the two are integrated to the point that they are seen as a unified whole. In Figure 12.1B, we see only the type, which seems more abundant when isolated than it does when combined with the underlying graphics (the mapped area). This type is responsible for a large proportion of the information contained in the map. Figure 12.1C reveals the relative simplicity of the underlying graphics. The strength of maps is realized when graphics and type work together in harmony.

The rules and guidelines for the use of type in cartography are derived from general rules of typography, but they have been modified over time to reflect the specific purposes of mapmaking. Fortunately, these rules and guidelines are relatively well defined. We encourage you to build a foundation of typographic skills by following the specific rules and guidelines presented here and then to consider alternative approaches. Always be prepared to explain or defend your typographical decisions.

12.3.1 CHARACTERISTICS OF TYPE

Type is commonly organized according to characteristics such as type family, type style, and type size. Variations in type family, style, size, and other attributes such as color, provide the ability to differentiate map features, to imply relative importance, and to distinguish between qualitative and quantitative information (Brath and Banissi 2017). **Type family** refers to a group of type designs that reflect common design characteristics and share a common base name—Palatino, for example (Figure 12.2A). Within a type family, type is differentiated by **type style**: roman (normal), bold, and italic are common type styles[1] (Figure 12.2B). Additional type styles include condensed,

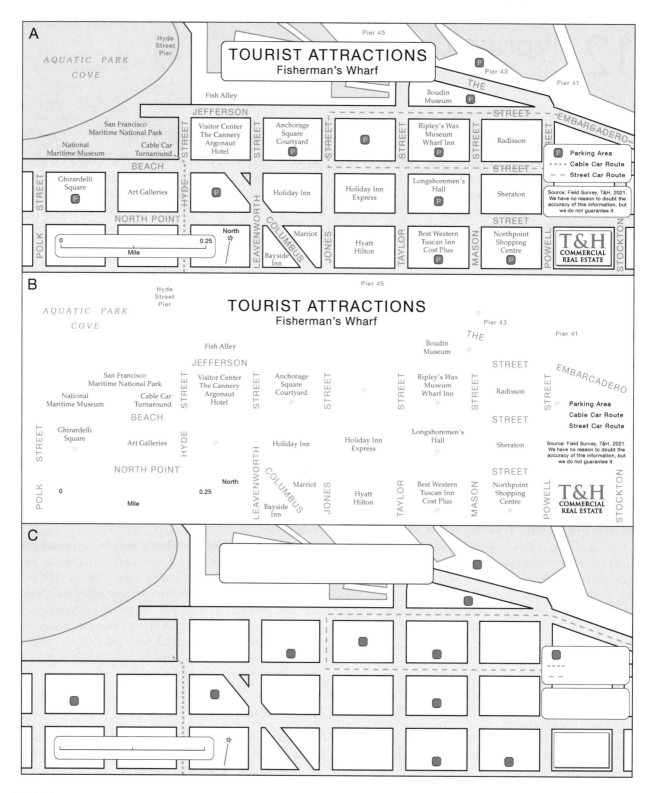

FIGURE 12.1   (A) Graphics and type working in harmony. (B) Type only. (C) Graphics only.

expanded, light, and extra bold. Type of a particular family and style is referred to as a **typeface**, such as Palatino Roman or Helvetica Bold (Figure 12.2C). A particular typeface is further differentiated by type size (Figure 12.2D). **Type size** (height) is measured in points (one point equals 1/72″). Although type sizes are described in points, the

actual height of a given character cannot be inferred from its point size; the point size refers to the height of the metal block on which type was created prior to the development of digital type. The term **font** also has roots in pre-digital typography. When type was set using metal blocks, a font referred to a set of all alphanumeric and special characters

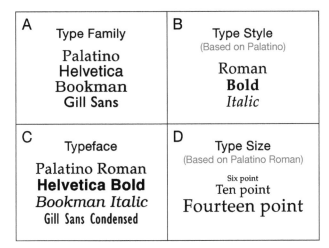

FIGURE 12.2   (A) Type family. (B) Type style. (C) Typeface (a particular type family and type style). (D) Type size measured in points.

of a particular type family, type style, and type size, such as Palatino Italic, 12 point. Palatino Italic, 10 point would represent a separate font because of its different type size. With the advent of digital type, the term font has become size independent, as a single typeface can be scaled to any size. As a result, the terms typeface and font have become somewhat interchangeable, with both representing the combination of a *particular type family and type style*. Many typographical terms are misunderstood and misused. For example, the term *font* is commonly (and inappropriately) used to refer to a type family.

Uppercase and lowercase letters are used in cartography, but lowercase letters have proved to be easier to read. This is because lowercase letters are less blocky, and they provide more detail that helps differentiate one letter from another. The majority of type on a map should be set in title case, as illustrated in Figure 12.3A. **Title case** is composed of lowercase letters with the first letter of each word

set in uppercase. Conjunctions and other "linking words" (*in, on, or, of, per, by, for, with, the, and, over*, etc.) are set in lowercase. Title case is appropriate for use in titles, subtitles, legend headings, legend definitions, labels for point and line features, and so on. **Sentence case** is composed of lowercase letters with the first letter of each sentence set in uppercase and is appropriate when formal sentences are used, such as in textual explanations or descriptions appearing on a map. Words set in all **uppercase** are used as labels for areal features, as described later.

Serifs are short extensions at the ends of major letter strokes, as illustrated in Figure 12.3B. Type families with serifs are termed serifed; type families without serifs are termed sans serif ("sans" means "without"). Serifed type is preferred in the context of written documents because the serifs provide a horizontal guideline that helps to tie subsequent letters together, reducing eye fatigue. In cartographic applications, where type is used primarily as short labels and descriptions, both serifed and sans serif type is used; neither has proved to be more effective than the other. In certain situations, a serifed type family can be used for one category of features (e.g., natural), and a sans serif type family can be used for another category of features (e.g., cultural).

Letter spacing, or character spacing, refers to the space between each letter in a word, and **word spacing** refers to the space between words. Default letter and word spacing (Figure 12.4A) results in compact type that is often easier to place on complex maps (because the words occupy less horizontal space). Slightly increased letter and word spacing results in type that appears to be less "cramped" and is easier to read. You should employ slightly increased spacing if possible. The blocky nature of all-uppercase type normally requires greater letter and word spacing than lowercase type to prevent it from looking cramped. Exaggerated letter and word spacing are often employed in conjunction with all-uppercase type when labeling areal features, as described later. Letter and word spacing should be kept consistent within individual blocks of type and among labels that are otherwise similar. For example, each line of type in a three-line data source should have the same letter and word spacing, as should every label that identifies a city on a particular map (Figure 12.4B). Figure 12.4C illustrates inconsistent letter and word spacing for the same data source and city labels. Related to letter and word spacing is *tracking*, which refers to the simultaneous alteration of space between both letters and words. Tracking should be used with caution, however, because it does not allow for the independent adjustment of letter and word spacing.

Kerning refers to the variation of space between two adjacent letters, as illustrated in Figure 12.4D. Different combinations of adjacent letters require different amounts of kerning if spaces between letters are to be visually consistent. For example, the letter pair WA requires the removal of space between the letters for the pair to look consistent with MN. Digital type includes preset "kerning pairs" that automatically set the space between various

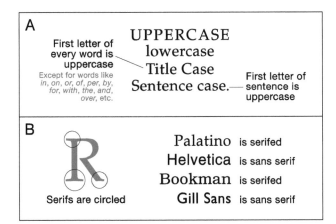

FIGURE 12.3   (A) Uppercase, lowercase, title case, and sentence case. The majority of type on a map should be set in title case. (B) Serifs are extensions at the ends of letter strokes.

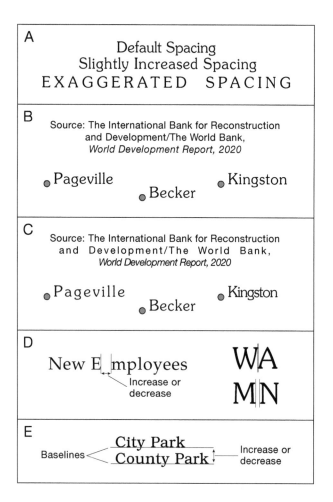

FIGURE 12.4 (A) Letter spacing and word spacing can be altered for different situations. (B) Consistent letter and word spacing. (C) Inconsistent letter and word spacing. (D) Kerning adjusts the space between two individual adjacent letters. (E) Leading is the space between lines of type, from baseline to baseline.

letter pairs. Kerning can also be performed manually, after letter and word spacing has been specified, to adjust the space between letter pairs that still appear to be too close together or too far apart. Kerning is measured according to a unit called an *em*, which is equal to the point size of the type being used.

Leading (pronounced like *heading*), or line spacing, refers to the vertical space between lines of type according to their *baselines* and is altered to place lines of type closer together or farther apart (Figure 12.4E). Leading should be great enough to allow multiple lines of type to be read easily but not so great as to result in wasted space between lines.

12.4 GENERAL TYPOGRAPHIC GUIDELINES

The following is a list of general guidelines for the use of type in cartography.

1. Avoid the use of decorative type families and use bold and italic styles sparingly. Script, cursive, and otherwise fancy and ornate styles are unnecessarily difficult to read. They should be avoided in favor of more practical type families, such as those used as examples in this chapter. The overuse of bold styles can overshadow other type and map elements, and bold is often not necessary if appropriate type sizes are chosen. If possible, italic type should be reserved for two primary applications: to label hydrographic (water) features and to identify publications in the data source, as illustrated in Figure 12.5. Italics are appropriate for hydrographic features because their slanted form resembles the flow of water. (It is also conventional to use the color cyan for hydrographic labels and features.) The use of italics for publications is standard bibliographic practice. An additional application of italic type is to identify the species of an organism on a map, such as *Pelecanus occidentalis*. The wide variety of features on general reference maps might require that you use bold and italic styles outside of these guidelines.

2. Avoid using more than two type families on a given map. Simpler maps can be limited to one type family. For the sake of consistency, map elements such as the title, subtitle, legend heading, legend definitions, data source, and scale should all employ the same type family. If two type families are required (e.g., to label a wide variety of map features), choose families that are distinctly different—one serifed and one sans serif, for example. As mentioned earlier, a serifed family can be used for one category of features (e.g., natural), and a sans serif family can be used for another category of features (e.g., cultural). This is illustrated in Figure 12.5, where the rivers, forest, and lake are labeled with a serifed type family, and the cities are labeled with a sans serif family. Notice that Lake Isabella is treated as a natural feature here. It could also be treated as a cultural feature formed by the construction of Isabella Dam.

3. Choose a realistic lower limit for type size; all type needs to be readable by the intended audience. Factors for consideration include the age and visual acuity of the map user, map reproduction method, anticipated lighting conditions, and the map user's physical proximity to the map. Mark Monmonier (1993) recommended a lower size limit of 7 points for lowercase type, but this value is conservative. Type as small as 4 points (usually uppercase, sans serif) is commonly used on congested paper street maps where space is at a premium. Readability is ultimately tied to the typeface used, crispness of reproduction, and other factors. The only way to ensure the readability of small type is to provide a sample to members of the intended audience.

4. Generally speaking, type size should correspond with the size or importance of map features. For example, the type representing the names of large

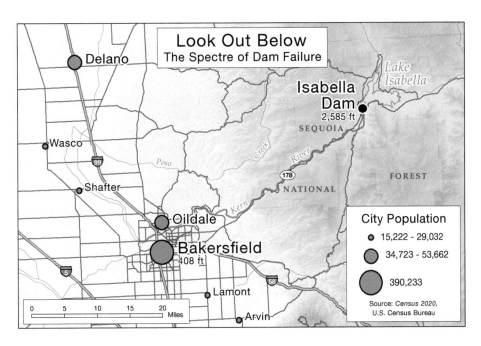

FIGURE 12.5 General typographic guidelines appropriately applied, including the use of distinctively different type families to represent natural and cultural features.

cities should be noticeably larger than the type used to represent small cities (Figure 12.5). Type size is also partially dictated by the relative importance of map elements, as described in sections 11.4 and 13.3. Because map users are not sensitive to slight differences in type size, avoid differences of less than 2 points if possible (Shortridge 1979).

5. Critically evaluate and apply type specifications such as type family, type style, type size, letter spacing, word spacing, kerning, and leading. Do not passively accept the default settings provided by software applications. Instead, consider the purpose of each unit of type in the context of the map, and apply type specifications accordingly.

6. All type should be spell-checked. Spelling errors interfere with cartographic communication and undermine the credibility of a map. Special attention should be focused on the most current spelling of place names, which change over time and are often controversial. Also, be aware that certain older place names are considered to be offensive or derogatory by today's standards (Monmonier 2006).

12.5 SPECIFIC TYPOGRAPHIC GUIDELINES

12.5.1 ALL FEATURES (POINT, LINEAR, AND AREAL)

The following is a list of specific guidelines for the placement of type associated with point, linear, and areal features.

1. Orient type horizontally, as illustrated in Figure 12.6A. One exception is when labeling a map that includes a graticule with curved parallels,

in which case the type should be oriented with the parallels (Figure 12.6B). Another exception is when labeling diagonal or curved linear and areal features, in which case the type should reflect the orientation of the features, as illustrated in Figure 12.6C.

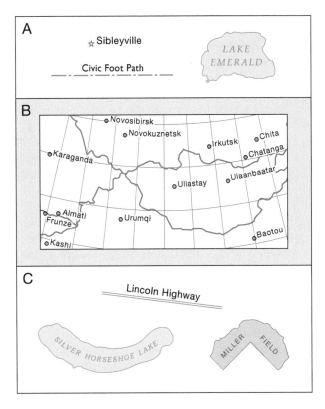

FIGURE 12.6 (A) Type placed horizontally. (B) Type oriented with a graticule. (C) Diagonal and curved type.

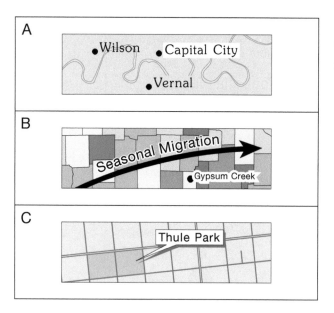

FIGURE 12.7 (A) Masks placed underneath Capital City and Vernal. (B) Halos around type. (C) A callout combines a mask with a leader line.

2. Avoid overprinting and, when unavoidable, minimize its effects. **Overprinting** is a phenomenon that occurs when a block of type is placed on top of a map feature (e.g., a river), obscuring the type and making it difficult to read, as illustrated by "Wilson" in Figure 12.7A. The effects of overprinting can be minimized through the use of a mask, halo, or callout. A **mask** is a subtle bounding box (e.g., a white rectangle) that is placed underneath type but above the mapped area, as illustrated by "Capital City" in Figure 12.7A. As seen in this example, masks can sometimes obscure too much of the mapped area, and thus they should be used with caution. Masks can also be specified with the same color as the underlying area, allowing them to blend in better, as illustrated by "Vernal" in Figure 12.7A. A **halo** is an extended outline of letters (Figure 12.7B). Halos cover less of the underlying mapped area than masks while still allowing the type to be read. Care should be exercised to ensure that halos are not so thick that they attract undue attention or partially obscure the point symbol, as illustrated by "Gypsum Creek" in Figure 12.7B. **Callouts** are a combination of mask and leader line (Figure 12.7C). Callouts are effective but should be used with caution because they are visually dominant and can overshadow other map features.

3. Ensure that all type labels are placed so that they are clearly associated with the features they represent. In pursuit of this goal, it is often useful to place larger type labels first, followed by intermediate and then smaller labels (Imhof 1975).

12.5.2 POINT FEATURES

The following is a list of specific guidelines for the placement of type associated with point features.

1. When labeling point features, select positions that avoid overprinting underlying graphics, according to the **sequence of preferred locations** illustrated in Figure 12.8A. This sequence is based on the work of Pinhas Yoeli (1972) but is modified by the authors according to the idea that, if possible, the symbol should be placed on the left and defined to the right (as in a legend). Notice, also, that the least preferred locations for a label are directly to the right and directly to the left of the symbol.

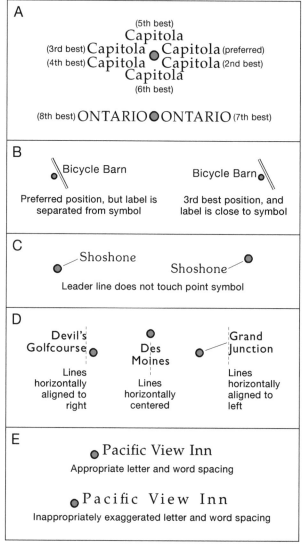

FIGURE 12.8 (A) Sequence of preferred locations for labeling point features. (B) A road is shown coming between a point symbol and its label. (C) Leader lines are used if necessary. (D) Stacked labels (multiple-line labels) are horizontally aligned or centered to imply an association with the symbol. (E) Letter and word spacing for point symbols.

This results in an "unfavorable optical coincidence" (Imhof 1975, 132), in which the point symbol might be misinterpreted as a type character in the label.

2. Do not allow other map features to come between a point symbol and its label (Figure 12.8B). Emphasize the association between the label and symbol by placing the label close to the symbol, even if it means choosing a less preferred location.

3. If the sequence of preferred locations does not provide a suitable option, consider using a mask, halo, or callout. Another option is to use a simple leader line, as illustrated in Figure 12.8C. Leader lines should be very thin (e.g., 0.25 point), not include an arrowhead, and point to the center of the point symbol without actually touching it.

4. Stacked labels (multiple-line labels) should be placed according to the sequence of preferred locations, and individual lines of type should be horizontally aligned or centered to emphasize the association between the label and point symbol (Figure 12.8D).

5. Do not exaggerate letter or word spacing when labeling point features. Exaggerated spacing weakens the association between a point symbol and its label and tends to emphasize the *areal* extent of features (and point features have no areal extent) (Figure 12.8E).

6. Point symbols on land that are close to coastlines should be labeled entirely on land if possible. Point symbols that touch coastlines should be labeled either entirely on land or entirely on water, depending on which option offers greater legibility, as illustrated in Figure 12.9A (Wood 2000). Avoid overprinting the coastline with type (Figure 12.9B).

7. These guidelines should be followed as closely as possible. In practice, however, it is often impossible to adhere to all guidelines simultaneously. Figure 12.9A illustrates how these guidelines might be applied to a map of restaurant locations in a coastal region. Notice in Figure 12.9A that thematic symbols (restaurant locations) are labeled with a sans serif type family, while base information such as streets and the bay are identified with a serifed type family. This is another example of how two distinctly different type families can be used to represent different categories of map features. Figure 12.9B illustrates how poorly labeled thematic symbols can result in a less attractive map that is more difficult to interpret.

12.5.3 LINEAR FEATURES

The following is a list of specific guidelines for the placement of type associated with linear features.

1. Place labels for linear features above the features, close to but not touching them, as illustrated in Figure 12.10A. Descenders, such as the lower

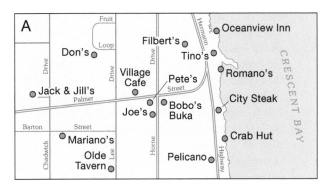
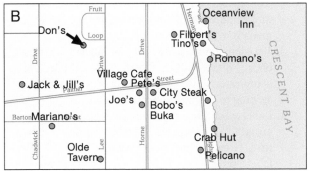

FIGURE 12.9 (A) Point features appropriately labeled. (B) Point features inappropriately labeled.

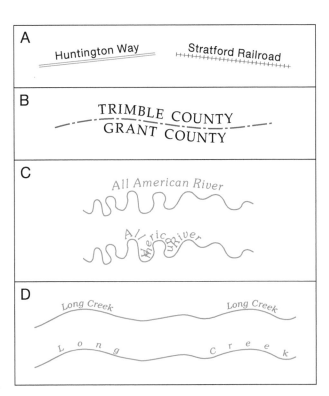

FIGURE 12.10 (A) Type labels placed above and oriented with linear features. (B) Labels on both sides of a boundary between two areas. (C) Labeling the general trend of a complex curve. (D) Labeling a long feature more than once is preferred to exaggerating letter and word spacing.

extensions of "g" and "y" in "Huntington Way," should just clear the line symbol. Type is placed above linear features because it appears to rest on the feature instead of hanging below it and because the bottom edge of lowercase type is normally less ragged than the upper edge, resulting in a more harmonious relationship between label and symbol. One exception is when labeling areas on both sides of a boundary, in which case type appears above and below the line (Figure 12.10B).

2. When labeling linear features that have complex curves, follow the general trend of the feature (Figure 12.10C) because type that curves too much is difficult to read.

3. Very long linear features can be labeled more than once (Figure 12.10D). The use of multiple labels is preferred to the exaggeration of letter and word spacing to emphasize linear extent.

4. Labels for linear features should be placed upright, not upside down. Upright labels read from left to right, whereas upside-down labels read from right to left, as illustrated in Figure 12.11. By convention, type that is absolutely vertical should be readable from the right side of the map (preferred), not the left side (not preferred). These rules actually apply to all type that appears on a map but are most closely associated with linear features. It is sometimes impossible to place an entire type label upright. For example, a long label might start out upright and then end upside down after following a curved feature. In this case, try to have the majority of the label upright, with the remainder upside down.

12.5.4 AREAL FEATURES

The following is a list of specific guidelines for the placement of type associated with areal features.

1. When labeling areal features that are large enough to fully contain a label, visually center the label within the feature, as illustrated in Figure 12.12A. Do not allow the label to crowd the areal symbol—if possible, allow a space of at least one and one-half times the type size (1.5 ems) between the ends of the label and the boundary of the feature (Imhof 1975). As when labeling linear features, follow the *general* trend of areal features that have complex curves.

2. If an area is large enough to contain a label, use all uppercase type (Figure 12.12A). The blocky nature of uppercase type can help to emphasize the areal extent.

3. Exaggerated letter and word spacing can be used to emphasize areal extent and should only be applied to all-uppercase type (Figure 12.12B); lowercase type tends to look disjointed when exaggerated spacing is applied (Figure 12.12C). Caution should be exercised when exaggerating letter and word spacing, as individual letters can become so far apart that the map user would have trouble seeing the label as anything other than individual letters.

FIGURE 12.12 (A) A type label visually centered within an areal feature. (B) Uppercase type with exaggerated letter and word spacing, emphasizing the areal extent of the feature. (C) Lowercase type is not as well-suited to exaggerated spacing. (D) Exaggerated leading. (E) A river labeled as an areal feature, a small areal feature (reflection pool) labeled as a point feature, and a leader line used to help identify an areal feature.

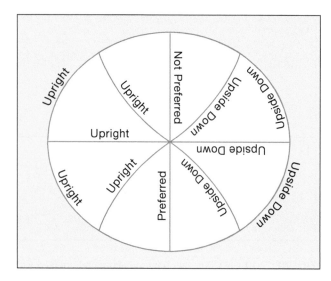

FIGURE 12.11 Type should be placed upright and should read from left to right. Vertical type should be readable from the right side of a map.

An extreme upper limit of four times the type size (4 ems) should be observed for letter spacing. Maximum word spacing may exceed this value, but care should be exercised to ensure that the relationship between words is clear.

4. Leading can also be exaggerated to emphasize an areal extent, as illustrated in Figure 12.12D. Leading should not be so great that the relationship between lines of type is lost.

5. Features typically thought of as being linear (e.g., rivers) that are represented at such a large scale that they appear as areas should be labeled like areal features, as illustrated by the river in Figure 12.12E.

6. If necessary, leader lines can be used with areal features. Leader lines should be very thin (e.g., 0.25 point), not include an arrowhead, and just enter the areal symbol, as illustrated by the Mollisol soil polygon in Figure 12.12E.

7. Areal features that are too small to contain a label can be labeled as if they were point symbols according to the sequence of preferred locations (Figure 12.8A), as illustrated by the reflection pool's label in Figure 12.12E.

8. Several additional approaches can be taken when labeling areal features that are too small to fully contain a label. The large lake in Figure 12.13A can easily accommodate the "LAKE" label at the given type size and letter spacing, so the label is appropriately placed by visually centering it within the area. As indicated previously, an areal feature that cannot accommodate a label can be treated as if it were a point symbol and labeled accordingly. Notice that the "Lake" label in Figure 12.13A is placed in the preferred location (upper right) and is set in title case without exaggerated letter spacing, just as a point symbol would be labeled. A second option is to retain the original type size and letter spacing and handle overprinting with halos (Figure 12.13B). A third option is to retain the original type size but reduce the letter spacing (Figure 12.13C). A fourth option is to use a "condensed" or "narrow" type style (Figure 12.13D). This can only be accomplished if a given type family has a condensed or narrow style. Do not attempt to modify an existing typeface by reducing its horizontal scale, as this essentially deforms the shapes of letters so that they occupy less horizontal space—something the type designer did not intend. A fifth option is to abbreviate the word or phrase (Figure 12.13E). Try to ensure that the abbreviation will be easily understood by the intended audience. A sixth option is to reduce the type size while retaining letter/word spacing relative to the full-sized label (Figure 12.13F). Finally, the areal feature can be treated like a point symbol, as described previously, but with an

area-style leader line that just enters the feature (Figure 12.13G). Regardless of the approach you take, you should try to be as consistent as possible. For example, if you are inclined to retain the original type size and letter spacing but overprint using halos, try to employ that same technique as often as possible.

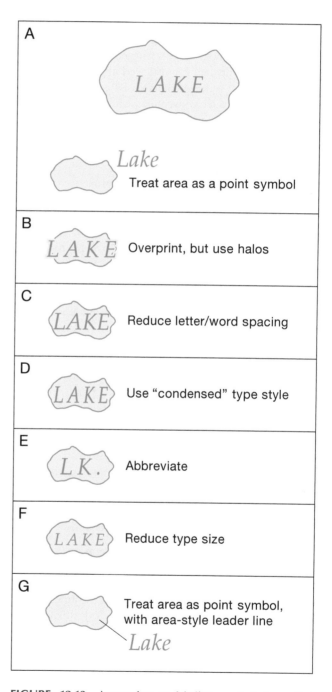

FIGURE 12.13 Approaches to labeling areal features that are too small to contain a label. (A) Labels placed inside and outside of areal symbols. (B) Overprinting with use of halos. (C) Reduction of letter/word spacing. (D) Condensed type style. (E) Abbreviation. (F) Reduced type size. (G) Use of a leader line for an areal symbol.

12.6 AUTOMATED TYPE PLACEMENT

Throughout this section, we have assumed that the cartographer is making all the decisions regarding type placement. When using graphic design software such as Adobe Illustrator, this is almost certainly the case. However, specialized **labeling software** has been developed for automatically positioning type, often within the context of a GIS. Labeling software focuses primarily on the placement of type associated with map features (e.g., streets and lakes) as opposed to positioning map elements that are composed of type, such as the title. Development of this software has been the focus of both computer scientists and cartographers, who have created sophisticated algorithms (including heuristics, or "rules of thumb") based on established rules, guidelines, and conventions of cartographic type placement. Most labeling software applications incorporate aspects of cartographic **expert systems**, which make decisions based on rules and guidelines obtained from cartographic experts (Zoraster 1991).

Labeling software is designed to approach or achieve "optimal" placement of type labels, which avoids both overprinting underlying map features and conflicts among type labels. Two general approaches have emerged: (1) placement of each label in its preferred position, followed by an iterative reorganization of labels to avoid or minimize conflict among labels (Freeman 1995) and (2) casual, suboptimal placement of labels, followed by an iterative reorganization of labels until the combined placement of all labels approaches an optimal state (Edmondson et al. 1996).

The biggest potential advantage of labeling software is its ability to save time, as manual type placement remains one of the most time-consuming aspects of map construction, particularly when labeling linear and areal features. One problem associated with labeling software is the fact that optimal solutions are often computer-intensive when a considerable amount of type must be positioned. Another problem is that the wide variety of maps, together with variations in map scale and complexity, make it difficult to achieve satisfactory results in all situations. The finished product normally requires some interactive editing to arrive at a solution that is visually acceptable.

Two leading examples of labeling software are Label-EZ, distributed by MapText, Inc., and the Maplex Labeling Engine, which is incorporated into GIS applications distributed by Environmental Systems Research Institute (ESRI). Both examples allow the user to specify rules that dictate how type should be placed for individual feature classes (logical groupings of points, lines, polygons, etc.). For example, the user is allowed to specify exactly how labels should be placed for point features representing spot elevations versus how labels should be placed for soil polygons. The software applies these specifications to GIS map layers and attempts to place each type label in its "best" position, usually saving vast amounts of time in comparison with manual type placement.

In addition to the two examples of labeling software just described, a wide range of other computer algorithms has been developed using different methods, with the goal of increasing the efficiency and speed of calculations (Ooms et al. 2012). Calculation speed becomes increasingly important in the realm of interactive maps in which panning and zooming require frequent re-calculation of optimal label placement.

12.7 SUMMARY

In this chapter, we have presented aspects of typography and have identified it as being central to the utility and attractiveness of a map. The cartographer needs to ensure that type is legible and select appropriately from type families, type styles, and type sizes. Lowercase type is more legible than uppercase, and type set in title case (a combination of uppercase and lowercase type) is appropriate for most cartographic applications. Serifed and sans serif type are both used in cartography, as neither type has proved to be more effective. Blocks of type can occupy more or less space by altering letter spacing, word spacing, and leading; spaces between letter pairs can be fine-tuned through kerning. General guidelines for the use of type were described, including the following: avoid decorative type styles and minimize the use of bold and italic; limit a map to two type families; select a minimum type size that will be readable by members of the intended audience; size the type to correspond with the relative size and importance of map features; critically evaluate and specify all aspects of type; and spell check type that appears on a map. Specific guidelines were provided for the use of type associated with point, linear, and areal features, and an overview of automated type placement was given.

12.8 STUDY QUESTIONS

1. Define typography and describe the special role it plays in cartography.
2. Describe characteristics of type including type family, type style, type face, type size, font, case, and serifs.
3. Compare letter spacing, word spacing, kerning, and leading.
4. Explain several general typographic guidelines.
5. Summarize several specific typographic guidelines for point features, linear features, and areal features.
6. Describe automated type placement and provide one advantage and one disadvantage of related labeling software.

NOTES

1 When applying type styles, it is best to use a member of a type family that has been specifically designed with that style (e.g., Bookman Bold). Many software applications allow the roman type to be crudely modified into italic or bold, resulting in type that is unsuitable for high-quality printing or display.

REFERENCES

Brath, R., and Banissi, E. (2017) "Multivariate label-based thematic maps." *International Journal of Cartography 3*, no. 1:45–60.

Edmondson, S., Christensen, J., Marks, J., et al. (1996) "A general cartographic la-belling algorithm." *Cartographica 33*, no. 4:13–23.

Freeman, H. (1995) "On the automated labeling of maps." In *Shape, Structure and Pattern Recognition*, ed. by D. Dori and A. Bruckstein, pp. 432–442. Singapore: World Scientific.

Imhof, E. (1975) "Positioning names on maps." *The American Cartographer 2*, no. 2:128–144.

Monmonier, M. (1993) *Mapping It Out: Expository Cartography for the Humanities and Social Sciences.* Chicago: University of Chicago Press.

Monmonier, M. (2006) *From Squaw Tit to Whorehouse Meadow: How Maps Name, Claim, and Inflame.* Chicago: University of Chicago Press.

Ooms, K., De Maeyer, P., Fack, V., et al. (2012) "Investigating the effectiveness of an efficient label placement method using eye movement data." *The Cartographic Journal 49*, no. 3:234–246.

Romano, F. J. (ed.). (1997) *Delmar's Dictionary of Digital Printing and Publishing.* Albany, NY: Delmar.

Shortridge, B. G. (1979) "Map reader discrimination of lettering size." *The American Cartographer 6*, no. 1:13–20.

Wood, C. H. (2000) "A descriptive and illustrated guide for type placement on small scale maps." *The Cartographic Journal 37*, no. 1:5–18.

Yoeli, P. (1972) "The logic of automated map lettering." *The Cartographic Journal 9*, no. 2:99–108.

Zoraster, S. (1991) "Expert systems and the map label placement problem." *Cartographica 28*, no. 1:1–9.

13 Cartographic Design

13.1 INTRODUCTION

This chapter introduces **cartographic design** as a process in which the cartographer conceptualizes and creates maps according to the needs of the intended map user. Topics in this chapter expand upon those presented in Chapters 11 and 12 (Map Elements and Typography); therefore, we strongly recommend that you read Chapters 11 and 12 first. Aspects of cartographic design have been derived from the results of **map design research** and have been influenced by principles of graphic design and **Gestalt principles** of perceptual organization.

The design process is distilled into seven procedures, including determination of how the map will be reproduced, selection of appropriate scale and map projection, identification of appropriate data classification and symbolization methods, selection and appropriate implementation of map elements, the establishment of an **intellectual hierarchy**, creation of rough **sketch maps**, and construction of the map using a software application. The design process culminates in an evaluation of the map by members of the intended audience. Specific aspects of cartographic design are **visual hierarchy**, **contrast**, **figure-ground**, and **balance**.

A case study is presented that illustrates how an effective, attractive thematic map can be designed according to the information in this chapter and in Chapters 11 and 12. General steps include consideration of what the real-world distribution of the phenomenon might look like, definition of the map's purpose and its intended audience, and data collection. Specific steps include the list of seven procedures just mentioned that range from the determination of how the map will be reproduced to the construction of the map using a software application.

This chapter was not written with specific software applications in mind. Graphic design applications such as Adobe Illustrator typically provide the greatest control over graphics and type, but recent advances in the design capabilities of Geographic Information Systems (GIS) have narrowed the gap, allowing you to produce high-quality maps in a user-friendly GIS environment. The examples in this chapter reflect simple thematic maps, although the principles set forth can also be applied to more complex thematic maps and general reference maps.

13.2 LEARNING OBJECTIVES

- Describe cartographic design in terms of the process, goals, map design research, and Gestalt principles.

- Enumerate the seven cartographic design processes that comprise step 4 in the map communication model.
- Establish an appropriate intellectual hierarchy for a thematic map that includes thematic symbols, base information, and map elements.
- Differentiate between the intellectual hierarchy and the visual hierarchy.
- Describe the roles of contrast, figure-ground, and balance in the design process.
- Provide an overview of how the contents of Chapters 11–13 have been applied to the Real Estate Site Suitability case study map.

13.3 ELEMENTS OF CARTOGRAPHIC DESIGN

Cartographic design is a partly mental and partly physical process in which maps are conceived and created. The design process appears formally as step 4 of the map communication model that was introduced in Chapter 1 (Figure 13.1), but it encompasses aspects of all five steps, from imagining the real-world distribution to evaluating the resulting map. The design process does not end until the final map has been completed. The word *design* also describes the appearance of a map; a map can have a particular design, but design in that sense is only the end result of the design *process*. Successful cartographic design results in maps that effectively communicate geographic information. Maps do not necessarily communicate *knowledge* but, rather, stimulate and suggest knowledge through the transmission of information (Montello 2002).

Cartographic design involves the conceptualization and visualization of the map to be created and is driven by two goals: (1) to create a map that appropriately serves the map user based on the map's intended use, and (2) to create a map that communicates the map's information in the most efficient manner, simply and clearly. Edward Tufte (1990, 53) echoed this second goal, eloquently stating that "Confusion and clutter are failures of design, not attributes of information." The physical act of placing, modifying, and arranging map elements is often referred to as an activity separate from cartographic design: as map construction or "layout." Because of the holistic nature of the design process, however, we consider map construction to be largely integrated with the cartographic design process.

Cartographic design is directed in large part by rules, guidelines, and conventions, but it is also relatively unstructured. A single, optimal solution to a given mapping problem generally does not exist. Instead, several acceptable solutions are usually possible. "Good design is simply

DOI: 10.1201/9781003150527-14

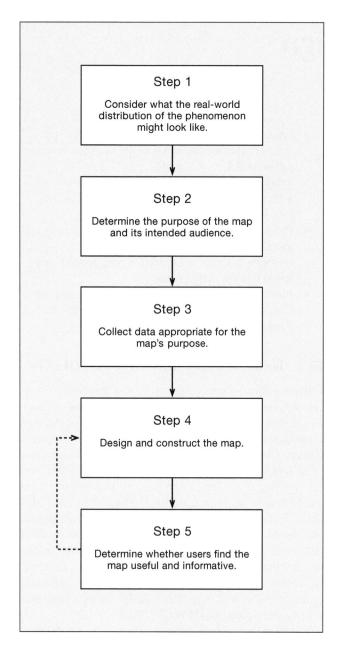

FIGURE 13.1   The map communication model.

the best solution among many, given a set of constraints imposed by the problem" (Dent et al. 2009, 208). If map design were easy and straightforward, this chapter would be unnecessary, as cartographic design expert systems would be used to design most maps. Expert systems are defined in Chapter 12 as software applications that make decisions based on rules and guidelines obtained from cartographic experts. Cartographic expert systems, however, are currently limited to narrow aspects of cartographic design and thus lack the ability to completely and consistently design top-quality maps.

As was the case with map elements and typography, we believe that it is important to build a foundation of cartographic design skills by following the specific rules and guidelines presented here, and then to consider alternative

approaches. Always be prepared to explain or defend your design decisions. In addition, the following points from Chapter 11 bear repeating here: (1) Think carefully about every action you take when designing a map—*you* are responsible for the outcome, (2) Do not implement something simply because you saw it on another map—the world is flooded with examples of poor cartographic choices, and (3) Do not passively accept default settings that are built into software applications—defaults are often created by computer programmers with little or no training in cartographic design.

Many aspects of cartographic design have been guided by the results of **map design research**. Arthur Robinson (1952) sparked enthusiasm for this research with *The Look of Maps*, in which he emphasized the importance of a map's function over its form and called for objective experimentation with regard to map design. Much of this research has focused solely on determining which mapping techniques are most effective, without regard for why they are effective (a behaviorist view). In contrast, *cognitive map design research* focuses on why certain techniques are effective by applying knowledge structures to the ways that people perceive maps. This research has been driven by the idea that understanding cognitive processes and applying their principles to map design can result in more effective maps (Montello 2002). Figure 13.2 illustrates the results of eye-movement studies performed by George Jenks in 1973. These studies revealed variations in eye-scan paths followed by different map users and remain as seminal examples of map design research. Notice in Figure 13.2 that the map user's eyes "entered" the map at the title; passed through the mapped area, subtitle, and legend; returned to the mapped area for a better look; and then exited at the title.

Although most map design research represents a scientific approach to understanding how maps work, the "art" of maps also plays an important role in cartographic communication. The artistic aspect of maps is guided less by experimentation and more by intuition and critical examination (MacEachren 1995). It is difficult to anticipate the map user's sensitivity to the artistic aspects of a map. However, it seems likely that a map that has been created

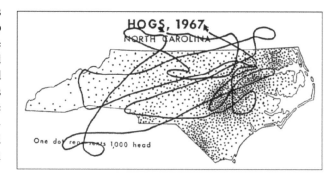

FIGURE 13.2   An eye-movement study as an example of map design research. (From Jenks, G. F. (1973) "Visual integration in thematic mapping: Fact or fiction?" *International Yearbook of Cartography* 13:27–35.)

with an artistic synthesis of contrast, balance, color, and so on has a greater chance of communicating information than a map that has been created in the absence of an artistic sensibility.

The link between cartographic design and graphic design is strong. *Graphic design* has been described as "problem solving on a flat two-dimensional surface…to communicate a specific message" (Arntson 2003, 2). Both cartographic design and graphic design emphasize the communication of information through graphical means—the primary difference being that graphic design is mainly oriented toward advertisements and packaging, as opposed to maps. While most maps are produced by cartographers, GIS specialists, and other geoscientists, graphic designers are responsible for a significant proportion of maps produced for print and online publication.

Both cartographic design and graphic design incorporate **Gestalt principles** of perceptual organization. Gestalt is a theory of visual perception developed in the 1920s that attempts to describe the manner in which humans see the individual components of a graphical image and then organize the components into a *unified whole*. These principles represent the theoretical underpinning for many cartographic design rules, guidelines, and conventions (MacEachren 1995). *Closure* allows us to complete an image even when parts are missing, as illustrated in Figure 13.3A. The administrative boundary is seen as a complete polygon as opposed to individual dashes. *Common fate* allows us to group elements that share the same moving direction, as illustrated by the tornado path arrows in Figure 13.3B. *Continuity* allows us to move our eyes from one object to another. In Figure 13.3C, our eyes "continue" the river, even though it is broken by the text label. *Figure-ground* allows us to perceive certain objects as being closer to us than other objects and, therefore, more important. In Figure 13.3D, New Zealand acts as a figure to the Pacific Ocean, which acts as ground. Simultaneously, the city of Wellington acts as a figure to New Zealand. Figure-ground is described further in Section 13.3.4. *Proximity* allows us to view objects that are close together as a group, as illustrated by the cluster of coyote sightings in Figure 13.3E. *Similarity* allows us to group objects that are similar in size, shape, color, etc., as illustrated by the brown land cover areas in Figure 13.3F. *Smallness, or Area*, allows us to view smaller areas as figures and larger areas as ground. This is illustrated in Figure 13.3G by the relatively small Hawaiian Islands in the context of a larger study area. *Symmetry* allows us to view symmetrical objects as whole figures that form around their centers (as opposed to individual marks), as illustrated by the point symbols in Figure 13.3H. These symbols are symmetrical because each can be bisected into two identical halves. This symmetry helps us to group the individual components of each point symbol into a unified whole.

It is important for you to build a mental inventory of designs and design possibilities from maps and other graphical images. Borden Dent et al. (2009) referred to this mental inventory as an *image pool*, which can be built by critically viewing as much art, graphic design, and maps as possible. Many examples of well-designed maps appear in this textbook. Use these maps to help build your image pool.

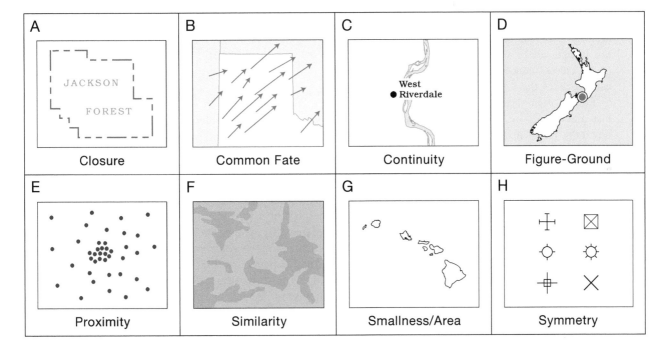

FIGURE 13.3   Cartographic examples of Gestalt principles.

13.3.1 The Design Process

The design process can be distilled into the following list of procedures. This list is an expanded version of step 4 of the map communication model (Figure 13.1); it assumes that steps 1 through 3 of the map communication model have been completed and that step 5 will be completed toward the end of the design process. Be aware that the procedures in this list are iterative and need to be repeated until the map has been completed. They will sometimes need to be executed simultaneously or even out of the pre-scribed order.

1. Determine how the map will be reproduced. Reproduction considerations, such as the printing method to be used, will impact almost every aspect of the design process and need to be resolved first. See Chapter 14 for a discussion of map reproduction.
2. Select a scale and map projection that are appropriate for the map's theme. See Chapter 6 for a discussion of scale and generalization. See Chapter 9 for a discussion on selecting an appropriate projection.
3. Determine the most appropriate methods for data classification and symbolization. See Chapter 5 for a discussion of data classification and Chapter 4 for an introduction to symbolization. See Chapters 15 through 22 for discussions of specific mapping techniques.
4. Select which map elements to employ and decide how each will be implemented. See Section 11.4 for a discussion of map elements. You must also decide how to implement type, as discussed in Chapter 12.
5. Establish a ranking of symbols and map elements according to their relative importance. This ranking is referred to as an **intellectual hierarchy**, or a "scale of concepts" (Monmonier 1993), and usually takes the form of a list. The intellectual hierarchy often varies depending on the type of map and its purpose, but the following is a general hierarchy for thematic maps (from most to least important):
   • Thematic symbols and type labels that are directly related to the theme
   • Title, subtitle, and legend
   • Base information such as boundaries, roads, place names, and so on
   • Scale and north arrow
   • Data source and notes
   • Frame line and neat line
6. Create one or more sketch maps. A **sketch map**, also called a thumbnail sketch, is a rough, generalized hand drawing that represents your developing idea of what the final map will look like, as illustrated in Figure 13.4. The sketch map should

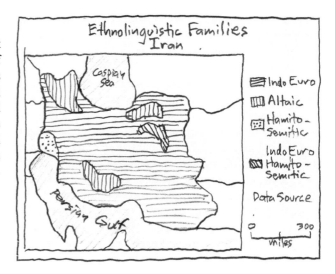

FIGURE 13.4   A rough sketch map. The map resulting from this sketch is illustrated in Figure 13.5.

include all selected map elements and should reflect the intellectual hierarchy established in the previous procedure. Methods for graphically emphasizing and deemphasizing map elements are described in Sections 13.3.3 and 13.3.4. You should experiment with various arrangements of map elements, striving for balance between them. Methods for establishing good balance are discussed in Section 13.3.5.

7. Construct the map with your chosen software application. Place, modify, and arrange map elements according to your sketch map and in the order that was recommended in Section 11.4. Print or display **rough drafts** that will allow you to reevaluate and refine the evolving map. Figure 13.5 illustrates the end result of this procedure, based on the sketch map of ethnolinguistic families (Figure 13.4).

13.3.2 Visual Hierarchy

Visual hierarchy refers to the graphical representation of the intellectual hierarchy described earlier, in which symbols and map elements are ranked according to their relative importance. When implementing the visual hierarchy, thematic symbols are graphically emphasized while base information is deemphasized. Similarly, more important map elements such as the title and legend are graphically emphasized, and less important elements such as the bar scale and data source are deemphasized. An effective visual hierarchy attracts the map user's eyes to the most important aspects of the map first and to less important aspects later.

Visual hierarchy is implemented by applying *contrast* to map features, as described in Section 13.3.3. The **visual weight** of map features refers to the relative amount of attention they attract and can be manipulated to emphasize

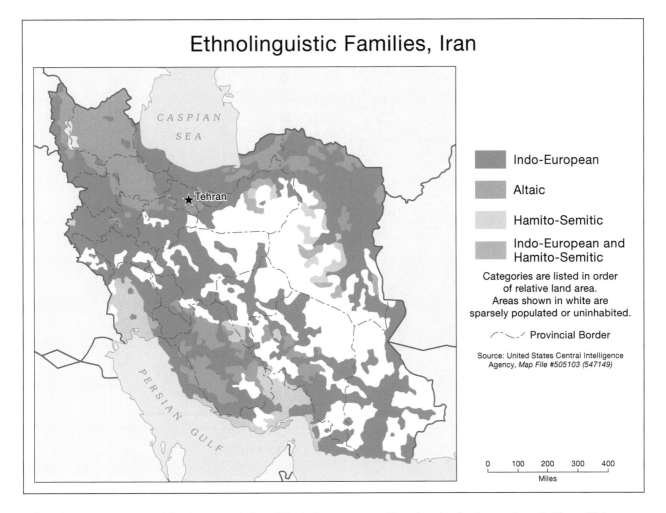

Ethnolinguistic Families, Iran

Indo-European

Altaic

Hamito-Semitic

Indo-European and
Hamito-Semitic

Categories are listed in order
of relative land area.
Areas shown in white are
sparsely populated or uninhabited.

Provincial Border

Source: United States Central Intelligence
Agency, *Map File #505103 (547149)*

0   100   200   300   400
Miles

FIGURE 13.5   The map resulting from completion of the design process, and based on the sketch map shown in Figure 13.4.

or deemphasize features. The map in Figure 13.6A reflects an *inverted* (incorrect) visual hierarchy based on the general intellectual hierarchy listed earlier. Thematic symbols (i.e., symbols representing elementary, junior high, and high schools) have been deemphasized in favor of base information such as roads, universities, and the river. More important map elements such as the title and legend have also been deemphasized, whereas less important map elements such as the bar scale and data source have been given far too much visual weight. Try looking at this map while squinting (the "squint test") and identify the map features that stand out. The bar scale, data source, universities, river, and roads should certainly *not* be the most noticeable features on a map that focuses on public schools. The map in Figure 13.6B has a visual hierarchy that appropriately reflects the intellectual hierarchy listed earlier. Thematic symbols are visually dominant, as are the title and legend; they are dominant because of their greater visual weight, achieved through the manipulation of contrast (described next). Base information (universities, river, roads, etc.) is subdued, as are the data source and bar scale. An effective visual hierarchy results in a map that clearly reflects the relative importance of symbols and map elements. A map

with an appropriate visual hierarchy is easier to interpret and is also more attractive.

13.3.3 Contrast

Contrast refers to visual differences between map features that allow us to distinguish one feature from another. Contrast adds interest to a map by providing graphical variety. It can be used to differentiate features, to imply their relative importance, or to represent differences in quantitative magnitudes. Several techniques can be used to create contrast, including manipulation of the visual variables that were presented in Chapter 4: spacing, size, perspective height, orientation, shape, arrangement, and all aspects of color.

The map in Figure 13.7A lacks appropriate contrast in four respects: type size, lightness and size of thematic symbols (circles), size of lines (line width), and the difference between the mapped area and the background. Type on this map is insufficiently differentiated by increments of 0.5 point. Remember that in Section 12.4, a minimum difference of 2 points was recommended. The title (10 points), legend heading (9.5 points), legend definitions (9 points),

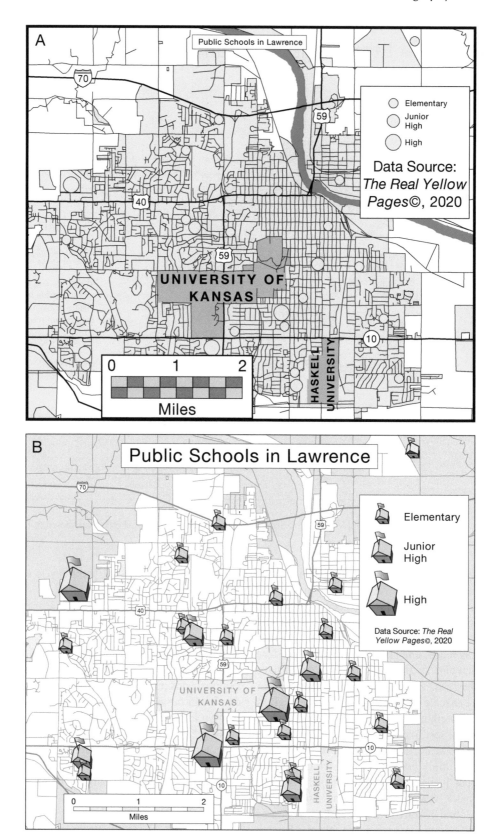

FIGURE 13.6   A map showing the distribution of schools in Lawrence, KS, with (A) an inverted (incorrect) visual hierarchy. (B) A correctly applied visual hierarchy, appropriately reflecting the intellectual hierarchy.

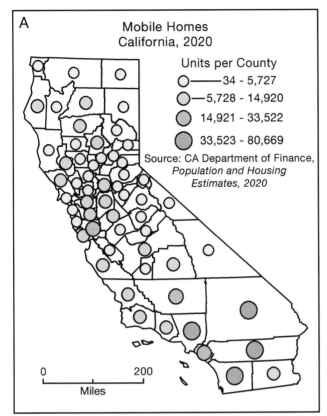
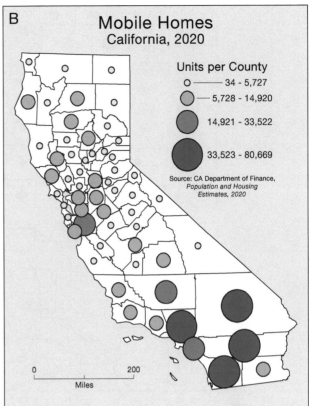

FIGURE 13.7 (A) Insufficient contrast in type size, lightness and size of thematic symbols (circles), line width, and the difference between the mapped area and the background. (B) Sufficient contrast in all respects.

bar scale type (8.5 points), and data source (8 points) are difficult to differentiate by size. This lack of contrast contributes to a monotonous design in which the relative importance of map elements is unclear at best and misleading at worst. The data source in this map appears almost as large as the title, even though it is far less important. A lack of contrast between the lightness of the circles, together with a lack of contrast in circle size, contributes to a dull design that makes it difficult to differentiate between classes. (See Chapter 10 for a discussion of contrast in grays and Chapter 18 for a discussion of contrast in the size of point symbols.) The lines in Figure 13.7A also lack contrast; the frame line, circle outlines, county boundaries, bar scale, and leader lines in the legend are all 1 point wide. Again, this lack of contrast contributes to a monotonous design in which the relative importance of map features is unclear. Bar scales and leader lines are less important than thematic symbols or base information (county boundaries in this map) and should appear so. Finally, the map in Figure 13.7A lacks sufficient contrast between the mapped area and the background, resulting in the mapped area appearing to blend in with the background.

The map in Figure 13.7B exhibits sufficient contrast in each of the four respects just described. It is easier to interpret, is more visually stimulating, and is more attractive. Notice the significant difference the gray background makes—it enhances the mapped area and makes it seem

more important than the background. This technique creates a special type of contrast called figure-ground that has already been introduced but deserves special attention.

13.3.4 Figure-Ground

Figure-ground was described previously as a Gestalt principle of perceptual organization. It refers to methods of accentuating certain chosen objects over others by making the chosen objects appear closer to the map user. Map design research has failed to produce guidelines for figure-ground that are guaranteed to work in every situation (MacEachren and Mistrick 1992), but we have found that the following guidelines work well in most cases: When accentuating points, the figure-ground relationship is established by making the points darker than their surroundings. In Figure 13.8A, a lack of contrast results in a situation in which no features advance easily as figures, and none recede as ground. In Figure 13.8B, the base information has been lightened, allowing the point symbols to emerge as figures. **Screening** is a term that describes the lightening of graphics to reduce their visual weight, as illustrated by the administrative boundaries in Figure 13.8B, which have been reduced from black to a 30% tint of black (a 30% gray).

When accentuating lines, the figure-ground relationship is established using the same technique as just described for points: lines are made to be darker than their surroundings.

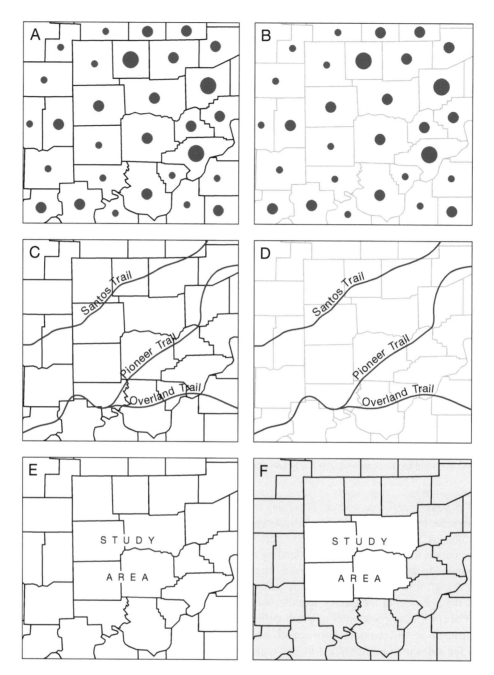

FIGURE 13.8 (A) Ambiguous figure-ground relationship. (B) Darker points emerge as figures; lighter administrative boundaries recede as ground. (C) Ambiguous figure-ground relationship. (D) Darker lines and type labels emerge as figures. (E) Ambiguous figure-ground relationship. (F) Lighter study area emerges as a figure.

In Figure 13.8C, a lack of contrast results in a situation in which no features advance easily as figures, and none recede as ground. In Figure 13.8D, the base information has been lightened, allowing the line symbols to emerge as figures. The same is true for the type labels that identify the trails.

When accentuating an area, the figure-ground relationship is established through opposite means—by making the area *lighter* than its surroundings. In our experience, making an area lighter than its surroundings causes it to stand out while simultaneously setting the stage for the establishment of figure-ground for point and line features, which should

be darker than their surroundings. The extent of the study area in Figure 13.8E is ambiguous and is only suggested by the "STUDY AREA" label. In Figure 13.8F, the study area is strongly emphasized and clearly defined due to its relative lightness. The figure-ground relationship accentuating Western Europe in Figure 13.9A is similar but is actually a special case, one in which land is the figure and water is the ground: the classic *land–water contrast*. Making an area lighter than its surroundings effectively emphasizes the area but is not always appropriate. For example, in situations when the mapped area is dense with areal thematic symbols (Figure 13.9B), it would be inappropriate to apply a

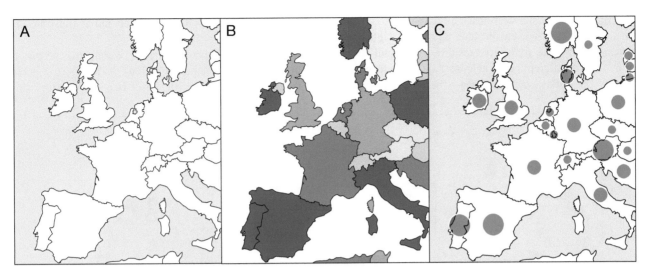

FIGURE 13.9 (A) Land–water contrast as a special case of the figure-ground relationship. (B) Situation in which the application of a darker color in the surrounding area would be inappropriate. (C) Three levels of information established through figure-ground relationships.

dark color to the surrounding area. This is because the areal thematic symbols provide enough contrast between the mapped area and its surroundings. Notice in Figure 13.9C that there are actually three levels, or layers, of information represented. The circles act as figures in relation to the countries (which act as ground)—the circles are dominant and appear closer to the map user. The countries, although subordinate to the circles, act as figures in relation to the water (which acts as ground)—the countries appear to be farther away than the circles, yet closer than the water.

Figure 13.10 represents three alternative methods of establishing a figure-ground relationship that accentuates areas. In Figure 13.10A, the British Isles are established as figures (although less strongly than in previous examples) because they appear to stand in front of the graticule, which acts as ground—an effect known as interposition. Stylized effects such as the vignette (boundary shading) surrounding the British Isles in Figure 13.10B can also enhance the

figure-ground relationship but should be used with caution because they can become visually dominant. Figure 13.10C illustrates the "waterlines" technique of establishing a figure-ground relationship of land on water. The waterlines technique incorporates a series of lines that parallel the coastline—the lines increase in spacing and decrease in visual weight as they move away from shore. Although more widely used in the nineteenth century, the waterlines technique can have a dynamic effect by evoking waves washing up on shore (Huffman 2010). Be aware that waterlines are purely cosmetic—they should not be confused with bathymetric contour lines, which quantitatively represent water depths.

13.3.5 BALANCE

Balance refers to the organization of map elements and empty space that results in visual harmony and equilibrium.

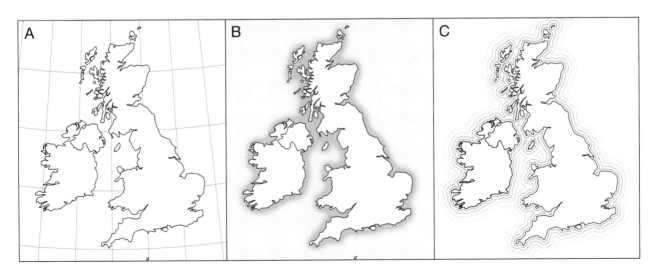

FIGURE 13.10 Figure-ground established using (A) a graticule, (B) a vignette, and (C) waterlines.

The map elements in a well-designed map tend to complement one another, whereas those in a poorly designed map appear to compete for space, resulting in visual disharmony. Before attempting to achieve balance, you need to identify the initial available space—the area the map will occupy—as defined by the frame line and as illustrated in Figure 13.11A.

Once the initial available space is defined, the placement of larger map elements such as the mapped area and inset can be considered (the inset is excluded from this

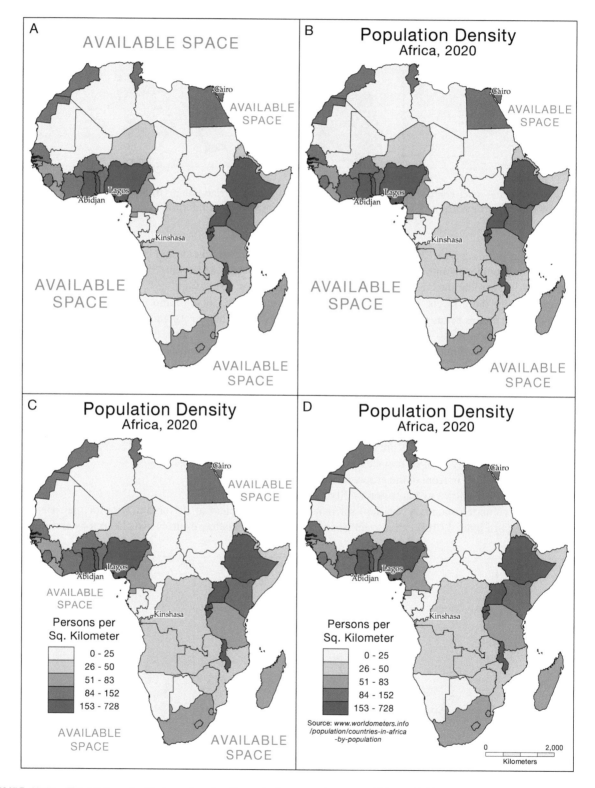

FIGURE 13.11 Establishment of balance by visually centering map elements within available spaces. (A) The frame line defines the initial available space and the mapped area is approximately centered within this frame line. (B) The title and subtitle are placed. (C) The legend is placed. (D) The data source and bar scale are placed.

example). The mapped area should be as large as possible within the available space while leaving ample room for the remaining map elements. The mapped area should also be visually centered within the available space if possible. The concept of visually centering objects has been discussed but warrants repetition: Visually centering objects is accomplished through visual approximation, not precise measurement; what matters most is that an object *looks* centered within the available space, as Africa does in Figure 13.11A. With the mapped area placed, the title (which is intermediate in size) can be positioned at the top center (Figure 13.11B). Notice that the title appears to be visually centered (vertically) within the space above the mapped area, as well as being horizontally centered within the frame line.

After the mapped area and title have been placed, the evolving map needs to be reevaluated for available space (Figure 13.11B); locations must be identified for the remaining map elements based on the guidelines for each map element presented in Section 11.4. Like the title, the legend is normally intermediate in size—larger than the data source, bar scale, and north arrow—and should be placed next. Visually center the legend within a larger area of available space (Figure 13.11C), reevaluate the remaining available space, and then place the smaller map elements such as the data source and bar scale, as illustrated in Figure 13.11D. These smaller map elements should also be visually centered within the available spaces. Notice in Figure 13.11D that available space remains, even after all map elements have been placed. It is not necessary to fill all available spaces, but an effort should be made to occupy the larger areas with well-placed map elements.

The addition of each map element alters the preexisting balance and available space. Map elements will probably need to be rearranged several times to achieve good balance. Useful questions for the cartographer to ask are whether a map looks left-heavy, right-heavy, top-heavy, bottom-heavy, or whether certain areas appear cramped or barren. Top-heavy designs are of particular concern because they tend to make people uncomfortable; humans are intrinsically aware of gravity and tend to feel more comfortable with objects that are closer to the ground or the bottom of a page (Arntson 2003). If the answer to any of the previous questions is yes, then the map elements should be rearranged with the goal of visually centering them within the available spaces. Consideration should also be given to the different visual weights of the individual map elements. Do not place too many "heavy" objects in the same area, but, rather, try to balance heavier objects with lighter ones.

Instead of allowing map elements to crowd one another, try to use them to balance one another. For example, the bar scale in Figure 13.12A crowds the legend, competing for its space, whereas the bar scale in Figure 13.12B (along with the data source) helps to counterbalance the legend. The map in Figure 13.12A is poorly balanced in many respects, but most of the problems are rooted in the fact that the map

elements were not visually centered within the available spaces. Figure 13.12B represents a well-balanced design in which all map elements are in harmony and equilibrium. Map elements were placed according to the same sequence described in Figure 13.11 (the Africa population density map). Certain individuals are intrinsically better than others at judging balance, but experience can improve one's skills.

13.4 CASE STUDY: REAL ESTATE SITE SUITABILITY MAP

In this section, we will apply the concepts, rules, and guidelines presented in this and the previous two chapters to a real-world map design problem. The goal will be to create an efficient, attractive map that represents the relative suitabilities of residential building parcels (thematic symbols), together with appropriate base information. The map will be provided to a residential real estate agent, who will distribute the map to potential clients. The source materials for this map are derived from county-level GIS data and are limited in extent to an area slightly smaller than an urban zip code region. These materials are described below.

1. Thematic symbols are building parcels (polygons) within two neighborhoods that are:
   a. Zoned for single-family residences
   b. Not within 500 feet of a freeway or freeway onramp/offramp
   c. Not within 250 feet of an active railway
2. Thematic symbols have been ranked for suitability according to the following criteria. Higher suitability parcels are characterized by:
   a. Relatively low crime density.
   b. Relatively high percentage of college-educated residents (Bachelor's degree or higher). Be aware that this criterion is highly subjective and suggests that people with degrees are somehow associated with more desirable neighborhoods. In fact, the opposite might be true. It is assumed here that either the real estate agent or the clients have established this criterion.
   c. Relatively high elevation (the region has a great potential for flooding). These three criteria have been weighted equally and represent the same level of importance in the analysis. Details of the suitability analysis model are not discussed here.
3. Base information includes:
   a. Unranked parcels
   b. Freeways and ramps
   c. Railways
   d. River
   e. Descriptive type labels

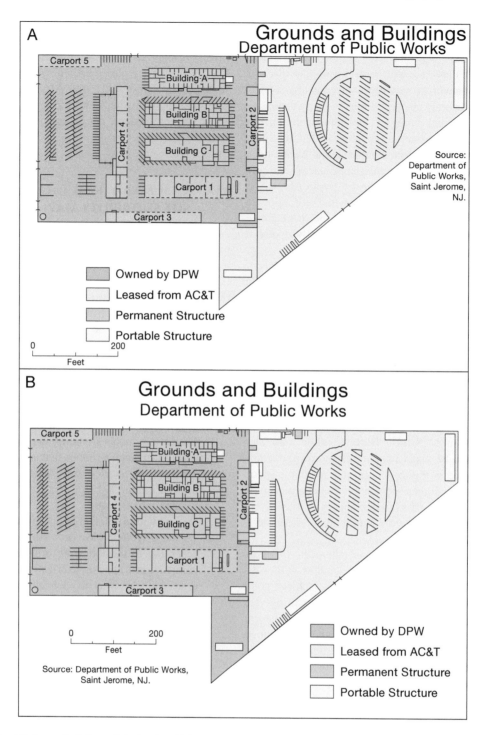

FIGURE 13.12 (A) A poorly balanced design in which map elements compete for space. (B) A well-balanced design in which map elements exist in relative harmony and equilibrium.

4. Additional symbols—crime hot spots—will also be included due to the immediate and present concern of crime to the potential homeowner. They represent areas of relatively high crime density: areas with a relatively high number of crimes per square mile. The crime hot spots are hybrid symbols; they are part thematic symbol and part base information. They are similar to thematic symbols because they are directly related to the map's theme—crime was used as an input to the site suitability ranking. They will be used here as a visual reference to areas of relatively high crime and will provide a frame of reference for the thematic symbols, much like base information does.

We will base this case study on the map communication model described previously (Figure 13.1), with an emphasis on step 4: Design and Construct the Map.

13.4.1 STEPS 1–3 OF THE MAP COMMUNICATION MODEL

Step 1 of the map communication model requires that we consider what the real-world distribution of the phenomenon might look like. We are fairly well acquainted with the study area and have general impressions of which areas are most and least suitable for residential property ownership. However, the specific distribution and ranking of parcels require the design and execution of spatial analysis within the realm of a GIS. Spatial analysis of this sort is beyond the scope of this text but is fairly common. It incorporates various geoprocessing operations that are linked within the construct of a graphical modeling application.

Step 2 of the map communication model requires us to determine the purpose of the map and its intended audience. The intended audience is defined as clients of a residential real estate agent. The purpose of the map will be to provide the clients with a graphical representation of the geographic distribution of parcels, which are ranked according to the criteria described above. The map will focus attention on the most suitable parcels and will include related base information but will not provide specific information on the parcels or the homes that occupy them.

Step 3 of the map communication model involves the collection of appropriate data. It is assumed that this step has been completed as part of the GIS analysis process. We are now ready to proceed to step 4 of the map communication model, where we will describe in detail each procedure in the design process. This textbook is available in both print and eBook form, but the emphasis here is on print reproduction.

13.4.2 STEP 4 OF THE MAP COMMUNICATION MODEL: DESIGN AND CONSTRUCT THE MAP

Procedure 1 of the design process, as outlined in Section 13.3.1, requires us to consider how the map will be reproduced. As will be described in Chapter 14, the chosen reproduction method will influence almost every aspect of the cartographic design process. Since our map will ultimately be printed in this textbook, we can assume that the reproduction methods will involve either high-resolution inkjet printing or offset lithography (Section 14.8). These methods result in high-resolution, high-quality printed images in full color and place few restrictions on the design process.

Procedure 2 of the design process involves the selection of a scale and map projection that is appropriate for the map's theme. The scale of our map will be strictly dictated by two factors: our need

to represent two neighborhoods and our need to do so within the confines of roughly seven inches of width in this textbook. Experimentation has resulted in an optimal scale of roughly 1:20,000. Because of the very small geographic extent of this region (in comparison with the extent of the entire Earth), our choice of map projection is virtually inconsequential. Distortions resulting from the map projection process will be insignificant. For reference, the data used in the GIS analysis phase of this project are stored in the State Plane Coordinate System, incorporating the Lambert Conformal Conic projection. This projection correctly represents the shapes of small areas but distorts the relative sizes of areas. In an area as small as two neighborhoods, it is assumed that this distortion will be of no consequence to the map user.

Procedure 3 in the design process dictates that we choose appropriate data classification and symbolization methods. The classification method will be quantile. This approach will appropriately represent the quantitative "low-to-high" suitability ranking that was incorporated in the GIS analysis and will ensure that the map has roughly equal numbers of parcels in each class. Data standardization will not be performed on the raw suitability values; we intend instead to represent a simple ranking of raw suitability values. Because our thematic symbols are polygons that represent the variation in magnitude of a single attribute, a choropleth-style symbolization approach will be taken. Each suitable polygon will be symbolized with a color that gets darker in relation to increased suitability. The map user should be able to scan the map and differentiate between various levels of suitability based solely on the polygon symbolization; specific magnitudes will be defined by the legend.

Procedure 4 in the design process involves the selection of appropriate map elements, consideration of how they should be implemented, and decisions related to typography. While it is important to make preliminary decisions in regard to these topics, the map designer typically needs to progress further along in the design process in order to make final decisions. For example, the process of creating a sketch map (Procedure 6) often presents problems and reveals solutions that have a direct bearing on map elements and how they should be implemented. This procedure (Procedure 4) is relatively lengthy and time consuming compared with the other six procedures, and because of this, we will simply list the appropriate map elements here and then return to a more comprehensive treatment of how the map elements and typography should be implemented in Section 13.4.3. We have decided to include a frame line and/or a neat line, the mapped area, an inset (a locator map), a title and subtitle, a legend, a data source, and a bar scale. An indication of orientation (north arrow) might be required depending on the orientation of the mapped area.

Procedure 5 in the design process requires us to establish an intellectual hierarchy. We will use the hierarchy presented in Section 13.3.1 because, in our experience, it has proved to be effective for most simple thematic maps. This hierarchy is repeated below (from most to least important).

- Thematic symbols and type labels that are directly related to the theme
- Title, subtitle, and legend
- Base information such as boundaries, roads, place names, and so on
- Scale and north arrow
- Data source and notes
- Frame line and neat line

In a subsequent procedure, we will establish the visual hierarchy by graphically expressing the intellectual hierarchy.

Procedure 6 in the design process involves the creation of one or more sketch maps that can help an evolving design take shape. The creation of sketch maps can be especially useful when the map designer is not completely comfortable with the software application that will be used to create the final map. A map can be drawn by hand without any of the distractions or problems associated with computers and software applications. Sketch maps become less valuable when the map designer is completely comfortable with the software. Under that condition, the designer can use the software as a sketch pad of sorts, quickly experimenting with various design options. A hybrid approach can also be taken, in which the designer hand-sketches on top of computer-generated rough drafts. We will create both hand-drawn and hybrid sketch maps.

Procedure 7 in the design process is where the final map will be created using a software application. While graphic arts applications offer the greatest amount of control over graphic elements and type, we estimate that most maps are currently produced using GIS applications. As a result, we have chosen to create this map using ESRI ArcGIS Pro, and we will subsequently export the map to Adobe Illustrator for "finishing" and preparation for print reproduction. Procedure 7 also includes the creation of rough drafts representing the evolving map design. Rough drafts can act as a surrogate for hand-sketched maps or, as described previously, can be used in conjunction with hand-sketching. Whichever approach is taken, sketch maps and rough drafts allow the designer to evaluate the current state of the map and to explore alternative approaches to the design problem.

13.4.3 RETURN TO PROCEDURE 4: IMPLEMENTATION OF MAP ELEMENTS AND TYPOGRAPHY

As indicated previously, Procedure 4 of the design process is relatively lengthy and time consuming compared with the other six procedures. At this point, we return to Procedure 4 in order to consider how each of the selected map elements should be implemented. We will also make decisions related to typography. We will place the following map elements in the order that was recommended in Section 11.4—from largest to smallest. Justifications for including these map elements, together with implementation details, are provided below:

1. Frame line and neat line
2. Mapped area
3. Inset
4. Title and subtitle
5. Legend
6. Data source
7. Scale
8. Orientation

13.4.3.1 Frame Line and Neat Line

The first map element placed will be the frame line. It is included because it can help focus the map user's attention on what is within it. The frame line also establishes the initial available space within which all other map elements will be placed. This frame line will also act as a neat line, cropping the mapped area, as described below. A solid thin black line (0.5 point) will serve these purposes without attracting undue attention. The size (width and height) of the frame line is restricted in part by the maximum printable width of this textbook (7 inches).

13.4.3.2 Mapped Area

The mapped area is then placed within the frame line. It will be sized as large as possible without cramping the frame line and will leave ample room for the remaining map elements. It will be visually centered (both horizontally and vertically) within the frame line. At this point, a distinction needs to be made between the two components of the mapped area: thematic symbols and base information. As you recall, the mapped area always consists of thematic symbols and sometimes includes base information. In this case, thematic symbols are the residential building parcels representing various levels of suitability—choropleth-style area symbols. Base information includes unranked parcels, freeways and ramps, railways, a river, street names, and place names.

Thematic symbols will appear only in the neighborhoods of interest; it is this component of the mapped area that will be sized and positioned according to the guidelines stated above. Base information will extend beyond the neighborhoods of interest and will act as a geographic frame of reference for the thematic symbols. As a consequence, the thematic symbols will exist completely

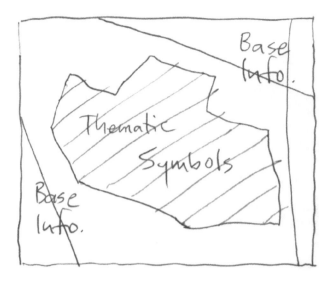

FIGURE 13.13 A sketch map representing thematic symbols centered within the frame line and base information cropped by it.

within the frame line (without touching it), whereas the base information will extend as far as the frame line and will be cropped by it, as illustrated by the sketch map in Figure 13.13. In this case, the frame line also acts as a neat line, cropping or limiting the extent of the base information. We will continue to refer to this line as the frame line. In comparison with a floating mapped area, which is removed from its geographic context, this cropped mapped area will leave relatively little available space for the remaining map elements. Placement of subsequent map elements will require careful consideration, together with the use of masks to cover the underlying base information. The size and position of the mapped area will likely need to be adjusted simultaneously with the placement of additional map elements.

The intellectual hierarchy dictates that thematic symbols appear to be most important when represented on this map—they should have the greatest visual weight in the visual hierarchy. Accordingly, the parcels will be dominantly symbolized with black outlines and a part-spectral color scheme. Higher suitabilities will correspond with darker symbols, lower suitabilities with lighter symbols. Base information will be subdued through screening, providing the map user with a frame of reference without challenging the thematic symbols for attention. Before adding thematic symbols, we will focus on the appropriate symbolization of base information. It is often useful to symbolize base information first, as it helps to set the context for thematic symbols.

The various layers of base information will need to be symbolized so that each component is easily identifiable. We will accomplish this by applying contrast to each component—not necessarily to imply relative importance but to differentiate one component from another. Figure 13.14A illustrates the unranked parcels, freeways and ramps,

railways, and river (type labels will be added later) in the absence of contrast; the visual weight of these symbols is excessive, making it difficult to differentiate the various components. Figure 13.14B illustrates the base information after contrast has been applied, primarily through the use of screening (lightening), coloring, and the manipulation of line thickness and line style. Appropriate symbolization allows us to differentiate the individual components of base information and prepares the map for the addition of thematic symbols.

Figure 13.15 illustrates the yellow-orange-red thematic symbols dominantly superimposed on the base information. Compare Figure 13.15 with the sketch map in Figure 13.13. In both cases, the thematic symbols are visually centered within the frame line and the base information is cropped by the frame line. Also, notice in Figure 13.15 that the crime hot spots (irregular polygons with horizontal line patterns) have been superimposed on the thematic symbols and base information. As described earlier, the crime hot spots are hybrid symbols: part thematic symbol and part base information. Be aware that line and dot patterns can be irritating to the eye, so use them sparingly and apply them subtly. The line pattern here is being used to allow for transparency: the underlying mapped area can be seen through the hot spot symbols. The final component of base information—descriptive type labels—will be added toward the end of the design process.

The figure-ground relationship has been established as follows: thematic symbols act as "figure," appear closer to the map user than the base information and are interpreted as being more important; base information acts as "ground," appears farther from the map user, and is interpreted as being less important. The hybrid crime hot spots are, technically speaking, closer to the map user because they appear to rest on top of the underlying symbols, but they attract only slightly more attention than the base information because they are partly see-through. In Section 13.3.4, separate guidelines were provided for implementing figure-ground contrast to accentuate points and lines and, alternatively, to accentuate areas. It was stated that in order to accentuate an area, the area needs to be lighter than its surroundings. It was also stated that this approach is not always appropriate, especially when the mapped area is "dense with areal thematic symbols," as is the case here. Because the building parcels are represented by colors that get progressively darker, they naturally emerge as a figure when superimposed on a lighter background.

At this point, it will be useful to evaluate the available space and consider where the remaining map elements might be placed. Figure 13.15 represents the mapped area with the thematic symbols visually centered within the frame line. Available space appears both above and below the thematic symbols, with slightly more existing below. We will now identify potential locations for the remaining map elements: inset, title/subtitle, legend, data source, scale, and orientation. We will hand-sketch each map element on top of a computer-generated rough draft.

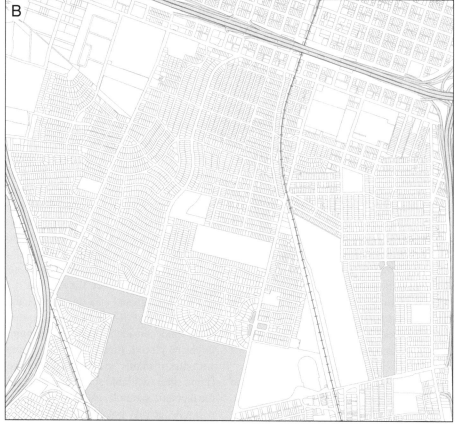

FIGURE 13.14   (A) Base information in the absence of contrast. (B) Base information differentiated by means of contrast.

FIGURE 13.15   Thematic symbols superimposed on base information; inclusion of crime hot spot symbols.

13.4.3.3   Inset

The geographic extent of the mapped area is quite small—smaller than a single zip code region—and might not be recognized without the use of an inset (a locator map). This inset will represent the primary mapped area at a much smaller scale within the context of the city boundary and major highways/freeways. Because an inset can serve various purposes, it does not appear in the intellectual hierarchy listed previously. In this case, the inset is for reference purposes only and will be treated much like base information. Accordingly, it will seem to have an intermediate level of importance in the visual hierarchy. The inset will contain a minimal amount of type, such as the name of the city and "Area of Detail," but a discussion of type characteristics will be withheld until the title is placed. The inset will be placed in a relatively large portion of available space, as dictated primarily by the frame line and thematic symbols. Examination of Figure 13.15 reveals that the largest portion of available space appears in the lower left. This location might be appropriate for the inset, provided that a mask is placed over the underlying base information. Figure 13.16 shows the inset sketched on top of a computer-generated rough draft (a hybrid sketch map). The finished inset is shown in Figure 13.17A. It contains just enough information to allow the map user to locate the area of detail within the larger geographic region. Be aware that the inset and all subsequent map elements are shown at their actual

sizes—they need to be small in order to fit within the confines of our map.

13.4.3.4   Title and Subtitle

The title will succinctly reflect the map's theme as "Suitability of Residential Parcels." The subtitle will identify the geographic region and date. (It is assumed that the map user will not otherwise recognize the region.) Assuming that we mask the base information, there appears to be sufficient space for the title and subtitle at the top center—we will use this space accordingly (Figure 13.16). The title and subtitle will initially be placed at the top center but, as with all map elements, can subsequently be moved if necessary.

At this point, we are prompted to make our first typographical decisions. We have selected a simple sans serif type family (Helvetica Neue) over fancier, ornate families. The decision to use a sans serif family instead of a serifed family is somewhat arbitrary. The use of sans serif type families often results in cleaner, more modern-looking maps, but sans serif families have no proven advantage over serifed families. The type style for the title and subtitle will be normal (roman). Careful choice of type size will avoid the need to use a bold style, and italics will be reserved for the identification of the publication in the data source and for hydrographic features. The title will be the largest type on the map, followed by the subtitle and legend heading.

FIGURE 13.16   A hybrid sketch map representing possible locations for map elements and type.

To increase readability, letter spacing and word spacing will be slightly increased.

Remembering that top-heavy designs tend to make people uncomfortable (Section 13.3.5), we will attempt to place the remaining map elements toward the bottom of the page, in the available space below the mapped area. We will also mask the underlying base information where necessary.

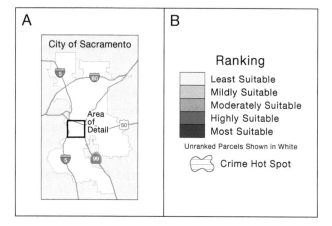

FIGURE 13.17   (A) The inset designed for use as a locator map. (B) The legend defining parcel suitability ranking and crime hot spots.

13.4.3.5   Legend

The legend will define the suitability ranking of thematic symbols by means of rectangles and definitions ranging from "Least Suitable" to "Most Suitable." The rectangles will be connected because the parcels they represent are mostly connected and because they emphasize a gradation of suitabilities as opposed to unique qualitative categories. The legend heading is a good place to provide information such as the unit of measure, the type of enumeration unit (parcels in this case), and the data collection date. Because this information will be provided in the title and subtitle, the legend heading will simply be "Ranking." It is assumed that the map user will be able to identify and use the legend without the inclusion of the word "Legend" in the heading.

While thematic symbols are virtually always defined in the legend, base information is either included or omitted, depending on various factors. The thematic symbols are represented as residential building parcels and are identified as such in the title. Because of this, the unranked parcels used as base information could probably be omitted from the legend. We could assume that the map user will intuitively identify the white polygons as being unranked parcels because they look like parcels but lack thematic symbology. However, in order to prevent any ambiguity, we have chosen to identify the unranked parcels in the legend with the label

"Unranked Parcels Shown in White." Freeways, railways, and the river are assumed to be self-explanatory and will be omitted from the legend. Crime hot spots—areas of relatively high crime—are not self-explanatory and will need to be defined in the legend. Because the crime hot spots are irregular in shape, they will be identified in the legend by an irregular polygon. For consistency, all type characteristics in the legend will be identical to those used in the title, subtitle, and inset, with the exception of type size. The legend heading will be visibly smaller than the subtitle, and the legend definitions will be smaller than the legend heading.

Like the inset and title, the legend will be placed in a relatively large portion of available space, as dictated primarily by the frame line, thematic symbols, and inset (Figure 13.16). A mask will be used to cover the underlying base information. The finished legend is shown in Figure 13.17B.

13.4.3.6 Data Source

The data source will indicate the three sources of the thematic data. The heading "Data Sources:" will be included to avoid confusion with map authorship, and publication names will be italicized. Because the data source appears toward the bottom of the intellectual hierarchy, it will be visually deemphasized through the use of a very small type size: 5 points. It is assumed that the average reader of this textbook will be able to discern type of this size from a distance of roughly 12 inches. Individual lines of type will be horizontally centered to establish a self-balancing block of type. While the most appropriate location for the data source is directly below the legend, limitations of available space dictate that the data source be placed elsewhere, most likely in a portion of available space that can accommodate the wide but "short" dimensions of the data source. Examination of Figure 13.16 reveals a wide, short portion of available space in the extreme lower right. This space might also accommodate the bar scale—a map element with similar dimensions. With the exception of type size and the use of italics for publications, type characteristics will be identical to those used in the title, subtitle, inset, and legend. The finished data source is shown in Figure 13.18A, together with the bar scale.

13.4.3.7 Scale

A bar scale will be included because it will allow the map user to assess neighborhood characteristics, such as block lengths and proximity to crime hot spots. The scale will begin with zero (no extension scale), and the maximum distance will be 0.5 mile—a value that is round and easy to work with. A maximum value of 1 would also be suitable, but the resulting bar scale would be prohibitively wide. One intermediate tick and a center value of 0.25 will be included. Like the data source, the bar scale will be designed to be subtle—it should not attract attention. A skeletal style incorporating a very thin line width (0.25 point) is more appropriate than bulkier checkerboard styles. Again, like the data source, the bar scale needs to be placed in a portion of available space that can accommodate its wide but short dimensions. Both the data source and the bar scale can probably fit in the available space in the lower right (Figure 13.16). Placement of the data source and bar scale in this location will also leave a portion of available space just to the left. This space can be used for notes related to the site suitability ranking criteria. The finished bar scale is shown with the data source in Figure 13.18A.

13.4.3.8 Orientation

The inclusion of a north arrow must be carefully considered. This map is oriented to geographic north and is not intended for navigational, surveying, or orienteering purposes. Thus, a north arrow is not warranted under these conditions. However, because of the irregular orientation of streets and freeways (many are oriented to magnetic north, roughly 17° E), we have decided to include a subtle north arrow in order to clarify an ambiguous orientation. The north arrow will include an "N" for north and a five-pointed star representing Polaris, the North Star. It will be placed in a tall, thin portion of available space, as seen in the right-hand margin in Figure 13.16. The finished north arrow is shown in Figure 13.18B.

13.4.4 Final Procedures

One of the final procedures in the design process, while not officially presented in the list in Section 13.3.1, is the placement of descriptive type labels and other type, such as explanatory notes. In this case, descriptive type labels will identify base information, primarily streets and place names. These labels are for reference purposes only and should not attract undue attention. They will be designed and placed according to the typographical rules and guidelines presented in Section 12.5. Labels for linear features (roads) will be set in title case; areas will be labeled in all uppercase with exaggerated letter and word spacing; italic style will be applied to the hydrographic feature (the river); and all descriptive type labels will be oriented to reflect the features they represent. Masks and halos will be used to cover the underlying mapped area. Explanatory notes will also be included in order to provide basic information

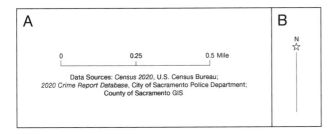

FIGURE 13.18   (A) The data source and bar scale. (B) The north arrow.

related to the suitability ranking. The relatively small size of this map prevents the inclusion of a greater number of street labels—a larger version would allow for more.

In the process of placing map elements in order from largest to smallest and visually centering each in appropriate portions of available space, the map should already reflect a general sense of balance. After all map elements and type have been placed, however, the overall balance of the map needs to be reconsidered. Map elements might need to be repositioned so that they complement rather than crowd one another, and a sense of equilibrium between map elements and empty space needs to be attained. This can be achieved through the careful examination of rough drafts and thoughtful editing. A second set of eyes can also be helpful. At this stage in the design process, the map designer often loses a degree of objectivity and can benefit from the comments and opinions of others.

Figure 13.19 shows our map with all map elements placed and with descriptive type labels and notes added. Compare this map with the hybrid sketch map in Figure 13.16. The map in Figure 13.19 should look similar to the sketch map, which acted as a template during the design process.

Now that we have completed the seven procedures of the map design process, we return to step 5 of the map communication model. This step involves the participation of the map's intended audience (the clients of the residential real estate agent) in order to evaluate its effectiveness. Feedback solicited from the map users can be incorporated into the final map design.

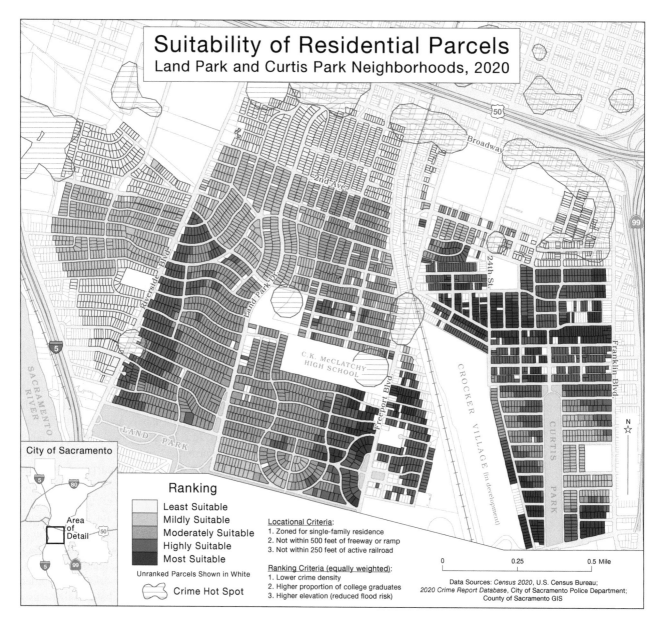

FIGURE 13.19   The final real estate site suitability map, resulting from the case study in Section 13.4.

Completion of this map required several iterations of editing, fine-tuning, and critiquing based on rough drafts. The map is a relatively efficient and attractive communicator of geographic information, but it represents only one possible solution to the problem at hand. As we stated earlier in this chapter, we believe that it is important to build a foundation of cartographic design skills by following the specific rules and guidelines presented here and then to consider alternative approaches. Always be prepared to explain or defend your design decisions.

13.5 SUMMARY

In this chapter, cartographic design was introduced as a holistic process, both mental and physical, that results in the creation of efficient maps that clearly communicate geographic information. Cartographic design is strongly guided by the needs of the map user and the purpose of the map. It is also guided by rules, guidelines, and conventions, but there is often no optimal result of the design process; the cartographer thus crafts an appropriate solution from among many possibilities. Cartographic design has been guided by the results of map design research and is strongly influenced by graphic design and Gestalt principles of perceptual organization. The design process was distilled into seven procedures, which are incorporated into the map communication model introduced in Chapter 1: (1) determine how the map will be reproduced; (2) select an appropriate scale and map projection; (3) decide on data classification and symbolization methods; (4) select appropriate map elements and determine how to implement them; (5) establish an intellectual hierarchy that represents the relative importance of symbols and map elements; (6) create rough sketch maps that represent the evolving map design; and (7) construct the map in your chosen software application based on your sketch maps, and print rough drafts that will allow you to reevaluate the evolving map.

Visual hierarchy was described as being the graphical representation of the intellectual hierarchy. Contrast is applied to map features in order to manipulate their visual weights—the relative amount of attention they attract. Contrast refers to the visual differences between map features that allow us to tell one feature from another and that sometimes allow certain map features to appear to be more or less important. We learned that figure-ground is a special application of contrast that allows certain map features to be emphasized by appearing closer to the map user. Balance was described as the organization of map elements and empty space, resulting in a map that appears to be in a state of visual harmony and equilibrium. Good balance can be obtained, in part, through the establishment, identification, and use of available space. Map elements are placed in order from largest to smallest and are visually centered within appropriate areas of available space.

Finally, the concepts, rules, and guidelines set forth in this chapter and in Chapters 11 and 12 were employed to design a thematic map representing the suitability of residential building parcels. This case study was organized according to the map communication model and the seven procedures of the cartographic design process. It included consideration of how the map will be reproduced; selection of an appropriate scale and map projection; determination of appropriate data classification and symbolization methods; selection and appropriate implementation of map elements; establishment of an intellectual hierarchy; creation of sketch maps; and creation of the map using software applications. The final map effectively communicated its theme by means of an appropriate visual hierarchy, contrast, figure-ground, and balance.

13.6 STUDY QUESTIONS

1. Explain what cartographic design is, what its goals are, and how it has been influenced by map design research and art.
2. List four Gestalt principles of perceptual organization and give cartographic examples of each.
3. Establish an appropriate intellectual hierarchy by placing the following in order from most important to least important, top-to-bottom. Also, provide an explanation of why you have placed them in that order.
   - Base information such as boundaries, roads, place names, and so on
   - Data source and notes
   - Frame line and neat line
   - Scale and north arrow
   - Thematic symbols and type labels that are directly related to the theme
   - Title, subtitle, and legend
4. Use the map in Figure 13.5 to identify and describe three types of contrast that have been applied appropriately.
5. Use the map in Figure 13.6B to identify and describe figure-ground contrast that has been applied appropriately.
6. Provide several tips for achieving good balance on a map.
7. Review the seven procedures in Step 4 of the map communication model and comment on how they were implemented in the design of the residential site suitability map that was presented in the cartographic design case study (Section 13.4).

REFERENCES

Arntson, A. E. (2003) *Graphic Design Basics* (4th ed.). Fort Worth, TX: Harcourt Brace College Publishers.

Dent, B. D., Torguson, J. S., and Hodler, T. W. (2009) *Cartography: Thematic Map Design.* (6th ed.). Boston, MA: McGraw-Hill.

Huffman, D. P. (2010) "On waterlines: arguments for their employment, advice on their generation." *Cartographic Perspectives*, no. 66:23–30.

MacEachren, A. M. (1995) *How Maps Work: Representation, Visualization, and Design.* New York: Guilford.

MacEachren, A. M., and Mistrick, T. A. (1992) "The role of brightness differences in figure-ground: Is darker figure?" *The Cartographic Journal 29*, no. 2:91–100.

Monmonier, M. S. (1993) *Mapping It Out: Expository Cartography for the Humanities and Social Sciences.* Chicago: University of Chicago Press.

Montello, D. R. (2002) "Cognitive map-design research in the twentieth century." *Cartography and Geographic Information Systems 29*, no. 3:283–304.

Robinson, A. H. (1952) *The Look of Maps.* Madison, WI: University of Wisconsin Press.

Tufte, E. R. (1990) *Envisioning Information.* Cheshire, CT: Graphics Press.

14 Map Reproduction

14.1 INTRODUCTION

This chapter presents topics related to **map reproduction** in terms of printing: **print reproduction**. Print reproduction is used to create identical copies of maps in small-to-large quantities on specific materials (e.g., paper, cloth, mylar), using pigments (ink or toner), with specific page sizes and possible fold patterns. We encourage you to plan ahead for map reproduction to avoid problems associated with specific map reproduction methods, and we introduce **map editing**—a critical activity that can save you time and money during map reproduction.

The process of printing a digital map is distilled into four steps and involves raster image processing. **Screening** methods are introduced as methods for creating tints from base colors. **Process colors**, **spot colors**, **high-fidelity process colors**, and **color management systems** are described as aspects of color printing.

Offset lithography is identified as the dominant method for high-volume print reproduction and is characterized by excellent print quality, high speed, and volume discounts. The **prepress** phase of offset lithography is centered on the **service bureau** and involves the production of **printing plates** and **proofs**. Portable document formats are introduced for the delivery of digital maps to the service bureau. The offset lithographic printing press transfers images from printing plates to print media via a series of rolling cylinders. **Registration** refers to the proper alignment of colors produced by an offset press, and **trapping** is used to correct for improper registration.

14.2 LEARNING OBJECTIVES

- Explain the importance of planning ahead for print reproduction and provide specific factors to consider.
- Provide strategies for effectively editing a map.
- Describe raster image processing in the context of printing a digital map.
- Define screening and compare halftone with stochastic screening.
- Contrast process colors with spot colors and describe their typical uses in the printing of maps.
- Describe the prepress phase of offset lithographic printing and provide examples of prepress products and file formats.
- Explain the offset lithographic printing process.

14.3 PLANNING AHEAD

Reproduction methods can be quite specific—each provides certain benefits but also imposes strict limitations. The reproduction method you choose will influence almost every aspect of the cartographic design process. If you design a map without a specific reproduction method in mind, you could very well end up with a map that cannot be practically or economically reproduced. For example, if you design a color map using the RGB color model and then attempt to print to a CMYK device, you will encounter problems with color accuracy that will take time (and probably money) to resolve. Another example involves the reproduction of folded maps. Many large-format maps are designed to be folded in specific patterns so that, for example, the title and cover art appear on the front and the legend appears on the back when the map is folded. If you design a map with a special fold pattern in mind without ensuring that an appropriate folding mechanism is available, you will be forced to redesign your map. Another potential problem with folding is cracking: the flaking of ink or toner at creases. Thicker paper is typically more prone to cracking than thinner paper (Joyce 2015). If you hire a second party to reproduce a map, choose carefully, and make sure to establish a good relationship with this party before making major design decisions.

The following is a list of questions that need to be answered in the early stages of the design process. If you take the time to answer these sorts of questions carefully, the map reproduction process will be less likely to present problems and roadblocks that will prevent you from successfully completing a mapping project.

1. Who is the intended audience, and what is the purpose of the map? As in the design process, the answer to this question influences almost every aspect of the reproduction process.
2. What is your budget? Different reproduction methods vary greatly in cost.
3. When is your deadline? Make sure to allow enough time for reproduction, considering the speed of the method you have chosen. Consider that problems often arise that delay the reproduction of a map.
4. What material will be used? Paper? Cloth? Plastic? Different materials and different grades of material absorb ink and toner differently.
5. Will the map be limited to black, white, and grays, or will it be color? How many colors will be used? Printers have strict limitations regarding color, and

DOI: 10.1201/9781003150527-15

the cost of print reproduction generally increases when more colors are used.

6. What size will the map be? Printing devices have strict limitations on size and prices increase with size.

7. How many copies will be required? Certain reproduction methods offer a significant decrease in the cost per unit as the number of copies increases; other methods do not.

8. Will the map be folded? What will the fold pattern be? Folding mechanisms are limited to certain patterns.

9. Will printing occur on one side or on both sides? Certain materials are opaque, while others allow some of the image on the opposite side to show through.

10. What level of print quality is acceptable? Different reproduction methods provide various levels of quality.

11. Will you copyright the map? Will the map infringe on an existing copyright? Copyright information can be found at www.copyright.gov.

14.4 MAP EDITING

A map must be carefully edited before it is reproduced (Figure 14.1). **Map editing** is the critical evaluation and correction of every aspect of a map; it begins the first time the cartographer views the map in its early stages and culminates with a final edit that occurs just before reproduction begins. Editing could easily be treated as an aspect of cartographic design, but it is presented in this chapter because of its critical role in map reproduction. Because reproduction methods consume time and money, you simply cannot

afford to reproduce a map that contains errors. The following is a list of general questions that should be addressed when editing:

1. Map design: Does the design appropriately serve the map user based on the map's intended use? Does it communicate the map's information in the most efficient manner, simply and clearly?

2. Completeness: Are any features, map elements, or type labels missing?

3. Accuracy: Are features, map elements, and type labels correctly placed? Are words spelled correctly? Are numeric values correct?

Overfamiliarity and fatigue are problems faced by cartographers who edit their own maps. The cartographer spends so much time working on a map that he or she loses a certain degree of objectivity and often misses errors that are obvious to others. One solution to this problem is to have someone else (or a group of people) edit the map. Other tips for editing include the following:

1. Edit with "fresh eyes" at the beginning of a work session.

2. Edit large maps in sections.

3. View maps upside down or sideways—this forces you to see maps in new ways, possibly illuminating errors.

4. Read type out loud.

5. If possible, edit several days after completing a map.

14.5 RASTER IMAGE PROCESSING FOR PRINT REPRODUCTION

Virtually all modern print reproduction methods involve **raster image processing**: the conversion of a digital map into a raster image that can be processed directly by a raster-based printing device. The resulting raster image is representative of the **raster data model** (a general category of computer data consisting of rows and columns of square pixels), as illustrated in Figure 14.2A. In contrast to raster

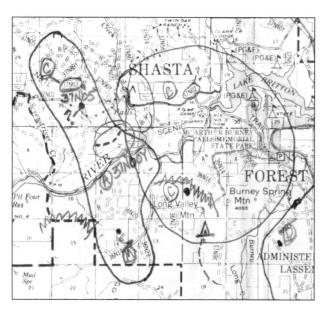

FIGURE 14.1 A Forest Visitor Map in the process of being edited. Carefully executed corrections, additions, and deletions save time and money during reproduction.

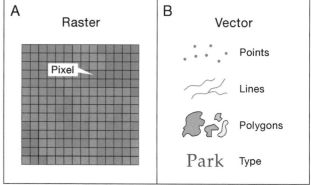

FIGURE 14.2 (A) Raster and (B) vector data models.

is the **vector data model**, which is composed of precisely defined points, together with the lines that connect them. The vector data model closely resembles what is drawn by hand and allows the representation of discrete points, lines, and polygons, as illustrated in Figure 14.2B. High-quality type is also represented using the vector data model. Most thematic maps are created using the vector data model because of its ability to represent clearly defined features such as wells, roads, administrative boundaries, and so on. The raster data model is commonly used when representing continuous phenomena such as terrain or when representing the results of raster-based GIS analysis or satellite image classification. Many maps incorporate both raster and vector data models. Regardless of the data models used in the creation of a map, the map is converted into a raster image for printing.

14.5.1 PRINTING THE DIGITAL MAP

Several steps are involved in printing a digital map, as illustrated in Figure 14.3. First, the cartographer creates the map in digital form using **application software**. A description of the wide range of application software capable of producing maps is beyond the scope of this text, but general categories include graphic design, GIS, and remote sensing applications; examples include Adobe Illustrator, ESRI ArcGIS Pro, and ERDAS Imagine, respectively. The digital map file produced by the application software is converted via the **printer driver** into **page description data** that consist of a set of printing instructions describing every graphical and textual component of the map. The page description data are written in a **page description language** (PDL) such as Adobe's **PostScript** or Hewlett Packard's **Printer Control Language** (PCL). PDLs are device-independent to a certain extent, allowing the user to print from, or to, almost any device equipped to interpret a particular PDL. PostScript has proved to be more device-independent and better suited to high-end printing than other PDLs, making it the de jure standard PDL; PostScript is the only PDL to be recognized by the International Organization for Standardization (ISO) (Adobe Systems Incorporated 1997).

After the page description data have been created by the printer driver, they are interpreted by a **raster image processor (RIP)**. A RIP is software or firmware that interprets page description data and converts them into a raster image that can be processed directly by a printing device. For example, a PostScript-compatible RIP interprets PostScript page description data and converts them into a raster image for printing. RIPs exist inside printing devices such as laser printers or as software applications that operate inside computers. Many less expensive printing devices are incompatible with PDLs and do not possess a RIP. These devices depend on RIP software inside the computer to create the raster image, which is subsequently sent to the printer.

The RIP produces raster images of a specific resolution that normally coincides with the maximum resolution supported by the printing device that will be used. Resolution refers to the number of pixels per inch or dots per inch (dpi) that a device can print. After the RIP has created the raster image, the image is processed by the printing device, which in turn creates the printed map (the last step in Figure 14.3).

14.6 SCREENING FOR PRINT REPRODUCTION

Screening is a technique used in most print reproduction methods in which colors are made to appear lighter by reducing the amount of ink or toner[1] applied to the *print medium* (e.g., paper). Screening is also referred to as *dithering*, but the term *screening* is more commonly used in the printing industry. Screening is used to create tints of a particular base color, and to represent continuous tone surfaces. A **tint** is a lighter version of a *base color*: Gray is a tint of the base color black, and pink is a tint of the base color red. A continuous tone surface is composed of tints that continuously vary in lightness within a given area, as illustrated by the terrain beneath the "REGIONAL PARK" label in Figure 14.4.

A standard (monochromatic) laser printer is only capable of printing black because it uses black toner. Through the application of screening, however, the same laser printer is capable of producing what *appear* to be grays even though black is the only color used (Figure 14.4). Screening can produce what appear to be lighter versions of *any* base color, not just black. In the screening process, ink or toner of a particular color, say, dark green, is applied to the print medium in a pattern of individual dots, allowing the color of the underlying print medium (normally white) to show through. If the pattern is fine enough, the human eye is capable of combining the color of the individual dark green dots with the white of the underlying print medium. This allows the

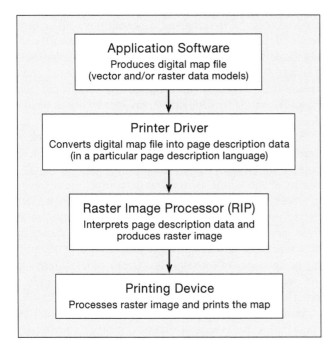

FIGURE 14.3 General steps involved in printing a digital map.

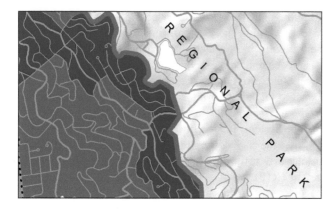

FIGURE 14.4 Black is the only color used in this map; the illusion of grays is produced through screening.

map user to perceive light green instead of dark green; the pattern itself typically goes unnoticed. Most color printing devices work on this principle; each color that the device can produce can be lightened through screening, allowing for the creation of far more colors than the number of inks or toners would imply.

14.6.1 HALFTONE AND STOCHASTIC SCREENING

Several methods of screening have been developed, but two have emerged as the most widely used: halftone and stochastic.[2] **Halftone screening** is used in almost all print reproduction methods, with the exception of inkjet printing, which normally employs stochastic screening. Halftone screening involves the application of ink or toner in a pattern of equally spaced dots of *variable size*, as illustrated in Figure 14.5A. The size of each dot determines the degree of lightness that is achieved; the smaller the dot, the lighter the result because more of the underlying print medium will show through. Halftone dots are composed of pixels produced by a RIP, with the pixel size normally corresponding with the maximum resolution of the printing device to be used (Figure 14.6). As a result, halftone dots are always coarser than the maximum resolution of the

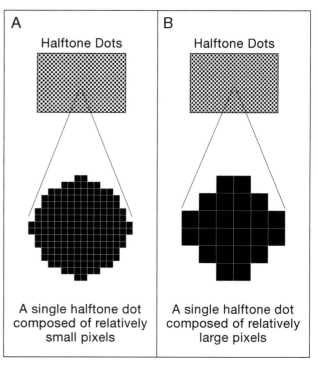

FIGURE 14.6 Halftone dots composed of pixels: (A) output from a higher resolution device; (B) output from a lower resolution device.

printing device. Smaller pixels (compare parts A and B of Figure 14.6) allow for the creation of finer, better defined halftone dots with more size variation, allowing a larger number of tints to be derived from a base color.[3]

Halftone screening is sometimes referred to as *amplitude modulation* (AM) screening because variations of lightness are achieved through the alteration of the size, or amplitude, of each dot. Halftone dots increase in size until a 50 percent tint is surpassed, at which point all color is removed from the dots and is applied to the area surrounding the dots, essentially creating white dots on a color background (70 percent and 90 percent in Figure 14.7A). As tints greater than 50 percent are specified, the background area expands, eventually filling the areas formerly occupied by dots.

Stochastic screening involves the application of ink or toner in a pattern of very small, pseudorandomly[4] spaced dots of *uniform* size, as illustrated in Figure 14.5B. The density of dots within a given area determines the degree of lightness that is achieved; dots spaced farther apart produce lighter results. Stochastic screening is also referred to as *frequency modulation* (FM) screening because variations of lightness are achieved through the alteration of the spacing or frequency of each dot. The term *stochastic* refers to the pseudorandom approach used to place dots within a given area.

Halftone screening has been practiced and refined for about 200 years, whereas stochastic screening is a relatively new method. While the halftone method is well defined, reliable, and relatively forgiving, the stochastic method is

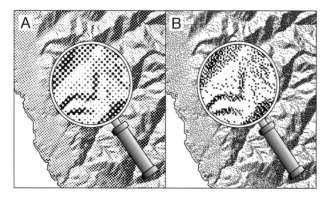

FIGURE 14.5 (A) Halftone screening, produced by equally spaced dots of variable size. (B) Stochastic screening produced by pseudorandomly spaced dots of uniform size.

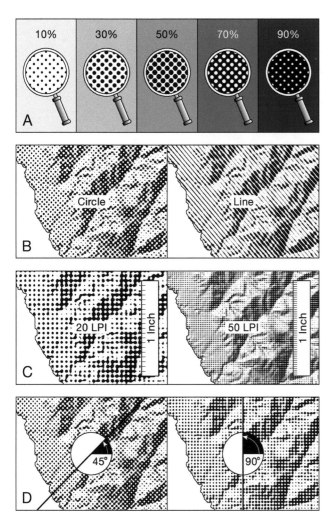

FIGURE 14.7 Halftone screening parameters: (A) tint percentage; (B) cell type; (C) screen frequency, measured in lines per inch (lpi); (D) screen angle, measured in degrees, counterclockwise from horizontal.

still being refined and is relatively unforgiving of inaccuracies in the reproduction process. Very small dots are difficult to print correctly, requiring that the entire reproduction process be tightly controlled. The stochastic method, however, offers several improvements over halftone screening. For example, it can result in clearer images with greater contrast because individual dots can be placed closer together and because there is less overprinting of inks when colors are mixed on the page. It can also eliminate the possibility of undesirable moiré patterns (described later) that are inherent in halftone screening (Joyce 2015). Although stochastic screening has been predicted to eclipse halftone screening, this has not yet occurred—both methods are still widely used.

14.6.2 HALFTONE SCREENING PARAMETERS

Halftone screening is controlled by four parameters: tint percentage, cell type, screen frequency, and screen angle. The **tint percentage** controls the degree to which the

appearance of ink or toner is lightened, as illustrated in Figure 14.7A. Tint is specified as a percentage, with lower values resulting in lighter colors. The **cell type** refers to the shape of each individual mark of the halftone pattern; the circle (dot) is the most widely used shape, but the line and other shapes are also used (Figure 14.7B).

Screen frequency, or screen ruling, refers to the spacing of halftone dots within a given area or, more specifically, the spacing of the *lines* that the dots are arranged in (Figure 14.7C). The screen frequency parameter controls how coarse or fine the halftone pattern is. Lower frequencies produce a coarse halftone pattern in which individual dots are clearly visible, which is typically employed only when a lower resolution printing device is used. Higher frequencies produce a fine halftone pattern in which it is more difficult to discern individual dots. This pattern is easier for the human eye to interpret as representing a solid but lighter color. Screen frequency is measured in lines per inch (lpi). Lpi is related to, but should not be confused with, dpi (dots per inch). Dpi refers to the maximum resolution of a printing device, independent of the lpi setting of a halftone screen. Halftones with higher lpi settings consist of smaller dots that are most accurately produced by high-resolution printing devices.

The **screen angle** parameter controls the angle at which the lines of halftone dots are oriented, as illustrated in Figure 14.7D. Screen angle becomes important when multiple halftone patterns are printed in the same area. For example, imagine one halftone pattern that represents a 30 percent tint of black, and another that represents a 30 percent tint of cyan. To mix the two colors, both halftone patterns need to be printed in the same area, allowing black and cyan to "mix" on the page. If both screen angles are identical (and if all other halftone parameters are equal), the black halftone dots will cover and partly block out the cyan dots. To prevent this, the screen angle of the black halftone would need to be offset from the cyan halftone, preventing the direct overlapping of dots, as illustrated in Figure 14.8. The concept of mixing colors on the page is described further in section 14.7.1. Screen angles need to be precisely specified in order to prevent the creation of a **moiré pattern** (Figure 14.9), which is an unwanted print artifact resulting from the interplay of dots in overlying halftones (Johansson et al. 2011). Most software applications and printing devices have been programmed to use screen angles that prevent the creation of moiré patterns.

14.6.3 STOCHASTIC SCREENING PARAMETERS

Stochastic screening is typically controlled by one parameter only: tint percentage. As with halftone screening, the tint percentage controls the degree to which the appearance of ink or toner is lightened. Tint is specified in percentage, with lower values resulting in lighter colors. The additional parameters associated with halftone screening do not apply to stochastic screens. The cell type is always a very small dot, usually representing the smallest pixel mark a particular

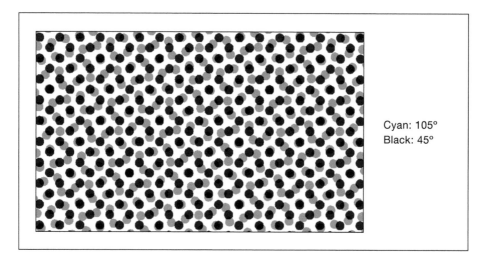

Cyan: 105°
Black: 45°

FIGURE 14.8   Different halftone screen angles allow tints of black and cyan to mix on the page.

printing device can produce. Because these dots are placed pseudorandomly within a given area, screen frequency and screen angle become meaningless. The irregular spacing of dots eliminates the potential problem of moiré patterns inherent in halftone screening.

14.7   ASPECTS OF COLOR PRINTING

14.7.1   Process Colors

In contrast with display devices, which generate color through the addition of red, green, and blue light (the additive primary colors—see Chapter 10), most printing devices generate color using inks and toners based on the **subtractive primary colors**: *cyan* (C), *magenta* (M), and *yellow* (Y). In theory, a mixture of pure C, M, and Y (without screening) will result in pure black. In reality, this mixture produces a "muddy" dark brown that lacks crisp detail, and the grays produced by mixing C, M, and Y are not pure.

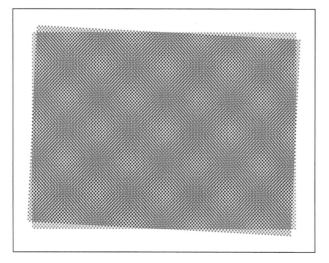

FIGURE 14.9   A moiré pattern produced from the incorrect specification of cyan and magenta halftone screen angles.

As a result, black (K) is normally used in conjunction with C, M, and Y, creating CMYK, or the **process colors**. Most color printing devices mix the process colors on the page by applying them, in sequence, to the same area. Inks and toners based on the process colors are semi opaque or translucent, allowing them to combine on the page; new colors are created where the process colors overlap. This, together with screening techniques, allows for the creation of a wide variety of colors and is referred to as **four-color process printing**. When process colors are mixed on the page, tints of each base color are represented by halftone patterns, each with a unique screen angle to avoid moiré patterns. This results in a special pattern of inks or toners called a **rosette** (Figure 14.10). The human eye interprets fine rosettes (i.e., ones that result from halftones with high screen frequencies) as representing a solid color. Four-color process printing is an efficient method of producing many colors from four base colors. However, the actual number of colors that are possible is fewer than the number possible using the additive primary colors (RGB). Specifically, vibrant colors such as orange, red, green, and blue are difficult to reproduce using the process colors (McCue 2014).

14.7.2   Spot Colors

An alternative to mixing process colors on the page is to use solid colors, or **spot colors**. Spot colors take the form of opaque inks that are *premixed* before they reach the printing device. As with process colors, tints can be created from spot colors through screening. Exact color matches are easier to achieve using spot colors because they do not rely on the printing device for mixing—a situation that often results in color inconsistencies from device to device. The *Pantone Matching System* (PMS) is the most universally accepted spot-color system, but TOYO and other systems are also popular. Spot colors can be selected on-screen or (more appropriately) from printed color swatches. Spot inks are mixed according to specific formulas, such as those

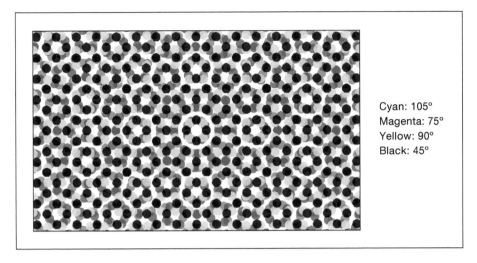

Cyan: 105°
Magenta: 75°
Yellow: 90°
Black: 45°

FIGURE 14.10 A rosette pattern produced from halftone screens of the process colors, each at a different screen angle.

found in the *Pantone Color Formula Guide* (Joyce 2015). A disadvantage of using spot colors is the cost of reproduction when a map has many colors. Unlike the process colors, which can produce a wide variety of colors through mixing, spot colors normally require separate inks to be used for each color, increasing the cost of reproduction.

14.7.3 HIGH-FIDELITY PROCESS COLORS

Newer methods of color printing have been developed in response to the disadvantages of process and spot colors, specifically the limited variety and vibrancy of process colors and the expense of spot colors when they are used in large numbers. **High-fidelity process colors** are based on the traditional process colors (CMYK) but include two or three additional colors. This allows for dramatic increases in the variety and vibrancy of colors and also allows for accuracy in color matching that rivals the Pantone Matching System.

Various approaches to high-fidelity color have been taken. Many of these incorporate CMYK and two additional colors such as orange and green, or light cyan and light magenta. The additional colors result in prints with more variety, vibrancy, and contrast than traditional process-color systems but are usually more expensive.

14.7.4 COLOR MANAGEMENT SYSTEMS

A common problem confronting cartographers is matching the colors on a graphic display with those produced by a printing device. Without deliberate intervention, on-screen colors will rarely, if ever, truly match colors that are printed. This is partly due to differences between the color models employed by graphic displays and the color models employed by most printing devices (RGB vs. CMYK). It is also partly due to differences between software applications and display and printing technologies. Each device in the workflow, including scanners, graphic displays, and printers, introduces subtle variations in color. Even devices that are supposedly identical—two identical printers, for example—can produce different results due to variations in calibration, operating conditions, and so on. All of these factors contribute to differing color **gamuts** (ranges of colors produced) among devices. **Color management systems** (CMSs) attempt to provide consistency and predictability of color by identifying differences in gamuts and correcting the variations in color introduced by each device (Johansson et al. 2011).

CMSs are centered on the **color profile**, which is an electronic file that describes the manner in which a particular device introduces color variations. Before a profile is created, the device—a color printer, for example—needs to be warmed up to normal operating temperature and calibrated so that it operates within the manufacturer's specifications. A sample page is printed, which consists of a series of strictly defined colors. Colors on the printed sample page are measured using a *colorimeter* or *spectrophotometer* (both devices precisely measure color). The measured values are then sent to the CMS, which creates the device's color profile by comparing the colors on the sample page with the colors as they were originally defined. Once color profiles have been created for every device in the workflow, the CMS is able to correct for variations in color, resulting in printed colors that are *as similar as possible* to those seen on the graphic display, captured by the scanner, and so on. It is impossible to match all colors in every situation, however, because the color gamuts of electronic devices can be substantially different.

Color management has become partly standardized with the formation of the **International Color Consortium** (ICC). The ICC has developed a standard for a vendor-neutral, cross-platform color profile called the *ICC color profile*. Color profiles based on this standard can be implemented at the operating-system level on Macintosh and Windows computers (Green 2010).

14.8   HIGH-VOLUME PRINT REPRODUCTION

When large numbers of maps are required (e.g., more than 200), issues of cost and time become critical. Methods for low-volume print reproduction such as laser and ink-jet printing become too costly or time-consuming when reproducing large numbers of maps, especially when full-color output is required. Low-volume methods also present strict limits on page size. Instead of providing an in-depth discussion of low-volume methods, we will focus on high-volume print reproduction. While low-volume reproduction methods are plentiful and varied, high-volume reproduction is dominated by a single method: offset lithography. *Lithography* is a printing process in which ink is made to stick only to certain areas of a printing surface (through chemical or electronic means) and is subsequently transferred to a print medium (e.g., paper). Off**set lithography** is a form of lithography in which ink is transferred to an *intermediate printing surface* before being transferred to the print medium. Virtually all mass-produced maps are the result of offset lithography, which is characterized by excellent print quality, high printing speed, and a significant decrease in the cost per unit as the number of copies increases—that is, the last copy costs significantly less than the first. The cost-per-unit decrease occurs because the majority of expenses are incurred in preparation for printing; the cost of paper and ink for additional copies is minimal in comparison with the costs of prepress and press setup, activities that occur before printing begins.

14.8.1   THE PREPRESS PHASE

The **prepress phase** of high-volume map reproduction consists of various technologies and procedures that make offset lithographic printing possible. The press in "prepress" refers to the **offset lithographic printing press** (Figure 14.11), which is further described in Section 14.8.4. The

FIGURE 14.11   A four-color offset lithographic printing press. (This file is licensed under the Creative Commons Attribution-Share Alike 4.0 International license. https://commons.wikimedia.org/wiki/File:Offset_press.jpg.)

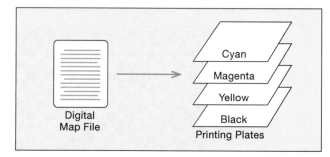

FIGURE 14.12   Printing plates for a full-color map to be printed using process colors.

abbreviated terms *offset press* and *offset printing* are often used to describe this device. Many aspects of prepress revolve around the **printing service bureau**, a business that specializes in the creation of products such as printing plates and proofs, and provides print-related services. The service bureau often plays a key role in the reproduction process, intermediate between the cartographer and the press operator (the person who operates the offset press).

The **printing plate** is a sheet of aluminum (or polyester) that is ultimately mounted on a roller on an offset lithographic printing press. It is laser-engraved to receive a positive, latent (invisible) image on a **platesetter**. One printing plate is created for each color that will be used on the offset press, as illustrated in Figure 14.12. After it is mounted on the offset press, the printing plate is the first representation of the digital map to come into contact with ink, which is transferred to the print medium shortly afterward. Plates are produced by service bureaus or by press operators, depending on the capabilities of each.

The **proof** is a representation of what the final, reproduced map will look like. It is an essential component of the prepress phase and is used in conjunction with editing to ensure that the map will be reproduced just as the cartographer intends. Cruder, less expensive proofs (rough drafts) are created repeatedly during the map design process, but higher-quality (and more expensive) proofing methods are required in the prepress phase. Proofs are created using various technologies and are produced by individuals, service bureaus, and press operators. Proofing methods are described in detail in Section 14.8.3.

14.8.2   FILE FORMATS FOR PREPRESS

A digital map can be delivered to the service bureau in a variety of formats. For example, it can be delivered in the native format of the application software that created it (an ESRI ArcGIS Pro Project file, an Adobe Illustrator file, etc.), but this can be a risky option for several reasons. First, the service bureau might not have the exact software application or fonts that you used. Second, related data such as linked images or GIS data sets might not transfer well to the service bureau's computer, making it difficult or impossible for the service bureau to properly open the file. Finally, application files are easily editable, making it possible for the service bureau to edit your file, either intentionally or accidentally. Delivering a digital map in the application's native file format can yield good results, but only if the service bureau can assure you that these issues will not pose problems.

Another option is to deliver the digital map as a file that contains page description data (e.g., a PostScript file). Page description files can be written by choosing the Print to File option when printing a map. This can be a practical option because related data and fonts get *embedded* into page description data and because the service bureau has the ability to download page description data directly to RIPs and printing devices without the software application that was used to create the map. Downsides to this approach include the facts that page description files can be very large, they need to contain specific information regarding the printing device that will be used (information you might not have), and their contents can be viewed only by a handful of software applications.

A more attractive option is to deliver the digital map in a *portable* document format such as **Encapsulated PostScript (EPS)** or **Portable Document Format (PDF)**. An EPS file is a subset of the PostScript PDL that allows digital maps and other documents to be transported between software applications and between different types of computers. It consists of PostScript code for high-resolution printing and an optional low-resolution raster image for on-screen display. Because they are written in PostScript, EPS files can be very large, but they do not require the specific printing device information required by PostScript page description files. EPS files also have the ability to embed related data and fonts. A PDF file is similar to an EPS file in that it is related to PostScript but is more flexible and more efficient. In addition to being able to embed related data and fonts, PDF files also have the ability to embed, or encapsulate, features such as hyperlinks, movies, and keywords for searching and indexing. Zooming and panning capabilities

for on-screen viewing are also provided. PDF files can be viewed using Adobe Acrobat Reader (a free download) and can be created from virtually any application via Adobe's Acrobat and Acrobat Distiller. Acrobat compresses and optimizes digital maps for printing, Web display, and so on. Although the PDF format was originally intended for on-screen display, it has become a format of choice for the delivery of maps and other documents to service bureaus for high-end printing (Joyce 2015).

Portable document files are created by converting digital maps from their native file format into EPS or PDF files. This is normally accomplished either by selecting the Export command within the application that was used to create the map or by "printing" the map to a PDF file using Acrobat. Be aware that, on occasion, certain graphical elements of your map (e.g., custom line patterns) will be altered or lost during the conversion process. If you choose to deliver your digital map to the service bureau in a portable document format, it is especially important that you proofread your map based on output from the EPS or PDF file so that you can identify any changes resulting from the file conversion.

14.8.3 Proofing Methods

Proofing methods range from low-cost, low-quality techniques to high-cost, high-quality techniques, as illustrated in Figure 14.13. The proofing method that most closely resembles the ultimate reproduction method is usually the most reliable. Proofs are useful in both low-volume and high-volume reproduction, but the emphasis here is on proofing for high-volume reproduction using an offset press. In print reproduction, the correction of errors during editing is generally cheapest when working with crude proofing methods and becomes increasingly expensive with more sophisticated proofing methods.

Soft proofing (or on-screen display) is the least expensive proofing method (assuming a computer was used to create the map); it involves viewing a digital map on a

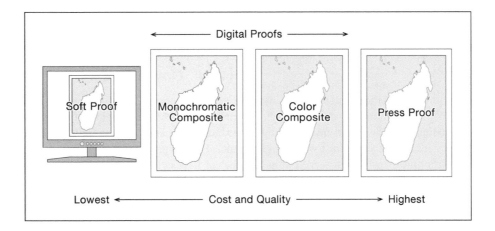

FIGURE 14.13 Proofing methods for print reproduction. The most reliable method is the one that most closely resembles the intended reproduction method.

computer screen. Maps are typically displayed using application software or document-viewing software such as Adobe Acrobat Reader, which displays PDF files. Soft proofing is the primary proofing method for nonprint reproduction but is the crudest of methods for print reproduction. On-screen display allows you to perform basic editing and spell-checking but should not be trusted entirely when editing for print reproduction for three reasons:

1. Because computer screens are often smaller than the map being edited, the entire map cannot be seen at full size.
2. Because the typical computer screen resolution is far lower than the resolution of printing devices, fine detail will not be visible on-screen unless the cartographer zooms in on a portion of the map.
3. On-screen colors rarely, if ever, truly match colors that are printed. CMSs can improve the accuracy of on-screen colors, but accurate color proofing is still beyond the scope of on-screen display.

Despite the drawbacks of on-screen proofing, the PDF file format provides various features that enhance on-screen proofing. When viewed in Adobe Acrobat Reader, PDF files can be zoomed and panned, searched, spell-checked, and commented on. Adobe Acrobat Reader also provides proofing simulation tools that allow the user to see what the map might look like when printed on paper. Because of these features, PDF files are often accepted as contract proofs (described later) in noncritical printing situations (McCue 2014).

Creation of **monochromatic composites** is the second least expensive proofing method. This involves printing a map in black, white, and grays, typically on a laser printer. Monochromatic composites are classified as **digital proofs** because they are created on digital printers rather than an offset press. The term *composite* refers to the fact that all color information is applied to the page at virtually the same time and, in this case, is represented as black, white, and grays. Monochromatic proofs created by laser printers are sufficiently detailed for general editing tasks, but color cannot be discerned, making monochromatic proofing unsuitable for color editing. Laser prints are typically limited to page sizes of 11″ × 17″ or smaller.

Color composites represent a middle ground between low- and high-quality proofing methods and are also classified as digital proofs. Various printing devices can produce color composites including color laser and inkjet printers, although inkjet devices are probably the most widely used. Inkjet printers range from inexpensive letter-sized units to expensive, large-format devices developed primarily for producing color composite proofs. Inexpensive inkjet printers cannot reproduce fine detail as well as a monochromatic laser printer, but they normally have the ability to represent color. However, the color accuracy of inexpensive inkjet printers is not good enough to warrant using them for critically evaluating colors. High-end, large-format inkjet printers allow large maps to be printed on one sheet and, when used in conjunction with a CMS, can produce colors that are far more accurate than those produced by inexpensive, small-format inkjet printers. Color laser printers produce high-quality images but are usually limited to page sizes of 11″ × 17″ or smaller.

The highest quality, most expensive proofs available (before the offset press is run) can act as *contract proofs*: contractually binding documents. Contract proofs are produced by very high-resolution inkjet printers that are calibrated to simulate the output from a specific offset lithographic printing press. If the cartographer is satisfied with a high-quality proof, and if the print operator assures the cartographer that the final, printed map will look *very similar* to the proof, then the proof can act as part of the contract between the two parties.

The ultimate (and most expensive) proof is the **press proof**, which is printed on the offset press. The press proof can be produced in the prepress phase, or it can be produced at the **press check**. At the press check, printing plates are mounted on the offset press, the press is inked and calibrated, and sample prints are made. If the cartographer is satisfied with the samples, the press run can begin. As the printing press runs, the press operator periodically compares the prints with the approved samples and makes adjustments to the press if necessary. Cartographic errors discovered during the press check will be extremely expensive to correct because prepress and press setup activities will need to be repeated. Because of this, the press check should be viewed as an opportunity to identify issues that the press operator can address rather than an opportunity to identify cartographic errors. For example, the press operator is capable of producing slight variations in the lightness of particular colors by altering the density of ink that is applied to the page without increasing the cost of the print job.

Regardless of the proofing method, it is best to view a proof under the same conditions that the final map will be viewed. This includes lighting conditions (outdoors in sunlight, indoors with fluorescent lighting, etc.) and more specific locational conditions such as inside a helicopter or on a boardroom table.

14.8.4　Offset Lithographic Printing

Once the prepress phase is complete, the offset lithographic printing process can begin. Offset presses are mechanical printing devices that incorporate aspects of digital technologies. Presses are categorized, in part, according to the number of base colors they can produce. The simplest presses are limited to one base color and its tints, whereas others are capable of printing ten or more base colors and their tints. Tints are typically created through halftone screening. The offset press contains one or more **printing units** (Figure 14.14) composed of many cylinders that ultimately transfer ink to the print medium. Each printing unit is

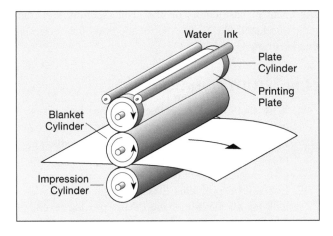

FIGURE 14.14 A printing unit (simplified) of an offset lithographic printing press.

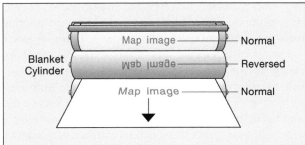

FIGURE 14.15 A blanket cylinder transferring the map image from the printing plate to the print medium.

capable of printing one base color. There are three primary cylinders in each printing unit: the *plate cylinder*, to which a printing plate is mounted; the *blanket cylinder*, which receives an image from a printing plate and subsequently transfers it to the print medium; and the *impression cylinder*, which helps move the print medium through the press. There are also several smaller cylinders that come into contact with the plate cylinder; these are used to apply ink and water to the printing plate. For simplicity, they are represented in Figure 14.14 as one cylinder for water and one for ink.

As stated earlier, the printing plate is the first representation of a digital map to come into contact with ink. The ink used in offset printing is oil-based—a key characteristic in determining the manner in which ink adheres to the printing plate. The printing plate contains image areas that are *oleophilic* (oil-liking) and nonimage areas that are *oleophobic* (oil-fearing). As the offset press runs, water and ink are applied to the printing plate by the series of smaller cylinders near the top of the printing unit; ink adheres only to the image areas and is washed away from nonimage areas.

Once the ink is applied to the printing plate, the image on the plate could, in theory, be transferred directly to the print medium. If this were to occur, however, the resulting image would be reversed, as illustrated by the image on the blanket cylinder in Figure 14.15. To print a normal, non-reversed image, it is necessary to transfer the image from the printing plate onto the blanket cylinder, which subsequently transfers the image onto the print medium. The *offset* in "offset lithography" refers to this process—the printing plate is offset from the print medium by the blanket cylinder. The rubber coating on the blanket cylinder allows it to perform two additional duties. First, it acts as a buffer between the easily worn printing plate and the print medium. Second, it applies ink to porous print media more effectively than the rigid printing plate would be able to. As the ink is transferred to the blanket cylinder and then to the print medium, it is forced to spread out due

to pressures exerted between the cylinders. This effect is termed **dot gain**. It tends to deteriorate fine detail and make tints slightly darker as halftone dots are increased in size. Dot gain is sometimes compensated for by reducing the size of halftone dots when printing plates are produced (McCue 2014).

The procedure just described accounts for the application of a single base color to the print medium. Multicolor print jobs are produced in the same manner, except that multiple printing units are employed, one for each base color, as illustrated in Figure 14.16. As the print medium passes through the press, it receives a different color ink from each successive printing unit. The printing units are spaced at a particular distance from one another to allow each successive color to print on the same region of the print medium. This allows all base colors to overlay correctly and mix on the page.

Registration refers to the alignment of colors in multicolor printing. When using the process colors, for example, a map feature composed of a mixture of cyan, magenta, and yellow *should* have crisp, well-defined edges, indicating that all three inks have been applied to the exact same area of the page. In reality, misalignment often prevents

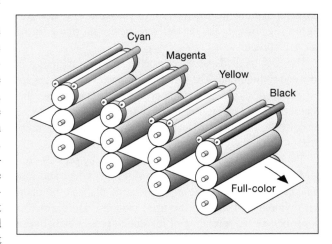

FIGURE 14.16 Multicolor printing achieved through the use of multiple printing units. In this example, process colors are mixed on the page, resulting in full-color output.

the perfect placement of colors, and misregistration occurs. Misregistration can result from poorly produced printing plates or a poorly calibrated printing press. **Trapping** refers to a series of techniques used to minimize the effects of misregistration. It involves the manipulation of ink placement to improve the appearance of areas where inks overlap incorrectly or where gaps occur (i.e., where inks should print but do not). The trapping process has become largely automated through the use of special software employed by service bureaus, relieving the cartographer of what used to be a tedious and often difficult procedure (McCue 2014).

14.9 SUMMARY

In this chapter, we have presented topics related to print map reproduction. We have encouraged you to plan ahead for map reproduction and have stressed the importance of map editing, both of which can help you avoid difficulties when reproducing a map.

Raster image processing was described as being central to printing digital maps, with the printing process distilled into four steps: creation of the digital map using application software, conversion into page description data, interpretation by a RIP and creation of a raster image, and printing of the image on a printing device. Halftone and stochastic screening methods were presented as ways to create tints from base colors. Aspects of color printing were described, including process colors, which mix on the page; spot colors, which are premixed; high-fidelity process colors, which increase color variety and vibrancy; and color management systems, which identify and correct for variations in color introduced by electronic devices.

Offset lithography was identified as the dominant method for high-volume print reproduction, characterized by high quality, high speed, and volume discounts. Prepress involves the creation of printing plates that are mounted on the offset press, each of which represents one color on an offset press, and proofs that facilitate map editing. Prepress is centered on the printing service bureau. We learned that portable document formats such as EPS and PDF are preferred file formats for the delivery of digital maps to the service bureau.

Proofing methods were described on a scale ranging from low-end to high-end. The soft proofing method is the cheapest and crudest, followed by monochromatic composites normally produced on laser printers; color composites produced on color laser printers, inkjet printers, and so on; and the press proof—the highest quality and most expensive proof.

We learned that offset lithographic printing presses contain printing units composed of rolling cylinders and that images are transferred from the printing plate to the blanket cylinder and then to the print medium. Dot gain was described as the spreading of ink due to pressures between cylinders on the press. Registration was described as the perfect alignment of base colors on the print medium, and trapping was identified as a series of methods for correcting misregistration.

14.10 STUDY QUESTIONS

1. Why should map reproduction issues be considered early in the cartographic design process? Give three specific examples of map reproduction problems that can arise if the designer does not plan ahead sufficiently.
2. List several problems that a cartographer should look for when editing a map. Also, provide several tips that can be helpful in the editing process.
3. Describe the process of printing a digital map in terms of application software, the print driver, page description data, and raster image processing.
4. Compare and contrast halftone and stochastic screening and the contexts in which each screening method is typically used.
5. Define process colors and spot colors and provide examples of why each is typically used.
6. Describe the prepress phase of the offset lithographic printing process, including the preferred file formats for delivery, printing plates, and proofs.
7. Explain the offset lithographic printing process in terms of the various cylinders and printing units that are employed.

NOTES

1 Toner is a powder that is used to form images in laser printers and copy machines.
2 Certain screening methods represent a combination of halftone and stochastic approaches, also known as second-order screening (Joyce 2015).
3 The number of tints that can be derived from a base color can be calculated using this formula: (printer resolution/screen frequency)^2+1.
4 Computing devices are incapable of producing truly random numbers.

REFERENCES

Adobe Systems Incorporated (1997) *The Adobe PostScript Printing Primer*. San Jose, CA: Adobe Systems Incorporated.

Green, P. (2010) *Color Management: Understanding and Using ICC Profiles*. Hoboken, NJ: Wiley.

Johansson, K., Lundberg, P., and Ryberg, R. (2011) *A Guide to Graphic Print Production* (3rd ed.). Hoboken, NJ: John Wiley and Sons.

Joyce, M. P. (2015) *Designing for Print*. Santa Ana, CA: Inez D. Inc.

McCue, C. (2014) *Real World Print Production with Adobe Creative Cloud*. San Francisco, CA: Peachpit/Pearson Education.

Part II

Mapping Techniques

15 Choropleth Mapping

15.1 INTRODUCTION

The choropleth map is arguably the most commonly used (and abused) method of thematic mapping. This chapter covers several topics that will help you create a well-designed choropleth map and thus avoid these abuses. One common abuse is mapping raw counts such as numbers of people. Section 15.3 considers several approaches for converting these raw counts into standardized data, which is essential for choropleth mapping. Another common abuse is using inappropriate colors to represent each class of data. Ideally, an appropriate **color scheme** should logically represent the quantitative nature of the data depicted on choropleth maps. Section 15.4 covers ten factors that can be considered in selecting appropriate color schemes.

One of the challenges in choropleth mapping is determining whether you will create a classed or an unclassed map. Classed maps are far more commonly used and recommended, but we will see that unclassed maps have their advantages; for instance, if there is a positive skew in the data, then this positive skew can be reflected in the mapped distribution. Section 15.5 covers several approaches (including web-based tools) for creating classed and unclassed maps, and Section 15.6 summarizes issues that you should consider when deciding whether you should create a classed or unclassed map.

Another challenge in choropleth mapping is designing an appropriate legend, as an effective color scheme may be useless if the legend is poorly designed. Section 15.7 responds to this challenge by illustrating a range of legend design options. The bulk of this chapter deals with traditional choropleth mapping in which enumeration units are symbolized with visual variables related to color (e.g., hue and lightness). In the concluding section of the chapter, we consider the **illuminated choropleth map** in which traditional choropleth symbology is combined with an illuminated 3-D prism view, which can produce an eye-catching graphic.

15.2 LEARNING OBJECTIVES

- Utilize various approaches for standardizing raw-total data that are often collected for choropleth mapping.
- Explain the difference among sequential, diverging, and spectral color schemes and indicate how the kind of data (unipolar or bipolar) can play a role in selecting a color scheme.
- Consider the broad range of factors that can play a role in selecting a color scheme, such as color

vision impairment, simultaneous contrast, map use tasks, color associations, and color aesthetics.
- Learn about web-based tools for selecting color schemes for classed and unclassed choropleth maps.
- Decide when it is appropriate to use classed and unclassed choropleth maps.
- Design legends appropriate for choropleth maps.
- Contrast the illuminated choropleth map with the traditional choropleth map and indicate any advantages of the illuminated method.

15.3 SELECTING APPROPRIATE DATA

Ideally, the choropleth technique is most appropriate for a phenomenon that is uniformly distributed within each enumeration unit, changing only at enumeration unit boundaries. For instance, state sales tax rates are appropriate for choropleth mapping because they are constant throughout each state, changing only at state boundaries. Because this ideal is seldom achieved in the real world, you should be cautious in using the choropleth map. If you wish to focus on "typical" values for enumeration units, you can utilize the choropleth map, but you should realize that error might be present in the resulting map. Instead of using the choropleth map because you have data readily available for enumeration units, you should consider whether some of the other symbolization methods discussed in this book might be more appropriate. (See Section 4.6 for further elaboration of this issue.)

Of particular concern in choropleth mapping is the size and shape of enumeration units. Ideally, the method works best and is most accurate when there is no significant variation in the size and shape of the units. The problem can be understood by considering the county-level map of U.S. population density shown in Figure 1.6. Imagine trying to determine the population density for one of the large counties in California. Because people are unlikely to be uniformly spread throughout a county, the color shown for that county is unlikely representative of the population density throughout the county. In contrast, the color for a very small county, such as one of the five counties of New York City, would be more representative because the population density in those counties is relatively uniform. The net result is that there is considerable variation across the choropleth map in terms of our ability to properly represent population density. Also, note that the largest counties provide the biggest visual impact but also have potentially the largest error of representation. Although the choropleth technique is more appropriately used when enumeration units are similar in size, such as for counties of Iowa, in reality, cartographers cannot control the size and shape of enumeration units, and so appropriate standardization is essential.[1]

Another important issue in selecting data for choropleth maps is ensuring that raw-total data have been standardized to account for varying sizes of enumeration units. In Chapter 4 (see Figure 4.9), we standardized by dividing an area-based raw total (i.e., acres of wheat harvested for Kansas counties) by the areas of those counties. Because the numerator and denominator were both in the same unit of measurement (acres), a proportion (or percentage) resulted. We now consider additional approaches for standardizing data. One approach is to divide an area-based raw total by some other area-based raw total. For example, for acres of wheat harvested, we might divide by acres of wheat planted, with the resulting ratio providing a measure of success of the wheat crop (Figure 15.1A). As an alternative, we might divide acres of wheat harvested by acres harvested for all major crops, producing a map illustrating the relative importance of wheat to the agricultural economy of each county.

A second approach for standardizing data is to create a density measure by dividing a raw total not involving area by either the areas of enumeration units or some area-based raw total. For example, if the number of bushels of wheat produced is divided by the area of the county (in acres), then the result is bushels of wheat per acre. This approach might not be meaningful, however, if the raw-total data occur only within a portion of the enumeration unit. For example, with the wheat data, it makes more sense to divide the bushels produced by the acres of wheat harvested, yielding bushels of wheat produced per acre of wheat (Figure 15.1B). A third approach for standardizing data is to compute the ratio of two raw totals, neither of which involves area. For

example, we might divide the value of wheat harvested (in dollars) by the value of all crops harvested (in dollars). The resulting proportion would indicate the relative value of wheat in each county. Although the area is not included in the formula, this approach indirectly standardizes for the area because larger areas tend to have larger values of both attributes.

When computing the ratio of two raw totals not involving area, it is common to express the result as a *rate*. Although we are all familiar with rates, we often aren't aware of how they are computed. To illustrate, consider how we compute cancer death rates for counties if we wish to express the death rate as the number of deaths per 100,000 people. First, we establish a simple proportional relationship as follows:

$$\frac{\text{Cancer deaths for county}}{\text{Population for county}} = \frac{\text{Number of cancer deaths}}{100,000 \text{ people}}$$

We then solve for the number of cancer deaths by rewriting the equation as follows:

$$\text{Number of cancer deaths} = \frac{\text{Cancer deaths for county}}{\text{Population for county}} \times 100,000.$$

The resulting number of cancer deaths is termed the cancer death rate. More generally, the formula for rates is:

$$\text{Rate} = \frac{\text{Magnitude for category of interest}}{\text{Maximum possible magnitude}} \times \text{Units of the rate,}$$

where the units of the rate are some power of 10 such as 1,000 or 1,000,000.

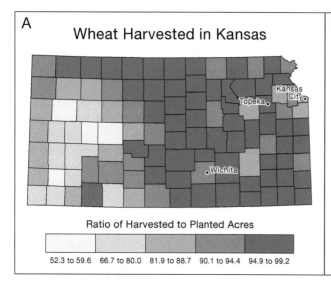

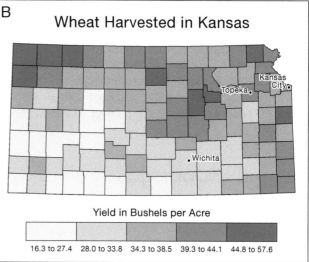

FIGURE 15.1 Examples of approaches for standardizing data. (A) Dividing acres of wheat harvested by acres of wheat planted; both are area-based raw totals and thus proportions (or percentages) result. (B) Dividing bushels of wheat harvested by acres of wheat harvested; the result is a density measure (bushels per acre). (From National Agricultural Statistics Service, U.S. Department of Agriculture, http://www.nass.usda.gov/Quick_Stats/.)

A fourth standardization approach is to compute a summary numerical measure (e.g., mean or standard deviation) for each enumeration unit. For example, we could compute the average size of farms in each county by dividing the acreage of all farms by the number of farms. Note that this approach accounts for a larger acreage in a larger county by dividing by a greater number of farms.

In general, the standardization approaches discussed thus far involve ratios. As a result, one might suggest that computing ratios are the key to standardization. A simple example illustrates that this is not the case. Imagine that for each county, you computed the mean number of acres of wheat harvested over a ten-year period. Clearly, the values would be ratios (the numerator would be the sum of acreages over ten years, and the denominator would be 10), but the data would not be standardized because the denominator in each case would be 10. No adjustment would have been made to account for the fact that larger counties tend to have larger numbers of acres harvested each year.

It is important to recognize that the various standardization approaches lead to quite different maps of the phenomenon being investigated. For example, Figure 4.9B (representing the percentage of land area from which wheat was harvested) reveals that higher values tend to occur along a north–south line extending through the center of the state. In contrast, Figure 15.1A (illustrating the percentage of planted wheat actually harvested) reveals that higher values occur in the eastern two-thirds of the state. The map of wheat yield (Figure 15.1B) produces yet another pattern, with higher values occurring in northern and eastern sections of the state. Clearly, different methods of standardization produce different views of the spatial pattern of wheat harvested in Kansas. It is your responsibility as a mapmaker to think carefully about which standardization procedure is most appropriate, given your data and the message that you wish to communicate.

15.4 FACTORS FOR SELECTING A COLOR SCHEME

In this section, we consider numerous factors for selecting a color scheme for choropleth maps, focusing largely on the work of Cynthia Brewer and colleagues. For illustrative purposes, we will map the percentage of the adult population in the 48 contiguous U.S. states that did not graduate from high school (Table 15.1). Note that this dataset is unipolar as there is no natural dividing point in the data. We have chosen to display only classed maps in this section because classed choropleth maps are much more common than unclassed choropleth maps. You should realize, however, that the factors described here also apply to unclassed maps. We have chosen five classes for these classed maps so that readers can easily discriminate the colors for differing classes on all of the maps.

TABLE 15.1

Percentage of the Adult Population, in the U.S. States, That Did Not Graduate from High School (Based on 2011 Data)

State	Not Graduating (%)
Alabama	17.3
Arizona	14.3
Arkansas	16.2
California	18.9
Colorado	9.8
Connecticut	10.9
Delaware	13.0
Florida	14.1
Georgia	15.7
Idaho	11.4
Illinois	12.8
Indiana	12.7
Iowa	9.4
Kansas	10.0
Kentucky	16.9
Louisiana	17.5
Maine	9.1
Maryland	11.1
Massachusetts	10.8
Michigan	11.2
Minnesota	8.0
Mississippi	18.9
Missouri	12.4
Montana	7.7
Nebraska	9.0
Nevada	16.0
New Hampshire	8.6
New Jersey	11.9
New Mexico	16.8
New York	15.0
North Carolina	15.3
North Dakota	9.3
Ohio	11.7
Oklahoma	13.7
Oregon	10.6
Pennsylvania	11.4
Rhode Island	15.2
South Carolina	15.8
South Dakota	9.4
Tennessee	15.8
Texas	18.9
Utah	9.7
Vermont	8.2
Virginia	12.2
Washington	9.9
West Virginia	15.8
Wisconsin	9.6
Wyoming	8.0

Source: American Community Survey, https://www.census.gov/programs-surveys/acs/.

15.4.1 KIND OF DATA

In her early work, Brewer (1994) suggested that the *kind of data* (e.g., unipolar vs. bipolar) should play an important role in selecting a color scheme. For unipolar data, Brewer recommended that the sequential steps in the data should be represented by sequential steps in lightness; Brewer termed this a **sequential color scheme**. The most obvious example would be different lightnesses of gray, although, in general, a sequential scheme can be achieved by holding hue and saturation constant and varying lightness (e.g., progressing from light orange to dark orange). Concurring with earlier cartographers' recommendations, Brewer advocated using light colors for low data values and dark colors for high data values, respectively.[2]

Brewer argued that lightness differences should predominate for sequential schemes, but she stressed that visual contrast between individual colors could be enhanced if saturation differences also are used. Although subtle, this increased contrast can be seen in Figure 15.2, in which a pure lightness scheme is compared with a combined lightness–saturation scheme. In Figure 15.2A, only lightness changes—from light to dark green; in Figure 15.2B, lightness again changes from light to dark green, but saturation also changes, increasing from the first to the third class and then decreasing for the latter two classes. Although, in this case, the increase and decrease in saturation do not correspond logically to the continual increase in the data, Brewer (1994, 137) indicated that this is acceptable "if high saturation colors do not overemphasize unimportant categories."

Brewer suggested that hue differences could also be used for sequential schemes but that such differences should be subordinate to lightness differences. Figure 15.3A illustrates the middle five classes of a yellow to green to a purple scheme that she advocated. Brewer did not recommend a greater range of hues, although she noted that "sequential schemes may be constructed that use the entire color circle" (as in Figure 15.3B). In our opinion, the latter implies qualitative differences and therefore should not be used for numerical data.[3]

For bipolar data, Brewer (1994, 139) recommended a **diverging color scheme**, in which two hues diverge from a common light hue or neutral gray. An example converging on a light hue is a dark green–greenish yellow–yellow–orange–dark orange scheme, and an example converging on a neutral gray is a dark blue–light blue–gray–light red–dark red scheme (Figure 15.4 illustrates the latter). The argument could be made that it is inappropriate to display the high school graduation data with a diverging scheme as the data are unipolar. We do so here to enable us to contrast the various color schemes with one another. Also, as we will see below, in one experiment, Brewer did apply a broad range of color schemes to a unipolar dataset.

Brewer's (1994) early work was based on "personal experience, cartographic convention and the writings and graphics of others" (124). In later work, Brewer and colleagues (1997) performed experiments with people to determine the effectiveness of various color schemes; in particular, they applied sequential, diverging, and spectral schemes to unipolar data. The logic of applying a diverging scheme to unipolar data is that if a quantiles classification

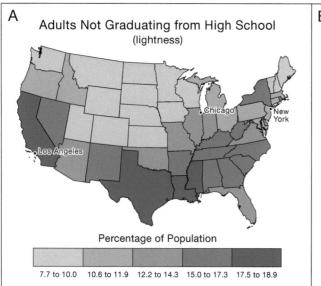
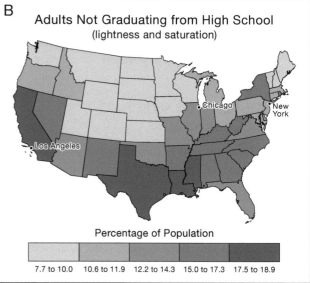

FIGURE 15.2 A comparison of two sequential color schemes: (A) only lightness varies; (B) both lightness and saturation vary, with saturation increasing from the first to the third class and then decreasing for the latter two classes. (Colors based on CMYK values shown in Brewer 1989.)

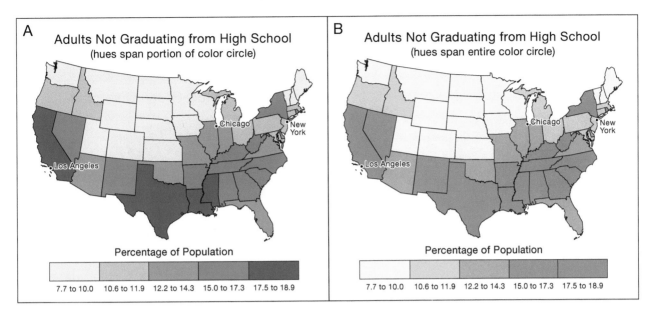

FIGURE 15.3 A comparison of two sequential color schemes in which differing hues are used: (A) the hues span a portion of the color circle; (B) the hues span the entire color circle. (CMYK values for maps provided by Cynthia Brewer.)

is used, the median will fall in the middle class and serve as a logical dividing point in the data.[4] A **spectral color scheme** (or **rainbow color scheme**) is based on the electromagnetic spectrum (red, orange, yellow, green, blue, indigo, and violet, or ROYGBIV) and is widely used in the data visualization community. Traditionally, cartographers opposed using spectral schemes because yellow is inherently a light color and thus seems out of place if shown in the middle of a spectral scheme. Brewer and

colleagues, however, argued that a satisfactory spectral scheme could be developed by progressing from *dark* red to *bright* yellow to *dark* blue. Interestingly, Brewer and colleagues found that sequential, diverging, and spectral schemes yielded very similar visualizations of data, in part because they took great care in developing these schemes. Figure 15.5 illustrates some of their schemes applied to the high school education data.

A subset of the spectral scheme is the **traffic light color scheme**, which like a traffic light, extends from red to yellow (or orange) to green. This has been a popular scheme because it is easier to order than a full spectral scheme, and it is readily interpretable (green means go or good, yellow means caution, and red means stop or danger).[5] We will see shortly, however, that like the spectral scheme, the traffic light scheme is problematic for some color deficient readers because of the confusion of green and red.

15.4.2 COLOR NAMING

In addition to working with *kind of data*, Brewer and her colleagues utilized several other factors in developing their color schemes. One important factor for diverging schemes is selecting colors that cannot be confused with one another and thus are readily "named" as distinct colors. For instance, if red and pink were assigned to opposite ends of a diverging scheme, a user might have difficulty associating a particular color on the map with the proper end of the diverging scheme in the legend. In contrast, blue and red would not be confused, as blue colors look quite different from red colors and thus are readily named as distinct colors.

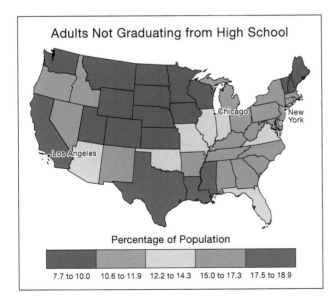

FIGURE 15.4 An example of a diverging scheme in which two sequential schemes converge on a neutral gray.

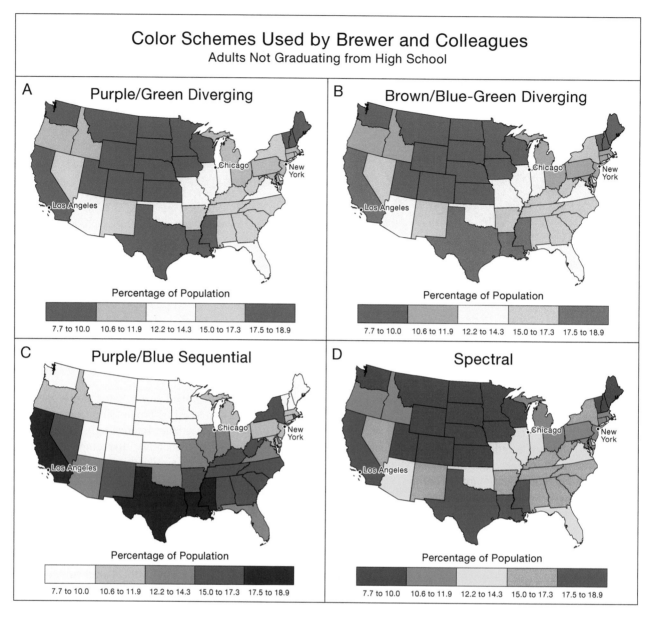

FIGURE 15.5 Some of the effective color schemes tested in Brewer et al.'s (1997) work: (A) a purple/green diverging scheme, (B) a brown/blue–green diverging scheme, (C) a purple–blue sequential scheme, and (D) a spectral scheme. (CMYK values for maps provided by Cynthia Brewer.)

The second column of Table 15.2 indicates which color pairs Brewer (1996) indicated are easily named (an "OK" appears in the column).

15.4.3 COLOR VISION IMPAIRMENT

The various forms of color vision impairment can be represented in the CIE color model by a series of *confusion lines* or lines along which colors are confused with one another (see Chapter 10 for an introduction to color vision impairment and the CIE color model). For example, Figure 15.6 illustrates confusion lines for protanopes and deuteranopes (two subgroups of dichromats, people who are missing red and green cones, respectively, and cannot distinguish red from green). Colors running along the confusion lines

will be confused by these groups, whereas colors running roughly perpendicular to the lines should be distinguishable. Using CIE diagrams like the one shown in Figure 15.6, Judy Olson and Cynthia Brewer (1997) developed sets of confusing and accommodating color schemes. Not surprisingly, Olson and Brewer found that color-impaired readers interpreted accommodating schemes more easily.

The third column of Table 15.2 indicates (with an "OK") which diverging color pairs are acceptable for the color vision–impaired. If you are showing a diverging scheme in a printed publication that will be widely disseminated, you should consider using these color pairs. Alternatively, if the maps are to be shown in an interactive environment, then you might wish to provide an option for viewers to indicate whether they are color vision–impaired. Those without impairment would

TABLE 15.2

Color Pairs Appropriate for Diverging Color Schemes

	Confusions		
Color Pair	Color Naming	Color Vision Impairment	Simultaneous Contrast
Pink–Red	*	*	OK
Pink–Orange	*	*	OK
Pink–Brown	*	*	OK
Pink–Yellow	OK	*	OK
Pink–Green	OK	*	*
Pink–Blue	OK	*	OK
Pink–Purple	*	*	OK
Pink–Gray	*	*	OK
Red–Orange	*	*	OK
Red–Brown	*	*	OK
Red–Yellow	OK	*	OK
Red–Green	OK	*	*
Red-Blue	OK	OK	OK
Red–Purple	*	OK	OK
Red–Gray	OK	*	OK
Orange–Brown	*	*	OK
Orange–Yellow	*	*	OK
Orange–Green	OK	*	OK
Orange–Blue	OK	OK	OK
Orange–Purple	OK	OK	OK
Orange–Gray	OK	*	OK
Brown–Yellow	*	*	OK
Brown–Green	*	*	OK
Brown–Blue	OK	OK	*
Brown–Purple	*	OK	OK
Brown–Gray	*	*	OK
Yellow–Green	*	*	OK
Yellow–Blue	OK	OK	*
Yellow–Purple	OK	OK	OK
Yellow–Gray	*	OK	OK
Green–Blue	*	*	OK
Green–Purple	OK	*	*
Green–Gray	*	*	OK
Blue–Purple	*	*	OK
Blue–Gray	*	OK	OK
Purple–Gray	*	*	OK

Source: After Brewer (1996, 81).

Note: An asterisk indicates that the color pair is inappropriate for the parameter listed in the associated column.

then be able to take advantage of the broader set of easily named colors shown in column 2 of the table.

To see how a color scheme you have created will appear to the color vision–impaired, we encourage you to experiment with the Color Oracle software tool (https://colororacle.org/). With this tool, you simply display a color map on your computer screen, specify the type of color vision impairment that you wish to simulate (e.g., protanopia or deuteranopia), and your screen will be updated to reflect what the colors will look like with the corresponding

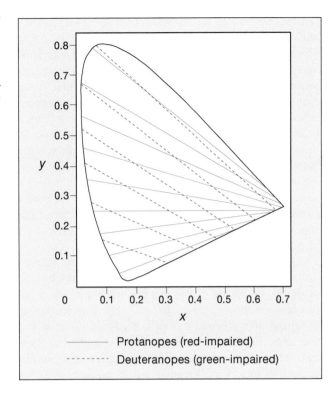

FIGURE 15.6 Confusion lines for protanopes and deuteranopes drawn on a *Yxy* 1931 CIE diagram. Colors of similar lightness will be difficult to discriminate if they are placed along the same confusion line. (After Olson and Brewer 1997, 108.)

color impairment. Before using Color Oracle, you may wish to review Section 10.3.5, which introduces the basic terminology for color vision impairment.

15.4.4 Simultaneous Contrast

As discussed in Section 10.3.4, simultaneous contrast causes the color of an area to shift toward the opponent color of a surrounding color (e.g., a gray surrounded by green will appear reddish). Fortunately, most color pairs are not affected by simultaneous contrast ("OK" predominates in the fourth column of Table 15.2). Unfortunately, though, note that many of those color pairs affected by simultaneous contrast coincide with easily named color pairs. The fact that a color pair is unacceptable from the standpoint of simultaneous contrast, however, does not necessarily prevent it from being used on a particular map. Rather, you need to examine map patterns to see whether situations arise in which simultaneous contrast is likely to be problematic (the ColorBrewer program discussed in Section 15.5.1.3 facilitates this process).

15.4.5 Map Use Tasks

Map use tasks refers to whether a map will be used to present specific or general information and whether the user will acquire this information while looking at the map or by recalling it from memory. To examine the role that color schemes play in map use tasks, Janet Mersey (1990) had map readers interpret a spatial distribution using a variety of color schemes. For specific acquisition tasks, Mersey found that unordered hues (e.g., hues that do not follow the spectral sequence and have the same lightness and saturation) worked best. This is not surprising, given that colors for ordered schemes are often difficult to discriminate and that this discrimination is complicated by simultaneous contrast. For general acquisition tasks, Mersey found that ordered schemes (such as a sequential scheme) performed better, thus supporting traditional cartographic thinking on using lightnesses of a single hue. Mersey focused just on general tasks in her discussion of results for memory, finding that lightness-based schemes outperformed hue-based schemes, with a hue–lightness scheme being the best overall.

Although Mersey's study illustrates that certain map use tasks are more effectively accomplished with particular color schemes, you should realize that it is often difficult to predict how a map will be used. It is probably more appropriate to utilize color schemes that permit a broad range of map use tasks. In Mersey's case, she found that the hue–lightness scheme worked best overall (see Figure 15.7), scoring highest on four of the ten tasks tested and scoring a close second or third on the remaining tasks. In their subsequent research, which we mentioned previously, Brewer and colleagues (1997) found that a broad range of carefully selected color schemes works well for a wide variety of map use tasks.

15.4.6 Color Associations

The fact that certain colors are often associated with particular phenomena might allow you to pick a logical color scheme (e.g., in the United States, people associate the color green with money). You should be aware, however, that these associations are cognitive and cultural in nature and thus are

likely to vary with geographic location, may change over time, and be inconsistent among map users. A good illustration of the potential for change over time is the use of blue and red for cool and warm temperatures. In an early study, David Cuff (1973) concluded that these associations were not effective, but a later study by Dale Bemis and Kenneth Bates (1989) found the associations to be effective. Bemis and Bates suggested that the different results might be a function of the more common use of this scheme since the time of Cuff's study and thus users' greater familiarity with it.

15.4.7 Aesthetics

The aesthetics of a color scheme is an important consideration, regardless of how effective that scheme might be otherwise. As an example, consider the study of Terry Slocum and Stephen Egbert (1993), in which participants compared traditional static maps (in which the map is viewed all at once) with sequenced maps (in which the map is built by the computer while the reader views it). Their study was split into two major parts: a formal one in which participants viewed one of the two types of maps (static or sequenced) and performed various map use tasks, and an informal one in which participants viewed both types of maps and were asked to comment on them. Slocum and Egbert used a yellow–orange–red scheme similar to Mersey's hue–lightness scheme in the formal portion and a lightness-based blue scheme in the informal portion. Although the major purpose of the informal portion was to have participants compare the method of presentation (static or sequenced), they often commented on the color schemes, indicating their preference for the blue scheme over the supposedly more appropriate yellow–orange–red scheme.

Interestingly, studies of color preference outside the discipline of cartography have also found blue colors appealing. For example, using people's rating of Munsell color chips, J. P. Guilford and Patricia Smith (1959) found that colors were preferred in the order blue, green, purple, red, and yellow. In a later study, I. C. McManus and colleagues (1981) found that people's preferences were more variable, but that blue and yellow were still the most and least preferred, respectively. One limitation of such studies is that they tend to focus on particular groups of people. For example, McManus and colleagues studied only "undergraduate members of the University of Cambridge" (653). What might the results have been if the study had been done in other areas of the world or with other age groups? Because color preference varies among individuals, and because there are likely to be differences among cultures, a key point to remember is that the color schemes *you* find attractive might not be attractive to others.

15.4.8 Age of the Intended Audience

The age (and presumably experience) of the intended audience is an important consideration in selecting colors. For example, Marcel Zentner (2001) found that three and

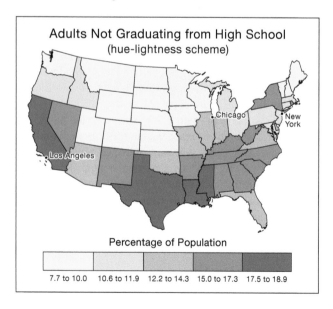

FIGURE 15.7 A hue–lightness color scheme that Mersey found scored highest on four of the ten tasks that she tested in experiments with map readers. (After Mersey 1990.)

four-year-old children preferred a red color, which is in direct contrast to the results for older individuals described above. Presumably, older readers' dislike for red is that over time we associate negative things (such as blood) with the color red. As another example of the effect of age, it is common knowledge that young children prefer saturated colors.[6]

It is important to bear in mind that our color vision capability is not consistent throughout our lifetimes. Research has shown that as much as 20 percent of those over the age of 72 may have difficulty distinguishing blue from green and yellow from violet.[7] You can see the effect of this deficiency by examining color maps using the tritanopian option within Color Oracle.

15.4.9 PRESENTATION VS. DATA EXPLORATION

From Chapter 1, recall that maps can be used in two basic ways: to *present* information to others (for *communication* purposes) and to *explore* data. In selecting color schemes, we have assumed that the former is our goal. If exploration is instead our goal, then the previously recommended schemes can still be used, but a greater variety of schemes also might be useful, even those that might be considered "inappropriate." For example, John Berton (1990, 112) argued:

> While it is usually critical that a palette not "double-back" on itself … certain kinds of banding effects which violate the traditional order of the color wheel can provide excellent markers for transitional areas in the data.

15.4.10 ECONOMIC LIMITATIONS AND CLIENT REQUIREMENTS

In an ideal world, there would be no economic limitations, and thus the cartographer would have sole responsibility for selecting colors. In the real world, of course, this usually is not the case. One obvious economic limitation is the expense of color reproduction in book or journal form: Although color might communicate information more effectively than black and white, it might not be feasible from an economic standpoint. As an example, academic journals sometimes require authors to pay for color reproduction in the printed version of the journal.

If you are employed in a cartographic production laboratory, you generally do not design a map completely on your own, but rather, you must respond to client requirements or desires. Although you might suggest an ideal, or optimal, set of colors, clients might reject these because they find other colors more pleasing or have traditionally used other colors. For example, the director of a cartographic production laboratory told us, "the school district made us make a boundary map for attendance areas in hot pink and blue to 'match their old map.'" Although mapmakers can try to dissuade clients from choosing such schemes, they must bear in mind that it's the clients who are keeping the laboratory in business.

15.5 SYSTEMS FOR SPECIFYING COLOR SCHEMES

In the previous section, we considered numerous factors involved in selecting color schemes. Our emphasis was on the general nature of color schemes, as opposed to specifying the specific colors making up each scheme. Here we consider approaches for specifying the particular colors that are used. We have split this section into approaches for classed and unclassed maps, respectively because the approaches differ for each.

15.5.1 APPROACHES FOR CLASSED MAPS

For classed maps, cartographers generally have argued that areal symbols for classes should have maximum contrast with one another (i.e., that colors for each class should appear to be equally spaced from one another and thus be perceptually uniform). To illustrate, consider how we might specify areal symbols for a printed five-class choropleth map in which achromatic grays are used. If we assume that 0 percent and 100 percent area inked are used for the lowest and highest classes, respectively, then you might argue that the intermediate classes should be 25 percent, 50 percent, and 75 percent area inked, as these values would have maximum contrast with 0 percent and 100 percent. However, as we will see in Section 15.5.1.2, it turns out that our perception of the area inked is not a simple linear function of the actual area inked, and so the correct intermediate values would be 14 percent, 35 percent, and 63 percent area inked. Let's now consider some of the approaches that have been used to create colors for intermediate classes on choropleth maps.

15.5.1.1 Color Ramping and HSV Systems

The red, blue, and green (RGB) color model (see Section 10.4.1) is commonly used to specify colors when working with digital displays. The specific RGB values utilized in this model are oftentimes selected using **color ramping**. In this approach, users select the endpoints for a desired color scheme from a color palette, and the computer automatically interpolates values between the endpoints on the basis of RGB values. As an example, imagine that a user selects endpoints of white (RGB values of 255, 255, 255) and black (RGB values of 0, 0, 0) and wishes to produce a five-class map. If we space the values evenly between 0 and 255, we find the following RGB values (rounding to the nearest whole number): 191, 191, 191; 128, 128, 128; and 64, 64, 64. The problem with this approach, as discussed in Section 10.4.1, is that a constant increase in RGB values does not correspond to a constant increase in perceived lightness, and thus the resulting interpolated grays will not *appear* equally spaced. Although interpolation algorithms might be modified to account for this discrepancy, they generally are not modified. As a result, color ramping must be used with caution.

A similar problem arises when working with the HSV color model shown in Figure 10.10. Although the terms hue, saturation, and value suggest that visual perception has been dealt with, the HSV model has not been adjusted to deal with the manner in which we perceive individual colors. Thus, colors linearly interpolated between, say, a red having a V value of 1.0 and a gray with a V value of 0.2 would not necessarily appear to be equally spaced, and so you should use the HSV color model with caution.

15.5.1.2 The Munsell Curve

One approach that does account for visual perception is the **Munsell curve** shown in Figure 15.8.[8] Note that the Munsell curve is plotted above a simple linear relation between the percent area inked and perceived blackness. For low percent area inked values on the Munsell curve, a small difference in percent area inked equates to a large difference in perceived blackness, while for large percent area inked values, a large difference in percent area inked equates to a small difference in perceived blackness. The Munsell curve was created using **partitioning**, in which a user places a set of grays of differing lightnesses between white and black such that the resulting grays appear equally spaced. It can be argued that the grays resulting from such an approach will have maximum contrast with one another and thus be appropriate for a classed choropleth map.

The Munsell curve can be utilized in a four-step process:

Step 1. Pick the smallest and largest perceived blackness that you desire. Let's assume that you select values of 12 and 100 as initial choices. (A value of 12 equates to 6 percent area inked and thus will produce a light gray that can be seen as a figure against the white background typically used on printed maps. A value of 100 will produce a solid black.)

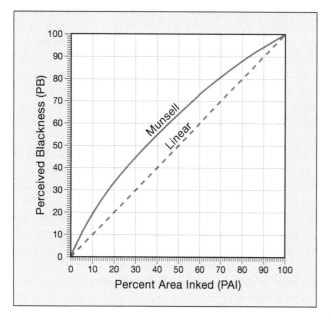

FIGURE 15.8 Munsell curve for converting desired perceived blackness to percent area inked. (After Kimerling 1985, 137.)

Step 2. Determine the contrast between each pair of perceived blackness values, assuming the values are equally spaced from one another. To accomplish this, divide the range of perceived blackness values by the number of classes minus 1. Assuming five classes, and the preceding perceived blackness values, the result is:

$$\frac{PB_{max} - PB_{min}}{NC - 1} = \frac{100 - 12}{4} = \frac{88}{4} = 22,$$

where PB_{max} and PB_{min} are the perceived maximum and minimum blackness, and NC is the number of classes.

Step 3. Interpolate the intermediate perceived blackness values. This involves simply adding the contrast value derived in step 2 to each perceived blackness value, beginning with the lowest. For perceived blackness values of 12 and 100, and a contrast value of 22, we have: 12 + 22 = 34; 34 + 22 = 56; 56 + 22 = 78; and 78 + 22 = 100.

Step 4. Determine the percent area inked corresponding to each perceived blackness value. Using Figure 15.8, this can be accomplished by drawing a horizontal line from the perceived blackness value to the Munsell curve and, from this point, drawing a vertical line to the proper percent area inked value. For perceived blackness values of 12, 34, 56, 78, and 100, you should find percent area inked values of approximately 6, 20, 41, 67, and 100.

Traditionally, cartographers promoted using the Munsell curve solely for achromatic grays of varying lightness (e.g., Kimerling 1985). We have, however, used the approach successfully for chromatic schemes that vary solely in lightness (as for the blue areal symbols shown in the choropleth maps in Chapter 5).

15.5.1.3 ColorBrewer

Mark Harrower and Cynthia Brewer (2003) developed the program ColorBrewer to assist novice map designers in selecting appropriate color schemes.[9] Although ColorBrewer was not designed explicitly to create maximum contrast colors, Brewer did use "her knowledge of the relationships between CMYK color mixture and perceptually ordered color spaces, such as Munsell" (Harrower and Brewer 2003, 30). One must also bear in mind that many of the schemes included in ColorBrewer vary in hue, saturation, and lightness, and thus the process of creating maximum contrast colors would be much more complicated than when only lightness varies. In any case, in our review of the cartographic literature and data visualization literature more generally, we find that ColorBrewer is widely recommended.[10]

The most recent version, ColorBrewer 2.0 (http://colorBrewer2.org), provides three basic schemes: sequential, diverging, and qualitative. Sequential schemes "are suited to ordered data that progress from low to high," whereas diverging schemes "put equal emphasis on mid-range critical values and extremes at both ends of the data range." Qualitative schemes are not relevant to this chapter because they are intended for qualitative (or nominal) data. A total of 18 sequential and 9 diverging schemes are possible, with sequential schemes ranging from 3 to 9 classes and diverging schemes ranging from 3 to 11 classes. Using a hypothetical spatial pattern, ColorBrewer 2.0 allows users to see whether individual class colors can be easily discriminated on a map. The program also allows users to determine whether the color scheme will be problematic for the color deficient and provides an indication of whether the color scheme will work in various situations (e.g., whether it can be printed or photocopied). ColorBrewer provides the specifications for color schemes in a variety of formats, including RGB, CMYK, and hexadecimal. Given the unipolar character of the high school education data, we have chosen to create a "Reds" sequential scheme using ColorBrewer (Figure 15.9); you should consider which of the factors in the preceding section would justify using a "Reds" color scheme.

15.5.2 Approaches for Unclassed Maps

Specifying colors for unclassed maps is more complicated than for classed maps because the large number of possible colors means that we cannot use a simple manual interpolation as we did with the Munsell curve. In this section, we consider two approaches, one that uses a mathematical equation to describe the Munsell curve and one developed by Peter Kovesi that attempts to specify colors in the CIE color model.[11]

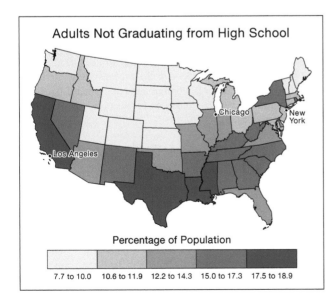

FIGURE 15.9 A "Reds" sequential scheme created using ColorBrewer (Colors from https://colorbrewer2.org/.)

15.5.2.1 Applying the Munsell Curve

To apply the Munsell curve, we digitized the curve shown in Figure 15.8 and used curvilinear regression to fit the digitized data (Davis 2002, 207–214). The following is the equation that resulted:

$$PAI = 0.287029 + 0.329157PB + 0.008753PB^2$$
$$-6.24654 \times 10^{-7}PB^4 + 4.16010$$
$$\times 10^{-11}PB^6,$$

where PB represents a desired perceived blackness and PAI represents the percent area ink necessary to achieve that level of blackness. The equation produces an excellent fit as the maximum error between predicted and digitized values for the percent area inked is 0.3 percent.

Although this equation appears complex, it can be incorporated into a spreadsheet in the following three-step procedure:

Step 1. Determine the proportion of the data range represented by each data value. To accomplish this, subtract the minimum of the data from each raw data value, and divide the result by the range of the data:

$$Z_i = \frac{X_i - X_{min}}{X_{max} - X_{min}},$$

where X_i is a raw data value, X_{min} is the minimum of the raw data, X_{max} is the maximum of the raw data, and Z_i is the proportion of the data range. The resulting Z values will range from 0 to 1.
Step 2. Convert the values calculated in step 1 to perceived blackness. This involves computing

$$PB_i = Z_i(PB_{max} - PB_{min}) + PB_{min}$$

where Z_i, PB_{min}, and PB_{max} are defined as before, and PB_i is the perceived blackness for an individual data value.
Step 3. Insert the perceived blackness values into the equation representing the Munsell curve.

The unclassed maps shown in Figure 15.11 were created using this approach. We'll consider these maps in detail in Section 15.6.

15.5.2.2 Kovesi's Approach

Various researchers have attempted to create color schemes for unclassed data (sometimes referred to as *continuous data*). Generally, this has involved use of the CIE color model because, as we indicated in Section 10.4.5, variants of the CIE model can be used to produce perceptually uniform color schemes. We will utilize an approach developed by Peter Kovesi because (1) he took great care to develop a wide variety of perceptually uniform color schemes; (2) he provides a detailed discussion of the color scheme creation process (see Kovesi 2015); and (3) he makes the resulting color schemes available for download in a wide variety of formats appropriate for various software that you might be using to create maps (https://peterkovesi.com/projects/colourmaps/index.html).[12]

The idea behind a perceptually uniform color space is that if a line is drawn through this color space and points are equally spaced along this line, then the resulting colors associated with these points should appear to be logically perceptually ordered and have uniform perceptual contrast with one another. Kovesi, however, notes that the traditional CIELAB uniform color space is limited because it was based on experiments with human subjects in which a relatively large portion of the visual field was covered, in contrast to small portions of the visual field that are often examined in scientific images, such as maps. To solve this problem, Kovesi modified the standard CIELAB process by accumulating the lightness change along a line through the color space, dividing the cumulative lightness into equal intervals, and then back calculating to determine the CIE specifications that would produce those color values. Although Kovesi did not test the resulting schemes with human subjects, he did examine the schemes using a carefully constructed test image, which revealed whether a color scheme properly represented the data and did not introduce undesirable artifacts.[13]

Kovesi termed his color schemes **color maps**, and we will use that terminology to be consistent with his paper that introduced the schemes (and because the term is commonly used in the data visualization literature). Kovesi introduced five general types of color maps, three of which are relevant to choropleth mapping: linear, diverging, and rainbow. A **linear color map** is similar to the sequential scheme that we introduced for classed maps. The term "linear" is used to emphasize that the lightness gradient is constant across the different colors composing the color map. A **diverging color map**, like a diverging scheme for classed maps, is appropriate for data that have a natural or meaningful dividing point. A **rainbow color map** is analogous to the spectral schemes that we introduced for classed maps. Like Brewer, Kovesi notes the limitations of rainbow color maps and takes care to develop rainbow color maps that produce minimum confusion for the reader.

To illustrate the use of Kovesi's color maps, Figure 15.10 portrays an unclassed map of the high school education data using a linear blue-green-yellow-white color map.[14] Figure 15.10A portrays the entire color map, with the white (light) end used for low data values and the blue (dark) end used for high data values. One problem with the resulting color map is that the white for the lowest data value (for the large state of Montana) has no contrast with the white background surrounding the map. To solve this problem, we eliminated the three lightest colors and mapped the data for all 48 states using only the remaining 253 colors—the result is shown in Figure 15.10B.[15] To assist you in evaluating the usefulness of the resulting map, in the next section,

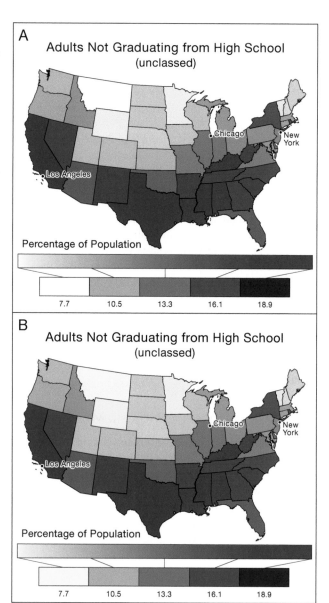

FIGURE 15.10 Maps created using Kovesi's (2015) linear blue–green–yellow–white color map: (A) the entire color map (256 colors) is used, (B) the three lightest colors are not used. (Colors are from Peter Kovesi, Good Colour Maps: How to Design Them, https://peterkovesi.com/projects/colourmaps/index.html, licensed under Creative Commons License (CC BY-Attribution 4.0 International).)

we consider several criteria that you might consider in deciding whether to use a classed or an unclassed map.

15.6 CLASSED VS. UNCLASSED MAPPING

Prior to the early 1970s, classed choropleth maps were the norm.[16] Cartographers argued that classed maps were essential because of the limited ability of the human eye to discriminate shades for areal symbols; also, practically speaking, classed maps were the only option because of the time and effort required to produce unclassed maps using traditional photomechanical procedures. In 1973, the latter constraint was eliminated when Waldo Tobler introduced a method for creating unclassed maps using a line plotter (Figure 4.11C could have been produced using such a device; Tobler 1973). Today, unclassed maps can be created using a wide variety of hardware devices.

The development of unclassed mapping led to a hotly contested debate on its merits and demerits (e.g., Dobson 1973, 1980; Muller 1980) and numerous experimental studies, which we will consider shortly. Although the results of the experimental studies can be helpful in selecting between classed and unclassed maps, two criteria should be considered first: (1) whether the cartographer wishes to maintain the numerical relations among data values and (2) whether

the map is intended for presentation or exploration. In this section, we consider these two criteria and then appraise the results of some experimental studies.

15.6.1   MAINTAINING NUMERICAL DATA RELATIONS

To illustrate the notion of maintaining numerical data relations, consider the classed and unclassed maps for foreign-owned agricultural land and high school graduation shown in Figure 15.11. The optimal method was used to create the classed maps because the optimal method does the best job of minimizing classification error and so, in theory, will produce classed maps that are most similar to the unclassed maps.

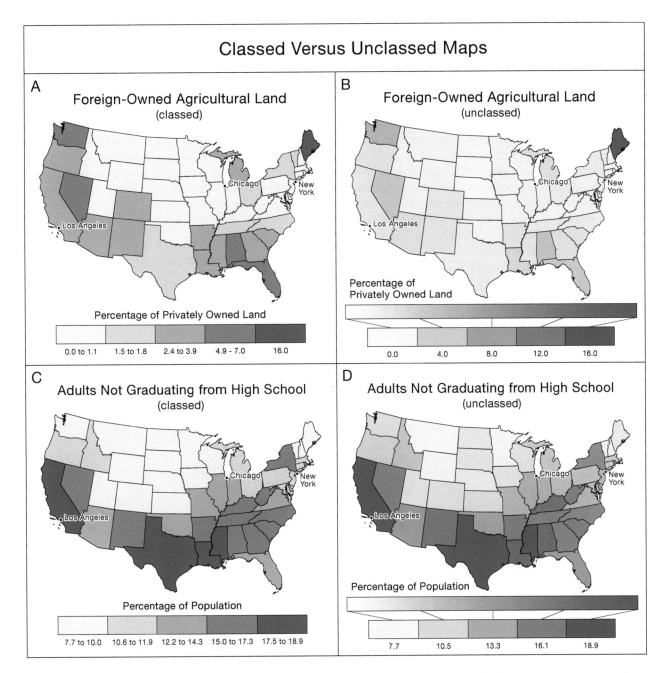

FIGURE 15.11   Optimally classed (A and C) and unclassed (B and D) maps of two attributes: foreign-owned agricultural land and high school education.

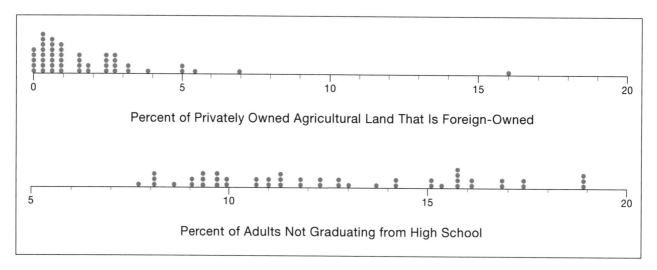

FIGURE 15.12   Dispersion graphs for the attributes shown in Figure 15.11.

Areal symbols for the classed maps were selected using the Munsell curve described in the preceding section and a conventional maximum-contrast approach in which the lightnesses for symbols are perceptually equally spaced from one another. In contrast, the lightnesses of symbols on the unclassed maps were made directly proportional to the values falling in each enumeration unit, thus maintaining the numerical relations among the data, with an adjustment for visual perception made using the same Munsell curve. We have chosen to vary only lightness on these maps (hue and saturation are held constant) because the bulk of experimental studies that we describe below varied only lightness.

Clearly, the maps within each pair look different, with the difference most distinct for the agricultural maps. The difference for the agricultural maps is a function of the severe skew in the data (examine Figure 15.12 and note the concentration of data on the left, with Maine a notable outlier at 16.0). The unclassed agricultural map is a spatial expression of this skew, as we see numerous light blue states and one darker blue state (Maine). In contrast, on the classed map, we see only five different lightnesses of blue, with Maine not appearing quite so distinct. Although unclassed maps do a better job of portraying the actual data relations, a disadvantage is that for skewed distributions, the ordinal relations in much of the data might be hidden. For example, in the case of foreign-owned agricultural land, it is difficult to determine whether states in the northcentral U.S. differ from many of those in the southern U.S. On the classed map, these differences are more obvious.

15.6.2 PRESENTATION VS. DATA EXPLORATION

When a map is intended for presentation, normally, only one view of the data can be shown, so the mapmaker must make a choice between classed and unclassed maps. If, however, the intention is to explore data, then numerous options are possible. One is to compare a variety of classification approaches visually, as we did in Figure 5.2. With the appropriate software, such a comparison might also

include an unclassed map. Another option in data exploration is to apply unclassed symbolization to only a portion of a data set. For example, for the agricultural data, Maine can be assigned a unique symbol, with the remainder of the data displayed using the unclassed method (Figure 15.13). Note how this approach makes it easier to contrast the pattern of lighter blues for northcentral states with the pattern of darker blues for southern states.[17]

15.6.3 SUMMARIZING THE RESULTS OF EXPERIMENTAL STUDIES

This section summarizes the results of some experimental studies that might assist you in determining whether classed or unclassed maps are appropriate.[18] One problem

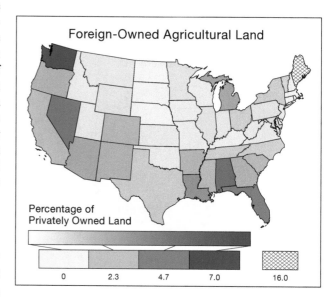

FIGURE 15.13   An unclassed map in which continuous-tone shading is applied to a portion of the data set (in this case, Maine is separated from the remaining data and represented by a cross-hatched shade).

in summarizing the studies is that some of them did not use unclassed maps but rather varied the number of classes (e.g., from 3 to 11 classes); thus, in the following discussion, it is useful to substitute "maps with few classes" for classed maps and "maps with many classes" for unclassed maps. The results of the studies can be summarized conveniently under the types of information that readers acquire and recall from maps: specific and general.

15.6.3.1 Specific Information

For the *acquisition* of specific information, classed maps generally are more effective. This result might seem surprising because unclassed maps are usually touted as being more accurate (because data are not grouped into classes). The high accuracy of unclassed maps, however, is mathematical, not perceptual. Visually matching a shade on an unclassed map with a shade in the legend is difficult because there are so many shades and because their appearance is affected by simultaneous contrast. Note, however, that for some data sets, unclassed maps can be useful for visualizing ordinal relations. For example, on the unclassed education map shown in Figure 15.11D, note that Montana and Wyoming appear to have a lower percentage of not graduating than Utah and Colorado. On the classed map, such ordering is impossible to determine because these states all fall in the same class. Although ordinal relations can be obtained from unclassed maps, the task can be difficult if enumeration units are far apart (and thus likely appear in different contexts) or if a distribution is highly skewed (as with the foreign-owned agricultural land data).

15.6.3.2 General Information

For *acquisition* of general information, studies generally have revealed no significant difference for classed and unclassed maps. Only Katherine Mak and Michael Coulson (1991) found any significant difference, and this occurred only when individual regions were analyzed; they concluded that "for more complex classed maps (six to eight classes) classification may have a distinct advantage over unclassed maps" (121). For *recall* of general information, Alan MacEachren (1982) found no significant relationship between the number of classes and subjects' ability to recall data, and Janet Mersey (1990) found that a greater number of classes decreased recall effectiveness. Based on his findings, MacEachren suggested that classification might be unnecessary for general tasks. In contrast, Mersey noted that "A large number of map symbols, even on a map with regular surface trends, may create a notion in the mind of the user that the distribution is more complex than in fact it is" (p. 125).

15.6.3.3 Discussion

One problem with interpreting the results of these studies is that usually, there was no direct comparison between classed and unclassed maps. For example, Mak and Coulson asked subjects to divide each map into five ordinal-level regions (ranging from "low" to "high"). Rather than comparing the resulting regions directly on classed and unclassed maps, they analyzed the consistency for each type of map (classed and unclassed). As a result, it is possible that the consistency was similar for both but that the locations of regions differed.

Another problem is that only one of the studies (Gilmartin and Shelton 1989) measured processing time. Measuring processing time is important because classed and unclassed maps could provide the same information, but one map could take less time to process. (The more rapidly processed map clearly would be more desirable.)

As we have indicated, one limitation of these studies is that they often did not utilize truly unclassed maps. Thus, it would be interesting to evaluate classed vs. unclassed maps using color schemes such as those developed by Kovesi that come closer to creating a truly unclassed map (his schemes permit 256 different colors). In Chapter 24, we will discuss a study by Mark Harrower (2007) that contrasted the effectiveness of classed and unclassed choropleth maps for animated mapping.

15.7 LEGEND DESIGN

General principles of legend design were presented in Section 11.4.5. Here we consider some finer points of legend design that pertain to choropleth maps. Although the design of the legend might seem a minor point, a map intended for presentation could fail to communicate if the legend is poorly designed. Our focus here is on legend design for classed choropleth maps, as they are more common than unclassed maps. One issue in legend design is whether a horizontal or vertical legend should be used (Figure 15.14). An argument for a horizontal legend is that its orientation matches the traditional number line; as such, values should increase from left to right (Figure 15.14A would be appropriate, whereas Figure 15.14B would be inappropriate).

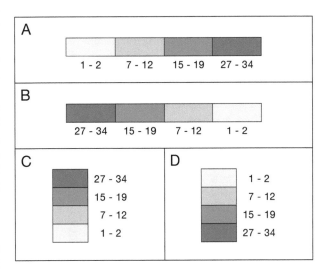

FIGURE 15.14 Some horizontal and vertical legends. (A), (C), and (D) are appropriate, whereas (B) is inappropriate because the legend limits decrease from left to right (unlike on a number line).

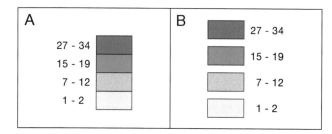

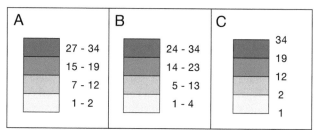

FIGURE 15.15 Inappropriate legend designs: (A) the numeric values for classes precede the legend boxes, and (B) the legend boxes are not contiguous.

FIGURE 15.16 Methods for indicating class limits: (A) the actual range of data falling in each class is shown, (B) class limits are expanded to eliminate gaps, (C) only the minimum and maximum data values and the upper limit of each class are shown.

For a vertical legend, high values can be shown either at the top or bottom (Figure 15.14C and D). The logic of showing high values at the top is that people associate "up" with "higher"—think of climbing a mountain: High elevations are at the top. One problem, however, with having high values at the top is that we normally read from left to right and from top to bottom; as such, it might seem awkward to find values increasing from left to right but decreasing from top to bottom. The decision to use a horizontal or vertical legend likely will depend on available map space; for instance, we chose a horizontal design for most choropleth maps in this book because the design seemed to fit well with the geographic shapes and column widths we were dealing with.

Numeric values should be placed either at the bottom of legend boxes (for a horizontal legend) or to the right of boxes (for a vertical legend); thus, the design in Figure 15.15A should be avoided. The logic is that we read from left to right and from top to bottom; we normally see an areal symbol on the map and want to know the associated range of values. Legend boxes should be placed directly next to one another, as opposed to being separated by spaces (avoid Figure 15.15B). The logic is that enumeration units are contiguous to one another, so legend boxes also should be contiguous. (See Section 11.4.5 for some exceptions to this rule.)

Several approaches are possible for specifying class limits, including (1) indicating the range of data actually falling in each class, which produces numeric gaps between classes (e.g., the gap between "1–2" and "7–12" in Figure 15.16A); (2) eliminating these gaps by expanding classes (Figure 15.16B); and (3) indicating the minimum and maximum data values and the upper limit

of each class (Figure 15.16C). The advantage of the first approach is that the reader will know precisely the values falling in each class. The latter two approaches avoid the problem of gaps, but they do not show the range of data actually occurring in each class on the map. An advantage of the last approach is that the reader will have fewer numbers to work with; on the other hand, there might be confusion concerning the bounds of each class (e.g., in Figure 15.16C, which class is "19" in?). As explained in Chapter 11, a hyphen can be used to separate numeric values, as we have done with the figures here. If negative values are included in the legend, then the word "to" is frequently used instead to separate values. Because some of the data sets in this book include negative values, we generally have used "to" for consistency.

It might also be desirable to integrate a graphical display with the legend. For instance, Figure 15.17 shows two approaches for integrating a dispersion graph. In Figure 15.17A, the legend boxes are sized to reflect the range of the data in each class, whereas, in Figure 15.17B, the legend boxes are equal in size. A problem with the former approach is that the size of the small legend box for the first class might make it difficult to match the shade with a shade on the map.

15.8 ILLUMINATED CHOROPLETH MAPPING

Conventional choropleth maps are typically created by using visual variables related to color (e.g., hue and lightness in a hue-lightness color scheme), as we have done throughout the bulk of this chapter. James Stewart and Patrick Kennelly

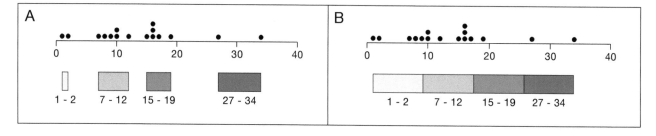

FIGURE 15.17 How a graphical display can be integrated with the legend: (A) legend boxes are scaled to reflect the range of data in each class, and (B) legend boxes are constant.

(2010) have developed an intriguing alternative termed an **illuminated choropleth map** in which they symbolize a dataset redundantly using both a color scheme and a planimetrically correct 3-D prism view (Figure 15.18).[19]

Traditionally, prism maps (such as the one shown in Figure 4.11A) are viewed at an oblique angle, which can result in considerable blockage of smaller prisms. The illuminated choropleth map largely eliminates this problem by providing a top-down view while simultaneously displaying classed and unclassed versions of the same data set. In Figure 15.18, the classed version is represented by five classes via a sequential color scheme and the unclassed version is depicted using the 3-D prisms, where individual states are raised to a height proportional to their data values. Normally, when a traditional 3-D prism map is viewed directly from above, it is not possible to see differences in the heights of prisms. The illuminated choropleth map attempts to solve this problem by using an illumination model for prisms that produces soft shadows, which allows the color of an enumeration unit to be seen even when a shadow is cast by an adjacent enumeration unit. In the resulting visualization, we cannot make precise estimates of heights, but we can generally see which enumeration units are higher than other units (although this task can be difficult when the units are not adjacent).

Based on an extensive user study, Stewart and Kennelly argued that an advantage of their approach is that is possible to determine which of two adjacent enumeration units is greater (even when those units are in the same choropleth class) and still be able to perform many of the same functions that we would expect with a classed choropleth map (e.g., match a shade on the map with one in the legend). For instance, in Figure 15.18, we can see that the state of

Mississippi has a higher percentage of not graduating from high school than the surrounding states that also fall in the highest data class. The illuminated choropleth method provides a good illustration of how modern computer-based techniques can change the look and function of traditional cartographic products.

15.9 SUMMARY

Raw-total data need to be standardized prior to choropleth mapping. Representative examples include dividing two area-based raw totals (e.g., dividing the number divorced by the number married), calculating a rate (e.g., the number of suicides per 100,000 people), and calculating a summary numerical measure (e.g., the median value of homes as opposed to mapping the total value of homes).

A key factor in selecting an appropriate color scheme is whether the data being mapped are unipolar or bipolar. For unipolar data (where there is no natural dividing point), a sequential scheme is most appropriate as the sequential steps in the data are represented by sequential steps in lightness of the scheme (e.g., a series of green shades that become darker as the data values increase). For bipolar data (where there is a natural dividing point), a diverging scheme is logically appropriate, in which two hues diverge from a common light hue or neutral gray (an example diverging from a neutral gray would be a dark orange–orange–light gray–purple–dark purple scheme).

Other factors involved in selecting an appropriate color scheme include color vision impairment, simultaneous contrast, map use tasks, color associations, and color aesthetics. Spectral schemes (such as ROYGBIV) illustrate the problem of color vision impairment, as approximately 8 percent of the population (mostly males) will confuse red and green in this scheme. Although a color scheme can be selected that will enhance certain map use tasks (e.g., major hue differences can enhance specific information), the schemes in ColorBrewer are desirable because they have been designed to work with a wide variety of tasks. Utilizing color associations (e.g., using green for a map of forest growth because green is commonly associated with vegetation) can be desirable, but remember that some color associations are culturally determined and may vary from person to person.

Classed choropleth maps are more commonly used because they enable accurate and rapid matches of shades in the legend with those depicted on the map. Unclassed maps, however, can be useful if you wish your mapped distribution to reflect the statistical character of the data (e.g., if the data have a positive skew, you may want to see how this is reflected in the mapped distribution).

If you have made sound decisions in standardizing your data and in selecting an appropriate color scheme, you also need to be certain that you have created a well-designed legend. Some factors that you need to consider include creating a balanced map (e.g., determining if there is a convenient

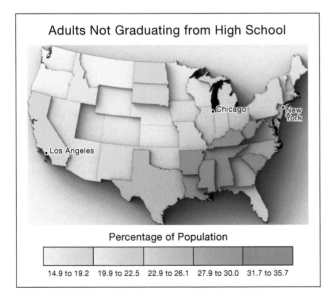

FIGURE 15.18 An example of Stewart and Kennelly's illuminated choropleth mapping approach using 1990 education data for states in the United States. (Courtesy of James Stewart and Patrick Kennelly.)

open space for the legend, as on our Florida maps), taking advantage of how we normally read things (e.g., we read from the left to right, and so it makes sense to have numbers to the right of legend boxes in a vertical legend), and making legend boxes contiguous (because enumeration units on the map are contiguous).

15.10   STUDY QUESTIONS

1. Presume that you have collected 2020 U.S. Presidential election data for each county in the state of Georgia. Indicate whether each of the following datasets would be appropriate for a choropleth map and briefly explain why in each case: (A) the number of people in each county voting for President Biden; (B) the percentage of all Presidential votes in each county voting for President Biden; and (C) the number of people in each county voting for President Biden divided by the population of the entire state.

2. Compute a COVID-19 death rate (per 100,000) for a country that had 25,000 deaths due to COVID-19 and a population of 50,000,000.

3. Indicate whether a sequential or a diverging color scheme would be most appropriate for each of the following datasets and explain why in each case: (A) the percent increase (or decrease) in voter participation in census tracts in a city in Presidential elections over a 20-year period; and (B) the percent of loss in forest cover in tropical rainforest areas.

4. Use Table 15.2 to determine which diverging color schemes would avoid both color naming and color vision impairment problems.

5. Which of the ten factors discussed in Section 15.4 would be most relevant if you were creating a county-level map of the entire U.S. showing the percent voting Republican in a Presidential election?

6. Using ColorBrewer (https://colorbrewer2.org/), choose "9" for "Number of data classes" and select "sequential" for "Nature of your data." Then select the "Greens" scheme under "Single hue." Using the resulting map, explain which of the ten color scheme factors discussed in Section 15.4 is being demonstrated by this map.

7. Go to Peter Kovesi's website (https://colorcet.com/gallery.html) and scroll down to where the "Linear Colour Maps" is displayed. Select one of these color maps that you think would provide a useful alternative to the scheme displayed in Figure 15.10 and discuss the logic of your choice.

8. Contrast the ease of acquiring specific information from the two high school education maps shown in Figure 15.11.

9. Contrast the nature of the spatial patterns (general map information) that a map reader might acquire

from the classed and unclassed foreign-owned agricultural maps shown in Figure 15.11.

10. Describe (and illustrate if you like) how you would design a legend for a six-class choropleth map of the state of Oklahoma.

NOTES

1 See Boscoe and Pickle (2003) for an overview of criteria that ideally should be considered in selecting enumeration units for choropleth mapping.

2 Brewer's suggestion is supported by a recent study by Karen Schloss and colleagues (2019, 817), who found color schemes are most effective when darker, more opaque colors represent high data values on a light background.

3 Colors that appear qualitatively different are more appropriate for *chorochromatic maps*, in which areas are colored to represent nominal level data, such as on a land use/land cover map.

4 We classed the data for these maps using the optimal method. As a result, the median of the data (11.0) does not fall in the middle of the third class but rather in the second class.

5 In a survey of the use of color in academic journals, White et al. (2017) found that traffic light schemes comprised 7.7 percent of all maps examined.

6 Although children may prefer saturated colors, it should be noted though that they do not necessarily perform map tasks significantly better with saturated colors (Buckingham and Harrower 2007).

7 See Hobbins (2019) for a discussion of a range of issues associated with designing maps for the color deficient.

8 The Munsell curve is part of the Munsell color model described in Section 10.4.4.

9 Although ColorBrewer uses only choropleth maps for illustration, the program could be utilized to select colors for any map containing areal symbols (e.g., an isarithmic map).

10 Another web-based approach for specifying colors for classed maps is the Sequential Color Scheme Generator, available at http://eyetracking.upol.cz/color/.

11 Some cartographers have recommended using an alternative curve, the Stevens curve, for unclassed maps (e.g., see Kimerling 1985, 141).

12 For an alternative approach, see Morse et al. (2019).

13 For a detailed discussion of the special image, see https://peterkovesi.com/projects/colourmaps/colourmaptestimage.html.

14 Linear_bgyw_20-98_c66_n256 is the code Kovesi uses.

15 An alternative approach for increasing the contrast would be to use a color other than white for the background of the map.

16 Interestingly, the first choropleth map (produced by Charles Dupin in 1827) was unclassed, but after the 1830s, classed maps were much more common (Robinson 1982, 199).

17 In Chapter 25, we will discuss how this can be achieved using the CommonGIS software.

18 The experimental studies include Muller (1979), MacEachren (1982), Gilmartin and Shelton (1989), Mersey (1990), and Mak and Coulson (1991).

19 We have used 1990 education data rather than the 2011 data used for earlier maps because the larger range of data produced a more interesting illuminated choropleth map.

REFERENCES

Bemis, D., and Bates, K. (1989) "Color on temperature maps." Unpublished manuscript, Department of Geography, Pennsylvania State University, University Park, PA.

Berton, J. A. J. (1990) "Strategies for scientific visualization: Analysis and comparison of current techniques." *Extracting Meaning from Complex Data: Processing, Display, Interaction. Proceedings, SPIE*, Volume *1259*, Santa Clara, CA, pp. 110–121.

Boscoe, F. P., and Pickle, L. W. (2003) "Choosing geographic units for choropleth rate maps, with an emphasis on public health applications." *Cartography and Geographic Information Science 30*, no. 3:237–248.

Brewer, C. A. (1989) "The development of process-printed Munsell charts for selecting map colors." *The American Cartographer 16*, no. 4:269–278.

Brewer, C. A. (1994) "Color use guidelines for mapping and visualization." In *Visualization in Modern Cartography*, ed. by A. M. MacEachren and D. R. F. Taylor, pp. 123–147. Oxford, England: Pergamon.

Brewer, C. A. (1996) "Guidelines for selecting colors for diverging schemes on maps." *The Cartographic Journal 33*, no. 2:79–86.

Brewer, C. A., MacEachren, A. M., Pickle, L. W. et al. (1997) "Mapping mortality: Evaluating color schemes for choropleth maps." *Annals of the Association of American Geographers 87*, no. 3:411–438.

Buckingham, B., and Harrower, M. (2007) "The role of color saturation in maps for children." *Cartographic Perspectives* no. 58: 28–47.

Cuff, D. J. (1973) "Colour on temperature maps." *The Cartographic Journal 10*, no. 1:17–21.

Davis, J. C. (2002) *Statistics and Data Analysis in Geology* (3rd ed.). New York: Wiley.

Dobson, M. W. (1973) "Choropleth maps without class intervals?: A comment." *Geographical Analysis 5*, no. 4:358–360.

Dobson, M. W. (1980) "Perception of continuously shaded maps." *Annals of the Association of American Geographers 70*, no. 1:106–107.

Gilmartin, P., and Shelton, E. (1989) "Choropleth maps on high resolution CRTs/the effect of number of classes and hue on communication." *Cartographica 26*, no. 2:40–52.

Guilford, J. P., and Smith, P. C. (1959) "A system of color-preferences." *The American Journal of Psychology 72*, no. 4:487–502.

Harrower, M. (2007) "Unclassed animated choropleth maps." *The Cartographic Journal 44*, no. 4:313–320.

Harrower, M., and Brewer, C. A. (2003) "ColorBrewer.org: An online tool for selecting colour schemes for maps." *The Cartographic Journal 40*, no. 1:27–37.

Hobbins, D. (2019) "Map design for the color vision deficient." In *Handbook of the Changing World Language Map*, ed. by S. Brunn and R. Kehrein, pp. 1–13. Cham, Switzerland: Springer.

Kimerling, A. J. (1985) "The comparison of equal-value gray scales." *The American Cartographer 12*, no. 2:132–142.

Kovesi, P. (2015) "Good colour maps: How to design them." arXiv:1509.03700[cs.GR], https://arxiv.org/abs/1509.03700, accessed September 29, 2020.

MacEachren, A. M. (1982) "The role of complexity and symbolization method in thematic map effectiveness." *Annals of the Association of American Geographers 72*, no. 4:495–513.

Mak, K., and Coulson, M. R. C. (1991) "Map-user response to computer-generated choropleth maps: Comparative experiments in classification and symbolization." *Cartography and Geographic Information Systems 18*, no. 2:109–124.

McManus, I. C., Jones, A. L., and Cottrell, J. (1981) "The aesthetics of color." *Perception 10*, no. 6:651–666.

Mersey, J. E. (1990) "Colour and thematic map design: The role of colour scheme and map complexity in choropleth map communication." *Cartographica 27*, no. 3:1–157.

Morse, P. E., Reading, A. M., and Stal, T. (2019) "Well-posed geoscientific visualization through interactive color mapping." *Frontiers in Earth Science 7*, no. 274:1–17.

Muller, J.-C. (1979) "Perception of continuously shaded maps." *Annals of the Association of American Geographers 69*, no. 2:240–249.

Muller, J.-C. (1980) "Perception of continuously shaded maps: Comment in reply." *Annals of the Association of American Geographers 70*, no. 1:107–108.

Olson, J. M., and Brewer, C. A. (1997) "An evaluation of color selections to accommodate map users with color-vision impairments." *Annals of the Association of American Geographers 87*, no. 1:103–134.

Robinson, A. H. (1982) *Early Thematic Mapping in the History of Cartography*. Chicago: University of Chicago Press.

Schloss, K. B., Gramazio, C. C., Silverman, A. T. et al. (2019) "Mapping color to meaning in colormap data visualizations." *IEEE Transactions on Visualization and Computer Graphics 25*, no. 1:810–819.

Slocum, T. A., and Egbert, S. L. (1993) "Knowledge acquisition from choropleth maps." *Cartography and Geographic Information Systems 20*, no. 2:83–95.

Stewart, J., and Kennelly, P. J. (2010) "Illuminated choropleth maps." *Annals of the Association of American Geographers 100*, no. 3:513–534.

Tobler, W. R. (1973) "Choropleth maps without class intervals?" *Geographical Analysis 5*, no. 3:262–265.

White, T. M., Slocum, T. A., and McDermott, D. (2017) "Trends and issues in the use of quantitative color schemes in refereed journals." *Annals of the American Association of Geographers 107*, no. 4:829–848.

Zentner, M. R. (2001) "Preferences for colours and colour: Emotion combinations in early childhood." *Developmental Science 4*, no. 4:389–398.

16 Dasymetric Mapping

16.1 INTRODUCTION

In the previous chapter, we stressed that a choropleth map ideally should be used only when a phenomenon is uniformly distributed within enumeration units composing the choropleth map. The dasymetric map is an alternative to the choropleth map when the phenomenon is not uniformly distributed within enumeration units. Like the choropleth map, a **dasymetric map** displays standardized data using areal symbols, but the bounds of the symbols do not necessarily match the bounds of enumeration units (e.g., a single enumeration unit might have a full range of gray tones representing differing levels of population density). Population is the attribute most typically portrayed using a dasymetric map.

Although the dasymetric map ultimately portrays standardized data (e.g., population density), you will need raw totals for each enumeration unit (e.g., the number of people residing there) so that you can redistribute these raw totals within the enumeration units. In a fashion similar to dot mapping (see Chapter 19), redistribution is accomplished through the use of **ancillary information**, such as a land use/land cover map. For instance, to create a map of population density, you would likely set the population density of a water land cover category to 0, presuming that people do not live on the water (although in some areas of the world, this would not be true!). Section 16.3 provides further discussion of the data used in dasymetric mapping.

In Section 16.4, we consider several basic approaches for assigning total counts associated with enumerated units to a set of zones defined by ancillary information, including the binary method, the three-class method, and a limiting variable method. In Section 16.5, we examine the effectiveness of these three methods using a study by Eicher and Brewer (2001). In Section 16.6, we examine a more sophisticated approach developed by Mennis and Hultgren (2006) known as intelligent dasymetric mapping (IDM), which utilizes both the cartographer's domain knowledge and the relationship between the enumeration units and the ancillary information to distribute raw totals.

In Section 16.7, we present two approaches for creating dasymetric maps of population density. For comparison, we first create a simple choropleth map of population density. We then incorporate limiting ancillary information such as land cover, zoning, and other areas in which people typically do not live, such as parks, schools, airports, etc. The resulting choropleth map and the two dasymetric maps are compared, revealing that the dasymetric maps represent more realistic images of where people live and in what densities.

In the last two sections of the chapter, we discuss some applications of dasymetric mapping. In Section 16.8, we describe the development and use of SocScape (http://150.254.124.68/socscape_usa/), a web application that was developed using Mennis and Hultgren's approach and that enables you to examine racial diversity across the entire coterminous United States at a 30-meter resolution. In Section 16.9, we introduce various approaches for mapping global population density and discuss a web application (http://luminocity3d.org/WorldPopDen/#3/52.32/31.11) that enables you to visualize global population densities associated with the Global Human Settlement Layer project.

16.2 LEARNING OBJECTIVES

- Define source zones, ancillary zones, and target zones and relate them to the notion of dasymetric mapping.
- Define ancillary information and indicate how it can be utilized in dasymetric mapping.
- Summarize the following three basic dasymetric mapping approaches: the binary method, the three-class method, and the limiting variable method.
- Summarize Mennis and Hultgren's intelligent dasymetric mapping (IDM) approach.
- Provide examples of limiting ancillary information that can be used to produce dasymetric maps of population density.
- Describe common assumptions regarding where populations are concentrated and provide exceptions to those assumptions.
- Discuss five advantages of converting vector-based enumeration units to a uniform raster-based dataset when mapping population data.
- Describe the character of the Gridded Population of the World (GPW) and explain how the Global Human Settlement Layer (GHSL) can be utilized to enhance GPW.
- Describe the nature of the population data provided by LandScan and the ancillary information used to generate that data.

16.3 SELECTING APPROPRIATE DATA AND ANCILLARY INFORMATION

Like the choropleth map, a dasymetric map uses areal symbols to represent presumed zones of uniformity, but unlike the choropleth map, the bounds of zones need not match enumeration unit boundaries (Figure 16.1). Also, like the choropleth

DOI: 10.1201/9781003150527-18

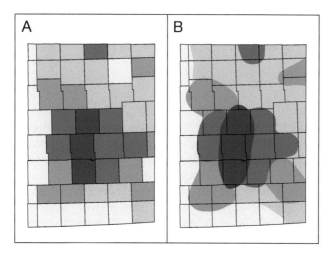

FIGURE 16.1 A hypothetical comparison of (A) choropleth and (B) dasymetric mapping.

map, the dasymetric map is used to display standardized data. Traditionally, population density has been the most common form of standardized data used, but a wide variety of standardized data are possible. For instance, we might create a dasymetric map of the density of cattle, of percent forest cover, or of the rate for a specific type of cancer. Although a standardized map is the end result, raw-total data (unstandardized data) are required for each enumeration unit so that the raw totals can be redistributed within enumeration units.

Redistribution of raw totals is accomplished utilizing **ancillary information**. Typically, this has involved using a land use/land cover map or remotely sensed imagery. For instance, assume that you have data available on the distribution of horses and that these data are available only at the county level. Let's presume further that you acquire a land use/land cover map that provides considerable detail at the county level and consists of the classes water, urbanized areas, grassland, cropland, and forest. Theoretically, you should be able to use this land use/land cover map to distribute horses within each county. The number of horses that you assign to each land use/land cover class will be a function of the relative proportion of area covered by the class and the likelihood that horses will occur in that class. In some cases, the likelihood that horses will occur within a class will be obvious (e.g., we generally don't expect horses to appear in the water), but in others, it may not be so obvious (e.g., the density of horses in a forested region would be a function of the percent forest cover). In the following section, we will learn about some different approaches that have been used to determine the likelihood of a phenomenon occurring within particular ancillary classes.

16.4   SOME BASIC APPROACHES FOR DASYMETRIC MAPPING

In considering various approaches for dasymetric mapping, it is useful to consider Figure 16.2, which depicts a set of *source zones*, *ancillary zones*, and *target zones*. Typically, one has available a set of source zones containing population values and a set of ancillary zones defined by land use/land cover. The basic problem is how to distribute population to the target zones formed by the overlay of the source and ancillary zones.

Traditionally, a fair amount of subjectivity was involved in assigning population to the various land use/land cover classes. The simplest of these subjective approaches is known as the **binary method**, in which each land use/land cover category is considered either habitable or uninhabitable. For example, if the categories are water, forest, grassland, and urban, you might consider water, forest, and grassland uninhabitable and urban habitable. This approach is obviously problematic because some people likely live in forested and grassland regions. We might instead assign forest and grassland to the habitable category, but then we cannot distinguish between the population density of forested/grassland and urban areas.

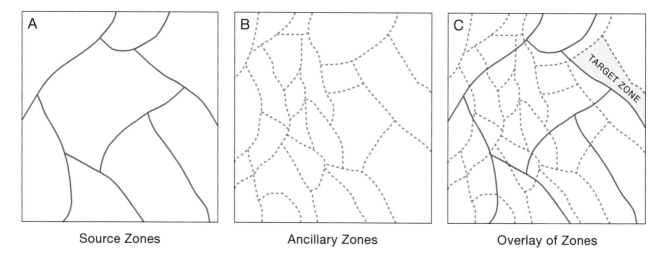

FIGURE 16.2 Nature of source, ancillary, and target zones in dasymetric mapping: (A) source zones; (B) ancillary zones based on land use/land cover; and (C) an overlay of source and ancillary zones forms target zones.

A somewhat more involved subjective approach is the **three-class method**, where we consider three habitable classes; in our case, these would be forest, grassland, and urban. For each of these classes, we would assign a certain percentage of the population to each class. For instance, based on our knowledge of the geography of a region, we might assign 15 percent of the population to forest, 30 percent to grassland, and 65 percent to urban. Although an improvement over a simple binary approach, one problem with the three-class approach is that no consideration is made for the area falling in each category. Thus, a source zone with very little area in the urban category would still have 60 percent of its population allotted to that category, which could produce an unrealistically high population density for the associated target zone.[1]

A solution to the area problem associated with the three-class method is to utilize the **limiting variable method**, which is often attributed to John K. Wright (1936).[2] To illustrate some of the computations involved in this method, we'll consider a simple hypothetical example in which only two categories are present (forested and urban), the population to be assigned is 1,000, the area of a region is 1 square kilometer, and the forested area accounts for 15/16 or 0.9375 square kilometers. The density of population in the overall region is (1,000 people)/(1 sq km) = 1,000. In Figure 16.3A, we have assigned the same density value of 1,000 to both the forested and urban areas, initially presuming that we do not differentiate these areas.

Obviously, the population density should be much higher in the urban area. Let's assume that a prior study suggests a forested threshold of 15 people per square kilometer. With this information, we can calculate the density for the urban category using the following limiting variable formula (Wright 1936):

$$D_n = \frac{D - D_m a_m}{1 - a_m},$$

where a region has been divided into two areas n and m, D is the overall density of the region, D_m is the estimated density of subregion m, a_m is the fractional area of region m (relative to the entire region), and D_n is the density of region n. For the example illustrated in Figure 16.3, the overall density, D, is 1,000, the forested and urban regions would be equivalent to regions m and n, $D_m = 15$, and $a_m = 0.9375$. Plugging these numbers into the formula for D_n, we have:

$$D_n = \frac{1,000 - (15)(0.9375)}{1 - 0.9375} = 15,775.$$

The resulting value is shown for the urban category in Figure 16.3B. In Section 16.5, we will consider a study by Cory Eicher and Cynthia Brewer that contrasts this limiting variable method with the binary and three-class methods.

An alternative approach for dasymetric mapping is to create a statistical model that specifies the relationship between population data and land use/land cover information using a regression approach (see Section 3.6.3 for an introduction to correlation and regression). One regression equation that has been used is

$$P_s = \alpha + \left(\sum_{c=1}^{C} \beta_c A_{sc} \right) + \varepsilon_s,$$

where P_s is the population of a source zone s; α is an intercept term; β_c is the coefficient for land cover c; A_{sc} is the area of land cover c within zone s; C is the number of populated land cover types occurring within the study region; and ε_s is a random error term. Input to the model consists of population values for each source zone (the P_s values) and the area of each land cover type within each source zone (the A_{sc} values). When the equation is solved for α and the β_c, the resulting β_c may be loosely interpreted as the population density for each land cover class (Langford 2006).

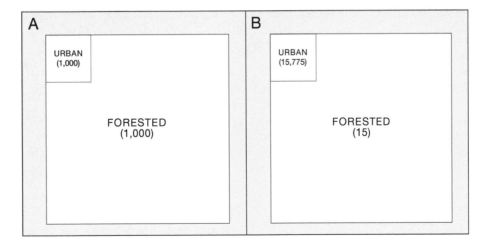

FIGURE 16.3 Computing population density figures for regions using the limiting variable method: (A) if the forested and urban regions are not differentiated, the same density (1,000 people per square kilometer) is assigned to both regions; (B) if a maximum value of 15 people per square kilometer is presumed for the forested region, then a density value of 15,775 is computed for the urban region. See text for associated formula.

The above regression may be applied to the entire study area of interest (this is termed a *global regression*) or separate regression equations may be developed for subregions (this is termed a *regional regression*) (Langford 2006). Rather than attempting to define the regional regression themselves, researchers can use a more sophisticated quantile regression (Cromley et al. 2012). Also, a more sophisticated autoregressive approach may also be used to deal with spatial autocorrelation, which is characteristic of geographic data that violates the independence assumption of standard regression (Qiu et al. 2010).

Although numerous researchers have used land use/land cover information to create dasymetric maps, the results have often been less than satisfactory because land use/land cover classes are not necessarily a direct expression of population. Thus, a number of alternative approaches have been explored, including road networks (Reibel and Bufalino 2005), cadastral and parcel data (Maantay et al. 2007; Tapp 2010), building volumes based on LIDAR (Qiu et al. 2010), the relationship with a physical process (e.g., the desire to live near the coast—Kar and Hodgson 2012), and address information (Tapp 2010; Zandbergen 2011).

The various dasymetric methods that we have discussed in this section are also often contrasted with an **areal weighting** approach for interpolation, in which each source zone contributes to the target zone in direct proportion to the percentage of the source zone's area that is found in the target zone (and thus no ancillary information is utilized). If we denote a source zone as s, and a generic map zone as z (which may overlap more than one source zone), then the target zone $t = z$, and the general formula for areal weighting is as follows (Mennis and Hultgren 2006, 180):

$$\hat{y}_t = \sum_{s=1}^{n} \frac{y_s A_{s \cap z}}{A_s},$$

where:

$\hat{y}_t$ = the estimated count for the target zone

y_s = the count of a source zone, which overlaps the target zone

$A_{s \cap z}$ = the area of intersection between the source and target zones

A_s = the area of the source zone

n = the number of source zones with which z overlaps

If each target zone intersects only one source zone, then this formula can be simplified to (Mennis and Hultgren 2006, 180):

$$\hat{y}_t = \frac{y_s A_t}{A_s},$$

where A_t is the area of the target zone. As a simple example of the latter formula, imagine that we wish to compute the population counts (as opposed to densities) for the areas shown in Figure 16.3A. Our source zone, A_s, is the entire one-kilometer region; the count of this source zone, y_s, is 1,000; the area of intersection for the source and forested category (target zone) is 0.9375 square kilometers. Thus, we

can compute $\hat{y}_t = (1{,}000 \times 0.9375)/1 = 938$ people for the forested region. Similarly, for the urban region we compute $\hat{y}_t = (1000 \times 0.0625)/1 = 62$ people.

16.5 EICHER AND BREWER'S STUDY

Cory Eicher and Cynthia Brewer (2001) were among the first to compare several methods for dasymetric mapping in an automated context. They created dasymetric maps of several population-related attributes (e.g., population density and density of homes of a specified value) for a 159-county region in four states—Pennsylvania, West Virginia, Maryland, and Virginia—and the District of Columbia. For ancillary information, they utilized a land use/land cover data set from the United States Geological Survey that produced 330 zones for the four-state area. When these 330 zones were overlaid with the 159-county region, the result was 580 target zones that fell into either urban, agricultural, woodland, forested, or water categories (the boundaries of the zones are shown in Figure 16.4). For the discussion to follow, we assume that population density is the attribute to be mapped using the dasymetric method.

To allot population from the counties to dasymetric zones, Eicher and Brewer compared three methods: binary, three-class, and limiting variable. Per the definition of the

FIGURE 16.4 The 580 target zones resulting from overlaying the county boundaries and land use/land cover zones utilized by Eicher and Brewer in their study of dasymetric mapping. (After Eicher and Brewer 2001. First published in *Cartography and Geographic Information Science* 28, no. 2:127. Reprinted with permission from the Cartography and Geographic Information Society. Courtesy of Cory Eicher and Cynthia Brewer.)

binary method (see Section 16.4), the land use/land cover categories were split into two groups: habitable and uninhabitable. The habitable group included the urban, agricultural, and woodland categories, and the uninhabitable group consisted of the water and forested categories.

For the three-class method, Eicher and Brewer combined the agricultural and woodland categories into one category they termed "agricultural/woodland," and they presumed that population would be assigned to only three categories: urban, agricultural/woodland, and forested (the water category was still considered uninhabitable). Based in part on prior research by Holloway et al. (1996), Eicher and Brewer assigned percentages of population to the three categories as follows: 70 percent for urban, 20 percent for agricultural/woodland, and 10 percent for forested.

In applying the limiting variable method, Eicher and Brewer assigned population so that the density of the three habitable categories just mentioned was identical. Thus, if a county consisted of 1,000 people and was 10 square kilometers in area, and if only habitable categories were found in the county, a density of 1,000/10 or 100 was assigned for the entire county. Next, thresholds were set for the maximum density allowed in each habitable category: these were 15 and 50 people per square kilometer for the forested and agricultural/woodland categories, respectively. If the overall density figure exceeded these thresholds, then the densities would be set to the maximum threshold, and the remaining population would be assigned to other categories on the map. In our example, because the value 100 exceeds both thresholds, the excess people all would have to be assigned to the urban category.

Eicher and Brewer evaluated the effectiveness of the three approaches for allotting county populations to the 580 target zones by comparing the population values computed for these zones with those for the U.S. Bureau of the Census block groups composing the zones. This constituted a reasonably precise check because the 580 target zones were composed of more than 13,000 block groups. Using this approach, Eicher and Brewer found that the limiting variable method produced significantly better results (from a statistical standpoint) than the other two methods. The results were also visualized using maps such as those shown in Figure 16.5. Figure 16.5A is a dasymetric map of population density resulting from the binary method. Obviously,

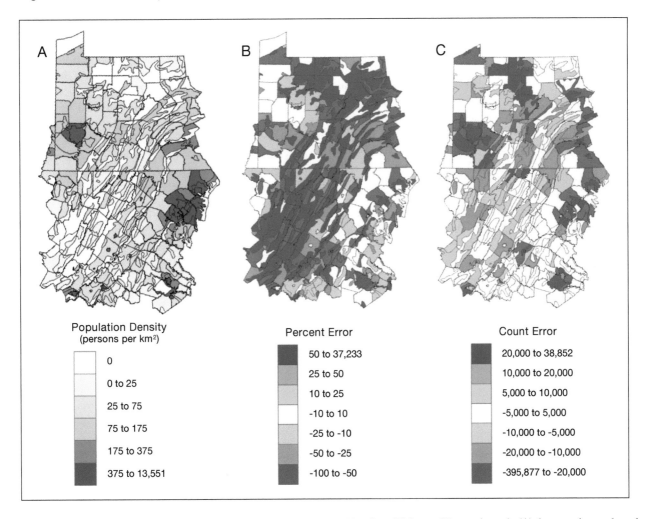

FIGURE 16.5 Example of dasymetric and associated error maps resulting from Eicher and Brewer's work: (A) dasymetric map based on the binary method; (B) percent error map; (C) count error map. (After Eicher and Brewer 2001. First published in *Cartography and Geographic Information Science* 28, no. 2:134. Reprinted with permission from the Cartography and Geographic Information Society. Courtesy of Cory Eicher and Cynthia Brewer.)

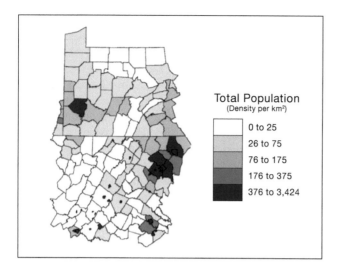

FIGURE 16.6 A choropleth map of population density for the 159-county region utilized by Eicher and Brewer. (After Eicher and Brewer 2001. First published in *Cartography and Geographic Information Science* 28, no. 2:128. Reprinted with permission from the Cartography and Geographic Information Society. Courtesy of Cory Eicher and Cynthia Brewer.)

such a map portrays considerably more detail than a simple choropleth map of the data (Figure 16.6). Parts B and C of Figure 16.5 are error maps associated with the population density map. These error maps were computed by comparing the estimated values for each zone with the actual values for these zones based on the block groups composing each zone. The percent error map shows error relative to the total population of a zone, and the count error map simply shows the difference between the estimated and true values. Note the linear patterns running northeast to southwest (in this ridge-and-valley physiographic region) on the percent error map. Eicher and Brewer indicated that such patterns are a function of the binary method assigning zero population to forested areas (which is unrealistic). The limiting variable method did not exhibit such patterns.

16.6 MENNIS AND HULTGREN'S INTELLIGENT DASYMETRIC MAPPING (IDM)

Responding to the subjectivity of traditional dasymetric mapping approaches, Jeremy Mennis and Torrin Hultgren (2006) developed a new approach, which they termed **intelligent dasymetric mapping (IDM)**.[3] This approach utilizes both the cartographer's domain knowledge and the relationship between source zones and ancillary information. If the cartographer has considerable knowledge about the phenomenon being mapped, then it may be possible to specify appropriate density values for particular ancillary classes. If, however, the cartographer does not have that knowledge, the IDM approach will determine appropriate density values as a function of the relationship between the source zones and the ancillary classes.

As with other dasymetric mapping approaches, IDM deals with how the raw totals for source zones should be

redistributed to the target zones (Figure 16.2). If we consider a source zone *s* and an ancillary zone *z* associated with ancillary class *c*, then our basic problem is to determine the estimated count for a target zone *t* defined by the area of overlap of *s* and *z*. The basic formula for calculating the estimated count for the target zone *t* is as follows:

$$\hat{y}_t = y_s \left[\frac{A_t \hat{D}_c}{\sum_{t \in s} \left(A_t \hat{D}_c \right)} \right],$$

where:

$\hat{y}_t$ = the estimated count for the target zone *t*

y_s = the count of a source zone, which overlaps the target zone

A_t = the area of the given target zone

$\hat{D}_c$ = the estimated density of ancillary class *c*

Note that the numerator of the formula within parentheses deals with the target zone for which an estimate is being calculated, while the denominator considers all target zones that overlap the source zone.

The cartographer may use their domain knowledge to specify the value of $\hat{D}_c$ or use one of the following three methods to compute the value: containment, centroid, or percent cover. In the containment method, IDM selects all source zones that are wholly contained within the individual ancillary class associated with the target zone. In the centroid method, IDM selects those source zones that have their centroids contained within the individual ancillary class. In the percent cover method, IDM allows the cartographer to specify a threshold percentage and then selects those source zones whose percentage of coverage by the ancillary class equals or exceeds that threshold. Once the source zones have been selected, the estimated density for the ancillary class may be calculated as follows:

$$\hat{D}_c = \sum_{s=1}^{m} y_s \bigg/ \sum_{s=1}^{m} A_s,$$

where:

$\hat{D}_c$ = the estimated density of ancillary class *c*

y_s = the count of a source zone

A_s = the area of a source zone

The resulting density is thus a ratio of the sum of the raw totals for all sampled source zones divided by the sum of the areas of all those source zones.

The above formulas are intended to illustrate the basic concepts involved in IDM. Mennis and Hultgren note that in the first of the equations, the term A_t applies only to those target zones associated with non-zero values (i.e., the cartographer has not specified a density of 0). Similarly, the term A_s in the second equation applies only to source zones with densities other than 0. Mennis and Hultgren also describe a more complex approach in which an additional set of "region zones" are specified so that a separate density estimate can be made for each ancillary class in each region. For example, you might be using census tracts for source zones and wish to also utilize county-level information.

In a fashion similar to Eicher and Brewer, Mennis and Hultgren evaluated the effectiveness of their approach by working with a region of 373 census tracts along the Front Range of Colorado. The 373 census tracts served as source zones, and a land use/land cover scheme available from the United States Geological Survey was used to create the following ancillary categories: high-density residential, low-density residential, non-residential developed, vegetated, and water. The attributes examined included total population, the Hispanic population, number of children (people under the age of 21), and number of households. A total of 19 different dasymetric maps were created for each of the four attributes. The 19 maps contrasted a wide variety of options available in IDM; for instance, maps were created using the containment, centroid, and percent cover methods (with various threshold values for the latter), and some included preset values for certain classes based on domain knowledge, while others did not include preset values. The maps also included the binary approach similar to that utilized by Eicher and Brewer and a traditional areal weighting approach.

For all 19 maps, Mennis and Hultgren analyzed the results by comparing the estimated values for target zones with actual values at the U.S. Bureau of the Census block level. In analyzing the results, Mennis and Hultgren noted the importance of domain knowledge in producing a satisfactory dasymetric map: "Those methods using preset density estimates generally performed better than conventional and IDM methods that did not use presets, particularly for the total population variable" (Mennis and Hultgren 2006, 189). They stressed, however, that when domain knowledge is not available, the sampling approaches (containment, centroid, and percent cover) can perform as good or better than the classic binary and areal weighting methods. The best results were obtained when presets and sampling were combined together.

16.7 TWO APPROACHES FOR PRODUCING DASYMETRIC MAPS OF POPULATION DENSITY

As we described previously, a dasymetric map is a modified choropleth map in which ancillary data are incorporated in order to more accurately represent relative densities—typically population densities. Because dasymetric maps are based on choropleth maps, it makes sense to produce an initial choropleth map to compare with the dasymetric maps that we wish to create. For illustrative purposes, we chose to map population density for the U.S. Bureau of the Census block groups in the city of Sacramento, California, as shown in Figure 16.7.[4] For the sake of clarity, the only other geographic features shown on this map are the rivers and other large bodies of water—features that help to define

the region. Note that population densities range from 109 to 30,589 persons per square mile, with darker red representing higher population densities. The choropleth technique has produced a useful representation of population densities, but it also paints an unrealistic picture, as it gives the impression of internal homogeneity within block group polygons when in reality, population densities vary within each block group. Although it is not apparent at this point, this choropleth map gives the impression that people live in areas that people do not normally live, such as parks, airports, golf courses, schools, colleges, cemeteries, and arenas. We will describe two approaches for modifying the choropleth map using ancillary data such as the parks, airports, etc., plus two additional data sources: land cover and zoning. The result will be two dasymetric maps that represent more realistic versions of population density.

16.7.1 APPROACH ONE: USING LAND COVER AND LIMITING ANCILLARY DATA SETS

The National Land Cover Database (NLCD; https://www.mrlc.gov/national-land-cover-database-nlcd-2016) is a seamless, nationwide raster data set that represents 20 land cover categories and subcategories that are derived from classified Landsat satellite images. The spatial resolution is 30 meters (each cell measures 30 meters by 30 meters). The Anderson Classification System that has been applied to the NLCD includes land cover categories such as open water, barren land, three subcategories of forest, and four subcategories of developed land: Developed, High Intensity; Developed, Medium Intensity; Developed, Low Intensity; and Developed, Open Space. The four subcategories of the developed land will be used to refine the choropleth map by removing parts of block group polygons that exist outside of developed areas. It is important to realize that some people do live outside of developed areas, but for simplicity, we assume that they do not. The resulting polygon map layer will then be further refined by removing all developed areas that do not act as traditional residences such as parks, airports, golf courses, schools, colleges, cemeteries, and arenas. Population densities will then be calculated using raw population counts for the block group polygons (and partial block group polygons) using a weighted approach that accounts for the relative areas of the polygons. This is an example of the binary method of dasymetric mapping, in which areas are determined to be either inhabited or uninhabited. It also incorporates areal weighting, in which populations are allocated to block groups and partial block groups (target zones) in direct proportion to their areas.

A spatial model was built in ModelBuilder, which is part of ESRI ArcGIS Pro, to perform the necessary geoprocessing. The model takes the form of a flowchart diagram: it takes data sets as inputs to geoprocessing tools, which in turn, produce output data sets in a daisy chain fashion. Instead of showing the actual model, a simplified version is shown in Figure 16.8. We will use the model diagram as a

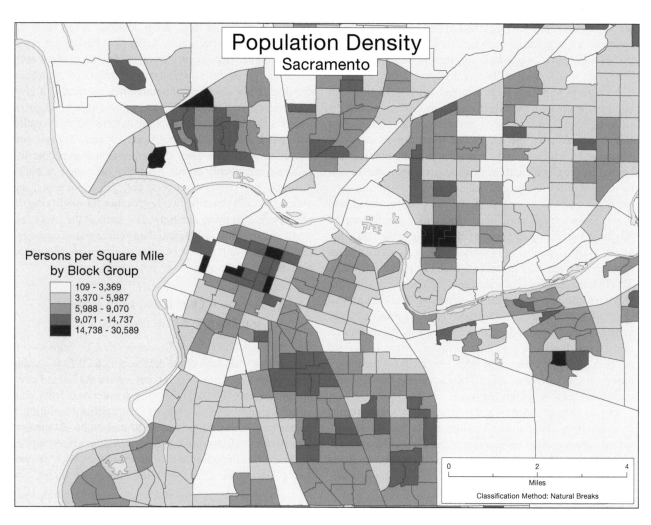

FIGURE 16.7   A choropleth map of population density, symbolized at the U.S. Bureau of the Census block group level. The map gives the incorrect impression that people are evenly distributed within the block group polygons.

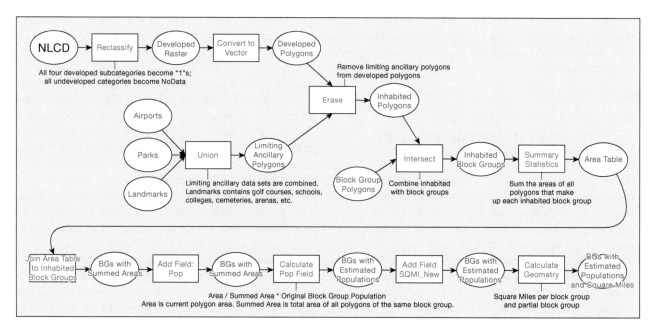

FIGURE 16.8   A simplified diagram of the model that was used in ModelBuilder to create the first dasymetric map. It includes reclassification, data conversion, map overlay, and other geoprocessing operations.

guide in describing the creation of the first dasymetric map. The model begins in the upper-left of Figure 16.8 with the NLCD grid, which has been clipped to an area that corresponds with the extent of the block group polygons that will provide the population data. In Figure 16.9A, we see that the clipped NLCD grid displays the four subcategories of developed land as various tints of red, which make up most of the area, but not all. The light gray area to the west is Yolo County, which has been excluded from this analysis. All other non-red grid cells represent undeveloped land such as herbaceous, cultivated crops, and open water, which, for simplicity, we assume is uninhabited.

Returning to the model diagram in Figure 16.8, we see that in the first step, the NLCD grid is reclassified in order to create a new raster that contains all four developed land cover subcategories but none of the other (undeveloped) categories. More specifically, the four developed subcategories are reclassified as "1"s, and the other categories are reclassified to NoData. The result is the Developed Raster (Figure 16.9B), which defines the developed areas that we assume people live in while ignoring the undeveloped categories. Notice the blank areas to the northwest and southeast and the blank area that meanders from east to west. These areas appear in the NLCD grid (Figure 16.9A) as yellows, tans, and browns—land cover categories considered to be undeveloped and uninhabited. Also, notice the very thin red linear features that appear in the northwest and southeast. These are collections of grid cells that represent developed areas along major roads—they are considered to be inhabited.

Since the block group polygons and all of the limiting ancillary data sets are in the vector data model (polygons), it is necessary to convert the Developed Raster into vector polygons that can be combined with the other vector data sets using the Union, Intersect, and Erase geoprocessing tools, which accept only vector inputs. The result is the Developed Polygons vector data set shown at the end of the first row of the model diagram and seen in Figure 16.9C. It represents the same developed areas but as vector polygons.

Looking at the model diagram shown in Figure 16.8, we see that the Erase tool is now used to remove areas from the Developed Polygons data set that correspond with the various limiting ancillary data sets (Airports, Parks, and Landmarks). Before the Erase tool can be run, the Union tool is used to combine the Airports, Parks, and Landmarks, creating the Limiting Ancillary Polygons data set shown in Figure 16.10A. Notice the large blue area to the southwest—this is the municipal airport for Sacramento. The meandering area that runs from east to west is a wooded area that borders the American River. The carrot-shaped area in the northeast is a golf course. All of these places are examples of areas in which people do not live. At this point, both inputs to the Erase tool are ready to be processed.

The Erase tool (see Figure 16.8) creates the Inhabited Polygons data set, which appears as the pink region in

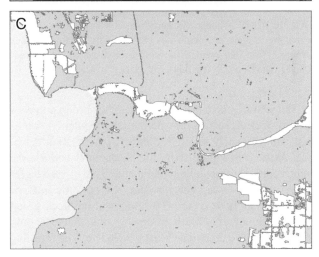

FIGURE 16.9 (A) The National Land Cover Database (NLCD) grid for the Sacramento area, most of which is classified as developed. (B) The Developed Raster which isolates four subcategories of developed land and eliminates all undeveloped land. (C) The Developed Polygons, which are vector polygons derived from the Developed Raster.

Figure 16.10B. We are presuming that people live within this pink region but not within the white areas that have been defined by the developed land and the limiting ancillary data sets.

FIGURE 16.10 (A) The Limiting Ancillary Polygons—areas in which we assume people do not live. (B) The Inhabited Block Groups, which represent all developed land except those areas found in the Limiting Ancillary Polygons data set.

FIGURE 16.11 (A) The block group polygons that contain the raw population counts that will be used to estimate population density. (B) The Inhabited Block Groups, which represent block group polygons (and partial block group polygons) that exist on developed land, but not in the limiting ancillary features.

The next step is to incorporate the block group polygons (Figure 16.11A), which provide the raw population counts that will ultimately be used to estimate population densities. As seen in Figure 16.8, this is accomplished via the Intersect tool by combining the Inhabited Polygons data set with the Block Group Polygons data set, but only in areas in which they both occur. The result will be block group polygons (and partial block group polygons) that exist only in areas we have determined that people live. The resulting Inhabited Block Groups data set (Figure 16.11B) provides all of the raw data we will need to estimate population densities, but several more processing steps will be required.

First, the Summary Statistics tool is used to calculate the sum of the areas of all polygons that make up each inhabited block group. At this point in the model, some block group polygons have maintained their original areas, while others have lost some of their areas as a result of being combined with the developed land and the limiting ancillary data sets. This can be seen if you compare Figure 16.11A with Figure 16.11B—in Figure 16.11B, there appear to be

holes in many of the block groups that represent uninhabited areas. Calculation of population density for those block group polygons that have retained their original areas will be straightforward, as we can simply divide the raw population count by the block group area.

Calculation of population density will be more involved for the partial block group polygons. First, we need to capture the sums of all the inhabited polygons that exist in each block group. In Figure 16.12A, there are two rectangular block group polygons bounded with black lines: Block Group 1 and Block Group 2, each of which has raw population counts of 1,000 and 1,500, respectively. Each block group polygon has been separated into inhabited and uninhabited areas. Figure 16.12B shows the same areas after the uninhabited regions have been removed. Adding Area A to Area B gives us the total area of the inhabited polygons in Block Group 1: 5 km². The single inhabited polygon in Block Group 2 has an area of 6 km². These values will be integral to the estimation of population densities

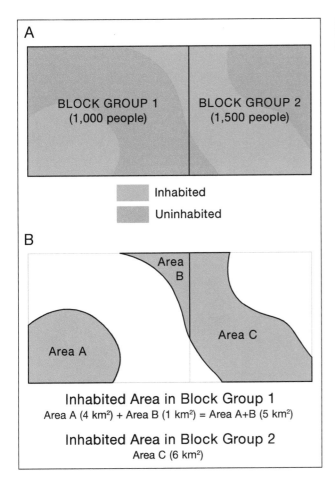

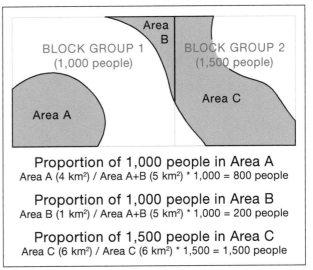

FIGURE 16.13 Calculation of population estimates based on areal weighting and raw population counts.

FIGURE 16.12 (A) Block group polygons with raw population counts, separated into inhabited and uninhabited regions. (B) Inhabited polygons for each block group have their areas summed in preparation for areal weighting and the estimation of population density.

in a subsequent step. The actual sums are stored in the Area Table at the far-right of the model diagram in Figure 16.8.

At the bottom-left of the model diagram, the Join tool is used to join the newly created Area Table to the Inhabited Block Groups table. We call the result the BGs with Summed Areas data set (BG = block group). Now we have, in a single location, two area fields that will allow us to assign weights to each polygon in the Inhabited Block Groups data set (Figure 16.11B), which, in turn, will be used to assign a proportion of each block group's original raw population count to each inhabited polygon. The first area field holds the current area of each polygon in the Inhabited Block Groups data set; the second holds the summed areas for all inhabited polygons in each block group (see Figure 16.12B).

The rest of the model involves adding fields to the Inhabited Block Groups attribute table and then calculating values that will populate those fields. The Pop field is populated with a ratio of the two area fields just described (current area/summed area) times the POP2015 field, which contains raw population counts for each block group. This

procedure is illustrated in Figure 16.13, where Area A is assigned an estimated population of 800, Area B is assigned 200, and Area C is assigned 1,500. The SQMI_New field is populated with the polygon areas in square miles (instead of square meters, which is what the data are stored in). The result is the BGs with Estimated Populations and Square Miles data set, which is the final result of the model.

The final data set, BGs with Estimated Populations and Square Miles, is then symbolized just as a choropleth map would be symbolized using the estimated populations and standardized (normalized) by the new SQMI_New field. A sequential color scheme is applied that emphasizes higher population density areas with darker colors.

The final dasymetric map is shown in Figure 16.14, complete with various map elements. Compare this map with the choropleth map in Figure 16.7. You should see that the dasymetric map is more geographically specific than the choropleth map because the thematic symbols that represent population density only appear in developed areas and not within the limiting ancillary areas such as parks, airports, etc. The population density values in the dasymetric map are also slightly higher in many cases, which makes sense because the density values from the choropleth map are distributed in smaller polygons in many cases—in block group polygons that had areas removed by the geoprocessing that occurred in the model. The minimum density value in the choropleth map is 109 persons per square mile, while the minimum value in the dasymetric map is 153, indicating that the lowest density block group in the choropleth map lost some area in geoprocessing, resulting in a higher density value. The highest density values, however, are the same for both the choropleth and the dasymetric map: 30,589 persons per square mile. This indicates that the highest density block group polygon in the choropleth map was unaffected by the geoprocessing and did not lose any of its original area.

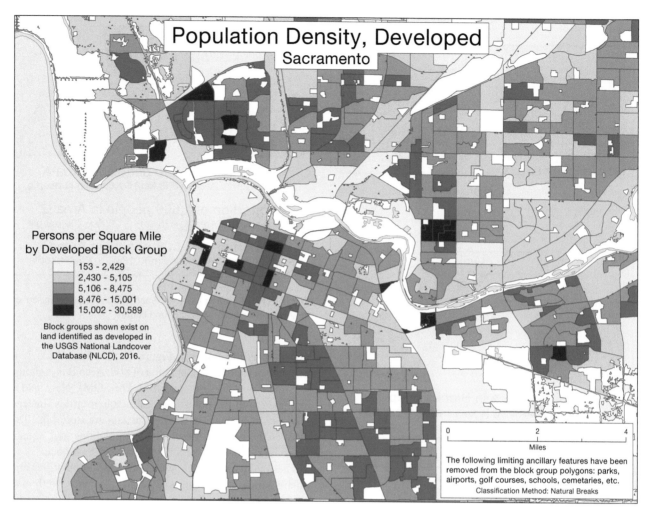

FIGURE 16.14 The final dasymetric map based on raw block group population counts, developed areas, and the limiting ancillary features. In comparison with the choropleth map in Figure 16.7, this is a more realistic representation of where people actually live and in what densities.

16.7.2 APPROACH TWO: USE ZONING POLYGONS AND LIMITING ANCILLARY DATA SETS

This second approach is almost identical to the first, with the exception that zoning polygons will be used to determine which areas are inhabited and which are uninhabited.[5] More specifically, instead of using the NLCD land cover raster to identify developed areas, we will isolate only those zoning polygons that have a zoning code that begins with R for residential (there are over 20 varieties of residential zones) and those that have a zoning code that begins with AR for agricultural residential (there are five varieties). The use of zoning polygons to differentiate between inhabited and uninhabited areas will limit populations to even more specific areas than the NLCD-derived developed lands did. This, together with the incorporation of the same limiting ancillary data sets used in the previous approach will produce a highly geographically specific representation of population density. Of course, as with the previous dasymetric approach, some people surely live outside of residential zoning areas, but for this exercise, we assume that they do not.

A spatial model was built that is almost identical to the model that was described previously, with the exception of the first few processes. In the upper-left of Figure 16.15, the Zoning Polygons data set is the initial input to the model. The zoning polygons are shown in Figure 16.16A. Many of these polygons are very small, especially in the central core of the city, making them difficult to distinguish. A larger scale version is shown in Figure 16.16B. In the previous approach, the NLCD grid used as the initial input was reclassified to isolate the developed areas using the Reclassify tool. Because the Zoning Polygons data set is vector, a different tool is used to isolate the residential areas—a tool that selects all of the residential and agricultural residential zoning polygons and then exports them to a new data set. This model actually has one step fewer than the previous model because we begin with a vector data set; there is no need to perform a raster to vector conversion.

As just described, the first process in the model uses the Select and Export tool (as shown in Figure 16.15) to create the Residential Polygons data set, which represents the inhabited areas, zoned as residential. The Erase tool is then

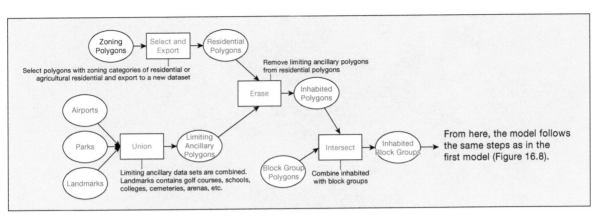

FIGURE 16.15 A partial diagram of the model that was used to produce the second dasymetric map. Zoning polygons were used as the initial input instead of the NLCD grid that was used as the initial input to the first model.

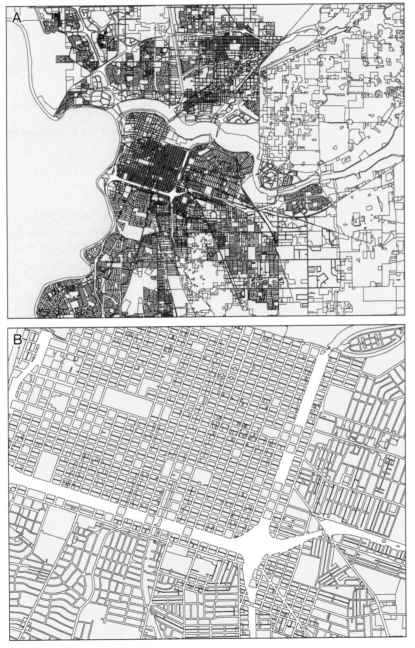

FIGURE 16.16 (A) The zoning polygons that were obtained from the City and County of Sacramento. (B) A larger scale map of the same zoning polygons. We are interested only in those polygons that are zoned as residential or agricultural residential.

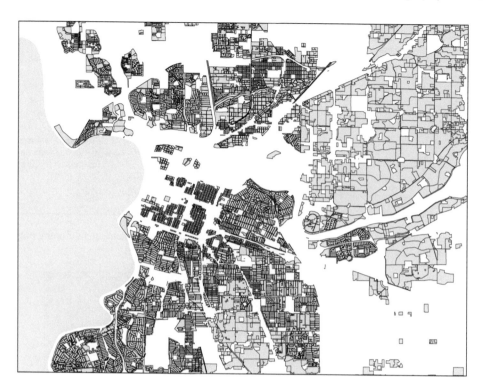

FIGURE 16.17   The Inhabited Block Groups, which represent block group polygons (and partial block group polygons) that are zoned as residential or agricultural residential but not in the limiting ancillary features.

used to remove the Limiting Ancillary Polygons from the Residential Polygons, producing the Inhabited Polygons. The Inhabited Polygons are then combined with the Block Group Polygons using the Intersect tool. The result is the Inhabited Block Groups as seen in Figure 16.17. From here, the model follows the same steps as in the first model, resulting in the BGs with Estimated Populations and Square Miles data set, which will be symbolized just as a choropleth map would be symbolized using the estimated populations, and standardized (normalized) by the SQMI_New field. A sequential color scheme is applied that emphasizes higher population density areas with darker colors, and, for the sake of clarity with such small polygons, the polygon outlines have been removed. The final dasymetric map is shown in Figure 16.18.

The overall distribution of population densities in the final dasymetric map is similar to both the choropleth map (Figure 16.7) and the first dasymetric map (Figure 16.14), but the inhabited areas are far more geographically specific, with people constrained to zoning polygons that are much smaller than the original block group polygons. A result of this is far higher population density values. The choropleth and the first dasymetric map had a maximum density value of 30,589 persons per square mile. This map has a maximum value of 123,494 persons per square mile. All three maps are based on the same block group raw population counts; the far higher density values in this map are the direct result of "placing" people in much smaller areas, areas in which they actually live, as opposed to being evenly distributed in the much larger original block group polygons.

16.7.3   Discussion

It is important to remember that in both of these approaches, important assumptions were made that we felt were sound but were also somewhat subjective. We assumed that people do not live in undeveloped areas, in areas such as parks, airports, schools, etc., or outside of areas zoned for residential use. Surely, some people do live in those areas, but for the purposes of demonstrating dasymetric techniques, we assumed they do not. This is a reflection of the binary nature of this type of dasymetric mapping: people live in certain areas, but they do not live in others. Various other approaches to dasymetric mapping have been presented in this chapter, some of which are less rigid in their view of whether or how many people might live in areas that we have deemed to be uninhabited.

The overall appearance of the three maps is partly dependent on the data classification methods that were used. The choropleth map (Figure 16.7) and first dasymetric map (Figure 16.14) use the natural breaks method, as it provides a more balanced visual impression than quantile or equal interval, especially on the first dasymetric map where there are lots of very small polygons. The second dasymetric map (Figure 16.18) uses the quantile classification method because it looks more "comparable" to the first two. A general rule for map comparison is that the same classification method should be used for every map, and that natural breaks is not a good choice, because it essentially creates a custom classification for each data set, making direct comparisons difficult. We feel that these maps are different, and they allow us more flexibility in choosing classification methods. All

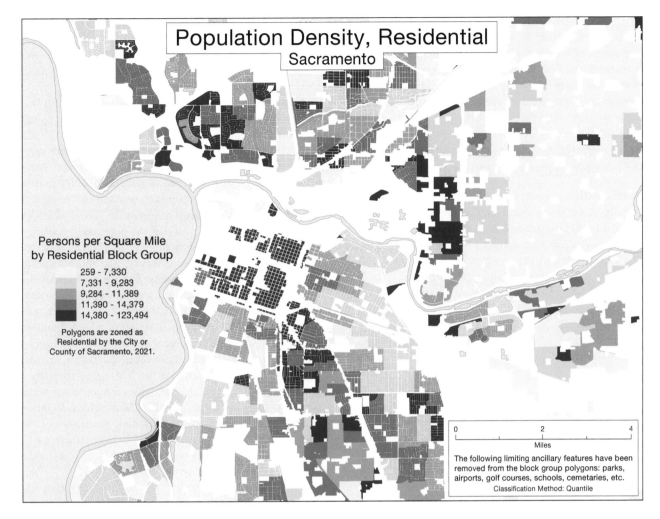

FIGURE 16.18   The final dasymetric map based on raw block group population counts, residential zoning, and the limiting ancillary features. In comparison with the choropleth map and the first dasymetric map shown in Figure 16.14, this is a more realistic representation of where people actually live and in what densities. Density values are far higher than on the two previous maps, as populations are limited to much smaller areas.

three maps are based on the same raw population counts, but the enumeration units are so different in size, shape, and number, and the density values are so different, especially on the second dasymetric map, that using the same classification method for all three maps seems less important than providing visually balanced impressions from map-to-map.

Upon close comparison of the three maps, we discovered an interesting issue related to the assumptions just described. There is a block group polygon that appears in its entirety on the choropleth map (Figure 16.19A), appears in partial form on the first dasymetric map because parts were determined to be developed and inhabited (Figure 16.19B), but does not appear at all on the second dasymetric map (Figure 16.19C). The block group has a significant population value of 2,666. We wondered why it did not appear on the second dasymetric map, and more importantly, what happened to those 2,666 people? We learned that the block group polygon is in an area that is not zoned for residential use, so it was eliminated under the assumption that people do not live outside of residential areas. It turns out that 2,666 people *do* live in that block group and all in the same building: the County

Jail (the small yellow feature in Figure 16.19). In this case, our assumption broke down, and we were forced to make a decision. Should we force the jail into the residential zoning category and show it on the map, or should we leave the jail off the map and leave the population "unallocated"? We adopted the second approach for two reasons. First, we chose to be consistent with our assumption that people do not live outside of residential areas. Second, we determined

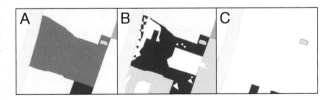

FIGURE 16.19   (A) A block group polygon (the red area) with 2,666 people on the choropleth map. (B) The same block group polygon with areas removed on the first dasymetric map. (C) Absence of the same block group polygon on the second dasymetric map. The small yellow feature is the County Jail, which is not zoned as residential, yet holds 2,666 people.

that prisoners at the jail do not live there permanently, in fact, many stay there for very short periods. Because the jail population is transient, we decided to leave the 2,666 people unallocated and off the map. A decision to include the jail population might be equally reasonable. Either way, we hope to emphasize the subjective nature of dasymetric mapping.

16.8 SOCSCAPE: A WEB APP FOR VISUALIZING RACIAL DIVERSITY

In this section, we introduce a dasymetric approach developed by Anna Dmowska and colleagues (2017) that enables racial diversity information normally depicted for enumeration units (such as census tracts) to be shown at a 30-meter resolution. The end product that we will ultimately focus on is the web application SocScape (http://150.254.124.68/socscape_usa/), which enables you to examine racial diversity patterns in the coterminous United States for the years 1990, 2000, and 2010.

Dmowska et al. noted numerous disadvantages to the traditional use of enumeration units for depicting population data such as race, including:

- Large vector-based files are difficult to work with
- The spatial resolution of related census data varies considerably between urban and rural areas
- The mapped attribute appears to be distributed uniformly within enumeration units (when in fact, it is not)
- Statistics (such as the correlation coefficient described in Section 3.6.3) depend on the size of enumeration units (this is the **modifiable areal unit problem**)
- The boundaries of enumeration units may change over time, which complicates temporal comparisons (Dmowska et al. 2017, 4)

Regular gridded data solve these problems as the resulting raster files are easier to process than vector files, the cell size is consistent between urban and rural areas, the fine resolution enables much greater detail to be shown, the consistent cell size eliminates the modifiable areal unit problem, and temporal data are not problematic as the cell size is constant over time.

To create a highly detailed racial diversity map for the United States for 2010, Dmowska et al. began with the highest resolution U.S. census data available for 2010 (the *block* level). For dasymetric modeling, they utilized both the National Land Cover Database for 2011 (https://www.mrlc.gov/data/nlcd-2011-land-cover-conus-0) and a National Land Use Dataset (Theobald 2014) to create six ancillary classes at a 30-meter resolution: one uninhabited class and five inhabited classes (open space, low intensity, medium intensity, high intensity, and vegetation).[6] Population totals were distributed from the block level to the 30-meter cells using Mennis and Hultgren's

IDM approach introduced in Section 16.6.[7] Since race-based subpopulations are not necessarily correlated with the ancillary information, the race-based attributes were distributed in the same fashion as the overall population totals (Dmowska et al. 2017, 6).

Using previous research by Holloway et al. (2012) on ethnic diversity as a basis, Dmowska et al. chose to map the racial diversity categories shown in Figure 16.20A. Here you can see that six major racial groups were specified: White, American Indian, Black, Other races, Asian, and Hispanic. A 30-meter cell is associated with a particular racial group if that group is most dominant in the cell. Note that each racial group in Figure 16.20A is either low or medium in diversity and that the population density of each group varies from low to high. In general, less saturated colors represent low diversity, and more saturated colors represent higher diversity. For a particular level of diversity, a darker color represents a higher population density. The High diversity category shown at the bottom of Figure 16.20A represents a situation where the dominant race is less than 50 percent of the total population.[8]

To illustrate the use of these racial categories, consider Figure 16.20B, which displays the 2010 racial diversity in SocScape centered on the borough of Manhattan within New York City. Note that the bulk of Manhattan falls largely in the White racial category, with high population density and either medium or low diversity.[9] In the northern part of Manhattan, however, we see two distinctly different areas, one that is largely medium diversity, high-density Black, and one that is medium diversity, high density Hispanic. Across the Hudson River in New Jersey, we see another interesting pattern. Along the river shore, there is a White high-density area that is largely medium diversity, whereas inland, there is an obvious change as Hispanic high density predominates, with a mix of medium and low diversity. We encourage you to use the SocSpace app to explore this area and other metropolitan areas in the United States.

16.9 MAPPING THE GLOBAL POPULATION DISTRIBUTION

Creating a detailed map of the global population distribution is critical given the potential of natural and manmade disasters (e.g., the effects of earthquakes or tsunamis such as in the Indian Ocean in 2004 and Japan in 2011 or the recent COVID-19 pandemic). Global population maps are also important given concerns about global climate change, which may lead to increased sea levels, extreme weather events, and a modification of global circulation patterns. If we are to determine the effect of these events on populations, we need to have detailed maps available of these populations.

16.9.1 GRIDDED POPULATION OF THE WORLD

The Gridded Population of the World (GPW), developed by Waldo Tobler and colleagues (1995) at the National Center for Geographic Information and Analysis, was arguably the first

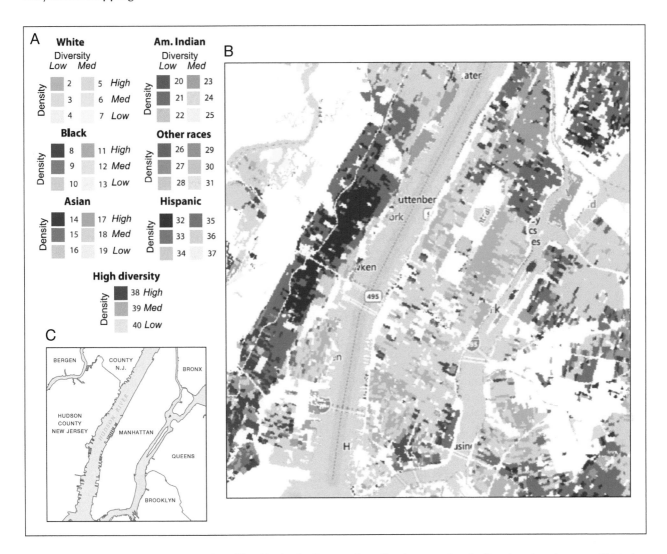

FIGURE 16.20 (A) The racial categories utilized in the SocScape web application; note that the first six categories are split into low and medium diversity and that each diversity category varies from low to high population density. (B) A screen capture from SocScape showing 2010 racial diversity centered on Manhattan within New York City. (C) An inset map for the New York City area. (Courtesy of Anna Dmowska and Tomasz Stepinski.)

major attempt to create a consistent spatial global population dataset (Rose and Bright 2014). GPW has been progressively updated and is now available through NASA's Socioeconomic Data and Applications Center (SEDAC) at https://sedac.ciesin.columbia.edu/data/collection/gpw-v4. GPW converts population data collected for enumeration units around the world to an approximately 1-km resolution for the entire globe using the areal weighting method described in Section 16.3.[10] Thus, no ancillary information is used; although this may seem undesirable, we will see shortly that GPW data can be combined with land use/land cover data to create more spatially accurate population distribution maps.

16.9.2 LANDSCAN

LandScan (https://landscan.ornl.gov/), developed at Oakridge National Laboratory in 1998, was a pioneer in combining population data with dasymetric techniques to create a global population database at an approximately

1-kilometer resolution. The earliest version of LandScan focused on utilizing landcover (derived from remote sensing), slope, location relative to roads, populated places, and *nighttime lights* as dasymetric parameters (Dobson et al. 2000). As described in the online documentation (https://landscan.ornl.gov/documentation/#inputData), the parameters used in today's LandScan include "land cover, roads, slope, urban areas, village locations and high resolution imagery analysis … [with the models] … tailored to match the data conditions and geographical nature of each individual country and region."

An important concept in LandScan is distinguishing between *residential* and *ambient populations*. Normally, censuses determine population counts based on where people reside, and thus the result is a residential population. For instance, the 2000 U.S. census indicated that there were only 55 people in the block that used to house the World Trade Center (Dobson et al. 2003, 267). In the case of natural and manmade disasters, we are instead interested in where people are apt to be located at an instant in time—this is termed the ambient population. In the case of LandScan, the ambient population is considered to be an average over a 24-hour period. Since other population databases generally utilize residential populations, care should be taken in comparing LandScan data with other population databases.

In addition to the global 1-kilometer resolution, LandScan also provides a higher resolution 90-meter database for the United States (https://geoawesomeness.com/landscan-usa-population-data-free/). An intriguing feature of this database is the availability of both daytime and nighttime population information, which is arguably critical in understanding the population density characteristics of cities that have large numbers of commuters. To illustrate, Figure 16.21 portrays the daytime and nighttime populations for San Francisco, California. The images show LandScan grids draped over a grayscale satellite image of the city. In the daytime image, populations are greatest in the Financial District and parts of the adjacent Downtown and South of Market neighborhoods, which appear as the darkest red. This is where high concentrations of people work during the day, with concentrations decreasing with distance from this central business core. The outlying lower population areas are represented primarily with orange and yellow in the daytime image. During the nighttime, the

pattern is reversed. The central business core appears to be almost uninhabited, with much of the underlying gray satellite image showing through. Outlying residential and mixed use neighborhoods such as the Western Addition, Pacific Heights, and Russian Hill, appear to have higher nighttime populations as people who work during the day return home from work. These neighborhoods appear primarily as red and orange in the nighttime image.

16.9.3 GLOBAL HUMAN SETTLEMENT LAYER

The mission of the Global Human Settlement Layer project (GHSL; https://ghsl.jrc.ec.europa.eu/about.php) is to produce "global spatial information about the human presence on the planet over time, [including] … built-up maps, population density maps, and settlement maps." One major GHSL effort has utilized *machine learning* remote sensing technology to create both 1-kilometer and 250-meter resolution built-up areas for the globe.[11] These built-up areas are expressed as a proportion of building footprint area within a grid cell, and the associated database is referred to as GHS-Built (https://ghsl.jrc.ec.europa.eu/ghs_bu2019.php). GHSL has treated the GHS-Built database as ancillary data and used a dasymetric technique to combine GHS-Built with the GPW population data described in Section 16.9.1. Population totals for the vector-based enumeration units in

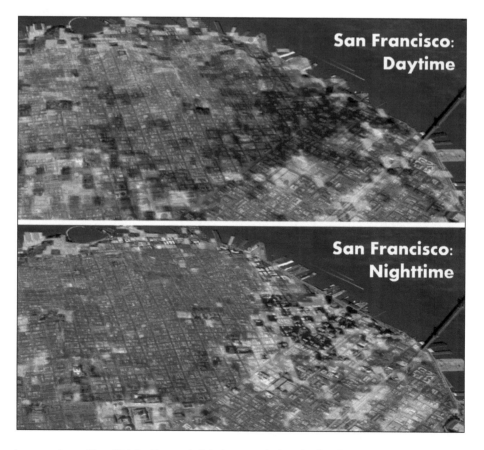

FIGURE 16.21   A comparison of LandSat daytime and nighttime populations for San Francisco, California. (Courtesy of Oak Ridge National Laboratory/U.S. Department of Energy.)

GPW were assigned to 250-meter cells of the GHS-Built database as follows:

A. If the enumeration unit generated 250 m cells and contained built-up areas (BU), the population was disaggregated in proportion to the density of the BU.
B. If the polygon generated 250 m cells and did not contain BU, population was disaggregated using areal weighting.
C. If the polygon did not generate its own 250 m cell, the polygon was converted to a point (based on its centroid), which was then associated with a particular cell (Freire et al. 2016, 3).[12]

The resulting database is termed GHS-Pop (https://ghsl.jrc.ec.europa.eu/ghs_pop2019.php).

Duncan Smith (2017) has created an online interactive application for visualizing the population density data associated with GHS-Pop (http://luminocity3d.org/WorldPopDen/#3/12.00/10.00). If you access this application, you will see that the legend includes a larger number of classes and a larger number of colors than cartographers normally recommend for thematic maps (see the legend in Figure 16.22A). Smith, however, presents several arguments in favor of this unusual classification and symbolization. First, we need to realize that global values of population density vary considerably, from less than one person per square kilometer in some wilderness areas to more than

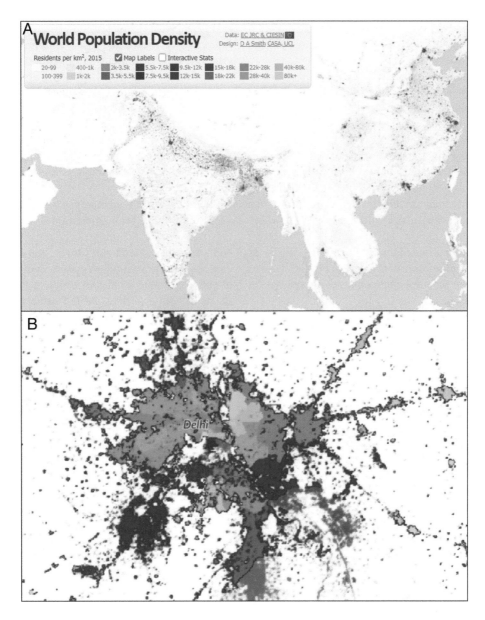

FIGURE 16.22 Images from Smith's (2017) interactive World Population Density map, which merges population data from the Gridded Population of the World with built-up areas defined by the Global Human Settlement Layer. (A) The legend for the map and a small-scale view of southern Asia; (B) A large-scale view of the city of Delhi, India. (Cartography: Duncan A. Smith, UCL. Data: EC JRC & CIESEN.)

50,000 people per square kilometer in dense urban areas. Smith argues that we would like to portray numerous transition densities within this large range, including wilderness to rural, rural to suburban, suburban to urban, and urban to super-urban. Moreover, if we examine settlement patterns across the globe, we find that there are different density levels for urban and suburban in different countries. The high number of classes used (16) allows us to see the transition densities and to utilize the application in varied areas of the globe.

In a similar fashion, Smith found that conventional methods of data classification were not suitable for mapping the varied population densities—particularly problematic was the high end of the distribution where urban land uses represented a small portion of the area, but a large portion of the population. As a result, Smith created a manual modification of the Jenks (optimal) classification approach so that more classes were included at the high end of the scale.

Given a large number of classes and the desire to clearly see the various population density transitions, Smith created a unique color scheme by combining a light-to-dark green-to-blue transition for the lowest eight classes with a multispectral blue to red to high saturation orange and yellow for the highest eight classes. Although cartographers often argue against using so many hues, this approach enables considerable detail to be seen in heavily urbanized areas, as in Delhi, India, in Figure 16.22B. Contrast this view with that of Southern Asia in Figure 16.22A, where the map appears to be almost entirely covered by the lowest eight classes. Smith evaluated the legibility of the 16 colors for those people with a color deficiency and found that the map was legible for the most common types of color vision deficiency.

16.10   SUMMARY

At several points in this book, we stressed that although choropleth maps are commonly used, they are often inappropriate because phenomena generally are not uniformly distributed within the enumeration units composing a choropleth map. Like the choropleth map, the dasymetric map uses areal symbols to represent presumed zones of uniformity, but the bounds of these zones need not match enumeration unit boundaries. The dasymetric map thus provides a potential alternative to choropleth mapping.

The key to dasymetric mapping is the availability of ancillary information at a finer resolution than the enumeration units for which you have attribute data available. Most typically, ancillary information is available as a land use/land cover map, although we saw that other sources of ancillary information are possible, such as remotely sensed imagery (utilized in the GHSL project) and zoning information provided by City and County governments (utilized for one of our Sacramento dasymetric maps). The ancillary information is used to define a set of ancillary zones, which are then overlaid with source zones containing raw totals (most typically population) to create a set of target zones that raw totals are distributed to.

There are a wide variety of approaches for distributing raw totals to target zones. We focused on an areal weighting method, a binary method, a three-class method, a limiting variable method, and intelligent dasymetric mapping (IDM). The latter was the most sophisticated approach as it allowed the cartographer to either use their domain knowledge to select estimated densities for ancillary classes associated with target zones or have these values automatically determined by the software. You can find additional dasymetric techniques by exploring the Further Reading section. One question you may pose is whether there is a "best" or "optimal" approach for dasymetric mapping. A study by Jose Pavía and Isidro Cantarino (2017) is useful in this respect as it compares a wide variety of approaches. Interestingly, they found that 3-D approaches that considered the vertical dimension of buildings were most useful in their study area of Barcelona, Spain. Although such comparative studies are useful, we argue that the choice of a dasymetric technique should be based on several factors, including the purpose of your dasymetric map, the time and cost needed to acquire the necessary source and ancillary data, your ability to understand the more sophisticated dasymetric techniques, and your knowledge of suitable software.

We noted several advantages for converting traditional vector-based files for enumeration source units into a regular raster grid, including faster processing, a consistent cell size in urban and rural areas, the capability to show data at a finer resolution (in greater detail), the elimination of the modifiable areal unit problem, and ease of working with temporal data. At the same time, we also saw with the Sacramento dasymetric maps that a vector-based approach can be useful too. That is because certain data sets, such as zoning, are only available as vector polygons, and the very small polygon sizes will not convert well to raster. The use of vector-based data also makes common geoprocessing tools available that only work with vector inputs such as Union, Intersect, and Erase. Vector also provides for better comparisons with traditional, vector-based choropleth maps.

Finally, we examined several approaches for examining the global population distribution. The GPW provides detailed population density for the world at a one-kilometer resolution. A limitation of this approach, however, is that it uses simple areal weighting to allocate population from source zones to the 1-km grid. The Global Human Settlement Layer (GHSL) provides a solution to this limitation by merging detailed urban built-up information with the GPW. The LandScan system is especially useful for evaluating the effect of natural and manmade disasters on the global population as it utilizes a wide range of ancillary information and considers the ambient population averaged over a 24-hour day rather than the simpler residential population.

16.11 STUDY QUESTIONS

1. Presume that you are provided the number of elephants estimated to exist in each country of Africa in 2022, along with land use/land cover information that would enable you to portray the distribution of elephants within these countries. Discuss the use of the terms source zones, ancillary zones, and target zones in this context.

2. (A) Based on the relative areas of the urban and forested regions shown in Figure 16.3A, use the limiting variable method to compute a population density for the urban category assuming that the total population of the region is 5,172 and the population density threshold for the forested region is 329 people per square kilometer. (B) Compute a population count for the urban category assuming that simple areal weighting is used.

3. (A) Discuss the terms source zones, ancillary zones, and target zones in the context of the Eicher and Brewer study. (B) In general, how did Eicher and Brewer compare the effectiveness of the three dasymetric approaches that they tested?

4. Describe how limiting ancillary information can be used to produce dasymetric maps of population density based on raw block group population counts.

5. Provide common assumptions regarding where populations are concentrated and provide exceptions to those assumptions.

6. Discuss five advantages of converting vector-based enumeration units to a uniform raster-based dataset when mapping population data.

7. Using the SocScape (http://150.254.124.68/socscape_usa/) application, contrast the racial diversity in 1990 (select the "Racial diversity 1990 MYC" data layer) with the racial diversity in 2010 (select the "Racial diversity 2010 MYC" data layer) in the Kansas City Metropolitan Statistical Area. (Note that it will be easiest to make this comparison if you open your web browser as two separate windows that can be viewed simultaneously.)

8. Discuss the difference between residential populations and ambient populations.

9. Discuss why it makes sense to combine the GHS-Built data with the GPW population data rather than use the GPW data alone.

10. Access the World Population Density application (http://luminocity3d.org/WorldPopDen/#3/52.27/31.11). (A) Zoom as far as you can into Houston, Texas. Describe the character of the population density in this metropolitan area. (Note that you can get some statistics on population density for Houston if you set Interactive Stats on in the legend and then hover over Houston.) Now zoom into Surat, India (on the west coast of India) and contrast its character with that of Houston. (B) Read the Analysis tab and indicate what map projection issues concern you when interpreting this interactive application.

NOTES

1 For a more sophisticated three-class method, see Langford (2006).

2 Although Wright is often recognized with developing the limiting variable method, Benjamin Semenov-Tian-Shansky utilized the method in the 1920s, and the origins of the method precede that time; for a discussion of the history of dasymetric mapping, see Petrov (2012).

3 Mennis and Hultgren were not the first to use the term "intelligent" in association with dasymetric mapping. For example, Eicher and Brewer (2001, 127) used the term to describe any areal interpolation that utilized ancillary land use/land cover data.

4 The block group is the smallest U.S. Census enumeration unit for which data are collected by the American Community Survey (ACS), which is performed every year. Census blocks are geographically smaller, but block-level data are only collected every ten years as part of the Decennial Census. We have chosen to use block group data over block-level data to allow for the use of more current population counts, which we will convert to population densities.

5 Zoning data sets are provided by certain City and County governments, typically those that provide GIS data online. We acquired two zoning data sets free of charge, one from the City of Sacramento (https://data.cityofsacramento.org) and one from the County of Sacramento (https://data-sacramentocounty.opendata.arcgis.com/). When combined, they provided us with zoning polygons for the City and the unincorporated County.

6 Details of this process can be found in Dmowska and Stepinski (2017, 15).

7 For details of the dasymetric modeling, see Dmowska and Stepinski (2017, 15–16).

8 Additionally, the so-called *standardized information entropy* is high (Dmowska et al. 2017, 8).

9 If you are accessing the Web app, you can use the Query data layers tool to make certain that you are interpreting the colors correctly. Be sure that you have set the transparency slider to the extreme right so that the colors in the app come close to matching those shown in the textbook.

10 For details on the areal weighting process, see Doxsey-Whitfield et al. (2015).

11 For details on this process, see Pesaresi et al. (2016).

12 For a more complete explanation, see Freire et al. (2016).

REFERENCES

Cromley, R. G., Nanink, D. M., and Bentley, G. C. (2012) "A quantile regression approach to areal interpolation." *Annals of the Association of American Geographers 102*, no. 4: 763–777.

Dmowska, A., and Stepinski, T. F. (2017) "A high resolution population grid for the conterminous United States: The 2010 edition." *Computers, Environment and Urban Systems 61*, Part A:13–23.

Dmowska, A., Stepinski, T. F., and Netzel, P. (2017) "Comprehensive framework for visualizing and analyzing spatio-temporal dynamics of racial diversity in the entire United States." *PLoS ONE 12*, no. 3:e0174993.

Dobson, J. E., Bright, E. A., Coleman, P. R. et al. (2000) "LandScan: A global population database for estimating populations at risk." *Photogrammetric Engineering and Remote Sensing 66*, no. 7:849–857.

Dobson, J. E., Bright, E. A., Coleman, P. R. et al. (2003) "LandScan: A global population database for estimating populations at risk." In *Remotely Sensed Cities*, ed. by V. Mesev, pp. 267–279. London: Taylor & Francis.

Doxsey-Whitfield, E., MacManus, K., Adamo, S. B. et al. (2015) "Taking advantage of the improved availability of census data: A first look at the Gridded Population of the World, Version 4." *Papers in Applied Geography 1*, no. 3:226–234.

Eicher, C. L., and Brewer, C. A. (2001) "Dasymetric mapping and areal interpolation: Implementation and evaluation." *Cartography and Geographic Information Science 28*, no. 2:125–138.

Freire, S., MacManus, K., Pesaresi, M. et al. (2016) "Development of new open and free multi-temporal global population grids at 250 m resolution." *Proceedings of the 19th AGILE Conference on Geographic Information Science*, Helsinki, Finland, pp. 1–6.

Holloway, S. R., Schumacher, J. and Redmond, R. (1996) *People and place: Dasymetric Mapping Using Arc/Info*. Missoula, MT: Wildlife Spatial Analysis Lab, University of Montana.

Holloway, S. R., Wright, R., and Ellis, M. (2012) "The racially fragmented city? Neighborhood racial segregation and diversity jointly considered." *Professional Geographer 64*, no. 1:63–82.

Kar, B., and Hodgson, M. E. (2012) "A process oriented areal interpolation technique: A coastal county example." *Cartography and Geographic Information Science 39*, no. 1:3–16.

Langford, M. (2006) "Obtaining population estimates in non-census reporting zones: An evaluation of the 3-class dasymetric method." *Computers, Environment and Urban Systems 30*:161–180.

Maantay, J. A., Maroko, A. R., and Herrmann, C. (2007) "Mapping population distribution in the urban environment: The Cadastral-based Expert Dasymetric System (CEDS)." *Cartography and Geographic Information Science 34*, no. 2:77–102.

Mennis, J., and Hultgren, T. (2006) "Intelligent dasymetric mapping and its application to areal interpolation." *Cartography and Geographic Information Science 33*, no. 3:179–194.

Pavía, J. M. and Cantarino, I. (2017) "Can dasymetric mapping significantly improve population data reallocation in a dense urban area?" *Geographical Analysis 49*, no. 2:155–174.

Pesaresi, M., Ehrlich, D., Ferri, S. et al. (2016) *Operating Procedure for the Production of the Global Human Settlement Layer from Landsat Data of the Epochs 1975, 1990, 2000, and 2014*. JRC Technical Report, European Union.

Petrov, A. (2012) "One hundred years of dasymetric mapping: Back to the origin." *The Cartographic Journal 49*, no. 3:256–264.

Qiu, F., Sridharan, H. and Chun, Y. (2010) "Spatial autoregressive model for population estimation at the census block level using LIDAR-derived building volume Information." *Cartography and Geographic Information Science 37*, no. 3:239–257.

Reibel, M. and Bufalino, M. E. (2005) "Street-weighted interpolation techniques for demographic count estimation in incompatible zone systems." *Environment and Planning A 37*, no. 1:127–139.

Rose, A. N. and Bright, E. A. (2014) "The LandScan global population distribution project: Current state of the art and prospective innovation." *Population Association of American Annual Meeting*, Boston, MA, pp. 1–21.

Smith, D. A. (2017) "Visualising world population density as an interactive multi-scale map using the global human settlement population layer." *Journal of Maps 13*, no. 1:117–123.

Tapp, A. F. (2010) "Areal interpolation and dasymetric mapping methods using local ancillary data sources." *Cartography and Geographic Information Science 37*, no. 3:215–228.

Theobald, D. M. (2014) "Development and applications of a comprehensive land use classification and map for the US." *PLoS ONE 9*, no. 4: e94628.

Tobler, W., Deichmann, U., Gottsegen, J. et al. (1995) *The Global Demography Project*, Technical Report TR-95-6. Santa Barbara, CA: National Center for Geographic Information and Analysis.

Wright, J. K. (1936) "A method of mapping densities of population: With Cape Cod as an example." *The Geographical Review 26*, no. 1:103–110.

Zandbergen, P. A. (2011) "Dasymetric mapping using high resolution address point datasets." *Transactions in GIS 15*, no. s1:5–27.

17 Isarithmic Mapping

17.1 INTRODUCTION

Isarithmic maps (the most common form being the **contour map**) depict smooth, continuous phenomena, such as rainfall, barometric pressure, depth to bedrock, and, of course, Earth's topography. After the choropleth map, the isarithmic map is probably the most widely used thematic mapping technique and is certainly one of the oldest, dating to the eighteenth century. Data for isarithmic mapping may be available as either a continuous grid of attribute values (such as the Copernicus Digital Elevation Model (DEM); https://registry.opendata.aws/copernicus-dem/) or as a set of irregularly spaced **control points** with associated attribute values (such as temperature data for weather stations). In this chapter, we focus on the latter kind of data and introduce approaches for interpolating attribute values between control points.

There are two basic kinds of data for control points: true and conceptual. **True point data** actually occur at point locations (such as oil well production), whereas **conceptual point data** are conceived to occur at points (such as the centroids of census tracts). A wide variety of techniques have been developed for interpolating true point data—we will introduce four of these that are commonly used: triangulation, inverse-distance weighting, ordinary kriging, and thin-plate splines. Although these techniques might also be used to interpolate conceptual point data, it is more appropriate to consider the area over which the data actually occur. This is accomplished using **pycnophylactic interpolation** in which we conceive of raising each enumeration unit to a height proportional to its associated data value and then gradually smooth this 3-D surface, keeping the volume within each enumeration unit constant.

In the last section of the chapter, we discuss several symbolization techniques that have been commonly used for isarithmic mapping, including contour lines, the application of color schemes to areas between contour lines, continuous-tone maps (or unclassed isarithmic maps), and the use of the perspective height visual variable (e.g., to create a fishnet map). An intriguing possibility is to view spectral color schemes with special glasses that enhance the **color stereoscopic effect** (in which long-wavelength colors appear nearer than short-wavelength colors).

It is important to recognize that we are only providing an introduction to interpolation techniques. Entire books have been written on this topic (e.g., Watson 1992), with many of these focusing on kriging (e.g., Isaaks and Srivistava 1989, Webster and Oliver 2007). We encourage you to explore these and other references found in the list of references at the end of the chapter and in the Further Reading section at www.routledge.com/9780367712709.

17.2 LEARNING OBJECTIVES

- Distinguish between true point data and conceptual point data, and indicate which interpolation methods are ideally appropriate for each type of data.
- Define the terms Thiessen polygon and Delaunay triangles, and indicate how the triangulation interpolation method is related to these terms.
- List the three basic steps involved in inverse-distance weighting, and explain how a value at a grid point is estimated as a function of surrounding control points.
- Define the terms semivariance and semivariogram, and indicate why the semivariogram is a key element of ordinary kriging interpolation.
- Explain why ordinary kriging is often said to produce an optimal interpolation and what characteristics might prevent the interpolation from being truly optimal.
- Describe the general process of creating thin-plate splines, and indicate what sort of spatial distribution would likely be appropriate for splining.
- Define cross-validation and describe three measures of error that are typically used as a measure of cross-validation.
- Explain how color can assist in the interpretation of isarithmic maps and why a broader range of color schemes may be possible on isarithmic maps (as opposed to choropleth maps).
- Define the color stereoscopic effect and explain how this effect can be enhanced using specialized viewing glasses.

17.3 SELECTING APPROPRIATE DATA

In considering the kind of data appropriate for isarithmic mapping, it is important to remember that the underlying phenomenon generally is presumed to be *continuous* and *smooth*. Thus, the phenomenon is presumed to exist throughout the geographic region of interest and to change gradually between individual point locations (as opposed to abruptly). For instance, think of the mean yearly rainfall for the state of Oregon. There is potentially a value for every location, and the change between locations is relatively gradual. Phenomena that are largely continuous and smooth but that have some discontinuities can be handled with the isarithmic approach, but you must specify the discontinuities and have software available that will handle them. For

DOI: 10.1201/9781003150527-19

instance, a geologist might be interested in creating an isarithmic map showing the height of the Dakota sandstone in Kansas based on a series of sample wells. If any faults in the bedrock were known, these would have to be specified; otherwise, a smooth transition between elevation values would be presumed along the fault zone.

Data for isarithmic mapping are sampled from an underlying phenomenon at point locations termed **control points**. As with proportional symbol maps, two forms of data can be collected at control points: true and conceptual. For **true point data**, values are actually measured at a point location; for instance, weather data, such as pressure and insolation, are measured at individual weather stations. In contrast, **conceptual point data** are collected over an area (or volume) but are presumed to occur at point locations; for instance, we might collect population data for counties and presume that the population is located at the centroids of counties. Maps produced using true and conceptual point data are termed **isometric** and **isopleth maps**, respectively. You should be especially cautious when working with conceptual point data, as the assumption of continuity might not be met. For instance, population-related data might be characterized by sharp discontinuities that would be more appropriately mapped by the dot or dasymetric methods.

As with the choropleth map, the isopleth map requires that data be standardized to account for the area over which conceptual point data are collected. Thus, it makes no sense to create an isopleth map of the raw number of people in counties because a larger county tends to have a higher population. Methods of standardization for isopleth maps are identical to those we have discussed for choropleth maps (see Chapter 15).

In a fashion similar to dasymetric and dot maps, it is also possible to utilize ancillary information to enhance the interpolation of the main phenomenon of interest in isarithmic mapping. For example, imagine that you have collected data on rainfall at weather stations located in a mountainous region (such as Chile). It is well known that precipitation tends to increase with elevation due to the orographic effect of elevation, and so it makes sense to include elevation as an ancillary attribute in the interpolation procedure (Goovaerts 2000, 115). Since our focus in this chapter is on basic concepts of interpolation and the visualization of the resulting maps, we will not delve into ancillary information in detail. For now, you should recognize that not all interpolation methods (and associated software) include an option to handle ancillary information.

17.4   MANUAL INTERPOLATION

A key problem in isarithmic mapping is that data often are available only at irregularly spaced control points, whereas the phenomenon is presumed to exist throughout the region of interest. Therefore, it is necessary to determine the values of intermediate points through the

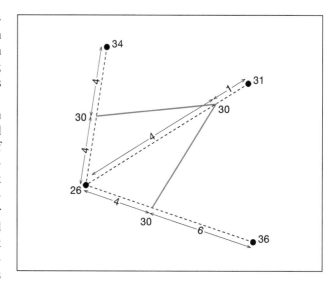

FIGURE 17.1   Manual interpolation involves a linear interpolation between control points.

process of **interpolation**. **Manual interpolation**, or interpolation by "eye," was the only approach available prior to the development of computer-based methods, which did not become popular until the 1970s. Although not commonly practiced today, it is important that you be aware of manual interpolation because it is frequently used as a yardstick against which automated methods are compared. Manual interpolation is accomplished by mentally connecting neighboring control points with straight lines and then linearly interpolating along these lines to create contours (or isolines) of equal value. For instance, in Figure 17.1, you can see how a contour line of 30 would be drawn through a set of four control points. *Linear* interpolation refers to the idea that the contour line is positioned proportionally between the control points (e.g., in Figure 17.1, the contour line of 30 is 4 units from 26 and 1 unit from 31, and so is positioned 4/5 of the distance between 26 and 31).

When control points are positioned near the corners of a square, it might not be clear where to draw contours. For instance, in Figure 17.2A and B, we see a situation in which the area in the middle of the four control points can be either below or above the 25-contour line. This problem is handled by averaging the four control points to create an additional control point (Figure 17.2C).

A strict definition of linear interpolation produces the angular contour line shown in Figure 17.1. Because real-world phenomena generally do not exhibit such angularity, contour lines are smoothed by considering the trend of control point values within the region. This is illustrated in Figure 17.4A for a larger data set. In Figure 17.4A, also note that when neighboring control points are connected, a set of triangles is formed. These triangles are also utilized in *triangulation*, one of several automated interpolation methods that we consider in the next section.

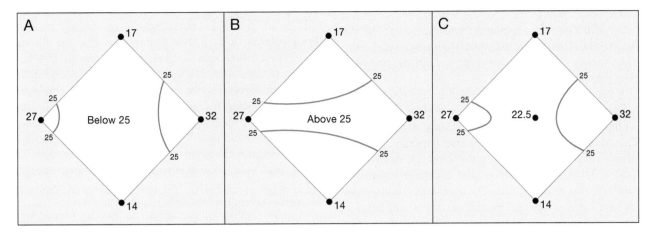

FIGURE 17.2 When the control points to be manually interpolated form a square, a fifth control point is created by averaging the four original control points.

17.5 AUTOMATED INTERPOLATION FOR TRUE POINT DATA

In this section, we consider four automated interpolation methods appropriate for true point data: triangulation, inverse-distance weighting, thin-plate splines, and ordinary kriging. The first three of these are considered *deterministic methods* because the interpolated points are determined by mathematical functions of the data. In contrast, ordinary kriging is considered a *geostatistical method* because it includes a measure of *geography* (the spatial autocorrelation in the data) and uses *statistical* principles (e.g., the data are often assumed normally distributed). To assist in comparing the four methods, we utilized the software ArcMap to create isarithmic maps of all four methods (Figure 17.12).[1] Within ArcMap, we used the Geostatistical Analyst for inverse-distance and kriging, the Spatial Analyst for thin-plate splines, and the 3-D Analyst for triangulation.[2] The resulting interpolated maps are based on mean annual snowfall for 1981–2010 for 563 weather stations within Wisconsin (151 stations) and surrounding states (412 stations). Although our focus is on Wisconsin, we need to include adjacent states so that interpolated values are accurate along the state boundary of Wisconsin. We will discuss the four maps shown in Figure 17.12 once we have introduced all four interpolation methods.

Prior to creating mapped interpolations of the data, it is wise to explore the data in map and graphical form. For this purpose, we created a proportional symbol map of the control point data for Wisconsin (Figure 17.3), and we made use of an Explore Data tool available within the Geostatistical Analyst. The proportional symbol map reveals that there appears to be an increase in the mean snowfall as we go north, particularly in extreme northern Wisconsin. This trend was also apparent in the Explore Data tool, as it revealed a linear trend in the snowfall data when it was projected onto a graph extending from south to north. The Explore Data tool also revealed a slight increase in the data from west to east (as reflected in a linear trend

when the data were projected onto a graph extending from west to east). The general increase in snowfall as we move north through the state can be explained to some extent by a movement northward (away from warmer regions to the south) and the higher elevations generally found in northern Wisconsin. The very high amounts of snowfall in northern Wisconsin, however, are largely due to the *lake effect* snow associated with Lake Superior. Although Lake Michigan

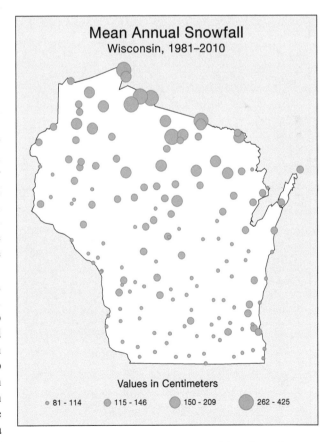

FIGURE 17.3 Proportional symbol map of the Wisconsin snowfall data. (Data source: Midwestern Regional Climate Center; https://mrcc.purdue.edu/.)

bounds Wisconsin to the east, the lake effect snows associated with that lake are primarily to the east of the lake (https://thumbwind.com/2021/03/07/great-lakes-lake-effect-snow/). This is why we don't see a distinctive linear trend running west to east.

A histogram of the data provided by the Explore Data tool also revealed a distinctive positive skew in the snowfall data (the data were not normal). Since normal data are often recommended for kriging, we experimented with various data transforms (using the approach described in Section 3.5.2) and concluded that a natural logarithmic transform was most effective. For consistency, and because we were curious what effect a transform would have, we decided to utilize both the raw and transformed data with all four interpolation methods.

The fact that there were potential trends in the data and that elevation might play a role suggests that we might want to utilize interpolation methods that consider such issues. For now, though, we will presume that we wish to interpolate just the snowfall data. We will return to these other issues in Section 17.5.5.

17.5.1 TRIANGULATION

Triangulation is logical to address first because it emulates how contour maps are constructed manually. A key step in triangulation is connecting neighboring points to form a set of triangles that are analogous to those employed in manual contouring. John Davis (2002, 375) indicated that one of the challenges for those developing triangulation was determining a "best" set of triangles, given that "simply entering the [control] points in a different sequence could result in contour lines with a conspicuously different appearance." Fortunately, this is no longer a problem, as **Delaunay triangles** provide a unique solution regardless of the order in which control points are specified. Delaunay triangles are closely associated with **Thiessen polygons**, which are formed by drawing boundaries between control points such that all hypothetical points within a polygon are closer to that polygon's control point than to any other control point. For example, in Figure 17.4B, hypothetical point S is closer to control point C than to any other control point, so point S is part of the Thiessen polygon associated with

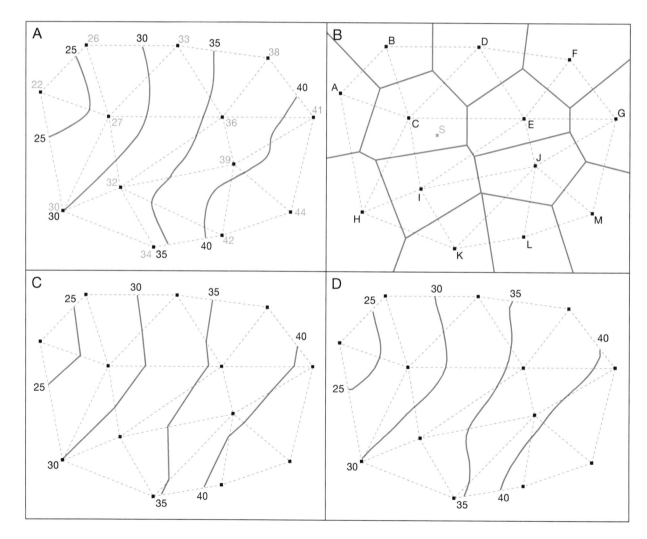

FIGURE 17.4 Hypothetical data showing (A) manual contouring, (B) Thiessen polygons and Delaunay triangles, (C) simple linear interpolation along Delaunay triangle edges, and (D) smoothed interpolation along Delaunay triangle edges.

control point C. Delaunay triangles are created by connecting control points of neighboring Thiessen polygons. For example, in Figure 17.4B, triangle ABC is formed because the Thiessen polygons associated with control points A, B, and C are all neighbors of one another.

Once Delaunay triangles are formed, contour lines can be created by interpolating along the edges of triangles in a fashion similar to manual interpolation. Delaunay triangles are desirable for this purpose because the longest side of any triangle is minimized, and thus the distance over which interpolation must take place is minimized. As with manual interpolation, a strict linear interpolation along triangle edges leads to angular contour lines, as shown in Figure 17.4C. Smoothed contours can be created using a **natural neighbors interpolation**. In this approach, the Thiessen polygons already described are first created for the given data set. To interpolate a value for any arbitrary point, a new Thiessen polygon is computed for that point, and the proportion of overlap of that new polygon with the original polygons is used to determine the weight assigned to each of the control points associated with the original polygons (https://tinyurl.com/how-natural-neighbor-works).

Triangulation is commonly used in DEMs because the density of Thiessen polygons (and associated Delaunay triangles) can be varied as a function of the variability in the landscape. An important characteristic of triangulation is that it *honors* the control point data (it is an **exact interpolator**), as estimated values for control points will be identical to the raw values originally measured at those control points. We can see this in Figure 17.4D, where the contour line for 30 passes directly through control point H, which has a raw value of 30.

17.5.2 INVERSE-DISTANCE WEIGHTING

Inverse-distance weighting (or simply inverse-distance) involves three steps: (1) laying a grid on top of the control points, (2) estimating values at each grid point (or grid node) as a function of the distance to control points, and (3) interpolating between the grid points to actually create a contour line. There is no analogy to this method in manual contouring—a grid is used because "it is much easier to draw contour lines through an array of regularly spaced grid nodes than it is to draw them through the irregular pattern of the original points" (Davis 2002, 380).

The term *inverse-distance* is used because control points are weighted as an inverse function of their distance from grid points: Control points near a grid point are weighted more than control points far away. The basic formula for estimating a value at a grid point is:

$$\hat{Z} = \frac{\sum_{i=1}^{n} Z_i / d_i^k}{\sum_{i=1}^{n} 1 / d_i^k},$$

where:

$\hat{Z}$ = estimated value at the grid point
Z_i = data values at control points

d_i = Euclidean distances from each control point to a grid point
k = power to which distance is raised
n = number of control points used to estimate a grid point

To illustrate, Figure 17.5 depicts a portion of a hypothetical grid laid on top of four control points $(n = 4)$ along with calculations for the distances from each control point to the central grid point. If we assume $k = 1$ (for simplicity in manual computation), then the central grid point value is:

$$\hat{Z} = \frac{\left(Z_1/d_1^1\right) + \left(Z_2/d_2^1\right) + \left(Z_3/d_3^1\right) + \left(Z_4/d_4^1\right)}{\left(1/d_1^1\right) + \left(1/d_2^1\right) + \left(1/d_3^1\right) + \left(1/d_4^1\right)}$$

$$= \frac{(40/2.24) + (60/1.00) + (50/1.00) + (40/1.41)}{(1/2.24) + (1/1.00) + (1/1.00) + (1/1.41)}$$

$$= 49.5$$

Oftentimes, however, k is set to 2, which cancels the square root computation in distance calculations and thus saves computer time (Isaaks and Srivastava 1989, 259). We chose to use an option within the Geostatistical Analyst that optimized the selection of k by using a cross-validation approach similar to what we will describe in Section 17.5.5—the result was a k value of 1.6 for the raw data and 2.3 for the transformed data.

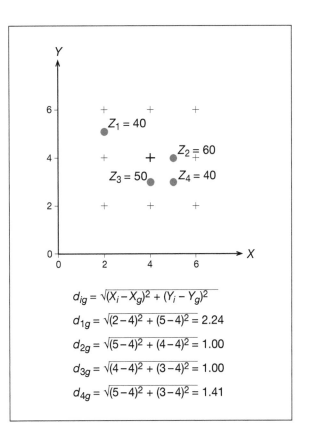

$$d_{ig} = \sqrt{(X_i - X_g)^2 + (Y_i - Y_g)^2}$$

$$d_{1g} = \sqrt{(2-4)^2 + (5-4)^2} = 2.24$$

$$d_{2g} = \sqrt{(5-4)^2 + (4-4)^2} = 1.00$$

$$d_{3g} = \sqrt{(4-4)^2 + (3-4)^2} = 1.00$$

$$d_{4g} = \sqrt{(5-4)^2 + (3-4)^2} = 1.41$$

FIGURE 17.5 Computation of distances for inverse-distance contouring. Four control points are overlaid with a hypothetical grid; distance calculations are shown from each control point to the central grid point.

One should bear in mind that although Euclidean distance is the metric most often used, Euclidean distance is inappropriate for small-scale maps that encompass a substantial portion of Earth's surface. For small-scale maps, it is necessary instead to use spherical distance, which measures distance along the arc of a great circle (Willmott and Matsuura 2006, 91). Simple Euclidean distance computations also may be modified to account for the landscape characteristics of subregions on the map; for example, Steve Lyon and colleagues (2010) used the slope to modify distance computations when interpolating the pH of soils.

With real-world data, the number of control points obviously will be greater than the simple hypothetical case shown in Figure 17.5. A larger number of control points leads one to wonder which control points should be used to estimate each grid point. One approach is to use *all* control points; in this case, n in the formula for $\hat{Z}$ is simply the number of control points. This approach seems unrealistic, however, because control points far away from a grid point are unlikely to be correlated with that grid point. This sort of thinking, and the need to minimize computations, led software developers to create various strategies for selecting an appropriate subset of control points, as shown in Figure 17.6.

The *simple (no sectors)* strategy requires that control points fall within an ellipse (typically, a circle) of fixed size; normally, only a subset of all control points is used to make an estimate (here, we have assumed that eight control points will be used). *Quadrant* and *octant* strategies also require that control points fall within an ellipse, but the ellipse is divided into four and eight sectors, respectively, with a specified number of control points used within each sector. For this example, we have again assumed a total of eight control points, and so two points are located in each sector in the quadrant strategy and one point is located in each sector in the octant strategy. For these hypothetical data, it

appears that either the quadrant or octant strategy is preferable to the simple strategy because the latter does not use any control points southwest of the grid point.

For the Wisconsin snowfall data mapped in Figure 17.12B, we assumed a total of 12 control points and a quadrant strategy; thus, we stipulated a maximum of three points in each sector. ArcMap also permits the specification of a minimum number of control points that must be found within a sector. This specification is potentially relevant to the Wisconsin snowfall data because of the two Great Lakes which bound the state of Wisconsin. We specified a minimum number of two control points within each sector.

In discussing inverse-distance weighting, we have focused on grid point calculations because of the considerable number of options involved. Remember, however, that after grid point calculation, an interpolation must also take place between grid points. As with triangulation, a simple linear interpolation leads to angular contour lines; angularity can be avoided either by increasing the number of grid points (which increases computations and memory requirements) or by generalizing the resulting grid. In ArcMap, generalization is accomplished using bilinear interpolation, in which a value for a point on the generalized grid is a weighted average of the original grid points (https://tinyurl.com/bilinear-interpolation). This leads to a smoother surface, but as we see in Figure 17.12B, it does not necessarily lead to smooth contours between data classes. We will consider this issue again when we contrast the four maps in Section 17.5.5.

One problem with inverse-distance weighting is that it cannot account for trends in the data. For example, a visual examination of z values for control points might suggest a value for a grid point higher than any of the control points, but the inverse-distance method cannot produce such a value. This problem exists because the formula for inverse-distance is a weighted *average* of z values for control points: An average of values cannot

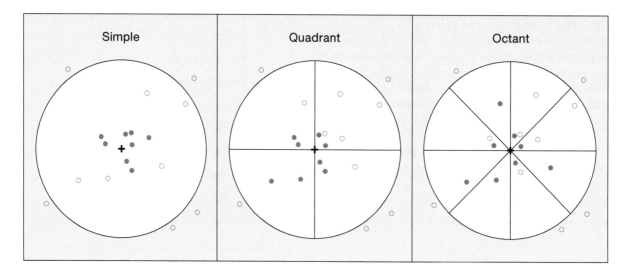

FIGURE 17.6 Various strategies for selecting an appropriate subset of control points. The cross in the middle represents a grid point location to be estimated, and the large circle represents the search radius within which control points must be located. Actual control points selected in each case are shown in red.

be lesser or greater than the range of input values. Like triangulation, inverse distance is an exact interpolator, so when a grid point coincides with a control point, the distance to the control point will be zero, and that control point has a weight of 1/0 or infinity.

17.5.3 ORDINARY KRIGING

The term **kriging** comes from Daniel Krige, who developed the method for geological mining applications (Cressie 1990). Kriging is similar to inverse-distance interpolation in that a grid is overlaid on control points, and values are then estimated at each grid point as a function of surrounding control points. Kriging, however, considers not only the distance from surrounding control points to a grid point but also the distances between control points. Understanding how kriging works requires an understanding of semivariance and the semivariogram.

17.5.3.1 Semivariance and the Semivariogram

Consider the simplified graphic shown in Figure 17.7A, which portrays attribute data for five hypothetical, equally spaced control points. Normally, of course, control points are not equally spaced, but it is easier to understand semivariance if initially, we assume that they are. **Semivariance** is defined as:

$$\gamma_h = \frac{\sum_{i=1}^{n-h}\left(Z_i - Z_{i+h}\right)^2}{2(n-h)},$$

where:

Z_i = values of the attribute at control points
h = multiple of the distance between control points
n = number of sample points

In Figure 17.7A, the distance between control points is 10; h can take on values of 1, 2, 3, and 4 (that is, distances of 10, 20, 30, and 40, respectively, are possible); and n is 5. Note that the largest possible multiple is 4 because the maximum possible distance between the control points is 40.

To illustrate computations for semivariance, we can insert the data from Figure 17.7A into the preceding formula for $h = 1$.

$$\gamma_1 = \frac{\sum_{i=1}^{5-1}\left(Z_i - Z_{i+1}\right)^2}{2(5-1)}$$

$$= \frac{\left(Z_1 - Z_2\right)^2 + \left(Z_2 - Z_3\right)^2 + \left(Z_3 - Z_4\right)^2 + \left(Z_4 - Z_5\right)^2}{8}$$

$$= \frac{\left(20-30\right)^2 + \left(30-35\right)^2 + \left(35-40\right)^2 + \left(40-45\right)^2}{8}$$

$$= 21.88$$

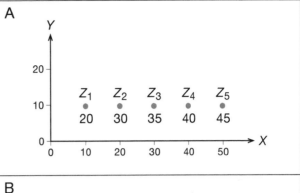

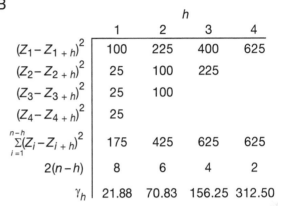

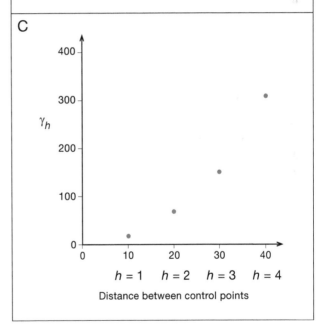

FIGURE 17.7 Computation of semivariance using equally spaced, hypothetical control points: (A) the equally spaced points, (B) semivariance computations, and (C) semivariogram.

A summary of these calculations for all values of h is shown in Figure 17.7B. You should compute the column for $h = 2$ to ensure that you understand the method of computation.

Real-world computation of semivariance requires two modifications to this approach. First, because control points are not normally equally spaced, a tolerance must

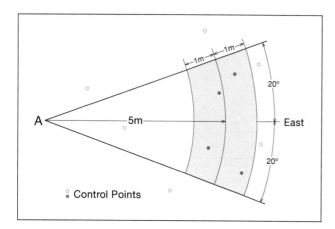

FIGURE 17.8 Determining which control points will be used in semivariance computations. A fixed distance yields no points directly 5 meters east of point A. A tolerance of 1 meter and 20°, however, yields four control points. (After Isaaks and Srivastava, 1989, 71.)

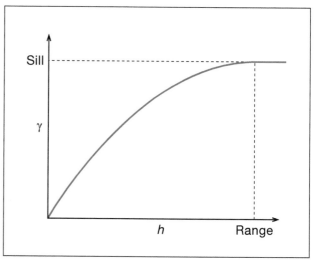

FIGURE 17.9 Idealized semivariogram illustrating a flattening in the semivariance values. The distance at which this occurs is known as the *range* and the associated semivariance value is the *sill*.

be allowed in both distance and direction between control points. For example, consider the control points shown in Figure 17.8. If we assume the distance between control points is 5 meters, note that no control point occurs exactly 5 meters east of point A. However, if we permit a distance tolerance of 1 meter and an angular tolerance of 20°, then four control points (the solid red ones) are east of point A. The second modification is to calculate semivariance in a variety of directions; thus, in addition to computing along the *x*-axis (east–west direction), computations should be made in north–south, northwest–southeast, and northeast–southwest directions. For now, we assume that such computations are combined to create a single semivariance value.

The **semivariogram** is a graphical expression of how semivariance changes as *h* increases. For example, Figure 17.7C is a semivariogram for the hypothetical data shown in Figure 17.7A. Clearly, as *h* increases (i.e., as control points become more distant), the semivariance also increases. This basic feature is characteristic of most data and should not surprise us: We expect nearby geographical data to be more similar than distant geographical data. The behavior of semivariograms with larger data sets is characterized by the idealized curve shown in Figure 17.9. Note that the semivariance increases as it did in Figure 17.7C but eventually reaches a plateau. The value for semivariance at which this occurs is known as the *sill*, and the distance at which it occurs is known as the *range*. The plateau normally indicates that the data are no longer similar to nearby values but rather that the semivariance has approached the variance in the entire data set.

When using the semivariogram in kriging, it is necessary to make an estimate of the semivariance for some arbitrary distance between points. For example, in Figure 17.7C, we might wish to estimate the semivariance for a distance of 26. This is normally accomplished by fitting a curve (or model) to the set of points comprising the semivariogram, a process known as *modeling the semivariogram*. Once an

equation for the model is determined, a value for distance can be inserted into the equation, and a value for semivariance computed. The simplest model is a straight line (linear) one; other common models include the spherical and exponential (Figure 17.10). For the snowfall data, we chose a *K-Bessel* model, which appeared to work reasonably well for both the raw and transformed data.

17.5.3.2 Kriging Computations

There are a wide variety of approaches used in kriging. At an introductory level, it is common to focus on **ordinary**

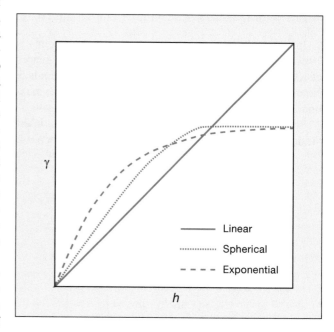

FIGURE 17.10 Models commonly used to summarize the points comprising the semivariogram: linear, spherical, and exponential. (After Olea 1994, 31.)

kriging, in which the mean of the data is assumed to be constant throughout geographic space (i.e., there is no trend or drift in the data). To simplify the discussion, we assume that only three control points are used to estimate a grid point; later, we will relax this constraint.

In a fashion similar to inverse-distance, kriging uses a weighted average to compute a value at a grid point. For three control points, the equation is

$$\hat{Z} = w_1 Z_1 + w_2 Z_2 + w_3 Z_3,$$

where:

$\hat{Z}$ = estimated value at a grid point

Z_1, Z_2, and Z_3 = data values at the control points

w_1, w_2, and w_3 = weights associated with each control point

The w_i weights are analogous to the $1/d$ values used in inverse-distance computations; the formula appears simpler for kriging because the weights are constrained to sum to 1.0.

In kriging, the weights are chosen to minimize the difference between the estimated value at a grid point and the true (or actual) value at that grid point. This is analogous to the situation in regression analysis in which we minimize the difference between the estimated value of a dependent attribute and its true value (see Section 3.6.3.2). In kriging, minimization is achieved by solving for the w_i in the following simultaneous equations:

$$w_1 \gamma\left(h_{11}\right) + w_2 \gamma\left(h_{12}\right) + w_3 \gamma\left(h_{13}\right) = \gamma\left(h_{1g}\right)$$

$$w_1 \gamma\left(h_{12}\right) + w_2 \gamma\left(h_{22}\right) + w_3 \gamma\left(h_{23}\right) = \gamma\left(h_{2g}\right)$$

$$w_1 \gamma\left(h_{13}\right) + w_2 \gamma\left(h_{23}\right) + w_3 \gamma\left(h_{33}\right) = \gamma\left(h_{3g}\right)$$

$$w_1 + w_2 + w_3 = 1.0,$$

where $\gamma\left(h_{ij}\right)$ = the semivariance associated with the distance between control points i and j (e.g., $\gamma\left(h_{12}\right)$ = the semivariance for control points 1 and 2), and $\gamma\left(h_{ig}\right)$ is the semivariance associated with the distance between the ith control point and a grid point (e.g., $\gamma\left(h_{1g}\right)$ = the semivariance for control 1 and a grid point) (Davis 1986, 385).[3] Note that these equations consider not only the distance from control points to the grid point (used in calculating $\gamma\left(h_{ig}\right)$) but also the distances between the control points themselves (used in calculating $\gamma\left(h_{ij}\right)$); contrast this approach with the inverse-distance method, which considers only distances between the grid point and the control points.

To illustrate the nature of these equations, assume that we are given three hypothetical control point values and wish to estimate a value for a grid point, as shown in Figure 17.11A. Figure 17.11B lists the X and Y coordinates for all points, and Figure 17.11C lists the distances between

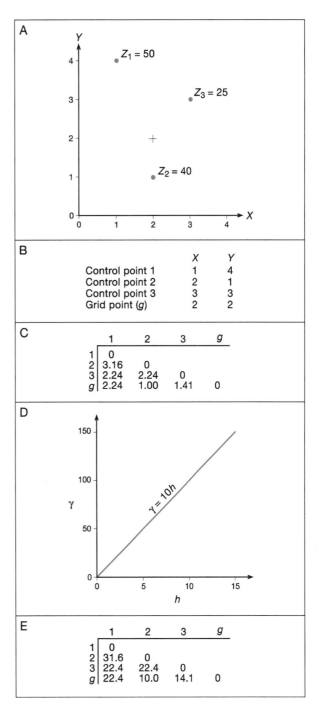

FIGURE 17.11 Semivariance computations for a grid point and three control points: (A) graph of the grid point and control points; (B) X and Y coordinates for the points; (C) distances between points; (D) linear model of a semivariogram for a data set from which the points were presumed to have been taken; (E) semivariances associated with distances between points. (After Davis 1986, 387.)

pairs of points. Also, assume that Figure 17.11D is an appropriate linear model for a semivariogram associated with a larger data set from which the three points were taken: The model $\left(\gamma = 10h\right)$ indicates that the semivariance is simply 10 times the distance (when h is 5, γ is 50).

Determining a semivariance for use in the simultaneous equations requires plugging distance values into the model. For example, for control points 1 and 2, the distance is 3.16, and the semivariance is 31.6. Repeating this process for each pair of points produces the semivariance values shown in Figure 17.11E. Inserting these values into the simultaneous equations, we have

$$w_1 0.00 + w_2 31.6 + w_3 22.4 = 22.4$$

$$w_1 31.6 + w_2 0.00 + w_3 22.4 = 10.0$$

$$w_1 22.4 + w_2 22.4 + w_3 0.00 = 14.1$$

$$w_1 + w_2 + w_3 = 1.0.$$

Davis (1986, 385–388) provided details of how these equations can be solved using matrix algebra. We did so using the statistical package SPSS and found the following values for w: $w_1 = 0.15$, $w_2 = 0.55$, and $w_3 = 0.30$. Inserting the values for w and the attribute values for the control points into the basic weighted-average formula for kriging, we have

$$\hat{Z} = 0.15(50) + 0.55(40) + 0.30(25) = 37.$$

Using kriging in real-world situations (e.g., with the Wisconsin snowfall data) differs from this simplified example in two important respects: (1) more than three control points normally are used to estimate each grid point and (2) more than one semivariogram might be appropriate. For the number of control points, a cartographer normally specifies a search strategy similar to that used for the inverse-distance method. For comparative purposes, we chose the same parameters as for the inverse-distance map: a total of 12 control points and a quadrant strategy; thus, there was a maximum of three points in each sector and a minimum of two control points within each sector.[4] The resulting map is shown in Figure 17.12C (for the raw data).

More than one semivariogram might be appropriate for a particular data set because semivariance can vary as a function of orientation (e.g., the semivariance for an east–west orientation may differ from that for a north–south orientation). With the Wisconsin snowfall data, it is reasonable to expect the semivariance values to be more similar as one moves east-west than north-south because snowfall is related to temperature, which is more similar east-west at this latitude. This notion is handled in ArcMap using an anisotropy option, which automatically accounts for directional influences in the computation of the semivariance.

Although kriging is admittedly more complex than many other methods of interpolation, it can produce a more accurate map; as such, it is often said to produce an **optimal interpolation**. It must be stressed that this is true only if one has properly specified the semivariogram(s) and associated semivariogram models. This is important to recognize because the software for kriging might provide simplified defaults (e.g., a single semivariogram and linear model) that

will not produce an optimal kriged map. Another advantage of kriging is that it provides a measure of the error associated with each estimate, known as the *standard error of the estimate*. This error measure can be used to establish a *confidence interval* for the true value at each grid location, assuming a normal distribution of errors. For example, if the kriged estimate at a grid point is 50 cm of rainfall and the corresponding standard error is 7 cm, then a 95 percent confidence interval for the true value at that grid point would be $50 \pm 2(7)$. This is equivalent to saying we are 95 percent certain that the true value is within the range 36–64.[5]

17.5.4 THIN-PLATE SPLINES

Thin-plate splines involve fitting a mathematical surface to the data such that the roughness of the surface is minimized. Depending on the options chosen for interpolation, the surface will either pass directly through the values for control points or be close to those values. The term thin-plate is suggestive of how you should attempt to visualize the process—imagine bending a thin plate of flexible material (such as rubber) and trying to distort it to fit the values of the control points in three dimensions. Note that this approach is quite different from the weighted average of surrounding control points used in inverse distance and kriging. Since surrounding points are not averaged, it is possible for the resulting surface to extend beyond the limits of the input data (i.e., to predict values outside the range of the data). We were especially interested in experimenting with thin-plate splines because the technique is commonly utilized with climatic data (e.g., Tatalovich et al. (2006), Snehmani et al. (2015), Plouffe et al. (2015)).

Since the mathematics of thin-plate splines are beyond the scope of this textbook (see Lloyd 2010 for some of the details of the mathematics), we will focus on selecting software parameters. The Spline routine within Spatial Analyst in ArcMap has two basic types of thin-plate splines: regularized and tension. The regularized type is considered appropriate for creating smooth, gradually changing surfaces, and an associated weight specifies the degree of smoothing. The least amount of smoothing occurs with a weight of 0, in which the surface will pass directly through values for control points. As the weight is increased, a smoother surface results, although the surface will no longer necessarily pass directly through the points. For the tension type of thin-plate spline, the opposite is the case, as higher values of the weight result in coarser (less smooth) surfaces.

Given the fact that our snowfall data is an average computed over a 30-year period, we expected the resulting surface would be relatively smooth, although we also expected there could be rapid changes in some locations (e.g., in the northern region where lake effect snows are common). Thus, we felt that we should explore both the regularized and tension types. We created interpolated maps using a variety of weight values and evaluated the maps using the cross-validation procedure described in the next section, in which true snowfall data values are compared with interpolated

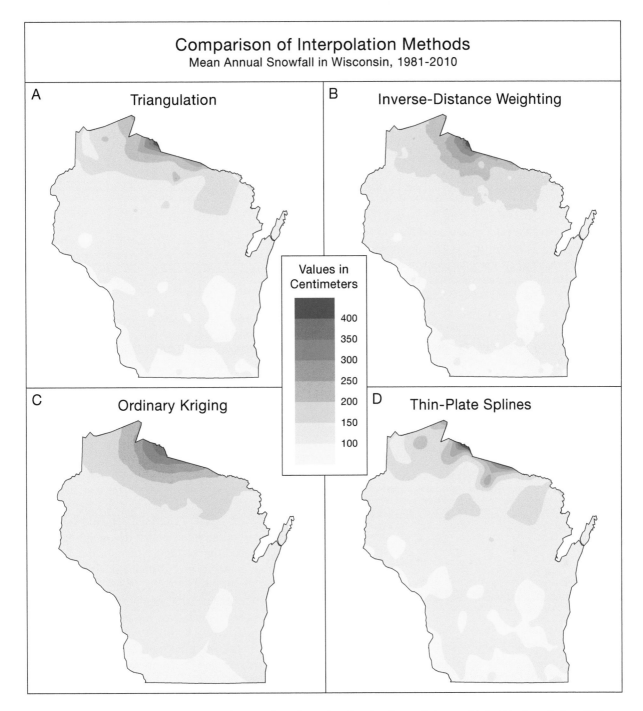

FIGURE 17.12   Isarithmic maps of the snowfall data for Wisconsin using four interpolation methods: (A) triangulation, (B) inverse-distance weighting, (C) ordinary kriging, and (D) thin-plate splines.
(Data source: Midwestern Regional Climate Center; https://mrcc.purdue.edu/.)

snowfall values at corresponding locations. Ultimately, we chose the map shown in Figure 17.12D, which was created using the regularized type with a weight of 0.

17.5.5   Choosing among the Interpolation Methods

Given the similarity of the maps resulting from applying the four methods of interpolation to the Wisconsin snowfall

data (Figure 17.12), how does one decide which method should be used? A common approach to choosing among a subset of interpolation methods is to conduct a **cross-validation** to see how well an interpolated map predicts the true values of observations. There are two basic approaches for cross-validation. One approach is termed **leave-one-out (LOO)**, in which we: (1) remove a control point from the data to be interpolated, (2) create an interpolated map using all other control points, (3) estimate the value of the attribute (snowfall in our case) at the location of the removed point, and (4) compute the **residual**, the difference between the known and estimated control point values. These steps are repeated for each control point in turn, and the resulting residuals are analyzed using the set of measures described below. An obvious advantage of this approach is that virtually all data are retained in the analysis each time that a residual is computed. A disadvantage is that there is greater computational time because the interpolation must be run *n* times, where *n* is the number of control points in the data set.

An alternative approach for cross-validation is **data splitting**, in which the original data set is split into two groups, one to create the isarithmic map and one to evaluate it. Attribute values (snowfall in our case) for points in the evaluation group are the known data values—these values are compared with estimated values on the interpolation surface at the longitude and latitude values corresponding to the known data values. The resulting residuals are then analyzed using the measures described below. One problem with data splitting is that it does not test the actual data set that will ultimately be used to create the interpolated map but rather a subset of it. Also, the results may vary as a function of which points are chosen for the two groups of data. We chose to utilize the data splitting approach, however, because the LOL approach was not available for two of the interpolation methods that we considered in ArcMap.

We computed three common measures of cross-validation for data splitting: the **mean bias error (MBE)**, the **root mean square error (RMSE)**, and the **mean absolute error (MAE)**:[6]

$$MBE = \frac{\sum_{i=1}^{m}\left(Z_i - \hat{Z}_i\right)}{m} \qquad RMSE = \sqrt{\frac{\sum_{i=1}^{m}\left(Z_i - \hat{Z}_i\right)^2}{m}}$$

$$MAE = \frac{\sum_{i=1}^{m}\left|Z_i - \hat{Z}_i\right|}{m},$$

where:

$\hat{Z}_i$ is the estimated value for the ith control point in the evaluation group

Z_i is the true value of the ith control point in the evaluation group

m is the number of control points in the evaluation group

As the name suggests, MBE measures whether there is a bias in the errors. Ideally, positive and negative residuals should cancel one another, and MBE should be close to 0. RMSE is a common measure of the total error for residuals. Note that each of the residuals $\left(Z_i - \hat{Z}_i\right)$ is squared in the formula for RMSE. This is problematic because large residuals will produce very large values when squared (Willmott and Matsuura 2006, 93). MAE is an alternative total error measure that computes the absolute value of the residuals and is thus not adversely impacted by large errors.

TABLE 17.1

Cross-Validation Error Measures for the Four Interpolation Methods for the Wisconsin Snowfall Data

	Raw Data	Transformed Data
MBE		
Triangulation	−5.98	−1.40
Inverse distance	−2.15	2.62
Ordinary kriging	−3.98	−0.19
Thin-plate splines	−4.37	−1.52
RMSE		
Triangulation	30.86	30.51
Inverse distance	31.77	28.97
Ordinary kriging	31.68	31.16
Thin-plate splines	39.47	39.33
MAE		
Triangulation	21.58	20.32
Inverse distance	22.03	20.27
Ordinary kriging	21.79	21.40
Thin-plate splines	27.24	26.00

Table 17.1 shows the results of cross-validation for each of the three error measures for the four interpolation methods for both the raw and transformed data. Comparing the raw and transformed data, we can see that the RMSE and MAE were lower in all cases for the transformed data. With the exception of inverse distance, the absolute value of the mean bias error was also smaller for the transformed data. Comparing the transformed results with one another, we can see that triangulation, inverse-distance, and ordinary kriging had similar RMSE and MAE values (with inverse-distance having the smallest errors) and that thin-plate splines clearly had the largest error (e.g., the MAE for thin-plate splines, 26.00, was approximately 21 percent more than the MAE for ordinary kriging, 21.40).

The maps showing the results of interpolation for the four methods in Figure 17.12 provide an interesting complement to the cross-validation. Here we have mapped the results of the transformed data for all but ordinary kriging. Although the cross-validation error for the transformed data was smaller for ordinary kriging, the resulting map did not reflect the rapidly changing snowfall in the extreme north (the two highest classes did not appear) and so we chose the interpolation based on the raw data (though even here the top class did not appear).

Another obvious difference among the maps is the smoothness of the contour lines between classes. Thin-plate splines appear to have the smoothest lines, whereas inverse-distance has more irregular contour lines. Given that our snowfall data are a thirty-year mean and that snowfall tends to be smooth and continuous, the smoother contour lines for thin-plate splines seem desirable. Unfortunately, the poorer performance of thin-plate splines on cross-validation is problematic. Triangulation, however, has relatively smooth contour lines and scored quite well on the cross-validation. The irregular contour lines of the inverse-distance method are problematic as they do not appear to reflect the smooth and continuous nature of the underlying phenomenon of snowfall. Inverse-distance also has a number of island-like areas, which it turns out often coincide with the location of control points—this is likely a function of the nature of the underlying computations and the fact that it is an exact interpolator.[7]

It is important to stress that we have only covered basic procedures for creating an interpolated map. In introducing the Wisconsin snowfall data, we noted two important characteristics: a general trend for increased snowfall as one moves north (with a notable jump in the extreme north) and the notion that snowfall likely increases as elevation increases. Trends in the data can be accounted for in **universal kriging**, and ancillary information (such as elevation) can be accounted for in **cokriging**.[8] We encourage you to explore these other approaches for interpolation.

In addition to conducting a cross-validation, we encourage you to review the literature to determine which approaches others have used to examine the particular phenomenon that you are interested in. In this respect, both the References at the end of the chapter and the Further Reading section for this chapter at www. routledge.com/9780367712709 contain several references that compare interpolation methods. Other factors that you should consider when selecting an interpolation method include the potential difficulty of understanding the technique, the amount of effort needed to select appropriate parameters for a technique, and the execution time (which can be an important consideration for large data sets).

17.6 TOBLER'S PYCNOPHYLACTIC INTERPOLATION

Although the interpolation methods discussed in Section 17.5 were developed to handle true point data, they also have been commonly used for conceptual point data (and thus for isopleth mapping). For example, we used ordinary kriging to create the isopleth map of wheat harvested in Kansas counties described in Chapter 4 (the map is repeated in Figure 17.14B). We accomplished this by assigning the percentage of wheat harvested to the centroids of each county, which served as control points for interpolation.

Waldo Tobler's (1979) **pycnophylactic** (or volume-preserving) **interpolation** is a more sophisticated approach for handling conceptual point data. To visualize this method, consider the standardized data associated with enumeration units as a clay model in which each enumeration unit is raised to a height proportional to the data (as in Figure 4.3B). The objective of the pycnophylactic method

is to "sculpt this surface until it is perfectly smooth, but without allowing any clay to move from one [enumeration unit] to another and without removing or adding any clay" (Tobler, 520).[9] Relating this concept to the other interpolation approaches we have considered, we can think of volume preservation as a form of *honoring* the data associated with each *enumeration unit*.

To illustrate the pycnophylactic method in more detail, we use a simplified algorithm developed by Nina Lam (1983, 148–149). This algorithm illustrates the pycnophylactic character of Tobler's method but not the smoothness constraints that Tobler specified; for the latter, see his original paper. To illustrate Lam's algorithm, presume that we are given raw counts $\left(RC_i\right)$ for the three hypothetical enumeration units shown in Figure 17.13A. These raw counts might be the number of people or the acres of wheat harvested in each enumeration unit. In step 1 of Lam's algorithm, a set of square cells is overlaid on top of the enumeration units, and it is determined which cells fall in each enumeration unit (Figure 17.13B). A cell is considered part of an enumeration unit if its center falls within that unit—this can be determined by a so-called *point-in-polygon test*. Note that four cells fall within the enumeration unit E_1. In step 2 of the algorithm, a raw count for each cell is determined by dividing the raw count for each enumeration unit by the number of cells in that unit; for example, cells within enumeration unit 1 receive a value of 25/4, or 6.25 (Figure 17.13C). This step essentially standardizes the data by computing a density measure.

Steps 3–5 of Lam's algorithm are executed in an iterative fashion. The steps are as follows:

Step 3. Each cell is computed as the average of its nondiagonal neighbors. For example, cell (2,2) becomes

$$\frac{6.25 + 6.25 + 7.00 + 5.00}{4} = 6.13.$$

The results for all cells for the first iteration are shown in Figure 17.13D. This step accomplishes the smoothing portion of the algorithm.

Step 4. The cell counts within each enumeration unit at the end of step 3 (Figure 17.13D) are added to obtain a total smoothed count value, SC_i, for each enumeration unit. For example, the total for enumeration unit 1 is

$$SC_1 = 6.25 + 6.50 + 5.83 + 6.13 = 24.71.$$

The results for enumeration units 2 and 3 are 32.40 and 39.39, respectively.

Step 5. All cell values are multiplied by the ratio $\frac{RC_1}{SC_1}$. For cell (2,2), the result is

$$6.13 \times \left(\frac{RC_1}{SC_1}\right) = 6.13 \times \left(\frac{25}{24.71}\right) = 6.20.$$

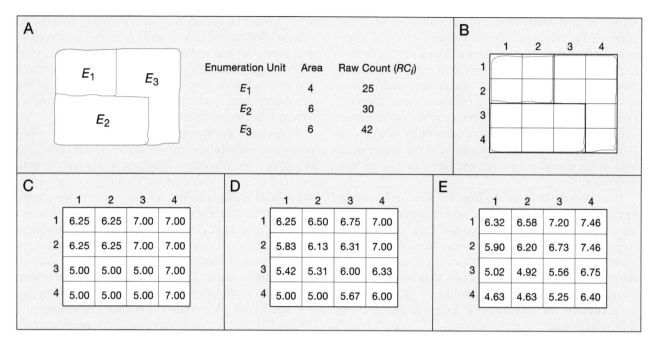

FIGURE 17.13 Basic steps of pycnophylactic (volume-preserving) contouring: (A) three hypothetical enumeration units; (B) square cells overlaid on the enumeration units; (C) initial density values for each cell (computed by dividing the raw count for each enumeration unit by the number of cells in that unit); (D) smoothed cell values (achieved by averaging neighboring cells); (E) smoothed values adjusted so that the sum within an enumeration unit equals the original total sum for that enumeration unit (the volume is preserved). (After Lam 1983, 148–149.)

The results for all cells for the first iteration are shown in Figure 17.13E.

Note that if the counts in each cell of an enumeration unit in Figure 17.13E were added, the resulting sum would be equal to the original raw count for that unit; for example, for enumeration unit 1, we have

$$6.32 + 6.58 + 5.90 + 6.20 = 25.00.$$

As a result, steps 4 and 5 enforce the pycnophylactic (volume-preserving) constraint. Remember that steps 3 to 5 are executed in an iterative fashion. For the second iteration, the results shown in E would be placed where C currently is depicted in Figure 17.13, and steps 3 to 5 would be executed again. Iteration continues until there is no significant difference between the raw and smooth counts for each enumeration unit or there is no significant change in the cell values compared with the last iteration.

One issue stressed by Tobler and not dealt with in this algorithm is how the boundary of the study area is handled. His computer program for pycnophylactic interpolation, PYCNO, provided two options: one in which zeros are presumed to occur outside the bounds of the region and one in which a constant gradient of change is presumed across the boundary. The former would be appropriate if the region is surrounded by water, as when mapping population along the coasts of the United States. The latter would be appropriate when mapping a phenomenon that is presumed to have similar characteristics in surrounding enumeration units, as when contouring wheat harvest data for the state of Kansas.

To contrast pycnophylactic interpolation with methods intended for true point data, Figure 17.14 portrays contour maps of the percentage of wheat harvested in counties of Kansas, using both the pycnophylactic and ordinary kriging methods. The maps obviously look different from one another, largely because of the emphasis that the pycnophylactic method places on honoring the data for enumeration units. To illustrate this, consider the southcentral portion of the map where there is a contour line with a value of 45 on the pycnophylactic map, but the highest value for a contour line on the kriged map is only 40. Percentage harvested values for the two counties (Harper and Sumner) through which the 45 contour line passes are 43.9 and 43.5, respectively. To show a smooth transition to the lower-valued surrounding counties, Harper and Sumner's edges must be beveled off, but to retain the same volume (remember the notion of a clay model), their center portions must be built up, thus resulting in a value above 45 in their interior. In contrast, the kriging method we used cannot produce a value higher than any control point because each grid point is a weighted average of surrounding control points.[10]

Although the pycnophylactic method is arguably more appropriate for isopleth mapping than point-based interpolation methods, the method should be used with caution if the data are not truly continuous. For example, Jie Lin and colleagues (2017) show how the pycnophylactic method can be modified to map population density when there are known areas that have no population, and so ancillary information should be utilized.

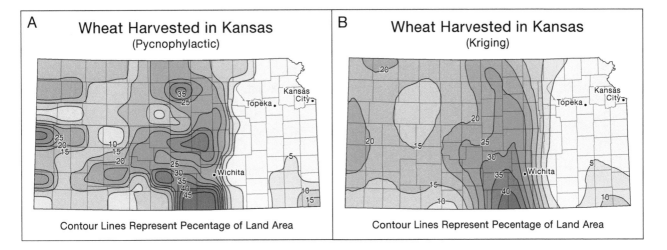

FIGURE 17.14 Comparison of interpolation methods for data collected for enumeration units: (A) the pycnophylactic approach, which expressly deals with the fact that the data were collected from enumeration units; (B) the ordinary kriging approach, which treats the areal data as conceptual point locations.

17.7 SYMBOLIZATION

The last step in isarithmic mapping is to symbolize the interpolated data. When mapping Earth's topography, a variety of specialized symbolization techniques are possible—we consider these in Chapter 23. In this chapter, we consider symbolization approaches that are appropriate for either topographic or nontopographic phenomena.

17.7.1 SOME BASIC SYMBOLIZATION APPROACHES

A range of basic symbolization approaches is shown in Figure 17.15. **Contour lines** (Figure 17.15A) frequently have been used to depict smooth, continuous phenomena, particularly when maps were produced manually because little production effort was involved. An obvious problem with contours, however, is that visualizing the 3-D surface requires careful examination of the position and values of individual contour lines—that is, the surface does not simply pop out at you.

The application of a color scheme to the areas between contour lines (Figure 17.15B) clearly enhances the ability to visualize a 3-D surface because light and dark colors can be associated with low and high values, respectively. The color schemes that we discussed in Chapter 15 for choropleth maps can also be utilized on isarithmic maps. Note, however, that a broader range of color schemes is possible on isarithmic maps because symbols of similar value must occur adjacent to one another (i.e., because the order of symbols on the map matches the order in the legend, the legend needs to be consulted less frequently). Still, we recommend caution when using spectral (rainbow) color schemes (such as those often used on TV weather maps), as they will be problematic for the color deficient and the lightnesses of the colors do not logically relate to the numeric data.[11]

One problem with applying a color scheme to the areas between contour lines is that the limited number of colors (or classes) suggests a stepped surface rather than the smooth one that occurs in reality. This problem can be ameliorated by creating a **continuous-tone map** (Kumler and Groop 1990), in which each point on the surface is colored proportional to the value of the surface at that point (Figure 17.15D). This approach is analogous to unclassed choropleth mapping (see Section 15.6), in which enumeration units are colored proportional to the data value in that unit. One problem with interpreting a continuous-tone map is that it is difficult to associate numbers in the legend with particular locations, but this problem can be solved by overlaying continuous tones with traditional contour lines (Figure 17.15E).

Another approach that assists in interpreting smooth, continuous phenomena is to utilize the perspective height visual variable (for example, in Figure 17.15C, we use a **fishnet map**). With this approach, not only does the surface change gradually, but we can also actually "see" that certain points are higher or lower than others. Although a perspective height symbolization is useful in isarithmic mapping, it has the same disadvantages as a prism map (discussed in Section 4.7), including the blockage of low points by higher ones and the fact that rotation might produce a view unfamiliar to readers who normally see maps with north at the top.

When choosing among the various symbolization methods shown in Figure 17.15, it is useful to consider a study by Mark Kumler and Richard Groop (1990) that compared classed and continuous-tone isarithmic maps. People were asked to complete the following tasks while looking at maps: "locat[e] surface extrema, interpret ... slope directions between points on the surface, estimat[e] relative surface values, and estimat[e] absolute values at points on the surface" (282). Based on the accuracy with which people completed these tasks, Kumler and Groop found that continuous-tone maps were more effective than classed maps. Also, when asked to pick their favorite approach, the

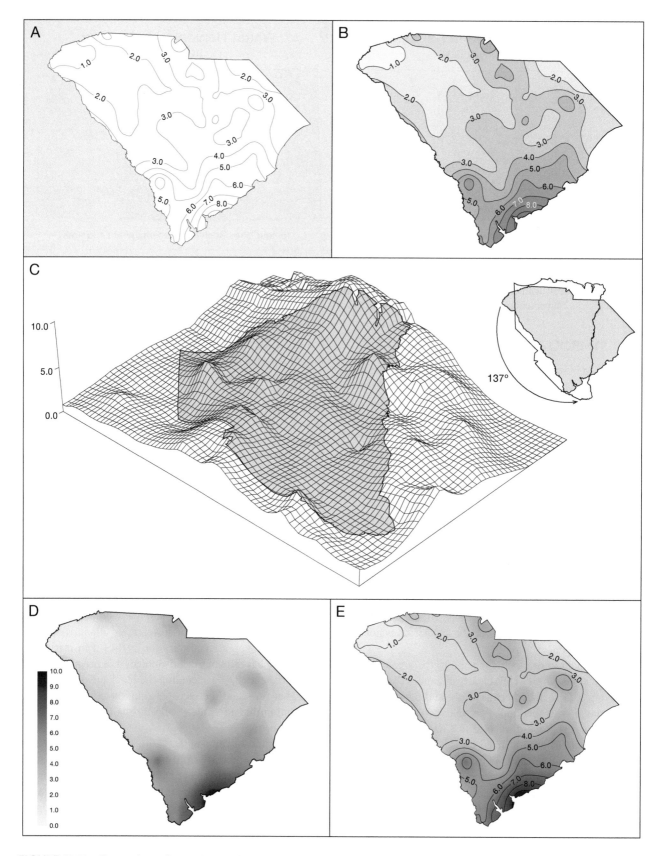

FIGURE 17.15 Comparison of several methods for symbolizing smooth, continuous phenomena (in this case, precipitation associated with Hurricane Hugo in South Carolina in 1989): (A) using only contour lines; (B) combining contour lines and hypsometric tints; (C) a fishnet map; (D) continuous tones analogous to those used on unclassed choropleth maps; (E) continuous tones combined with contour lines. (Data Source: Southeast Regional Climate Center; https://sercc.com/.)

majority of people selected one of the continuous-tone maps (a full-spectral scheme). Although Kumler and Groop's study suggests that continuous-tone maps are effective, the study was limited because it did not include classed color maps (only black-and-white classed maps were used). Thus, an interesting follow-up to their study would be a comparison of continuous-tone and classed color maps.

The 3-D appearance of the fishnet map in Figure 17.15C can also be achieved by using stereo pairs and anaglyphs, both of which permit stereo views. With **stereo pairs**, two maps of the surface are viewed with a stereoscope, which you might be familiar with from examining 3-D views of aerial photographs. We will examine the use of stereoscopy in the realm of virtual environments in Chapter 28. With **anaglyphs**, two images are created, one in green or blue and the other in red; these images are viewed with anaglyphic glasses, which use colored lenses to produce a 3-D view. In their book *Infinite Perspectives: Two Thousand Years of Three-Dimensional Mapmaking*, Brian and Jeffrey Ambroziak (1999) described a specialized anaglyph approach—the Ambroziak Infinite Perspective Projection (AIPP). In contrast to traditional anaglyphs, in which the image is viewed from only a single vantage point, AIPP permits viewing from multiple positions. Additionally, AIPP is "self-scaling" (that is, vertical exaggeration increases as one moves away from the image), permits features that both rise above and descend below the image plane, and permits the inclusion of additional data that are not part of the main anaglyph image (Ambroziak and Ambroziak 1999, 86–87). Images viewed using the AIPP produce a striking 3-D appearance—to see this, we encourage you to examine the images found in the Ambroziaks' book.

17.7.2 Color Stereoscopic Effect

The **color stereoscopic effect** refers to the notion that colors from the long wavelength portion of the electromagnetic spectrum (e.g., red) appear slightly nearer to the viewer than colors from the short wavelength portion (e.g., blue), primarily because the lens of the eye refracts light as a function of wavelength (Travis 1991, 135–139). This effect can be used as an argument for utilizing a spectral (rainbow) color scheme (red, orange, yellow, green, blue, indigo, and violet). J. Ronald Eyton discussed several approaches for enhancing the color stereoscopic effect. In his early work, Eyton (1990) focused on how spectral colors should be generated and how they might be combined with contours and hill shading to enhance the color stereoscopic effect. He indicated that, ideally, the subtractive CMYK process should not be used to create spectral colors because the process is not purely subtractive, as "subtractive ink dots … adjacent to each other produce desaturated colors that are formed additively" (21). As one solution to this problem, Eyton created maps on a computer graphics display and then produced hard copies by recording the images onto film. As another solution, he printed maps using **fluorescent inks**, which produced "brilliant, intense color" (23–24). Maps

resulting from his use of fluorescent ink can be found in Plates 1 to 4 of his paper.

In spite of taking these careful approaches, Eyton found little color stereoscopic effect when spectral colors were used alone. Eyton did find, however, that when contour lines were added, users perceived a distinct color stereoscopic effect. Although Eyton could find no explanation for this enhancement in the literature, he hypothesized that the rapid change in slope associated with closer contours might provide depth cues.

Eyton achieved a particularly dramatic color stereoscopic effect when he combined spectral colors with a hill-shaded display. (**Hill shading** or **shaded relief** is a process in which terrain is shaded as a function of its orientation and slope relative to a presumed light source, typically from the northwest; see Chapter 23 for more details.) Eyton indicated that many viewers found a combination of spectral colors, contours, and hill shading to be most effective for enhancing the color stereoscopic effect.

In later work, Eyton (1994) described how the color stereoscopic effect could be enhanced via special viewing glasses. Originally developed by Richard Steenblik (1987), these glasses are now available through https://chromatek.com/. The glasses can produce dramatic 3-D images when spectral colors are combined with contours or hill shading. We encourage you to acquire a set of glasses and examine Figure 17.16, which was created using hill shading and a spectral scheme developed by Eyton (2017, 37–42). We also encourage you to examine Kenneth Field's color stereoscopic map *Over the Edge: Death in the Grand Canyon*, which provides a visual representation of the more than 900 people who have lost their lives in the canyon (https://tinyurl.com/over-the-edge).

17.8 SUMMARY

In this chapter, we have focused on isarithmic maps, which are used to depict smooth, continuous phenomena. Two basic forms of data are utilized for isarithmic mapping: true point data and conceptual point data. True point data are actually measured at point locations (e.g., the hours of sunshine received over the course of the year), whereas conceptual point data are collected over areas (or volumes) but conceived as being located at points (e.g., birthrates for census tracts within an urban area). The terms isometric map and isopleth map are used to describe maps resulting from using true and conceptual point data, respectively.

A fundamental problem in isarithmic mapping is interpolating values between known control points (such as weather stations). We covered four interpolation methods appropriate for true point data: triangulation, inverse-distance weighting, ordinary kriging, and thin-plate splines. Triangulation fits a set of triangles to control points and then interpolates along the edges of these triangles in a fashion analogous to manual contouring. Triangulation is commonly used in association with DEMs as it is useful when the density of control points is variable. Inverse-distance

FIGURE 17.16   A combination of hill shading and a spectral color scheme developed by Eyton (2017) are applied to Crater Lake, Oregon. When viewed with special glasses (http://www.chromatek.com/), the color stereoscopic effect is exaggerated, producing a striking 3-D image. (Courtesy of J. Ronald Eyton.)

weighting estimates values on an equally spaced grid of points as an inverse function of distance from control points to a grid point (thus, nearby control points are weighted more than distant control points). Inverse-distance weighting is computationally efficient but can lead to a bulls-eye appearance.

Ordinary kriging estimates values at grid points by considering not only the distance to control points but also the distances between control points using the semivariance (which provides information on how spatial data varies over distance) and the semivariogram (a graphical expression of the semivariogram). Kriging approaches are sometimes referred to as optimal interpolators because they minimize the difference between observed and predicted data values. However, they are optimal only if an appropriate model has been used to represent the semivariogram. Thin-plate splines fit a mathematical surface to the control point data such that the roughness of the surface is minimized. A distinct advantage of thin-plate splines is that they produce a smooth representation of a phenomenon, which is arguably appropriate as isarithmic maps are intended to represent smooth, continuous phenomena.

The effectiveness of interpolation methods can be evaluated using cross-validation procedures in which predicted values on an interpolated surface are compared with known data values. Cross-validation can be accomplished by either leave-one-out (LOO) or data splitting. In LOO, a known observation is removed from the data set, the data are interpolated, and a predicted value based on the interpolation is compared with the known data value. This process is repeated for each value in the data set. In data splitting, the complete data set is split into two parts, one to produce an interpolation and a second to evaluate its effectiveness. An advantage of LOO is that it uses essentially the entire data set to evaluate the effectiveness of interpolations. In addition to conducting cross-validation, it is important to compare maps of the interpolated data to see whether the maps appear to properly represent the underlying phenomena (e.g., is the smooth character of the phenomenon properly reflected in the resulting map?).

We considered one method for handling conceptual point data: Tobler's pycnophylactic interpolation. The pycnophylactic method begins by assuming that each enumeration unit is raised to a height proportional to the value of its associated control point. This 3-D surface is then gradually smoothed, keeping the volume within each individual enumeration unit constant; the smoothing is accomplished using a cell-based generalization process analogous to procedures used in image processing. Although pycnophylactic interpolation is appropriate for conceptual point data (and thus isopleth maps), it should be used only if the assumption of a smooth, continuous phenomenon seems reasonable.

Methods for symbolizing isarithmic maps include contour lines, the application of color schemes to areas between the contour lines, continuous-tone maps (a form of unclassed map), and the use of the perspective height visual variable (such as a fishnet map). An intriguing possibility is to view a spectral color scheme with special glasses that enhance the color stereoscopic effect (in which long-wavelength colors will appear to be nearer than short-wavelength colors). In Chapter 23, we will look at additional symbolization methods that are especially appropriate for topography.

17.9 STUDY QUESTIONS

1. For each of the following data sets, indicate whether the data are true or conceptual and whether the data would be appropriate for isarithmic mapping: (A) rainfall recorded by volunteer weather observers as a hurricane passes over the state of Florida; (B) the median value of homes in each census tract of Dallas, Texas; (C) the depth to the bottom of Lake Superior recorded at 1,000-meter intervals by a research vessel as it traversed the lake; and (D) population for each country of the world.

2. There are a variety of ways of connecting control points to form a set of triangles. Why is the Delaunay triangulation considered to be the most desirable approach?

3. Define the term "exact interpolator" and explain whether thin-plate splines are exact interpolators.

4. Compute a value for inverse-distance weighting for $\hat{Z}$ for the data shown in Figure 17.5 assuming that distance is raised to a power of 2.

5. Presume that you are given precipitation values at weather station locations for 2020 for the entire continent of Asia, that you create an inverse-distance weighted map using the inverse-distance formula for $\hat{Z}$ given in the textbook, and that you use the simple (no-sector) approach with 12 control points. Discuss any problems that you see with this approach.

6. Using the formula for semivariance, show how to compute a value for semivariance for the data in Figure 17.7A when $h = 2$.

7. Using the formula for $\hat{Z}$ for ordinary kriging, explain whether this technique can produce an estimate beyond the range of the data.

8. Discuss two reasons why ordinary kriging might be favored over the inverse-distance weighting approach.

9. What is a distinct advantage of thin-plate splines when compared to the other interpolation methods covered in this chapter?

10. Assume that you have measured nitrogen dioxide levels at 50 locations within an urban area. You use 45 locations to create an isarithmic map and 5 locations to cross validate. (A) Given that the following are known and estimated values for nitrogen dioxide (in micrograms per cubic meter), compute the mean bias error (MBE) and the MAE (mean absolute error) for a cross-validation.

z	80	110	190	250	400
$\hat{Z}$	85	92	175	290	200

11. Explain why a wider range of color schemes can be utilized on classed isarithmic maps than on choropleth maps.

12. Describe the "color stereoscopic effect" and explain why specialized viewing glasses are necessary to truly appreciate the effect.

NOTES

1 Other software commonly used for interpolation include Surfer (https://www.goldensoftware.com/products/surfer), R (e.g., https://rspatial.org/raster/analysis/4-interpolation.html), Global Mapper (https://www.bluemarblegeo.com/products/global-mapper.php), and ANUSPLIN (https://fennerschool.anu.edu.au/research/products/anusplin for thin-plate splines).

2 Similar functions are available within ArcGIS Pro.

3 We reference Davis's (1986) second edition here because we feel it is more appropriate for the introductory student. Please be aware, however, that more recent information on kriging is available in Davis's (2002) third edition.

4 These are similar to parameters recommended by Isaaks and Srivastava (1989, Chapter 14).

5 For a more sophisticated approach to providing a measure of error associated with kriging, see Chainey and Stuart (1998).

6 The terminology for these three terms is taken from Willmott and Matsuura (2006), who provide more complete formulas in which the size of each grid cell can be utilized. The latter approach is critical for global data sets.

7 When the contour lines of an interpolated map appear irregular (as in the inverse-distance case here), you should consider using a smoothing operator to remove some of this irregularity. For instance, in ArcMap, you could use Smooth Line in the Spatial Analyst.

8 For a discussion of universal kriging, see Davis (2002, 428–437). A variety of approaches have been developed for handling ancillary information; for example, see Goovaerts (2000) and Meng et al. (2013).

9 See http://www.ncgia.ucsb.edu/pubs/gdp/pop/pycno.html for a visual representation of pycnophylactic interpolation.

10 Universal kriging can estimate values beyond the range of the data, but it would not necessarily honor the data associated with the enumeration unit.

11 For a discussion of the use of color on weather maps, see Monmonier (1999).

REFERENCES

Ambroziak, B. M., and Ambroziak, J. R. (1999) *Infinite Perspectives: Two Thousand Years of Three-Dimensional Mapmaking*. New York: Princeton Architectural Press.

Chainey, S., and Stuart, N. (1998) "Stochastic simulation: An alternative interpolation technique for digital geographic information." In *Innovations in GIS 5*, ed. by S. Carver, pp. 3–24. London: Taylor & Francis.

Cressie, N. (1990) "The origins of kriging." *Mathematical Geology 22*, no. 3:239–252.

Davis, J. C. (1986) *Statistics and Data Analysis in Geology* (2nd ed.). New York: Wiley.

Davis, J. C. (2002) *Statistics and Data Analysis in Geology* (3rd ed.). New York: Wiley.

Eyton, J. R. (1990) "Color stereoscopic effect cartography." *Cartographica 27*, no. 1:20–29.

Eyton, J. R. (1994) "Chromostereoscopic maps." *Cartouche* Autumn/Winter Special Issue:*15*.

Eyton, J. R. (2017) *3D Globes from G.Projector*. https://www.blurb.com/b/8091137-3d-globes-from-g-projector.

Goovaerts, P. (2000) "Geostatistical approaches for incorporating elevation into the spatial interpolation of rainfall." *Journal of Hydrology 228*, no. 1/2:113–129.

Isaaks, E., and Srivastava, R. M. (1989) *An Introduction to Applied Geostatistics*. New York: Oxford University Press.

Kumler, M. P., and Groop, R. E. (1990) "Continuous-tone mapping of smooth surfaces." *Cartography and Geographic Information Systems 17*, no. 4:279–289.

Lam, N. S.-N. (1983) "Spatial interpolation methods: A review." *Cartography and Geographic Information Systems 10*, no. 2:129–149.

Lin, J., Hannik, D. M., and Cromley, R. G. (2017) "A cartographic modeling approach to isopleth mapping." *International Journal of Geographical Information Science 31*, no. 5:849–866.

Lloyd, C. D. (2010) *Spatial Data Analysis: An Introduction for GIS Users*. New York: Oxford University Press.

Lyon, S. W., Sorensen, R., Stendahl, J., et al. (2010) "Using landscape characteristics to define an adjusted distance metric for improving kriging interpolations." *International Journal of Geographical Information Science 24*, no. 5:723–740.

Meng, Q., Liu, Z., and Borders, B. E. (2013) "Assessment of regression kriging for spatial interpolation—Comparisons of seven GIS interpolation methods." *Cartography and Geographic Information Science 40*, no. 1:28–39.

Monmonier, M. (1999) *Air Apparent: How Meteorologists Learned to Map, Predict, and Dramatize Weather*. Chicago: University of Chicago Press.

Olea, R. A. (1994) "Fundamentals of semivariogram estimation, modeling, and usage." In *Stochastic Modeling and Geostatistics*, ed. by J. M. Yarus and R. L. Chambers, pp. 27–35. Tulsa, OK: American Association of Petroleum Geologists.

Plouffe, C. C. F., Robertson, C., and Chandrapala, L. (2015) "Comparing interpolation techniques for monthly rainfall mapping using multiple evaluation criteria and auxiliary data sources: A case study of Sri Lanka." *Environmental Modelling & Software 67*:57–71.

Snehmani, A. B., Singh, M. K., Gupta, R. D. et al. (2015) "Modelling the hypsometric seasonal snow cover using meteorological parameters." *Journal of Spatial Science 60*, no. 1:51–64.

Steenblik, R. A. (1987) "The chromostereoscopic process: A novel single image stereoscopic process." *True Three-Dimensional Imaging Techniques and Display Technologies. Proceedings, SPIE 761*:27–34.

Tatalovich, Z., Wilson, J. P., and Cockburn, M. (2006) "A comparison of Thiessen polygon, kriging, and spline models." *Cartography and Geographic Information Science 33*, no. 3:217–231.

Tobler, W. R. (1979) "Smooth pycnophylactic interpolation for geographical regions." *Journal of the American Statistical Association 74*, no. 367:519–536.

Travis, D. (1991) *Effective Color Displays: Theory and Practice*. London: Academic.

Watson, D. F. (1992) *Contouring: A Guide to the Analysis and Display of Spatial Data*. Oxford, England: Pergamon.

Webster, R., and Oliver, M. A. (2007) *Geostatistics for Environmental Scientists*. (2nd ed.). Chichester: Wiley.

Willmott, C. J., and Matsuura, K. (2006) "On the use of dimensioned measures of error to evaluate the performance of spatial interpolators." *International Journal of Geographical Information Science 20*, no. 1:89–102.

18 Proportional Symbol Mapping

18.1 INTRODUCTION

The **proportional symbol map** is used to represent numerical data associated with point locations, whether those be true point locations (such as oil wells) or conceptual point locations (such as the centroids of countries when mapping global data). Circles are the most commonly used proportional symbol, although a wide variety of geometric and pictographic symbols are possible. *Unclassed* proportional symbol maps are common (in contrast to the classed maps common in choropleth mapping), in part because prior to the development of digital cartography, approaches were available to create a wide variety of symbol sizes for geometric symbols. A key challenge that you will face is deciding whether unclassed symbols should be mathematically scaled to reflect the actual relations in the data or whether the symbols should be perceptually scaled to reflect the tendency of readers to underestimate the size of larger symbols. Classed proportional symbol maps (also known as **range-graded** or **graduated symbol maps**) do not reflect the actual data relations but rather use a set of symbols that clearly differentiate each class from one another; as such classed maps can enhance the perception of spatial pattern. Other challenges that we will cover in proportional symbol mapping include determining whether it is necessary to standardize data, designing an appropriate legend, and handling symbol overlap.[1]

One purpose of this book is to introduce novel approaches that broaden our thinking about potential symbolization options. The **necklace map** is one such novel approach that eliminates the problem of symbol overlap by displaying proportional symbols on a one-dimensional curve (the necklace) that surrounds the mapped region of interest. Although the necklace map complicates the perception of spatial pattern, we will see that it can be used to show additional information (e.g., flows between symbols).

18.2 LEARNING OBJECTIVES

- Decide whether data for a proportional symbol map need to be standardized.
- Decide whether abstract geometric symbols (e.g., circles) or mnemonic pictographic symbols (e.g., oil derricks) should be used.
- Determine whether a classed or an unclassed proportional symbol map is appropriate.
- Utilize computational formulas to determine the size of unclassed proportional symbols relative to one another.

- If a classed proportional symbol map is desired, specify an appropriate set of range graded (graduated) symbols.
- Design legends appropriate for proportional symbol maps.
- Determine an appropriate amount of overlap for a proportional symbol map and select symbols that enhance figure-ground contrast when overlap occurs.
- Avoid the problem of symbol overlap entirely by creating a necklace map.

18.3 SELECTING APPROPRIATE DATA

Proportional symbol maps can be used to depict two forms of point data: true and conceptual. **True point data** are actually measured at point locations; examples include the number of calls made from telephone booths and temperatures at weather stations. Although we generally use isarithmic maps to display continuous phenomena such as temperature (see Chapter 17), we can use proportional symbols to focus on the raw data as they are collected at point locations. **Conceptual point data** are collected over areas (or volumes), but the data are conceived as being located at points for purposes of symbolization. An example is the number of microbreweries and brewpubs in each state, where the center of each state is the conceptual point location (Figure 18.1A). Some data are not easily classified as either true or conceptual. For example, data associated with cities are collected over the areal extent of each city, but the data normally are treated as occurring at point locations because, at typical mapping scales, cities are depicted as points. An example is the number of out-of-wedlock births to teenagers in major U.S. cities (Figure 18.2A).

The concept of data standardization discussed for choropleth maps (see Chapters 4 and 15) is also applicable to proportional symbol maps. As an example, consider the data for out-of-wedlock births. The unstandardized map (Figure 18.2A) is useful for showing the sheer magnitude of out-of-wedlock births, but care must be taken in interpreting any spatial patterns on this map because cities with large populations are apt to have a large number of out-of-wedlock births. (Note the strong visual correlation between parts A and B of Figure 18.2)

One method of standardization is to compute the ratio of two raw-total attributes. In the case of the birth data, we can compute the ratio of out-of-wedlock births to the total number of births, obtaining the proportion of out-of-wedlock births in each city (this is shown in percentage form in Figure 18.2C). This map has a markedly different

DOI: 10.1201/9781003150527-20

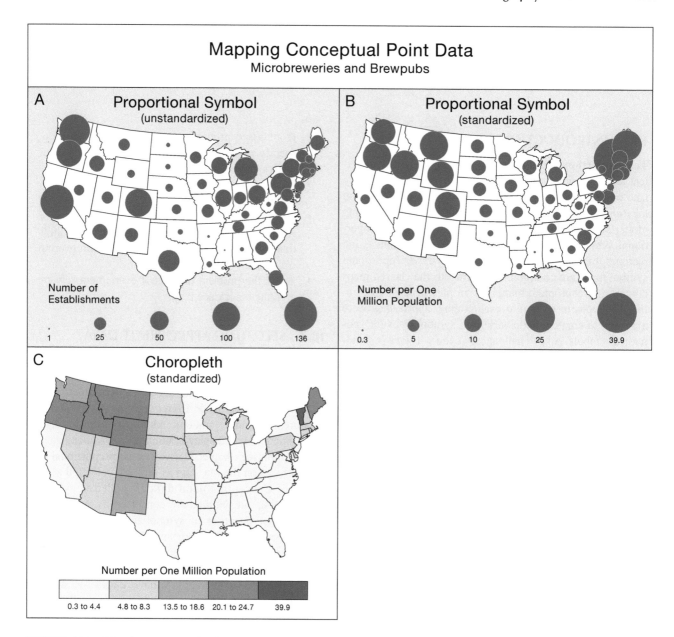

FIGURE 18.1 Mapping conceptual point data: (A) proportional circles represent the raw number of microbreweries and brewpubs in each state; (B) proportional circles represent standardized data—the number of microbreweries and brewpubs per 1 million people; (C) a choropleth map representing the standardized data shown in B. (For similar data on microbreweries and brewpubs, see https://www.brewersassociation.org/statistics-and-data/state-craft-beer-stats/.)

appearance than the unstandardized map, in large part because of a much narrower range of data: 3.3 to 22.7 percent, as opposed to 102 to 11,236 births. This map also illustrates the difficulty of using proportional circles when the data range is relatively narrow—here, the largest value is more than six times the smaller, but the map does not immediately suggest this. In Section 18.5.2, we discuss how formulas for circle sizes might be modified to handle this problem.

One can also argue that the microbrewery and brewpub data should be standardized, given that the number of microbreweries and brewpubs is likely to be greater in more populous states. In this case, a useful standardization is the number of microbreweries and brewpubs per 1 million population (Figure 18.1B). Although standardized conceptual point data can be represented with proportional symbols, a choropleth map is more commonly used (Figure 18.1C) because conceptual point data are associated with areas.

18.4 KINDS OF PROPORTIONAL SYMBOLS

Proportional symbols can be divided into two basic groups: geometric and pictographic. **Geometric symbols** (e.g., circles, squares, spheres, and cubes) generally do not mirror the phenomenon being mapped, whereas **pictographic symbols** (heads of wheat, caricatures of people,

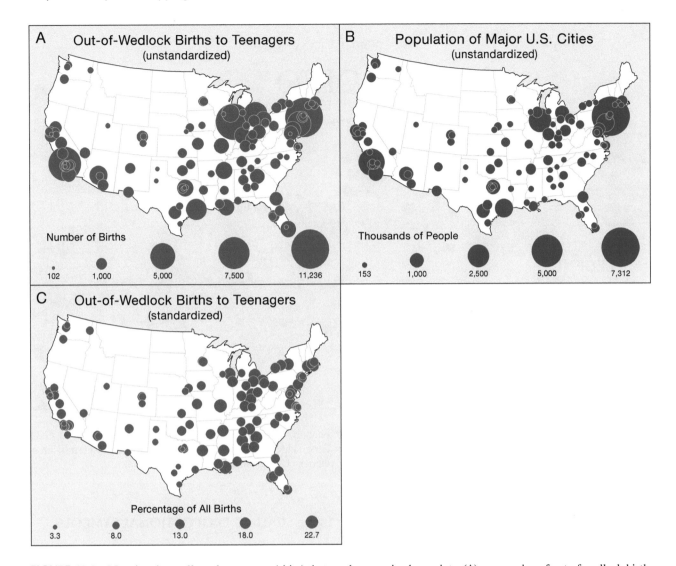

FIGURE 18.2 Mapping data collected over areas (cities) that can be conceived as points: (A) raw number of out-of-wedlock births for U.S. cities with a population of 150,000 or more; (B) the population of those cities (note that this map correlates highly with A); (C) a standardized map obtained by dividing the number of out-of-wedlock births by the total number of births. (Data source: National Center for Health Statistics, https://www.cdc.gov/nchs/data_access/Vitalstatsonline.htm.)

and diagrams of barns) do. Prior to the development of digital mapping, geometric symbols predominated because templates were readily available for manually constructing common geometric shapes. Today, digital mapping has eased the development of pictographic symbols: One can create the basic design for a symbol by hand, scan the symbol into a computer file, and then use design software to duplicate the symbol in various sizes. The map of beer mugs shown in Figure 4.6 was created using this approach. Alternatively, one can find numerous pictographic symbols already in digital form in **clip art** files.

The ease with which readers can associate pictographic symbols with the phenomenon being mapped, their eye-catching appeal, and their ease of construction in the digital realm suggest that these symbols may become more common in the future. Pictographic symbols, however, are not without problems. One problem is

that when symbols overlap, they may be more difficult to interpret than geometric symbols (compare the northeastern portions of Figures 4.6 and 18.1A). Another problem is that it may be more difficult to judge the relative sizes of irregular pictographic symbols (e.g., judging size relations among beer mugs is arguably more difficult than judging size relations among circles).

Circles have been the most frequently used geometric symbol. Arguments offered for using circles include the following:

1. Circles are visually stable.
2. Users prefer circles over other geometric symbols.
3. Circles (as opposed to, say, bars) conserve map space.

Traditionally, cartographers recommended against using 3-D geometric symbols (spheres, cubes, and prisms)

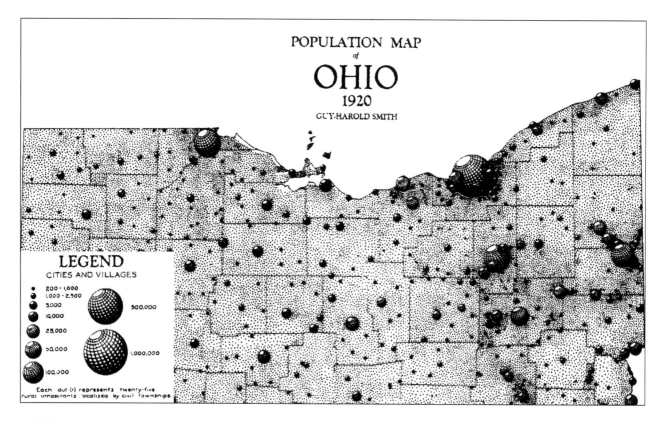

FIGURE 18.3 An eye-catching map created using 3-D geometric symbols. (From "A population map of Ohio for 1920" by G.-H. Smith, *The Geographical Review* 1928, copyright © the American Geographical Society of New York, reprinted by permission of Taylor & Francis Ltd, http://www.tandfonline.com on behalf of the American Geographical Society of New York.)

because of the difficulty of both estimating their size and constructing them. Like pictographic symbols, however, 3-D symbols can produce attractive, eye-catching graphics (Figure 18.3). Additionally, 3-D symbols can be useful for representing a large range in data; for example, in Figure 18.4, the small 3-D symbol is easily detected, whereas the corresponding two-dimensional symbol nearly disappears.

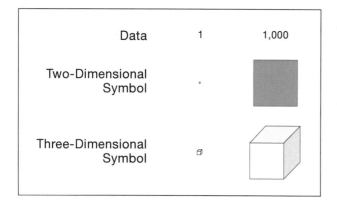

FIGURE 18.4 Attempts to portray a large range of data (note that the smallest 3-D symbol is more easily seen than the smallest two-dimensional symbol).

18.5   SCALING PROPORTIONAL SYMBOLS

18.5.1   Mathematical Scaling

Mathematical scaling sizes areas (or volumes) of point symbols in direct proportion to the data; thus, if a data value is 20 times another, the area (or volume) of a corresponding point symbol will be 20 times as large (Figure 18.5). We now consider some formulas for calculating symbol sizes; we deal with circles first because of their common use.

Remembering from basic math that the area of a circle is equal to πr^2, we can establish the relation

$$\frac{\pi r_i^2}{\pi r_L^2} = \frac{v_i}{v_L},$$

where:
r_i = radius of the circle to be drawn
r_L = radius of the largest circle on the map
v_i = data value for the circle to be drawn
v_L = data value for the largest circle

Note that this formula specifies circle areas in direct proportion to corresponding data values. The relation uses the largest radius (and largest data value) for one of the circles because proportional symbol maps are often constructed by

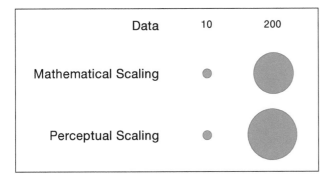

Data	10	200
Mathematical Scaling		
Perceptual Scaling		

FIGURE 18.5 Mathematical vs. perceptual scaling. In mathematical scaling, the areas of the circles are constructed directly proportional to the data; in perceptual scaling, the size of the larger circle is increased (using Flannery's exponent) to account for underestimation.

beginning with a largest symbol size to minimize the effect of symbol overlap.

Because the values of π cancel, this equation reduces to

$$\frac{r_i^2}{r_L^2} = \frac{v_i}{v_L}$$

Taking the square root of both sides, we have

$$\frac{r_i}{r_L} = \left(\frac{v_i}{v_L}\right)^{0.5}$$

Finally, solving for r_i, we compute

$$r_i = \left(\frac{v_i}{v_L}\right)^{0.5} \times r_L.$$

To apply this formula, consider how a radius for the circle representing Los Angeles, California, was computed for Figure 18.2A. After some experimentation, we decided that the largest circle (New York) should have a radius of 0.2125 inches (0.5398 cm). To determine the radius for Los Angeles, we inserted the number of out-of-wedlock births for Los Angeles and New York (8,507 and 11,236, respectively) into the formula, along with the largest radius as follows:

$$r_{\text{Los Angeles}} = \left(\frac{v_{\text{Los Angeles}}}{v_{\text{New York}}}\right)^{0.5} \times r_{\text{New York}}$$

$$r_{\text{Los Angeles}} = \left(\frac{8,507}{11,236}\right)^{0.5} \times 0.2125 = 0.1849.$$

Formulas for other geometric symbols can be derived in a similar fashion. The results are as follows:

$$\text{SQUARES: } s_i = \left(\frac{v_i}{v_L}\right)^{0.5} \times s_L,$$

where s_i is the length of a side of a square to be drawn and s_L is the length of a side of the largest square;

$$\text{BARS: } h_i = \left(\frac{v_i}{v_L}\right) \times h_L,$$

where h_i is the height of a bar to be drawn and h_L is the height of the tallest bar;

$$\text{SPHERES: } r_i = \left(\frac{v_i}{v_L}\right)^{\frac{1}{3}} \times r_L,$$

where r_i is the radius of a sphere to be drawn and r_L is the radius of the largest sphere;

$$\text{CUBES: } s_i = \left(\frac{v_i}{v_L}\right)^{\frac{1}{3}} \times s_L,$$

where s_i is the length of a side of a cube to be drawn and s_L is the length of a side of the largest cube.

In using these formulas, there are several issues that you need to consider. First, although these formulas are sometimes embedded in mapping software, you may have to enter them into spreadsheets along with the raw data. Second, because these formulas typically involve taking either the square or cube root of the ratio of data (i.e., raising the ratio to either the 1/2 or 1/3 power), mathematical scaling is sometimes referred to as *square root* or *cube root scaling*. We term the power to which the ratio of data is raised the *exponent for symbol scaling*.

Third, it is important to recognize that these formulas produce unclassed maps, as differing data values are depicted by differing symbol sizes. Classed or **range-graded maps** (see Section 18.5.3) are created by classing the data and letting a single symbol size represent a range of data values, but *unclassed* proportional symbol maps are more common. This might seem surprising given the frequency with which *classed* choropleth maps are used. The difference stems, in part, from the ease with which unclassed proportional symbol maps could be created in manual cartography (either an ink compass or a circle template could be used to draw circles of numerous sizes). Also, it is easier to differentiate proportional symbols from one another (especially when the range of symbols is large) than it is to differentiate shades on a choropleth map (e.g., the range of lightnesses for a hue is fixed).

Fourth, you will find that some software permits arbitrarily specifying the smallest and largest symbol sizes, with other symbols scaled proportionally between these symbols. For symbols based on area, the formulas are

$$z_i = \frac{v_i - v_S}{v_L - v_S}$$

$$A_i = z_i\left(A_L - A_S\right) + A_S,$$

where:

v_S and v_L = smallest and largest data values

z_i = a proportion of the data range associated with the data value v_i

A_S and A_L = smallest and largest areas desired

A_i = area of a symbol associated with the data value v_i

This approach produces an unclassed map, but the symbols are not scaled proportional to the data (i.e., a data value twice another does not have a symbol twice as large). An advantage of the approach, however, is that it can enhance the map pattern, just as do an arbitrary exponent and range grading, as described in the following sections.

18.5.2 PERCEPTUAL SCALING

Numerous studies have shown that the perceived size of proportional symbols does not correspond to their mathematical size; rather, people tend to underestimate the size of larger symbols. For example, in viewing the larger mathematically scaled circle in Figure 18.5, most people would estimate it to be less than 20 times as large as the smaller circle. If larger symbols are underestimated, it seems reasonable to assume that formulas for mathematical scaling might be modified (or "corrected") to account for underestimation; this process is known as **perceptual** (or psychological) **scaling**.

18.5.2.1 Formulas for Perceptual Scaling

To develop formulas for perceptual scaling, it is useful to consider how researchers have summarized the results of experiments dealing with perceived size. The relation between actual and perceived size typically has been stated as a *power function* of the form

$$R = cS^n,$$

where

R = response (or perceived size)

S = stimulus (or actual size)

c = a constant

n = an exponent

To differentiate the exponent n from the one for symbol scaling introduced earlier, we term it the **power function exponent**.

The power function exponent is the key to describing the results of experiments involving perceived size. If the size is estimated correctly, the exponent will be close to 1.0. Underestimation and overestimation are represented by exponents appreciably below and above 1.0, respectively. For example, an oft-cited study by James Flannery (1971) found that for circles, $R = (0.98365)S^{0.8747}$; here, the exponent of 0.8747 is indicative of underestimation.

The power function equation states what response arises from a certain stimulus. For constructing symbols, we need to know the reverse: what stimulus must be shown to get a certain response. Therefore, the power function must be transposed by dividing each side by c and then by raising each side to the $1/n$ power. The result is

$$S = c_1 R^{1/n},$$

where c_1 is a constant equal to $(1/c)^{1/n}$. In Flannery's study, the transpose was $S = (1.01902)R^{1.1432}$.[2] To simplify computations, the value of the constant (1.01902) in the transposed equation can be ignored because it is close to 1.0; thus, we have $S = R^{1.1432}$. Because this equation expresses the relation between the areas of circles, and because circles are constructed on the basis of radii, we need to take the square root of both sides, producing $S^{0.5} = R^{0.5716}$. Again for simplicity, the value 0.5716 has normally been rounded to 0.57. As a result, a perceptual scaling formula for circles is

$$r_i = \left(\frac{v_i}{v_L} \right)^{0.57} \times r_L.$$

The result of using an exponent of 0.57 in the circle scaling formulas can be seen for the perceptually scaled circles in Figure 18.5; for most readers, the larger circle should now appear closer to 20 times larger than the smaller circle.

The magnitude of the power function exponent varies depending on the symbol type. For squares, Paul Crawford (1973) derived an exponent of 0.93 (which yields an exponent of 0.54 for symbol scaling), indicating that squares are estimated better than circles. For bars, Flannery found that underestimates were balanced by overestimates and thus recommended no corrective formula. Although these results suggest that squares and bars should be used when precise estimates are desired, we recommend using squares and bars with caution. A study by Slocum and colleagues (2004) revealed that squares were not aesthetically pleasing (when compared with several geometric and pictographic symbols) and that bars were difficult to associate with a point location.

Power function exponents for 3-D symbols generally have been appreciably lower than for two-dimensional symbols, indicating severe underestimation. For example, for spheres and cubes, Ekman and Junge (1961) derived power function exponents of 0.75 and 0.74 (corresponding to exponents of 0.44 and 0.45 for symbol scaling). Interestingly, research by Ekman and Junge also indicated that when truly 3-D cubes were used (they were "made of steel with surfaces polished to a homogeneous, dull silvery appearance," 2), the exponent was 1.0. This result suggests that the manner in which 3-D symbols are portrayed might have an impact on the exponent. For example, if an interactive graphics program gives the impression of traveling through 3-D space, we might expect an exponent closer to 1.0. In more recent research dealing with actual physical representations of spheres (as might be created by a 3-D printer), Yvonne Jansen and Kasper Hornbæk (2016) found that the surface area of spheres (as opposed to their volumes) approximated the perceived size of the spheres.

18.5.2.2 Problems in Applying the Formulas

Unfortunately, there are numerous problems in applying formulas for perceptual scaling: (1) the value of a power function exponent can be affected by experimental factors; (2) using a single exponent might be unrealistic because of the variation that exists between subjects and within an individual subject's responses; (3) the formulas might fail to account for the spatial context within which symbols are estimated; and (4) experimental studies for deriving exponents have dealt only with acquiring specific map information (as opposed to considering memory and general information). We briefly consider each of these problems.

Kang-tsung Chang (1980) summarized the various experimental factors that can affect a power function exponent. One factor is whether subjects are asked to complete a ratio or magnitude estimation task. **Ratio estimation** involves comparing two symbols and indicating how much larger or smaller one symbol is than the other (e.g., noting that "this symbol appears to be five times larger than this one"); on a map, this involves comparing symbols without consulting the legend. Flannery's power function exponent of 0.87 for circles was developed on this basis. **Magnitude estimation** involves assigning a value to a symbol on the basis of a value assigned to another symbol; for example, if a single circle is included in a map legend, values for other circles on the map can be estimated by comparing them with the legend circle. Using this approach, Chang (1977) found that the use of a moderately sized circle in the legend led to a power function exponent of 0.73 for circles.

T. L. C. Griffin (1985) is one of several researchers who have noted that reporting a single exponent for an experiment hides the variation among and within subjects. Although overall, Griffin found a power function exponent of 0.88 for circles (nearly identical to Flannery's 0.87), he stressed that the exponents for individuals varied from approximately 0.4 to 1.3. He also noted that "perceptual rescaling was shown to be inadequate to correct the estimates of poor judges, while seriously impairing the results of those who were more consistent" (35).

One solution to the problems noted by Chang and Griffin is to apply no correction and to stress the importance of a well-designed legend. For example, Chang recommended including "three standards [in the legend]—small, medium, [and] large"; he based this recommendation, in part, on a magnitude estimation study in which he found an exponent of 0.94 when three legend circles were used (Chang 1977). Chang (1980, 161) also recommended including the statement that "circle areas are made proportional to quantities" to encourage users to make estimates based on area.

Another solution to the problems noted by Chang and Griffin is to train users how to read proportional symbol maps. For example, Judy Olson (1975) found that when readers were asked to make estimates of circle size and then were given feedback on the correct answers, the power function exponent was closer to 1.0. Olson also found, however, that only when the practice was combined with perceptual scaling did the dispersion of errors also decrease,

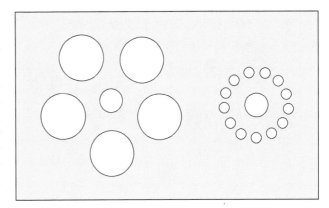

FIGURE 18.6 The Ebbinghaus illusion: The two circles in the middle of the surrounding circles are identical in size, but the one surrounded by larger circles appears smaller.

suggesting that training and perceptual scaling should be used in concert.

The importance of spatial context in developing a perceptual scaling formula can be seen by considering the **Ebbinghaus illusion** shown in Figure 18.6: The two circles in the middle of the surrounding circles are identical in size, but the one surrounded by larger circles appears smaller. Applying this principle to a map, a small circle surrounded by larger circles should appear smaller than it really is, whereas a large circle surrounded by smaller circles should appear larger than it really is. In an experimental study with maps, Patricia Gilmartin (1981) found that spatial context actually had these effects. On this basis, one could argue for a formula in which each circle is scaled to reflect its local context. Gilmartin argued against this, however, indicating it would "have the undesirable effect of weakening the overall pattern perception" (162).

Another limitation of perceptual scaling is that experiments for deriving an appropriate exponent have focused solely on the specific map information. In Chapter 1, we argued that the portrayal of general information is a more important function of maps. Thus, it seems that general tasks should be considered in developing an exponent, or at least that the effect of the exponent on the overall look of the map should be considered. To illustrate this point, consider Figure 18.7, which compares mathematically and perceptually scaled maps of the microbrewery and brewpub data. Here the largest circle size on each map has been held constant to minimize the effects of circle overlap on the two maps. Although a larger range of circle sizes appears on the perceptually scaled map, the patterns on the two maps are similar, suggesting that perceptual scaling has little effect on general information.

Rather than basing the exponent on specific circle estimates, Judy Olson (1976, 155–156) suggested that an arbitrary exponent might be used to enhance the recognition of spatial pattern. To illustrate, Figure 18.8 compares a mathematically scaled map (the exponent in the circle-scaling formula is set to 0.5) with a map having an arbitrary exponent (the exponent is set to 1.0) for the standardized birth data. The narrow range of data on the mathematically scaled map (Figure 18.8A) makes it difficult to detect any pattern.

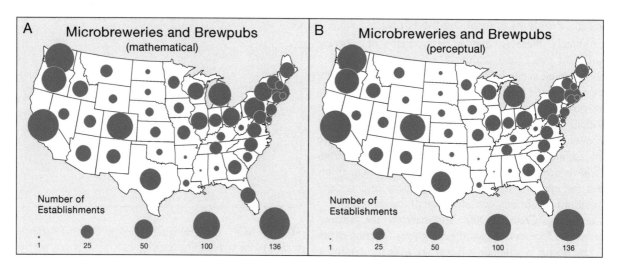

FIGURE 18.7 Effect of mathematical vs. perceptual scaling on map pattern: (A) mathematically scaled map; (B) perceptually scaled map based on Flannery. Note that perceptual scaling appears to have little effect on the overall pattern when the largest circle on both maps is the same size, as is the case here.

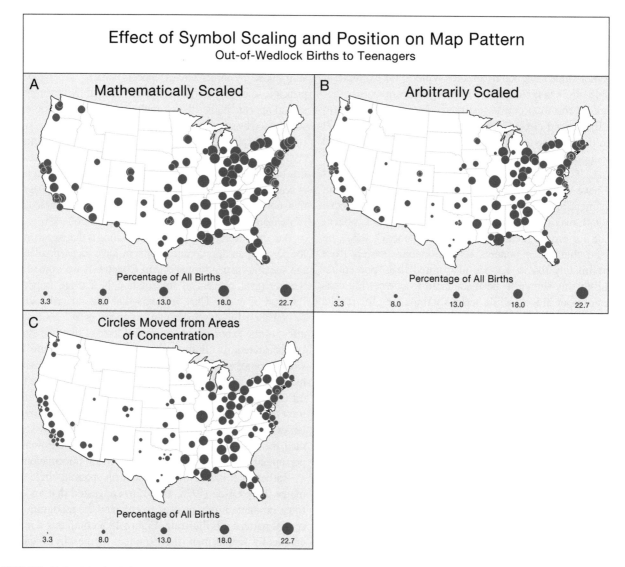

FIGURE 18.8 Manipulating symbol scaling and symbol position to enhance map pattern: (A) a mathematically scaled map (the exponent in the circle-scaling formula is set to 0.5); (B) an arbitrarily scaled map (the exponent in the circle-scaling formula is set to 1.0); (C) circles are moved away from congested areas.

In contrast, on the arbitrarily scaled map (Figure 18.8B), it is easier to detect a pattern; for example, note the lower percentage of out-of-wedlock births in the extreme south central part of the United States (in Texas).

A final limitation of perceptual scaling is that most experiments have been based only on information *acquisition*, as opposed to *memory* for that information. Studies by psychologists (e.g., Kerst and Howard 1984) have revealed that underestimation of larger sizes occurs twice: once when acquiring information and once again when recalling that information. As a result, the exponent for memory is appreciably lower than that for acquisition (i.e., values obtained are approximately the square of that for acquisition). Although the exponent in the perceptual scaling formulas might be adjusted to account for this, we suspect that most cartographers would not make this adjustment because map readers are not expected to remember precise, specific information. It is, however, interesting to note that the lower exponent resulting from this approach would have an effect similar to that just suggested for enhancing spatial pattern.

18.5.3 RANGE-GRADED SCALING

An alternative to mathematical and perceptual scaling is **range-graded scaling** (or **graduated-symbol scaling**), in which data are grouped into classes, and each class is represented with a different-sized symbol (e.g., if the data are grouped into five classes, then five symbol sizes are used). Three basic decisions must be made in range grading: the number of classes to be shown, the method of classification to be used, and the symbol sizes to be used for each class. The first two decisions are standard in any classification of numerical data and were discussed extensively in Chapter 5.

The sizes of range-graded symbols normally are selected to enhance the visual discrimination among classes. Figure 18.9 portrays two sets of symbols that cartographers have developed for this purpose. The first (Figure 18.9A) was developed by Hans Meihoefer (1969) in a visual experiment. Meihoefer indicated that people "were able generally to differentiate" (112) among all these circles in a map environment. Arguing that some of Meihoefer's circles were difficult to discriminate, Borden Dent (1999, 181–183) used his personal experience in designing maps to develop the set shown in Figure 18.9B. For both sets, the mapmaker simply selects *n* adjacent circles, where *n* is the number of classes to be depicted on the map.

Range grading is considered advantageous because readers can easily discriminate symbol sizes and thus readily match map and legend symbols; another advantage is that the contrast in circle sizes might enhance the map pattern in a fashion similar to the use of an arbitrary exponent described in the preceding section. To illustrate the latter, consider Figure 18.10, which compares range-graded and mathematically scaled maps for both the standardized out-of-wedlock birth data and the raw microbrewery and brewpub data.[3] In the case of the microbrewery and brewpub data, we see that range grading has relatively little effect on the spatial pattern because the range of circle sizes on the range-graded and mathematically scaled maps are similar. For the birth data, however, the maps are dramatically different, with range grading resulting in considerable overlap.

Because specific range-graded sizes have been recommended only for circles, individual mapmakers must determine appropriate sizes for other symbol types. For example, to create the range-graded proportional-square map shown in Figure 18.11, distinctly different small and large squares were selected, and then three intermediate squares were specified so that their areas were evenly spaced between the smallest and largest squares.

Range grading is particularly desirable for pictographic symbols because their unusual shapes often make precise relationships between symbols awkward to compute (for the mapmaker) and awkward to estimate (for the map reader). Thus, it makes sense to select intermediate-sized symbols by eye, as opposed to spacing them regularly on the basis of area or volume. This was the approach taken to construct the pictographs of beer mugs shown in Figure 4.6.

A disadvantage of range grading is that readers might misinterpret specific information if they do not pay careful attention to the legend. For example, a reader failing to examine the legend might say that the value for circle T in Figure 18.10A is considerably larger than the value for circle S (say, approximately 30 times as large), but the numbers specified in the legend indicate that the values differ by only a factor of about 4. Another disadvantage is that range grading creates clear differences in circle size and thus potentially creates a pattern when there might not really be a meaningful pattern. For example, for a data set with minimum and maximum values of 10 and 11 percent, respectively, range grading would create obvious differences among circles, even though a 1 percent difference might be of no interest or significance to the reader.

18.6 LEGEND DESIGN

There are two basic problems in designing legends for proportional symbol maps: deciding how the symbols should be arranged and determining which symbols should be included.

18.6.1 ARRANGING SYMBOLS

Two basic legend arrangements are used on proportional symbol maps: nested and linear. In the **nested-legend arrangement**, smaller symbols are drawn within larger

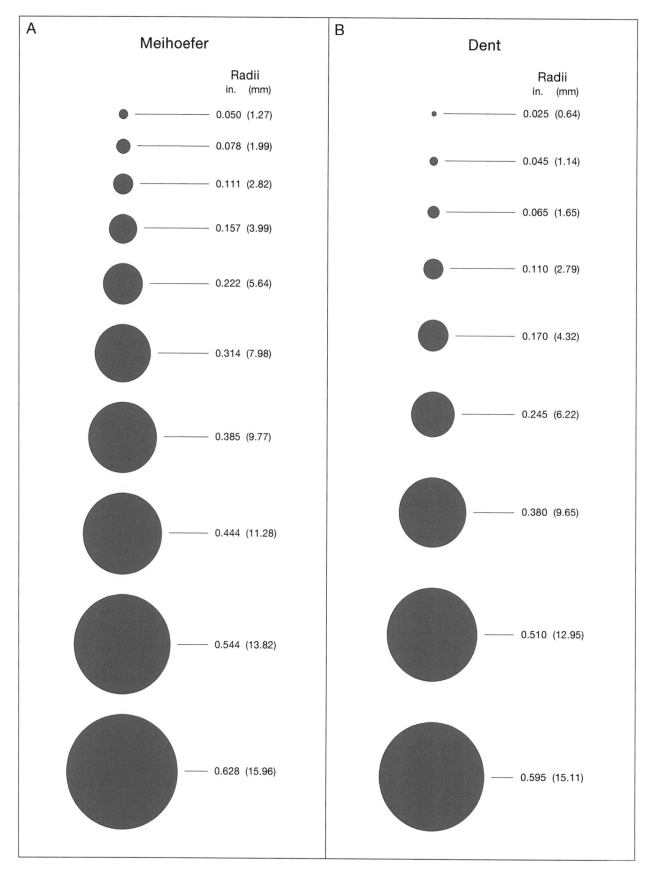

FIGURE 18.9 Potential circle sizes for range grading: (A) a set developed by Meihoefer in a visual experiment (after Meihoefer 1969, Figure 18.4); (B) a set developed by Dent based on practical experience (after Dent 1999, 183).

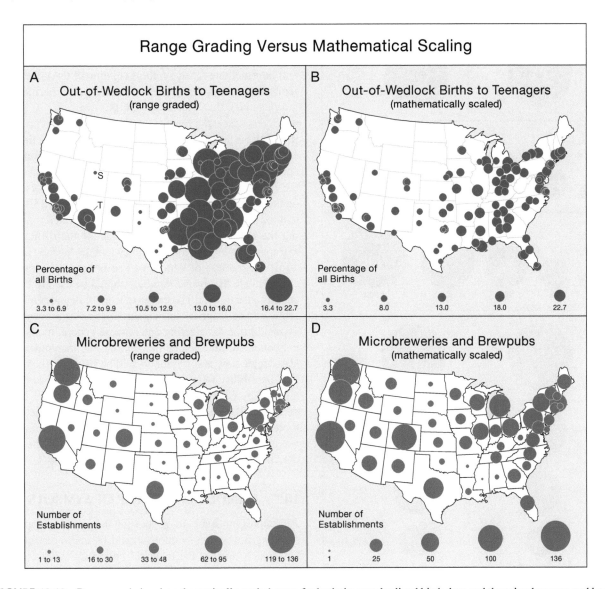

FIGURE 18.10 Range-graded and mathematically scaled maps for both the standardized birth data and the microbrewery and brewpub data. Maps A and C are range graded based on Dent's five smallest circles. Maps B and D are mathematically scaled.

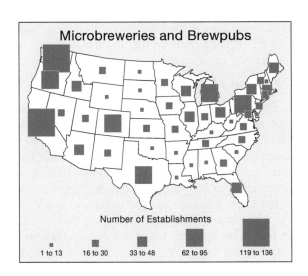

FIGURE 18.11 A range-graded map using squares as a point symbol.

symbols (Figure 18.12A and B), whereas in the **linear-legend arrangement**, symbols are placed adjacent to each other in either a horizontal or a vertical orientation (Figure 18.12C–F). An advantage of the nested arrangement is that it conserves map space. Note, however, that this arrangement might make it difficult to compare a symbol in the legend with a symbol on the map because symbols in the legend (with the exception of the smallest) are covered by other symbols.

When a linear horizontal arrangement is used, you must decide whether the symbols should be ordered with the smallest on the left and the largest on the right or vice versa. Displaying larger symbols on the right is most desirable, given that the traditional number line progresses from left to right. When a linear vertical arrangement is used, you similarly must decide whether symbols should be ordered with the largest at the top or at the bottom. As with choropleth maps (see Section 15.7), we might use the argument that "people associate 'up' with 'higher' and 'higher' with

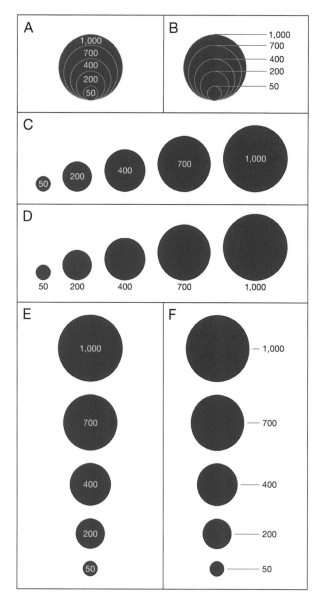

FIGURE 18.12 Various legend arrangements for unclassed proportional symbol maps: (A) and (B) are nested; (C) and (D) are linear with a horizontal orientation; and (E) and (F) are linear with a vertical orientation.

larger data values" to justify placing larger symbols at the top. Alternatively, we could argue that larger symbols are visually "heavier" and thus are more stable if placed at the bottom. With range-graded maps, it makes sense to place smaller symbols at the top so that data values increase from left to right and from top to bottom, matching the order in which we normally read text. Ultimately, the choice of a particular linear legend design is likely to depend on available map space.

18.6.2 WHICH SYMBOLS TO INCLUDE

With range grading, legend symbols are a function of the classes displayed on the map (e.g., a five-class map yields five legend symbols). For mathematical and perceptual

scaling, there are two general methods for selecting legend symbols. One is to include the smallest and largest symbols actually shown on the map and then interpolate several intermediate-sized symbols (Figure 18.13A). Bernhard Jenny and colleagues provide software (http://berniejenny. info/legend/) that can assist in this process. A second method is to select a set of symbols that are most representative of those appearing on the map, which should minimize estimation error. The latter method can be implemented by applying Jenks's optimal classification (see Section 5.8) to the raw data and then constructing legend symbols based on the median (or mean) of each class (Figure 18.13B). This method might also be refined to include circles representing the extremes in the data because of the difficulty of extrapolating beyond symbols shown in the legend (Dobson 1974). Regardless of which method is used, the argument can be made that round numbers should be used in the legend because they will be easier to work with and remember. Jenny and colleagues' software assists in this process too.

In addition to selecting one of these general methods for mathematical and perceptual scaling, the mapmaker must also decide how many symbols will be shown in the legend. One possibility would be to use three symbols, as recommended by Chang. Although three symbols might be sufficient for data with a small range, it makes sense to use more than three when the range is large. As a rule of thumb, we suggest using as many symbols as appear to be easily discriminated.

18.7 HANDLING OVERLAP OF SYMBOLS

An important issue in proportional symbol mapping is deciding how large symbols should be and, consequently, how much overlap there should be. Small symbols cause little or no overlap, and thus a map potentially devoid of spatial pattern; in contrast, large symbols create a cluttered map, making it difficult to interpret individual symbols. This section considers two issues: deciding how much overlap there should be and how the overlap should be symbolized. In the next section, we also will consider a novel method for avoiding circle overlap, the necklace map.

18.7.1 HOW MUCH OVERLAP?

Unfortunately, there are no rules regarding the appropriate amount of overlap. Rather, cartographers have suggested subjective guidelines; for instance, Arthur Robinson and colleagues (1984) indicated that the map should appear "neither 'too full' nor 'too empty'"(294). Examples of improper amounts of overlap are shown in parts A and B of Figure 18.14. Although most cartographers would agree that such extreme cases should be avoided, there would be disagreement as to which amount of overlap between the extremes is more proper. (Parts C and D of Figure 18.14 are two possibilities.)

The role that overlap plays is determined to some extent by whether the map is to be used for communication or for

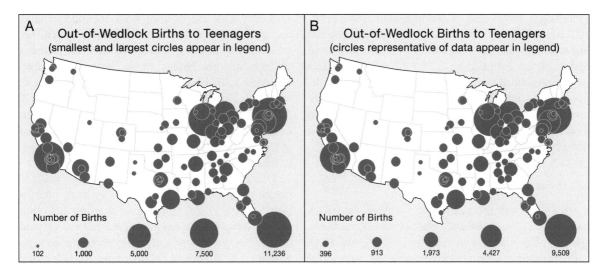

FIGURE 18.13 Approaches for selecting circles for the legend on an unclassed proportional symbol map: (A) the smallest and largest symbols actually shown on the map are used, along with several intermediate-sized symbols; (B) the raw data are classified, and a representative value of each class is used (in this case, the median).

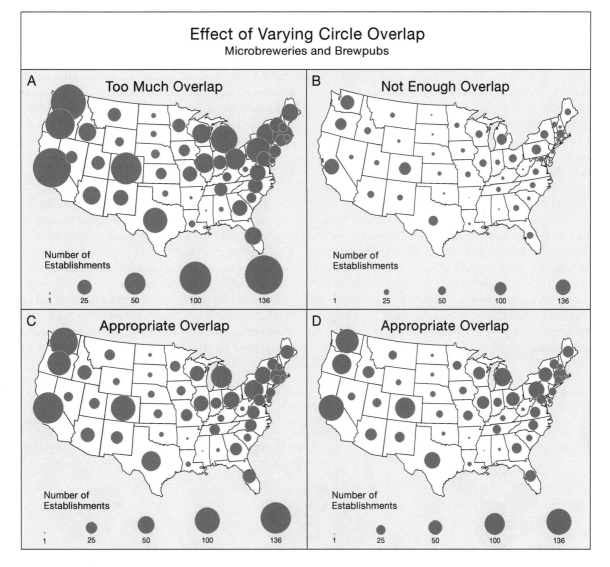

FIGURE 18.14 Effect of varying the amount of overlap: (A) too much overlap—the map appears crowded, particularly in the northeast; (B) not enough overlap—the map appears empty; (C) and (D) are examples of maps having an appropriate amount of overlap.

data exploration. In the case of communication, you probably will want to manipulate circle overlap to enhance the spatial pattern. For example, Figure 18.14C might be a more appropriate choice than Figure 18.14D if you wish readers to note the concentration of microbreweries in the northeast part of the United States. In the case of data exploration, mapping software ideally should provide an option to easily change symbol sizes, such as the interactive option described by Slocum and Yoder (1996).

18.7.2 SYMBOLIZING OVERLAP

There are a variety of approaches for handling overlap. A key parameter is the amount of opacity (or alternatively transparency) associated with each symbol. Figure 18.15 shows three options: opaque, fully transparent, and semi-transparent. Opaque (Figure 18.15A) symbols enhance figure-ground contrast because the symbols tend to appear as a figure against the background of the rest of the map. A downside, however, is that symbols blocked by other symbols will be more difficult to estimate. Also, note that it is difficult to associate a circle with a particular state because the state boundaries are often blocked. In contrast, fully transparent symbols (Figure 18.15B) allow the bounds of all symbols to be seen (along with the state boundaries), thus making it easier to estimate symbol sizes and associate symbols with the appropriate state. A downside of this approach though is that the figure-ground contrast is weakened.

A semi-transparent approach provides a compromise between the opaque and fully transparent approaches. Each proportional symbol in Figure 18.15C is filled with the same color as those in Figure 18.15A but has been made semi-transparent by applying a transparency level of 25 percent (or an opacity level of 75 percent). Semi-transparent proportional symbols act like transparent symbols in that they allow readers to see underlying features. In addition, the semi-transparency used in symbol overlap produces progressively darker colors, which enhances the figure-ground relationship. A possible third advantage is that the progressively darker colors give the impression of increased density of the variable being represented.

A key characteristic of both the opaque and semi-transparent approaches described above is that the circles are drawn from the largest to the smallest so that smaller symbols are not hidden by larger ones. An alternative approach is to develop algorithms that try to maximize the total visible boundaries of all circles (e.g., Cabello et al. 2010, Kunigami et al. 2014). Although we have not made a full evaluation of these approaches, an illustration in Cabello et al. suggests that the result is similar visually to the sorted approach that we use here.

Other solutions for handling overlap include **inset maps**, which portray a congested area at an enlarged scale (see Section 11.4.3); the use of a **zoom function** in an interactive graphics environment; and the possibility of displacing symbols slightly away from the center of congested areas. Figure 18.8C illustrates the effect of moving symbols away from congested areas. In comparing parts B and C of Figure 18.8, we can see that in Figure 18.8C, it is easier to compare the sizes of individual circles and the spatial pattern is potentially more obvious. Recent research by Tomasz Opach and colleagues (2019), however, indicates that approaches other than displacement may be more appropriate when users are able to easily zoom in and out on a proportional symbol map. In contrasting the displacement approach with the aggregation of enumeration units and several forms of modified symbolization (e.g., volume-scaling and color-coded classes), they found that displacement was the least effective of the approaches.

18.8 NECKLACE MAPS

The **necklace map** developed by Bettina Speckmann and Kevin Verbeek (2010) provides a thought-provoking alternative to the traditional proportional symbol map. Rather than placing proportional symbols at conceptual point locations, symbols are placed on a one-dimensional curve that surrounds the mapped region of interest (Figure 18.16). An attempt is made to place symbols near associated enumeration units, and the association between a symbol and an enumeration unit is clarified by showing the same color within both the

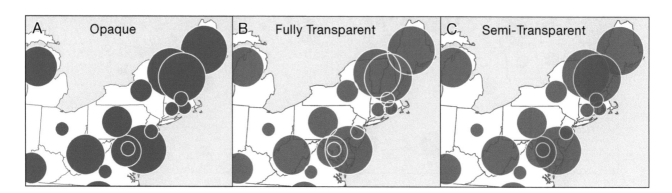

FIGURE 18.15   A comparison of opaque (A), fully transparent (B), and semi-transparent (C) proportional symbols.

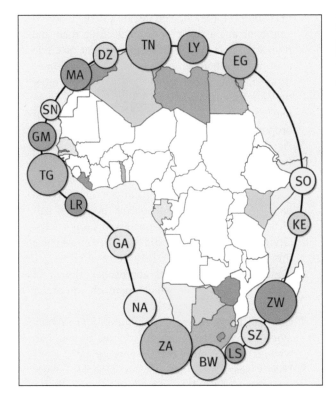

FIGURE 18.16 An example of a necklace map: Internet usage per thousand inhabitants in African countries in 2002. No legend is shown here because Speckmann and Verbeek did not implement legends in their algorithm. (© 2010 IEEE. Reprinted, with permission, from Figure 1 of Speckmann, B. and Verbeek, K. (2010) "Necklace maps." *IEEE Transactions on Visualization and Computer Graphics* 16, no. 6:881–889.)

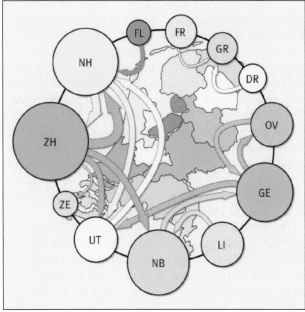

FIGURE 18.17 A necklace map in which flows are shown between symbols (in this case, the circles represent the population for provinces in the Netherlands in 2005, and the flows represent relocations between those provinces in 2005). (© 2010 IEEE. Reprinted, with permission, from Figure 1 of Speckmann, B. and Verbeek, K. (2010) "Necklace maps." *IEEE Transactions on Visualization and Computer Graphics* 16, no. 6:881–889.)

symbol and the enumeration unit. Abbreviations for the enumeration units typically are shown within the symbols (note the two-letter codes for countries of Africa in Figure 18.16).

As implemented by Speckmann and Verbeek, a distinct advantage of necklace maps is that proportional symbols will not overlap one another, a characteristic that can be especially useful when you wish to display large symbols within small enumeration units. The obvious disadvantage is that the perception of the spatial pattern of symbols is compromised because symbols are not shown at their true geographic locations. Speckmann and Verbeek argue that this problem might be handled by creating an interactive system in which one continuously morphs between the necklace map and a traditional proportional symbol map.

Other advantages of the necklace map are that flows can be shown between symbols on the necklace and additional information can be clearly depicted within the symbols. Figure 18.17 illustrates the use of flows; here, the circles represent the population of provinces in the Netherlands in 2005, and the arrows represent relocations between those provinces in that year. In Chapter 21, we will see how the necklace map concept can be extended to create effective interactive flow maps.

18.9 SUMMARY

Proportional symbol maps are used to display numerical data associated with point locations, whether those be true point locations (such as water wells associated with center-pivot irrigation) or conceptual point locations (such as the number of fatalities due to drunk driving in each state). Although raw totals are frequently symbolized on proportional symbol maps, care should be taken that these totals are not simply reflecting another variable (such as population). If this is the case, then the data should be standardized by, say, computing rates or percentages.

On an unclassed proportional symbol map, symbols can be scaled mathematically in direct proportion to the data or perceptually to account for the tendency to underestimate larger symbols. When circles are used as the proportional symbol, the Flannery correction is commonly used to scale circles, as shown by an exponent of 0.57 in the following formula:

$$r_i = \left(\frac{v_i}{v_L}\right)^{0.57} \times r_L.$$

Arbitrarily increasing this exponent can also enhance the perception of spatial pattern. When creating a classed map (a range-graded map), symbol sizes for classes are selected that enable readers to easily discriminate the different symbol sizes (e.g., using Dent's circles); this classed approach can also enhance the perception of the spatial pattern.

Basic problems of legend design include the arrangement of symbols (nested or linear) and the number and size of symbols. A nested arrangement conserves map space but may make it more difficult for the map user to compare a symbol in the legend with a symbol on the map. Symbol overlap is a common problem in proportional symbol mapping, as symbol size estimation and spatial pattern perception become more difficult as the overlap increases. You should attempt to have some overlap (to enhance the spatial pattern), but not so much overlap that the map appears cluttered. The semi-transparent option for showing overlap is desirable because it enhances the perception of spatial pattern but also enables the perception of the bounds of individual symbols and the underlying base map.

The necklace map avoids the problem of circle overlap by placing proportional symbols on a one-dimensional curve surrounding a region of interest; this approach is potentially useful when large symbols are associated with small enumeration units. A disadvantage of the necklace map is that the perception of the spatial pattern of symbols is compromised because symbols are not shown at their true geographic locations. In Chapter 21, we will see how the necklace map concept can be extended to create effective interactive flow maps.

18.10   STUDY QUESTIONS

1. The following are some datasets that might be mapped using proportional symbols. For each dataset, indicate whether you feel that you should standardize the dataset, the method of standardization, and justify your approach: (A) for each county in Michigan, the number of households that have an income in excess of $100,000; (B) for each metropolitan statistical area (MSA) in the United States, the percentage of teenagers who indicate that they consume alcohol on a regular basis; and (C) for each country around the world, the number of coronavirus cases during the recent pandemic.

2. Imagine that you wished to create a county-level proportional symbol map of the distribution of covered bridges in the New England region of the United States. Discuss: (A) how you would choose between a geometric symbol (circles) and a pictographic symbol (covered bridges) to represent the number of covered bridges in each county and (B) whether you would create a classed or an unclassed map for the pictographic symbol.

3. Presume that you wish to create a proportional circle map of the covered bridge data and that you have the following parameters: the maximum number of bridges in a county is 15, and the largest radius is 0.62 cm. Compute the radius for a county with four bridges assuming: (A) a mathematically scaled circle and (B) a perceptually scaled circle.

4. Imagine that you have a dataset with a relatively large range, extending from a minimum of 27 to a maximum of 10,000. You decide to create a conventional unclassed proportional circle map and find that if you try to limit the amount of circle overlap, the smallest circles on the map seem to disappear. (A) What geometric symbology other than circles might you use to try to solve this problem? (B) Do you see any problems in using your proposed symbology?

5. According to the U.S. Bureau of Labor Statistics, the unemployment rates for U.S. metropolitan areas in September 2020 varied from 2.8 to 23.6 percent. (A) What problem can you envision happening if you try to visualize the spatial pattern of this data with an unclassed proportional circle map? (B) How would you attempt to solve this problem?

6. Contrast the two legend approaches shown in Figure 18.13. Indicate which approach you prefer and discuss why you prefer it.

7. Contrast maps A and C in Figure 18.14. Which amount of overlap seems appropriate to you, and why is that amount of overlap appropriate?

8. Use a search engine such as Google images (https://www.google.com/imghp?hl=en) to find a proportional symbol map that you find interesting. Answer the following questions: (A) What sort of symbology is used? Is it geometric or pictographic? (B) Which method appears to be used to scale proportional symbols? (C) Which legend design approach is used? (D) How has symbol overlap been handled? (E) Is the figure-ground relationship clear? (F) Overall, does the map seem well designed?

NOTES

1   We have used the term "proportional symbol" to cover both unclassed and classed point symbol maps, as this seems common in the cartographic literature. Similarly, we have used the term "graduated symbol" for classed maps, as this is also common in recent cartographic literature. Historically, however, the situation is complicated by the fact that the terms proportional symbol and graduated symbol have been used interchangeably.

2   Flannery did not raise $1/k$ to $1/n$, so his constant differed slightly from the value reported here.

3   For the range-graded maps, the data were classed using Jenks's optimal approach and Dent's five smallest circles were utilized.

REFERENCES

Cabello, S., Haverkort, H., van Kreveld, M. et al. (2010) "Algorithmic aspects of proportional symbol maps." *Algorithmica 58*, no. 3:543–565.

Chang, K. (1977) "Visual estimation of graduated circles." *The Canadian Cartographer 14*, no. 2:130–138.

Chang, K. (1980) "Circle size judgment and map design." *The American Cartographer 7*, no. 2:155–162.

Crawford, P. V. (1973) "The perception of graduated squares as cartographic symbols." *The Cartographic Journal 10*, no. 2:85–88.

Dent, B. D. (1999) *Cartography: Thematic Map Design* (5th ed.). Boston, MA: McGraw-Hill.

Dobson, M. W. (1974) "Refining legend values for proportional circle maps." *The Canadian Cartographer 11*, no. 1:45–53.

Ekman, G., and Junge, K. (1961) "Psychophysical relations in visual perception of length, area and volume." *The Scandinavian Journal of Psychology 2*, no. 1:1–10.

Flannery, J. J. (1971) "The relative effectiveness of some common graduated point symbols in the presentation of quantitative data." *The Canadian Cartographer 8*, no. 2:96–109.

Gilmartin, P. P. (1981) "Influences of map context on circle perception." *Annals of the Association of American Geographers 71*, no. 2:253–258.

Griffin, T. L. C. (1985) "Group and individual variations in judgment and their relevance to the scaling of graduated circles." *Cartographica 22*, no. 1:21–37.

Jansen, Y., and Hornbæk, K. (2016) "A psychophysical investigation of size as a physical variable." *IEEE Transactions on Visualization and Computer Graphics 22*, no. 1:479–488.

Kerst, S. M., and Howard, J. H. J. (1984) "Magnitude estimates of perceived and remembered length and area." *Bulletin of the Psychonomic Society 22*, no. 6:517–520.

Kunigami, G., de Rezende, P. J., de Souza, C. C. et al. (2014) "Optimizing the layout of proportional symbol maps: Polyhedra and computation." *INFORMS Journal on Computing 26*, no. 2:199–207.

Meihoefer, H.-J. (1969) "The utility of the circle as an effective cartographic symbol." *The Canadian Cartographer 6*, no. 2:105–117.

Olson, J. M. (1975) "Experience and the improvement of cartographic communication." *The Cartographic Journal 12*, no. 2:94–108.

Olson, J. M. (1976) "A coordinated approach to map communication improvement." *The American Cartographer 3*, no. 2:151–159.

Opach, T., Korycka-Skorupa, J., Karsznia, I. et al. (2019) "Visual clutter reduction in zoomable proportional point symbol maps." *Cartography and Geographic Information Science 46*, no. 4:347–367.

Robinson, A. H., Sale, R. D., Morrison, J. L. et al. (1984) *Elements of Cartography* (5th ed.). New York: Wiley.

Slocum, T. A., Sluter, R. S., Kessler, F. C. et al. (2004) "A qualitative evaluation of MapTime, A program for exploring spatiotemporal point data." *Cartographica 39*, no. 3:43–68.

Slocum, T. A., and Yoder, S. C. (1996) "Using Visual Basic to teach programming for geographers." *Journal of Geography 95*, no. 5:194–199.

Smith, G.-H. (1928) "A population map of Ohio for 1920." *The Geographical Review 18*, no. 3: 422–427.

Speckmann, B. and Verbeek, K. (2010) "Necklace maps." *IEEE Transactions on Visualization and Computer Graphics 16*, no. 6:881–889.

19 Dot Mapping

19.1 INTRODUCTION

Dot maps are utilized when you have collected raw total data for enumeration units and wish to show that the underlying phenomenon is not uniform throughout the enumeration units. As introduced in Section 4.6.4, we create a dot map by letting one dot equal a certain amount of some phenomenon and then place dots where that phenomenon is most likely to occur. For instance, imagine that you have data on the number of elephants living in each country of Africa. You could make a proportional symbol map of such data, but this map would be misleading because each country would have a single symbol, suggesting no variation in the spatial distribution of elephants within a country. A dot map would provide a more realistic representation of the distribution of elephants. In Section 19.3, we return to the wheat harvested in Kansas dot map introduced in Chapter 4 and consider the following key issues involved in dot mapping: (1) how ancillary information can be utilized to select regions where dots will be placed, (2) the selection of **dot size** (how large each dot is) and **unit value** (the raw total represented by each dot), and (3) approaches for placing dots within regions.

Dot size is normally held constant throughout a dot map. Using equal-sized dots is useful for examining the detailed spatial pattern of a phenomenon, but extracting specific information (such as the raw total of a phenomenon for an area) is awkward due to the large number of dots that may need to be visualized. In Section 19.4, we examine one solution to this problem, the **graduated dot map**, which replaces densely packed dots with dots of larger sizes.

Traditionally, dot maps have been viewed in a static map environment. With the continued proliferation of web mapping in the last ten years or so, we have seen an increase in the development of interactive web maps. In Section 19.5, we discuss the capability to examine large data sets (so-called Big Data) in an interactive dot mapping environment and consider some map design issues associated with this capability.

19.2 LEARNING OBJECTIVES

- Select appropriate data for dot mapping.
- Identify appropriate ancillary information to be used when creating a dot map.
- Select an appropriate dot size and unit value for dot mapping.
- Recognize that dot mapping software often places dots randomly, which leads to undesirable clusters and gaps in the dot pattern.
- Be aware of dot algorithms that avoid the problems of placing dots randomly.
- Consider the possibility of creating a graduated dot map, in which sizes of dots are varied to ease the extraction of specific information.
- Understand the advantages and disadvantages of an interactive dot mapping web environment.

19.3 KEY ISSUES INVOLVED IN DOT MAPPING

In Chapter 4, we illustrated a dot map of wheat harvested in Kansas (Figure 4.10D is repeated in the lower right portion of Figure 19.1). This section considers three issues relevant to creating such a dot map: (1) determining regions within which dots should be placed, (2) selecting dot size and unit value, and (3) placing dots within selected regions. Although we consider these issues in the context of the wheat harvested example, they are generic to most dot-mapping problems.

19.3.1 DETERMINING REGIONS WITHIN WHICH DOTS SHOULD BE PLACED

Dot maps (like dasymetric maps; see Chapter 16) use **ancillary information** to create a detailed map of a phenomenon. Ancillary information is commonly split into limiting and related attributes. *Limiting attributes* place absolute limits on where dots can be placed. For example, in general, it does not make sense to place dots representing population within water bodies. *Related attributes* are those that are correlated with the phenomenon being mapped but that do not place absolute limits on the location of dots. For example, as the views of a distant mountain range become more prominent, we might expect to find more houses in those desirable viewing regions.

In constructing the dot map of wheat harvested in Kansas, we first considered several limiting attributes. Obvious limiting attributes included the location of water bodies (in the case of Kansas, there are a number of reservoirs), the location of urban areas (e.g., Wichita and Kansas City), and slope (e.g., tractors cannot be used on very steep slopes). Such limiting attributes might be found on paper maps and entered into a GIS as layers via *scanning* or *digitizing* (or they might already be available in digital form). The resulting layers can then be overlaid, and dots placed only in areas not constrained by the limiting attributes.

There are also a variety of related attributes that might be relevant to mapping the distribution of wheat harvested. One potential related attribute would be precipitation because it varies from a high of about 90 centimeters (35 inches) in

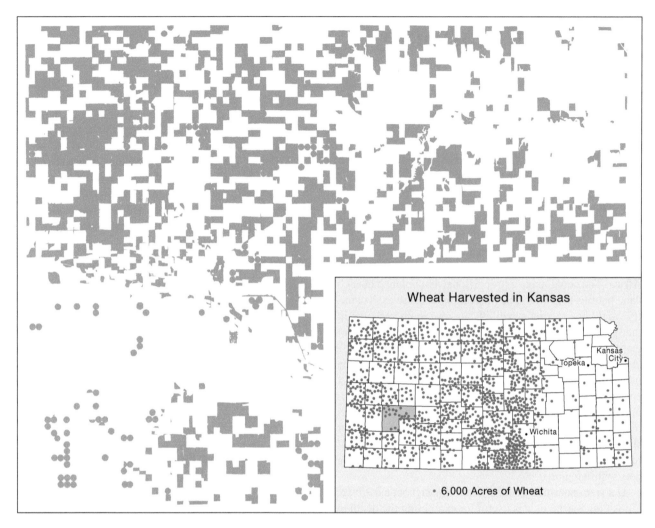

FIGURE 19.1 Distribution of wheat for Finney County on the 2005 Kansas land use/land cover map (upper left) and the dot map resulting from utilizing this land use/land cover map, with Finney County highlighted in gray.

eastern Kansas to only about 40 centimeters (15 inches) in western Kansas. Presuming an ideal amount of precipitation for wheat, a mathematical function could be developed that would place a higher probability on locating dots near the ideal precipitation area. Such an approach also could consider other complicating attributes, such as the timing of the precipitation. Another attribute related to precipitation would be major hail events, which are known to damage or destroy a wheat crop. Hail events could be treated as either limiting or related attributes, depending on the severity of damage.

Another potentially related attribute would be irrigated cropland. The United States Department of Agriculture provides data on the acres of wheat harvested from irrigated land (http://quickstats.nass.usda.gov/), so if the location of irrigated cropland could be determined, a percentage of dots could be placed in such areas equivalent to the percentage of wheat irrigated in that area. Remote sensing could be used for this purpose, particularly in the case of center-pivot irrigation, which traces a distinctive circle on the landscape (Astroth et al. 1990).

Still another possible related attribute would be the distribution of soils within each county. One might suspect that

certain soils would be more conducive to wheat production than to other crops. In discussing this possibility with those knowledgeable about Kansas agriculture (a farmer and a county extension agent), we found that the selection of crops at the county level was more likely a function of farm policy and associated programs rather than of soil type. For example, wheat would continue to be planted at a particular location because a farm program specified that the same crop must be planted for financial support to be retained.

As with the limiting attributes, the resulting related attributes could be entered into a GIS (or existing digital data could be utilized). Since the related attributes provide a likelihood of finding wheat harvested at particular locations (as opposed to indicating the presence/absence of wheat harvested), we would need to establish some sort of weighting system to combine the various layers.

Rather than combining limiting and related attributes in an overlay approach, we utilized a 2005 Kansas land use/land cover map (https://tinyurl.com/KansasLandCoverMaps) developed by Dana Peterson and colleagues (2009). This 25-class land use/land cover map provided not only the location of limiting attributes (e.g., urban areas and water bodies) but also showed the detailed location for wheat fields in 2005. Figure 19.1 illustrates the kind of detail that was provided for wheat on the land use/land cover map. Peterson and her colleagues indicated that the accuracy of identifying wheat fields using remotely sensed data was in excess of 90 percent. Although these images recorded wheat growing rather than wheat harvested, we expected that wheat growing areas would be a strong indicator of wheat harvested areas. Additionally, although our wheat data were from 2011, our experience with mapping wheat data suggests that there would be little change in the pattern of wheat production between 2005 and 2011.

Creating land use/land cover maps can be time-consuming and expensive; for example, Peterson et al. described the creation of the Kansas land use/land cover map as an 18-month endeavor that involved numerous people. Thus, you should not assume that such a detailed map will be available for a phenomenon that you wish to map. In a previous edition of this textbook, we used a less-detailed land use/land cover map, which split the landscape into ten classes: five urban (residential, commercial-industrial, grassland, woodland, and water) and five rural (cropland, grassland, woodland, water, and other). In that case, we located dots largely in the cropland class but also placed some dots in the grassland class because of known errors of land use/land cover classification (Whistler et al. 1995, 773).

19.3.2 Selecting Dot Size and Unit Value

Dot size (how large each dot is) and **unit value** (the total represented by each dot) are important parameters in determining the look of a dot map. Cartographers have argued that very small dots produce a "sparse and insignificant" distribution and that very large dots "give an impression of excessive density." Similarly, a small unit value produces a map that "gives an unwarranted impression of accuracy," whereas a large unit value results in a map that "lacks pattern or character" (Robinson et al. 1995, 498). Generally, it has been argued that dots in the densest area should just begin to coalesce (that is, merge with one another), although some cartographers have argued that dots should not overlap so that individual dots can be counted if desired.

J. Ross Mackay (1949) developed a graphical device known as the **nomograph** (Figure 19.2) to assist in selecting dot size and unit value. To use the nomograph, we must first calculate the number of dots per square centimeter for a sample area on our map. For example, imagine that we found our densest region to be an area with 500,000 acres of wheat and an area of 6 square centimeters (at the mapped scale). If we presume a unit value of 1 dot equals 1,000 acres, then the number of dots required would be 500 (500,000/1,000), and the number of dots per square centimeter would be 83.3 (500/6). (The dashed vertical line in Figure 19.2 represents the value of 83.3.) If we now construct

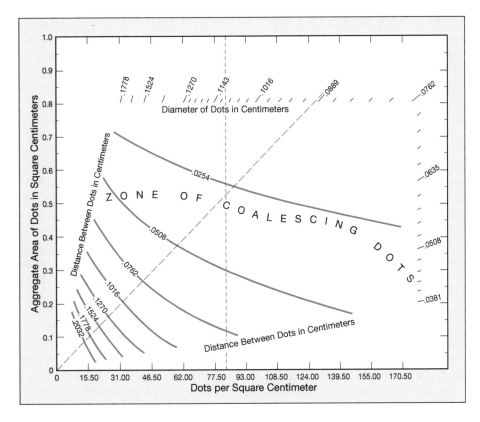

FIGURE 19.2 The nomograph—a device used for computing appropriate dot size and unit value. A vertical line is drawn representing a relatively dense area on the map, and a second line is drawn from the origin to a desired dot size. The distance between dots can be determined by where the two lines intersect. (After Mackay 1949. First published in *Surveying and Mapping* 9, p. 7. Reprinted with permission from the Cartography and Geographic Information Society.)

a line from the origin to a desired dot diameter, and note where this line intersects the vertical dashed line, we will have an indication of the approximate distance between the edges of dots. For instance, in Figure 19.2, we can see that a dot diameter of approximately 0.0889 centimeters has been selected, and thus the distance between dots will be slightly greater than 0.0254 centimeters. Note that this distance falls in the "zone of coalescing dots."

It should be noted that calculations using the nomograph presume that dots are distributed in a relatively uniform hexagonal pattern and that dots do not overlap one another. Jon Kimerling (2009) indicated that the latter constraint is problematic because dots on dot maps often do overlap one another (if the objective is to have the dots coalesce). Kimerling also noted that the nomograph was developed during an era when maps were drawn by hand using pen-and-ink techniques, which limited the smallest size of dots that could be created consistently. As a result, the nomograph does not include the smaller dot sizes that are sometimes used in digital cartography.

In practice, some experimentation is generally required to select an appropriate dot size and unit value, regardless of whether the nomograph is used. For our wheat map, we selected several counties that had both a high percentage of cropland in wheat and a large value for acres of wheat harvested. Selecting counties on the basis of the percentage of cropland in wheat ensured that dense areas of harvested wheat would be considered while having a large value for acres harvested ensured that the dense area would consist of a relatively large number of dots. Using a GIS (ArcMap), we experimented with various dot sizes and unit values until we arrived at a solution that resulted in the coalescence of some of the dots (our goal was to have the closest dots touch one another but not overlap).

One problem with the approach we have described is that it fails to consider the difficulty that map readers have in correctly estimating the number and density of dots within subregions of a map. Analogous to the underestimation of proportional symbol size (see Section 18.5.2), readers also underestimate the number and density of dots. Judy Olson (1977) indicated that it is possible to correct for this underestimation, but the effect on the overall pattern is small and thus of questionable utility.

It is interesting to note that mapmakers can ignore the issue of dot size and unit value when the phenomenon is located explicitly via remote sensing (as in the Finney County example shown in the upper left of Figure 19.1); in this case, the darkened pixels tell us precisely where the phenomenon (wheat in this case) is located. One disadvantage of this approach, however, is that it might be difficult to make a visual estimate of the magnitude of production as individual pixels merge together. Of course, because the map is automatically generated, the computer can be used to make a precise estimate of the acres of wheat grown in any area (albeit with the estimate a function of the ability to correctly interpret the phenomenon

via remotely sensed imagery). Another disadvantage is that the map arguably provides excessive detail (imagine trying to interpret the Finney County large scale image when it is reduced to the scale of the dot map for the entire state). Clearly, some degree of generalization would be required in this instance.

19.3.3 Placing Dots within Regions

It is important to distinguish how dots were placed manually (when pen-and-ink cartography was common) and how they are placed using digital methods.

19.3.3.1 Placing Dots Manually
Mackay (1949, 5) described the manual placement process as "placing the first dot near the center [of a region] and then each successive dot in the largest remaining space." Perfect uniformity is not desirable because patterns in nature are not perfectly uniform. Pure randomness also is not desirable because this could lead to unrealistic clustering and large gaps between some dots. Thus, the result might better be termed *uniform with a random component*.

Figure 19.3 illustrates the pattern of dots that might result from a manual placement approach for soybeans in Kansas. In Figure 19.3A, the areas to be mapped are counties of Kansas and no ancillary information was used. The dots were distributed relatively uniformly by hand in each county and the result is essentially a choropleth map, as we can see the effect of county boundaries in the eastern part of the state. Figure 19.3B shows a variation of this dotting approach, where again, no ancillary information is used, but the dots are not only plotted relatively uniformly but are geographically weighted toward neighboring higher density counties; this is logical as we do not expect the density of wheat harvested to change rapidly at county boundaries. The result is essentially an isopleth map, based on the assumption of smooth, continuous changes between enumeration units. In this case, the effect of county boundaries nearly disappears. Figure 19.3C shows a map in which dots have been placed relatively uniform in areas defined by ancillary information. This map could be further enhanced by also weighting dots within these areas toward denser adjacent areas. The geographically based approach is, of course, the preferred approach as it takes advantage of geographic knowledge and produces the most detailed dot map.

19.3.3.2 Placing Dots Digitally
Unfortunately, present-day software for dot mapping often does not include satisfactory approaches for placing dots within desired regions. The most common approach is to utilize so-called *dot-density routines* that place dots randomly within geographic areas specified by the user. This is accomplished by overlaying the area of interest with an *xy* grid, computing the desired number of points to be plotted, and then repeatedly plotting points at random *x* and *y* coordinates. The problem with this approach is that

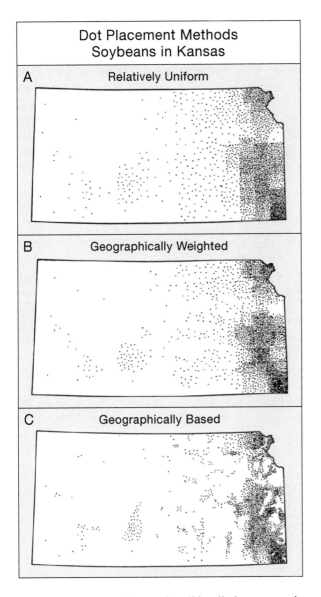

FIGURE 19.3 Approaches used traditionally by cartographers to place dots manually on a dot map: (A) relatively uniform—dots are placed uniformly but with a random component, (B) geographically weighted—dots are plotted relatively uniformly but weighted toward denser areas, and (C) geographically based—dots are plotted relatively uniformly within areas specified by ancillary information.

plotted dots tend to form clusters, irregular gaps often appear between dots, and dots may overlap one another, as shown in Figure 19.4A for both low- and high-density situations. An advantage of this technique, however, is its simplicity and speed of execution. We used a basic random dot density routine within a GIS to initially place dots for the dot map shown in Figure 19.1 and then used a graphic design program to move dots in order to avoid what we deemed inappropriate clusterings, gaps, and overlaps.

Jon Kimerling (2009) developed a *pseudo-random* algorithm that improves upon the basic random dot-density routine. In this approach, each randomly computed dot is checked to see whether it overlaps a previously plotted dot.[1] If overlap does not occur, the computed dot is plotted. If overlap does occur, the computed dot is eliminated, a new random dot is computed, and the test for overlap is repeated. This process is repeated until all desired dots are plotted. Figure 19.4B shows the result of the pseudo-random approach. For the high-density case, there clearly is an improvement when using the pseudo-random approach, as the dot pattern appears more uniform than in the random case (albeit with the desired random component). For the low-density case, however, there is still undesirable clustering of dots and, in this particular image, irregular gaps between dots.

To see whether we could avoid the undesirable clustering and gaps, we modified Kimerling's approach by also requiring that dots be no closer than a specified distance. Since we were unsure what the minimum distance between dots should be, we experimented with several values. We started with the mean distance between points for a theoretically random distribution of points as defined by mathematicians (Clark and Evans 1954) and found that we had to increase that by about 40 percent before the dots exhibited a pattern closer to what cartographers traditionally created manually. The result is shown in Figure 19.4C, where there is a distinct improvement for the low-density case. In the next section on graduated dot maps, we will introduce a more sophisticated approach that begins with a random distribution of points and disperses these dots while maintaining the density of the original dots—the result is termed a blue-noise pattern.

A quite different approach for dot placement was developed by Annette Hey and Ralf Bill (Hey 2012; Hey and Bill

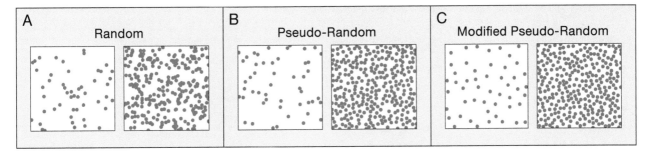

FIGURE 19.4 Approaches for placing dots digitally: (A) random—dots are plotted using randomly selected x and y coordinates, (B) pseudo-random—dots are plotted randomly, but a dot is not plotted if it overlaps another dot, (C) modified pseudo-random—dots are plotted randomly and without overlap, but a dot is not plotted if it is less than 40 percent of the average distance to the nearest dot in a theoretically random distribution.

2014). Their approach consists of two major steps, assuming a desired dot density within a particular area. First, dot positions are defined by a series of logarithmic spirals using the centroid of the area to be dotted as a reference point. The result is a regular pattern of dots that would seem to violate the manual principle of "relative uniformity" of dots. To provide a pattern of dots that comes closer to approximating traditional manual approaches, Hey and Bill randomly move individual dots within a circle surrounding each dot. Since they use the *x* coordinate of the original dot to seed the random number generator, the resulting moved dot will be in the same position from one computer run to another. This is desirable as one disadvantage of the random approaches described above is that a different dot pattern within the same enumeration unit will arise with each computer run.

19.3.4 DESIGNING A LEGEND

If you examine Figure 19.1, you will note that the legend for the dot map consists of a single dot with an associated unit value of 6,000. We deemed this simple legend appropriate because our primary purpose in creating this dot map was to illustrate the details of the spatial pattern of wheat harvested in Kansas in 2011. Although readers could conceivably either count or estimate the number of dots occurring within an area, that was not our intention.

If you do wish readers to make estimates of the number of dots occurring within an area, it is advisable to create a legend that shows at least three representative densities, as shown in Figure 19.7A. Robert Provin (1977) showed that including such a legend nearly eliminated the difference between the perceived number and the actual number of dots. Provin also found that the estimates were improved when smaller dots were used (the dots did not coalesce), suggesting that overlap of dots should be avoided if you wish to have people make correct estimates of the number of dots.

19.4 GRADUATED DOT MAPPING

Although a legend with representative dot densities can ease the burden of making estimates of the number of dots occurring within an area, this task is still challenging when the number of dots is large and the density of dots is high. Therefore, to further ease the burden of making estimates of the number of dots, Nicholas Arnold and colleagues (2017) promote the use of the **graduated dot map** in which dots of varying sizes are used. Arnold and colleagues develop a digital approach for creating such a map and test the effectiveness of the approach with map readers. An example of a graduated dot map taken from the 1977 *Atlas of Switzerland* is shown in Figure 19.5. Here we see

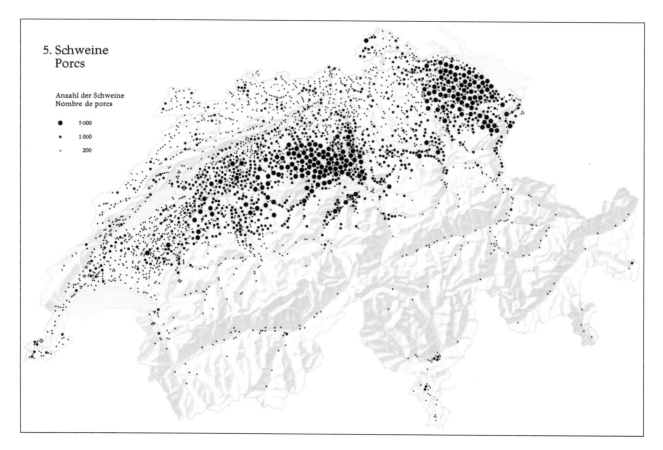

FIGURE 19.5 An example of a graduated dot map showing the distribution of schweine/porcs (swine) in Switzerland. (© Atlas of Switzerland (1977), sheet 51; www.atlasofswitzerland.ch.)

that three sizes of dots are used to represent the distribution of schweine/porcs (swine) in Switzerland. The smallest dot sizes occur in the least dense areas, and the two larger dot sizes occur in the denser areas, thus arguably easing the burden of making estimates of the number of dots.

As we have pointed out, random dot approaches produce awkward gaps and clusters not warranted in the underlying data. Ideally, we would like to have dots placed relatively uniformly but with a random component. This is essentially what computer scientists term a **blue-noise pattern**: point distributions "… with large mutual distances between points and no apparent regularity artifacts" (Balzer et al. 2009, 1). To create a blue-noise pattern, Arnold and colleagues use a capacity-constrained Voronoi tessellation (CCVT) algorithm developed by Balzer et al. The CCVT algorithm "disperses dense groups of dots such that their distribution is uniform but randomized, while maintaining the density distribution of the original dots" (Arnold et al. 2017, 2529). An example of applying the algorithm is shown in Figure 19.6, where Arnold and colleagues started with a random dot-density pattern (Figure 19.6A),[2] applied the CCVT algorithm, and ended up with a blue-noise pattern (Figure 19.6B).

Viewed broadly, Arnold and colleagues' digital approach for producing graduated dot maps consists of three components: (1) creating a traditional digital dot map using a random dot approach; (2) eliminating some of the gaps and clusters resulting from the random dot approach by rearranging them in a blue-noise pattern; and (3) combining densely clustered dots to form the larger dot sizes found on the graduated dot map. The second and third components are used iteratively to produce the graduated dot map. Arnold and colleagues argue that three dot sizes "… is a good number to avoid confounding users' ability to differentiate between dot sizes" (2531).

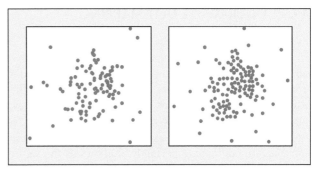

FIGURE 19.6   A comparison of two dot placement approaches: (A) random dot-density and (B) a blue-noise pattern resulting from applying the CCVT algorithm to (A). (After Arnold, N. D., Jenny, B. and White, D. (2017) "Automation and evaluation of graduated dot maps." *International Journal of Geographical Information Science* 31, no. 12:2524–2542, Taylor & Francis Ltd, https://www.tandfonline.com/journals/tgis20.)

Arnold and colleagues tested the effectiveness of graduated dot maps by comparing map readers' ability to estimate values on traditional dot maps and on graduated dot maps.[3] Figure 19.7 illustrates an example of a pair of maps that were tested. Note that on the traditional dot map, a legend with a single dot (representing the unit value) and three representative densities (values) is shown, whereas, on the graduated dot map, just the three sizes of dots are shown in the legend. Arnold and colleagues found that values were estimated more accurately on graduated dot maps than on traditional dot maps and that readers found graduated dot maps "clearer, more legible, and more visually appealing" (2540). One limitation of their study was that readers did not perform spatial pattern tasks, and so it would be interesting to conduct additional user studies to determine the nature of spatial patterns perceived on traditional and graduated dot maps.

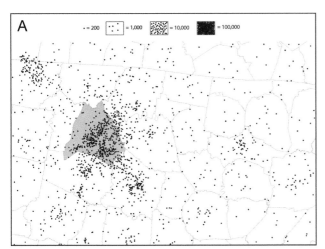

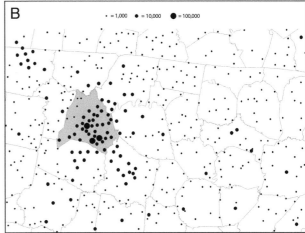

FIGURE 19.7   An example of maps utilized in the user study by Arnold and colleagues (2017): (A) a traditional dot map in which the size of dots is constant (note the legend consists of the unit value for a single dot and three boxes of differing densities), (B) a graduated dot map, with dots of three different sizes (here the legend indicates the unit value for each of the three dot sizes). In the user study, the graduated dot map was rotated 180 degrees to reduce potential learning effects. (After Arnold, N. D., Jenny, B. and White, D. (2017) "Automation and evaluation of graduated dot maps." *International Journal of Geographical Information Science* 31, no. 12:2524–2542, Taylor & Francis Ltd, https://www.tandfonline.com/journals/tgis20.)

19.5 INTERACTIVE DOT MAPPING ON THE WEB

In the first two sections of this chapter, we focused on approaches for creating static dot maps. In this section, we consider the potential of interactive dot maps, especially in a web mapping environment. A key advantage of interactive dot maps is the ability to zoom in and pan the map. Not only does this allow us to examine dense dot patterns in more detail, but more importantly, it allows us to examine much larger data sets than are normally viewed in a static map environment. To illustrate the capability of interactive dot mapping, we will examine an interactive map on educational attainment that Kyle Walker (2018) developed for the Web (http://personal.tcu.edu/kylewalker/maps/education/#5/22.791/-158.794). More specifically, Walker

examines the level of educational attainment for the U.S. population aged 25 and up using data obtained from the 2011–2015 American Community Survey. Figure 19.8 is a screenshot of the San Francisco, California, area from the web application. Here we can see that different colored dots are used to represent differing levels of educational attainment, extending from less than high school (in red) to graduate degree (in blue).

A key decision that Walker needed to make for the educational attainment application was the level of geography that should be examined. If you were making a static dot map of the entire United States at a typical page size, you probably would not use enumeration units any finer than the county level. But the capability to zoom and pan permits a much finer level of viewing. Walker chose to map at the *census tract* level because finer levels (*block groups* and

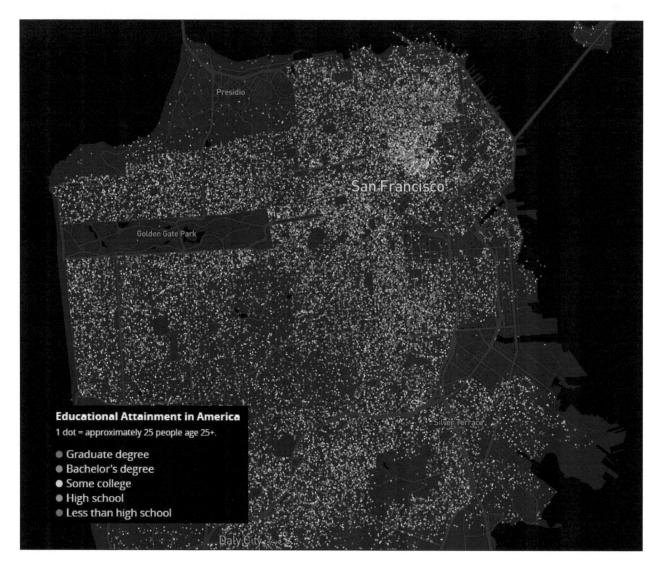

FIGURE 19.8   A screen capture of San Francisco, California, from Kyle Walker's interactive web mapping application for examining educational attainment throughout the United States at the census tract level. A notable region of higher educational attainment (and lower population density) can be seen in a roughly circular area in the center of the city (blue and green dots). Coincidentally, many of these neighborhoods sport a "higher" motif in their names: Diamond Heights, Twin Peaks, and Forest Hill. Regions of lower educational attainment form a ring around the center in the lower elevations of the city, with higher concentrations on the eastern side of the city. (Courtesy of Kyle E. Walker.)

blocks, respectively) have greater uncertainty in the data for educational attainment. At the time of data collection for the map, there were approximately 75,000 census tracts in the United States. This large number of enumeration units moves us into the realm of Big Data, a topic that we will discuss in greater depth in Chapter 26. For now, we focus on several issues in dot mapping related to examining this large dataset.

One issue is the nature of ancillary information used for placing dots within census tracts. Since each dot represents the number of people with a certain level of educational attainment, we want to make sure that dots are placed only where people live (e.g., you generally would avoid bodies of water when mapping population). To solve this problem, Walker chose to use census *blocks* as a limiting attribute: no dots were placed within blocks that had zero population (approximately 5,000,000 of the more than 11,000,000 blocks composing census tracts had a population of zero).[4]

A second issue is the color scheme used to depict the various levels of educational attainment shown in Figure 19.8. Ideally, Walker wanted readers to merge colors for the various levels of educational attainment in a fashion similar to the pointillist technique used in nineteenth-century painting (Jenks 1953; see Section 22.4.2.2). For the default color scheme, Walker chose to treat the different levels of attainment as categorically different and so used a qualitative scheme from ColorBrewer (https://colorbrewer2.org/#type=qualitative&scheme=Set1&n=6). Since the resulting scheme may be inappropriate for the color deficient (e.g., as discussed in Chapter 15, the green and red in Figure 19.8 are problematic), Walker also provides a diverging color scheme (https://tinyurl.com/Dot-map-colorsafe). The latter scheme could also be argued more appropriate if we treat the data as ordinal in nature (as *increasing* levels of attainment extending from "less than high school" to "graduate degree"). However, as Walker indicates (181), it is more difficult to distinguish neighboring categories with this diverging color scheme. Although Walker expected readers to use these color schemes to view all five attainment categories simultaneously, each category can be toggled on and off if a viewer struggles with the simultaneous view or wishes to focus on a particular category.

A third issue relevant to dot mapping with this large dataset is determining the appropriate unit value and dot size. Since users need to be able to zoom in and out to see the level of educational attainment at different scales (e.g., a national view vs. the variation within a city), Walker used different dot sizes at different scales and rounded these values for ease of interpretation (unit values of 25, 50, 100, 200, and 500 were used). Dot size was automatically modified using Mapbox's zoom-dependent styling so that smaller dot sizes were shown at smaller zoom levels (175).

For those readers interested in developing their own interactive Web applications, Walker covers several issues encountered while designing and distributing the application. For instance, one issue is that a developer must realize that the application will likely be used in a variety of viewing environments (e.g., phone vs. desktop), and requires a different interface because of the drastic difference in screen size. Another issue is that if you make an application generally available, you cannot anticipate the knowledge of a potential user about cartography. In this respect, Walker noted that users may be "upset that their neighborhoods are missing dots that are representative of their own educational achievements" (182). The problem here is that users need to realize that the data are based on a sample and, therefore, may not reflect their own personal situation.

19.6 SUMMARY

A dot map is appropriate when you have collected raw totals for enumeration units and wish to show that the underlying phenomenon is not uniform throughout the enumeration units. For example, if you are given data on the number of bears occurring in each county of Montana, you may be interested in showing the distribution of bears throughout each county. Ancillary information is critical to providing an effective dot map. With our bear example, we would need to study the literature on bears to determine which attributes are apt to be correlated with their location on the landscape (for example, bears may favor locales in which berries are plentiful in the summer months). Oftentimes, already existing maps and databases can ease the collection of ancillary information, as the Kansas land use/land cover map did in our case.

Important issues in dot mapping include selecting the dot size (how large each dot is) and the unit value (the total represented by each dot). A nomograph can be used for this purpose, although you will generally find that some experimentation with potential values is necessary. Once ancillary information has been used to determine areas where dots should be placed, the dots have typically been placed using a purely random procedure. It is more desirable to account for dot overlap, as in Kimerling's pseudo-random procedure. Ultimately, dots should be placed relatively uniformly, albeit with some random displacement—this might be achieved by using some variation of the modified pseudo-random procedure we described, where dots are purposely separated to provide a relatively uniform pattern mimicking traditional manual procedures.

One problem with traditional dot maps is that the large number and high density of dots sometimes complicate estimates of dot number. One solution to this problem is the graduated dot map, in which dots of more than one size are used. Such a map can be created by rearranging the dots to create a blue-noise pattern and then clustering the resulting dots to create larger dot sizes. It will be interesting to see to what extent this relatively novel approach is utilized and whether it can be effective for both specific and general (pattern) information.

Developing an interactive web application for dot mapping is intriguing because of the potential to visualize large quantities of spatial data at multiple scales, as we saw with Walker's educational attainment application. The ability to portray the data at multiple scales, however, raises challenges regarding the choice of appropriate ancillary information and suitable dot size and unit value. The large pool of users of the Web also raises questions about how the mapped information will be interpreted, as users unfamiliar with the dot mapping technique and the underlying data may misinterpret the meaning of the dot symbology used.

19.7 STUDY QUESTIONS

1. For each of the following datasets, indicate whether it would be appropriate for dot mapping and justify your decision (assume a static page-size map): (A) the number of electoral votes in each state of the United States and the District of Columbia (DC); (B) the location of organic crop farms in the state of Iowa (https://www.iowaorganic.org/find_farms); (C) the number of murders in each of the 100 largest cities (in terms of population) of the United States; (D) the acreage of agricultural land dedicated to grapes for wine in each department of France (departments are major subregions of France).

2. Presume that you wish to create a dot map of the distribution of corn harvested in Iowa and are given maps of the following attributes that might be used as ancillary information. Indicate whether each attribute is limiting, related, or not relevant, and briefly explain your choice. (A) "urban land," as designated on a land use/land cover map; (B) likelihood of hail damage during the growing season for corn; (C) reservoirs; (D) the degree of the slope (https://tinyurl.com/Degree-of-slope); and (E) distance to a storage facility for corn once it is harvested.

3. Let's assume that you've combined the appropriate ancillary information to define the regions within which dots will be placed for the corn map of Iowa. Also, assume that the densest region on your map has 300,000 acres of corn harvested in an area of 8 square centimeters (at the mapped scale). (A) Presuming that 1 dot on the map equals 2,000 acres of corn, compute the number of dots to be plotted in the densest region and the associated number of dots per square centimeter on the map. (B) Plot the dots per square centimeter value on the nomograph in Figure 19.2, and describe what the resulting densest region will look like on the map if you use a dot diameter of 0.0889 centimeters.

4. (A) Describe the basic legend designs that are possible for a static corn map of Iowa. (B) Select a legend design that you would consider most appropriate and explain your choice. (C) How might the choice of legend design vary if you developed an interactive web map?

5. (A) Describe how dots were plotted on a dot map using pen-and-ink cartography. (B) Discuss the digital dot mapping approaches illustrated in Figure 19.4, and indicate which comes closest to the pattern resulting from pen-and-ink cartography.

6. (A) Discuss why the graduated dot map might be used as an alternative to the traditional dot map. (B) What is one potential limitation of the graduated dot map?

7. Select a major metropolitan area of the United States and use Walker's interactive educational attainment application to evaluate the nature of educational attainment in that area. Remember that you can use either the default color scheme or the diverging color scheme. Select whichever color scheme you deem appropriate. Be sure to include one or more screenshots supporting your discussion.

NOTES

1 Kimerling also illustrated the possibility of varying the amount of overlap from 0 to 100 percent.
2 Arnold et al. used the term "pseudo-random" for the initial random dot pattern, but they did not use pseudo-random in the manner that Kimerling (2009) did; rather, they used this terminology to indicate that the points were not truly random, as is often the case with random number generators.
3 Proportional circle maps were also included in the evaluation, but we do not mention them here because of our focus on traditional vs. graduated dot maps.
4 Walker used the term "dasymetric mapping" to describe the use of ancillary information. We chose not to use that term here because we reserve the term dasymetric mapping for areal symbols that cover an entire enumeration unit, as discussed in Chapter 16.

REFERENCES

Arnold, N. D., Jenny, B. and White, D. (2017) "Automation and evaluation of graduated dot maps." *International Journal of Geographical Information Science 31*, no. 12:2524–2542.
Astroth, J. H. J., Trujillo, J., and Johnson, G. E. (1990) "A retrospective analysis of GIS performance: The Umatilla Basin revisited." *Photogrammetric Engineering and Remote Sensing 56*, no. 3:359–363.
Balzer, M., Schlömer, T., and Deussen, O. (2009) "Capacity-constrained point distributions: A variant of Lloyd's method." *ACM Transactions on Graphics 28*, no. 3:Article 86, 1–8.
Clark, P. J., and Evans, F. C. (1954) "Distance to nearest neighbor as a measure of spatial relationships in populations." *Ecology 35*, no. 4:445–453.
Hey, A. (2012) "Automated dot mapping: How to dot the dot map." *Cartography and Geographic Information Science 39*, no. 1:17–29.
Hey, A., and Bill, R. (2014) "Placing dots in dot maps." *International Journal of Geographical Information Science 28*, no. 12:2417–2434.
Jenks, G. F. (1953) "Pointillism as a cartographic technique." *The Professional Geographer 5*, no. 5:4–6.

Kimerling, A. J. (2009) "Dotting the dot map, revisited." *Cartography and Geographic Information Science 36*, no. 2:165–182.

Mackay, J. R. (1949) "Dotting the dot map." *Surveying and Mapping 9*, no. 1:3–10.

Olson, J. M. (1977) "Rescaling dot maps for pattern enhancement." *International Yearbook of Cartography 17*: 125–136.

Peterson, D., Whistler, J., Bishop, C. et al. (2009) "The Kansas next-generation land use/land cover mapping initiative." *ASPRS 2009 Annual Conference*, Baltimore, Maryland, pp. 1–12.

Provin, R. W. (1977) "The perception of numerousness on dot maps." *The American Cartographer 4*, no. 2:111–125.

Robinson, A. H., Morrison, J. L., Muehrcke, P. C. et al. (1995) *Elements of Cartography* (6th ed.). New York: Wiley.

Walker, K. E. (2018) "Scaling the interactive dot map." *Cartographica 53*, no. 3:171–184.

Whistler, J. L., Egbert, S. L., Jakubauskas, M.E. et al. (1995) "The Kansas State Land Cover Mapping Project: Regional scale land use/land cover mapping using Landsat Thematic Mapper data." *1995 ACSM/ASPRS Annual Convention & Exposition*, Volume *3* (Technical Papers), Charlotte, NC, pp. 773–785.

20 Cartograms

20.1 INTRODUCTION

When creating thematic maps, cartographers generally avoid distorting spatial relationships. For example, an equivalent (or equal-area) projection is normally used for a dot map so that the density of dots is solely a function of the underlying phenomenon (and not the map projection). Sometimes, however, cartographers purposefully distort space based on values of an attribute; the resulting map projection (Tobler 2004) is known as a **cartogram** (or **value-by-area map**). Probably the most common type of cartogram used in everyday life is the **distance cartogram**, in which real-world distances are distorted to reflect some attribute, such as the time between stops on subway routes: Here, cartograms are appropriate because the time between (and order of) stops is more important than the actual distance between stops.[1]

In the geographic literature, the **area cartogram** is the most common form of cartogram that is discussed, and it will be our focus in this chapter. Area cartograms are created by scaling (sizing) enumeration units as a function of the values of an attribute associated with the enumeration units; for example, sizes of countries might be made proportional to the population of each country. It is common to contrast the area cartogram with an equivalent projection, as enumeration units on an equivalent projection are shown in their proper size relations (as they would appear on a globe). Although cartograms often greatly distort these size relations, a distinct advantage of cartograms is that small enumeration units with high attribute values that would be hidden on an equivalent projection can be readily seen. Population is a typical attribute portrayed on area cartograms, but any data based on raw totals (unstandardized data) can be utilized. As we will see in this chapter, a common application of cartograms is the display of election results.[2]

Two basic forms of symbolization are possible with cartograms. When a single attribute is mapped with a cartogram (i.e., the sizes of enumeration units are varied as a function of that attribute, say population), the result is a **univariate cartogram**. Here the emphasis is on visualizing the sizes of the enumeration units relative to one another. It is also possible to create a **bivariate cartogram** in which the sizes of enumeration units are a function of a raw total (typically population), and a symbol is then placed within each enumeration unit representing a second attribute. For example, we might create a cartogram showing the population of counties within a state, and place a choropleth symbolization within each county representing the median income of each county. We will consider several examples of bivariate cartograms in this chapter; for additional examples, see http://bivariate.weebly.com/.

A wide variety of methods have been developed for creating cartograms. One way of dividing these methods is to consider whether they attempt to preserve the shape of enumeration units. Section 20.3 considers methods that attempt to preserve the shape of enumeration units; here, we consider noncontiguous, contiguous, and mosaic cartograms. Section 20.4 considers methods that use geometric symbols (e.g., circles or squares) to represent enumeration units and thus do not preserve their shape; here, we consider rectangular, Dorling, and Demers cartograms.

As we discuss methods for creating cartograms, we will make use of various aspects of accuracy commonly utilized to contrast cartogram approaches.[3] **Statistical accuracy** refers to whether the numerical relations among attribute values are properly represented by the relative sizes of enumeration units representing those attribute values.[4] **Topological accuracy** refers to whether the spatial adjacencies of enumeration units normally found on an equivalent projection are maintained on the cartogram. **Shape accuracy** refers to how well the shapes of enumeration units are maintained on the cartogram relative to their true shape on Earth's surface (as depicted on a globe). **Positional accuracy** refers to how well the geographic position of enumeration units is maintained (e.g., if an enumeration unit is west of another enumeration unit on the untransformed map, then it should be west of that unit on the cartogram). **Perceptual accuracy** refers to how easy and accurately people can make estimates of the size relations among enumeration units. In Section 20.5, we contrast major cartogram methods with one another using these aspects of accuracy. In Section 20.5, we also consider a user study that examines the effectiveness of major cartogram methods from the user's perspective.

Although cartograms are sometimes utilized because of their dramatic visual effect, they are often criticized because of the extreme shape distortion that can result. In the last section of the chapter, we consider three methods that have been utilized as alternatives to conventional cartograms: a combined choropleth/proportional symbol map, value-by-alpha mapping, and the balanced cartogram. We use the term *conventional cartogram* to refer to cartograms in which enumeration units are scaled in direct proportion to the values of an attribute. We will see that this is not the case for the balanced cartogram.

DOI: 10.1201/9781003150527-22

20.2   LEARNING OBJECTIVES

- Contrast the univariate cartogram with a bivariate cartogram.
- Describe the differences among cartograms that attempt to preserve the shape of enumeration units: the noncontiguous, contiguous, and mosaic cartograms.
- Describe the differences among cartograms that do not preserve the shape of enumeration units: the rectangular, Dorling, and Demers cartograms.
- Define the following aspects of accuracy pertaining to cartograms: statistical, topological, shape, positional, and perceptual.
- Contrast major cartogram methods in terms of the above aspects of accuracy.
- Explain how the following approaches can be utilized to provide an alternative to conventional cartograms: the combined choropleth/proportional symbol map, the value-by-alpha map, and the balanced cartogram.

20.3   METHODS THAT ATTEMPT TO PRESERVE THE SHAPE OF ENUMERATION UNITS

In this section, we consider methods for creating cartograms that attempt to preserve the shape of enumeration units. These can be divided into noncontiguous cartograms, contiguous cartograms, and mosaic cartograms.

20.3.1   NONCONTIGUOUS CARTOGRAMS

With the **noncontiguous cartogram**, the positions and shapes of enumeration units are retained, and the areas of enumeration units are downscaled relative to the enumeration unit having the largest value. In Figure 20.1, we see an example in which each state in the United States has been scaled based on the number of persons 65 years of age and over. The large gaps between enumeration units (as seen in Figure 20.1) may make it difficult to develop a notion of the spatial pattern of the distribution, but the full retention of shape arguably eases map interpretation (except when enumeration units become very small).[5] An interactive example of the noncontiguous cartogram known as Bouncy Maps can be found at https://www.bouncymaps.com/.

There are two characteristics of area cartograms that you should be aware of. One is that area cartograms oftentimes do not include a legend. In Figure 20.1, a legend is included, but this is not common. In this case, the legend tells us that a square area of the given size contains 100,000 people. We can thus use this square area to make estimates of the elderly population of each state, but this becomes difficult because of the varied shapes of the states. This difficulty is why mapmakers often do not include legends, as is true

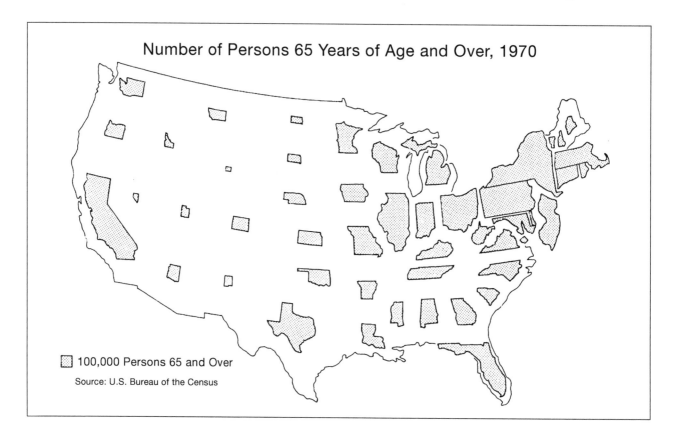

FIGURE 20.1   A noncontiguous area cartogram; the shape of enumeration units is retained, but gaps appear between units. (After Olson, J. M. (1976) "Noncontiguous area cartograms." *The Professional Geographer* 28, no. 4:372; original source U.S. Bureau of the Census.)

for the Bouncy Maps. Since the maps in Bouncy Maps are interactive, you can argue that a legend is unnecessary, as you can hover over a country and determine the precise data value associated with that country or scroll down and look at the table of population numbers.

Another characteristic of area cartograms is that data for the different sized areas generally are not classed (i.e., these are unclassed maps). This is arguably appropriate as the range in the size of areas represented is generally quite large, and so it is relatively easy to generate a large number of differently sized areas that are perceptually differentiable.

20.3.2 Contiguous Cartograms

With the **contiguous cartogram,** topological relations between enumeration units are retained (i.e., enumeration units adjacent to (or contiguous to) one another on an equivalent projection will also be adjacent on the cartogram), and there is an attempt to preserve shape. As an example, Figure 20.2 illustrates states and territories of Australia as they might appear (A) on an equivalent map projection and (B) on a contiguous cartogram when scaled in direct proportion to the population of each state or territory. In comparison with the noncontiguous cartogram, Borden Dent argued that the contiguous cartogram arguably best represents a "true" map because there is an attempt to maintain boundary relationships between enumeration units, there are no gaps between enumeration units, and the shape of the total study area is easier to maintain (Dent 1999, 209). A downside of the contiguous cartogram, however, is that the shapes of enumeration units may make them difficult to identify.

Traditionally, the digital construction of contiguous cartograms was complicated by two problems: the inability to retain correct topological relationships between enumeration units and slow execution times.[6] Michael Gastner and Mark Newman (2004, 2005) developed a novel approach that maintains topological relationships and is efficient (i.e., execution times are on the order of seconds).[7] Reflecting their physics background, Gastner and Newman (2005, 4) described their approach as follows:

> … one way to create a cartogram … is to allow population to "flow away" from high-density areas into low-density ones until the density is equalized everywhere. This … brings to mind the diffusion process of elementary physics….

Although the mathematics of their approach is beyond the scope of this text, Gastner and Newman presented numerous applications illustrating its effectiveness. Figure 20.3 illustrates one of these applications. Figure 20.3A is a dot map of lung cancer cases among males in New York State from 1993 to 1997 based on 1,598 zip codes on an equivalent map projection. It can be argued that this map is not particularly useful because the areas of concentration do not necessarily represent an increased likelihood of cancer but rather are indicative of higher population densities

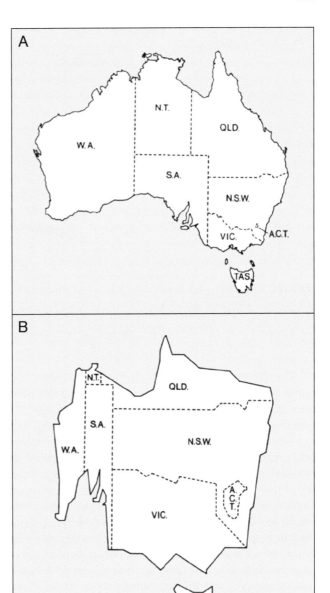

FIGURE 20.2 Creating a cartogram: (A) States and territories of Australia as they might appear on an equivalent projection; (B) a contiguous cartogram in which states and territories are scaled on the basis of 1976 population. (From Griffin 1983, 18; courtesy of T. L. C. Griffin.)

(note the heavy concentration in the New York City area in the southeast portion of the map). In contrast, Figure 20.3B is a cartogram created using the same data—in this case, the zip codes have been scaled to represent the underlying population, and the dots representing cancer cases have been plotted on the transformed map—this is thus an example of a bivariate cartogram. Now we can see that the dots are relatively uniformly distributed, suggesting that the incidence of lung cancer is not a function of environmental factors but rather individual behavior (Gastner and Newman 2005, 10).

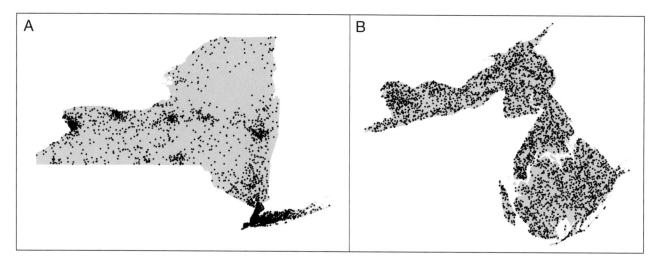

FIGURE 20.3 (A) A dot map of lung cancer cases for males in New York State from 1993 to 1997 based on 1,598 zip codes on an equivalent map projection; (B) a cartogram of the same data in which zip codes are scaled based on population. (From Gastner, M. T., and Newman, M. E. J. (2004) "Diffusion-based method for producing density-equalizing maps." *Proceedings of the National Academy of Sciences* 101, no. 20:7499–7504; Copyright (2004) National Academy of Sciences, U.S.A.)

The Worldmapper site (https://worldmapper.org/) provides an extensive collection of maps that further illustrates the usefulness of Gastner and Newman's approach. The purpose of Worldmapper is "to communicate to the widest possible audience … how parts of the world relate to each other and the implications of these relationships for society" (Dorling et al. 2006, 757).[8] This is accomplished by displaying a wide variety of data sets in cartogram form using the Gastner/Newman algorithm. Figure 20.4 illustrates an example map showing the projected population increase from 2015 to 2050. Colors on this map have been chosen to represent various continents and countries around the world; for example, the continent of Africa is shown in green, with different subregions of Africa shown in shades of green (for a summary of colors used for various world regions, see the *Colour Key* at https://tinyurl.com/WorldMapperPopIncrease).[9]

Figure 20.4 also illustrates what some have argued is the eye-catching appeal of cartograms (e.g., the drastic difference between the cartogram and a conventional equivalent projection of countries illustrates that major population increases are likely to occur in Africa), but we also see some of the difficulty of interpreting the map, as you may find it awkward to identify particular countries within Africa due to the distortion of geographic shape. The latter difficulty led Dorling et al. (2006) to argue that an interactive capability may ultimately need to be added to "allow users to 'brush' over the cartograms and see the names of territories change under their eyes … and morph from one projection to another to better illustrate just how

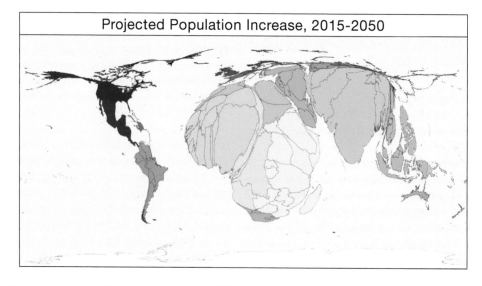

FIGURE 20.4 An example of a contiguous cartogram from Worldmapper that portrays projected population increases from 2015 to 2050. The map was created using the Gastner/Newman algorithm. (Courtesy of Worldmapper; www.worldmapper.org.)

unevenly peoples, goods and livelihoods are distributed over the surface of this planet" (762).[10] Other approaches for enhancing the interpretation of such contiguous cartograms include retaining characteristic features along boundaries of enumeration units (e.g., distinctive meanders of a river), providing an inset map depicting actual areal relations of enumeration units (or, alternatively, the units should be labeled), and consider not using contiguous cartograms if the anticipated readership is unfamiliar with the region depicted.

20.3.2.1 Gridded Cartograms

Although contiguous cartograms, such as those found in Worldmapper, can provide intriguing visualizations, they are limited in that they do not provide details about individual countries or territories. In response to this limitation, Benjamin Hennig (2014, 2019) developed the **gridded cartogram**. To create a gridded cartogram, a raster grid of data values is overlaid on a desired country or territory, and the resulting grid cells are then sized as a function of the values occurring within each grid cell using the Gastner/Newman algorithm. Although any quantitative attribute could be utilized to size the grid cells, a common approach is to map population, which results in a *gridded population cartogram*. Once the gridded population cartogram has been created, another attribute can then be mapped within the deformed grid cells to create a bivariate cartogram.

To illustrate the gridded cartogram, we will examine the U.S. presidential election results for 2020 shown in Worldmapper (https://worldmapper.org/maps/usa-politics-election-2020/) and depicted in Figure 20.5. In the upper right of this figure, you see the results of the election presented as a conventional choropleth map at the county level. Here you see that much of the United States is covered by a dark red, indicating that between 70 and 90 percent of the vote in these areas was for former President Donald Trump. Below the choropleth map is the gridded cartogram. Here you can see individual gridded cells, where each cell has been scaled to reflect the total number of people living in that cell. Each of the cells has then been colored using the same legend values used on the choropleth map, producing the bivariate cartogram. Obviously, the visual result for the gridded cartogram is quite different than for the choropleth map, as there appears to be a much greater amount of blue on the gridded cartogram, reflecting the larger number of votes that President Joe Biden received and that he ultimately won the election.[11] Hennig (2019, 76–77) notes that a gridded cartogram can be more difficult to read but that it "enables a more sophisticated understanding of the topic mapped." Since the distortion on a gridded cartogram is greater than on a nongridded cartogram, it is useful to add the names of key areas to the resulting map—thus, in this case, you can see that the names of major U.S. cities have been identified.

20.3.3 Mosaic Cartograms

As the name suggests, a **mosaic cartogram** is created by displaying a *mosaic* of small uniformly shaped geometric symbols (typically squares or hexagons) as a function of data that either consist of (or can be cast into) small integer units. A common example is the portrayal of electoral votes

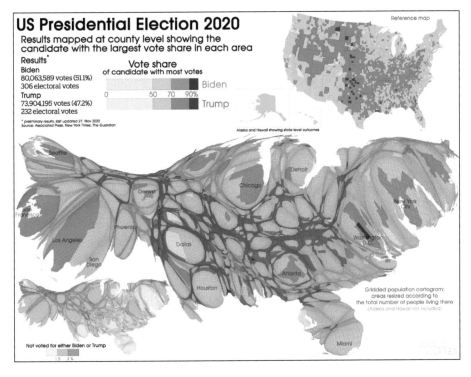

FIGURE 20.5 An illustration from Worldmapper that portrays the results of the 2020 U.S. presidential election using both a conventional choropleth map (upper right) and a gridded cartogram (bottom portion). (Courtesy of Worldmapper; www.worldmapper.org.)

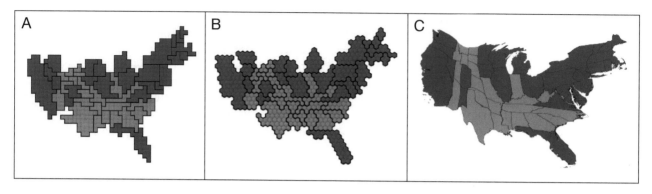

FIGURE 20.6 Comparison of square and hexagonal mosaic cartograms (A and B, respectively) with a contiguous cartogram (C) of electoral votes in the 2012 U.S. presidential election. States won by the Democrat (Barack Obama) are shown in blue, and states won by the Republican (Mitt Romney) are shown in red. (From Figure 1 of Cano, R. G., Buchin, K., Castermans, T., et al. (2015) "Mosaic drawings and cartograms." *Computer Graphics Forum* 34, no. 3:361–370, © 2015 The Author(s) *Computer Graphics Forum* © 2015 the Eurographics Association and John Wiley & Sons Ltd. Published by John Wiley & Sons Ltd.; map C courtesy of Mark Newman, University of Michigan.)

for U.S. presidential elections, as shown in Figure 20.6A and B.[12] Here we can see that the electoral votes for each state have been represented by a mosaic of either squares (Figure 20.6A) or hexagons (Figure 20.6B), with blue for states won by Barack Obama, a Democrat, and red for states won by Mitt Romney, a Republican. Note that, if desired, we can count the number of squares or hexagons (and thus the number of electoral votes) in each state and that some effort has been made to maintain the shape of each state (and thus our inclusion of this method in Section 20.3). Rafael Cano and colleagues (2015) describe algorithms for creating mosaic cartograms (from which Figure 20.6 was taken). Their algorithms produce no topological error, very low statistical error, and attempt to preserve the shapes of the input enumeration units. Although mosaic cartograms do not preserve shape as well as contiguous cartograms (Figure 20.6C), they seem particularly appropriate when one wishes the reader to be able to count the integer units making up individual enumeration units.

20.4 METHODS THAT DO NOT PRESERVE THE SHAPE OF ENUMERATION UNITS

In this section, we consider methods for creating cartograms that do not preserve the shape of enumeration units. These can be divided into rectangular cartograms, Dorling cartograms, and Demers cartograms based on the geometric symbol used to create each (rectangles, circles, and squares, respectively).

20.4.1 Rectangular Cartograms

A **rectangular cartogram** is created by sizing rectangles representing enumeration units as a function of the attribute being mapped. As with contiguous cartograms, an attempt is made to maintain the topology of the enumeration units. A classic example created by Erwin Raisz (1934) is shown in Figure 20.7, where the population of the United States by

state has been mapped for 1930. Raisz made the following argument for creating a rectangular cartogram:

> If a way could be found to increase the scale of the northeastern region and reduce that of the west, distribution could be shown more clearly. Simple distortion of the map would be misleading, but, if we go a step farther, discard altogether the outlines of the country, and give each region a rectangular from of size proportional to the value represented, we arrive at the rectangular statistical cartogram. (From p. 292 of "The rectangular statistical cartogram" by E. Raisz, *The Geographical Review* 1934, copyright © the American Geographical Society of New York, reprinted by permission of Taylor & Francis Ltd, http://www.tandfonline.com on behalf of the American Geographical Society of New York.)

Note that by using rectangles rather than the actual shape of enumeration units, it is easier to make detailed comparisons between enumeration units. For example, we can consider a *specific task* in which we compare the state of Minnesota ("MINN") with North Dakota ("N.D."), and readily see that Minnesota has about four times the population (this task obviously would be more difficult on a contiguous cartogram because of the different shapes involved). Raisz chose to include the actual data values on the map, but such comparisons could be made without the data values. Moreover, this argument for the ease of a specific task could be extended to a pattern task in which you might wish to compare the population of one region with another region. Note that Raisz also chose to make the size of urban areas proportional in size to their percent of a state's population (note the prominence of the "New York metropolitan district"). A downside of the rectangular cartogram is that it is difficult to associate the distribution with our mental map of the shape and distribution of U.S. states that we have formed from examining conventional map projections.

In recent years, a variety of researchers have developed digital methods for creating rectangular cartograms. One of the difficulties with developing a suitable algorithm is that it is not possible to simultaneously maintain both the statistical

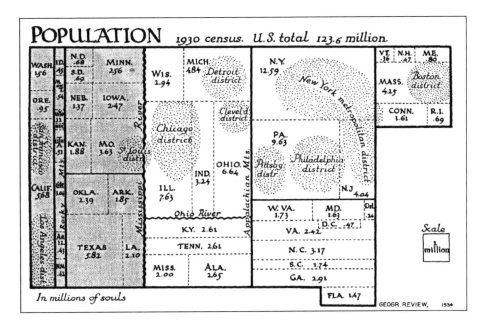

FIGURE 20.7 A rectangular cartogram created by Irwin Raisz that portrays the population of U.S. states in 1930. (From "The rectangular statistical cartogram" by E. Raisz, *The Geographical Review* 1934, copyright © the American Geographical Society of New York, reprinted by permission of Taylor & Francis Ltd., http://www.tandfonline.com on behalf of the American Geographical Society of New York.)

accuracy of the area relations between enumeration units and the topological accuracy of the spatial relations among enumeration units. For instance, based on an algorithm by Heilmann et al. (2004), Christian Panse (2018) developed an R package (https://cran.r-project.org/web/packages/recmap/index.html) that eliminates statistical error but can sometimes produce awkward topology. As a solution to this problem, Kevin Buchin and colleagues (2012) developed

an algorithm that produces more acceptable topologies and generally provides statistical error near zero. Figure 20.8 is an example that Buchin et al. created portraying electoral votes for the 2008 U.S. presidential election. This is an example of a bivariate cartogram as the size of each state is a function of the number of electoral votes, and the color within each state indicates whether the state was won by Barack Obama, a Democrat (in blue) or John McCain, a

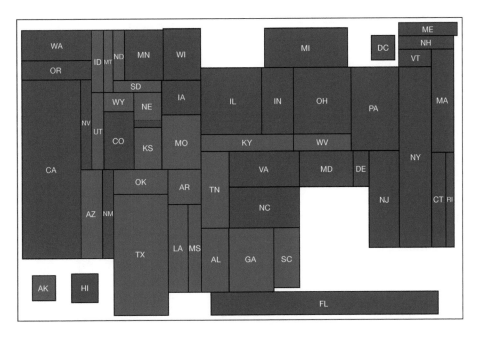

FIGURE 20.8 A rectangular cartogram portrays the electoral votes for the 2008 U.S. presidential election; the cartogram was created using an algorithm developed by Buchin et al. (2012). Blue and red represent those states won by Barack Obama and John McCain, respectively; Obama won the election. (Adapted by permission from Springer Nature: Springer eBook. *Geographic Information Science. GIScience* 2012 by N. Xiao, M.-P. Kwan, M. F. Goodchild, et al. (eds) COPYRIGHT (2012).)

Republican (in red). Although their approach generally produces rectangular cartograms that have acceptable topology and very low statistical error, Buchin et al. (2012) noted that when the attribute data being mapped has extremely high variability, it is difficult to create rectangular cartograms that simultaneously have correct topology, low statistical error, and a reasonable aspect ratio (40).

20.4.1.1 Rectilinear Cartograms

A **rectilinear cartogram** is a generalization of the rectangular cartogram in which enumeration units are represented by rectilinear polygons. A *rectilinear polygon* can be defined in two ways: (1) the line segments forming the polygon are all at right angles (the angle at each vertex is either 90° or 270°), or (2) the line segments forming the polygon are all parallel to either the *x* or *y* axis. If a rectilinear polygon is composed of only four line segments, then the resulting polygon is a rectangle.

Mark de Berg and colleagues (2010) described an algorithm for creating rectilinear cartograms; an example of a resulting map is shown in Figure 20.9. Here the gross domestic product (GDP) is shown for European countries (each country is labeled with a two-character abbreviation and areas of water are shown in blue). An advantage of rectilinear cartograms is that it is possible to have both 100 percent statistical accuracy and 100 percent topological accuracy (remember that rectangular cartograms cannot achieve this). In examining

Figure 20.9, we see that achieving these maximum levels of accuracy can be accomplished by using mostly rectangles. Of course, like rectangular cartograms, rectilinear cartograms exhibit extreme shape distortion, and the positions of enumeration units may appear awkward when compared to their true geographic positions (e.g., consider Slovenia (SI) and its position on an equivalent map projection).

20.4.2 DORLING CARTOGRAMS

When circles are used to represent enumeration units in a cartogram, the result is commonly termed a **Dorling cartogram** because of the digital techniques that Daniel Dorling (1993, 1995b) developed to create such maps. To create a Dorling cartogram, a uniformly shaped symbol (typically, a circle) is placed in the center of each enumeration unit, with the size of the symbol a function of an attribute (typically population). Initially, symbols overlap one another because small enumeration units can represent relatively large populations (Figure 20.10A). To eliminate overlap, an iterative procedure is executed in which symbols are gradually moved away from one another (Figure 20.10B–E). Wherever possible, points of contact between symbols should reflect points of contact between actual enumeration units, but sometimes it is not possible to meet this constraint (i.e., topology cannot necessarily be maintained). As a result, Dorling termed this a "noncontiguous form of cartogram," but the result

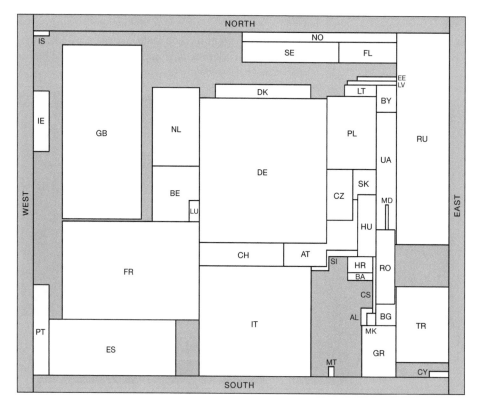

FIGURE 20.9 An example of a rectilinear cartogram depicting the gross domestic product (GDP) of European countries. Two-character abbreviations are utilized for countries of Europe, and water bodies are shown in blue. (Republished with permission of World Scientific Publishing Co., Inc., from "Optimal BSPs and rectilinear cartograms" by M. de Berg, E. Mumford, and B. Speckmann,. 20, no. 2, 2010; permission conveyed through Copyright Clearance Center, Inc.)

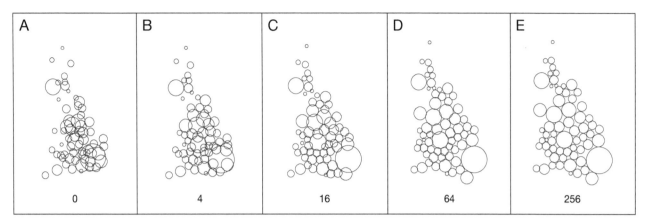

FIGURE 20.10 Dorling's algorithm for creating cartograms: (A) Uniformly shaped symbols (circles, in this case) are scaled on the basis of population and are placed in the center of enumeration units on an equivalent projection; (B–E) the circles are gradually moved away from one another so that no two overlap. The numbers beneath each illustration represent the number of iterations in the algorithm. (After Dorling 1995b, 274.)

is obviously quite different than the noncontiguous carto-grams described in Section 20.3.1.

The result of applying Dorling's algorithm to the population of 9,289 wards of England and Wales is shown in Figure 20.11. Figure 20.11A depicts the boundaries of wards as they would appear on a traditional equivalent projection, and Figure 20.11B displays the resulting cartogram. (The upper right portion of each figure portrays wards grouped into counties.) The basic difference between the maps is that small land areas with large populations are much more

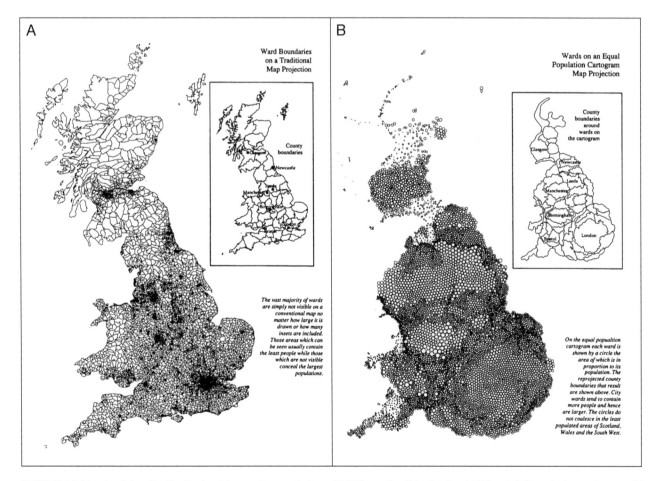

FIGURE 20.11   Applying Dorling's algorithm to the population of 9,289 wards of England and Wales: (A) Boundaries as they would appear on a traditional equivalent projection; (B) a cartogram resulting from applying Dorling's algorithm. (From "Map design for census mapping," by D. Dorling (1993), *The Cartographic Journal*, copyright © The British Cartographic Society, reprinted by permission of Taylor & Francis Ltd., http://www.tandfonline.com on behalf of The British Cartographic Society.)

apparent on Dorling's cartogram than they are on the equivalent projection. This is particularly apparent on the maps of county boundaries: Note that London is a mere dot on the traditional map but that London encompasses a substantial portion of the cartogram.[13]

A major purpose of Dorling's approach is not to map population per se but, rather, to serve as a base on which other attributes can be displayed (i.e., a bivariate cartogram is created). For example, Figure 20.12 portrays population density in Britain on both a traditional equivalent projection (the result is a choropleth map) and on Dorling's cartogram using the same data classification and color scheme.[14] Note the drastic difference in appearance between these two maps. The equivalent projection (A) suggests that most of the country is dominated by relatively low population densities (the blues and greens), whereas the cartogram (B) provides a detailed picture of the variations in population densities within urban areas. Dorling contends that his cartogram depicts the *human*

geography of a region rather than focuses on its *physical extent*, as depicted on the choropleth map.

One obvious difference between Dorling's cartograms and the noncontiguous and contiguous cartograms shown in Section 20.3 is that Dorling's cartograms provide no shape information for enumeration units. As a result, the addition of a basemap of the region on a conventional equivalent projection (either as an inset or a comparison map) would seem essential when examining Dorling's cartograms, particularly if the map reader is unfamiliar with the region depicted. Another difference is that Dorling's cartograms typically show a large number of enumeration units and thus considerable detail. As a result, readers must study these maps meticulously. In fact, the complexity of Dorling's cartograms suggests that they might be examined most effectively in a data exploration environment. For example, imagine having both a traditional equivalent projection and a cartogram displayed on a computer screen at the same time. As the cursor is moved over an enumeration

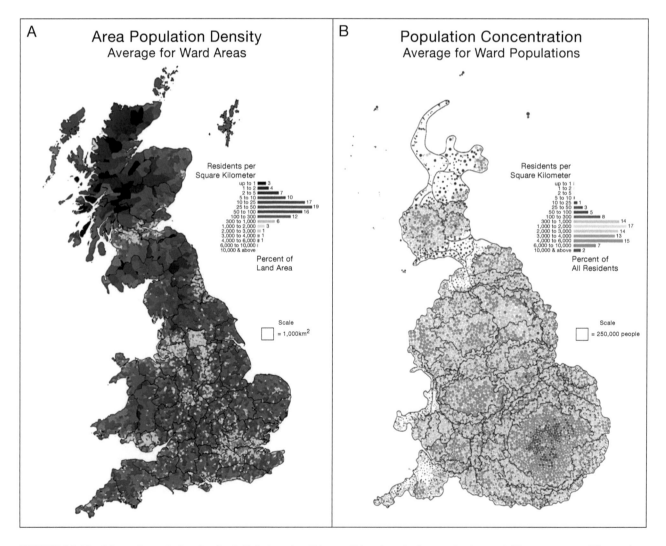

FIGURE 20.12 Maps of population density in Britain using (A) a traditional equivalent projection and (B) a cartogram. The equivalent projection suggests that most of Britain is dominated by relatively lower population densities (the blues and greens), whereas the cartogram provides a detailed picture of the variation in population densities within urban areas. (From Dorling 1995a, xxxiii. Courtesy of Danny Dorling.)

unit on the cartogram, this same enumeration unit could be highlighted and identified on the equivalent projection (or vice versa). Alternatively, the user could zoom in to examine subregions in greater detail on both maps.

20.4.3 DEMERS CARTOGRAMS

When squares are used to represent enumeration units on a cartogram, the result is commonly termed a **Demers cartogram**, as this technique was developed by Steve Demers.[15] An advantage of using squares is that they can be packed more compactly than circles and thus can capture the topology more effectively than circles (Nusrat and Kobourov 2016, 627). A potential disadvantage of using squares is that squares are not as aesthetically pleasing as circles (see Section 18.4). For an example of a Demers cartogram, see https://www.nytimes.com/elections/2012/ratings/electoral-map.html, where it is used to depict the results of the 2012 U.S. presidential election.

20.5 CONTRASTING VARIOUS CARTOGRAM METHODS

In this section, we contrast the various methods for creating cartograms that we introduced in the preceding two sections. First, we contrast the methods based on various aspects of accuracy (e.g., how well the method preserves the shape and topology of enumeration units). Then we consider a user study that evaluated the effectiveness of major cartogram methods from the user's standpoint.

20.5.1 CONTRASTING CARTOGRAM METHODS IN TERMS OF ASPECTS OF ACCURACY

Figure 20.13 provides a summary comparison of the major methods for creating cartograms based on five aspects of accuracy: statistical, topological, shape, positional, and perceptual. More effective approaches are depicted with a darker green. Note that the results for the contiguous and rectangular approaches will depend on the algorithm that is used to construct the cartogram. Here we assume that the Gastner/Newman and Buchin et al. algorithms described previously are utilized to construct the contiguous and rectangular cartograms, respectively.

Looking at *statistical accuracy* in Figure 20.13, the noncontiguous, Dorling, and Demers approaches all can represent the attribute values without error and so are most effective. The contiguous and rectangular methods, however, are only slightly less effective as they generally represent attribute values with little or no error. It should be stressed that at this point, we are considering the mathematical relations between areas, not their perceived relations—the latter is dealt with in the "perceptual accuracy" criterion discussed later. We see that *topological accuracy* is maintained on the contiguous, rectangular, and Demers approaches, but that there is some loss in topological accuracy with the Dorling method, and topology definitely is not maintained with the noncontiguous method (as there are distinct gaps between the resulting enumeration units).

For *shape accuracy*, the noncontiguous method is clearly most effective, as the shapes of enumeration units are not distorted, although the shapes may be difficult to determine when enumeration units become very small. In contrast, the shape clearly is not maintained with the rectangular, Dorling, and Demers methods, which all utilize geometric symbols. Note that although the shape of individual enumeration units is not maintained in the Dorling and Demers methods, the general shape of the overall region being depicted can often be recognized, and so we have depicted these as being slightly more effective than the rectangular method. The contiguous approach does provide some shape information, although, as we saw in Figure 20.4, some enumeration units may be distorted

	Noncontiguous	Contiguous (Gastner/ Newman)	Rectangular (Buchin et al. 2012)	Dorling	Demers
Statistical Accuracy	No Error	Minimal Error	Minimal Error	No Error	No Error
Topological Accuracy	Not Maintained	Maintained	Maintained	Some Loss	Maintained
Shape Accuracy	Maintained	Moderately Maintained	Not Maintained	Individual Units Not Maintained	Individual Units Not Maintained
Positional Accuracy	Maintained	Generally Maintained	Often Not Maintained	Sometimes Not Maintained	Sometimes Not Maintained
Perceptual Accuracy	Difficult	Most Difficult	Somewhat Difficult	Somewhat Difficult	Easiest

FIGURE 20.13 Comparison of major methods for creating cartograms using various accuracy criteria. More effective methods are shown with a darker green.

considerably, which can make it difficult to recognize the shape of individual units. The results for *positional accuracy* are similar to that for shape accuracy, with the noncontiguous method again clearly most effective.

As we indicated in the introduction, we utilize *perceptual accuracy* to refer to how easily and accurately people can make estimates of the size relations among enumeration units. Here we assume that one wishes to make ratio estimates (e.g., one area is three times as large as another) as opposed to ordinal estimates (e.g., one area is smaller or larger than another). The Demers method should be most effective because regular shapes (squares) are being compared, and there is little estimation error involved in estimating size relations between squares (see Section 18.5.2.1). The Dorling method utilizes regular shapes (circles), but there is a greater error in estimating circle sizes than squares (again, see Section 18.5.2.1). The rectangular approach uses regular shapes (rectangles), but these vary in length and width, which certainly will make estimates more difficult. The noncontiguous and contiguous approaches both involve comparing nongeometric shapes, which should be more difficult than geometric shapes, with the potentially highly irregular shapes of the contiguous approach the most difficult.

20.5.2   A User Study of Major Cartogram Methods

We based Figure 20.13 on a survey of the literature dealing with cartograms and on a visual examination of numerous cartograms. In determining the effectiveness of cartograms, it is also useful to consider user studies that have evaluated how effectively people can utilize them.[16] For our purposes, we focus on a recent study by Sabrina Nusrat and colleagues (2018) that evaluated four types of cartograms: noncontiguous, contiguous, Dorling, and rectangular. Thirty-three participants were asked to perform the following broad range of visualization tasks: locate, compare, detect change, find top-k, summarize, adjacency, and recognize. Participant responses were measured in terms of both accuracy and response time. *Locate* involved finding an enumeration unit in a cartogram given its location on an undistorted map. Nusrat et al. hypothesized that noncontiguous and contiguous cartograms would be more effective for this task because the position of enumeration units is better preserved in these approaches. Similarly, the Dorling method should be more effective than the rectangular approach. *Compare* involved comparing two enumeration units on the cartogram and indicating which unit was larger. Nusrat et al. hypothesized that contiguous cartograms would be more effective than either Dorling or rectangular cartograms, arguing that people have difficulty estimating circle sizes (for Dorling cartograms) and dealing with the changing aspects ratios (for rectangular cartograms). Note, however, that this ordinal task is simpler than the ratio task that we assumed for "perceptual accuracy" in

Figure 20.13. *Detect change* involved comparing the size of an enumeration unit on the cartogram with its size on an undistorted map and indicating whether the unit has grown or shrunk.[17] *Find top-k* involved finding either the highest or the second-highest enumeration unit on the map. As with the compare task, Nusrat et al. hypothesized the contiguous cartogram would perform best for both the detect change and find top-k tasks.

Summarize involved asking participants to find patterns and trends in the cartograms. Here Nusrat et al. hypothesized that the rectangular cartogram would be least effective because it distorts position, shape, and topology. In discussing Figure 20.13, we noted that topology is retained with the Buchin et al. rectangular algorithm, but note that the visualization of that topology appears different on a rectangular cartogram than on an undistorted map (e.g., the length of boundary between two units is not the same on the two maps). *Adjacency* involved determining which of four enumeration units was a neighbor of a highlighted enumeration unit on the cartogram. Contiguous and rectangular cartograms were hypothesized to be most effective as they preserve the topology of enumeration units. *Recognition* involved determining whether an enumeration unit on the cartogram could be matched with a unit on the undistorted map. This task was included only for the noncontiguous and contiguous approaches because the results for Dorling and rectangular cartograms were expected to lead to low accuracy and response times.

The results of the user study largely supported the above hypotheses. Overall, contiguous cartograms were most effective for almost all of the tasks. Dorling cartograms were effective in providing a "big picture" (e.g., participants scored well on the summarize task). Although noncontiguous cartograms were particularly effective on the recognition task, they performed poorly on the adjacency task. Rectangular cartograms were clearly the least effective approach.

In addition to completing the above tasks, participants were also asked to provide a subjective appraisal of the four cartogram techniques. There was a clear preference for both the contiguous and Dorling cartograms, although Nusrat et al. noted that this might be a function of their more common use.

Nusrat et al. stressed the potential that interactive graphics could provide for cartograms. In this respect, some of the difficulties that users had with utilizing the various approaches could be obviated if they were able to interact with the display. For instance, the attribute values for enumeration units could be provided through a simple mouse-over. Similarly, an option could be provided to show all adjacent neighbors when hovering over a particular enumeration unit. As we have discussed previously, the cartogram and undistorted map displays could also be linked with one another so that when an enumeration unit is selected in one display, the same enumeration unit could also be highlighted in the other display.

20.6 ALTERNATIVES TO CONVENTIONAL CARTOGRAMS

In this section, we consider three alternatives to conventional cartograms: combined choropleth proportional symbol maps, value-by-alpha maps, and balanced cartograms. Although these approaches have been utilized largely for bivariate cartograms, we will see that the principle used to create balanced bivariate cartograms could also be used to create balanced univariate cartograms.

20.6.1 COMBINED CHOROPLETH/ PROPORTIONAL SYMBOL MAPS

As an illustration of the use of a combined choropleth/ proportional symbol map as an alternative to a bivariate cartogram, we will consider a user study by Silvan Kaspar and colleagues (2011). For the bivariate cartogram, Kaspar et al. utilized the Gastner/Newman algorithm to create a contiguous cartogram. The size of each enumeration unit in the cartogram was a function of raw totals (e.g., the number employed), and the shading within the enumeration unit was a function of standardized data (e.g., the percent unemployed), as depicted in Figure 20.14A.[18] For the traditional mapping approach, a choropleth map was utilized to depict the standardized data and overlaid proportional symbols were utilized to depict the raw totals, as depicted in Figure 20.14B. In addition to comparing the two mapping techniques with one another, Kaspar et al. also examined the effect of using regularly shaped units (i.e., counties in Kansas) vs. irregularly shaped units (i.e., communes in a canton of Basel), and simple (i.e., specific information) vs. complex (i.e., general, or map pattern, information) questions posed to the user study

participants. Both accuracy and response time were measured for 50 participants in the study.

The results revealed that simple questions were answered with very high accuracy on both types of maps (there was in excess of 95 percent accuracy for both map types). For complex questions, however, there was a notable and significant difference, with the choropleth/proportional symbol approach having 68 percent correct answers and the cartogram 56 percent correct answers. Interestingly, the difference for the complex questions was even more dramatic when the shape of the enumeration units was considered. The percentage of correct answers for the complex questions was similar when irregular units were utilized (71 percent vs. 67 percent for the traditional and cartogram, respectively) but was very different when regular units were utilized (65 percent vs. 47 percent for the traditional map and cartogram, respectively). One potential reason for this difference is suggested by Figure 20.15, which shows the transformation of irregularly shaped (A) and regularly shaped (B) enumeration units. Clearly, the shape of the regularly shaped unit is changed dramatically. The results of this study suggest that the contiguous cartogram can be nearly as effective as a traditional choropleth/proportional symbol scheme when irregular enumeration units are utilized. However, when regular enumeration units are utilized, the choropleth/proportional symbol approach would be recommended.

Although Kaspar et al.'s study suggested that the choropleth/proportional symbol approach was overall more effective, we should realize that this combined symbology would be awkward when a large number of enumeration units are utilized, and the areas of the enumeration units vary greatly in size, thus producing some small enumeration units. If the small enumeration units are associated with an urban area,

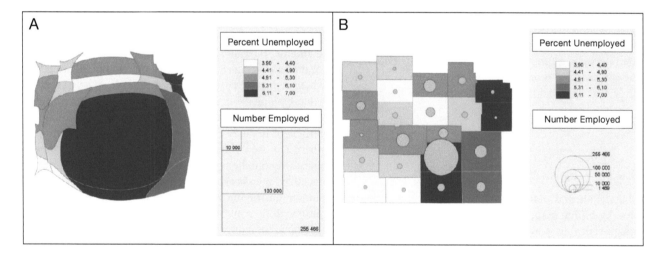

FIGURE 20.14 Comparison of a bivariate cartogram created using the Gastner/Newman algorithm (A) and a combined choropleth/ proportional symbol map (B). (After Figure 2 of Kaspar, S., Fabrikant, S. I., and Freckmann, P. (2011) "Empirical study of cartograms." *Proceedings of the 25th International Cartographic Conference*, Paris, France, pp. 1–8, Creative Commons Attribution 4.0 License, https://www.ica-conference-publications.net/licence_and_copyright.html; text for maps has been translated into English.)

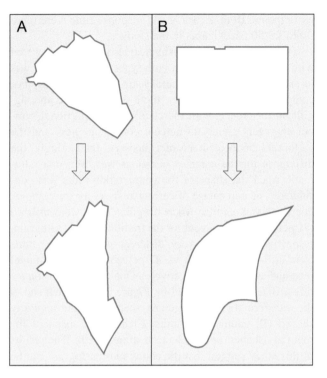

FIGURE 20.15 Illustration of the effect of the Gastner/Newman algorithm on irregular (A) and regular (B) shapes. Regular shapes clearly are more affected by the algorithm. (After Figure 12 of Kaspar, S., Fabrikant, S. I., and Freckmann, P. (2011) "Empirical study of cartograms." *Proceedings of the 25th International Cartographic Conference*, Paris, France, pp. 1–8, Creative Commons Attribution 4.0 License, https://www.ica-conference-publications.net/licence_and_copyright.html; colors of image were changed.)

then the raw count attribute would tend to be large, which would cause circles to overlap one another and hide the choropleth information.[19] Thus, the combined choropleth/proportional symbol approach will not work in many situations. An alternative traditional symbolization is to again utilize proportional symbols for raw totals but to shade their interior as a function of a standardized attribute rather than applying a choropleth symbology to each enumeration unit.[20] This approach, however, does not solve the problem of circles overlapping one another when enumeration units are small.

20.6.2 Value-by-Alpha Maps

Robert Roth and colleagues (2010) developed the novel **value-by-alpha map**, which utilizes the visual variable **transparency** to portray what they term the *equalizing attribute* (e.g., population) that is normally used to size enumeration units in a conventional cartogram. Enumeration units with low population values are made more transparent, whereas enumeration units with high population values are made less transparent. The attribute of interest (such as

election data results) is then mapped within enumeration units. The net result is the creation of a *spotlight effect* for those areas having low transparency.

Figure 20.16 is an example of a value-by-alpha map of U.S. presidential election results by county for 2008. Examining the legend, we see that counties with higher populations have lighter blues and reds (they are less transparent), whereas counties with lower populations have darker blues and reds (they are more transparent). The attribute of interest, in this case, is whether Obama or McCain had the majority of votes in the county (blue and red are used for counties where Obama and McCain had a majority of votes, respectively). Note that the background for this map is shown in black, which enhances the spotlight effect, as lighter counties tend to pop out against the dark background. In this case, a relatively large portion of the United States with a lower population tends to merge with the background spotlighting the more highly populated urban areas. Also, note that the value-by-alpha approach does not distort the shape of enumeration units, which is a common criticism levied against traditional cartograms.

The term *alpha* in *value-by-alpha map* comes from the usage of the *alpha channel* in computer graphics. In an interactive graphical display, individual picture elements (**pixels**) are composed of red, green, and blue primary colors. A separate alpha channel is used to determine the level of transparency (or alternatively *opacity*) of each pixel. In Figure 20.16, the values of the alpha channel would be varied as a function of the population values. Assuming that alpha values range from 0 to 100 percent, the background would be depicted with an alpha of 0 percent (completely opaque), the highest population class would be depicted with an alpha of 100 percent (completely transparent), and intermediate classes would have values between 0 and 100 percent (Roth et al. suggested alpha values of 10, 30, 55, and 80 percent for the intermediate classes).[21]

Roth et al. noted three limitations of the value-by-alpha technique. First, they noted that classified data are utilized for the equalizing variable (e.g., population in Figure 20.16), which reduces that data to an ordinal level. As such, the value-by-alpha technique should not be used when there is a desire to portray numerical information for the equalizing variable. This is in contrast to the conventional cartogram where the use of the visual variable size allows considerable differentiation in the areas of enumeration units (an unclassed map is typically created) and thus greater potential for portraying numerical information. For the same reason, Roth et al. do not recommend the value-by-alpha technique for a univariate cartogram, "as the resulting map would essentially be a choropleth map with a greyscale sequential color scheme" (p. 137).

A second limitation of value-by-area maps is that small enumeration units may be difficult to read even if the equalizing variable is completely transparent. This is especially

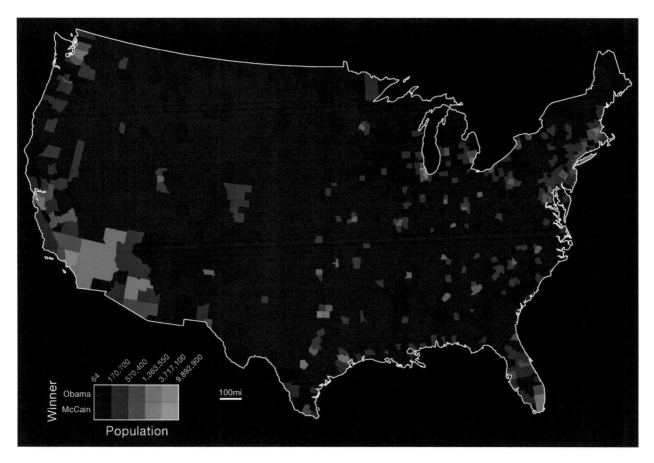

FIGURE 20.16   A value-by-alpha map of the 2008 U.S. presidential election by county. (From "Value-by-alpha maps: An alternative technique to the cartogram," by R. E. Roth, A. W. Woodruff, and Z. F. Johnson, *The Cartographic Journal*, copyright © The British Cartographic Society 2010, reprinted by permission of Taylor & Francis Ltd., http://www.tandfonline.com on behalf of The British Cartographic Society.)

problematic because small enumeration units are often found in urban areas where the population values are highest. In fact, this characteristic is the argument made for distorting areas as a function of population in conventional cartograms. A third limitation is that the visual variable of color used to depict the equalizing variable in value-by-alpha maps will not hold up well in different print and digital media, whereas the equalizing variable of size used in conventional cartograms will be unaffected by different print and digital media.

20.6.3   Balanced Cartograms

Richard Harris and colleagues (2017) provide an interesting critique of the conventional contiguous cartogram, stating:

> Most cartograms create a high level of geographical distortion, changing both the shape and locations of the areas on the map, potentially reversing the problem they are trying to solve: the areas of lowest population density become the smallest on the map and, worse, their illegibility is compounded by the distortion of their shape. (1945)

As a solution to this problem, they suggest a modified form of the contiguous cartogram in which all enumeration units are scaled based on their physical size, except that any place that would fall below a specified interpretability threshold is set to that threshold. The result is termed a **balanced cartogram**, as there is a *balance* between *distortion* and *geographic interpretation*—some distortion of enumeration units is permitted, but the distortion is not so great that the interpretation of the geography suffers. This approach is illustrated in Figure 20.17 for the 2015 United Kingdom General Election, where we see a traditional undistorted map on the left and a balanced cartogram on the right. In this case, Harris et al. set the threshold to 2 mm^2—thus, enumeration units that would normally be scaled to smaller than 2 mm^2 would simply be set to an area of 2 mm^2. Harris et al. argue that the resulting map appears much easier to interpret than a conventional cartogram yet still provides enhanced detail for urban areas (such as London within the area outlined in black). Given the difficulty that readers often have with interpreting conventional cartograms, the balanced cartogram seems like an idea worth exploring in greater depth.[22]

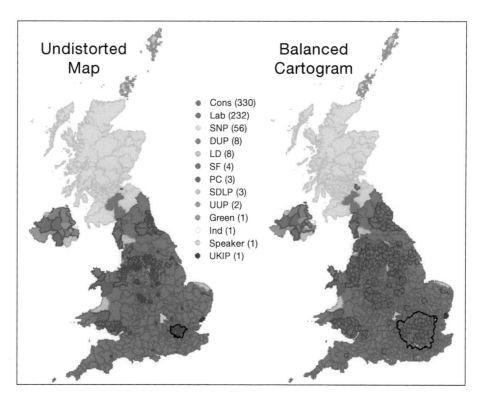

FIGURE 20.17   A comparison of a traditional undistorted map (A) and a balanced cartogram (B) of the 2015 United Kingdom General Election results. (After Figure 1 of Harris, R., Martin, C., Brunsdon, C., et al. (2017) "Balancing visibility and distortion: Remapping the results of the 2015 UK General Election." *Environment and Planning A* 49, no. 9:1945–1947, https://journals.sagepub.com/doi/full/10.1177/0308518X17708439.)

20.7   SUMMARY

In this chapter, we focused on the area cartogram, a map projection that sizes enumeration units as a function of an attribute, such that larger values of an attribute are represented by larger enumeration units. The resulting distortion of geographic area obviously moves us away from the equivalent projections that are commonly used with thematic maps. Cartograms are also symbolization methods that can use color schemes, like choropleth symbolization, to represent quantitative data. If only a single attribute is mapped (such as population), and no color scheme is used, then we term the result a univariate cartogram. If we wish to map a second attribute within each scaled enumeration unit, such as the percent voting for a particular candidate, then a color scheme is used, and the result is termed a bivariate cartogram.

One way of categorizing area cartograms is to contrast those that tend to maintain the shape of enumeration units (the noncontiguous, contiguous, and mosaic cartograms) with those that do not maintain the shape of enumeration units (the rectangular, Dorling, and Demers cartograms). Noncontiguous cartograms retain the positions and shapes of enumeration units by downscaling enumeration units within their original undistorted geographic locations. Retaining correct positions and shapes, however, eliminates a clear expression of topological relations, as there will be gaps between enumeration units. In contrast, contiguous cartograms retain the correct topology of enumeration units but often greatly distort the shape of individual enumeration units. The contiguous cartogram (typically created using the Gastner/Newman algorithm) has been much more commonly used than the noncontiguous cartogram, possibly because of the "shock value" provided by the resulting distortion. The mosaic cartogram is a highly specialized technique that is appropriate for integer-based data (such as electoral votes for U.S. states)—although it does attempt to show the shapes of individual enumeration units, these shapes are not nearly as clear as on the noncontiguous or contiguous cartogram.

The rectangular, Dorling, and Demers cartograms are distinguished on the basis of the geometric symbols used to represent enumeration units (rectangles, circles, and squares, respectively). The Dorling and Demers methods appear to do a better job of maintaining the overall shape of the region formed by a set of enumeration units. The Demers approach is arguably appropriate if one wishes to make numerical estimates of the sizes of enumeration units (squares are more accurately estimated than circles) and because the topology of enumeration units can be better retained with squares than circles. Circles are, however, more aesthetically pleasing, which possibly has been part of the explanation for the widespread use of the Dorling cartogram.

We contrasted major cartogram methods using the following aspects of accuracy: statistical, topological, shape, positional, and perceptual. An important concept is that it is not possible to simultaneously have statistical, topological, shape, and positional accuracy with cartograms. For instance, it is possible to create a Demers cartogram that maintains statistical and topological accuracy, but the result will likely distort shape and/or positional accuracy. Another key concept is that the statistical accuracy is mathematical in nature—if the relations between enumeration units are mathematically accurate, they may not be perceived accurately (e.g., circles are not estimated as accurately as squares, and thus there is a need to consider perceptual accuracy).

Given the extreme shape distortions that can result from using cartograms, we discussed three approaches as alternatives to the conventional cartogram: a combined choropleth/proportional symbol map, a value-by-alpha map, and a balanced cartogram. The value-by-alpha approach is intriguing because it eliminates the distortion found in traditional cartograms while providing an attractive spotlight effect that emphasizes high values associated with the equalizing variable (e.g., population) in a bivariate cartogram. The balanced cartogram reduces the extreme distortion found in conventional cartograms by setting a lower threshold on how small an enumeration unit can become. Although we illustrated a balanced *bivariate* cartogram, the approach could also be used to create a balanced *univariate* cartogram.

20.8 STUDY QUESTIONS

1. Assume that you have collected data on the total number of COVID-19 cases and the death rates due to COVID-19 for counties in New York State over the course of the COVID-19 pandemic. (A) Indicate whether these data sets are unstandardized or standardized, and if a data set is standardized, indicate how it was likely standardized. (B) Describe how you would use the contiguous cartogram technique to display these data. (C) What terminology would you use to describe your resulting cartogram or cartograms?

2. (A) Go to the Bouncy Maps (https://www.bouncymaps.com/) and examine the default map of "Population, 2019." Which three countries have the largest populations? (B) Now select Health and choose the Obesity attribute. Which three countries have the largest number of obese individuals? How do these results compare with part (A)? (C) Which parameters of *accuracy* described in this chapter are met by the maps in Bouncy Maps? (D) Which parameter of *accuracy* is clearly not met, and what is the ramification of this failure?

3. Go to Worldmapper (https://worldmapper.org/) and select Maps, People, Population, and "Population Year 2018." Click on the resulting map. (Note that you can also download a "poster" describing this attribute.) (A) Ignoring the slightly different date, how does your visualization of population in Worldmapper compare to your visualization of population in Bouncy Maps? (B) Which parameters of accuracy are clearly maintained (and not maintained) in Worldmapper? (C) Scroll through the various topics found in Worldmapper, pick one topic that you find interesting, and comment on the resulting map.

4. (A) Contrast the mosaic cartogram shown in Figure 20.6A with the contiguous cartogram shown in Figure 20.6C. How effective do you feel each map is? (B) Describe an interactive system that could assist in interpreting these two maps.

5. If you had to add a rectilinear cartogram to Figure 20.13, explain how the result would compare to the rectangular cartogram?

6. Using information presented in this chapter, contrast the Dorling and Demers cartograms with one another.

7. A classmate tells you that they find contiguous bivariate cartograms difficult to interpret because of their extreme distortion and that they prefer to utilize a combined choropleth/proportional symbol approach. Indicate a situation in which you suspect contiguous bivariate cartograms would perform nearly as well as the combined approach and also a situation in which the combined approach clearly would not work.

8. Discuss how a balanced cartogram is able to portray the character of highly populated urban areas that oftentimes are not properly displayed on an undistorted map (i.e., on an equivalent projection) and yet not produce the considerable distortion that is often found on cartograms.

NOTES

1. A classic example of a distance cartogram was Henry Charles Beck's 1930s cartogram of the London subway system (Garland 1994).

2. Although standardized data could conceivably be mapped using this technique, the narrow range of data values associated with standardized data would unlikely produce clear differences in the sizes of enumeration units—standardized data for enumeration units are best mapped using the choropleth technique.

3. The general notion of examining the accuracy of cartogram methods was taken from Nusrat and Kobourov (2016). They did not explicitly consider *perceptual accuracy*.

4. Another term for statistical accuracy would be "cartographic accuracy."

5. Olson (1976, 377) argued that the large gaps served to stress the difference in data values among the enumeration units separated by the gaps.

6. For a review of various approaches that have been used to create contiguous cartograms, see Tobler (2004).

7. An enhanced version of the Gastner/Newman algorithm can be found in Gastner et al. (2018).

8 For a more recent overview, see Hennig (2019).

9 Note that the different colors are used to differentiate regions of the world from one another—the colors have no quantitative meaning. The same color scheme is used across the bulk of the maps found in Worldmapper making it easier to determine the country or region portrayed at a particular location.

10 For examples of transforming between a conventional map projection and a cartogram, see http://metrocosm.com/how-to-make-cartograms-with-animation/.

11 Since the winner of the U.S. presidential election is determined by electoral votes rather than actual votes cast, care must be taken in interpreting the maps shown in Figure 20.5. For a discussion of the electoral vote process, see https://www.usa.gov/election.

12 Again, for a discussion of the electoral vote process, see https://www.usa.gov/election.

13 No theme is displayed on the traditional map because just the boundaries of wards appear; the dark areas represent where small wards bleed together.

14 You can see the latter if you examine the legend of each map carefully.

15 See http://www.ncgia.ucsb.edu/projects/Cartogram_Central/types.html.

16 For a summary of prior user studies, see Nusrat et al. (2018, 1105–1106).

17 The noncontiguous method was not considered for the change task due to the manner in which the sizes of enumeration units are computed (Nusrat et al. 2018, 1111–1112).

18 Note that we have translated the original maps to the English language.

19 The problem of circle overlap might be handled by using a carefully selected set of range graded circles.

20 For an example, see Gao et al. (2019).

21 For more detail on approaches for creating alpha-by-value map, see Roth et al. (2010, pp. 134–137).

22 For some of their more recent work with balanced cartograms, see Harris et al. (2018).

REFERENCES

Buchin, K., Speckmann, B., and Verdonschot, S. (2012) "Evolution strategies for optimizing rectangular cartograms." In *Geographic Information Science. GIScience 2012*, ed. by N. Xiao, M.-P. Kwan, M. F. Goodchild, et al., pp. 29–42. Berlin: Springer.

Cano, R. G., Buchin, K., Castermans, T., et al. (2015) "Mosaic drawings and cartograms." *Computer Graphics Forum 34*, no. 3:361–370.

de Berg, M., Mumford, E., and Speckmann, B. (2010) "Optimal BSPs and rectilinear cartograms." *International Journal of Computational Geometry & Applications 20*, no. 2: 203–222.

Dent, B. D. (1999) *Cartography: Thematic Map Design* (5th ed.). Boston, MA: McGraw-Hill.

Dorling, D. (1993) "Map design for census mapping." *The Cartographic Journal 30*, no. 2:167–183.

Dorling, D. (1995a) *A New Social Atlas of Britain*. Chichester, England: Wiley.

Dorling, D. (1995b) "The visualization of local urban change across Britain." *Environment and Planning B: Planning and Design 22*, no. 3:269–290.

Dorling, D., Barford, A., and Newman, M. (2006) "Worldmapper: The world as you've never seen it before." *IEEE Transactions on Visualization and Computer Graphics 12*, no. 5: 757–764.

Gao, P., Li, Z., and Qin, Z. (2019) "Usability of value-by-alpha maps compared to area cartograms and proportional symbol maps." *Journal of Spatial Science 64*, no. 2: 239–255.

Garland, K. (1994) *Mr Beck's Underground Map*. Harrow Weald, United Kingdom: Capitol Transport Publishing.

Gastner, M. T., and Newman, M. E. J. (2004) "Diffusion-based method for producing density-equalizing maps." *Proceedings of the National Academy of Sciences 101*, no. 20: 7499–7504.

Gastner, M. T., and Newman, M. E. J. (2005) "Density-equalizing map projections: Diffusion-based algorithm and applications." *GeoComputation 2005*, Ann Arbor, MI, pp. 1–23, http://www-personal.umich.edu/~mejn/papers/geocomp.pdf.

Gastner, M. T., Seguy, V., and More, P. (2018) "Fast flow-based algorithm for creating density-equalizing map projections." *Proceedings of the National Academy of Sciences 115*, no. 10: E2156–E2164.

Griffin, T. L. C. (1983) "Recognition of areal units on topological cartograms." *The American Cartographer 10*, no. 1: 17–29.

Harris, R., Charlton, M., and Brunsdon, C. (2018) "Mapping the changing residential geography of White British secondary school children in England using visually balanced cartograms and hexograms." *Journal of Maps 14*, no. 1:65–72.

Harris, R., Martin, C., Brunsdon, C., et al. (2017) "Balancing visibility and distortion: Remapping the results of the 2015 UK General Election." *Environment and Planning A 49*, no. 9:1945–1947.

Heilmann, R., Keim, D. A., Panse, C., et al. (2004) "RecMap: Rectangular map approximations." *IEEE Symposium on Information Visualization*, Austin, Texas, pp. 33–40.

Hennig, B. (2014) "Gridded cartograms as a method for visualising earthquake risk at the global scale." *Journal of Maps 10*, no. 2: 186–194.

Hennig, B. D. (2019) "Remapping geography: Using cartograms to change our view of the world." *Geography 104*, Part 2:71–80.

Kaspar, S., Fabrikant, S. I., and Freckmann, P. (2011) "Empirical study of cartograms." *Proceedings of the 25th International Cartographic Conference*, Paris, France, pp. 1–8.

Nusrat, S., Alam, M. J., and Kobourov, S. (2018) "Evaluating cartogram effectiveness." *IEEE Transactions on Visualization and Computer Graphics 24*, no. 2:1105–1118.

Nusrat, S., and Kobourov, S. (2016) "The state of the art in cartograms." *Computer Graphics Forum 35*, no. 3:619–642.

Olson, J. M. (1976) "Noncontiguous area cartograms." *The Professional Geographer 28*, no. 4:371–380.

Panse, C. (2018) "Rectangular statistical cartograms in R: The recmap package." *Journal of Statistical Software 86*, Code Snippet 1:1–27.

Raisz, E. (1934) "The rectangular statistical cartogram." *Geographical Review 24*, no. 2:292–296.

Roth, R. E., Woodruff, A. W., and Johnson, Z. F. (2010) "Value-by-alpha maps: An alternative technique to the cartogram." *The Cartographic Journal 47*, no. 2:130–140.

Tobler, W. (2004) "Thirty-five years of computer cartograms." *Annals of the Association of American Geographers 94*, no. 1:58–73.

21 Flow Mapping

21.1 INTRODUCTION

Flow maps are used to display the flow of some phenomenon between geographic locations. Examples of phenomena that we might illustrate with a flow map include the migration of people between countries, the transport of agricultural commodities (such as wheat and corn) from grain storage facilities to manufacturing sites, the movement of water along stream networks, and the flow of internet traffic along major communication routes. Traditionally, such flows have been represented on flow maps by lines of differing widths, with arrows frequently used at the end of lines to represent the direction of flow. For instance, Figure 1.2 illustrates common international flight routes in February of 2021, and Figure 9.13 illustrates migration to the United States from Europe and Asia for four different time periods. Although the visual variable size (i.e., the width of flow lines) has been common on flow maps, other visual variables have been experimented with (such as varying the lightness of flow lines as a function of associated data values).

Section 21.3 introduces three basic types of flow maps (origin-destination, trajectory-based, and continuous) and the nature of data associated with each type of flow map. We will see that the availability of *location-aware technologies (LATs)* such as Global Positioning System (GPS) data recorders, mobile phones, and radiofrequency identification (RFID) chips have greatly increased the amount of data that are available for flow mapping. Section 21.4 introduces a number of issues involved in designing flow maps. We will see that both unclassed and classed flow maps are possible. With unclassed flow maps, a common approach is to make the widths of flow lines directly proportional to the underlying data. Alternatively, on a classed map, you might purposely enhance the differences between flow line widths to stress the differences among low and high flows.

Sections 21.5 and 21.6 provide a historical perspective on the development of flow maps. In Section 21.5, we consider examples of flow maps prior to the advent of digital cartography. The creation of flow maps by pen-and-ink cartography required a strong sense of map design, but the result was often a particularly effective map in terms of communicating spatial information to the map reader. A classic example was Charles Joseph Minard's map illustrating Napolean's Russian Campaign of 1812. In Section 21.6, we consider early efforts to automate the process of flow mapping, focusing on the work of Waldo Tobler, who arguably provided a foundation for the flow mapping of Big Data.

In Section 21.7, we provide three examples of recent efforts in digital flow mapping. First, we examine the development and character of Daniel Stephen and Bernhard Jenny's interactive map of U.S. migration available at http://usmigrationflowmapper.com/. This software is illustrative of a broad range of approaches that can be utilized to create attractive flow maps of Big Data. Second, we examine the development and character of Caglar Koylu and colleagues' web-based software https://flowmapper.org which enables you to illustrate a variety of flows with your own data. Third, using research by Yalong Yang and colleagues (2019), we examine the potential of flow mapping in virtual environments. We will see that a 3-D virtual environment provides intriguing opportunities for handling the clutter that is common on flow maps traditionally depicted in a 2-D environment.

There has been a considerable amount of work in flow mapping in the realm of geovisual analytics. Although we do not fully define and illustrate geovisual analytics until Chapter 26, we feel that it is appropriate to point out some of the major efforts in geovisual analytics and flow mapping in the present chapter. We do so in Section 21.8.

21.2 LEARNING OBJECTIVES

- Define and contrast the three basic kinds of flow maps: origin-destination, trajectory-based, and continuous.
- List five essential design strategies for creating flow maps.
- Utilize formulas that will enable you to either make flow line widths proportional to the data or enhance the differences among flows.
- Describe Charles Joseph Minard's classic flow map illustrating Napolean's Russian Campaign of 1812.
- Explain why Waldo Tobler's early digital flow mapping efforts constitute pioneering work with Big Data.
- Explain why the design of flow maps in Stephen and Jenny's flow mapping software (http://usmigrationflowmapper.com/) is equal to or surpasses that of traditional pen-and-ink flow maps.
- Suggest some alternative designs for flow lines that might be utilized to reduce clutter when showing two-way flows between origins and destinations.
- Describe some flow mapping designs that might be effective when the 3-D space of a virtual environment is utilized.
- Discuss how the space-time cube can be utilized to visualize flows that have a temporal component.

DOI: 10.1201/9781003150527-23

21.3 BASIC TYPES OF FLOW MAPS AND ASSOCIATED DATA FOR FLOW MAPPING

A review of the literature on flow mapping suggests that there are three basic kinds of flow maps: origin-destination, trajectory-based, and continuous. **Origin-destination flow maps** are used to portray flows between geographic locations when the actual route of flow is unimportant. For instance, if we wish to create a map of the migration of people between countries of the world, we generally are not interested in the particular routes that each person took, but rather we desire to depict the total flow from one country (an origin) to another country (the destination). We can depict the flow as either a straight line or a curved line (the latter will tend to avoid overlap of flow lines and produce a visually pleasing graphic); Figures 1.2 and 9.13 are examples of origin-destination flow maps. Data for origin-destination flow maps can be associated with either enumeration units or point locations. An example of enumeration unit data would be the migration between countries of the world, and an example of point location data would be the bus traffic flow between major U.S. cities (assuming that the precise route of flow is not important).

Trajectory-based flow maps are used to depict flows when the actual route of flow is deemed important. Figure 21.1 is an example of a trajectory-based flow map that depicts

the routes of 55 turkey vultures (*Cathartes aura*) over periods ranging from 1 month to 11 years along seasonal migration routes. A complete understanding of the routes shown on this map would likely require interactive software, but such a map clearly raises questions about the particular routes that were taken. Harvey Miller and colleagues (2019, 856) argue that maps like Figure 21.1 are possible because of the "stunning advances in location-aware technologies (LATs) for moving objects data (MOD) collection, such as Global Positioning System (GPS) data recorders, mobile phones, radiofrequency identification (RFID) chips, geotags, radiolocation devices and georeferenced social media." The underlying data for Figure 21.1 consist of more than 500,000 GPS points.

Continuous flow maps are used to depict flows for spatially continuous data such as wind speed and direction and ocean current speed and direction. Traditionally, the data for such maps were based on a limited set of point locations (e.g., weather stations), which would then be interpolated to produce a continuous data set. Today, however, other technologies (e.g., radar) can be utilized to collect such data (Horstmann et al. 2015), or the data may be generated by weather forecasting models. For an intriguing continuous flow map, see Cameron Beccario's animation of global wind patterns (and other weather conditions) at https://earth.nullschool.net/.

Flow maps can be used to map either qualitative or quantitative data. An example of qualitative data would be the depiction of various linear features such as roads, streams, and railroads by, respectively, red, blue, and black lines (including small perpendicular black marks for the railroads). Here the intention would be to illustrate that the types of flows for these features differ qualitatively from one another. Such a map might also portray quantitative data; for instance, narrow and wide red lines might be used to represent interstate and U.S. highways, respectively. In the latter case, the intention would be to express an ordinal relationship between the two types of roads—consequently, no numerical information would be provided in the legend.

More commonly, quantitative flow maps depict numerical (interval-or ratio-level) data. For example, the flow maps in Figure 9.13 illustrate the *magnitude* of migration. By referring to the legend, you can see that the magnitude of migration from Asian countries was approximately three to four times greater between 1981 and 1990 than between 1961 and 1970.

Similar to proportional symbol maps, flow maps can be based on either unstandardized or standardized data. Unstandardized data (i.e., using raw counts) are by far more common. One must be careful, however, with using unstandardized data. For instance, a large raw migration from China is to be expected because of its large population, but one might wonder how the rate of migration from China compares with that from Japan. If the latter is of interest, then standardized data should be used.

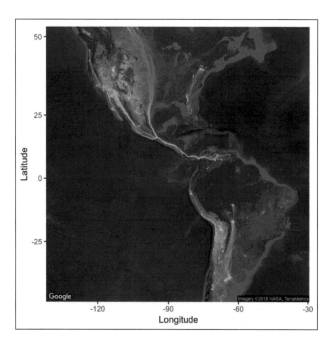

FIGURE 21.1 A trajectory-based flow map depicting the seasonal migration routes of turkey vultures. (Article Source: Graña Grilli M., Lambertucci S. A., Therrien J-F., and Bildstein K. L. (2016) "Wing size but not wing shape is related to migratory behavior in a soaring bird." *Journal of Avian Biology* 47: 1–10. **doi:10.1111/jav.01220**; Data Package Source: Bildstein K. L., Barber D., Bechard M. J., and Graña Grilli M. (2016) Data from "Wing size but not wing shape is related to migratory behavior in a soaring bird." Movebank Data Repository. doi:**10.5441/001/1.37r2b884**.)

21.4   ISSUES IN DESIGNING FLOW MAPS

Borden Dent and colleagues (2009, 194) provide an excellent summary of "essential design strategies" in creating flow maps. Below we summarize six of these:

- Flow lines have the highest intellectual importance and thus should be placed at the top of the visual hierarchy on the map.
- Smaller flow lines should appear on top of larger flow lines.
- Arrows should be used if the direction of flow is critical to map meaning.
- Care should be taken in selecting a map projection appropriate for flow mapping (for an example, see Section 9.4.3).
- If data permits, flow lines should be placed such that the map appears balanced (e.g., it should not appear top heavy or bottom heavy).
- Legends should be unambiguous and include units where necessary.

Since we have discussed similar design strategies elsewhere in this textbook, we will not expand on the first five of these. With respect to legend design, it is important to emphasize that some flow maps do not include a legend. The logic is that sometimes we are more interested in interpreting the general spatial pattern than in deriving precise numerical information.

As with proportional symbol maps, either unclassed or classed (range graded) flow maps are possible. On an unclassed flow map, each differing data value is represented by a differing flow line width, whereas on a range-graded flow map, the data are grouped into classes, and a differing line width is used for each class. Figures 1.2 and 9.13 are both examples of range-graded flow maps.

Once you have decided whether an unclassed or a classed map is appropriate, you need to determine how the widths of lines will be assigned as a function of the data. A common approach is to make the widths of lines *proportional* to the data. Thus, on an unclassed map, if a data value is twice as large as another data value, the line width should be twice as large. The formula is as follows:

$$w_i = \frac{v_i}{v_L} \times w_L$$

where:

w_i is the width of the flow line to be drawn
v_i is the data value for the flow line to be drawn
v_L is the largest data value
w_L is the largest flow line width desired

In the case of a range-graded map, we can use a representative value of the class (e.g., the midpoint of the class) and scale the line widths proportional to these representative values. This is the approach that was used for Figure 9.13.

If the data have a small range, and you wish to enhance the differences among the data on an unclassed map, you can use the following formulas to adjust the flow line widths so that differences among flows are more easily seen:

$$z_i = \frac{v_i - v_S}{v_L - v_S}$$

$$w_i = z_i \left(w_L - w_S \right) + w_S$$

where:

v_S and v_L = smallest and largest data values
z_i = a proportion of the data range associated with the data value v_i
w_S and w_L = smallest and largest flow line widths desired
w_i = width of a flow line associated with the data value v_i

The above formulas could also be used on a range-graded map, where the data values would be representative values of each class. Alternatively, you could approximate this formal approach by picking the smallest and largest line width and then select line widths for intermediate classes "by eye"—this was the approach used to create Figure 1.2. When individual flow lines are a constant width, it generally has been argued that you do not need to worry about perceptual issues when selecting line widths, as a line width twice as large as another line width should be seen as approximately twice as large (McCleary 1970).

21.5   FLOW MAPPING PRIOR TO AUTOMATION

Arthur Robinson (1982, 147) indicated that the first flow maps were produced in 1837 by Henry Drury Harness to depict the flow of passengers and "traffic" via railway. Robinson (150) argued, however, that Charles Joseph Minard was a much more prolific flow map designer, as 42 of the 52 maps that he produced were flow maps. Figure 21.2 is an example of one of Minard's maps, which illustrates the volume of French wine exported by sea in 1864.[1] Although Minard did not have the precise ship tracking data that we have available today, he did know the general shipping routes that were used, and so the flow lines in Figure 21.2 follow these general routes (e.g., note the route around South America, as the Panama Canal did not yet exist). As such, this map can be considered an early form of a trajectory-based flow map.

It is interesting that Minard adjusted the design of the geographic base to place more emphasis on the shipping routes. Arthur Robinson (1982, 151) stated:

> The logic is unassailable: if a flow line of a given width is to portray ocean transport it would be ridiculous to have it cross land, such as the margins of the Channel or the Straits of Gibraltar.... The solution was simply to widen the Channel or the Strait.

Clearly, Robinson was impressed by Minard's flow maps, noting that Minard "... certainly brought that class

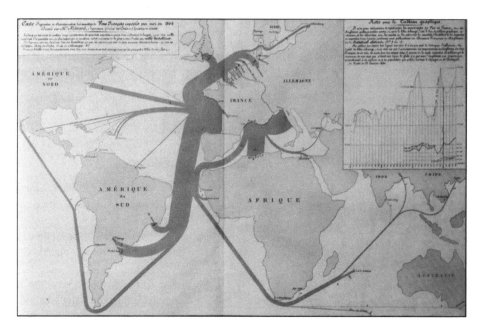

FIGURE 21.2 Minard's (1865) flow map illustrating the export of French wine by sea in 1864. This is arguably an example of a trajectory-based flow map in which there is an attempt to depict the actual route of flow. (From p. 151 of Robinson, Arthur H. (1982) *Early Thematic Mapping in the History of Cartography.* University of Chicago Press; original from l'Ecole Nationale des Ponts et Chaussées, Paris.)

of cartography to a level of sophistication that has probably not been surpassed" (Robinson 1967, 105).

A more popular example of Minard's work is the graphic shown in Figure 21.3, which illustrates the loss of troops associated with Napolean's Russian Campaign of 1812. We find this graphic frequently mentioned in the literature, often with the statement by Edward Tufte (1983) that it "may well be the best statistical graphic ever drawn." Although

some feel that too much attention has been placed on this graphic,[2] it does provide a captivating portrayal of the loss of troops associated with Napolean's campaign. Looking at Figure 21.3, we can see that the campaign began at Kowno (Kaunas in present-day Lithuania), went east to Moscou (Moscow) and then returned to an area to the south of the starting point. Movements east are shown in brown and movements west are shown in black. Although no legend

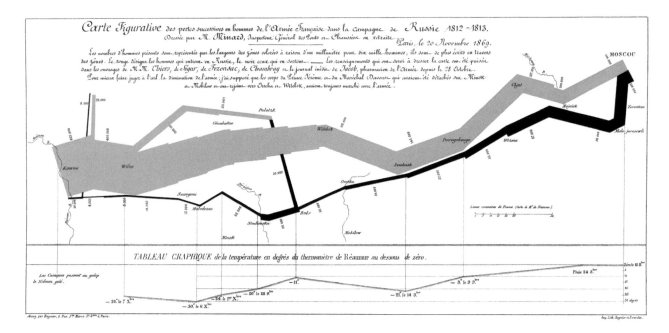

FIGURE 21.3 Minard's (1869) classic flow map illustrating changes in the number of troops in Napolean's Russian Campaign of 1812. For an interactive version of the map, see https://www.masswerk.at/minard/. (From Bibliothèque numérique patrimoniale des ponts et chaussées (https://heritage.ecoledesponts.fr/enpc/) and https://www.masswerk.at/nowgobang/2018/minard-morse-tufte.)

is shown on the map, a translation of the text on the map indicates that "The numbers of men present are represented by the width of the colored zones at a rate of one millimeter for every ten thousand men ..."[3] Thus, the decreases in line width are proportional to troop losses, which is confirmed by the numbers written alongside the flow lines. Overall, we can see that Napolean began with 422,000 troops and ended with only 10,000 troops.

Time is not shown along the flowlines, but instead, a second graphic is shown below the flow lines depicting the changes in temperature (along the y-axis) associated with changes in time along the x-axis for the return trip from Moscow. Here we can see that all temperatures during this portion of the campaign were below 0°R, which is equivalent to 0°C.[4] Although the cold temperatures were certainly an important consideration in the loss of troops on the return journey from Moscow, Menno-Jan Kraak (2014) discusses other reasons for these troop losses in his book *Mapping Time: Illustrated by Minard's map of Napoleon's Russian Campaign of 1812*. We will return to Kraak's work again in Section 24.10 when we discuss the space-time cube as an alternative to animation.

21.6 EARLY DIGITAL FLOW MAPPING EFFORTS BY WALDO TOBLER

Waldo Tobler is widely recognized for his early digital flow mapping efforts: In reviewing recent research on flow mapping for this book, we repeatedly saw Tobler's (1987) early work on migration referenced. Tobler argued that migration data are especially logical for *digital* portrayal because of the large number of movements that must be depicted. For example, for the 48 contiguous U.S. states, there are 2,256 possible movements (assuming that all pairs of states are considered); if we consider all 3,000+ U.S. counties, there are more than 9 million possible movements, and this does not even consider the attribute of time. Thus, we can argue that Tobler was starting to work with Big Data before the term became popular (Rey et al. 2020, 538)!

One of the simpler options in Tobler's software was the depiction of one-way migration to or from a particular state by arrows of varying widths; Figure 21.4 illustrates this for the state of California for the period 1965–1970. When one wishes to show the migration between all pairs of states simultaneously, a different approach is required because of the large number of arrows that result. The key is recognizing that many migration movements are small; deleting such movements will allow the map reader to focus on the more important movements. Tobler indicated that by deleting migration values below the mean, he generally was able to remove 75 percent or more of the flow lines while deleting less than 25 percent of the migrants. For example, Figure 21.5 illustrates this approach for net migration for U.S. states from 1965 to 1970.

Another interesting feature of Tobler's software was the ability to route data through states lying between the starting and ending points for migration, thus reflecting the

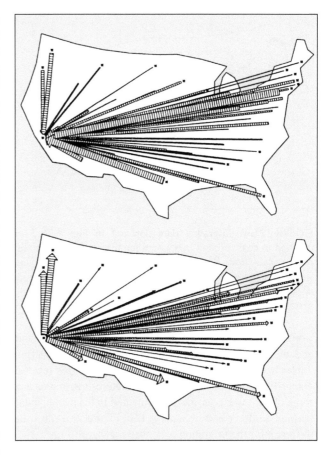

FIGURE 21.4 Migration to and from California, 1965–1970. (After Tobler 1987. First published in the *American Cartographer* 14(2), p. 160. Reprinted with permission from the Cartography and Geographic Information Society.)

route over which people were presumed to migrate. Thus, Tobler began to experiment with trajectory-based flows, even though he did not have data on the actual routes used. For example, migration data from New York to California

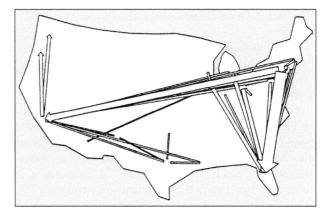

FIGURE 21.5 Net migration 1965–1970 for the 48 contiguous U.S. states, with flows below the mean net migration not shown. (Republished with permission of John Wiley & Sons—Books, from "A model of geographical movement" by W. R. Tobler, Vol. 13, no. 1, 1981; permission conveyed through Copyright Clearance Center, Inc.)

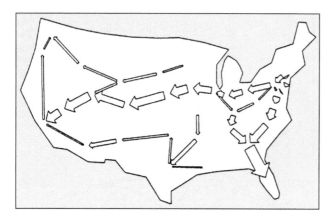

FIGURE 21.6 Migration data depicted in Figure 21.5 are rerouted to pass through states between the starting and ending points; again, flows below the mean are not shown. (Republished with permission of John Wiley & Sons—Books, from "A model of geographical movement" by W. R. Tobler, Vol. 13, no. 1, 1981; permission conveyed through Copyright Clearance Center, Inc.)

would ideally have to be routed through Pennsylvania, Ohio, and numerous other states. Although the details of how Tobler achieved this are beyond the scope of this text (see Tobler 1981, 7–8 for a summary), it is interesting that he used some of the same concepts implemented in his pycnophylactic method described in Section 17.6. Figure 21.6 illustrates the result for the migration data used for Figure 21.5.[5]

21.7 EXAMPLES OF RECENT DIGITAL FLOW MAPPING

In this section, we consider three examples of recent efforts in digital flow mapping. Additionally, you may wish to consider https://flowmap.blue/, a web-based software package developed by Ilya Boyandin that allows you to create flow maps of your own origin-destination data. You also may find it useful to examine Boyandin's website https://ilya.boyandin.me/, which illustrates numerous examples of approaches for digital flow mapping.

21.7.1 Stephen and Jenny's Interactive Web-Based Origin-Destination Flow Map

Daniel Stephen and Bernhard Jenny have created an interactive origin-destination flow map for visualizing the migration between U.S. counties from 2009 to 2013 (the resulting map is available at http://usmigrationflowmapper.com/). Here we consider a sequence of three steps that led to what we feel is a highly effective method for interpreting the more than 9 million possible migration movements.

In the first step, Jenny and colleagues (2018) began by conducting a *quantitative content analysis* (Riff et al. 2019) of 97 non-branching origin-destination flow maps that reflect traditional principles of good cartographic design (for instance, professional cartographers were asked to provide flow maps that they considered well designed).[6] The content analysis revealed eight design principles for these 97 maps.

Many of these principles were also supported by a review of the literature on graph drawing in the computer science literature (graph drawing uses nodes and edges, which are also found in flow maps: the starts and ends of flows are nodes, and the flows are connecting edges). For those design principles that were not supported by the graph drawing literature, a user study was conducted with 215 participants to determine which design approaches were most effective. The following were the key design principles that resulted from the combined results of the content analysis, the results found in the graph drawing literature, and the user study:

- Flow-on-flow and flow-on-node intersections must be minimized (Figure 21.7A).
- Sharp bends in flow lines should be avoided, and symmetric flows are preferred to asymmetric flows (Figure 21.7B).
- Avoid acute angles between crossing flows (Figure 21.7C).
- Flows must not pass under unconnected nodes (Figure 21.7D).
- Radial distribution of flow lines at nodes is preferred, thus avoiding narrow angles between flow lines at nodes (Figure 21.7E).
- Small flows should be placed on top of larger flows.
- Quantity is best represented by scaled flow width, although color brightness can also show quantity by using dark colors for higher values.
- Directions are best represented with arrowheads (as opposed to tapered flow lines).

In the second step, Jenny and colleagues (2017) used the results of the first step to develop a *force-directed layout method* for creating origin-destination flow maps. In the force-directed layout method, flow lines are modeled with quadratic Bézier curves because the resulting flow lines "… cannot have loops, are never S-shaped, and are included in common graphics libraries …" (1523). Conceptually, the control point of a quadratic Bézier curve is attached to a spring, with the opposing end of the spring attached to the midpoint of the line between the start and end points of the

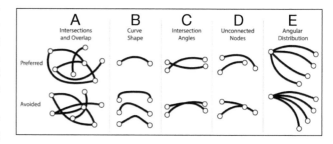

FIGURE 21.7 Summary of key design principles from Jenny et al.'s (2018) study; approaches in the top row are preferred. (From Jenny et al. 2018. First published in *Cartography and Geographic Information Science* 45(1), p. 73. Reprinted with permission from the Cartography and Geographic Information Society.)

flow, as shown in Figure 21.8. "An iterative process computes the equilibrium state between the retracting forces of the springs and the repulsing forces of other flows" (1523–1524). In this manner, Jenny and colleagues were able to consider the preferred layouts shown in Figure 21.7. A sample of nine expert cartographers found that flow maps created by the force-directed approach were similar in quality to manually produced flow maps.

In the third step, Daniel Stephen and Bernhard Jenny (2017) developed the software found at http://usmigrationflowmapper.com/. Stephen and Jenny noted that three approaches were utilized to generalize the more than 9 million flows between the 3,000+ counties: sub-setting, thresholding, and merging. *Sub-setting* refers to showing flows for only a selected geographic area. For instance, Figure 21.9A shows net flows at the state level for all 50 states (plus Puerto Rico and the District of Columbia), whereas

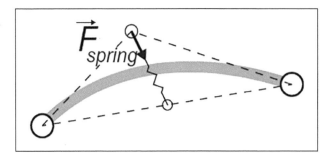

FIGURE 21.8 Jenny et al.'s (2017) use of the quadratic Bézier curve and associated spring to consider the preferred layouts shown in Figure 21.7. (From "Force-directed layout of origin-destination flow maps," by B. Jenny, D. M. Stephen, I. Muehlenhaus, et al., *International Journal of Geographical Information Science*, 2017, publisher Taylor & Francis Ltd., http://www.tandfonline.com, reprinted by permission of the publisher.)

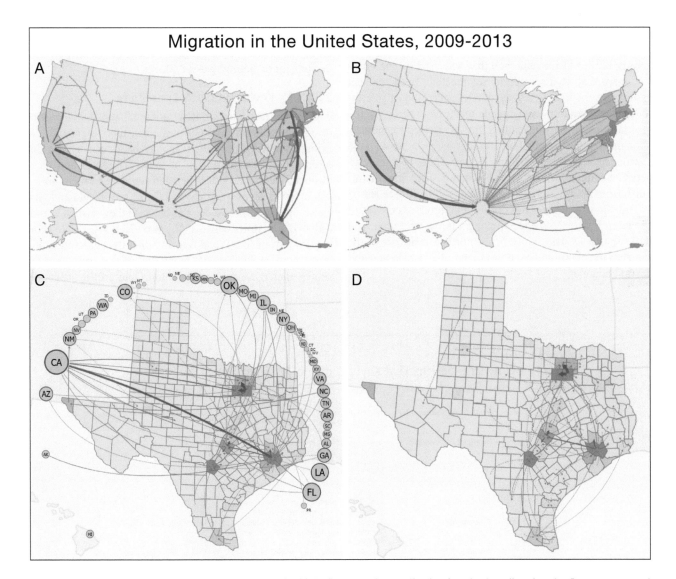

FIGURE 21.9 Maps created using Stephen and Jenny's (2017) flow mapping application found at http://usmigrationflowmapper.com/: (A) The top 50 net flows between pairs of states; (B) the net flows for just Texas relative to all other states; (C) the top 75 net flows for counties in Texas (flows for counties in other states are represented by the magnitude of circles surrounding the state of Texas); (D) the same information as (C), but the surrounding state information has been removed. (Courtesy of Daniel Stephen and Bernhard Jenny.)

Figure 21.9B shows the net flows for just Texas relative to all other states. *Thresholding* refers to showing the largest flows as a function of some threshold. Rather than use the mean of the data as a threshold (as Tobler did), Stephen and Jenny found that the clutter of flow lines is reduced by focusing on the *top n flows*, where the user can vary *n* from 10 to 100. For instance, in Figure 21.9A, we chose the top 50 flows (out of a maximum of 2,652 flows).

Merging involves combining enumeration units together when viewing flows from counties in one state to counties in other states. Merging is illustrated for net flows for counties in the state of Texas in Figure 21.9C, where the top 75 flows are shown. Here net flows to or from counties in Texas to other states are combined and represented by circles of varying sizes surrounding the state of Texas. The position of each state in the surrounding circles is a function of the geographic position of a state relative to the center of Texas; thus, we can see that California (CA) is located to the west. If you hover over the flow line for CA, you will see that Harris County in southeastern Texas makes the largest contribution with 5,880 net movers. The merging approach used here is similar to the approach that was used on the necklace maps that we discussed in Section 18.8. In the present case, however, the circles surrounding the area of interest refer to enumeration units outside that area, whereas in the necklace maps, the surrounding circles referred to enumeration units within the area of interest. One problem with the circular arrangement shown in Figure 21.9C is that flows between counties within the state can be difficult to distinguish from flows to or from other states. This is handled by a "Hide flows to other states option" that produces the result shown in Figure 21.9D. Here, using a hover operation, you will see that Dallas county in the northeastern part of the state has the maximum net flows to the adjacent Denton and Tarrant counties.

A distinct advantage of Jenny et al.'s flow mapping application is that alternative maps can be created in a matter of seconds and the user interface is easy to use; thus, we encourage you to experiment with the application. We utilized the net flows for illustration here, but you can also map total flows. A limitation of the application is that you cannot use it to visualize your own flow data. Stephen and Jenny have, however, developed a tool called Flox (https://github.com/OSUCartography/Flox) that enables you to experiment with a variety of parameters that were used to create the flow maps for http://usmigrationflowmapper.com/.

21.7.2 Koylu et al.'s Web-Based Software for Designing Origin-Destination Flow Maps

Caglar Koylu and colleagues (2021b) have developed web-based software that enables you to design your own origin-destination flow maps that focus on two-way flows between pairs of locations. Before discussing the capability of the software, we will consider a user study that Koylu and Guo (2017) conducted to examine the effectiveness of potential designs for depicting two-way flows.

21.7.2.1 Koylu and Guo's User Study

Table 21.1 provides an illustration of the five flow-line designs that Koylu and Guo tested in their user study. In initially viewing this set, you may be struck by the difference between the design of these flow lines and those ultimately utilized by Stephen and Jenny (e.g., there are no full arrow designs in Koylu and Guo's set). Koylu and Guo chose these alternative designs because they wished to portray two-way flows between any two locations (i.e., to show the flows from A to B and B to A, rather than either the total or net flow, as Stephen and Jenny did). If two-way flows are shown, then there are more flow lines intersecting at a node,

TABLE 21.1

The Five Flow Line Designs That Koylu and Guo Tested in Their User Study

Design	Name	Direction	Magnitude
	Monotone Arrowhead (MA)	Biased curvature, arrowhead, counter-clockwise orientation (right-hand traffic rule)	Line width, color value
	Divergent Arrowhead (DA)	Biased curvature, arrowhead, counter-clockwise orientation, varying thickness, gradual change of color hue from blue, to gray mid-break, to red	Line width
	Fading Arrowhead (FA)	Biased curvature, arrowhead, counter-clockwise orientation, gradual change of color hue and transparency from blue to, transparent white mid-break, to red	Line width
	Tapered (TA)	Biased curvature, counter-clockwise orientation, varying line width from wide to narrow	Line width, color value
	Teardrop (TD)	Biased curvature, counter-clockwise orientation, varying line width from narrow to wide, gradual change of color value	Line width, color value

Source: From p. 313 of Koylu, C. and Guo, D. (2017) "Design and evaluation of line symbolizations for origin–destination flow maps." *Information Visualization* 16, No. 4:309–331.

making it more difficult to use full arrow symbols; thus, there is a need to consider alternative designs, such as those shown in Table 21.1.

All of the designs in Table 21.1 are assumed to flow from left to right. In the Monotone Arrowhead (MA) design, a half arrowhead indicates direction, but direction is also indicated by the high curvature at the origin and the low curvature at the destination (i.e., there is a biased curvature). The latter is an approach that has been utilized effectively in graph drawing in the computer science literature. Note that this is a *counterclockwise orientation* (i.e., if we were to also show flows from right to left, the net result would be a counterclockwise direction of flow). The term Monotone is used to indicate that this flow line is the same color throughout; however, different magnitudes of flow are indicated by both line width and color value, with a lower flow represented by a narrower line and a lighter color (a lighter gray in this case).

The Divergent Arrowhead (DA) and Fading Arrowhead (FA) designs have an overall structure similar to MA, as they have the same biased curvature and half arrowhead, but they utilize a diverging color scheme, with blue at the origin and red at the destination. For both DA and FA, different magnitudes of flow are indicated by only different line widths (color value is not used). FA differs from DA in that the middle of the flow line becomes transparent for long flow lines, thus reducing clutter on the map. It can be hypothesized that the diverging color schemes should enhance map interpretation for tasks that involve determining a dominant direction of flows.

The Tapered (TA) design has the same biased curvature as the above methods, but direction is indicated by a tapering of the flow line, with the line width decreasing to 0 as it approaches the destination. A similar tapered approach has been found effective in graph drawing. In this case, different magnitudes are represented by both changing line width and color value, as was the case for MA. The final design, the Teardrop (TD), has an overall structure different than the others that we've examined, as it is narrow and relatively straight at the origin and then widens before finally narrowing at the destination. This technique also varies in color value from the starting point to the destination point, going from gray to black. In this case, different magnitudes are represented by both line width and color value. Ware et al. (2014) found that a variant of TD was effective in depicting continuous flows, such as wind flows and ocean currents.

Koylu and Guo analyzed the effectiveness of the above methods in a user study of approximately 500 participants. Since Koylu and Guo expected the results might be a function of the particular tasks that participants completed, care was taken to develop four distinctly different tasks (315):

- Task 1: Select the dominate direction that most flows are going to.
- Task 2: Select three flows that have the highest volume.

- Task 3: Select the circle (node) that receives the highest number of flows in the top 10 in length.
- Task 4: Select a cluster of blue circles that are near one another (blue circles represented net-exporters).

The latter task did not involve the perception of flows per se, but was included to see whether the flow symbolization had an impact on the perception of circles at nodes (the circles represented either the total flow or the net flow, depending on the task).

The results revealed that overall the Monotone Arrowhead (MA) was most effective, and thus Koylu and Guo recommend it when many different visualization tasks are required. Although it was hypothesized that the diverging scheme approaches (DA and FA) would be particularly effective for a direction task (e.g., Task 1), MA was most effective for such a task. However, the diverging schemes were more accurate than the TA and TD approaches for the direction task. The Fading Arrowhead (FA) approach was not as effective as was hoped, as the transparency "… made smaller magnitudes become more salient as compared to other designs" (327). Given Ware et al.'s promotion of The Teardrop (TD), it was expected to perform well, but it performed similarly to the Tapered (TA) approach, although it did lead to rapid response times for the direction task. TD and the diverging scheme methods all performed poorly on magnitude estimation (e.g., Task 2). In general, Koylu and Guo concluded that all of the designs have some potential but that their effectiveness will be a function of the tasks desired.

21.7.2.2 Koylu et al.'s FlowMapper Software

Koylu et al.'s software FlowMapper is available at https:// flowmapper.org and is described in Koylu et al. (2021b). As we have indicated, a key characteristic of this software is the ability to visualize your own flow mapping data. However, for illustrative purposes, we will utilize a tutorial data set on family migration between U.S. states for the period 1887–1924 that is distributed with FlowMapper.[7] Before reading the following paragraphs, you may find it useful to access the software using the directions shown in study question 7.

Figure 21.10 displays one of the maps that we created in FlowMapper with the assistance of Caglar Koylu. Here we mapped three attributes: the number of families migrating as two-way flows, the gross migration as circles, and the migration efficiency as a choropleth symbology. We discuss each of these symbologies below.

We have mapped the two-way flows of the number of families migrating between states using a "Curve Half Arrow" approach. Like the Monotone Arrowhead (MA), the Curve Half Arrow permits using both line width and color variation (e.g., lightness) to represent flows. We chose to vary just the line width (and not the lightness) in order to enhance the clarity of the smaller flows. Another parameter for the flow lines is the "Stroke color," which is the outline

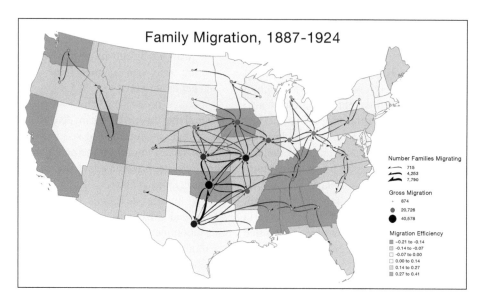

FIGURE 21.10 A multivariate map from Koylu and colleagues' FlowMapper (https://flowmapper.org) showing two-way migration flows between states (using the "Curve Half Arrow" approach, a variant of the Monotone Arrowhead), the gross migration for each state represented by a circle at the centroid of the state, and migration efficiency represented by a choropleth map with a diverging color scheme. (Courtesy of Caglar Koylu, Geng Tian, and Mary Windsor.)

used for the flow lines; here, we used a moderate gray. Koylu notes that the stroke color can be used "to increase the visual clarity of the parts of a flow map where there are too many overlapping flows."[8]

We chose to create unclassed flows by using a "Proportional" option within FlowMapper. A related FlowMapper option allows you to scale the flow line widths by entering desired "Min width" and "Max width" values.[9] To both avoid severe overlap of high flows and permit the visualization of small flows, we experimented with various values for Min width and Max width, ultimately choosing Min and Max values of 5 and 20 units, respectively; thus, the Min and Max flows differed by a factor of 4. If you wish to make the flow lines *directly* proportional to the data, you should use the following formula:

$$w_S = \frac{v_S}{v_L} * w_L = \frac{715}{7790} * 20 = 1.84$$

where:

  w_S is the minimum width of the flow lines to be drawn
  v_S is the smallest data value
  v_L is the largest data value
  w_L is the maximum width of the flow lines to be drawn

Here we have plugged in the smallest and largest number of families migrating (see the legend in Figure 21.10) and assumed a maximum flow width of 20 units (to be consistent with the Max width used above). Note that the resulting smallest and largest flows now differ by a factor of more than 10. If you examine the map in Figure 21.10 carefully, you will note that some pairs of states have only one flow appearing between them—this occurs because we chose an option to map only the top 80 flows, thus reducing the clutter that would appear with a larger number of flows.

Note that the gross migration associated with each state has been represented by a circle at the centroid of each state. FlowMapper permits a variety of options for scaling circle sizes. We chose a "proportional" (unclassed) approach within FlowMapper in which the user enters a desired "Min radius" and "Max radius." We experimented with various Min and Max radii values and ultimately chose radii values of 2 and 16, respectively, that we felt enhanced the overall readability of the map. If you instead wish to create Flannery-scaled circles, you should use the following formula from Chapter 18:

$$r_S = \left(\frac{v_S}{v_L}\right)^{0.57} \times r_L = \left(\frac{874}{40,578}\right)^{0.57} \times 16 = 1.79$$

where:

  r_S = radius of the smallest circle
  r_L = radius of the largest circle
  v_S = data value for the smallest circle
  v_L = data value for the largest circle

Here we have substituted the minimum and maximum gross migration values for the smallest and largest circle (see the legend in Figure 21.10) and assumed a Max radius of 16 units (to be consistent with the Max radius used above); the result of 1.79 would be the value for Min radius. In examining Figure 21.10, you will note that we chose to vary both the size and Fill color of circles (we used lightnesses of gray for the fill), as we wanted to experiment with the "look" of this redundant symbology. You can experiment with alternative designs in FlowMapper to determine whether you think this approach is effective.

Finally, we have shown a background choropleth map of the *migration efficiency*, which is computed as follows (Koylu et al., 2021b):

$$\text{Migration efficiency} = \frac{\text{Inmigration} - \text{outmigration}}{\text{Inmigration} + \text{outmigration}}$$

Migration efficiency measures the imbalance of migration flows. Values close to 0 indicate a balance between inflows and outflows, whereas values well above 0 indicate greater inmigration, and values well below 0 indicate greater outmigration. Since 0 constitutes a natural dividing point in the data (the data are bipolar), a diverging color scheme is appropriate here. For ease in interpreting the legend, we have chosen the value 0 as a dividing point and created equal interval classes below and above this point. Note that since we have mapped the flows, the gross migration, and the migration efficiency, the result is a **multivariate map**.

In addition to the Curve Half Arrow approach, FlowMapper permits "Straight Half Arrow," "Tapered," and "Teardrop" approaches for designing flow lines. Figure 21.11 utilizes the Teardrop approach to depict the family migration flow data. Here we have chosen a yellow-orange-red color scheme to depict the flows rather than the solid black flows used to create the map in Figure 21.10. To enable the flows to stand out against a gray background, we have not mapped the migration efficiency in this case. Koylu et al. indicate that they plan a variety of enhancements to FlowMapper, including adding the Divergent and Fading Arrowheads examined in the user study. You should keep in mind that FlowMapper provides several options for interacting with the data once you have designed a basic map. For instance, you can zoom in on cluttered areas to see the flows in greater detail; hover over locations to get detailed information about nodes, flows, and enumeration units; and easily change the total number of flows currently depicted.

21.7.3 FLOW MAPPING IN VIRTUAL ENVIRONMENTS

In this section, we consider research by Yalong Yang and colleagues (2019), which examined the utilization of flow maps in virtual environments. In Chapter 28, we will find that **virtual environments** enable a 3-D computer-based simulation of a real or imagined environment that users are able to navigate through and interact with. Yang et al. argue that such 3-D environments are appealing for flow mapping for several reasons, including:

- Lifting the flow lines off the map can reduce clutter.
- The height (or z-dimension) in 3-D space can be used to encode data (e.g., the height of a flow line can be made proportional to its associated data value).
- Occlusions between flow lines can be resolved by natural head movements or gesture manipulations (693).

Yang et al. conducted three user studies to examine the potential of various flow mapping designs in virtual environments. We will consider their first user study in detail and then briefly summarize the second and third user studies. In the first user study, five designs were examined:

- *2-D straight* (Figure 21.12A): A traditional 2-D flow map using straight lines to connect origins and destinations, which leads to the greatest clutter.
- *2-D curve* (Figure 21.12B): A traditional 2-D flow map using curves rather than straight lines, thus reducing clutter.
- *3-D constant height* (Figure 21.12C): Flows are represented by tubes raised to the same height above the surface. This should reduce some of the clutter of the 2-D maps, but clutter can still occur because the flows are at the same height.

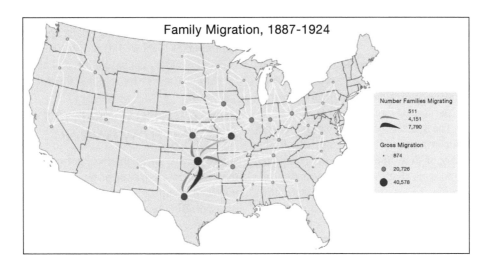

FIGURE 21.11 A map from FlowMapper showing the Teardrop approach for depicting flows with a yellow-orange-red color scheme. To enable the flows to stand out against a gray background, the migration efficiency has not been mapped. (Courtesy of Caglar Koylu, Geng Tian, and Mary Windsor.)

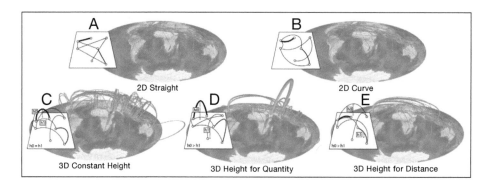

FIGURE 21.12   The five flow mapping designs that Yang and colleagues (2019) tested in their first user study. (Republished with permission of The Institute of Electrical and Electronics Engineers, Incorporated (IEEE), from "Origin-Destination Flow Maps in Immersive Environments," by Y. Yang, T. Dwyer, B. Jenny, et al., Vol. 25, no. 1, 2019; permission conveyed through Copyright Clearance Center, Inc.)

- *3-D height for quantity* (Figure 21.12D): The height is proportional to the flow magnitude, thus providing a redundant visual encoding of the data (flow thickness and height). The different heights should reduce clutter.
- *3-D height for distance* (Figure 21.12E): The height is proportional to the distance between origins and destinations. This is another approach for reducing clutter.

In examining Figure 21.12, note that a red-green color gradient was used to encode direction. Although this could conceivably create problems for the color deficient, Yang et al. (2019). noted that this was not a concern as the mapping tasks "… did not involve ambiguity regarding direction" (696). Twenty participants were asked to find and compare flows (and indicate which flow was greater) between two pairs of locations on maps of international migration flows between countries. Participants wore an HTC Vive head-mounted display that provided a 3-D virtual environment (see Section 28.4.3 for a discussion of head-mounted displays). Responses were evaluated in terms of both accuracy and speed of response.

A key finding from the first user study was that *3-D height for distance* produced the most accurate results, and this design was also the preferred visualization by participants. However, participants responded fastest to the *2-D straight* maps. The success of these two approaches led to their inclusion in a second user study, which also considered the display of flow line tubes on a *globe*, and a novel *MapsLink* approach that displayed origins on one flat map and destinations on a second flat map, with flows appearing as tubes in 3-D space above the two maps. For the *globe* case, the height of flow tubes was a function of the great circle distance between origins and destinations. For the *MapsLink* approach, the height of flow tubes was a function of the Euclidean distance between origins and destinations. The setup for this user study was similar to the first one, except that participants had two controllers available (one to position and rotate the map/globe and one to pick any location and drag it to a new position) (Yang et al. 2019, 698). The results of the second study revealed that the *globe* approach produced the most accurate results (the *3-D height for distance* approach was a close second) and that the *globe* approach was preferred by participants in terms of visual design. Both the *globe* and *3-D height for distance* approaches were preferred for their readability and led to more confident responses. *MapsLink* clearly led to the slowest responses.

Given the slow responses for *MapsLink*, the third user study contrasted just the three other approaches used in the second study, using denser flow data sets (which would produce greater clutter). For this user study, *globe* was both more accurate and faster than *2-D straight* and *3-D height for distance*, and participants preferred the *globe* approach. The effectiveness of the *globe* approach may seem surprising given that only one-half of the globe was viewable at one time. In general, Yang et al. (2019, 701) argued that the virtual environment head-mounted display technology allowed participants to resolve overlapping flows on 3-D flow maps by changing the relative position of the head and the map that is viewed. Overall, Yang et al. recommended using the *globe* approach for global flows and the *3-D height for distance* for regional flows.

21.8   GEOVISUAL ANALYTICS AND FLOW MAPPING

As we discuss in Chapter 26, geovisual analytics combines the visualization power of humans with the computational power of computers to make sense out of large geospatial datasets (Big Data). Data for flow mapping certainly fit into this framework because of the large data sets that can be collected in association with LATs such as GPS and mobile phones. For instance, in Section 26.6.1, we discuss Nivan Ferreira and colleagues (2013) efforts to understand the spatial patterns of the approximately 500,000 taxi trips each day in New York City. The focus of Chapter 26, however, is not on flow mapping per se, but rather on how geovisual analytics is utilized in a broad range of applications. Thus, in the present section, we point you to additional research that has focused explicitly on geovisual analytics and flow mapping.

Gennady and Natalia Andrienko have been leading proponents of the use of geovisual analytics to study spatial patterns of movement. You can find an overview of the many techniques that they have developed in Andrienko and Andrienko (2013) and a more detailed discussion in the book *Visual Analytics of Movement*, which they co-authored with other colleagues (Andrienko et al. 2013). Others who have made important contributions to geovisual analytics and flow mapping include Guo and Zhu (2014), Zhu et al. (2019), Wood et al. (2010), Yang et al. (2017), and Willems et al. (2013). For further contributions, see the Further Reading section of this chapter.

Space does not permit us to cover the breadth of computational approaches that have been utilized in applying geovisual analytics to flow mapping. For an introduction to the broad range of computational approaches that have been utilized, see Andrienko et al. (2020). We introduce these approaches in Section 26.4.

Visualization approaches used in geovisual analytics, to some extent, parallel those we have already discussed. Thus, the clutter resulting from mapping massive data sets is often handled with interactive methods similar to those that we discussed for Stephen and Jenny's (2017) flow mapper software discussed in Section 21.7.1. Additional visualization approaches that reduce clutter include *varying the opacity* of flows (in which lower magnitude flows are shown with less opacity, i.e., greater transparency) and **edge bundling**, in which nearby flows are bundled together (e.g., Holten and van Wijk (2009)).

Geovisual analytics data for flow mapping frequently include a temporal component. For instance, mobile phone data associated with traffic flow data typically contain not only geographic location coordinates but also a temporal time stamp for each location. There are three basic techniques that have been used for mapping flows over time: **animation**, the **small multiple**, and the **space-time cube**. We discuss all of these techniques in Chapter 24—we emphasize the animation of *movement and flows* in Section 24.5.1. Although these three techniques are all used to map flows over time, it is our impression that the space-time cube has received the greatest emphasis. To stress this point, Figure 21.13 displays the trajectory of a Galapagos Albatross on both a 2-D map (on the left) and a space-time cube (on the right). The *x* and *y* axes of the space-time cube correspond to longitude and latitude, and the *z*-axis displays the temporal aspect of the data (rather than the height of the bird's flight pattern). Figure 21.13 was taken from a paper by Somayeh Dodge and Evgeny Noi (2021) that summarizes various visualization approaches for mapping movement. The trajectory of the Albatross displayed in Figure 21.13 was based on GPS coordinates collected every 90 minutes from June to September 2008. In the space-time cube, we can see the path of the bird between Galapagos Island and mainland South America over time. The color of the dots in the image represents the magnitude of the tailwind (a value below 0 indicates a headwind). Although static space-time cubes such as this have assisted in exploring the behavior of migrating animals, they are generally best used in a highly interactive environment so that researchers can more easily interpret the 3-D space of the cube and filter the data when the data density becomes high. We examine the space-time cube in somewhat greater depth in Section 24.10 of the animation chapter.

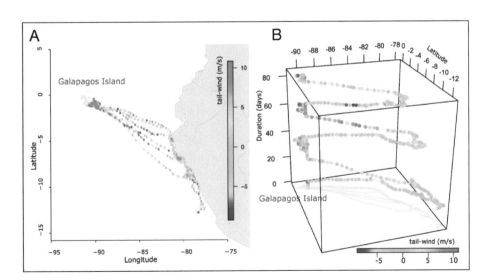

FIGURE 21.13 A 2-D map (A) and a 3-D space-time cube (B) depicting the flights of a Galapagos Albatross from June to September 2008. The 2-D map shows the various routes traveled by the bird over Earth's surface, while the space-time cube provides the temporal component of the travel. Both maps show the tailwind magnitude using a diverging spectral color scheme (note the location of the 0 point). (From Dodge and Noi 2021. First published in *Cartography and Geographic Information Science* 48(4), p. 356. Reprinted with permission from the Cartography and Geographic Information Society.)

21.9 SUMMARY

Flow maps are used to display the flow of some phenomenon between geographic locations. There are three basic kinds of flow maps: origin-destination, trajectory-based, and continuous. Origin-destination flow maps are utilized when the actual route of flow is unimportant (e.g., a flow map showing the total value of all agricultural commodities exported from the United States to other countries). Trajectory-based flow maps are utilized when the actual route of flow is important (e.g., a flow map showing the travel of a wolf pack over a three-month period). Continuous flow maps are utilized to display flows for spatially continuous data such as the speed and direction of ocean currents. Traditionally, origin-destination flow maps were the most common form of flow map, but trajectory-based flow maps are becoming popular because of the availability of LATs such as GPS data recorders.

Flows generally have been represented by lines of varying width, with arrows used at the end of lines to represent direction. More recently, researchers have begun to explore alternative flow line designs to handle the clutter that can result, especially when showing two-way flows. For instance, Koylu and Guo experimented with the Monotone Arrowhead, Diverging Arrowhead, Fading Arrowhead, Tapered, and Teardrop designs. Overall, they concluded that the Monotone Arrowhead worked best. Several of these alternative designs are available in the software https://flowmapper.org.

It is common to make flow line widths proportional to the data, thus honoring the relations among the data values. However, it is possible to exaggerate the differences among the data by arbitrarily selecting the smallest and largest line width and then scaling other line widths relative to this range. It is also possible to create either a classed or an unclassed flow map—your choice will likely depend on the sort of information you expect readers to extract from the map.

We explored some of the pre-digital flow maps (e.g., Minard's map of Napolean's Russian Campaign of 1812). Unfortunately, early digital efforts (such as Tobler's) did not produce the aesthetic quality of the smooth curves found on many traditional pre-digital flow maps. Jenny et al.'s careful development of digital flow mapping techniques (e.g., conducting a content analysis of previously produced flow maps and getting expert cartographers' opinions about flow maps) is a good example of how traditional cartographic design principles can be transferred into the digital realm. The resulting software (http://usmigrationflowmapper.com/) produces maps that are comparable in quality to traditional pen-and-ink drawn maps and provides a variety of approaches for displaying Big Data.

Conventionally, flow maps have been produced in 2-D space. We examined two ways in which 3-D space may be used to enhance the interpretation of flow maps. First, we reviewed a study by Yang and colleagues, which showed that flow maps could be effectively utilized in a 3-D virtual environment. For instance, their third user study revealed that the results for the 3-D *globe* approach were more accurate and faster than a 2-D approach. Second, in examining the use of geovisual analytics in flow mapping, we noted that the space-time cube is a common approach used to visualize change over time.

21.10 STUDY QUESTIONS

1. For each of the following data sets, indicate which type of flow map would result (origin-destination, trajectory-based, or continuous), and briefly explain your choice: (A) in-vehicle navigation systems that provide the GPS coordinates of routes of a set of rental cars in a city over the course of a week; (B) 1,000-meter grid cells that cover Lake Michigan, with each cell providing the wind speed and direction at the time a ship sunk; and (C) a spreadsheet showing the migration of people between U.S. states during the Dust Bowl of the 1930s.

2. Use the formulas in Section 21.4 to compute the following: (A) Presume that you wish to make the flow line widths directly proportional to the data on an unclassed map. Given that the smallest and largest migration values are 1,100 and 8,200, respectively, and the largest line width is 0.5 cm, compute the line width for a migration value of 3,300. (B) Use the same data as shown in (A), but presume that you wish to enhance the differences among the flows by making the largest line width 0.5 cm and the smallest line width 0.02 cm.

3. Using Minard's map displayed in Figure 21.3, answer the following questions: (A) Indicate whether you feel this is an origin-destination or a trajectory-based flow map and explain your choice. (B) Discuss whether you agree with the statement "the bulk of the troop losses appear to be due to the cold temperatures during the return trip from Moscow." (For an interactive version of the map, see https://www.masswerk.at/minard/.)

4. (A) On what basis can we say that Waldo Tobler was a pioneer in flow mapping with Big Data? (B) Give two examples of approaches that Tobler used to minimize the effect of clutter on his digital flow maps.

5. Access Stephen and Jenny's U.S. migration flow mapper (http://usmigrationflowmapper.com/) and answer the following questions: (A) What does this map tell you about the migration among U.S. states during the period 2009–2013? (Experiment with changing the number of flows shown.) (B) Select Florida by clicking on it. What is the total number of people moving in and out of Florida over the period 2009–2013? Click on the icon to "Show county-level flows for the selected state." Which were the top three counties in Florida in terms of net movers from New York State? Where are these counties located in Florida?

6. Describe the design of the Monotone Arrowhead (MA) studied by Koylu and Guo. How does this design differ from the design of traditional flow lines? Why might this design be more effective in some situations than a traditional design?

7. Access Koylu et al.'s FlowMapper (https://flowmapper.org), select Skip Tutorial, and then select File, Project, Load Project, and specify the file Fig 21.10.json, which can be downloaded from Chapter 21 at www.routledge.com/9780367712709. The resulting image should approximate Figure 21.10 in the textbook (although the resolution will be coarser). Click on Flows and you will note that the Min and Max width values of 5 and 20 described in Section 21.7.2.2 have been selected. Try to enhance the differences among the flow lines by specifying a larger Max width value and/or a smaller Min width value (select Map Flows to have a change take effect). Justify your choice of values.

8. Discuss how 3-D virtual environments can be useful for mapping flows that have traditionally been displayed in 2-D maps.

NOTES

1   Robinson (1982, 251) suggested a publication date of 1865, although no publication date was provided on the map.
2   As an example, in an article illustrating the use of a *Sankey map* (a combination of a Sankey diagram and a flow map), Barry Lehrman (2018, 56) states, "Minard's chart is the most frequently (but undeservedly) cited example of a flow map, thanks to the prolific graphic evangelizing of Tufte ..."
3   A translation of the map can be ordered from https://www.edwardtufte.com/tufte/posters.
4   At the time, France was using the Réaumur temperature scale, which had a slightly different meaning than the Celsius scale; for instance, −10°R equals −13°C (Kraak 2014, 21).
5   In 2003, David Jones updated Tobler's original software to a Windows-based environment (http://csiss.ncgia.ucsb.edu/clearinghouse/FlowMapper/). Although the updated software does not include all of the options described in Tobler's original work, it enables users to explore the sorts of flows that Tobler discussed.
6   Non-branching refers to flow maps in which the flows do not merge with one another.
7   These data were derived from a family tree data set compiled by Koylu et al. (2021a).
8   Personal communication, December 2021.
9   Since the line width of a flow line is not constant throughout, the line width specified for a particular flow line essentially refers to the point of maximum thickness along the line (ignoring the arrowhead) (Personal communication, Caglar Koylu, November 2021).

REFERENCES

Andrienko, N., and Andrienko, G. (2013) "Visual analytics of movement: An overview of methods, tools and procedures." *Information Visualization 12*, no. 1:3–24.

Andrienko, G., Andrienko, N., Bak, P., et al. (2013) *Visual Analytics of Movement*. Heidelberg: Springer.

Andrienko, N., Andrienko, G., Fuchs, G., et al. (2020) *Visual Analytics for Data Scientists*. Cham, Switzerland: Springer.

Dent, B. D., Torguson, J. S., and Hodler, T. W. (2009) *Cartography: Thematic Map Design*. (6th ed.). Boston, MA: McGraw-Hill.

Dodge, S., and Noi, E. (2021) "Mapping trajectories and flows: Facilitating a human-centered approach to movement data analytics." *Cartography and Geographic Information Science 48*, no. 4:353–375.

Ferreira, N., Poco, J., Vo, H. T., et al. (2013) "Visual exploration of Big spatio-temporal urban data: A study of New York City taxi trips." *IEEE Transactions on Visualization and Computer Graphics 19*, no. 12:2149–2158.

Guo, D., and Zhu, X. (2014) "Origin-destination flow data smoothing and mapping." *IEEE Transactions on Visualization and Computer Graphics 20*, no. 12:2043–2052.

Holten, D., and van Wijk, J. J. (2009) "Force-directed edge bundling for graph visualization." *Computer Graphics Forum 28*, no. 3:983–990.

Horstmann, J., Nieto Borge, J. C., Seemann, J., et al. (2015) "Wind, wave, and current retrieval utilizing X-band marine radars." In *Coastal Ocean Observing Systems*, ed. by Y. Liu, H. Kerkering and R. H. Weisberg, pp. 281–304. London: Academic Press.

Jenny, B., Stephen, D. M., Muehlenhaus, I., et al. (2017) "Force-directed layout of origin-destination flow maps." *International Journal of Geographical Information Science 31*, no. 8:1521–1540.

Jenny, B., Stephen, D. M., Muehlenhaus, I., et al. (2018) "Design principles for origin-destination flow maps." *Cartography and Geographic Information Science 45*, no. 1:62–75.

Koylu, C., and Guo, D. (2017) "Design and evaluation of line symbolizations for origin–destination flow maps." *Information Visualization 16*, no. 4:309–331.

Koylu, C., Guo, D., Huang, Y., et al. (2021a) "Connecting family trees to construct a population-scale and longitudinal geo-social network for the U.S." *International Journal of Geographical Information Science 35*, no. 12:2380–2423.

Koylu, C., Tian, G., and Windsor, M. (2021b) "FlowMapper.org: A web-based framework for designing origin-destination flow maps." *Journal of Maps*, https://arxiv.org/abs/2110.03662.

Kraak, M.-J. (2014) *Mapping Time: Illustrated by Minard's map of Napoleon's Russian Campaign of 1812*. Redlands, CA: ESRI Press.

Lehrman, B. (2018) "Visualizing water infrastructure with Sankey maps: A case study of mapping the Los Angeles Aqueduct, California." *Journal of Maps 14*, no. 1:52–64.

McCleary, G. (1970) "Beyond simple psychophysics: Approaches to the understanding of map perception." *Proceedings of the American Congress on Surveying and Mapping*, pp. 189–209.

Miller, H. J., Dodge, S., Miller, J., et al. (2019) "Towards an integrated science of movement: Converging research on animal movement ecology and human mobility." *International Journal of Geographical Information Science 33*, no. 5: 855–876.

Minard, C. J. (1865) "Carte figurative et approximative des quantités de vin français exportés par mer en 1864." Map. Paris.

Minard, C. J. (1869) "Carte Figurative des pertes succesives en hommes de l'Armée Française dans la campagne de Russie 1812–1813." Map. Paris.

Rey, S., Han, S. Y., Kang, W., et al. (2020) "A visual analytics system for space–time dynamics of regional income distributions utilizing animated flow maps and rank-based Markov chains." *Geographical Analysis 52*, no.4:537–557.

Riff, D., Lacy, S., and Watson, B. R., et al. (2019). *Analyzing Media Messages: Using Quantitative Content Analysis in Research*. (4th ed.). New York: Routledge.

Robinson, A. H. (1967) "The thematic maps of Charles Joseph Minard." *Imago Mundi 21*:95–108.

Robinson, A. H. (1982) *Early Thematic Mapping in the History of Cartography*. Chicago: University of Chicago Press.

Stephen, D. M. and Jenny, B. (2017) "Automated layout of origin–destination flow maps: U.S. county-to-county migration 2009–2013." *Journal of Maps 13*, no. 1:46–55.

Tobler, W. R. (1981) "A model of geographical movement." *Geographical Analysis 13*, no. 1:1–20.

Tobler, W. R. (1987) "Experiments in migration mapping by computer." *The American Cartographer 14*, no. 2:155–163.

Tufte, E. R. (1983) *The Visual Display of Quantitative Information*. Cheshire, CT: Graphics Press.

Ware, C., Kelly, J. G. W., and Pilar, D. (2014) "Improving the display of wind patterns and ocean currents." *Bulletin of the American Meteorological Society 95*, no. 10:1573–1581.

Willems, N., Scheepens, R., van de Wetering, H., et al. (2013) "Visualization of vessel traffic." In *Situation Awareness with Systems of Systems*, ed. by P. van de Laar, J. Tretmans and M. Borth, pp. 73–87. New York: Springer.

Wood, J., Dykes, J., and Slingsby, A. (2010) "Visualization of origins, destinations and flows with OD maps." *The Cartographic Journal 47*, no. 2:117–129.

Yang, Y., Dwyer, T., Goodwin, S., et al. (2017) "Many-to-many geographically-embedded flow visualisation: An evaluation." *IEEE Transactions on Visualization and Computer Graphics 23*, no. 1:411–420.

Yang, Y., Dwyer, T., Jenny, B., et al. (2019) "Origin-Destination Flow Maps in Immersive Environments." *IEEE Transactions on Visualization and Computer Graphics 25*, no. 1: 693–703.

Zhu, X., Guo, D., Koylu, C., et al. (2019) "Density-based multiscale flow mapping and generalization." *Computers, Environment and Urban Systems 77*:101359, 1–10.

22 Multivariate Mapping

22.1 INTRODUCTION

In the preceding chapters, we focused on univariate mapping—the display of *individual* attributes (or variables). Frequently, however, mapmakers need to display *multiple* attributes. For example, a climatologist might wish to simultaneously view temperature, precipitation, barometric pressure, and cloud cover for a geographic region. The cartographic display of such multiple phenomena is known as **multivariate mapping**; if only two attributes are displayed, the process is termed **bivariate mapping**. Bivariate mapping is covered in Section 22.3, and multivariate mapping involving three or more attributes is covered in Section 22.4.

A fundamental issue in multivariate mapping is whether individual maps should be shown for each attribute (in which case maps are compared) or whether all attributes should be displayed on the same map (in which case maps are combined). Thus, Sections 22.3 and 22.4 are divided into two subsections: one for map comparison and one for map combination. In the map comparison section for bivariate mapping, we consider the problem of selecting a method of data classification when comparing two choropleth maps; the notion of utilizing a proportional symbol map to show amounts and a choropleth map to show rates; and the possibility of comparing maps of the same attribute for two points in time (here we stress the importance of combining the two data sets before classifying the data).

The map combination section for bivariate mapping first describes how choropleth maps can be overlaid to create a **bivariate choropleth map**. Other symbols considered in this section include the **rectangular point symbol**, in which the width and height of a rectangle are made proportional to the values of two attributes being mapped, and the **bivariate ray-glyph**, in which the angle of rays (straight-line segments) pointing to the left and right of a small central circle represents two attributes.

Section 22.4 begins by considering **small multiples**—the simultaneous comparison of three or more maps. The difficulty of synthesizing information depicted via small multiples has led to numerous approaches for combining attributes onto one map. Techniques discussed include the **trivariate choropleth map** (in which three choropleth maps are overlaid), the **multivariate dot map** (in which different-colored dots are used to represent multiple phenomena), and **multivariate point symbols** (an intriguing example is the **Chernoff face**, in which various facial features are used to represent multiple attributes).

In Sections 22.3 and 22.4, we make the presumption that the mapmaker wishes to map the attributes directly. An alternative approach is to mathematically manipulate the attributes prior to mapping them. In this respect, we focus on a traditional approach known as **cluster analysis**, in which observations are grouped based on their scores on a set of attributes; for instance, we might group counties in California based on which counties have similar incomes and similar voting behavior. Thus, rather than mapping the raw data, we map the resulting clusters, which is relatively simple from the standpoint of symbolization. Cluster analysis is the topic of Section 22.5. In Chapter 26, we will consider additional mathematical approaches for manipulating multiple attributes.

This chapter focuses on static methods for displaying multivariate data. In Chapters 25 and 26, we introduce interactive techniques that can enable you to explore multivariate data. In addition to considering the techniques discussed there, you may wish to examine the Longitudinal Neighborhood Explorer (LNE) (http://su-gis.iptime.org/LNE), which enables you to interactively compare two or more choropleth maps of U.S. Census and American Community Survey data from 1970 to 2010.[1]

22.2 LEARNING OBJECTIVES

- Determine which data classification method is most appropriate when comparing two choropleth maps of differing attributes.
- When mapping a particular phenomenon (such as infant mortality), discuss the roles that proportional and choropleth symbology can play when comparing maps of the phenomenon.
- When comparing two maps of an attribute for differing time periods, discuss the importance of combining the data in order to create an appropriate symbology and legend.
- Describe various approaches for creating bivariate choropleth maps.
- Contrast the cross-hatched method for creating a bivariate choropleth map with the rectangular point symbol technique.
- Contrast a small multiple of dot maps with the multivariate dot map.
- Contrast the following multivariate point symbols with one another: a combined star and snowflake symbol with the Chernoff face.
- Describe how ring maps are designed.
- Summarize the eight steps utilized in a cluster analysis in which a contiguity constraint is not utilized.
- Contrast hierarchical cluster analyses with and without a contiguity constraint.

DOI: 10.1201/9781003150527-24

22.3 BIVARIATE MAPPING

22.3.1 COMPARING MAPS

In this section, we consider approaches for bivariate mapping in which individual maps are used to show each attribute (i.e., maps are compared). Initially, we focus on choropleth maps because they have been commonly used for map comparison.

22.3.1.1 Comparing Choropleth Maps

As with a single choropleth map, an important consideration in comparing two (or more) choropleth maps is deciding whether the data should be classed and, if so, which method of classification should be used. To begin, we assume that we wish to class the data, and thus we focus on the method of classification. Initially, we also assume

that we wish to compare two attributes for a single point in time (say, median income and percent of the adult population with a college education, both collected for 2000). In Section 22.3.1.3, we consider comparing maps for the same attribute collected over two time periods.

In selecting a method of classification, it is critical to consider the distribution of each attribute along the number line. If the attributes have differing distributions (e.g., if one is skewed and the other is normal), certain classification methods can lead to an inappropriate visual impression of correlation between the attributes. To illustrate, consider the hypothetical distributions shown in Figure 22.1A. Attribute 1 is clearly positively skewed, whereas attributes 2 and 3 appear to have normal distributions.[2] In Figure 22.1B, values of these attributes have been assigned spatially so that extremely high correlations result in each case (the correlation coefficients, r, appear in Figure 22.1C).[3]

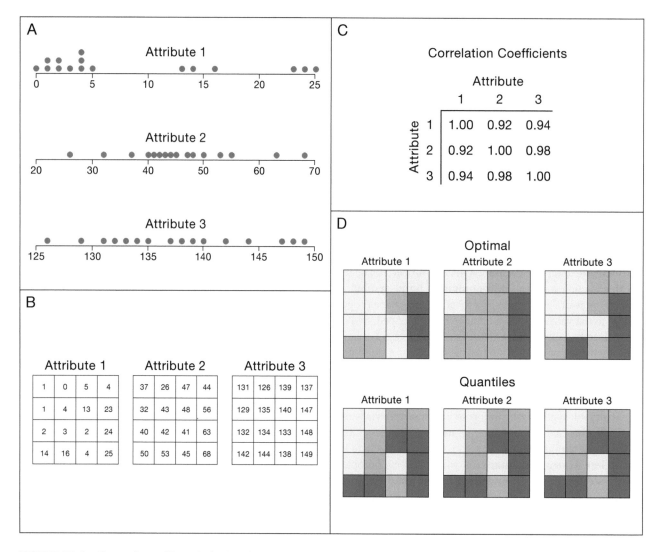

FIGURE 22.1 Comparison of hypothetical attributes using choropleth maps: (A) Three hypothetical attributes (attribute 1 is positively skewed, whereas attributes 2 and 3 are normal); (B) maps of the raw data for the three attributes; (C) correlation coefficients (r values) between each pair of attributes; (D) maps for optimal and quantiles classifications.

Recalling from Chapter 5 that the optimal method of classification is often recommended because it minimizes classification error, it seems natural to ask whether the optimal method might also be used for map comparison. If the optimal method is applied to all three attributes shown in Figure 22.1B, we obtain the maps shown in Figure 22.1D. Although these maps suggest positive associations between each pair of attributes, they do not support the high correlation coefficients found in Figure 22.1C (remember that 1.0 is the maximum possible value for *r*). The lack of a strong visual association between the skewed and normal distributions might not be surprising (compare the optimal maps for attributes 1 and 2), but note that there is also a lack of visual association between maps of the normal distributions (attributes 2 and 3). The optimal method fails to reflect the high correlations between the attributes because it focuses on the precise distribution of the individual attributes along the number line.

Recall from Section 5.5 that the quantiles method of classification places an approximately equal number of observations in each class based on the ranks of the data. Because the classes resulting from the quantiles method are unaffected by the magnitudes of the data, the method is arguably appropriate for comparing the differently shaped distributions shown in Figure 22.1A. This is illustrated in Figure 22.1D, where we can see that the quantiles method portrays high correlations between not only the normal distributions but also the skewed and normal distributions.

Our suggestion for using quantiles for choropleth map comparison is supported in formal studies done by cartographers. For instance, Robert Lloyd and Theodore Steinke (1976, 1977) found that the visual correlation of maps is affected by the amount of blackness on each map (assuming that lightnesses of gray are used for symbolization); in other words, if maps A and B and C and D have the same statistical correlation, maps A and B will be judged more similar if their blackness levels are more similar than the blackness levels of maps C and D. As a result, Lloyd and Steinke argued for using the **equal-areas** method of classification. The equal-areas method of classification is similar in concept to quantiles, but rather than placing an equal number of observations in classes, an equal portion of the map area is assigned (i.e., the desired area in each class is simply the area of the map divided by the number of classes desired). If enumeration units are equal in size (as in the hypothetical data), the equal-areas method produces a map identical to quantiles. A study by Cynthia Brewer and Linda Pickle (2002) is also relevant here, as they analyzed seven classification methods (including the optimal and quantiles methods) in terms of their effectiveness in comparing choropleth maps. Overall, they recommended the quantiles method.[4]

As with univariate choropleth maps, it is natural to ask whether the issue of selecting an appropriate classification method might be obviated simply by not classing the data. Studies by Michael Peterson (1979) and Jean-Claude Muller (1980) dealt with people's visual comparison of both classed and unclassed choropleth maps. These studies concluded that people perceive similar correlations on pairs of classed and unclassed maps; as a result, the authors raised questions about the need to class data for choropleth mapping. In contrast, in their study, Brewer and Pickle (2002) raised questions about the appropriateness of using unclassed maps, terming them "too-many-class maps,"[5] arguing that "the perception of differences across [such] maps is controlled partly by the quality of the perceptual scaling in the color system used to assign printer or display colors ..." (667). As we saw in Section 15.6, there has been debate among cartographers on the appropriateness of unclassed maps.

22.3.1.2 Comparing Miscellaneous Thematic Maps

Although it is common to hold the mapping method constant when comparing maps (e.g., showing two choropleth maps or two isopleth maps), useful information often can be acquired by comparing two different kinds of thematic maps. This is especially true when one map is used to show raw totals and another map is used to show standardized data. For example, consider Figure 22.2, which compares a proportional symbol map of the raw number of infant mortalities in New Jersey with a choropleth map of the infant mortality rate. The proportional symbol map suggests that the "problem" of infant mortality is in the northeastern part of the state (where the largest circles are). This map by itself, however, is not very meaningful because the pattern is likely a function of population (counties with more people are apt to have more infant deaths). In contrast, the choropleth map standardizes the raw mortality data by considering the number of deaths relative to the number of live births and suggests that the "problem" is found in three areas of the state (represented by the shades in the highest class). Unfortunately, a high rate on the choropleth map might not be meaningful if there are also few deaths. Only when the two maps are viewed together can the complete picture emerge: The northernmost high-rate area is most problematic because it is located where the raw number of deaths is high.

22.3.1.3 Comparing Maps for Two Points in Time

At the beginning of this section, we assumed that we wished to compare two attributes for a single point in time. It is also important to consider how we should compare maps of the same attribute for two points in time. To illustrate, let's return to the murder rate date that we introduced in Chapter 3. There we focused on the murder rate for U.S. cities in 1990 and related that data to a number of other attributes. You might wonder how the murder rate has changed over time, as you may have read about concerns with increasing murder rates in at least some U.S. cities. To examine this question, we compare unclassed proportional circle maps of murder rates for all U.S. cities with a population of 250,000 or more in both

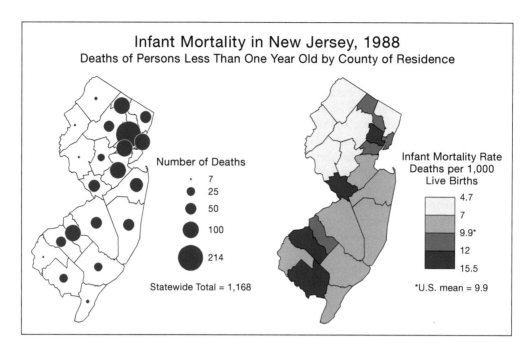

FIGURE 22.2   A comparison of proportional symbol and choropleth maps. The proportional symbol map provides information on the raw number of infant mortalities, whereas the choropleth map focuses on the rate of infant mortality. To understand the complete picture of infant mortality, both maps are necessary. (Republished with permission of University of Chicago Press—Books, from *Mapping It Out: Expository Cartography for the Humanities and Social Sciences* by M. Monmonier 1992; permission conveyed through Copyright Clearance Center, Inc.)

1995 and 2019 (Figure 22.3), the range of years for which the U.S. Federal Bureau of Investigation (FBI) provided data as of April 2021.[6] To make the data for the two years directly comparable, we combined the two data sets together and scaled the circles for both maps on the basis of the smallest data value found on the two maps (a value of 2.1 for Mesa, AZ in 2019). Thus, we show only one legend in Figure 22.3. In comparing the two maps, you can see that there is less red on the 2019 map (the circles are generally smaller), indicating

that murder rates have generally decreased, although you will note a few cases where the circles have increased in size over time. In Chapter 25, we discuss the notion of a **change map**, which will enable you to more clearly see these changes over time. Although we have utilized the proportional symbol map to illustrate the notion of comparing the change in an attribute over time, the principle of combining the two data sets to create a single legend would apply to other thematic mapping techniques, such as the choropleth map.

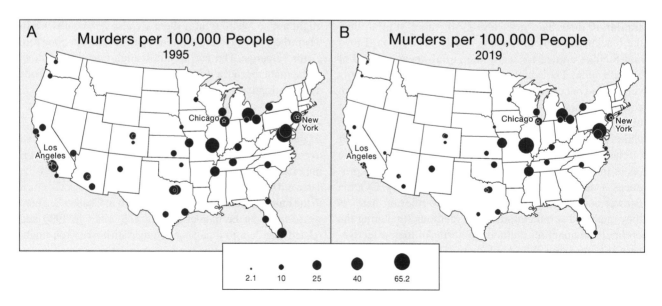

FIGURE 22.3   A comparison of murder rates for U.S. cities in 1995 and 2019. Note that the data for 1995 and 2019 were combined before symbolizing the data; thus, only one legend is used.

22.3.2 COMBINING TWO ATTRIBUTES ON THE SAME MAP

In this section, we consider approaches for bivariate mapping in which two attributes are combined on the same map. Again, we focus initially on choropleth maps because of their common use.

22.3.2.1 Bivariate Choropleth Maps

In the 1970s, the U.S. Bureau of the Census developed the **bivariate choropleth map**—a method for combining two colored choropleth maps.[7] Figure 22.4 depicts a bivariate choropleth map using a color scheme similar to the one

used by the Bureau of the Census, along with each of the univariate distributions used to create the bivariate map.[8] Although bivariate choropleth maps were deemed a success from a technical standpoint, they received considerable criticism for their presumed failure to communicate either information about individual distributions or the correlation between them. In response to these criticisms, Judy Olson (1981) conducted an experimental study using color schemes similar to those used by the Bureau of the Census. In contrast to the earlier criticism, Olson found that bivariate maps provided "information about regions of homogeneous value combinations" and that users, rather than being confused by these maps, actually had a positive attitude

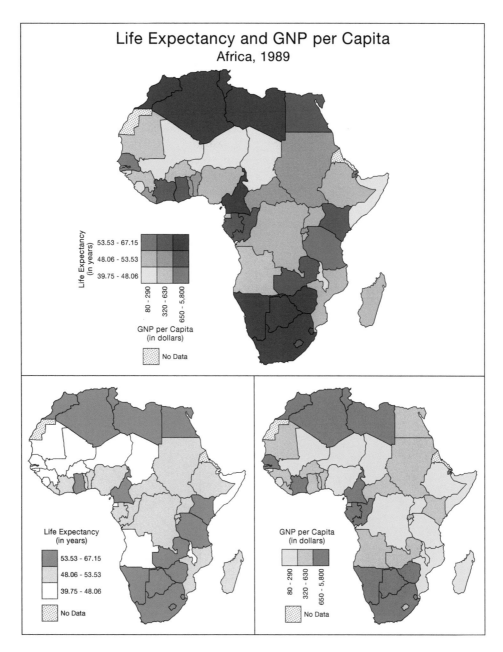

FIGURE 22.4 A bivariate choropleth map and its component univariate maps. The color scheme was taken from Olson (1981, 269) and is similar to those popular on U.S. Bureau of the Census maps in the 1970s. A quantiles classification was used for each map. (Data Source: ArcView 3.)

about them (275). Olson, however, stressed that a clear legend was critical to understanding bivariate maps, that both bivariate and individual maps should be shown,[9] and that an explanatory note should be included describing the types of information that could be extracted (see p. 273 of Olson's work for an example).

As an alternative to the Bureau of the Census approach, J. Ronald Eyton (1984) developed a bivariate method based on **complementary colors**, or colors that combine to produce a shade of gray.[10] Eyton chose red and cyan as the complementary colors for his maps, presumably because they produced an attractive map. Using the subtractive primary colors (CMY), Eyton created red by combining magenta (M) and yellow (Y), and he created cyan (C) directly from the cyan subtractive primary. Because overprinting colors in CMY does not produce a true gray, Eyton used black ink (K) for areas in which a true gray was desired. A bivariate map resulting from Eyton's process is shown in Figure 22.5.[11]

Eyton argued that most of the factors utilized by Olson were accounted for by his complementary method and that users appeared to understand the map more easily than one based on Bureau of the Census colors. A visual comparison of Figures 22.4 and 22.5 suggests that Eyton was correct. The Bureau of the Census color scheme implies nominal differences and requires careful examination of the legend, whereas Eyton's scheme appears more logically ordered and allows patterns on the map to be discerned more easily.

Note, for example, the ease with which the reddish-brown values (those above the white–gray–black diagonal) can be found using the Eyton scheme, compared with the corresponding values using the Bureau of the Census color scheme.

In addition to promoting complementary colors, Eyton also recommended that statistical parameters associated with the attributes be considered in bivariate mapping. Specifically, Eyton utilized the reduced major axis and bivariate normality. Recall from Section 3.6.3.3 that the reduced major axis method fits a regression line to a set of data such that the line bisects the regression lines of Y on X (Y is treated as the dependent attribute) and X on Y (X is treated as the dependent attribute; see Figure 22.6). The reduced major axis is thus appropriate when it is not clear which attribute should be treated as the dependent attribute.

A distribution is considered **bivariate-normal** if the Y values associated with a given X value are normal and if the X values associated with a given Y value are also normal. Generally, this condition is met if the individual distributions of X and Y are normal. Given a bivariate-normal distribution, it is possible to construct an **equiprobability ellipse** enclosing a specified percentage of the data. The equiprobability ellipse will be centered on the mean of the data and oriented in the direction of the reduced major axis (Figure 22.6). Using this ellipse and the means of the two attributes, Eyton created a bivariate map analogous to the

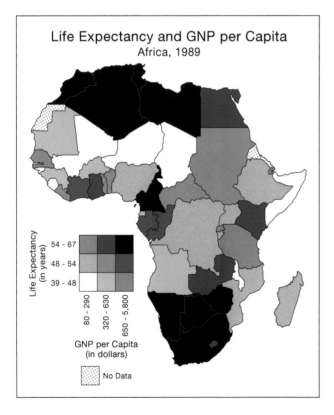

FIGURE 22.5 A bivariate choropleth map based on the complementary colors red and cyan. (After Eyton 1984.)

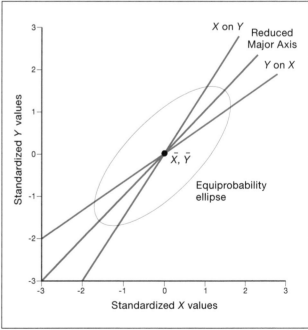

FIGURE 22.6 A reduced major axis and associated equiprobability ellipse. The reduced major axis is appropriate when one does not wish to specify a dependent attribute. The equiprobability ellipse can be used to enclose a specified percentage of the data associated with a bivariate normal distribution.

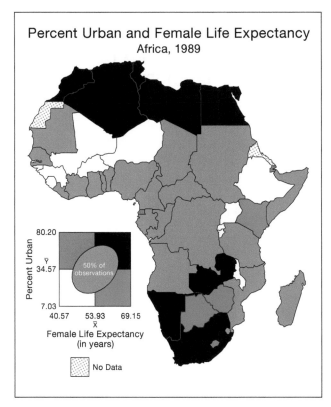

FIGURE 22.7 A bivariate choropleth map using complementary colors in which legend classes are based on the means of the variables and an equiprobability ellipse enclosing 50 percent of the data. Because the ellipse is based on a bivariate normal distribution, only normally distributed data should be mapped using this method. (After Eyton 1984; Data source: ArcView 3.)

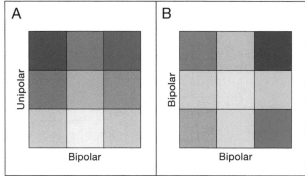

FIGURE 22.8 Some color schemes suggested by Brewer for bivariate maps: (A) Potential schemes for one unipolar and one bipolar attribute; (B) potential schemes for two bipolar attributes.

A potential concern in bivariate choropleth mapping is the number of classes used for each attribute. In the examples presented thus far, we have used three classes for each attribute. Our use of only three classes parallels Olson and Eyton's thinking; for instance, Eyton argued:

> The number of categories to be used should not exceed the number that can be dealt with by the reader. A 3 × 3 legend is both mechanically and visually simpler than a 4 × 4 arrangement and might actually convey more to the reader. (Eyton 1984, 480)

The approaches we have considered to this point for bivariate choropleth mapping have all been based on smooth (untextured) colors. As an alternative, both Laurence Carstensen (1982, 1986) and Stephen Lavin and J. Clark Archer (1984) created bivariate maps using **cross-hatched shading** consisting of horizontal and vertical lines (Figure 22.9). Interpretation of these maps focuses on the size and shape of the boxes formed by the cross-hatched lines (low values on both attributes are represented by large squares, whereas high values on both attributes are represented by small squares). A high positive correlation is represented on the map by a predominance of squares of varying sizes within enumeration units, and a high negative correlation is depicted by rectangles. In the case of Figure 22.9, we see a fair number of rectangles, which supports the correlation of $r = -0.71$.

Both Carstensen and Lavin and Archer stressed that *unclassed* cross-hatched bivariate maps should be used (thus, the legend in Figure 22.9 contains no class breaks). This suggestion contrasts with that of Eyton (1984, 485–486), who found smooth colored unclassed bivariate maps difficult to interpret. The reason that unclassed cross-hatched maps appear to be more effective than smooth colored unclassed maps is that the line-spacing attribute allows individual attributes to be seen. Carstensen (1982) found that cross-hatched bivariate maps are reasonably effective in communicating concepts about correlation but noted two problems: the difficulty of shading small enumeration units

one shown in Figure 22.7.[12] Eyton argued that the resulting map clearly contrasted observations near the means of the data (as defined by the 50 percent equiprobability ellipse) with extreme observations (i.e., those well above the mean on both attributes, those well below the mean on both attributes, and those high on one attribute and low on the other). For example, in Figure 22.7, we can clearly see that the countries of Botswana, Zimbabwe, Lesotho, and Kenya (those shown in cyan) have a low percent urban and high female life expectancy.

Cynthia Brewer (1994) argued that color schemes for bivariate maps should be a function of whether the attributes are unipolar or bipolar in nature. For two unipolar data sets (e.g., percent urban and female life expectancy), she recommended using either a complementary scheme or two subtractive primaries; the latter produces a diagonal of shades of a constant hue, as opposed to the grays resulting from complementary colors. (Brewer combined magenta and cyan to create a diagonal of purple blues.) If one attribute is unipolar and the other bipolar, she recommended using a sequential scheme for the unipolar data and a diverging scheme for the bipolar data (Figure 22.8A). If both are bipolar, she recommended using two diverging color schemes (Figure 22.8B).

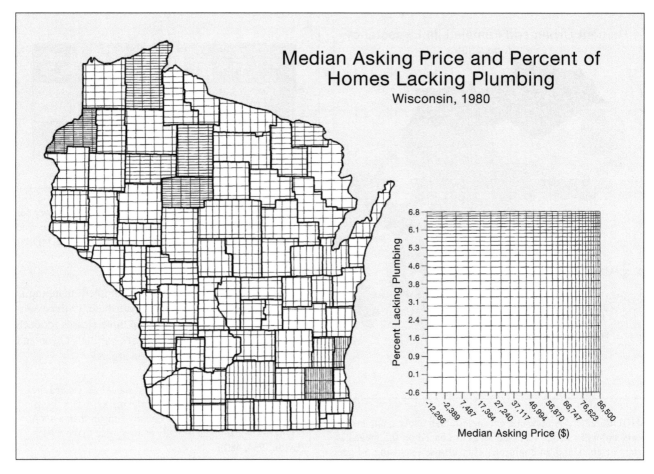

FIGURE 22.9 A bivariate choropleth map based on cross-hatching (the attributes are represented by horizontal and vertical lines of varying spacing). Note that each attribute is unclassed; negative values in the legend are a function of the major axis scaling used to fit the bivariate data. (From Carstensen 1986, p. 36; courtesy of Laurence W. Carstensen.)

and the unpleasant appearance of the symbology. These problems cause this method to be used less frequently than the smooth color methods that we have mentioned.

22.3.2.2 Additional Bivariate Mapping Techniques

In this section, we consider several additional approaches for bivariate mapping. The **bivariate point symbol** is one alternative. Figure 22.10 depicts one form of bivariate point symbol, the **rectangular point symbol**, in which the width and height of a rectangle are made proportional to each of the attributes being mapped. Both Stanton Wilhelm (1983) and Sean Hartnett (1987) proposed this method as an alternative to cross-hatched symbology; Hartnett argued that examining rectangular point symbols is much easier than inspecting the small boxes formed by cross-hatched lines and that the resulting map is more aesthetically pleasing. Because point symbols are more readily associated with point locations than with areas, bivariate point symbols might be particularly appropriate for mapping true point data (say, the number of out-of-wedlock births to teenagers and the number of hours of sex education for high school students for major cities in the United States).[13]

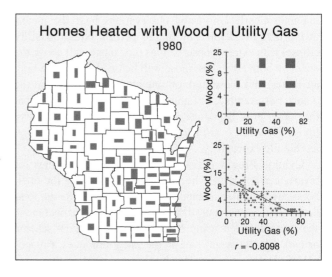

FIGURE 22.10 A bivariate map in which attributes are represented by the width and height of a rectangular point symbol. (Courtesy of Sean Hartnett.)

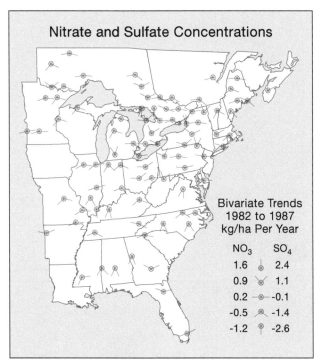

Nitrate and Sulfate Concentrations

Bivariate Trends
1982 to 1987
kg/ha Per Year

NO_3		SO_4
1.6		2.4
0.9		1.1
0.2		-0.1
-0.5		-1.4
-1.2		-2.6

FIGURE 22.11 A bivariate map based on a ray-glyph symbol. Rays (straight-line segments) pointing to the left and right represent changes in nitrate and sulfate concentrations, respectively, over the 1982 to 1987 time period. (After Carr et al. 1992. First published in *Cartography and Geographic Information Systems* 19(4), p. 234. Reprinted with permission from the Cartography and Geographic Information Society.)

Another form of bivariate point symbol is the **bivariate ray-glyph**, which Daniel Carr and his colleagues (1992) used to examine the trend in nitrate and sulfate concentrations from 1982 to 1987 in the eastern United States and Canada. In Figure 22.11, rays (the straight-line segments) pointing to the left represent nitrate concentrations, and rays pointing to the right represent sulfate concentrations. A maximum increase in concentrations occurs when the rays extend toward the top of the symbol, whereas a maximum decrease occurs when the rays extend toward the bottom of the symbol. An advantage of the ray-glyph is that the small symbols can be squeezed into a relatively restricted space. It seems likely, however, that the patterns represented by these symbols would be more difficult to interpret than those for the rectangular symbol.

A relatively common approach in bivariate mapping is to combine proportional and choropleth symbols with the size of the proportional symbol used for raw-total data and a choropleth shade within the symbol used for standardized data. For example, this approach could be used to depict on a single map the infant mortality data presented as two maps in Figure 22.2.

Space does not permit us to cover the full range of bivariate mapping techniques that have been developed. One technique that you might wish to examine is Colin Ware's (2009) use of *quantitative texton sequences*. In this approach, one smooth and continuous attribute is depicted using an areal color scheme, and another smooth and continuous attribute is depicted using a sequence of small graphical elements called *textons*, where each texton is associated with a particular numerical value. For example, this approach might be used to overlay maps of temperature and pressure. Ware found that the texton approach was more effective than traditional bivariate color schemes but that the approach is limited in that it cannot achieve the high resolution of smooth bivariate color schemes. For those with a strong interest in cartograms, Sabrina Nusrat and colleagues (2018) describe a variety of bivariate mapping approaches appropriate for cartograms. Various visualizations that they have created can be found online at http://bivariate.weebly.com.

22.4 MULTIVARIATE MAPPING INVOLVING THREE OR MORE ATTRIBUTES

22.4.1 COMPARING MAPS

If more than two attributes are symbolized, each as a separate map, the result is termed a **small multiple** (Figure 22.12). Edward Tufte (1990, 33) argued, "Small multiples, whether tabular or pictorial, move to the heart of visual reasoning—to see, distinguish, choose …. Their multiplied smallness enforces local comparisons within our eyespan, relying on the active eye to select and make contrasts." Although much can be gleaned from small multiples, they have their limitations. A general problem is that comparing two particular points or areas across a set of attributes can be difficult: Try using the dot maps shown in Figure 22.12 to describe the nature of agriculture in the state of Michigan. In the case of choropleth maps, one problem is the difficulty of discerning small enumeration units (e.g., a number of countries would disappear in a small multiple of choropleth maps of Africa). Problems such as these have led researchers to develop methods for combining multiple attributes on the same map.

22.4.2 COMBINING ATTRIBUTES ON THE SAME MAP

22.4.2.1 Trivariate Choropleth Maps

The notion of overlaying two choropleth maps can be extended to three choropleth maps, thus producing a **trivariate choropleth map**. Ideally, this approach should be used only for three attributes that add to 100 percent; examples include soil texture (expressed as percent sand, silt, and clay) and voting data for three political parties (e.g., percent voting Republican, Democrat, and independent). Colors can be assigned to the three attributes using a variety of approaches: CMY (Brewer 1994, 142; Metternicht and Stott 2003); RGB (Byron 1994); and red, blue, and yellow primaries (Dorling 1993, 172). The result of using RGB is shown

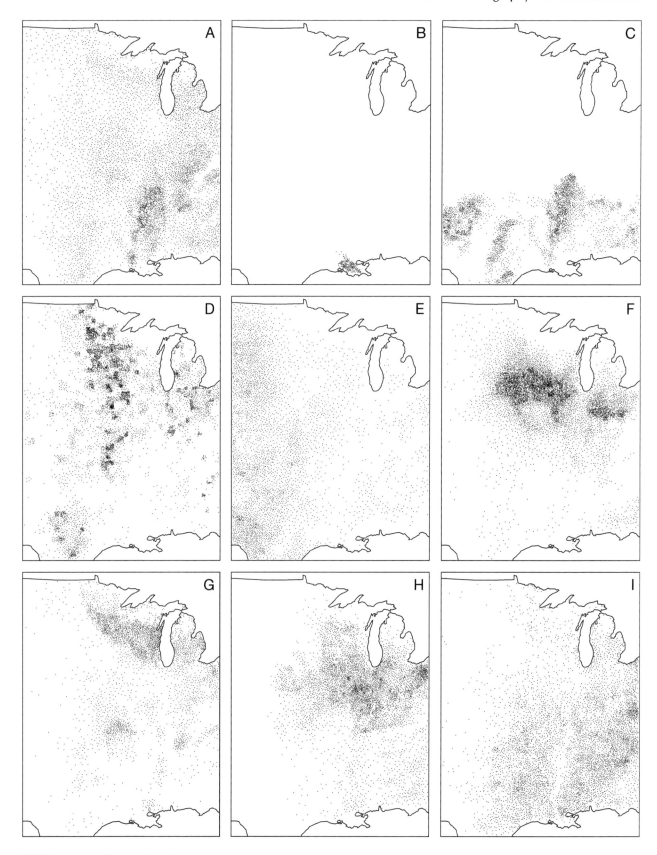

FIGURE 22.12 The small multiple: a method for multivariate mapping in which each attribute is depicted as a separate map. Dot maps depict the following agricultural data for 1954 for the central portion of the United States: (A) class V farms, (B) sugar cane, (C) cotton, (D) turkeys, (E) pasture, (F) hogs and pigs, (G) dairy, (H) expenditures for lime, and (I) residential farms. (After Lindberg 1987; courtesy of Mark B. Lindberg.)

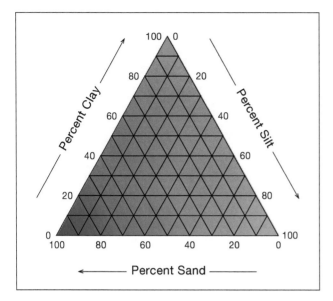

FIGURE 22.13 An RGB color scheme for creating a trivariate choropleth map. (After Byron 1994, p. 126.)

in Figure 22.13. The advantage of using attributes that add to 100 percent is that the resulting colors will be restricted to a triangular 2-D space. Brewer (1994) argued that if the attributes do not sum to 100 percent, a 3-D cube-shaped legend will be required and the resulting map will be difficult to interpret.

One potential solution to the difficulty of overlaying smooth colors is to *weave* the component colors with one another. Haleh Hagh-Shenas and colleagues (2007) accomplished this **color-weaving** approach by letting each color layer be represented by a sampling of independent pixels defined by a random noise function. In experiments involving the extraction of specific information from maps, Hagh-Shenas found that error rates for extraction remained low for up to four attributes and that the approach was significantly more effective than when color layers were simply blended together. Limitations of the color-weaving approach include the relatively coarse visual appearance and the fact that it would be awkward for small enumeration units.

22.4.2.2 Multivariate Dot Maps

The notion of univariate dot mapping (introduced in Chapter 19) can be extended to create a **multivariate dot map** if a distinct shape or color of symbol is used for each attribute to be mapped.[14] In the case of color, George Jenks (1953) introduced the notion of pointillism for multivariate dot mapping and developed two major maps based on it (Jenks 1961; 1962). Pointillism was a technique used by nineteenth-century painters to create various color mixtures by assuming the viewer would visually combine very small dots of selected colors. Jenks applied this principle by letting different-colored dots represent various crops or farm products; Figure 22.14;[15] he argued that viewers

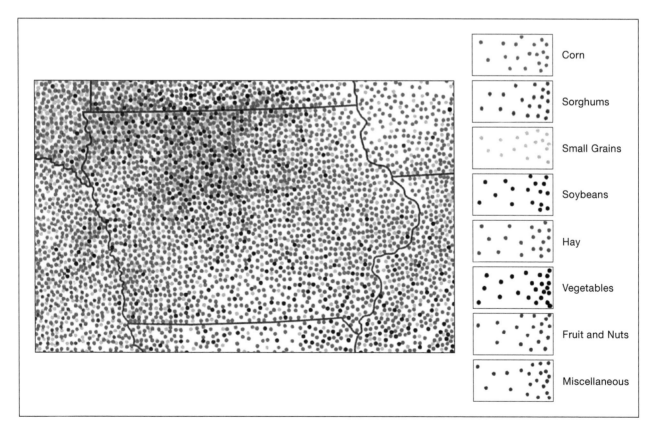

FIGURE 22.14 A portion of a multivariate dot map constructed on the basis of pointillism. Because this map was scanned from the original map, colors shown are approximations of the actual colors. (After Jenks 1961.)

could visually merge the separate colors to create mixtures, thus gaining a more realistic view of the transitional nature of cropping practices often found in the landscape. Furthermore, Jenks noted that if the dots were large enough, the map could provide detail regarding the location of individual crops in selected areas. Jenks (1953, 5) also provided some useful suggestions for dot colors:

1. Colors should remind the map reader of the crop that they represent.
2. High-value, low-acreage crops, such as tobacco or truck, should be of more intense hue than the more extensive and widely grown crops.
3. Selected minor crops, such as peanuts or soybeans, which tend to change the crop character of broader areas, should have colors of moderately high intensity.

Although considerable information can be obtained from multivariate dot maps, readers have difficulty determining the meaning of the areas of color mixture because the legend contains colors for only individual crops. This problem might be solved if the map were viewed in an interactive data exploration environment in which individual map areas could be focused on and enlarged, and individual categories of dots could be turned on and off so that the relative contribution of each category could be determined. Richard Groop (1992) developed an automated method for creating colored dot maps, but unfortunately, his method was not published.

A general question that can be asked of multivariate maps is how effective they are compared to a set of individual maps of each attribute (i.e., how would a multivariate colored dot map compare to a set of dot maps in small multiple format?). Jill Rogers and Richard Groop (1981) evaluated this question by having readers identify regions on both a multivariate dot map (consisting of three categories) and its component univariate dot maps. Readers were asked to identify both "homogeneous regions" in which one category appeared to predominate and "mixed regions" in which there appeared to be a mixture of two or three categories. Rogers and Groop found that the multivariate dot map "was slightly more effective in communicating perceptions of both homogeneous and mixed regions" (61). Their results, however, suggest that the perception of mixed regions on maps with more than three categories might not be effective.

22.4.2.3 Multivariate Point Symbol Maps

When multivariate data are depicted using a point symbol, the result is termed a **multivariate point symbol**. These symbols are obviously appropriate for point phenomena, but they generally must also be used for areal phenomena because of the difficulty of creating multivariate areal symbols (note the limitations of the trivariate choropleth map described previously when only three attributes are shown).

Two distinct forms of attributes are commonly mapped using multivariate point symbols: related (or additive) and nonrelated (or nonadditive). *Related attributes* are measured in the same units and are part of a larger whole. An example would be the percentages of various racial groups in a population: White, African American, Asian/Pacific Islander, Native American, and so on. Such attributes can be depicted using the familiar **pie chart**, in which a circle is divided into sectors representing the proportion of each attribute.

Nonrelated attributes are measured in dissimilar units and thus are not part of a larger whole (e.g., percent urban and median income). Multivariate point symbols used to represent nonrelated attributes are commonly termed **glyphs**, several of which are illustrated in Figure 22.15. The **multivariate ray-glyph** or **star** (Figure 22.15A) is constructed by extending rays from an interior circle, with the lengths of the rays proportional to the values of each attribute.[16] If rays composing the multivariate ray-glyph are connected, a **polygonal glyph** or **snowflake** is created (Figure 22.15B).

Donna Cox (1990) and her colleagues at the National Center for Supercomputing Applications at the University of Illinois at Urbana-Champaign developed several novel multivariate point symbols. One of these is 3-D bars, in which the bars are made to look 3-D, and the height of each bar is made proportional to the magnitude of various attributes (Figure 22.15C); in Cox's implementation, the individual bars representing attributes were shown in different colors, as opposed to using the different textures seen in Figure 22.15C. Another technique is the **data jack** (Ellson 1990), in which triangular spikes are extended from a square central area and made proportional to the magnitude of each attribute (Figure 22.15D). As with bars, the spikes of jacks can be distinguished most easily if they are displayed in different colors. An advantage of the 3-D structure of jacks is that they can be viewed from arbitrary positions in 3-D space.

A particularly intriguing multivariate point symbol is the **Chernoff face** (Figure 22.15E), developed by Hermann Chernoff (1973), in which distinct facial features are associated with various attributes. For example, fatness of the cheeks might represent one attribute, whereas the size of the eyes might represent another attribute. Figure 22.16, taken from Daniel Dorling's work with cartograms (see Section 20.4.2), is illustrative of Chernoff faces. In this case, the width of the face represents mean house price (fat cheeks indicate more expensive housing); mouth style represents the percent of adult employment (a smile for high employment); nose size represents the percent voting (a larger nose indicates a higher percentage); eye size and position represent the percent employed in services (large eyes located near the nose indicate a high percent employed in services); and total area of the face represents the number of voters in a parliamentary constituency.[17]

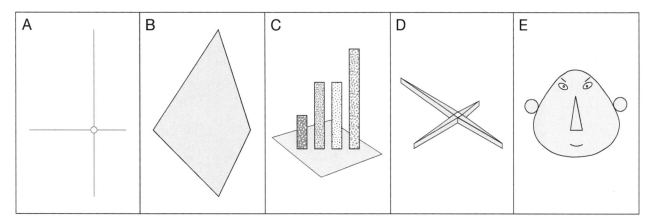

FIGURE 22.15 Examples of multivariate point symbols: (A) A multivariate ray-glyph or star: the lengths of rays are proportional to the values of attributes; (B) a polygonal glyph or snowflake: a polygon connects the endpoints of the rays shown in A; (C) 3-D bars: the height of bars is proportional to the magnitude of attributes; (D) data jacks: the spikes of the jack are proportional to the magnitude of each attribute; and (E) Chernoff faces: individual facial features (e.g., the size of the eyes) are associated with individual attributes.

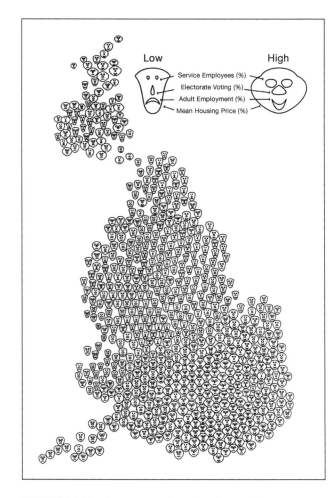

FIGURE 22.16 A cartogram in which Chernoff faces are used to display multivariate data. (From Dorling, D. (1995) "The visualization of local urban change across Britain," *Environment and Planning B: Planning and Design* 22, no. 3:269–290, https://journals.sagepub.com/doi/abs/10.1068/b220269?journalCode=epba.)

For simplicity, only four attributes are depicted by the symbols shown in Figure 22.15. Several of these symbols, however, can be modified to represent many more attributes. For example, numerous rays can be added to the star, and up to 20 different attributes can be represented by Chernoff faces (Wang and Lake 1978, 32–35).

The multivariate point symbols that we have considered thus far are often associated with points or areas in which there is some spacing between the point symbols. In contrast, several multivariate point symbols have been developed that are intended to depict numerical data that change continuously across geographic space (and thus, there may be little or no spacing between point symbols). An example of this approach is the "perceptual texture elements" (or **pexels**) developed by Christopher Healey and colleagues (Healey and Enns 1999; Healey 2001). Pexels are small 3-D bars (similar to those utilized by Cox and her colleagues) that can be varied in height, spacing, and color.[18] As an illustration, consider Figure 22.17, which portrays three attributes related to plankton in the northern Pacific Ocean (plankton are an important food source for salmon). The density of plankton is indicated by color, with low to high plankton values indicated by blue, green, brown, red, and purple, respectively. The speed of ocean current is indicated by the height of pexels (taller pexels indicate a stronger current), and the magnitude of sea surface temperature is indicated by the spacing of pexels (a tighter spacing indicates a warmer temperature). In examining Figure 22.17A, we see that in February, the density of plankton is relatively low (we see mostly blues and greens). At this time, ocean currents are relatively weak in the north central Pacific and along the south coast of Alaska (low bars are found at these spots), and most of the ocean is relatively cold (pexels are relatively widely spaced), although a warmer region can be seen in the south.

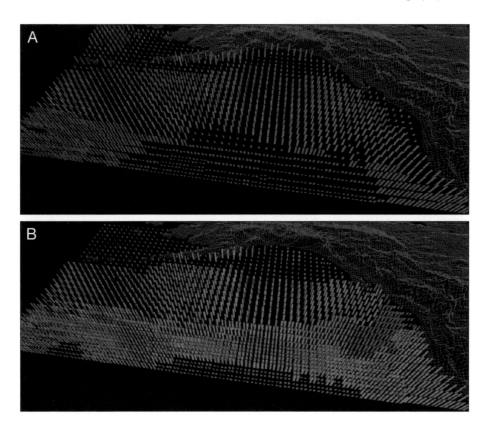

FIGURE 22.17 Using pexels to depict multivariate data for the northern Pacific Ocean (A) in February and (B) in June. Color of the pexels depicts the density of plankton (from low to high density, the scheme is blue, green, brown, red, and purple); height of the pexels depicts the ocean current (a higher pexel equals a stronger current); and spacing of the pexels depicts sea surface temperature (a tighter spacing indicates a warmer temperature). (Courtesy of Christopher Healey.)

In contrast, in June (Figure 22.17B), we see a higher density of plankton (reds and purples predominate), stronger currents seem to have shifted southward, and water is warmer farther north, although the ocean off the Alaskan and British Columbia coasts is still relatively cold. These sorts of summaries come not only from looking at static images such as those shown here but also from using interactive software that allows the user to view the 3-D image from different angles, as well as zoom in and pan around.[19]

22.4.2.4 Acquiring Specific and General Information from Multivariate Maps

From Chapter 1, you will recall that two kinds of information can be acquired from univariate thematic maps: specific and general. These same kinds of information can also be acquired from multivariate thematic maps, but the information can be acquired either for individual attributes or for combinations of attributes. Thus, we can conceive of the kinds of information acquired from a multivariate map as a two-by-two matrix, with specific–general on one axis and nature of the attributes (i.e., individual or combined) on the other axis (Nelson and Gilmartin 1996). As an illustration, consider Figure 22.18, which uses combined star and snowflake symbols to depict "quality of life" in South Carolina counties in 1992.[20] As indicated in the legend, each attribute on this map is symbolized by rays of three

lengths, indicating low, medium, and high desirability; for example, those counties with a high median family income are depicted as high desirability, whereas those counties with a high percentage of families below poverty level are depicted as low desirability. An example of specific information for an individual attribute would be the desirability associated with a particular county (for instance, the southernmost county is highly desirable in terms of attribute 2, median family income). A combined form of specific information would involve comparing the size of one symbol with the size of another (for instance, noting that the southernmost symbol is larger than the one to its immediate northwest). An example of general information for an individual attribute would involve examining the distribution of a single attribute across the state (e.g., examining the distribution of the attribute infant mortality); note that this is a difficult task for the combined star and snowflake symbol. A form of combined general information would involve examining the pattern of the size of symbols across the state (e.g., do you see anywhere in the state where the smaller symbols seem to concentrate?).

22.4.2.5 Ring Maps: An Alternative to Conventional Symbolization Approaches

Traditional approaches for multivariate mapping (such as trivariate choropleth maps and Chernoff faces) can sometimes be difficult to interpret because so much information

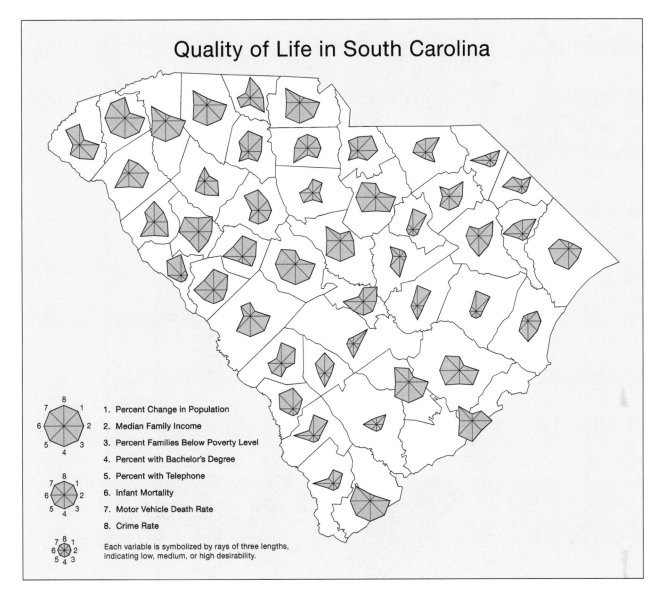

FIGURE 22.18   A multivariate map based on a combination of the star and snowflake symbols shown in Figure 22.15. (Data source: South Carolina State Budget and Control Board 1994.)

is placed in a single symbol. A potential solution to these traditional approaches is the **ring map** in which a single attribute is shown with a traditional mapping approach (e.g., the choropleth map in Figure 22.19), and surrounding concentric rings depict other attributes (as in the rings in Figure 22.19). Note also that each ring is split into *spokes*, with a leader line extending from each enumeration unit to its position on the ring. Sarah Battersby and colleagues (2011) note that ring maps have been largely used for temporal data but that they can also be used for non-temporal data, as illustrated in Figure 22.19.[21]

The choropleth map in Figure 22.19A depicts poverty rates for counties in South Carolina divided into three categories: low, medium, and high. The data were divided into quartiles for classification, with the low and high categories representing the lowest and highest quartiles and the medium category representing the two middle quartiles. The rings represent five health conditions that may be dependent on poverty, with each health condition divided into low, medium, and high categories (again based on the quartile approach just described). Note the clear association between those counties falling in the High poverty rate category and their tendency to fall in the High disease rate category on the five health conditions. For instance, the counties of Marlboro, Clarendon, and Bamberg all fall in the High poverty rate category, and these counties also fall in the High category on all five health conditions. This relationship has been highlighted by using drop shadows for counties in the highest category on the choropleth map and by also shading the associated county names. Note that the opposite tends to occur for counties in the Low poverty category.

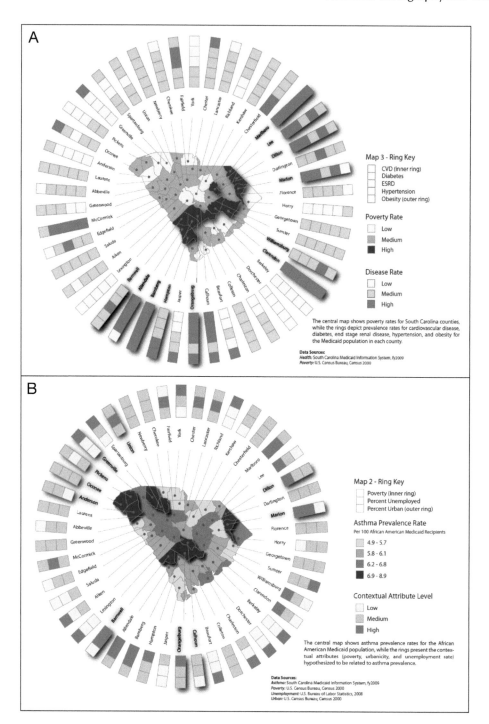

FIGURE 22.19 Examples of ring maps for South Carolina counties: (A) Examining the relationship between poverty and five health conditions; (B) examining the relationship between asthma prevalence rate and three potential explanatory attributes. (From Battersby, S. E., Stewart, J. E., Lopez-de Fede, A., et al. (2011) "Ring maps for spatial visualization of multivariate epidemiological data." *Journal of Maps*, copyright © Journal of Maps, reprinted by permission of Taylor & Francis Ltd., http://www.tandfonline.com on behalf of Journal of Maps.)

22.5 CLUSTER ANALYSIS

In Chapter 5, we considered classification methods appropriate for a single numeric attribute. Here we consider classification methods appropriate for multiple numeric attributes. As an example, we will cluster counties for the following attributes for counties in New York State (excluding counties for New York City): % population change (1990–2000), % unemployed in 2000, % African American in 2000, and infant mortality rate in 2000.[22] Our interest is in whether there are certain counties that have similar scores on one or more of these attributes. For instance, we might wonder whether some counties have a high percentage of African Americans, a high infant mortality rate, and a decreasing population (a negative percent population change). We might tackle such questions by visually comparing separate maps of the attributes, but this would require us to mentally overlay the maps to derive various combinations of attributes for particular counties. More useful would be a method that combines the maps and tells us which counties are similar to one another and how they are similar—this is the purpose of the mathematical technique of **cluster analysis**, which we describe in this section.

Cluster analysis techniques can be divided into hierarchical and nonhierarchical methods. *Hierarchical methods* begin by assuming that each observation (county, in the case of the New York data) is a separate cluster and then progressively combine clusters. The process is hierarchical in the sense that once observations are combined in a cluster, they remain in that cluster throughout the clustering procedure. Nonhierarchical methods presume a specified number of clusters and associated members and then attempt to improve the classification by moving observations between clusters. Because hierarchical methods are a bit easier to understand, we focus only on them here.[23]

22.5.1 Basic Steps in Hierarchical Cluster Analysis

To summarize the process of cluster analysis, we utilize an eight-step process that is a modification of six steps recommended by Charles Romesburg (1984).

Step 1. Collect an Appropriate Data Set. The first step is to collect a data set that the mapmaker wishes to cluster. Ultimately, we will cluster the New York State data mentioned previously, but for illustrative purposes, we also will work with the small hypothetical data set for livestock and crop production listed in Figure 22.20A and plotted in graphical form in Figure 22.20B.

Romesburg (1984, 38) indicates that cluster analysis can be used for three purposes:
- To create a scientific question.
- To create a research hypothesis that answers a scientific question.
- To test a research hypothesis to decide whether it should be confirmed or disproved.

Because we were relatively unfamiliar with the census data for New York, we wanted to use it to create a scientific question—to explore the patterns that might arise. Those knowledgeable about the geography of New York might use the resulting patterns to look for other attributes to explain the results or consider other attributes that might be suitable for a cluster analysis.

Step 2. Standardize the Data Set. The second step in cluster analysis is to standardize the data set, if necessary. Two forms of standardization are possible. One accounts for enumeration units of varying

sizes. Just as for choropleth maps (see Section 4.6.1), this standardization is appropriate because larger enumeration units are apt to have larger values of a countable phenomenon. For the New York State data, we standardized the raw number of African Americans living in each county by dividing by the total population of each county. For our small hypothetical agricultural data set, we will assume the data are already standardized.

The second form of standardization adjusts for different units of measure on the attributes. For instance, consider the case of median family income and percent voting in a presidential election. Median family income likely would be measured in thousands of dollars ranging from, say, 20,000 to 60,000, whereas percent voting in a presidential election is a percentage attribute ranging from, say, 30 to 50. If such data were clustered, the percentage values would have little impact on the cluster analysis; we would essentially be clustering family income. The most common approach for standardizing for different units of measure is to compute z scores:

$$z_i = \frac{X_i - \bar{X}}{s}$$

where z_i is the resulting z score for the ith observation, X_i is a raw data value for the ith observation, $\bar{X}$ is the mean of an attribute, and s is the standard deviation of an attribute. Applying this formula to our hypothetical data in Figure 22.20A, we would find respective mean and standard deviation values of 14.80 and 9.65 for the livestock attribute and a z score of $(5.00 - 14.80)/9.65 = -1.02$ for the first observation. Because we presume that our hypothetical data are in the same units of measure, however, this standardization is not necessary, and thus we will work with the unstandardized data. We chose, however, to standardize the New York State data using z scores because three of the attributes were measured in percent and the other (infant mortality) was measured in number of deaths per 1,000 births.

Step 3. Compute Initial Resemblance Coefficients. The third step of cluster analysis is to compute an initial set of **resemblance coefficients** that expresses the similarity of each pair of observations. Although numerous coefficients are possible, we will use a *Euclidean distance* coefficient because it is commonly used and readily interpreted graphically. The Euclidean distance for two attributes is

$$d_{ij} = \sqrt{\left(X_i - X_j\right)^2 + \left(Y_i - Y_j\right)^2}$$

where d_{ij} is the Euclidean distance between the ith and jth observations, X_i and X_j are the

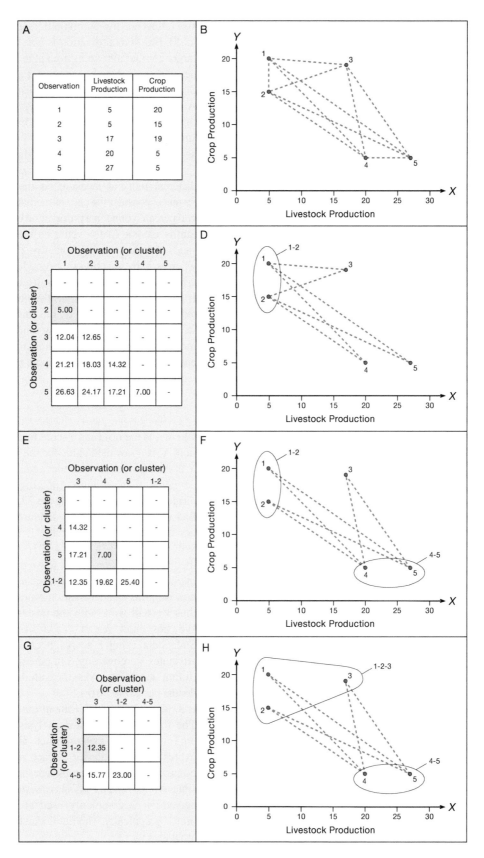

FIGURE 22.20 The basic cluster analysis process using the UPGMA method: (A) A hypothetical raw data set; (B) a graphical plot of each observation; (C) the initial resemblance matrix of Euclidean distance coefficients; (D) observations (or clusters) 1 and 2 combine because they have the smallest Euclidean distance coefficient; (E–H) the clustering process continues by combining clusters that have the smallest Euclidean distance coefficient.

values of observations i and j on attribute X, and Y_i and Y_j are the values of observations i and j on attribute Y. For instance, the Euclidean distance between the first two observations in Figure 22.20A is

$$d_{12} = \sqrt{(X_1 - X_2)^2 + (Y_1 - Y_2)^2}$$
$$= \sqrt{(5-5)^2 + (20-15)^2} = 5.00$$

Note that this value corresponds to the distance between points 1 and 2 in the graph in Figure 22.20B. Figure 22.20C provides a complete set of Euclidean distance coefficients for all pairs of observations listed in Figure 22.20A.

Step 4. Cluster the Data. The fourth step of cluster analysis is to actually cluster the data. In hierarchical cluster analysis, we begin by considering each observation as a separate cluster and then combining those clusters that are most similar to one another, as expressed by the initial resemblance coefficients. In our case, observations (clusters) 1 and 2 have the shortest Euclidean distance (see Figure 22.20C) and thus combine to form a new cluster, as shown in Figure 22.20D. For our purposes, we designate this as cluster 1–2.

When clusters are combined, we must recompute the resemblance coefficient between the newly formed cluster and all other existing clusters. Although numerous methods are possible for recomputing resemblance coefficients, we focus on one—the **unweighted pair–group method using arithmetic averages (UPGMA)**—because Romesburg (1984) indicated that it has been commonly used. In UPGMA, revised resemblance coefficients are computed by averaging coefficients between observations in the newly formed cluster and observations in existing clusters. For the newly formed cluster 1–2, we compute the following resemblance coefficients:

$$d_{3(1-2)} = \tfrac{1}{2}(d_{13} + d_{23}) = \tfrac{1}{2}(12.04 + 12.65) = 12.35$$

$$d_{4(1-2)} = \tfrac{1}{2}(d_{14} + d_{24}) = \tfrac{1}{2}(21.21 + 18.03) = 19.62$$

$$d_{5(1-2)} = \tfrac{1}{2}(d_{15} + d_{25}) = \tfrac{1}{2}(26.63 + 24.17) = 25.40$$

The results of these computations appear in the last row of the revised resemblance matrix shown in Figure 22.20E. Evaluating this revised resemblance matrix, we find that observations (clusters) 4 and 5 have the smallest Euclidean distance, and so they are grouped next to form cluster 4–5 (see Figure 22.20F). Again, we compute revised resemblance coefficients by averaging coefficients between observations in the newly formed cluster

and observations in existing clusters. For clusters 3 and 4–5, the result is

$$d_{3(4-5)} = \tfrac{1}{2}(d_{34} + d_{35}) = \tfrac{1}{2}(14.32 + 17.21) = 15.77$$

while for clusters 1–2 and 4–5, the result is

$$d_{(1-2)(4-5)} = \tfrac{1}{4}(d_{14} + d_{15} + d_{24} + d_{25})$$
$$= \tfrac{1}{4}(21.21 + 26.63 + 18.03 + 24.17) = 23.00$$

The resulting distances are shown in Figure 22.20G, where we see that the smallest Euclidean distance is 12.35 (between clusters 1–2 and 3). These clusters are combined in Figure 22.20H. Finally, we combine clusters 1–2–3 and 4–5 at an average Euclidean distance of 20.26 (the average of the lengths of the dashed lines shown in Figure 22.20H).

The results of a clustering process normally are summarized in the form of a **dendrogram**, which is a treelike structure that illustrates the resemblance coefficient values at which various clusters combined. For instance, for the hypothetical agricultural data, Figure 22.21 illustrates that clusters 1 and 2 combine at a Euclidean distance of 5 and that clusters 4 and 5 combine at a Euclidean distance of 7 (compare Figure 22.21 with Figure 22.20C and E). Figure 22.22 is a dendrogram of the results of UPGMA for the New York State data.

Step 5. Determine an Appropriate Number of Clusters. A key aspect of cluster analysis is determining an appropriate number of clusters. As we can see in the dendrograms, we begin with each observation as a separate cluster and ultimately combine all observations to create a single cluster. Clearly, the latter is not desirable as our goal is to create groups of observations, with each group being relatively homogeneous and different from the other groups. One common approach for selecting an appropriate number of clusters is to

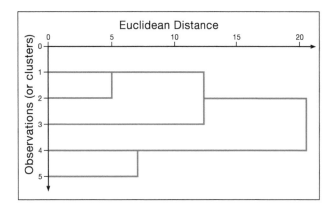

FIGURE 22.21 Dendrogram for the hypothetical data shown in Figure 22.20A using Euclidean distance as the resemblance coefficient.

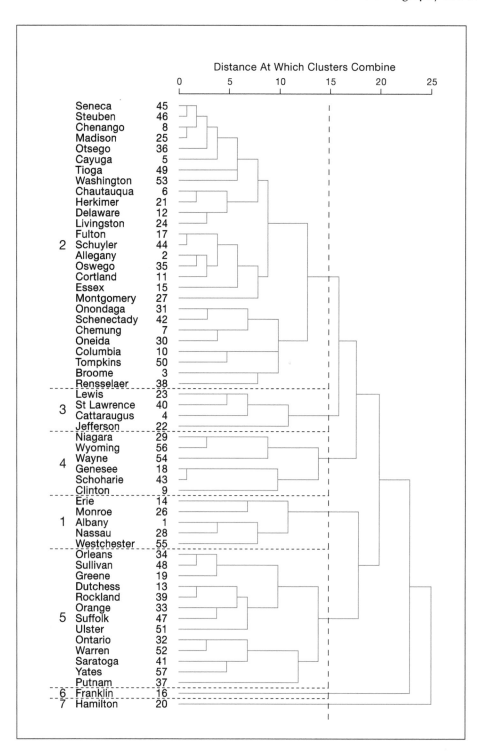

FIGURE 22.22 Dendrogram of the UPGMA method for the New York State data. Numbers on the left represent cluster numbers specified by the SPSS software used to generate the clusters and correspond to those shown in Table 22.1.

look for breaks in the dendrogram—places where combining two clusters would require making a fairly large jump in the resemblance coefficient. One simple way to do this is to cover the dendrogram with a piece of paper and then to gradually slide the paper along the dendrogram (in our case, from right to left). If you do this for the New York State data (Figure 22.22), you will find a relatively

large break for three clusters, but this is only the relatively uninteresting result that all but two observations fall in a single cluster. If you continue to slide the paper, you will find another relatively large break at seven clusters. Note that we have marked this solution in Figure 22.22 with a line of long dashes and indicated the composition of clusters with lines of short dashes.

TABLE 22.1

Mean *z* Scores for Attributes on Each Cluster Using the UPGMA Method with No Contiguity Constraint

Cluster	% Population Change	% Unemployed	% African American	Infant Mortality Rate
1	0.02	−0.86	2.19	−0.08
2	−0.56	−0.03	−0.40	−0.23
3	−0.40	1.95	−0.50	0.33
4	−0.47	0.31	−0.17	2.07
5	1.37	−0.67	0.26	−0.48
6	1.74	1.95	0.63	1.36
7	−0.04	2.35	−1.03	−2.26

Step 6. Interpret the Clusters. One challenge of cluster analysis is interpreting the resulting clusters. One approach is to compute, for each attribute, the mean z score for all observations falling in each cluster; for instance, Table 22.1 shows the mean *z* scores for the New York State data for the UPGMA method. We can name a cluster by using attributes with mean values that are relatively high or low (values near ±1.0 represent moderately high or low scores, whereas values near ±2.0 represent distinctly high and low scores). Thus, cluster 1 has moderately low unemployment and a high percentage of African Americans, whereas cluster 6 has a rapidly increasing population, high unemployment, and moderately high infant mortality. (Keep in mind that these statements are being made relative to other counties in New York State. A high value here might be low relative to some other areas of the country.)

Step 7. Map the Resulting Clusters. Once you have selected an appropriate set of clusters, you will want to map the results (Figure 22.23). In symbolizing clusters, you should bear in mind that clusters are likely to be qualitatively different, so symbols used to represent the clusters should reflect this qualitative difference. For the New York State data, we chose to use qualitatively different hues provided in ColorBrewer (https://colorbrewer2.org/#type=qualitative&scheme=Accent&n=7).[24]

Step 8. Determine Whether Clusters Are Really Appropriate. One should be cautious in interpreting the results of a cluster analysis because it is possible to cluster any data set, even a set of random numbers. One approach used to judge the effectiveness of a cluster analysis is the **cophenetic correlation coefficient**, which measures the correlation between raw resemblance coefficients (Euclidean distance values in the case of UPGMA) and resemblance coefficients derived from the dendrogram (normally referred to as the *cophenetic coefficients*). For our hypothetical data, these coefficients are depicted in

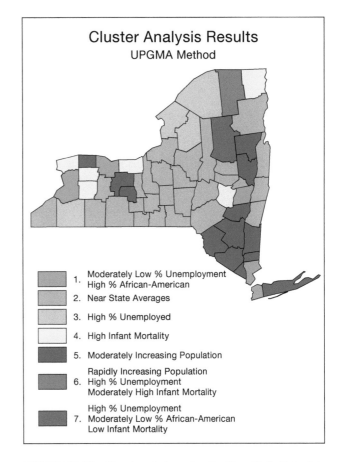

FIGURE 22.23   Results of clustering the New York State data using the UPGMA method. Numbers within the legend represent cluster numbers specified by the SPSS software used to generate the clusters.

Figure 22.24 as the *resemblance* and *cophenetic* matrices, respectively. Values for the resemblance matrix are taken from Figure 22.20C, and those for the cophenetic matrix are determined by considering the Euclidean distance at which observations combine in the dendrogram shown in Figure 22.21. For instance, cluster 1–2–3 does not combine with cluster 4–5 until a distance of 20.26, and so cophenetic values for observations 1 and 4, 2 and 4, and 3 and 4 are all 20.26.

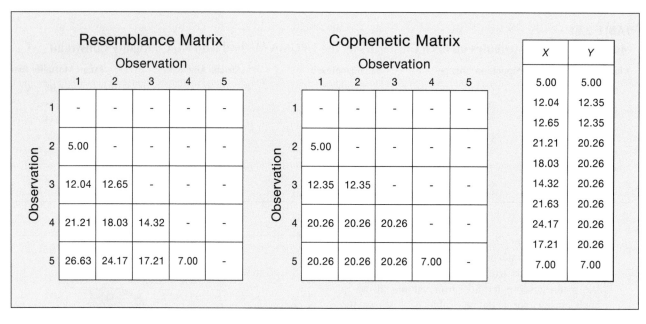

FIGURE 22.24 Computation of the cophenetic correlation involves relating corresponding values in the resemblance and cophenetic matrices.

Once the two matrices are determined, they are converted to two linear lists of X and Y values, as shown in Figure 22.24. A correlation coefficient r (see Section 3.6.3.1) is then computed between X and Y. For the matrices shown in Figure 22.24, r is equal to 0.91, which suggests that the clusters are reflecting the true relationships in the data, given that the maximum possible r is 1.0.[25] For the New York State data, we used the package R (http://www.r-project.org/) to compute the cophenetic correlation, deriving an r value of 0.83. This value also suggests that our clustered results are useful because Romesburg (1984, 27) indicates that cophenetic values of 0.80 or greater are desirable.

22.5.2 Adding a Contiguity Constraint to a Hierarchical Cluster Analysis

In Chapter 5, we noted that one limitation of traditional classification approaches is that they do not consider the spatial distribution of the data. In a similar fashion, we can argue that a limitation of the clustering method that we have described in Section 22.5.1 is that it does not explicitly consider the spatial distribution of the data. If your purpose is to create regions composed of enumeration units that are contiguous to one another, it can be argued that you should impose a contiguity constraint so that two enumeration units can fall in the same region only if they are contiguous to one another. A variety of approaches have been developed for imposing a contiguity constraint. We will utilize an approach developed by Diansheng Guo (2008) that is termed REDCAP (Regionalization with Dynamically Constrained Agglomerative Clustering and Partitioning).

REDCAP involves two basic steps: (1) a clustering of the data with a contiguity constraint to create a hierarchy of the data in the form of a *spatially contiguous tree*, and (2) a partitioning of this tree based on optimizing an objective function.

In the first step, Guo experimented with various forms of contiguity constraints and found that a so-called *full-order* method was most effective, in which the distance between two clusters is defined by all edges between enumeration units composing each cluster. Guo also experimented with various approaches for specifying the Euclidean distance between two clusters. One of these approaches is the average linkage that we described in Section 22.5.1. Others include *single linkage* and *complete linkage*. Single linkage defines the distance between two clusters as the shortest Euclidean distance between any pair of data points in each cluster. Alternatively, complete linkage defines the distance between two clusters as the longest Euclidean distance between any pair of points in each cluster.

In the second step of REDCAP, a measure of heterogeneity ($H(R)$) is computed for each potential new region as follows:

$$H(R) = \sum_{j=1}^{d}\sum_{i=1}^{n_r}\left(x_{ij} - \bar{x}_j\right)^2$$

where R is a region, d is the number of attributes, n_r is the number of objects in the region, x_{ij} is the attribute value for the jth attribute of the ith enumeration unit, and $\bar{x}_j$ is the mean of the jth attribute for all enumeration units in R. The goal is to divide the spatially contiguous tree such that the difference in heterogeneity between two resulting regions is as small as possible (the homogeneity of regions is maximized).

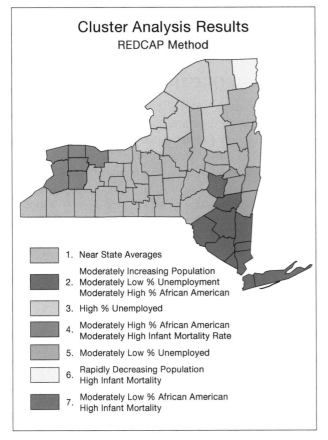

Cluster Analysis Results
REDCAP Method

1. Near State Averages

2. Moderately Increasing Population
   Moderately Low % Unemployment
   Moderately High % African American

3. High % Unemployed

4. Moderately High % African American
   Moderately High Infant Mortality Rate

5. Moderately Low % Unemployed

6. Rapidly Decreasing Population
   High Infant Mortality

7. Moderately Low % African American
   High Infant Mortality

FIGURE 22.25 Results of clustering the New York State data using the REDCAP method that includes a contiguity constraint. Numbers within the legend represent cluster numbers specified by the GeoDa software used to generate the clusters.

The GeoDa software (https://spatial.uchicago.edu/geoda) provides an implementation of the REDCAP approach.[26] Figure 22.25 illustrates the result of applying REDCAP to the New York cluster data set, with a full order average linkage for the contiguity constraint and seven regions (for comparison with the seven regions utilized in Figure 22.23).[27] Table 22.2 shows the mean z-scores for the resulting seven regions that were used to name each region. Obviously, the results for the contiguity-based approach (Figure 22.25) differ considerably from the approach in which contiguity is not considered (Figure 22.23). In designing Figure 25.25, we have used the same colors from ColorBrewer that were used to create Figure 22.23, and where possible, we have used the same colors for clusters with similar meanings in the two figures (e.g., purple is used for the category "Near State Averages" in both figures). In several cases, however, you will note that clusters with the same color in the two figures have quite different meanings.

When you decide whether or not you wish to impose a contiguity constraint, you need to consider carefully whether contiguity is important to your particular problem. For the New York State data, we note that the majority of counties in cluster 1 in Figure 22.23 (which are not contiguous to one another) have a major urban center: Buffalo in Erie County, Rochester in Monroe County, Yonkers in Westchester County, and Albany in Albany County. Note that these counties do not fall in the same cluster in Figure 22.25.

Space does not permit us to discuss the full range of issues associated with cluster analysis. For instance, one option available within GeoDa is the minimum number of enumeration units needed to create a cluster, which may be applicable in some situations. We encourage you to experiment with the wide range of clustering approaches and associated options that are available in GeoDa.

22.6 SUMMARY

In this chapter, we have covered a variety of methods for multivariate mapping involving the cartographic display of two or more attributes. When only two attributes are displayed, the process is commonly termed bivariate mapping. Multivariate mapping can be accomplished through either map comparison (in which a separate map is created for each attribute) or by combining all attributes on the same map. For bivariate map comparison, we stressed the importance of using an appropriate method of data classification. Although optimal data classification is often recommended for univariate mapping, it is generally inappropriate for bivariate mapping because it focuses on the precise distribution of individual data values along the number line. The quantiles method of classification is more appropriate for bivariate mapping.

TABLE 22.2
Mean z Scores for Attributes on Each Cluster Using the REDCAP Method with a Contiguity Constraint

Cluster	% Population Change	% Unemployed	% African American	Infant Mortality Rate
1	−0.26	−0.04	−0.53	−0.22
2	1.31	−0.88	1.02	−0.49
3	0.04	2.00	−0.42	−0.20
4	−0.13	0.26	0.99	1.03
5	−0.58	−0.98	0.67	0.19
6	−2.03	0.33	−0.19	2.88
7	−0.64	0.06	−0.81	2.78

When two choropleth maps are overlaid, the result is termed a bivariate choropleth map, a technique developed by the U.S. Bureau of the Census in the 1970s. Bivariate choropleth maps have been criticized because of their presumed failure to communicate information, although they can be effective if a clear legend is used, both bivariate and individual maps are shown, and an explanatory note is included. Eyton developed a logical color scheme for the bivariate choropleth map based on complementary colors (colors that combine to produce a shade of gray). Brewer noted that color schemes for bivariate choropleth maps should be a function of whether the attributes are unipolar or bipolar; for instance, if one attribute is unipolar and the other bipolar, Brewer recommended using a sequential scheme for the unipolar data and a diverging scheme for the bipolar data (Figure 22.8A). If both attributes are bipolar, she recommended using two diverging color schemes (Figure 22.8B).

Although choropleth maps have been the most frequently used symbology for combining two attributes on the same map, some interesting bivariate point symbols have been developed, including the rectangular point symbol (in which the width and height of a rectangle are varied) and the bivariate ray-glyph (in which straight-line segments point to either the right or left of a small central circle). It appears that the rectangular point symbol might be a suitable substitute for coarse cross-hatched symbology.

When more than two maps are displayed simultaneously, the result is termed a small multiple. Although small multiples can be useful for comparing patterns on multiple maps, they are difficult to interpret when comparing subregions within each map. A common alternative to the small multiple is the multivariate point symbol or glyph. Examples of glyphs include the star (in which multiple rays extend from a central circle), the snowflake (in which rays of the star are connected), 3-D bars (in which bars of varying height are placed alongside one another), data jacks (in which triangular spikes extend from a square central area), and Chernoff faces (in which distinct facial features are used). Although a considerable number of attributes can be represented by such methods, it is questionable whether map readers can fully understand the resulting symbols.

In this chapter, we also examined cluster analysis methods, which are appropriate when we wish to combine multiple numeric attributes on a single map; for example, we might want to know which census tracts are similar in terms of voting behavior, income, and attendance in private schools. In the hierarchical clustering process, we saw that there are numerous issues that must be tackled in cluster analysis, including standardizing the data, selecting a clustering method (we focused on one hierarchical method—UPGMA), selecting an appropriate number of clusters, interpreting those clusters, and symbolizing the clusters. Initially, we illustrated a cluster analysis in which a contiguity constraint was not imposed (i.e., enumeration units did not need to be contiguous (or adjacent) to one another in order to fall in the same cluster). We then introduced the REDCAP method in which enumeration units could only fall in the same cluster if they were contiguous to one another.

22.7 STUDY QUESTIONS

1. Assume that you wish to examine the relationship between the drug arrest rate and several potential attributes that might explain that rate. One of the attributes that you might use to explain the drug arrest rate would be the % of adults with either a two-year or four-year college degree. If the drug arrest rate has a positive skew and the % of adults with a college degree has a normal distribution, which of the following classification methods would you recommend using: optimal, quantiles, or mean-standard deviation? Explain your choice.

2. (A) Assume that you wish to compare abortion rates (the number of abortions per 1,000 women aged 15–44) for U.S. states for the years 1988 and 2017 using unclassed proportional circle maps. Would you utilize one legend or two legends, and how would you go about scaling circles and designing the legend(s)? The following are the minimum and maximum abortion rates: 9.4 for South Dakota and 44.9 for California in 1988; 4.5 for Utah and 28.0 for New Jersey in 2017. (Source: https://tinyurl.com/Guttmacher-Abortion-Data). (B) Make an argument for using a choropleth map rather than a proportional circle map for the abortion data. If you created two classed maps using the optimal method, would you utilize one legend or two legends, and how would you design the legend(s)?

3. Based on Figure 22.5, which countries of Africa seem to suggest a positive association between GNP per capita and Life expectancy in 1989?

4. If you wished to create a bivariate map of U.S. counties showing the % population change and % employment change, what sort of general color schemes would you recommend?

5. Contrast the cross-hatched shading approach for bivariate mapping with the rectangular point symbol approach for bivariate mapping. Which approach would you consider most effective?

6. Use the small multiple in Figure 22.12 to describe the nature of agriculture in Michigan in 1954. (Ignore the Upper Peninsula portion of Michigan.)

7. (A) How can the concept of pointillism be applied to create a multivariate dot map? (B) What is a potential limitation of this approach and how might we handle this limitation?

8. Describe (or use a screen capture program to draw) regions that you see in Figure 22.16.

9. Using Figure 22.18, (A) contrast the notion of specific information for an attribute with the notion of combined specific information for multiple attributes, and B) contrast the notion of general

information for an attribute with the notion of general information for multiple attributes.

10. Using Figure 22.19B, evaluate the relationship between the asthma prevalence rate for African American Medicaid recipients in South Carolina counties and the three attributes portrayed in the rings surrounding the state.

11. Would the following attributes be appropriate for a cluster analysis of African countries, or would you need to standardize one or more of the attributes? % urban, % of population with access to safe drinking water, GDP per capita.

12. Discuss, in general, how a cluster analysis with a contiguity constraint differs from a cluster analysis in which there is no contiguity constraint.

NOTES

1 For a discussion of the Longitudinal Neighborhood Explorer, see Han et al. (2019).

2 These visual assessments were confirmed by a Shapiro–Wilk test, an inferential test for normality (Stevens 1996, 245).

3 Kendall's rank correlations (Burt and Barber 1996, 396–398), which are arguably more appropriate for skewed data, were 0.98 between attributes 1 and 2 and 1 and 3 and 1.00 between attributes 2 and 3.

4 Brewer and Pickle found that a minimum boundary error method developed by Cromley (1996) that considered the spatial distribution of the data (see Section 5.11) also was quite effective.

5 From the work of Cromley (1995).

6 Data were acquired from https://ucr.fbi.gov/crime-in-the-u.s/1995 and https://ucr.fbi.gov/crime-in-the-u.s/2019/. Data for some cities is not included because the FBI indicated that in some cases, data for the different years was not comparable. We also did not include Oklahoma City as the large number of murders for 1995 was largely a result of the bombing of the Alfred P. Murrah Federal Building in that city.

7 See Meyer et al. (1975) for a summary of the technical methods used at that time.

8 Specifications for the colors used in Figure 22.4 were taken from Olson (1981, 269). For a discussion of the logic behind the selection of these colors, see Olson (1975).

9 Olson argued that the univariate maps should be shown in black and white so that the two distributions can be readily compared. In Figure 24.4, they are shown in color so that the colors composing the bivariate map are clear.

10 Complementary colors are opposite one another on the color wheel (see Section 10.5.1).

11 Eyton used cross-hatching (narrowly spaced horizontal and vertical lines) for intermediate classes, whereas we have chosen to use fine-textured (smooth) colors.

12 Given the requirement for normality, we chose to map two normal distributions: percent urban and life expectancy for females.

13 Ellipses would be an alternative to rectangles (MacEachren 1995, 95).

14 For a discussion of the use of shape on multivariate dot maps, see Turner (1977).

15 A scanned version of the map can be found at Jenks Dot Density.tif, which can be downloaded from Chapter 22 at www.routledge.com/9780367712709.

16 Several terms have been used to describe this symbol. The term *star* comes from Borg and Staufenbiel (1992). We use the term *multivariate ray-glyph* because of their similarity to the bivariate ray-glyph introduced earlier.

17 For an overview of the cartographic potential of Chernoff faces, see Nelson (2007).

18 Pexels can also vary in orientation, and the density can include a regularity parameter, but we do not illustrate these here.

19 Examples of other research that has utilized multivariate point symbols for continuous geographic data include Miller's (2007) work with *attribute blocks*, and Zhang and Pazner's work with *geo-icons* (Zhang and Pazner 2004; Zhang et al. 2012).

20 For research that examines various parameters associated with creating a combined star and snowflake symbol, see Klippel et al. (2009) and Fuchs et al. (2014).

21 Note that the approach used here is similar to the necklace maps described in Section 18.8.

22 The full data are available at NY_Cluster_Data.xls, which can be downloaded from Chapter 22 at www.routledge.com/9780367712709.

23 For an application of nonhierarchical clustering, see Polczynski and Polczynski (2014).

24 See Chapter 15 for a detailed discussion of ColorBrewer.

25 For more sophisticated approaches for evaluating cluster solutions, see Aldenderfer and Blashfield (1984, 62–74).

26 A version of REDCAP is also available through ZillionInfo (http://zillioninfo.com/).

27 Although Nassau County is technically not immediately adjacent to Westchester County, we modified the weights file in GeoDa so that these counties were considered contiguous; otherwise, GeoDa would not cluster the data (presumably because the counties of Nassau and Suffolk on Long Island could not be clustered with the rest of the state).

REFERENCES

Aldenderfer, M. S., and Blashfield, R. K. (1984) *Cluster Analysis*. Beverly Hills, CA: Sage.

Battersby, S. E., Stewart, J. E., Lopez-de Fede, A., et al. (2011) "Ring maps for spatial visualization of multivariate epidemiological data." *Journal of Maps 7*, no. 1:564–572.

Borg, I., and Staufenbiel, T. (1992) "Performance of snowflakes, suns, and factorial suns in the graphical representation of multivariate data." *Multivariate Behavioral Research 27*, no. 1:43–55.

Brewer, C. A. (1994) "Color use guidelines for mapping and visualization." In *Visualization in Modern Cartography*, ed. by A. M. MacEachren and D. R. F. Taylor, pp. 123–147. Oxford, England: Pergamon.

Brewer, C. A., and Pickle, L. (2002) "Evaluation of methods for classifying epidemiological data on choropleth maps in series." *Annals of the Association of American Geographers 92*, no. 4:662–681.

Burt, J. E., and Barber, G. M. (1996) *Elementary Statistics for Geographers* (2nd ed.). New York: Guilford.

Byron, J. R. (1994) "Spectral encoding of soil texture: A new visualization method." *GIS/LIS Proceedings*, Phoenix, AZ, pp. 125–132.

Carr, D. B., Olsen, A. R., and White, D. (1992) "Hexagon mosaic maps for display of univariate and bivariate geographical data." *Cartography and Geographic Information Systems 19*, no. 4:228–236, 271.

Carstensen, L. W. J. (1982) "A continuous shading scheme for two-variable mapping." *Cartographica 19*, no. 3/4:53–70.

Carstensen, L. W. (1986) "Bivariate choropleth mapping: The effects of axis scaling." *The American Cartographer 13*, no. 1:27–42.

Chernoff, H. (1973) "The use of faces to represent points in k-dimensional space graphically." *Journal of the American Statistical Association 68*, no. 342:361–368.

Cox, D. J. (1990) "The art of scientific visualization." *Academic Computing 4*, no. 6:20–22, 32–34, 36–38, 40.

Cromley, R. G. (1995) "Classed versus unclassed choropleth maps: A question of how many classes." *Cartographica 32*, no. 4:15–27.

Cromley, R. G. (1996) "A comparison of optimal classification strategies for choroplethic displays of spatially aggregated data." *International Journal of Geographical Information Systems 10*, no. 4:405–424.

Dorling, D. (1993) "Map design for census mapping." *The Cartographic Journal 30*, no. 2:167–183.

Ellson, R. (1990) "Visualization at work." *Academic Computing 4*, no. 6:26–28, 54–56.

Eyton, J. R. (1984) "Complementary-color two-variable maps." *Annals of the Association of American Geographers 74*, no. 3:477–490.

Fuchs, J., Isenberg, P., Bezerianos, A., et al. (2014) "The influence of contour on similarity perception of star glyphs." *IEEE Transactions on Visualization and Computer Graphics 20*, no. 12:2251–2260.

Groop, R. E. (1992) "Dot-density crop maps of the Midwest." Abstract, *Association of American Geographers Annual Meeting*, San Diego, CA, p. 89.

Guo, D. (2008) "Regionalization with dynamically constrained agglomerative clustering and partitioning (REDCAP)." *International Journal of Geographical Information Science 22*, no. 7:801–823.

Hagh-Shenas, H., Kim, S., Interrante, V., et al. (2007) "Weaving versus blending: A quantitative assessment of the information carrying capacities of two alternative methods for conveying multivariate data with color." *IEEE Transactions on Visualization and Computer Graphics 13*, no. 6:1270–1277.

Han, S. Y., Rey, S., Knaap, E., et al. (2019) "Adaptive choropleth mapper: An open-source web-based tool for synchronous exploration of multiple variables at multiple spatial extents." *International Journal of Geo-Information 8*, no. 11:1–20.

Hartnett, S. (1987) "Employing rectangular point symbols in two-variable maps." Abstract, *Association of American Geographers Annual Meeting*, Portland, OR, p. 39.

Healey, C. G. (2001) "Combining perception and impressionist techniques for nonphotorealistic visualization of multidimensional data." *SIGGRAPH 2001 Course 32: Nonphotorealistic Rendering in Scientific Visualization*, Los Angeles, California, pp. 20–52. https://www.csc2.ncsu.edu/faculty/healey/abstract/pubs.html#avi.02.

Healey, C. G., and Enns, J. T. (1999) "Large datasets at a glance: Combining textures and colors in scientific visualization." *IEEE Transactions on Visualization and Computer Graphics 5*, no. 2:145–167.

Jenks, G. F. (1953) "Pointillism as a cartographic technique." *The Professional Geographer 5*, no. 5:4–6.

Jenks, G. F. (1961) Crop patterns in the United States: 1959. Map, U.S. Bureau of the Census. (Scale 1:5,100,000.)

Jenks, G. F. (1962) Livestock and livestock products sold in the United States: 1959. Map, U.S. Bureau of the Census. (Scale 1:5,100,000.)

Klippel, A., Hardisty, F., Li, R., et al. (2009) "Color-enhanced star plot glyphs: Can salient shape characteristics be overcome?" *Cartographica 44*, no. 3:217–231.

Lavin, S., and Archer, J. C. (1984) "Computer-produced unclassed bivariate choropleth maps." *The American Cartographer 11*, no. 1:49–57.

Lindberg, M. B. (1987) "Dot map similarity: Visual and quantitative." Unpublished Ph.D. dissertation, University of Kansas, Lawrence, KS.

Lloyd, R. E., and Steinke, T. R. (1976) "The decisionmaking process for judging the similarity of choropleth maps." *The American Cartographer 3*, no. 2:177–184.

Lloyd, R., and Steinke, T. (1977) "Visual and statistical comparison of choropleth maps." *Annals of the Association of American Geographers 67*, no. 3:429–436.

MacEachren, A. M. (1995) *How Maps Work: Representation, Visualization, and Design*. New York: Guilford.

Metternicht, G., and Stott, J. (2003) "Trivariate spectral encoding: A prototype system for automated selection of colours for soil maps based on soil textural composition." *Proceedings of the 21st International Cartographic Conference*, Durban, South Africa, pp. 2341–2353.

Meyer, M. A., Broome, F. R., and Schweitzer, R. H. J. (1975) "Color statistical mapping by the U.S. Bureau of the Census." *The American Cartographer 2*, no. 5:100–117.

Miller, J. R. (2007) "Attribute blocks: Visualizing multiple continuously defined attributes." *IEEE Computer Graphics and Applications 27*, no. 3:57–69.

Muller, J.-C. (1980) "Visual comparison of continuously shaded maps." *Cartographica 17*, no. 1:40–51.

Nelson, E. S. (2007) "The face symbol: Research issues and cartographic potential." *Cartographica 42*, no. 1: 53–64.

Nelson, E. S., and Gilmartin, P. (1996) "An evaluation of multivariate quantitative point symbols for maps." In *Cartographic Design: Theoretical and Practical Perspectives*, ed. by C. H. Wood and C. P. Keller, pp. 191–210. Chichester, England: Wiley.

Nusrat, S., Alam, M. J., Scheidegger, C., et al. (2018) "Cartogram visualization for bivariate geo-statistical data." *IEEE Transactions on Visualization and Computer Graphics 24*, no. 10:2675–2688.

Olson, J. M. (1975) "The organization of color on two-variable maps." *AUTO-CARTO II, Proceedings of the International Symposium on Computer-Assisted Cartography*, pp. 289–294, 251, 264–266.

Olson, J. M. (1981) "Spectrally encoded two-variable maps." *Annals of the Association of American Geographers 71*, no. 2:259–276.

Peterson, M. P. (1979) "An evaluation of unclassed crossed-line choropleth mapping." *The American Cartographer 6*, no. 1:21–37.

Polczynski, M. and Polczynski, M. (2014) "Using the k-means clustering algorithm to classify features for choropleth maps." *Cartographica 49*, no. 1:69–75.

Rogers, J. E., and Groop, R. E. (1981) "Regional portrayal with multi-pattern color dot maps." *Cartographica 18*, no. 4:51–64.

Romesburg, H. C. (1984) *Cluster Analysis for Researchers*. Belmont, CA: Lifetime Learning.

Stevens, J. (1996) *Applied Multivariate Statistics for the Social Sciences* (3rd ed.). Mahwah, NJ: Lawrence Erlbaum Associates.

Tufte, E. R. (1990) *Envisioning Information*. Cheshire, CT: Graphics Press.

Turner, E. J. (1977) "The use of shape as a nominal variable on multipattern dot maps." Unpublished Ph.D. dissertation, University of Washington, Seattle, WA.

Wang, P. C. C., and Lake, L. T. G. E. (1978) "Application of graphical multivariate techniques in policy sciences." In *Graphical Representation of Multivariate Data*, ed. by P. C. C. Wang, pp. 13–58. New York: Academic.

Ware, C. (2009) "Quantitative texton sequences for legible bivariate maps." *IEEE Transactions on Visualization and Computer Graphics 15*, no. 6:1523–1529.

Wilhelm, S. D. (1983) "Two symbols for unclassed two-variable mapping: The bivariate box and the ratio bar." Unpublished MA thesis, University of North Carolina, Chapel Hill, NC.

Zhang, X., and Pazner, M. (2004) "The icon imagemap technique for multivariate geospatial data visualization: Approach and software system." *Cartography and Geographic Information Science 31*, no. 1:29–41.

Zhang, X., Wang, X., Li, J., et al. (2012) "IconMap-based visualisation technique and its application in soil fertility analysis." *Cartographic Journal 49*, no. 3:208–217.

Part III

Geovisualization

23 Visualizing Terrain

23.1 INTRODUCTION

Whereas Section 17.7 dealt with approaches for symbolizing a broad range of smooth, continuous phenomena, in this chapter, we consider approaches for symbolizing a particular type of smooth, continuous phenomenon—Earth's terrain. By "terrain," we mean Earth's elevation (both above and below sea level) and associated features found on Earth—its landscape. Specialized techniques have been developed for symbolizing terrain because of its importance in everyday life and in the burgeoning area of GIS.

Section 23.3 considers the kinds of data normally used to represent terrain. Fundamental data consist of elevation values above (or below) a particular vertical datum or starting elevation for individual point locations. These data are most commonly presented in a square gridded network (a so-called digital elevation model, or DEM). Data for depicting Earth's landscape can be acquired from mapping agencies such as the USGS.

Section 23.4 considers approaches for symbolizing terrain that have focused on a vertical (or plan) view, as though the map reader is viewing Earth from an airplane high above it. Methods covered include hachures, contour-based methods, Raisz's physiographic method, shaded relief, and various morphometric techniques (which focus on the structure of the terrain such as aspect and slope). Assuming that the topographic surface is illuminated (typically from the northwest), these methods utilize various approaches for distinguishing illuminated areas from areas in the shadows.

In Section 23.5, we consider approaches for symbolizing terrain that have focused on an oblique view of Earth, including block diagrams, panoramas (or bird's-eye views), and plan oblique relief. Although block diagrams and panoramas do not provide the geometric accuracy of a vertical view, their realism (and associated lack of abstraction) makes them an attractive alternative. Traditionally, only a single oblique view of a landscape was possible, and so some information was of necessity hidden. Now, however, interactive software permits users to create their own oblique views. Another limitation was the time needed to create realistic oblique views, but digital design tools are now making it easier for cartographers to create realistic-looking views.

Although maps based on an oblique view have a distinctive 3-D appearance, they are still created on a 2-D surface (either paper or a graphics display device). It can be argued that a more effective approach is a **physical model** that is truly 3-D (i.e., we can hold and manipulate small models in our hands). Such physical models are the topic of Section 23.6. Physical models have been around for more than 2,000 years, but only recently has technology enabled us to create such models from digital data.

In the last section of the chapter, we return to shaded relief techniques and focus upon them in depth, as they are the most commonly used technique for depicting the terrain. Issues covered include generalizing the terrain, selecting an appropriate orientation of the sun (known as the **azimuth**) and **sun elevation**, lighting model issues, the representation of Swiss-style rock drawing, and the use of color to depict the elevation of the terrain.

23.2 LEARNING OBJECTIVES

- Describe the nature of the data that is used to depict terrain including datums, point data sets, raster data sets, and a common source of elevation data.
- Contrast vertical views of terrain with oblique views of terrain and provide examples of each.
- Contrast Yoeli's and Samsonov's approaches for shadow hachures.
- Describe the difference between illuminated contours and shadowed contours, and indicate which is likely to be more effective for the map reader.
- Contrast the appearance of Raisz's map *Landforms of the United States* with Thelin and Pike's shaded relief map of the conterminous United States, and indicate what each would be useful for.
- Provide examples of visualizing the terrain that emphasize the morphometric structure of the terrain.
- Define the term panorama in the context of visualizing terrain, and describe representative examples.
- Define the term physical model (or physical visualization), and explain why this might be more effective than either a printed map or a map shown on an interactive display.
- Discuss the role that the following play in creating shaded relief: generalizing the terrain, selecting an azimuth for illumination, multiple light sources, and hypsometric tints.

23.3 NATURE OF THE DATA

Fundamental data for depicting terrain consists of elevation values above (or below) a particular vertical datum, or starting elevation, for individual point locations. Mean sea level has acted as a datum historically, but newer geoid-based[1] vertical datums such as the North American Vertical Datum of 1988 (NAVD88) are more commonly used today. Mean sea level is still used for small, local study areas in which mean sea level can be calculated with relative accuracy, but on a global scale, sea levels vary to the degree that calculation of an average is almost meaningless. Geoid-based vertical datums provide a more accurate and more globally consistent starting point for measuring most elevations, and they act as the basis for most digital elevation data sets.

Elevation data are available at irregularly spaced control points (as with other smooth, continuous phenomena) derived from field surveys, data collection stations, and remote sensing technologies such as LIDAR.[2] These point data sets are commonly converted to raster data sets. A raster consists of rows and columns of grid cells, each of which stores a numerical value that represents a real-world measurement (e.g., temperature) or a category (e.g., vegetation type). If the values represent elevations, then the raster is considered to be an elevation raster or, more specifically, a digital elevation model, or DEM. The DEM is analogous to the grid resulting from either the inverse-distance or kriging approaches discussed in Chapter 17. Elevation data are sometimes also provided as a triangulated irregular network (TIN), which is closely associated with the triangulation approach discussed in Chapter 17. The definitive source for DEMs in the United States is the USGS National Map Viewer Downloader (https://apps.nationalmap.gov/downloader/#/). This interactive web map displays seamless elevation data for the entire nation at various spatial resolutions (cell sizes) and several delivery formats such as GeoTIFF.[3] The user is able to download raster data sets for any region in the United States free of charge. The National Map Viewer also provides various types of "source" data that can be used to create DEMs such as LIDAR point clouds, in which each point consists of a longitude, a latitude, and an elevation that can be used to generate a DEM.

23.4 VERTICAL VIEWS

In this section, we consider approaches for symbolizing terrain that emphasize a vertical (or plan) view of the landscape, as opposed to an oblique view. A vertical view is often emphasized for two reasons. First, we may wish to visualize the landscape of a particular region (say, the Rocky Mountains in Colorado) in its entirety. If we choose an oblique view, that will, of necessity, hide some features. Second, we often wish to show features in their correct geographic location. In this respect, cartographers have distinguished between maps and paintings. For instance, Eduard Imhof (1982, 79–80) stated:

> The map is not only a picture; it is primarily, a means of *providing information* [italics in original]. It has other responsibilities than has painting …. Above all, it must include the depth of the valleys and the heights of mountains, not only in the visual sense but in the geometrical sense also.

To create the impression of a 3-D view, the bulk of the methods in this section assume that the surface is lit obliquely from the northwest (the elevation angle is set to 45° and the azimuth is 315° measured clockwise from the north). According to Imhof (1982, 178), a northwest light source is commonly used because of the preponderance of people who write with their right hand. When writing with the right hand, lighting is placed above and to the left to eliminate the possibility that a shadow will be cast over the written material. Our experience in using a light in this situation causes us to expect a light in a similar orientation in other situations such as relief depiction. If an alternate light direction is used, then the relief might appear to be inverted (e.g., valleys might seem to be mountains).

23.4.1 HACHURES

Hachures are constructed by drawing a series of parallel lines perpendicular to the direction of contours (and parallel to the slope). Two forms of hachures were traditionally used in manual cartography. In the first, known as *slope hachures*, the width and spacing of hachures were proportional to the steepness of the slope; thus, steeper slopes had thicker hachures that were closer together and darker than those on gentler slopes. Because slope hachures generally failed to create the impression of a third dimension, *shadow hachures* were developed in which an oblique light source was presumed. In this approach, the width of hachures was a function of whether the hachure was illuminated or in the shadow, with thicker lines used in less illuminated areas. Figure 23.1 illustrates shadow hachures using a computer-based approach developed by Pinhas Yoeli (1985).

Although hachures normally are associated with the bygone era of manual cartography, Imhof (1982, 229) argued that they should not be forgotten:

> Hachures alone, without contours, are more capable of depicting the terrain than is [relief] shading alone …. Hachures also possess their own special, attractive graphic style. They have a more abstract effect than [relief] shading, and perhaps for this reason are more expressive.

The map shown in Figure 23.1 was developed by Yoeli using the following rules recommended by Imhof for shadowed hachures:

1. Each hachure follows the direction of the steepest gradient.
2. Hachures are arranged in horizontal rows.
3. The density of hachures is constant except for the horizontal plane, which is blank.
4. The length of hachures corresponds to the local horizontal distance between assumed contours.
5. Wider hachures are used on steeper slopes.

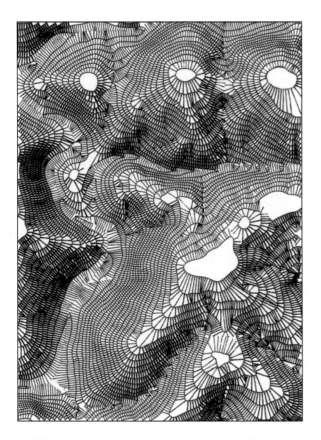

Recently, Timofey Samsonov (2014) noted several limitations of Yoeli's approach. One limitation is that Yoeli's hachures were straight lines, but ideally, they should be curved in order to conform to Imhof's rule that each hachure follows the direction of the steepest descent. Another limitation is that hilltops are not filled with hachures. Responding to these and other limitations, Samsonov developed a new algorithm based on the following requirements:

1. Contours are used as guides for placing hachures.
2. Hachures are derived as flowlines between contours, not straight segments.
3. Flowline divergence is treated by insertion of additional hachures.
4. Upslope hachures are derived along with downslope ones (65).

Figure 23.2 is an image of a volcanic area depicted using Samsonov's algorithm. Although Figures 23.1 and 23.2 do not permit a completely fair comparison of Yoeli's and Samsonov's approaches (as they are of different areas), we feel that Samsonov's approach does appear more representative of a real-world terrain because of the curved flowlines and the finer linework.

FIGURE 23.1 Shadow hachures created using Yoeli's computer-based procedure. (From "Topographic relief depiction by hachures with computer and plotter," by P. Yoeli (1985), *The Cartographic Journal*, copyright © The British Cartographic Society, reprinted by permission of Taylor & Francis Ltd., http://www.tandfonline.com on behalf of The British Cartographic Society.)

23.4.2 CONTOUR-BASED METHODS

Two basic methods for portraying terrain have been developed that involve manipulating contour lines: illuminated contours and shadowed contours. With **illuminated contours**,

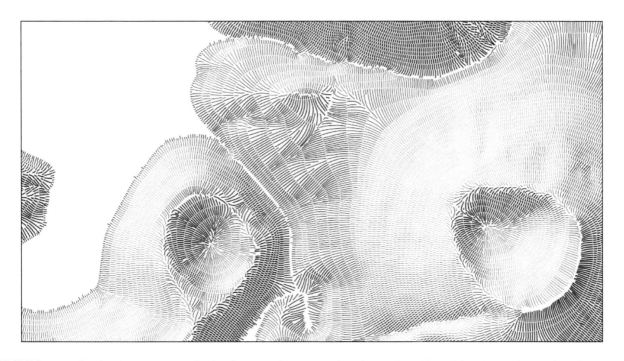

FIGURE 23.2 Shadow hachures created using Samsonov's computer-based procedure. (From "Morphometric mapping of topography by flowline hachures," by T. Samsonov (2014), *The Cartographic Journal*, copyright © The British Cartographic Society, reprinted by permission of Taylor & Francis Ltd., http://www.tandfonline.com on behalf of The British Cartographic Society.)

both the color and width of contour lines are varied as a function of their angular relationship to a presumed light source. Contour lines facing the light source normally are drawn in white (suggesting the illumination), whereas those in shadow are drawn in black, as shown in Figure 23.3. Lines perpendicular to the light source have the greatest width, whereas those parallel to the light source are the narrowest (again, examine Figure 23.3). Although illuminated contours were used as early as 1858 (Eynard and Jenny 2016, 1926), Kitiro Tanaka is often credited with originating the idea because of the attractive maps he produced (Figure 23.4) and the mathematical basis he provided.[4] In **Tanaka's method** (Tanaka 1950), the width of a contour line at a point is defined as the cosine of the angle formed between the aspect of that point and the azimuthal angle of illumination.

Like hachures, Tanaka's method originally was produced manually. Berthold Horn (1982) was one of the first to create a computer-based version of Tanaka's method. A visual analysis of Horn's maps suggests that Tanaka's method compares favorably with other computer-assisted methods for representing relief (see Horn 1982, 113–123). More recently, Patrick Kennelly and Jon Kimerling (2001) suggested improvements to Tanaka's method. One of the attractive maps that resulted from their work appears in Figure 23.5. Here we see that they drew contours in a fashion similar to Tanaka but added colors between the

FIGURE 23.3 A simplified representation of Tanaka's method for depicting terrain. The width of contour lines is a function of their angular relationship with a light source; contour lines facing the light source are drawn in white, whereas those in the shadow are drawn in black. (From "The relief contour method of representing topography on maps" by K. Tanaka (1950), *The Geographical Review*, copyright © the American Geographical Society of New York, reprinted by permission of Taylor & Francis Ltd., http://www.tandfonline.com on behalf of the American Geographical Society of New York.)

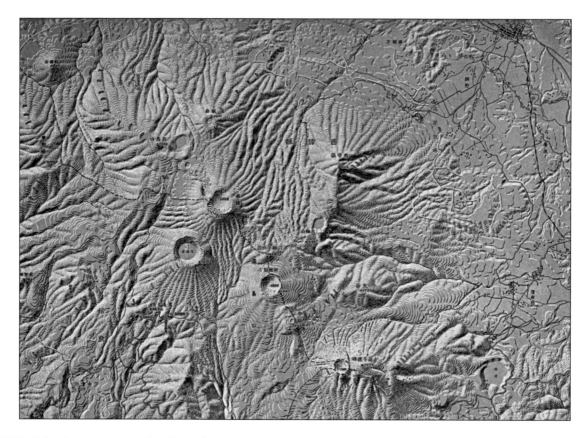

FIGURE 23.4 A map created using Tanaka's method for depicting terrain. (From "The relief contour method of representing topography on maps" by K. Tanaka (1950), *The Geographical Review*, copyright © the American Geographical Society of New York, reprinted by permission of Taylor & Francis Ltd., http://www.tandfonline.com on behalf of the American Geographical Society of New York.)

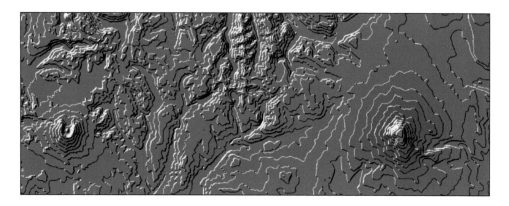

FIGURE 23.5 Tanaka's approach is modified by adding colors between the contours to give the map a more realistic appearance. (From Kennelly and Kimerling 2001. First published in *Cartography and Geographic Information Science* 28(2), p. 119. Reprinted with permission from the Cartography and Geographic Information Society.)

contours "to give the topography a somewhat realistic appearance" (117).

Shadowed contours are another approach for representing terrain by modifying contour lines. In this approach, only the width of the contour lines varies, with the thinnest lines facing the source of illumination and the thickest lines on the side opposite the source of illumination (where shadows are more likely). Figure 23.6 depicts a computer-based implementation of shadowed contours developed by Yoeli (1983).

23.4.2.1   Eynard and Jenny's Work

James Eynard and Bernhard Jenny (2016) evaluated the effectiveness of illuminated and shadowed contours, comparing these with both conventional contour lines and shaded relief (which we introduce in Section 23.4.4). They asked a total of 455 participants to perform two tasks on these four types of maps: (1) select the higher of two markers placed on the map, and (2) click on the location perceived as being the highest on the map. In addition, participants were asked to rate each map type on a Likert scale based on the following statement: "This map shows variations in elevation well and produces an appearance of the third dimension."

Overall, the illuminated contours method scored very well compared to the other techniques. For the relative height question, it received the highest percentage of correct responses (90.1 percent), and this result was significantly higher than for the other methods. In terms of speed of response, it was second on the relative height question (behind shaded relief, which was the least accurate on this question). For the maximum height question, it scored the highest (a mean of 79.5 percent correct), which was very similar to conventional contour lines (79.4 percent correct); shadowed contours scored somewhat lower (75.3 percent), and shaded relief was the poorest at 69.6 percent. The high percentage correct for illuminated contours is also noteworthy, as the conventional contour lines included labels, but the illuminated contours and shadowed contours did not. The speed of response for the maximum

FIGURE 23.6 A map created using shadowed contours. (After Yoeli 1983. First published in *The American Cartographer* 10(2), p. 109. Reprinted with permission from the Cartography and Geographic Information Society.)

height question was about the same for the three contour line approaches.

For the Likert scale question, the illuminated contours method received high scores similar to those of the highest-scoring shaded relief map, with 83.8 percent of those viewing the illuminated contour method indicating "agree" or "strongly agree" with the statement. One potential limitation of the study is that all contour-based methods used a gray background rather than the white background shown in Figure 23.6. As a result, the shadowed contours did not seem to pop out as readily as those shown in Figure 23.6. It would be interesting to repeat the experiment with various backgrounds for the three contour-based methods.[5]

23.4.3   RAISZ'S PHYSIOGRAPHIC METHOD

Based on the work of A. K. Lobeck (1924), which we consider in Section 23.5.1, Erwin Raisz (1931) developed the **physiographic method**, in which Earth's geomorphologic features are represented by a set of standard, easily recognized symbols. A sampling of Raisz's standard symbols is shown in Figure 23.7, and an interactive digital scan of Raisz's (1967) classic printed map *Landforms of the United States* is available through the David Rumsey Map Collection at https://tinyurl.com/David-Rumsey-Raisz-Map).[6] Raisz argued that his physiographic map "appeals immediately to the average [person]. It suggests actual country and enables [the individual] to see the land instead of reading an abstract location diagram. It works on the imagination" (Raisz 1931, 303). In the next section, we will contrast Raisz's approach with shaded relief.

23.4.4   SHADED RELIEF

Shaded relief (also known as **hill shading** or **chiaroscuro**) has long been considered one of the most effective methods for representing terrain. Swiss cartographers (Imhof in particular) have been widely recognized for their superb work. Like the preceding methods, hill shading presumes an oblique light source: Areas facing away from the light source are shaded, and areas directly illuminated

FIGURE 23.8   An example of a shaded relief map that Imhof created by hand. (Republished with permission of Walter de Gruyter and Company, from *Cartographic Relief Presentation* by E. Imhof, copyright 1981; permission conveyed through Copyright Clearance Center, Inc.)

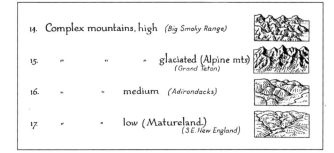

FIGURE 23.7   Examples of the standard symbols that Raisz used to create his physiographic diagrams. (From "The physiographic method of representing scenery on maps" by E. Raisz (1931), *The Geographical Review*, copyright © the American Geographical Society of New York, reprinted by permission of Taylor & Francis Ltd, http://www.tandfonline.com on behalf of the American Geographical Society of New York.)

are not shaded. Slope can also be a factor, with steeper slopes shaded darker. When produced manually, shading is accomplished using pencil and paper or an airbrush. Figure 23.8 is an example of one of the maps that Imhof created using manual methods.

Numerous people have contributed to automating shaded relief maps, with pioneering work done by Yoeli (1971), Brassel (1974), Hügli (1979), and Horn (1982). Today, shaded relief maps are a standard element of GIS software (e.g., ArcGIS) and software explicitly intended for mapping smooth, continuous phenomena (e.g., Surfer). In Section 23.7, where we discuss shaded relief in more depth, we introduce additional software.

The effectiveness of automated shaded relief maps is superbly illustrated by Gail Thelin and Richard Pike's (1991) map of the conterminous United States (a small-scale version of the complete map is shown in Figure 23.9, and an interactive digital version is available at https://pubs.usgs.gov/imap/i2206/usa_shade.pdf). Thelin

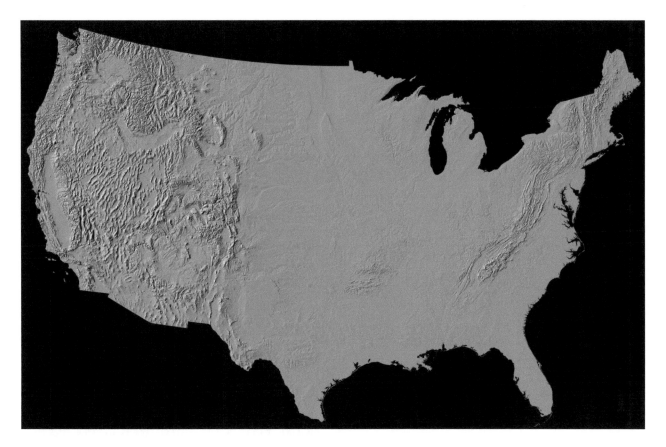

FIGURE 23.9 A small-scale version of the USGS map *Landforms of the Conterminous United States: A Digital Shaded-Relief Portrayal*, developed by Thelin and Pike (1991); for a PDF of this map, see https://pubs.usgs.gov/imap/i2206/.

and Pike (3) note that their map not only portrays well-known features such as the "structural control of topography in the folded Appalachian Mountains," but also less familiar features such as the Coteau des Prairies, a flat-iron shaped plateau in South Dakota associated with glaciation. Pierce Lewis (1992) described Thelin and Pike's map as follows:

> Let us not mince words. The United States Geological Survey has produced a cartographic masterpiece: a relief map of the coterminous United States which, in accuracy, elegance, and drama, is the most stunning thing since Erwin Raisz published his classic "… *Landforms of the United States.*" (289)

Because Raisz's map has been so widely used by geographers, it is natural to ask which is better, the USGS map or the Raisz map? Lewis argued that both maps have merit: The USGS map is "more like an aerial photograph" (298), allowing us to see the structure of the landscape as it actually exists (at least within the constraints of the computer program). In contrast, the Raisz map represents one knowledgeable geographer's view of the landscape, with certain features emphasized, as Raisz deemed appropriate.

23.4.5 MORPHOMETRIC TECHNIQUES

The techniques that we have described thus far are intended to provide an overall view of the terrain. Often, however, it is useful to focus on particular morphometric (or structural) elements of terrain. Aspect and slope are two of the most common structural elements that have been visualized. **Aspect** deals with the direction the land faces (e.g., north or south), whereas **slope** considers the steepness of the land surface. For example, a developer of a ski resort might desire north-facing aspects (to minimize snowmelt) and a diversity of slopes (to accommodate skiers' varied abilities). In this section, we consider a visualization technique that has been developed for handling aspect and slope simultaneously. We then consider a software package, LandSerf, that has been developed to visualize and analyze a broad range of morphometric elements.

23.4.5.1 Symbolizing Aspect and Slope: Brewer and Marlow's Approach

A common approach for symbolizing aspect and slope simultaneously is a technique developed by Cynthia Brewer and Ken Marlow (1993) that utilizes the **HVC** (hue, value, and chroma) **color model**, which was developed by

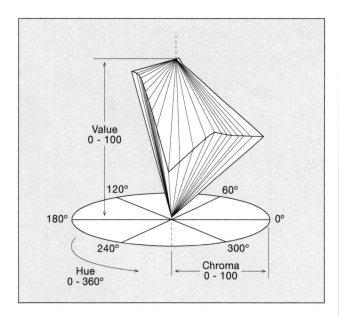

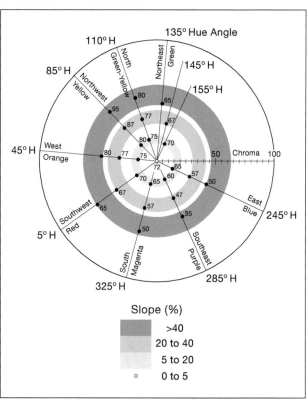

FIGURE 23.10 The HVC color model developed by Tektronix. (Taylor et al., 1991. First published in *Information Display*, 7 (4/5), p. 21. Courtesy of Tektronix, Inc.)

Tektronix to duplicate the Munsell color model (Section 10.4.4). Figure 23.10 provides an illustration of the HVC color model—here, we can see that hue is specified in degrees counterclockwise from 0°, the value varies vertically from 0 to 100, and chroma varies from 0 to 100 from the central axis to the edge of the model. The HVC model appears irregular in shape because, like the Munsell model, it is perceptually uniform, with equal steps in the model representing equal visual steps.

Using an earlier study by Moellering and Kimerling (1990) as a basis, Brewer and Marlow chose eight aspect hues, with yellow for the northwest aspect (where illumination is typically a maximum), as shown in Figure 23.11. For each aspect, they depicted three slope classes (5 to 20 percent, 20 to 40 percent, and >40 percent) with increasing levels of saturation for steeper slopes; they represented the <5 percent slope category as gray for all aspects. In total, they showed 25 classes of information (8 aspects × 3 slope classes, plus the <5 percent slope category). The colors selected by Brewer and Marlow for aspect and slope are shown graphically as HVC coordinates in Figure 23.11. Each black dot in the graph represents one of the 25 colors used. The eight lines extending from the interior to the edge of the circle represent the hues selected for the eight aspects. The hues do not appear to be equally spaced (as one might expect, given the perceptual uniformity of HVC) because a wide range of lightness and saturation values are not possible for all hues; for example, it is not possible to create a variety of lightnesses and saturations between green and blue. The saturation (or chroma) of a slope class within an aspect is represented by the distance of a black dot from the center of the diagram (these dots are equally spaced along the lines, reflecting the perceptual uniformity of

FIGURE 23.11 A plot of HVC colors used by Brewer and Marlow to depict aspect and slope; see the text for an explanation. (After Brewer and Marlow 1993, 331.)

the HVC model). The numerical values of distance can be determined using the scale labeled "Chroma" in the diagram; for example, for the blue hue, the third dot from the center of the diagram appears to have a saturation of approximately 45. The numbers printed alongside each dot represent the value (lightness) of each color. Note that higher lightness values tend to be found facing the direction of the light source (the northwest) and that lightness also varies with slope, with lightness increasing for steeper slopes facing the northwest and lightness decreasing for steeper slopes facing away from the northwest. HVC, *Yxy*, and RGB specifications for the 25 slope–aspect classes developed by Brewer and Marlow are shown in Table 23.1, and a map created using their color scheme is shown in Figure 23.12.[7]

23.4.5.2 Symbolizing Other Morphometric Parameters

Although aspect and slope are the most common parameters utilized in examining the morphometry of terrain, those interested in a detailed analysis (e.g., geomorphologists) utilize a richer set of parameters. For instance, two additional parameters are the *profile curvature*, which describes the rate of change of slope and *plan curvature*, which describes the rate of change of aspect. These four parameters can be used to define the following features: *pits*, *channels*, *passes*, *ridges*, *peaks*, and *planar regions*.

TABLE 23.1

HVC, *Yxy*, and RGB Specifications for Brewer and Marlow's Method for Representing Aspect and Slope

Aspect	H	V	C	Y	x	y	R	G	B
Maximum Slopes (>40 Percent Slope)									
0	110	80	56	55.3	0.339	0.546	132	214	0
45	135	65	48	35.2	0.252	0.486	0	171	68
90	245	50	46	19.8	0.180	0.191	0	104	192
135	285	35	51	10.3	0.240	0.135	108	0	163
180	325	50	60	19.6	0.350	0.198	202	0	156
225	5	65	63	33.7	0.442	0.319	255	85	104
270	45	80	48	53.8	0.421	0.421	255	171	71
315	85	95	58	81.1	0.395	0.512	244	250	0
Moderate Slopes (20 to 40 Percent Slope)									
0	110	77	37	50.6	0.319	0.451	141	196	88
45	145	67	33	37.0	0.253	0.400	61	171	113
90	245	57	31	25.7	0.221	0.238	80	120	182
135	285	47	34	17.4	0.262	0.209	119	71	157
180	325	57	40	25.1	0.332	0.240	192	77	156
225	5	67	42	34.3	0.390	0.316	231	111	122
270	45	77	32	47.3	0.377	0.384	226	166	108
315	85	87	39	63.7	0.359	0.445	214	219	94
Low Slopes (5 to 20 Percent Slope)									
0	110	75	19	45.4	0.305	0.381	152	181	129
45	155	70	16	39.4	0.270	0.348	114	168	144
90	245	65	15	32.7	0.260	0.284	124	142	173
135	285	60	17	27.5	0.280	0.272	140	117	160
180	325	65	20	32.6	0.311	0.282	180	123	161
225	5	70	21	37.2	0.339	0.317	203	139	143
270	45	75	16	41.8	0.328	0.348	197	165	138
315	85	80	19	49.7	0.321	0.375	189	191	137
Near-Flat Slopes (0 to 5 Percent Slope)									
None	0	72	0	39.0	0.290	0.317	161	161	161

Source: After Brewer and Marlow 1993. First Published in *Auto-Carto 11 Proceedings*. Reprinted with Permission from the Cartography and Geographic Information Society.

Note: Aspect is measured in degrees clockwise from north, which is 0°.

Jo Wood has developed an intriguing package, LandSerf (http://www.soi.city.ac.uk/~jwo/landserf/), that allows mapmakers to visualize and analyze these parameters and associated features. Figure 23.13 illustrates some of the capabilities of LandSerf. Figure 23.13A is a shaded relief map of the Columbia River area in Washington State in which the colors depict elevation using a green–brown–purple–white scheme. Figure 23.13B is a slope map (with the steepest slopes shown in yellow and red) overlaid on the shaded relief map. The slope map obviously shows quite different information than the shaded relief map shows, as the former clearly identifies the steepest slopes as being near major drainage basins. In Figure 23.13C, we see some of the key features found in this region: Channels are shown in blue, ridges in yellow, and planar regions in gray. At this scale, we cannot see them, but a blowup of the region would also show peaks (in red) and passes (in green). Finally, in Figure 23.13D, we see the effect of broadening the scale of analysis. In contrast to the images shown in Figure 23.13A–C, which were created by processing 3 × 3 windows of 30-meter DEM cells, Figure 23.13D was created using 65 × 65 windows of 30-meter cells. Obviously, the larger window size has generalized the landscape. (The border around the map appears because of the large window size and the lack of information surrounding the map.) We could continue the analysis by examining the characteristics of the features (e.g., channels and ridges) associated with this larger scale landscape. Because space does not permit us to illustrate the full potential of LandSerf here, we encourage you to explore the capability of this software on your own.

FIGURE 23.12   Brewer and Marlow's method for portraying aspect and slope. The image is for the San Francisco North 7 and one-half minute USGS quad. Caution should be taken in interpreting these colors because the RGB values recommended by Brewer and Marlow have been converted to CMYK values for printing.

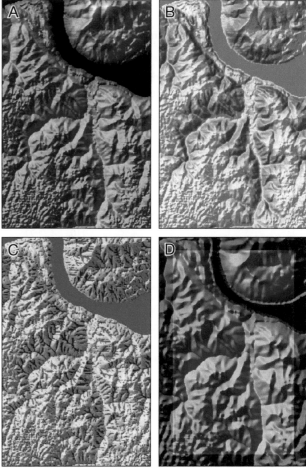

FIGURE 23.13   Some of the visualizations available in LandSerf. Shaded relief maps are overlaid with: (A) a green–brown–purple–white color scheme depicting elevation ranges; (B) a white–yellow–red color scheme depicting slope; (C) features found in the region, with channels shown in blue, ridges in yellow, and planar regions in gray; and (D) a generalization of (A) using a 65 × 65 cell window. (Courtesy of Jo Wood.)

23.5   OBLIQUE VIEWS

In Section 23.4, we introduced approaches for depicting terrain that utilize a vertical (top-down) view, which is arguably useful when we wish to show a large geographic area (a small-scale map) with a geographically accurate location. In contrast, in the present section, we introduce *oblique views*, in which the terrain is viewed at an angle (obliquely)—we will see that these approaches are generally useful for relatively large-scale maps and when geographic accuracy is not as critical. We consider three approaches: block diagrams, panoramas, and plan oblique relief.

23.5.1   BLOCK DIAGRAMS

In a fashion similar to the physiographic method described in the preceding section, **block diagrams** are useful for examining Earth's geomorphologic and geologic features, but with an oblique rather than a vertical view. Although block diagrams have their origins in the eighteenth century (Robinson et al. 1995), A. K. Lobeck (1958) is commonly considered to have originated them because of the engaging images that he created in

association with his book *Block Diagrams*—an example of one of these is shown in Figure 23.14. Lobeck indicated that block diagrams have a twofold function: "First, to present pictorially the surface features of the ground; and second, to represent the underground structure" (1). Obviously, a purely vertical view could not achieve this function. Lobeck recognized that cartographers might be concerned by the lack of accuracy in the block diagram (because certain features are purposely hidden, whereas others are exaggerated), but he felt that any inaccuracies were outweighed by the ease of understanding: "No conventions are needed to represent the topography. No explanation is needed to indicate the position of the geological cross-section. It carries its message directly to the eye and leaves a visual impression unencumbered by lengthy descriptions" (2).

A question that you may be wondering about is whether it is possible to automate the creation of the pen-and-ink approaches of Raisz's physiographic method and Lobeck's

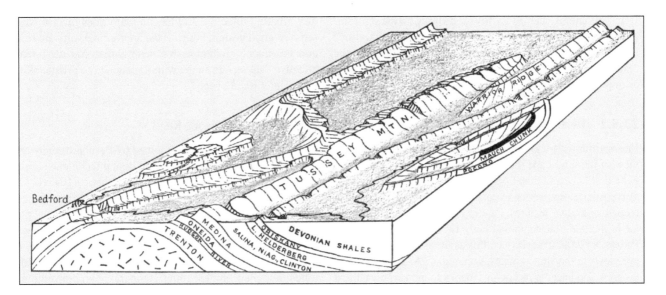

FIGURE 23.14 Example of a traditional block diagram. (From Figure 98, p. 69, of Lobeck, A. K. (1958) *Block Diagrams*. 2nd ed. Amherst, MA: Emerson-Trussel Book Company.)

block diagrams. Such automation falls under the rubric of **non-photorealistic rendering (NPR)**, which rather than attempting to depict realism, has other goals, such "… as imitating an artistic style, mimicking a look comparable to images created with specific reproduction techniques, or adding highlights and details to images" (Kennelly and Kimerling 2006, 35). Using NPR techniques, James Mower (2011) experimented with creating images that have the pen-and-ink look of Raisz and Lobeck's work and even created animations of such images (Mower 2016). For example, Figure 23.15 shows two images that Mower created of the West Temple feature of Zion National Park in Utah. Mower created such images using "a minimum number of strokes" by emphasizing three categories of lines:

- Silhouettes—lines representing borders between visible and invisible regions from the point of view of the observer.

- Creases—lines representing important features that must be rendered without regard to the viewer's position.
- Form lines—lines depicting overall surface curvature, independent of the point of view but unrelated to particular surface features (175).

Rather than try to recreate the pen-and-ink look of these traditional classic cartographic works, numerous software developers also have tried to create accurate representations of the underlying 3-D structure of block diagrams. This is common practice in the field of geology, where the structure of the underlying bedrock is of prime importance (e.g., in the search for oil). This is accomplished by collecting a set of sample points in the 3-D structure (e.g., via boreholes) and then interpolating a 3-D surface from these sample points. Figure 4.1 was created using this approach. A major advantage of automating block diagrams is that users are not restricted to a single

FIGURE 23.15 Images of the West Temple of Zion National Park created by Mower (2011) using NPR techniques: (A) Light is from the upper left corner at 45° above the horizon, with a maximum line width of 3 points for form lines; (B) The same lighting, but with a maximum line width of 2 points for form lines. Note how the relatively flatter regions are highlighted in B. (From Mower 2011. First published in *Cartography and Geographic Information Science* 38(2), p. 180. Reprinted with permission from the Cartography and Geographic Information Society.)

view—programs such as EarthVision (https://www.dgi.com/earthvision-software-for-3d-modeling-and-visualization/) developed by Dynamic Graphics allow users to rotate the diagram to different positions and to explore the data by, for example, slicing the diagram at different locations.

23.5.2 Panoramas and Related Oblique Views

Panoramas (also known as **bird's-eye views**) provide a view of the landscape that we might expect from a painting (which is how traditional panoramas were created), but they also pay careful attention to geography so that we can clearly recognize known features. Panoramas were especially popular in the late nineteenth and early twentieth centuries. Tom Patterson (2005, 61) described this as the golden era of birds-eye views in the United States, noting that "As towns and cities grew and became prosperous thanks to industrialization, it became fashionable and a matter of civic pride to advertise this newfound economic vitality in the form of oblique panoramic maps." Many of these are available from the U.S. Library of Congress (https://www.loc.gov/collections/panoramic-maps/). We have illustrated one of these for Anniston, Alabama, in Figure 23.16; if viewing this figure as an Ebook, you will find that you can zoom in on the image and see considerable detail, reflecting the considerable work that went into creating these panoramas.

One of the best-known panoramists was Heinrich Berann, whose work Patterson (2000) so ably described. Figure 23.17, a classic panorama of Denali National Park in Alaska, is an example of the fine work of Berann. Traditionally, creating panoramas such as Berann's required considerable artistic skill, but at the following website, Patterson provides numerous suggestions for designing static oblique 3-D terrain maps: http://shadedrelief.com/3D_Terrain_Maps/3dterrainmapsint.html.

Yes, creating effective 3-D terrain maps does involve some sense of good design, but with the appropriate software tools and Patterson's guidance, we believe that you can create effective 3-D terrain maps without having the artistic skills of a painter like Berann.[8]

23.5.3 Plan Oblique Relief

Plan oblique relief is a term coined by Bernhard Jenny and Tom Patterson (2007) to describe a digital technique similar to conventional shaded relief maps, but that causes relief to "stand up" and thus provide a more easily seen 3-D view, and yet preserve the geometric accuracy of conventional planimetric shaded relief maps. Jenny and Patterson argue that Raisz's maps were successful, in part because of his use of this technique, albeit in a manual world. Figure 23.18 illustrates how plan oblique relief compares with conventional shaded relief in terms of the projecting rays of light that are used to create a map image. Assuming that virtual rays of light are directed at a DEM of the terrain, Figure 23.18A shows the resulting projected rays of light that reach the image plane (e.g., a computer display) for the *orthographic* approach used in conventional shaded relief. In contrast, Figure 23.18B shows the result for the plan oblique approach, where α is the inclination angle between the image plane and the projected rays of light, with lower values of α producing greater vertical exaggeration.

One problem you may have with the plan oblique approach is that there may be a **terrain inversion effect** (or **relief inversion effect**), wherein concave structures (such as valleys) appear as convex structures (such as ridges), and vice versa (Biland and Çöltekin 2017). We found that we could eliminate this effect by utilizing a relatively low inclination angle. We will return to the terrain inversion effect

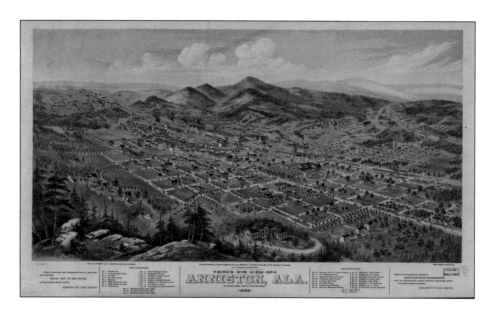

FIGURE 23.16 A classic bird's-eye view of Anniston, AL, in 1888. (From Glover, E. S. *Bird's eye view of Anniston, Ala.* 1888. Map. Shober & Carqueville Litho. Co., Chicago, c1888. Library of Congress, Geography and Map Division. *Map Collections: Cities and Towns, Panoramic Map Collection 1847–1929.* https://www.loc.gov/resource/g3974a.pm000020/ (accessed March 29, 2021).)

FIGURE 23.17   Berann's classic hand-drawn panorama of Denali National Park, Alaska. (Courtesy of National Park Service; artist Heinrich Berann.)

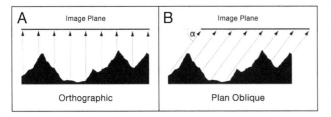

FIGURE 23.18   Projected light rays for (A) the orthographic approach (conventional shaded relief) and (B) the plan oblique approach. Smaller angles of α lead to greater vertical exaggeration. (From Jenny and Patterson 2007, 27; courtesy of *Cartographic Perspectives* (North American Cartographic Information Society, publisher), Bernhard Jenny, and Tom Patterson.)

in Section 23.7.2. Rather than use a conventional northwest illumination, Jenny and Patterson recommend an azimuth angle of 225° (a southwest illumination) for the plan oblique approach, arguing that a conventional northwest illumination "… places shadows on slopes facing the reader, which darken the map and decrease legibility" (33).[9]

23.6   PHYSICAL MODELS

In the preceding two sections, we covered approaches for depicting terrain that provide the illusion of a third dimension even though one does not actually exist. The associated maps were created either on a piece of paper or on a graphics display screen, both of which are 2-D. An alternative is to create a truly 3-D **physical model** (or **physical visualization**). Although physical models have been around since the time of Alexander the Great, cartographers traditionally used them infrequently because of the difficulty of obtaining terrain

information, the cost and time of creating them manually, and the difficulty of disposing of them. Today, such limitations no longer exist, as digital data are readily available, several technologies have been developed for converting the digital data into a solid model, and disposal is less of an issue, given the range of available technologies (Caldwell 2001).

Many of the early physical models that were created using digital technology were relatively large in scope and intended for viewing by several people at once. For instance, Solid Terrain Modeling (STM; http://www.stm-usa.com/) created 17 models for the National Geographic Explorer's Hall in Washington, DC, with the largest (the Grand Canyon) measuring 15.9 × 1.8 meters. Figure 23.19

FIGURE 23.19   Physical model for Hurricane Floyd created using technology developed by Solid Terrain Modeling. (Courtesy of Solid Terrain Modeling (https://www.solidterrainmodeling.com/).)

shows a model of Hurricane Floyd that was created for the Explorer's Hall. STM has created an even larger model (22.6 × 12.2 meters) of the province of British Columbia for the historic Crystal Garden in Victoria, British Columbia (http://www.solidterrainmodeling.com/british-columbia-experience/).

Although such large models are eye-catching and informative, it is also important to consider the potential of smaller physical models that we can hold and manipulate with our hands, especially given that the cost of **3-D printers** is diminishing. It is natural to ask how the effectiveness of such physical models would compare with maps viewed on a graphic display. For instance, we might ask whether a contour map depicted as a physical model (i.e., in true 3-D) is more easily understood and interpreted than a contour map displayed on a flat graphics display (essentially in 2-D). In this respect, Carlos Carbonell-Carrera and colleagues (2017) found that 3-D did enhance understanding of topographic concepts, but their study used both physical models and augmented reality (see Section 28.7.2)—it would be interesting to see what advantage there is to using physical models alone.

Although the focus of this chapter is on visualizing terrain and associated landscape features, it is important to realize that physical models also can be used to create a wide variety of maps, including thematic ones. In this respect, Figure 23.20 is a physical model that

FIGURE 23.20 Physical model depicting the average prices for building lots based on county-level data in Germany. (Reprinted by permission from Springer Nature Customer Service Centre GmbH: Springer Nature *True-3D in Cartography: Autostereoscopic and Solid Visualization of Geodata* by M. Buchroithner (ed.) Copyright 2011.)

Wolf-Dieter Rase (2012) generated depicting the average prices for building lots based on county-level data in Germany. Pycnophylactic interpolation (Section 17.6) was used to interpolate the point data for counties. The resulting isarithmic map could be depicted with just the yellow-to-red color scheme shown in Figure 23.20, but taking advantage of a 3-D printer, Rase chose to raise each point on the isarithmic map to a height proportional to its value. Rase argues that such physical models are advantageous in that "slight movements of the head or body suffice to compare heights, to solve viewing ambiguities or to reveal parts of the model that might be obscured in a fixed view" (120).

The physical models that we have discussed thus far are all static in the sense that their appearance does not change over time. It is possible, however, to project images onto a physical model and thus change the theme(s) that are displayed for a particular model of relief and to show changes over time. Gary Priestnall and colleagues (2012) have been leaders in this approach, which they call the **Projected Augmented Relief Model (PARM)**. The considerable potential of this approach is illustrated in a paper by Priestnall and Cheverst (2019), where they describe the use of a PARM to introduce visitors to a valley in the English Lake District, an area that is famous for its *fells* and its association with William Wordsworth and other poets (https://en.wikipedia.org/wiki/Lake_District). Priestnall and Cheverst utilized six projection effects for their application:

1. Animated illumination effects mimicking changing sun angle.
2. Static illumination effects mimicking a light source on the relief model surface…
3. Spotlight effects illuminating an area of land but with no hard boundary.
4. Highlighting map features where points, lines, or areas are emphasized through color, style, or brightness.
5. Animated map features to highlight movements including progress along a route.
6. Animated atmospheric or environmental conditions including flooding, snowfall, and cloud shadow effects. (p. 8, Creative Commons Attribution 4.0 International License.)

Some sense of the capability of these effects is provided in Figure 23.21, where we see examples of a lighting effect, highlighted map features, and a series of spotlights. A full appreciation of animated effects would, of course, require viewing the physical model live. We encourage you to examine Priestnall and Cheverst's paper and Priestnall's website (https://garypriestnall.wixsite.com/grandestviews/projection-models) for additional information on the PARM approach.

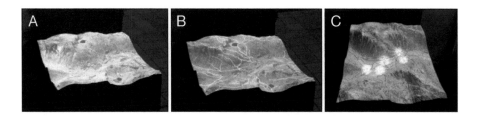

FIGURE 23.21 Images from a PARM used to introduce visitors to a valley in the English Lake District: (A) a lighting effect; (B) highlighted map features; (C) spotlights. (From Priestnall, G. and Cheverst, K. (2019), *Personal and Ubiquitous Computing*, https://link.springer.com/article/10.1007/s00779-019-01320-2, bottom half of Figure 8, Creative Commons Attribution 4.0 International License.)

23.7 ISSUES IN CREATING SHADED RELIEF

23.7.1 GENERALIZING THE TERRAIN

Terrain analysis typically requires high resolution, highly detailed elevation rasters and rasters derived from elevation, such as shaded relief, slope, and aspect. For cartographic representation, however, too much detail can detract from other features that appear on a map. In thematic mapping, in particular, shaded relief is often included as base information that is used in support of thematic symbols. In this case, generalized shaded relief images with less detail can draw attention to the thematic information, which is more important. Figure 23.22 shows two different shaded relief images for a portion of the Panamint Mountains in

FIGURE 23.22 (A) A detailed shaded relief image derived from a DEM with a sun azimuth of 315° (NW). (B) A generalized shaded relief image that was derived by applying a moving window filter that calculated the mean elevation values of surrounding cells.

southeastern California. The range separates Death Valley (to the east) from Panamint Valley (to the west). We chose this geographic region because of its dramatic topographic relief. Telescope peak, with an elevation of over 11,000 feet, stands in stark contrast with the floor of Death Valley, which in this image reaches a low elevation of 88 feet. Figure 23.22A was derived using the traditional sun azimuth of 315° (NW). The hillshade displays a high level of detail and reveals much about the complex topographic relief. The shaded relief image in Figure 23.22B used the same sun azimuth and was derived from the same DEM that was used to create Figure 23.22A, but the DEM was generalized using a moving window[10] that calculates the mean of all elevation values that fall within a circle with a radius of ten grid cells. This is probably the simplest way to produce a generalized shaded relief. The resulting image is noticeably blurred with much of the topographic detail removed. While providing less specific information about the terrain than Figure 23.22A, Figure 23.22B will compete less with any overlying thematic symbols and type, making it more suitable for use as base information on a thematic map.

An application named Terrain Sculptor (http://terraincartography.com/terrainsculptor/) has been developed by Anna Leonowicz and colleagues (2010) that provides a more sophisticated approach to generalizing terrain by producing shaded relief images that are not simply blurred but are altered in very specific ways. Terrain Sculptor uses a DEM as input and can be used to generalize shaded relief images by removing certain topographic details. It can also be used to enhance certain aspects of the terrain, but we focus here on its ability to generalize. Figure 23.23A shows a shaded relief image from Terrain Sculptor using the same DEM as in Figure 23.22A, using the default settings. It is slightly generalized and lighter in color than Figure 23.22A. Figure 23.23B is a shaded relief image based on the same DEM but with a reduced level of detail, achieved by increasing the "smoothing" factor. Terrain Sculptor allows the user to manipulate several other

FIGURE 23.23 (A) A generalized shaded relief image produced by the Terrain Sculptor application with the default settings. (B) An alternative shaded relief image produced by Terrain Sculptor with an increased "smoothing" factor.

variables that generalize the shaded relief in different ways, such as "ridges removal" and "valleys removal."

23.7.2 Selecting an Azimuth and Sun Elevation for Illumination

As we have noted, traditionally, it has been recommended that the azimuth be set to 315° (illumination from the northwest). Julien Biland and Arzu Çöltekin (2017) have examined this issue in depth, and noted that studies in psychology have shown that in the absence of cues to indicate the illumination position, people assume that light shines from *slightly above left*. Note that this concurs with Imhof's earlier suggestion for northwest lighting that we introduced in Section 23.4. As we also noted previously, Biland and Çöltekin indicated that if the azimuth differs considerably from 315°, the result can be a terrain reversal effect. Biland and Çöltekin examined this issue in a formal experiment by having 27 participants evaluate the relief inversion effect for 16 possible light directions

spanning the 360° of possible azimuths for eight different terrains. In general, the results revealed that illumination from the north led to the proper identification of landforms, and illumination from the south led to the strongest terrain inversion effect. The most accurate results were obtained with an angle of 337.5° (NNW) rather than the conventional 315° (NW), and results from 337.5° to 0° were more accurate than those from 315° to 337.5°.

Figure 23.24A was derived using Biland and Çöltekin's optimal azimuth of 337.5° (NNW). It is quite similar to Figure 23.22A, which was derived using the traditional azimuth of 315°, but it represents the landscape somewhat differently; we feel that it gives more emphasis to drainages that trend NNW to SSE, whereas the 315° azimuth in Figure 23.22A emphasizes drainages that run N to S. Figure 23.24B was derived with the opposite of Biland and Çöltekin's optimal azimuth: 157.5° (SSE). The inversion of the terrain is readily apparent—Telescope Peak appears to be the low point in a gigantic trough.

FIGURE 23.24 (A) A shaded relief image using Biland and Çöltekin's optimal sun azimuth of 337.5°. (B) An "inverted" shaded relief image using the opposite of Biland and Çöltekin's optimal azimuth: 157.5° (SSE).

In addition to azimuth, sun elevation is another variable that can be altered in the creation of shaded relief images. **Sun elevation** is the angle between the horizon and the height of the sun in the sky. Lower sun elevations produce darker images with greater contrast between the lightest and darkest grays in a shaded relief image (Figure 23.25A). Lower sun elevations can be used to emphasize the terrain in areas with little topographic relief as if the sun were lower in the sky. Higher sun elevations produce lighter images with less contrast between the lightest and darkest grays (Figure 23.25B) and can be used to reduce the visual impact of a shaded relief image without reducing the detail associated with generalization. Both shaded relief images in Figure 23.25 were created using an azimuth of 315°. (The shaded relief examples in Figures 23.22 and 23.24 were created using a sun elevation of 45°, which is a standard default value in GIS applications such as ArcGIS.)

23.7.3 OTHER LIGHTING MODEL ISSUES

The selection of the azimuth just described assumes that we wish to utilize a single light source and apply it throughout the entire shaded relief map. It is possible, however, to vary the nature of the light source locally and also to use multiple light sources. An example of multiple light sources is the work of Patrick Kennelly and James Stewart (2006), who developed a uniform sky illumination model in which the light source is presumed to be evenly distributed throughout the sky. A comparison of shaded relief images created using a single point source illumination and the uniform sky illumination model is shown in Figure 23.26. For the single point source (Figure 23.26A), a nontraditional northeast light source is used at a 45° angle above the horizon. In this case, intermediate slopes with aspects facing the northeast are brightest, and areas of steep slope

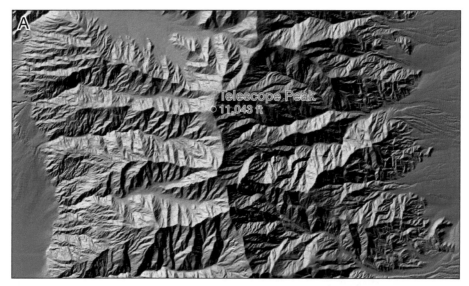

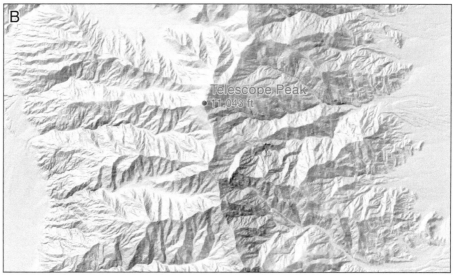

FIGURE 23.25   (A) A Shaded relief image using a sun elevation of 25°. (B) A shaded relief image using a sun elevation of 65°. Both images use an azimuth of 315°.

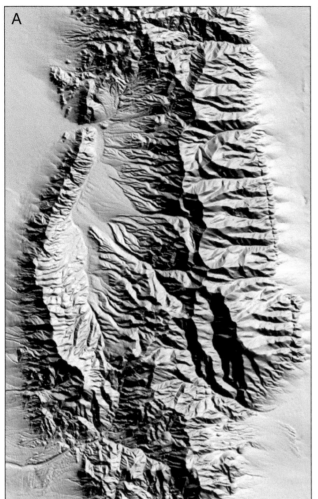
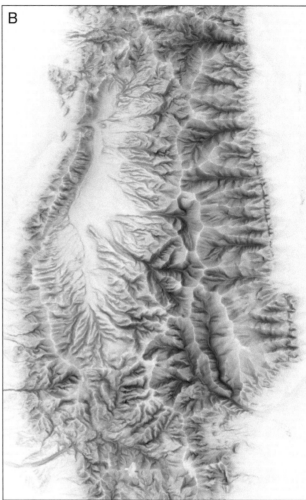

FIGURE 23.26 A comparison of shaded relief images using (A) a single point source illumination and (B) a uniform sky illumination model. The point source approach shows overall terrain structure and aspect, while the uniform approach shows subtle detail and relative elevation differences. (After Kennelly and Stewart 2006. First published in *Cartography and Geographic Information Science* 33(1), p. 30. Reprinted with permission from the Cartography and Geographic Information Society.)

facing the southwest are darkest. In the uniform model (Figure 23.26B), the brightest areas are either mountain ridge lines or relatively large, flat areas, whereas narrow valley bottoms are the darkest. Comparing the two images, it appears that the uniform model helps us visualize subtle terrain differences that are not apparent with the traditional point source approach. For instance, note the detail shown in the north–south trending valley in the southeastern portion of the map. The deepest understanding of the landscape, however, appears to be achieved when the two images are viewed together. The traditional approach appears useful for showing overall terrain structure and aspect, while the uniform model is useful for showing subtle detail and relative elevation differences.

23.7.4 Representation of Swiss-Style Rock Drawing

An application named Scree Painter (https://terraincartography. com/screepainter/) has been developed by Bernhard Jenny and colleagues (2010) that uses various digital inputs to simulate hand-drawn Swiss-style rock drawings. Scree is defined as a mass of rocks or stones that form a slope in a mountainous environment. Rock drawings depict bare rock, devoid of vegetation, that is commonly found in alpine environments. They emphasize how bare rock faces and scree might look from an observer on the ground (such as a rock climber) as opposed to how a landscape might be seen in an orthogonal or "birds-eye" view. When drawn by hand using the Swiss style, rock drawings consist of hachures that vary in width and density, which create a shading effect that also simulates the texture of actual rock faces. Scree is drawn as individual rocks of various sizes and shapes. Scree Painter requires four inputs, the first two being a DEM and a shaded relief image. The third input is a polygon data set that dictates where the scree will be placed. These polygons need to be derived through the investigation of landcover information via image interpretation and on-site visits. The fourth input consists of contours, spot elevations, and other map features upon which scree will not be placed. A most interesting aspect of Scree Painter is how the scree

FIGURE 23.27   A Swiss-style rock drawing including scree symbols produced in the Scree Painter application. (Courtesy of Bernhard Jenny.)

is depicted in digital form. In Scree Painter, the scree is rendered with scanned versions of actual hand-drawn scree stones taken from Swiss-style rock drawings. Figure 23.27 shows a sample map produced in Scree Painter. It includes the scanned scree stones, plus additional information such as contours, spot elevations, and water features.

23.7.5   COLOR CONSIDERATIONS

In this section, we consider three ways in which color can be utilized to enhance black-and-white shaded relief images. One approach is to utilize different colors (or **hypsometric tints**) to represent different elevation ranges. Figure 23.28 illustrates various color schemes for hypsometric tints that have been utilized historically, with low elevations represented by colors at the bottom of each scheme and high elevations represented by colors at the top of each scheme. In evaluating the potential effectiveness of these schemes, you should think about whether each scheme could logically represent increasing elevation. Note that schemes C and E to G were developed with the notion that the lightest colors should be used for the highest elevations to avoid darkening the shadows in the shaded relief for mountainous areas. Also, note that several of the schemes contain colors that could logically be associated with colors in the

environment; for instance, five of the schemes use green for lowland areas, which in many areas of the world would be appropriate for suggesting vegetation in those areas.

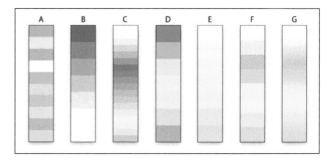

FIGURE 23.28   Various color schemes for hypsometric tints that have been utilized historically: (A) August Papen's polychromatic tints; (B) continuously progressing tints, with higher elevations represented by darker tints; (C) John Bartholomew Jr.'s tints; (D) Karl Peucker's tints, in which colors were supposed to provide a color stereoscopic effect; (E) Eduard Imhof's tints; (F) International Map of the World tints; (G) International Map of the World tints with a gradient. (From Patterson and Jenny 2011, 36; courtesy of *Cartographic Perspectives* (North American Cartographic Information Society, publisher), Tom Patterson, and Bernhard Jenny.)

Although using a color to suggest both elevation and the character of the environment may at first seem appropriate, we can obviously conceive of instances in which this does not work—think of those arid areas (such as Death Valley), which are at a low elevation, but are more likely to appear brown rather than green. Recognizing this limitation of hypsometric tints, Tom Patterson and Bernhard Jenny (2011) proposed **cross-blended hypsometric tints**, which combine traditional colors for hypsometric tints with colors that we might expect to find in the environment. Patterson and Jenny began with the scheme used for the 1962 International Map of the World shown in Figure 23.28G and modified that scheme to further emphasize the *higher is lighter* principle that we mentioned above. Using this scheme as a basis, they then created schemes for four separate environments with associated colors: arid (khaki brown), warm humid (rich yellow–green), cold humid (blue–green), and polar (cold gray),[11] as shown in Figure 23.29. As the term *cross-blended* suggests, these colors blend together across regions. The resulting cross-blended hypsometric tints and shaded relief are available for you to utilize in your maps at the Natural Earth website: https://www.naturalearthdata.com/.

Although cross-blended hypsometric tints are an improvement over raw hypsometric tints, a user study by Patterson and Jenny (2013) revealed that there were problems with properly interpreting the resulting tints. Thus, as an alternative, you may wish to consider using **Natural Earth** data, which combines satellite-derived land cover data (represented by different colors) with shaded relief. Two versions of the Natural Earth data are available (again at the Natural Earth website), one based on the actual land cover (Natural Earth 1) and one based on potential land cover, largely ignoring human influences (Natural Earth 2).[12]

23.8 SUMMARY

This chapter has examined methods for symbolizing Earth's terrain and associated landscape features. Basic data for depicting terrain are generally provided in an equally spaced gridded format (such as a digital elevation model or DEM). The definitive source for DEMs in the United States is the USGS National Map Viewer Downloader (https://apps.nationalmap.gov/downloader/#/).

We divided the methods for symbolizing terrain into vertical views, oblique views, and physical models. Methods that have emphasized a vertical view include hachures (e.g., drawing a series of parallel lines perpendicular to the direction of contours), contour-based methods (e.g., Tanaka's method), Raisz's physiographic method, and shaded relief (or hill shading). These methods assume that the topographic surface is illuminated, typically obliquely from the northwest. The various methods use different approaches to distinguish areas that appear illuminated from those that appear in the shadows. For example, in Tanaka's method, contour lines perpendicular to the light source have the greatest width, whereas those parallel to the light source are narrowest; furthermore, contour lines facing the light source are drawn in white, whereas those in shadow are drawn in black. We also considered morphometric methods as a special form of vertical view that emphasizes the structural elements of terrain. For instance, Brewer and Marlow developed an approach for symbolizing aspect and slope that utilizes the HVC color model, and Wood has developed the package LandSerf to visualize a variety of topographic features including pits, channels, passes, ridges, peaks, and planar regions.

Approaches for creating oblique views include block diagrams, panoramas, and plan oblique relief. Software for creating block diagrams (e.g., EarthVision) has been available for quite some time, whereas utilizing computers to create panoramas is a more recent phenomenon. Particularly interesting is the work of Patterson, who has provided numerous suggestions for designing oblique 3-D terrain maps (http://shadedrelief.com/3D_Terrain_Maps/3dterrainmapsint.html). Plan oblique relief is an intriguing technique that causes relief to "stand up" and yet preserves the geometric accuracy of conventional shaded relief maps.

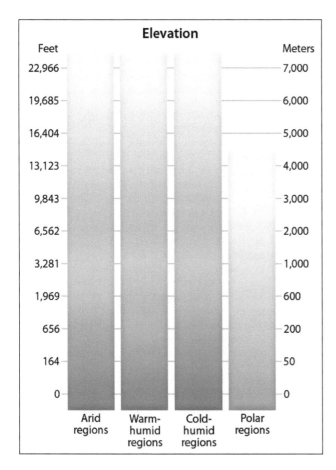

FIGURE 23.29 Color schemes for the cross-blended hypsometric tints developed by Patterson and Jenny. (From Patterson and Jenny 2011, 41; courtesy of *Cartographic Perspectives* (North American Cartographic Information Society, publisher), Tom Patterson, and Bernhard Jenny.)

Physical models are an attractive possibility for depicting terrain because the models are truly 3-D. Thus, users can work with these models as they would other 3-D objects in the real world; for instance, you can hold a small physical model in your hands and easily view it from different angles. Physical models have been around since the time of Alexander the Great, but only in the last few years has the technology become readily available to convert digital data into solid models.

Since shaded relief techniques are the most commonly used approach for depicting terrain, we focused on several related aspects, including generalizing the terrain, selecting an appropriate orientation of the sun (known as the azimuth) and sun elevation, lighting model issues (e.g., a model with multiple light sources), the representation of Swiss-style rock drawing, and the use of color to depict the elevation of the terrain.

23.9   STUDY QUESTIONS

1. Contrast the approaches used for shadowed hachures by Yoeli and Samsonov, and indicate which approach is likely to be most effective for representing the landscape.
2. (A) Describe the difference between illuminated contours and shadowed contours. (B) Using the results of the Eynard and Jenny study, indicate which of the methods in A is likely to be most effective. (C) Indicate any limitations that you see in the Eynard and Jenny study.
3. (A) Describe the appearance of Raisz's map *Landforms of the United States*, and contrast its appearance with Thelin and Pike's shaded relief map of the coterminous United States. (B) Using the interactive link for Raisz's map, zoom into the state of Wisconsin, and list the glacial features found on this map along with their approximate locations.
4. (A) Using Figure 23.11, describe how color is used in the Brewer and Marlow system for depicting aspect and slope. (B) Using Figure 23.12, contrast the character of the island located in the north-central part of the image with the relatively large island-like feature to its south.
5. Select a data set and/or a thematic mapping technique not discussed in the physical modeling section, and discuss why you think that data set/technique might be particularly appropriate for creating a physical model.
6. Assuming a single light source for illumination, contrast the effectiveness of the following azimuth values for the illumination of shaded relief: 180°, 315°, 337.5°, and 0°.
7. Explain why you might want to reduce the visual impact of a shaded relief image. Also, describe how a moving-window filter and the specification of sun elevation can be used to reduce the visual impact of a shaded relief image.

8. Using the knowledge of color schemes that you have developed in this book, and the information provided in the present chapter, discuss the usefulness of the hypsometric tints shown in A, D, and F of Figure 23.28 for representing elevation data.

NOTES

1   The geoid is the surface that would result on Earth if the oceans were allowed to flow freely over the continents. It is used to measure surface elevations with a high degree of accuracy and is further described in Section 7.4.2.3.
2   LIDAR, which stands for Light Detection and Ranging, is a remote sensing method that uses light in the form of a pulsed laser to measure distances to Earth, typically from an aircraft.
3   GeoTiff is a raster image format originally developed for graphics but has since been modified for use in GIS and Remote Sensing through the inclusion of geospatial coordinate system information in the header.
4   Tanaka did not use the term *illuminated contours*; rather, he referred to the technique as "the relief contour method," as specified in the title of the article in which the method was introduced.
5   Eynard and Jenny also developed open-source software (http://terraincartography.com/PyramidShader/) for those wishing to experiment with illuminated and shadowed contours.
6   A printed version of the map is available through Raisz Landform Maps (https://www.raiszmaps.com/).
7   For a slope/aspect method appropriate for hachuring, see Samsonov (2014).
8   Another option is Terrain Bender (https://www.terraincartography.com/terrainbender/), open-source software for creating panoramas.
9   For a full discussion of the advantages and disadvantages of plan oblique relief, see Jenny and Patterson (2007, 29–32). Jenny and Patterson noted that plan oblique relief was initially available in Natural Scene Designer (https://www.naturalgfx.com/products.html); the approach is also now available in ArcGIS's Terrain Tools (https://tinyurl.com/ArcGIS-Terrain-Tools). For related work on plan oblique relief, see Buddeberg et al. (2017).
10   A moving window is a type of filter that is applied to a raster. It is a collection of adjacent grid cells, typically in a 3-cell by 3-cell square, that is applied to an input raster. A calculation is made using the nine input cell values and the result is written to an output raster. The window then moves one cell over and the calculation is repeated with the new input values. The process continues until all input values have been processed and written to the output raster. A specific example involves calculating the mean value of the nine input values and writing the result to a single cell in the output grid (the cell location at the center of the square moving window). The window then moves to an adjacent cell, the mean is calculated for the new input values and is written to the output raster. Other calculations can be performed on the input cell values and the moving window itself can have different shapes and sizes.
11   Here we use color descriptors provided by Patterson and Jenny (2011, 39–40).
12   For a discussion of the development of the Natural Earth data using color, see Patterson and Kelso (2004) and Patterson (2012).

REFERENCES

Biland, J., and Çöltekin, A. (2017) "An empirical assessment of the impact of the light direction on the relief inversion effect in shaded relief maps: NNW is better than NW." *Cartography and Geographic Information Science 44*, no. 4:358–372.

Brassel, K. (1974) "A model for automatic hill shading." *The American Cartographer 1*, no. 1:15–27.

Brewer, C. A., and Marlow, K. A. (1993) "Color representation of aspect and slope simultaneously." *Auto-Carto 11 Proceedings*, Minneapolis, MN, pp. 328–337.

Buddeberg, J., Jenny, B., and Willett, W. (2017) "Interactive shearing for terrain visualization: An expert study." *GeoInformatica 21*, no. 3: 643–665.

Caldwell, D. R. (2001) "Physical terrain modeling for geographic visualization." *Cartographic Perspectives*, no. 38:66–72.

Carbonell-Carrera, C., Avarvarei, B. V., Chelariu, E. L., et al. (2017) "Map-reading skill development with 3D technologies." *Journal of Geography 116*, no. 5:197–205.

Eynard, J. D., and Jenny, B. (2016) "Illuminated and shadowed contour lines: Improving algorithms and evaluating effectiveness." *International Journal of Geographical Information Science 30*, no. 10:1923–1943.

Horn, B. K. P. (1982) "Hill shading and the reflectance map." *Geo-Processing 2*, no. 1:65–144.

Hügli, H. (1979) "Vom Geländemodell zum Geländebild. Die Synthese von Schattenbildern, Vermessung, Photogrammetrie." *Kulturtechnik 77*:245–249.

Imhof, E. (1982) *Cartographic Relief Presentation*. Berlin: Walter de Gruyter.

Jenny, B., Hutzler, E., and Hurni, L. (2010) "Scree representation on topographic maps." *The Cartographic Journal 47*, no. 2:141–149.

Jenny, B., and Patterson, T. (2007) "Introducing plan oblique relief." *Cartographic Perspectives*, no. 57:21–40, 88–90.

Kennelly, P., and Kimerling, A. J. (2001) "Modifications of Tanaka's illuminated contour method." *Cartography and Geographic Information Science 28*, no. 2:111–123.

Kennelly, P. J., and Kimerling, A. J. (2006) "Non-photorealistic rendering and terrain representation." *Cartographic Perspectives*, no. 54:35–54.

Kennelly, P. J., and Stewart, A. J. (2006) "A uniform sky illumination model to enhance shading of terrain and urban areas." *Cartography and Geographic Information Science 33*, no. 1: 21–36.

Leonowicz, A. M., Jenny, B., and Hurni, L. (2010) "Terrain Sculptor: Generalizing terrain models for relief shading." *Cartographic Perspectives*, no. 67:51–60.

Lewis, P. (1992) "Introducing a cartographic masterpiece: A review of the U.S. Geological Survey's digital terrain map of the United States." *Annals of the Association of American Geographers 82*, no. 2:289–304.

Lobeck, A. K. (1924) *Block Diagrams*. New York: Wiley.

Lobeck, A. K. (1958) *Block Diagrams* (2nd ed.). Amherst, MA: Emerson-Trussell.

Moellering, H., and Kimerling, A. J. (1990) "A new digital slope-aspect display process." *Cartography and Geographic Information Systems 17*, no. 2:151–159.

Mower, J. E. (2011) "Supporting automated pen and ink style surface illustration with B-spline models." *Cartography and Geographic Information Science 38*, no. 2:175–184.

Mower, J. E. (2016) "Concurrent drainage network rendering for automated pen-and-ink style landscape illustration." *Transactions in GIS 20*, no. 1:54–78.

Patterson, T. (2000) "A view from on high: Heinrich Berann's panoramas and landscape visualization techniques for the U.S. National Park Service." *Cartographic Perspectives*, no. 36:38–65.

Patterson, T. (2005) "Looking closer: A guide to making bird's-eye views of National Park Service cultural and historical sites." *Cartographic Perspectives*, no. 52:59–75, 96–106.

Patterson, T. (2012) Creating Natural Earth map data. *6th National Cartographic Conference (GeoCart 2012)*, Auckland, New Zealand, pp. 1–17, http://www.shadedrelief.com/new_zealand/natural_earth_nzcs.pdf.

Patterson, T., and Jenny, B. (2011) "The development and rationale of cross-blended hypsometric tints." *Cartographic Perspectives*, no. 69:31–46.

Patterson, T., and Jenny, B. (2013) "Evaluating cross-blended hypsometric tints: A user study in the United States, Switzerland, and Germany." *Cartographic Perspectives*, no. 76:5–17.

Patterson, T., and Kelso, N. V. (2004) "Hal Shelton Revisited: Designing and producing natural-color maps with satellite land cover data." *Cartographic Perspectives*, no. 47:28–55.

Priestnall, G., and Cheverst, K. (2019) "Understanding visitor interaction with a projection augmented relief model display: Insights from an in-the-wild study in the English Lake District." *Personal and Ubiquitous Computing*, https://link.springer.com/article/10.1007/s00779-019-01320-2, Creative Commons Attribution 4.0 International License.

Priestnall, G., Gardiner, J., Durrant, J., et al. (2012) "Projection augmented relief models (PARM): Tangible displays for geographic information." *Proceedings of Electronic Visualisation and the Arts*, London, England, pp. 180–187.

Raisz, E. (1931) "The physiographic method of representing scenery on maps." *The Geographical Review 21*, no. 2:297–304.

Raisz, E. (1967) *Landforms of the United States*. Map. (Scale approximately 1:4,500,000.)

Rase, W.-D. (2012) "Creating physical 3D maps using rapid prototyping techniques." In *True-3D in Cartography: Autostereoscopic and Solid Visualization of Geodata*, ed. by M. Buchroithner, pp. 119–134. Springer.

Robinson, A. H., Morrison, J. L., Muehrcke, P. C., Kimerling, A. J., and Guptill, S. C. (1995) *Elements of Cartography* (6th ed.). New York: Wiley.

Samsonov, T. (2014) "Morphometric mapping of topography by flowline hachures." *The Cartographic Journal 51*, no. 1:63–74.

Tanaka, K. (1950) "The relief contour method of representing topography on maps." *The Geographical Review 40*, no. 3:444–456.

Taylor, J., Tabayoyon, A., and Rowell, J. (1991) "Device-independent color matching you can buy now." *Information Display 7*, no. 4/5:20–22, 49.

Thelin, G. P., and Pike, R. J. (1991). Landforms of the conterminous United States—A digital shaded-relief portrayal, Map I-2206. Washington, DC: U.S. Geological Survey. (Scale 1:3,500,000.)

Yoeli, P. (1971) "An experimental electronic system for converting contours into hill shaded relief." *International Yearbook of Cartography 11*:111–114.

Yoeli, P. (1983) "Shadowed contours with computer and plotter." *The American Cartographer 10*, no. 2:101–110.

Yoeli, P. (1985) "Topographic relief depiction by hachures with computer and plotter." *The Cartographic Journal 22*, no. 2:111–124.

24 Map Animation

24.1 INTRODUCTION

In this chapter, we focus on **animated maps**, or maps characterized by continuous change while the map is viewed. You are probably most familiar with animated maps from watching the weather segment of the nightly news or The Weather Channel, but we will see that virtually any form of thematic symbology can be animated (albeit some more effectively than others). Although animation is commonly associated with recent technological advances, in Section 24.3, we will see that the earliest map animations were produced in the 1930s and that cartographers were discussing the potential of animation as early as the 1950s.

In Section 24.4, we consider visual variables for animation. From Chapter 4, you will recall that visual variables refer to the perceived differences in map symbols that are used to represent spatial data; examples include spacing, size, orientation, and shape. Animated maps can utilize these same visual variables, but they also use others, including duration, rate of change, order, display date, frequency, and synchronization.

Most animations depict changes over time, as changes in time can logically be represented by changes in the map symbolization associated with an animation (which also happens over time)—we focus on such **temporal animations** in Section 24.5, where we consider animations for a variety of thematic maps including flow, choropleth, proportional symbol, and isarithmic maps. **Nontemporal animations** have also been produced to animate spatial data that do not change over time. In Section 24.6, we consider some examples of these, including the animation of related attributes for multivariate data (e.g., various age groups within an urban area), the **sequencing** of portions of a single attribute, **fly-overs** (in which the user has the sense of flying over a landscape), and a *Wind Map* (in which **animated streamlets** provide a compelling sense of wind speed and direction).

A key issue involved in animation is the degree of interactivity that users are able to have with the animation. Generally, we have found that animations are most effective when users have considerable control over the animation (e.g., they are able to easily start and stop the animation and change its pace). In Section 24.7, we discuss several approaches for providing a greater degree of interaction than is found in most animations. Software for providing such potential is closely allied with the data exploration software that we describe in Chapter 25.

One alternative to animation is the **small multiple**, in which many small maps are displayed to show the change in an attribute over time or to compare many attributes for the same time period. In Section 24.8, we raise the question of whether animation works and consider some studies that have been done to evaluate the effectiveness of animation compared to small multiples. We will see that rather than seeing whether animation or small multiples are more effective, we should understand *how* readers utilize these techniques and *why* they are effective for certain tasks.

After viewing the various animations discussed in this chapter and exploring others available via the Web, we hope that you will consider designing your own animations. In this regard, Section 24.9 summarizes several guidelines for designing your own animations.

Normally, animations are displayed in 2-D space, and the third dimension (the *z*-axis) is used to display something that is 3-D (e.g., a mountain or a cloud). In Section 24.10, we consider the possibility of instead using the *z*-axis to depict time, which produces the **space-time cube**. This approach provides an alternative to animation that may allow us to detect information that either cannot be seen or is unlikely to be seen in animation. Since symbols in 3-D space frequently block one another, the space-time cube approach requires that users interact with the display.

24.2 LEARNING OBJECTIVES

- Describe how the traditional visual variables for static maps (e.g., location, size, and color) can be utilized on animated maps.
- Define and provide an example of usage for the following visual variables that are explicitly associated with animation: duration, rate of change, order, display date, frequency, and synchronization.
- Explain the congruence principle and describe three examples of its usage for temporal animations.
- Discuss two design parameters that illustrate why the design of animated maps often differs from the design of static maps.
- Explain how a bubble plot can be utilized to provide a temporal relationship among three attributes for countries of the world.
- Describe a nontemporal animation that makes usage of the congruence principle.
- Explain why it is often desirable to have users interact with the animations that they view.
- Define the terms small multiple and space-time cube, and indicate when each might be a useful alternative to animation.
- Identify four key guidelines to consider when designing your own animations.

DOI: 10.1201/9781003150527-27

24.3 EARLY DEVELOPMENTS

Although map animation is often viewed as a recent development, the notion has been around since the 1930s and was promoted by cartographers as early as the 1950s, most notably by Norman Thrower (1959, 1961), who clearly was enamored by its potential:

> By the use of animated cartography we are able to create the impression of continuous change and thereby approach the ideal in historical geography, where phenomena appear "as dynamic rather than static entities." Distributions which seem to be particularly well suited to animation include the spread of populations, the development of lines of transportation, the removal of forests, changing political boundaries, the expansion of urban areas, and seasonal climatic patterns. (From p. 10 of "Animated cartography," by N. J. W. Thrower, *The Professional Geographer*, copyright © 1959, reprinted by permission of Taylor & Francis Ltd., http://www.tandfonline.com on behalf of The Association of American Geographers www.aag.org.)

Because computers were only beginning to be developed in the 1950s, animations at that time were created using hand drawings ("cels") and a camera to record each frame of the animation.

Early computer-based animations were exemplified by Waldo Tobler's (1970) 3-D portrayal of a population growth model for the city of Detroit, Michigan, and Hal Moellering's (1976) representation of the spatiotemporal pattern of traffic accidents. Mark Harrower and colleagues (2000, 281) argued that work such as Moellering's illustrated the "power of map animation," as "the result was a clear representation of the daily cycle of accidents, with peaks during rush hour periods and troughs between those times, as well as the spatial pattern of weekday versus weekend accidents." Although Moellering continued to work with animation (we discuss his related work on data exploration in Section 25.5.1), there were few other efforts in cartographic animation prior to the 1990s. Craig Campbell and Stephen Egbert (1990, 24–25) argued that this was due to the lack of funds in geography departments, the lack of a "push toward the leading edges of technology" in these departments, a negative view of academics who develop and market software, and the difficulty of disseminating animations in standard print outlets. In the following sections, we will see that many of these limitations have been overcome.

24.4 VISUAL VARIABLES FOR ANIMATION

In 1991, David DiBiase and his colleagues created a videotape (now available at https://tinyurl.com/DiBiaseVideotape) in which they formulated visual variables unique to animation and categorized various types of animation. This videotape was supplemented by an article (DiBiase et al. 1992), and the visual variables were later expanded on by Alan MacEachren (1995).

DiBiase and colleagues described how the visual variables for static maps (see Section 4.5) could be applied to animated maps. For example, they animated a choropleth map of New Mexico population from 1930 to 1980 to show how the visual variable lightness (represented by gray tones) could represent changes over time, and they transformed the state of Wisconsin from its real-world shape to that of a cow to illustrate how the visual variable shape could be used in animation. More important, however, they showed how animation requires additional visual variables beyond those used for static maps, namely, duration, rate of change, and order.

In its simplest form, **duration** is defined as the length of time that a frame of an animation is displayed. More generally, a group of frames can be considered a *scene* (in which there is no change between frames), and we can consider the duration of that scene. Because the duration is measured in quantitative units (time), it can be logically used to represent quantitative data; for example, DiBiase et al. animated electoral college votes for president, varying the duration of each scene in direct proportion to the magnitude of the victory in each election.

Rate of change is defined as m/d where m is the magnitude of change (in position and attributes of entities) between frames or scenes, and d is the duration of each frame or scene. For example, Figure 24.1 illustrates different rates of change for the position of a point feature (A) and the attribute of circle size (B). The smoothness (or lack thereof) in an animation is a function of the rate of change. Either decreasing the magnitude of change between frames or increasing the duration of each frame will decrease

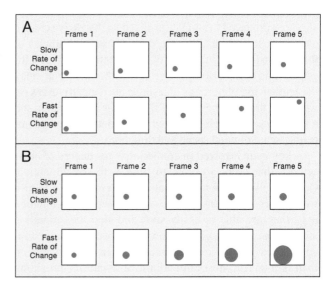

FIGURE 24.1 The visual variable rate of change is illustrated for (A) geographic position and (B) circle size. Rate of change is defined as m/d where m is the magnitude of change between frames or scenes, and d is the duration of each frame or scene. For these cases, duration is presumed to be constant in each frame. (After DiBiase et al. 1992, 205.)

the rate of change and hence make the animation appear smoother.

Order is the sequence in which frames or scenes are presented. Normally, of course, animations are presented in chronological order, but DiBiase et al. argued that knowledge also can be gleaned by reordering frames or scenes. For example, they modified not only the duration of frames associated with the presidential election data but also reordered the frames such that lower magnitudes of victory were presented first.

Alan MacEachren (1995) extended the visual variables for animation to include display date, frequency, and synchronization. MacEachren defined **display date** as the time some display change is initiated; for example, in an animated map of decennial population for U.S. cities, we might note that a circle first appears for San Francisco in 1850, during the California Gold Rush. **Frequency** is the number of identifiable states per unit time, or what MacEachren called *temporal texture*. To illustrate frequency, he noted its effect on *color cycling*, in which various colors are "cycled" through a symbol, as is often done on weather maps to show the flow of the jet stream. **Synchronization** deals with the temporal correspondence of two or more time series. MacEachren (285–286) indicated that if the "peaks and troughs of … [two] time series correspond, the series are said to be 'in phase,'" or synchronized. For example, if we examine animations of both precipitation and greenness (see Figure 1.13 for an example of a greenness map), we would expect them to be out of sync because the greenness will unlikely respond immediately to precipitation changes.

24.5 EXAMPLES OF TEMPORAL ANIMATIONS

In this section, we consider a range of temporal animations that have been developed by cartographers and other graphic designers. Because these animations are impossible to depict here (although we approximate a few with small multiples), it is important that you use the weblinks provided to view the animations online. We have divided these based on the thematic symbolization used.

24.5.1 ANIMATING MOVEMENT AND FLOWS

Temporal animations that depict *movements or flows* of a phenomenon arguably have great potential for animation, as the movements in the real world are represented by movements in the animation, thus making use of what Barbara Tversky and colleagues (2002, 247) termed the **congruence principle** ("the content and format of the graphic should correspond to the content and format of the concepts to be conveyed"). In the context of animating movement, an example of the congruence principle would be to depict the migration of birds via the movement of symbols representing birds. If the animation is implemented effectively, a viewer can actually see a phenomenon (in this case, birds) move from one location to another location. This notion

of the potential of animating movement was promoted by Daniel Dorling (1992) in his early work on animation:

> I suspect that animation needs to resemble natural visual experiences – in general, objects should not change color (even though that is easy to program), but should move, because we are good at registering movement. (218)

A relatively simple example that illustrates the use of movement in animation is a New York Times article by Rich Harris and colleagues (2020) that animates air traffic over China before and after flights were canceled due to the novel coronavirus in early 2020 (See the opening animation of https://tinyurl.com/NYTimesCoronaAnimation). In this case, each flight is shown by a small moving airplane icon (the visual variable location is used), and we can clearly see the dramatic decrease in flights as a result of the coronavirus by comparing the two animations. The animations primarily serve to attract our attention as we are not so concerned about where each plane is going but the differing concentration of planes. This is a temporal animation in the sense that each plane moves a certain distance over a period of time. There is no legend to indicate the passage of time, and there is no way to start and stop the animation, as these tasks are arguably not necessary, given that the emphasis is simply on the lower concentration and movement of planes subsequent to the spread of the coronavirus.

Another example that involves airplane movement is Michael Peterson and Jochen Wendel's animated atlas of air traffic over the United States, Canada, and the Caribbean (http://maps.unomaha.community/AnimatedFlightAtlas/).[1] This atlas illustrates a typical 24-hour cycle of air traffic for weekdays using airplane icons (when the distance across the display is less than 1,000 miles) and dots (when the distance is greater than 1,000 miles). Peterson and Wendel provide a broad range of options for animation, including all flights in either the United States, Canada, or the Caribbean; types of aircraft; major airlines; traffic associated with particular airports; major flight corridors (e.g., West Coast–Hawaii); regions (e.g., the Northeast); and individual states.

We were able to see distinctive patterns in several of the animations. Some of these patterns we saw ourselves, while others were prompted by text that is included with each animation. As an example, an animation of FedEx traffic showed very distinctive patterns associated with Memphis, Tennessee, which is a major hub for FedEx. Beginning around 4 PM, there was a distinct trend in outgoing flights from Memphis—this corresponds to planes leaving "to collect express mail and packages from airports across the country." Figure 24.2 shows a static snapshot around 4:30 PM (see the clock icon in the lower left of the figure); if you look carefully at this image, you can see that planes are generally moving away from Memphis (the front of the plane is pointed away from Memphis). In the animation, this is much more obvious, as we can *see* the planes moving away from Memphis. Around 10 PM, planes begin to return to Memphis, and this trend continues until about 1 AM.

FIGURE 24.2 A static frame from Peterson and Wendel's animation of FedEx traffic. Note that the time on the clock in the lower left is shortly after 4:30 PM and that planes appear to be moving away from Memphis, Tennessee. (Courtesy of Michael P. Peterson.)

The text associated with the animation indicates that from 11 PM to 3 AM, packages are sorted and distributed to the appropriate planes. You then see a major exodus of planes from Memphis between 3 and 5 AM. Although some of the animations in the atlas are fuzzy due to poor contrast, the animations certainly catch your attention and make you ask why particular spatiotemporal patterns exist. In contrast to the previous coronavirus example, in the atlas of air traffic, a clockface was absolutely essential to determine the spatiotemporal patterns that we have described. We also found it necessary to start and stop the animation and slow its speed (see the speed options in the lower right of the figure) in order to fully associate the patterns of planes with particular times.

A more complex example of animating movement is Jo Wood and colleagues (2014) efforts to animate the 16,000,000 bicycle trips associated with a public bicycle-sharing scheme in London, England, over a two-year period. It would be helpful if you view the following Tedx talk before reading our discussion of Wood et al.'s work with animation: https://tinyurl.com/WoodTedxTalk.[2] Although it might seem that an animation of 16,000,000 trips would be difficult to interpret, Wood et al. found that if careful decisions were made in the design, useful information could be gleaned from the animation; moreover, the animations "… resonated strongly with…policy makers… [and attracted] the attention of the popular press …" (2177). Some of the key design decisions were to represent trips by a Bezier curve with a relatively straight end at the starting point and a curved end at the destination, deleting the less frequent journeys, and categorizing the trips in a variety of ways. With respect to the latter, one animation contrasted the trips of ride-sharing members with those who are not

members (and pay by credit card), with members shown in blue and non-members shown in green. The resulting animation showed that the patterns of the two groups were quite different; for example, non-members concentrated around Hyde Park in the western portion of London.

24.5.2 ANIMATING CHOROPLETH MAPS

As we noted in Chapter 15, choropleth maps have been the most commonly used thematic mapping method, and so it is not surprising that there have been numerous efforts to animate them. We will begin with some basic examples of choropleth animation and then discuss some specialized approaches for animating them.

24.5.2.1 Some Basic Examples of Choropleth Animation

One simple example of a choropleth animation is Ed Stephan's animation of "The Sex Ratio" (specified as the number of males per 100 females) for the 48 contiguous states in the United States from 1790 to 1990 (http://www.edstephan.org/Animation/sexratios.html). This animation does not permit any user control, but it does illustrate distinctive changes over time. One of the keys to this animation is the red–blue diverging color scheme, which enables relatively higher ratios of males to females (in blue) to be contrasted with relatively lower ratios of males to females (in red). You must be careful in interpreting these ratios, however, as you will note that the dividing point for red and blue is not a ratio of 1 (100 males/100 females). Another key to the animation is that the sex ratios are a function of migration as the borders of the United States expanded over time. Thus, from 1790 to 1850, we see a largely red map with some blue in states added to the United States to the west, reflecting the fact that males tended to predominate as states were added to the United States. This pattern becomes even more dramatic from 1850 to 1920, as a large portion of the western United States appears in blue. Finally, from 1920 to 1990, we see a shift toward lower sex ratios and thus an entirely red map by the middle of the twentieth century.

Another more spatially complex example of a basic choropleth animation is the "Presidential Election Turnout" for the 3,000 plus counties in the United States from 1840 to 1972 (https://dsl.richmond.edu/voting/voterturnout.html).[3] We will focus just on the portion of the animation from 1912 to 1972, as the earlier portion is complicated by the fact that the boundaries of the United States were changing over that period. View the animation from 1912 to 1972, and try to summarize the overall spatial pattern over time (i.e., where, in general, are the low and high turnouts, and how does this spatial pattern change over time?). One of the difficulties that you may have in this process is associating a date with a particular pattern, as it is difficult to view the date and the spatial pattern at the same time (if you look at the date, you may miss the spatial pattern, and vice versa).

This is a common problem with many animations, not just choropleth animations. For this animation, you can handle this problem by controlling the pace of the animation by moving the small circle in the animation control located at the bottom of the screen—you can move the circle rapidly to greatly increase the pace of the animation or stop it to carefully associate a pattern with a particular date. One pattern that you should see is the distinctively lower turnout in the southeast United States (commonly termed the *South*) from 1912 to 1944; you may also note that this pattern is often associated with particular state boundaries. These lower turnouts were the result of voter suppression (of largely African Americans) through various means including poll taxes, literacy tests, and violence (https://www.britannica.com/topic/voter-suppression). The animation reveals that in 1948 the voter turnout in the South started to increase, and by 1968, the lower turnout in the southeast United States had largely disappeared. The key to the major change associated with the latter date was the passage of the Voting Rights Act in 1965, which outlawed many of the previous discriminatory voting practices (https://tinyurl.com/US-VotingRightsAct).

24.5.2.2 Should We Generalize Choropleth Animations?

It is our impression that many choropleth animations are difficult to follow, as we may become overwhelmed by the temporal changes taking place at multiple places on the map. This raises the question of whether it might be appropriate to generalize the animation in order to simplify it and thus more easily detect any temporal signals in the animation. Christoph Traun and Christoph Mayrhofer (2018) have developed a sophisticated mathematical approach for achieving such a generalization. Although the details of their approach are beyond the scope of the present textbook, the basic idea is to compute a generalized data value for each enumeration unit at each point in time as a function of that unit and neighboring enumeration units using both the time step being examined and adjacent time steps. An attempt is also made to modify data outliers that may distort temporal signals, although distinctive outliers are retained because they are considered important signals! Traun and Mayrhofer stress that their approach is most appropriate for highly autocorrelated data; thus, you should not expect their technique to assist in uncovering a pattern if your spatiotemporal data have a strong random component. In a user study of the technique, Traun et al. (2021) found that such generalization approaches did not lead to an improvement in trend detection, although users preferred these "noise reduced" animations. An intriguing finding was that global changes in a trend for highly autocorrelated distributions can be considered as movement, as users frequently mentioned terms that suggested movement (e.g., users described a "moving front" or the "movement of a cluster of low values") (17). Thus, for some choropleth maps, the notion of *movement* might be appropriate.

24.5.2.3 Should We Utilize Classed or Unclassed Maps?

As with static choropleth maps, a key question we can ask for animated choropleth maps is, "should we utilize classed or unclassed maps?" Our survey of animated choropleth maps suggests that classed maps are more common, but a user study by Mark Harrower (2007b) found that changes in the pattern were more effectively seen on unclassed animated choropleth maps. In the user study, 55 participants viewed animations of the monthly unemployment rates for the 3,000 plus U.S. counties and were asked to indicate which of four regions highlighted on the map exhibited the least stability and which of these four regions exhibited the greatest stability.[4] The key finding was that when the regions were relatively larger (individual states were highlighted as opposed to metro areas), participants were significantly better at identifying the lack of stability (i.e., change) on unclassed maps. Moreover, participants also felt that the unclassed maps looked smoother than the classed maps (this is something that we also noted when we viewed the animations).

Harrower explained the success of the unclassed animated maps using Figure 24.3. Here we see how three different locations (illustrated by the black lines) would be depicted on a three-class animated choropleth map (represented by light yellow, medium green, and dark green). Note that location 3 varies considerably over time, but the black line always falls in class 1 (light yellow), and location 2 increases over time, but the black line always falls in class 2 (medium green). In contrast, location 1 varies only slightly over time, but its class assignment (either medium green or dark green) varies considerably over time, thus producing a "jumpy" appearance in the animation. In contrast to the classed map, the unclassed map tends to avoid these sorts of problems, as proportional changes in data values are represented by corresponding proportional changes in symbolization. For instance, the

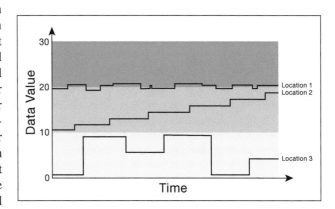

FIGURE 24.3 How changes in data values over time are depicted on a three-class animated choropleth map. (After "Unclassed animated choropleth maps," by M. Harrower, *The Cartographic Journal*, copyright © 2007 The British Cartographic Society, reprinted by permission of Taylor & Francis Ltd., http://www.tandfonline.com on behalf of The British Cartographic Society.)

increase in the data value for location 2 would be represented by a progressively darker green, and the variation in the data value for location 3 would be represented by a corresponding variation in the lightness of yellow.

24.5.3 ANIMATING PROPORTIONAL SYMBOL MAPS

As discussed in Chapter 4, proportional symbol maps are a common alternative to the choropleth map when one wishes to map data at either conceptual or true point locations. As such, numerous animations have been developed based on proportional symbols. Here we focus on what we argue is a particularly effective animation that shows job gains and losses in U.S. metropolitan areas. In Section 25.5.4, we will discuss the data exploration package MapTime, which contains a range of approaches for visualizing data for proportional circle maps, including map animation.

An animation based on proportional circles illustrating job gains and losses in U.S. metropolitan areas from 1999 to May 2021 can be found at https://tipstrategies.com/geography-of-jobs/. This animation was produced by TIP Strategies, Inc. (TIP, https://tipstrategies.com/about/) in association with Axis Maps (https://www.axismaps.com). The animation depicts a rolling 12-month net change in employment for each U.S. Metropolitan Statistical Area. Note that the animation uses two visual variables: the size of circles to depict change in employment and the color of the circles to depict gains (blue) and losses (orange).[5] Also, note the manner in which circle overlap is handled—it is similar to the semi-transparent approach that we utilized in Figure 18.15C, which allows you to see circle bounds even when circles overlap one another. Although the circles themselves do not move from their central locations, John Kostelnick and colleagues (2007, 41–42) argued that the increases and decreases in circle size on proportional circle maps can be seen as "apparent motion."

You can view the animation by selecting the "Play animation" button or you can control its pace (and select particular time elements) by dragging the rectangular box along the timeline. You can also use "Play animation" to start the animation and then hover over a metropolitan area to determine its precise employment gains or losses for each 12-month period. As you view the animation, you should see distinctive times in which there were notable job gains or losses and periods of relative stability. Remember that you should be cautious in interpreting the maps of gains and losses, as the resulting maps do not provide direct information on the current magnitude of employment; for example, many large blue circles at one point in time indicate many job gains, but these large circles do not necessarily indicate that the total magnitude of jobs has returned to a previous high level.

24.5.4 ANIMATING ISARITHMIC MAPS

As an example of animating isarithmic maps, we will consider Bernhard Jenny and colleagues (2016) efforts to map *A Year in the Life of Earth's CO_2* (http://co2.

digitalcartography.org/). Not only does this work provide an excellent example of the potential for animating smooth continuous phenomena such as the concentration of carbon dioxide in Earth's atmosphere, but it also illustrates the considerable interactive potential of modern computer technology. Jenny et al. describe their work as an *interactive video map*, although from our perspective, it meets our definition of *animation* as the resulting map is characterized by continuous change while the map is viewed.

Jenny et al. (2016) consider a *video* to consist "… of a stream of individual frames that are displayed in rapid sequence to create a movie" (36). In their case, frames for their video were based on a GEOS-5 computer model created by scientists at NASA. Jenny et al. use the term *interactive video map* to describe "… a map that contains at least one areally georeferenced video stream and offers interactive features" (37). The key point is that by georeferencing the video stream, Jenny et al. are able to interactively adjust the parameters of the map projection used to display the map (a plate carrée projection is used for the video stream), thus allowing you to center the animation on any desired location. Another key point is that by using HTML5 video and WebGL, most users of the video website should be able to manipulate the image in real-time over the web (the result will likely depend on your Internet speed and the computer you are using to access the video).

If you haven't already run the video, try doing so (note that you should also hear a voice-over describing what you are seeing in the video). The resulting video map gives the impression of motion, although the only visual variable that is being used is the changing color of each pixel location on the screen. We argue that the reason motion is apparent is that the concentration of CO_2 is a smooth continuous phenomenon, and we thus see changes between individual frames as motion. This will not necessarily be true of all smooth continuous phenomena, as a result will depend on how the data were sampled from the phenomenon and how the results are displayed. Experiment with changing the parameters of the map projection (the video map was cast using the Briesemeister equal-area modified azimuthal) by clicking on the map and dragging. You should be able to see the video continue to run as the center of the projection is changed.

24.5.5 OTHER TEMPORAL ANIMATIONS

If you explore the Web, you will no doubt find numerous additional examples of temporal animations. We encourage you to undertake such exploration! One temporal animation that we found particularly intriguing in our own exploration is the animated **bubble plot** (or **bubble chart**) used at https://tinyurl.com/BubblePlot. Figure 24.4 shows two frames taken from a bubble plot animation of the relation between Income and Life Expectancy for countries of the world from 1800 to 2018. Looking at the frame for 1945, we can see that Income (as represented by GDP per capita in inflation adjusted dollars) is plotted along the *x*-axis and Life Expectancy (in years) is plotted along the *y*-axis.

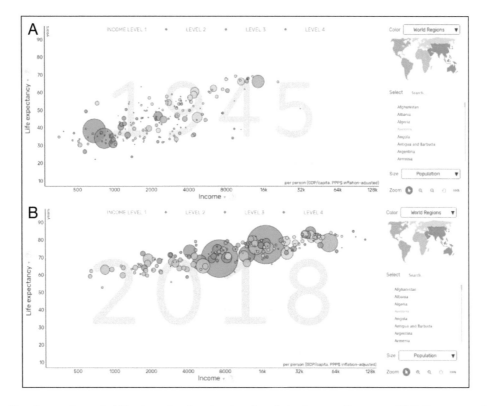

FIGURE 24.4   Two frames from a bubble plot animation showing the relation between Income and Life Expectancy for 1945 (A) and 2018 (B). (Showing free visualizations from gapminder.org, CC-by license.)

Each country is shown by a circle proportional in size to the population of that country (it is possible to change the attribute for circle size using the "Size" parameter in the lower right of the figure), with the colors of the circles taken from the map of World Regions shown in the upper right of the figure. Note that the date for this frame of the animation (1945) is represented by the large numbers in the background, an approach that allows you to easily see the date associated with the animation while it is running. If you run the animation, you will see that the circles move to new locations as a function of their values on Income and Life Expectancy for each year (the sizes of circles also change as a function of population, but this is not as obvious as the changes are slight for each year).

Comparing the graphs for 1945 and 2018 shown in Figure 24.4, we can see a dramatic improvement in life expectancy and the general tendency for a higher level of income over time. To see the changes for individual countries, however, it is useful to view the animation— note that individual countries can be highlighted in the animation to learn about their specific income and life expectancy by hovering on them or by selecting them from the menu on the right-hand portion of the screen. It is also interesting to note that like the examples discussed in Section 24.5.1, the key visual variable used is location, and the resulting movement is easily detectable. In this case, however, the movement takes place on a graph rather than on a map.

One limitation of the bubble plots in Figure 24.4 is that they do not show the attribute values on a map for each time step; rather, we just see a map of color-coded general world regions displayed in the upper right corner. Mikael Jern and colleagues at the National Center for Visual Analytics in Sweden developed several pieces of software that integrated bubble plots with maps and other graphical displays. Although this software is no longer available, you can get a feel for the approach by examining the link https://ncva.itn.liu.se/?l=en or by reading Ho et al. (2012).

24.6   EXAMPLES OF NONTEMPORAL ANIMATIONS

In this section, we consider examples involving animation in which spatial data do not change over time.

24.6.1   PETERSON'S EARLY WORK

Michael Peterson (1993) developed several early nontemporal animations for both univariate and multivariate data via choropleth symbolization. For univariate data, Peterson advocated animating both the classification method (say, mean–standard deviation, equal intervals, quantiles, and natural breaks) and the number of classes. For example, rather than displaying all four classification methods at once (as in Figure 24.5), Peterson displayed them in succession as an animation. For multivariate data, Peterson suggested using animation to examine the geographic trend of related attributes (i.e., those measured in the same units and part of a larger whole) and to compare unrelated attributes (i.e., those measured in dissimilar units and not part of a larger whole). With respect to related attributes, we might animate the percentage of population in various age groups (0–4 years old, 5–9 years old, etc.) within census tracts of a city. Peterson argued that such data "will usually show a clear regionalization in a city with older populations closer to the center and younger populations nearer the periphery. An animation … from low to high depicts high values 'moving' from the periphery to the center" (41–42). Two unrelated attributes would be median income and the percent of the adult population with high school education. Peterson argued that an animation of such attributes would detect similarities or differences in the two distributions (he recommended switching rapidly between the attributes).

Peterson (1999) also promoted the idea of "active legends," in which the user controls the order of the animation by selecting individual legend boxes on a map representing the elements to be animated, thus moving away from a "movie" to a more interactive form of animation. For instance, by moving the mouse over the upper four boxes shown in the lower right of Figure 24.6, a user could see various maps of health data displayed as an animation. Although Peterson's ideas for animating nontemporal data are intriguing, we found some of his animations difficult to follow. It seems that a small-multiples approach generally is more effective than animation for analyzing nontemporal data, primarily because the transitions between frames of

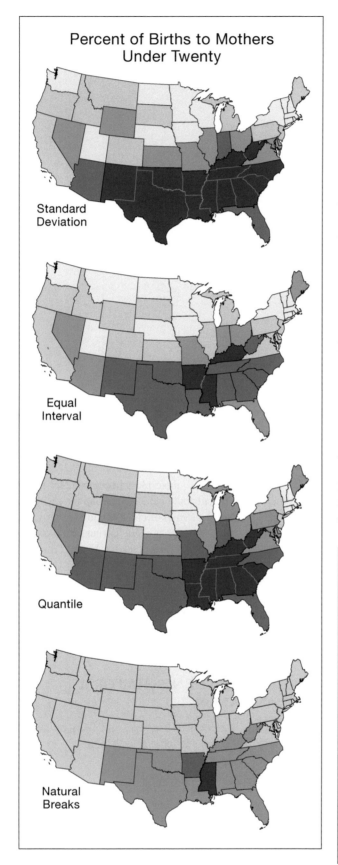

FIGURE 24.5 A small multiple illustrating four classification methods that Peterson animated. (After Peterson 1993. First published in *Cartography and Geographic Information Systems* 20(1), p. 41. Reprinted with permission from the Cartography and Geographic Information Society.)

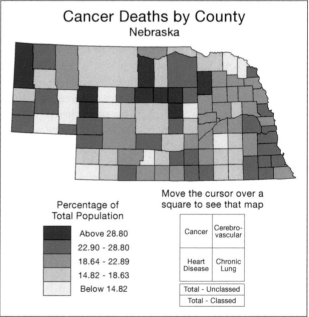

FIGURE 24.6 Peterson's (1999) notion of active legends: Users can control the order of frames within an animation by moving the mouse over the four legend boxes in the lower right representing various types of health data. (Courtesy of Michael P. Peterson.)

the animation are not smooth. In fact, Peterson (41) demonstrated some of his ideas in static form using small multiples (as in Figure 24.5 for the animation of the classification method).

24.6.2 Gershon's Early Work

Nahum Gershon (1992) also did some early experimental work with nontemporal animations. In one experiment, Gershon created an "animation of segmented components" in which users were shown an isarithmic map in a *sequenced presentation*: A high, narrow range of temperature appeared first, followed by a progressively wider range until the entire range to be focused on appeared; then the process was reversed with progressively narrower ranges of temperature displayed until just the highest narrow range appeared again. Gershon indicated that users "perceived the existence of structures, their locations and shapes better than in the static display of the data" (270). This result is a bit surprising, given the lack of success some have found with sequencing on choropleth maps (see Section 25.5.2). Gershon's superior results might be a function of either the particular sequencing strategy used, the data set used, or, possibly, sequencing is more appropriate for isarithmic than for choropleth maps.

Gershon (1992) also had people examine the correlation between two attributes as the two maps were quickly switched back and forth. This approach is analogous to Peterson's suggestion for animating more than one attribute. In contrast to the difficulty we had in understanding Peterson's animations, Gershon argued that the audience could "visually correlate structures appearing in both [maps]" (272). We suspect that Gershon's superior results were due to the similar structure of the maps: They were both of January sea surface temperatures but for different years. As a result, users could easily focus on the differences between the maps.

24.6.3 Fly-Overs

In a **fly-over** (or **fly-through**), a user is given the feeling of flying over or through a landscape or abstract thematic map. Respective examples include the *Grand Canyon Fly-through Animation* at https://www.nps.gov/grca/learn/photosmultimedia/fly-through.htm, and the fly-overs used to view election data depicted as 3D prism maps at https://www.esri.com/en-us/maps-we-love/gallery/election-results. Although fly-overs can produce eye-catching graphics, Mark Harrower and Ben Sheesley (2005) noted a number of problems with them, including their oblique perspective, their focus on realism, and thus information overload, visual occlusion, and user disorientation. The oblique perspective of fly-overs is problematic because it causes the scale to change across the image, making it difficult for viewers to estimate relative sizes, distances, and directions. One solution to this problem suggested by Harrower and Sheesley is to superimpose a grid on the scene. Although developers of

fly-overs have focused on creating realistic flights (likely because the power of computers has increased over time), Harrower and Sheesley argued that abstraction and generalization are often necessary in fly-overs. They stated:

> cartographic abstraction in 3-D fly-overs can be advantageous if it permits viewers to mentally organize and structure the landscape, for example, when roads are exaggerated, forest boundaries are highlighted, voice-overs are added, [and] visual hierarchy is established…. (6)

Visual occlusion in fly-overs is closely associated with the oblique perspective, as a low-angle view causes portions of the image to be obscured. One solution to this problem is to have the flight path circle key features in the landscape as opposed to simply flying by them. A second solution is to provide a static planimetric overview prior to showing the fly-over—the notion is that if the viewer can retain this planimetric overview in short-term memory while viewing the fly-over, their interpretation of the fly-over should be enhanced. A final solution is to provide both the planimetric overview and fly-over views simultaneously. In Section 28.5.2, we will see that this approach is also relevant in navigating virtual environments where the user has control over the flight.

User disorientation is often a problem for viewers of fly-overs, in part because viewers generally do not have control over the fly-over. Harrower and Sheesley suggested numerous solutions for disorientation, including the superimposed grid and overview maps mentioned previously, landmarks and labels, a jet contrail (which shows the flight path already completed), a spotlight path (which indicates the flight path that is ahead, and the flight path that has been completed), directional tick marks (or a floating compass), and heads-up display text. Figure 24.7 illustrates some of these solutions. In later work, Harrower and Sheesley (2007) tested the effectiveness of some of these approaches with users, finding that a horizon compass and monorail were effective orientation cues and that a grid was effective for directional tasks. The monorail was similar in concept to the spotlight path but served as a "physical representation of the actual flight path."[6] More recently, Richard Treves and Artemis Skarlatidou (2018) tested the nature of the flight path and the speed of the fly-over (which they called a **map tour**). Key findings were that a high flight path was most effective as it provides an overview of the map and that fast fly-overs produce an unacceptable cognitive load on the viewer.

24.6.4 Viégas and Wattenberg's *Wind Map*

A unique example of a nontemporal animation is Fernanda Viégas and Martin Wattenberg's *Wind Map* (http://hint.fm/projects/wind/), which depicts wind speed and direction for the current instant in time for the 48 contiguous U.S. states. Interestingly, Viégas and Wattenberg indicate that they created the *Wind Map* "… as an artistic exploration …," but it certainly can be used to get practical information about the wind. *Wind Map* was created using **animated streamlets**

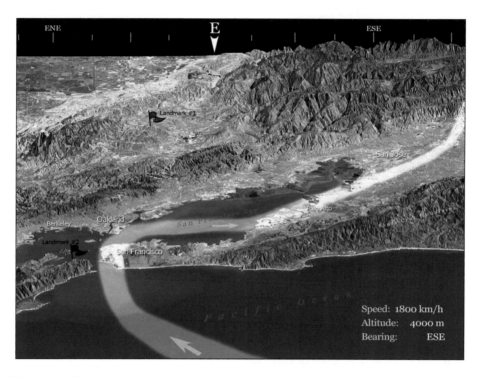

FIGURE 24.7 Solutions for disorientation in fly-bys: spotlight path (the light blue-shaded corridor); landmarks (the flags with the landmark labels); labels (e.g., the cities of Oakland and San Francisco); floating compass (the tic marks on the horizon); and heads-up display text (the text in the lower right). (Courtesy of Mark Harrower.)

in which small "comet-like trails" are used to show the direction and speed of the wind. Three redundant visual variables are used to depict the streamlets, as a greater wind speed is shown by a wider, brighter, and more rapidly drawn streamlet.[7] Direction is provided by drawing the streamlet in the direction of the wind flow.

Pyry Kettunen and Juha Oksanen (2019) tested the effectiveness of the animated streamlets used in *Wind Map* by altering several animation parameters, including the density, length, width, brightness interval, and overall color of the streamlets. Their major finding was that variation in these parameters did not affect the interpretation of maxima in the wind fields studied and that the values of these parameters thus can be chosen with respect to general map design considerations (e.g., color could be used for other map design parameters).

24.7 ENHANCING THE INTERACTIVITY IN ANIMATIONS

We have seen in the preceding two sections that most animations involve little or no interactivity (Jenny et al.'s interactive video map is a notable exception). We argue that the effectiveness of animations can often be enhanced if the user is able to interact with the representations of the phenomenon (or phenomena) under investigation and thus visualize the data in a variety of different ways. Below we describe some early efforts by Mark Harrower to achieve this and a recently developed application, CoronaViz, that enables users to visualize animations of data related

to the COVID-19 pandemic. Other examples of applications that permit considerable interactivity include The Refugee Project (http://www.therefugeeproject.org/#/2018) and Eruptions, Earthquakes & Emissions (https://volcano.si.axismaps.io/). In the subsequent chapter on data exploration, we will consider this notion in greater depth (MapTime as an example of data exploration that focuses on animation).

24.7.1 HARROWER'S WORK

Mark Harrower led the development of a range of approaches for interacting with animations; here, we consider three studies that Harrower directed. In the first, Harrower and colleagues (2000) developed the software EarthSystemsVisualizer, which was intended to facilitate undergraduate students' learning about global weather (Figure 24.8). Rather than simply showing "canned" animations, Harrower et al. wanted students to explore the underlying data using brushing and focusing tools. **Brushing** allows the user to highlight an arbitrary set of individual entities; for instance, the user might wish to see only animation frames associated with 6 AM. In contrast, **focusing** allows the user to focus on a particular subrange of data; for instance, you might wish to see an animation of only a 48-hour period, as opposed to an entire week. An important characteristic of EarthSystemsVisualizer was the attempt to represent different kinds of time (i.e., linear and cyclic) with corresponding linear and cyclic legends. *Linear time* refers to the notion of a general change over

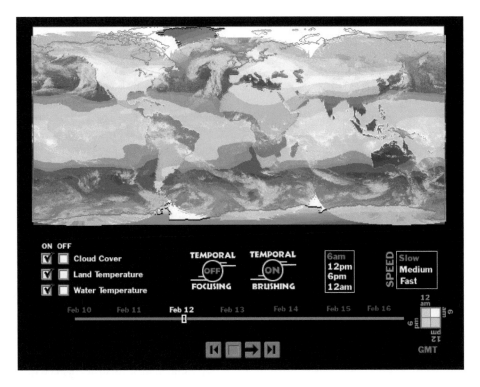

FIGURE 24.8   The interface for EarthSystemsVisualizer. (Courtesy of Mark Harrower.)

time, such as population growth for a city, whereas *cyclic time* refers to temporal changes that are repetitive in nature, such as diurnal changes in temperature. The bottom portion of Figure 24.8 shows examples of linear and cyclic legends (the date timeline and selected hours of the day, respectively). Important conclusions from Harrower et al.'s study were that students who understood the brushing and focusing tools performed the best and that those without access to the tools performed better than those who did not understand the tools, suggesting that instructors must take care in introducing such tools to students.

In a second study, Harrower (2001) promoted animation as a means for understanding satellite-based time-series data, stating:

> Current change-detection techniques are insufficient for the task of representing the complex behaviors and motions of geographic processes because they emphasize the outcomes of change rather than depict the process of change itself. Cartographic animation of satellite data is proposed as a means of visually summarizing the complex behaviors of geographic entities. (From Harrower 2001, 30; courtesy of *Cartographic Perspectives* (North American Cartographic Information Society, publisher) and Mark Harrower.)

To illustrate the potential of animation for satellite-based imagery, Harrower developed a prototype system known as VoxelViewer. Two key elements of VoxelViewer were tools that allowed a temporal resolution ranging from 10 to 80 days and a spatial resolution (pixel size) ranging from 10 to 250 km. Such tools were useful for reducing

spatial heterogeneity and short-term temporal fluctuations, ultimately producing smoother animations that captured general trends in the data. In an intriguing application, Harrower described how such tools can be used to examine the behavior of the Sahel, a semi-arid region in northern Africa. Using a spatial resolution of 10 km and a temporal resolution of 40 days, Harrower stated:

> The northern and southern boundaries do not move in unison. Rather there is a temporal lag in which the southern boundary moves northward first (since the monsoon rains arrive from the south) compressing the region before the northern boundary expands into the desert. The southern boundary is also the first to retreat southward with the onset of the dry season. (From Harrower 2001, 38; courtesy of *Cartographic Perspectives* (North American Cartographic Information Society, publisher) and Mark Harrower.)

Harrower noted that such a phenomenon could not be seen at other resolutions and would be missed if the animation were not used.

In a third study, Harrower (2002) developed the notion of **visual benchmarks** or reference points with which other frames of an animation can be compared. *Fixed* visual benchmarks remain constant from frame to frame; for example, Figure 24.9 illustrates fixed visual benchmarks for a proportional circle map—yellow and red circles represent, respectively, the minimum and maximum numbers of new AIDS cases for each country over the time period of the animation (1993–2000). *Dynamic* visual benchmarks are more complicated because they change from frame to frame; for example, a "ghost image" of a previous frame

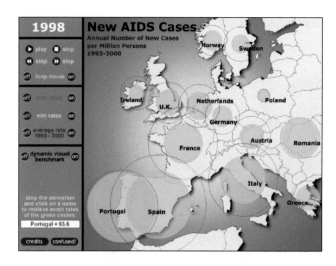

FIGURE 24.9 An illustration of Harrower's (2002) use of visual benchmarks (the yellow and red circles). (Courtesy of Mark Harrower.)

might appear superimposed on the current frame. Harrower implemented visual benchmarks for both proportional symbol and isarithmic maps, finding that benchmarks were more effective on proportional symbol maps. Although Harrower found that benchmarks were not as effective as he had hoped, he noted that their effectiveness might be enhanced through further use and training.

24.7.2 CoronaViz

Hanan Samet and colleagues (2020) developed CoronaViz (https://covidviz.umiacs.io/) to enable the dynamic visualization of a broad range of data related to the COVID-19 pandemic. Both unstandardized and standardized COVID-19 data can be visualized with CoronaViz. Unstandardized

data include raw totals for four attributes: confirmed cases, deaths, recoveries, and active cases. Standardized data include three attributes: the incidence rate (number of cases per 100,000 population), the mortality rate (the percentage of confirmed cases that led to death), and the recovery rate (the percentage of confirmed cases that have recovered).[8] Proportional circles are utilized to symbolize all seven attributes. In order to simultaneously display any subset of the attributes, hollow concentric circles are utilized, as shown in Figure 24.10 (Samet et al. term these *geo-circles*). Looking at the Control Panel on the right-hand side of the figure (note that the View option has been selected), we can see that the unstandardized and standardized data are symbolized with dashed and solid circles, respectively. Related attributes are shown in the same color (e.g., confirmed cases and incidence rate are both shown in black). Any combination of attributes can be selected, although, as we see in this figure, the resulting visualizations may be difficult to interpret if the number of attributes selected is large. Note also that the use of red and green circles on the same display may be problematic for the color deficient, as discussed in Section 15.4.3.

Since the ranges of each attribute may vary considerably from one another, and the unstandardized and standardized data are in different units, care must be taken in comparing circles for different attributes with one another. Care also must be taken in comparing circles of different attributes, as the scaling factor for each attribute can be adjusted by the user. By default, radii for circles are scaled as a function of the attribute values. The slider bar associated with each data attribute sets the circle radii. Using this default, our initial map for Figure 24.10 produced very large circles for the confirmed cases attribute, with the one for Brazil extending beyond the map boundary. Therefore, using the radii scaling sliders on the right-hand of the Control Panel,

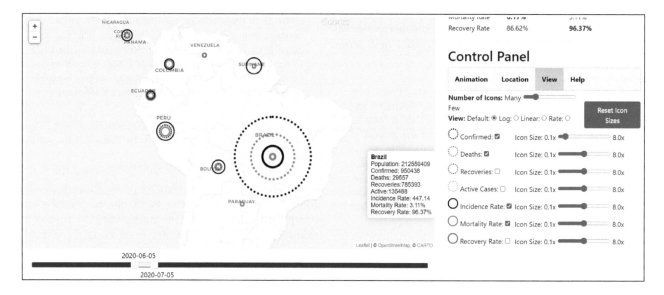

FIGURE 24.10 An illustration of the COVID-19 attributes that can be animated in CoronaViz. (After Samet, H., Han, Y., Kastner, J., and Wei, H. (2020) "Using animation to visualize spatiotemporal varying COVID-19 data." In 1st ACM SIGSPATIAL International Workshop on Modeling and Understanding the spread of COVID-19, Seattle, WA, pp. 53–62; courtesy of Hanan Samet.)

we chose to reduce the sizes of the confirmed cases circles. Also, note that since circles are scaled as a function of radii/attribute value relations (rather than area/attribute value relations), care should be taken in comparing the areas of circles. Rather than attempting to compare the magnitudes of areas, it is more appropriate to simply note the ordinal relations between circle sizes.

The underlying data for spatial entities can be found through a *hover* (or mouse over) operation, as is shown for Brazil in Figure 24.10. A highlighted country is indicated to the user through thicker lines used to draw a circle's perimeter. Accordingly, we see that the circles for Brazil have been drawn with a wider line, and we see the specific data values for each of the seven attributes displayed, along with the total population in the pop-up window. The temporal legend at the bottom of the display indicates that these values represent the total for the period from June 5, 2020 to July 5, 2020.

The flexibility of the animation approaches in CoronaViz is illustrated by the frame from a *Window* option shown in Figure 24.11. Rather than view an animation of the data for individual days, the Window option (in the Animation Control Panel tab) allows a user to combine the data for as many days as desired. For Figure 24.11, we chose to create a monthly window of thirty days for the number of deaths attribute, as an animation of the daily data for number of deaths produced an erratic animation with the circles rapidly changing in size. We chose to display the data for the number of deaths as a daily rate by selecting Daily Rate in the lower right of the Animation Control Panel. When viewing the animation, the changes in circle size were most obvious for Brazil, with a circle initially appearing there around the end of February 2020 (our dates refer to the middle of the monthly window), an initial high peak in

deaths from roughly June 15 to September 1, 2020 (with approximately 1,000 deaths per day), a drop in deaths to about 450 per day in late October 2020 and then another major increase in deaths, with a peak of about 2,900 deaths per day in early April 2021. Finally, deaths decreased to about 100 per day by mid-May 2021.

Note in the upper right of Figure 24.11 that it is possible to contrast the data for any two locations termed the Baseline Location and Focus Location. The baseline location is selected in the Location Control Panel, and the focus location is selected by simply clicking on that location. Another characteristic of CoronaViz is the automatic generalization of the COVID-19 data as the user zooms in or out. This may lead to some awkward descriptions of spatial entities provided by CoronaViz when hovering over a location, but it avoids the extreme overlap in circles that would result if generalization were not implemented. We encourage you to experiment with animations of the COVID-19 data; in their paper, Samet et al. provide a number of useful links to videos that can assist you in becoming familiar with the application.

24.8 DOES ANIMATION WORK?

Those developing animations are often enamored of their potential, but we (and others) often find the resulting animations difficult to understand. Is it possible that the capability of animation has been overhyped? Such thoughts were shared by Barbara Tversky and colleagues (2002), who undertook a meta-analysis of animation efforts that did not involve geospatial information. Although researchers suggested that the animations Tversky and colleagues evaluated were more effective than their static counterparts, Tversky et al. found that the animations often contained

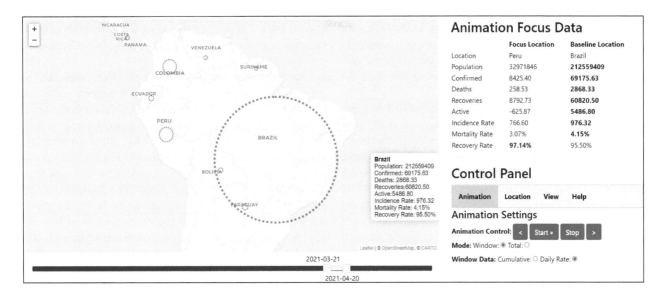

FIGURE 24.11   A frame from an animation of the number of deaths due to COVID-19 in a portion of South America illustrating the window option in CoronaViz. (After Samet, H., Han, Y., Kastner, J., and Wei, H. (2020) "Using animation to visualize spatiotemporal varying COVID-19 data." In 1st ACM SIGSPATIAL International Workshop on Modeling and Understanding the spread of COVID-19, Seattle, WA, pp. 53–62; courtesy of Hanan Samet.)

microsteps that were not available in the static displays, and thus they argued that a fair comparison was not possible. Tversky et al. also noted that although interactivity clearly has the potential to enhance an animation, studies had not yet been done that fully analyzed the effects of such interactivity relative to static maps. Finally, Tversky et al. made the following observation:

> ... Animations must be slow and clear enough for observers to perceive movements, changes, and their timing, and to understand the changes in relations between the parts and the sequence of events. This means that animations should lean toward the schematic and away from the realistic, an inclination that does not come naturally to many programmers, who delight in graphic richness and realism. (From p. 258 of Tversky, B., Morrison, J. B., and Betrancourt, M. (2002) "Animation: Can it facilitate?" *International Journal of Human–Computer Studies* 57:247–262; https://www.sciencedirect.com/science/article/abs/pii/S1071581902910177.)

These seem to be useful thoughts for those of us considering developing animations for others.

Experimental studies of the effectiveness of *map animations* have produced mixed results. A study often used to illustrate the benefits of map animation is the work of Koussoulakou and Kraak (1992), who found that although the percentage of correct answers did not differ, animated maps were processed significantly faster than static maps when depicted as a small multiple. Koussoulakou and Kraak stressed that users might have done even better if they could have interacted with the animated maps. A study by Patton and Cammack (1996) also revealed support for animation on sequenced choropleth maps (see Chapter 25). In contrast, studies by Slocum and Egbert (1993), Johnson and Nelson (1998), and Cutler (1998) found little difference between animated and static maps. For instance, in comparing an animated map with a small multiple, Cutler found that the animated version of a shaded isarithmic map resulted in a significantly lower percentage of correct answers and a slightly slower processing time than the small multiple.

Amy Griffin and colleagues (2006) responded to some of Tversky et al.'s concerns by conducting a carefully controlled study of animation vs. small multiples for the detection of spacetime clusters. Figure 24.12 is a small multiple illustrating an example of a spacetime cluster; these abstract hexagonal displays were used to control for the effect that the structure of a real geographic region might have on the results. Griffin et al. found that animated maps did enable "users to more often correctly identify whether a particular type of pattern was present than did the static small-multiple representation" (749). An interesting finding of their study was that for animated spacetime clusters, males were significantly more likely to find patterns than were females. Griffin et al. suggested that this difference might be a function of the enhanced "visual attention skills" of boys, resulting from their more common use of video games.

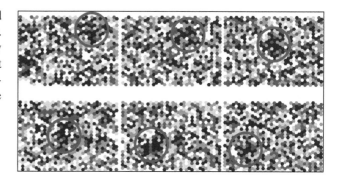

FIGURE 24.12 A small multiple illustrating an example of a spacetime cluster used in Griffin et al.'s experiment. (From "A comparison of animated maps with static small-multiple maps for visually identifying space-time clusters," by A. L. Griffin, A. M. MacEachren, F. Hardisty, et al. (2006) *Annals of the Association of American Geographers*, copyright © The Association of American Geographers, reprinted by permission of Taylor & Francis Ltd., http://www.tandfonline.com on behalf of The Association of American Geographers, www.aag.org.)

More recently, Sara Fabrikant and colleagues (2008) and Amy Lobben (2008) argued that it is not appropriate to simply ask whether animated maps or small multiples are more effective; we should instead consider a range of factors, including the properties of the data to be mapped, the symbology to be utilized, and the types of information that we wish to extract from maps in determining whether animation, small multiples, or some other mapping technique is appropriate. They also noted that in making a comparison between animation and small multiples, it is not necessary to make the two approaches informationally equivalent, as different kinds of information can be extracted from each form of display. Rather than seeing whether animation or small multiples are more effective, we should understand *how* readers utilize these displays and *why* they are effective for certain tasks. Answering such questions should help us make improvements in the design of both animation and small multiples.

One technique that Fabrikant et al. (2008) promoted for examining small multiples and animation is an *eye-movement* data-collection method. The eye-movement method enables the cartographer to determine where a map reader is looking (and for how long) while viewing either an animated map or a small multiple. For example, Figure 24.13 shows an individual map reader's eye-movement patterns for a small multiple for two different tasks. The graduated circles show where the reader's eye fixated (the larger the circle, the longer the fixation), and the connecting lines between circles represent eye movements between fixations. The task for Figure 24.13A was "to gain an overall impression of the [small multiple display] and verbally describe the patterns that are discovered during its visual exploration" (p. 205). In contrast, the task for Figure 24.13B was to compare the two small multiple maps for May and August. Obviously, these tasks led to dramatically different eye movements.

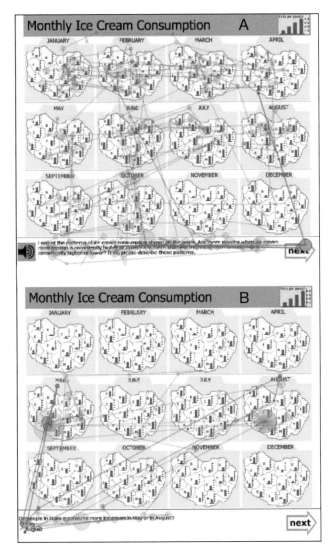

FIGURE 24.13 An individual map reader's eye-movement patterns for a small multiple for two different tasks. Graduated circles show where the reader's eye was fixated (the larger the circle, the longer the fixation), and connecting lines represent eye movements between the fixations. (From "Novel method to measure inference affordance in static small-multiple map displays representing dynamic processes," by S. I. Fabrikant, S. Rebich-Hespanha, N. Andrienko, et al., *The Cartographic Journal*, copyright © The British Cartographic Society 2008, reprinted by permission of Taylor & Francis Ltd., http://www.tandfonline.com on behalf of The British Cartographic Society.)

Several cartographers who have studied map animations have pointed out the difficulty of interpreting them, noting that information in an animation can be missed, especially when the animation is complex (e.g., when changes are happening at five or six different places on the map). In the psychological literature, this failure to detect changes in the visual field is known as **change blindness**.[9] In one famous experiment (Simons and Chabris 1999), participants were shown a video of two basketball teams, one wearing white shirts and the other wearing black shirts, and were asked to count how many times the ball was passed between members of one team. While the ball was being passed, a person in a gorilla suit walked among the players, stopped to face the camera and thump its chest, and then walked away. Only about 50 percent of the participants even noticed that the gorilla had appeared! Although we don't have gorillas walking across our maps, this humorous example illustrates the likelihood that we may miss changes in one portion of a map display while attending to others.

Carolyn Fish and colleagues (2011) conducted a change blindness user study in cartography, focusing on choropleth maps. They examined three issues: (1) how well map readers are able to detect the change, (2) the method of *tweening* used between a starting frame and an ending frame, and (3) users' confidence in their change detection abilities compared to their accuracy in detecting change. Users had difficulty detecting the full nature of change—they generally could detect that a change had taken place, but they had difficulty specifying the nature of that change. Maybe, more importantly, they overestimated their ability to detect the change, a phenomenon known as *change blindness blindness* (i.e., we are blind to our own change blindness). Three approaches were utilized to test tweening: (1) a gradual transition between a starting frame and an ending frame over a two-second period; (2) a one-second delay prior to a one-second transition period between the started and ending frame; and (3) no tweening, with an abrupt change from the starting frame to the ending frame after two seconds. The results were similar for the two tweening approaches, and these were clearly more effective than no tweening.

Clearly, more research is necessary if we are to determine whether an animation is truly effective. We suspect that the effectiveness of animation will be a function of numerous factors, including the type of symbology used (e.g., depicting temporal change on a proportional symbol map seems easier than depicting temporal change on a choropleth map); how the animation is viewed (a "movie" in which the user has no control is obviously quite different from an animation in which the user can move forward, backward, and jump to different frames); whether the user has expertise in the domain being mapped (understanding the animated map of changing CO_2 in the interactive video map may be easier for a scientist who fully understands the sources and sinks for CO_2); and experience with animation (does one who uses animation on a regular basis become more adept at using it?).

24.9 GUIDELINES FOR DESIGNING YOUR OWN ANIMATIONS

The following are some suggested guidelines that we hope will assist you in developing your own animations.

1. Don't create an animation just because it is lively. Just as with static maps, the key is whether users will understand the animation. Remember that there are several alternatives to animation,

including small multiples, and the space-time cube (discussed in the next section). Oftentimes, a spatiotemporal distribution can be best understood by using several visualization approaches.

2. Consider utilizing the congruence principle.

As we have stressed throughout this book, the symbology that you select should be a function of the data that you are mapping. In the case of animation, you also want to consider the congruence principle (whether the content and format of the graphic correspond to the content and format of the concepts to be conveyed).

3. Remember that some animations may be difficult to interpret.

If changes are happening rapidly in multiple places on an animated map, then it is likely that readers will have difficulty interpreting the animation due to change blindness. Consider testing your animation with a small sample of viewers to make certain that they are comfortable viewing the animation and that they are deriving useful information from it.

4. Shorter animations likely will be more effective.

Animations should be no more than about 30–60 seconds (Muehlenhaus 2014, 181), although longer animations may be appropriate when a voice-over is used to direct the reader's attention to key elements.

5. Carefully consider the pace of your animation.

A common problem with animations is that they move too quickly for the user to interpret the information presented. On the other hand, for some animations, a rapid pace may be required. Determine an appropriate pace through some informal experimentation with potential users. Provide controls that enable the user to stop and restart the animation, and adjust the pace if so desired.

6. If you are creating a temporal animation, consider the nature of time (is it linear or cyclic?).

For a linear time, a bar-like temporal legend is appropriate, while for cyclic time, a clock-like legend makes sense. Although textual temporal legends are common, the time to interpret them may detract from the user's ability to follow the animation.

7. Try to minimize the effort that users will require to move between the map and the temporal legend.

One alternative would be to allow the temporal legend to be moved so that it is close to the area that will be focused on in the animation. For some novel approaches for designing temporal legends, see Midtbø (2001) and Midtbø et al. (2007).

8. Consider using sound to enhance your animation.

Sound can be used in a variety of ways. Perhaps most obvious is a voice-over to direct the reader to key elements of an animation that you wish viewers to focus on. A second usage is when the phenomenon being mapped can logically be associated with a sound (e.g., the noise associated with a nuclear weapon). Another potential use is to enhance the interpretation of a temporal legend; for example, imagine hearing key dates spoken as you watch an animation of population growth. Such an approach should be effective because the visual and aural channels work independently of one another.

24.10 USING 3-D SPACE TO DISPLAY TEMPORAL DATA

The bulk of the animations we have described show geospatial changes over time by varying images in 2-D space. A potential disadvantage of this approach is that at any instant, we can see only one point in time. An alternative approach is to use the third dimension (the z-axis) to show the temporal changes, creating a **space-time cube**. In theory, the space-time cube should allow us to see not only spatial patterns for individual moments in time but also changes over time. Although the space-time cube is not technically a form of animation (i.e., it requires a great deal of interaction with the display), we include it in this chapter because it provides a distinct alternative to animation when wishing to examine changes over time.

Torsten Hägerstrand (1970) is generally credited with developing the idea of a space-time cube, but computer technology was not sufficiently developed at that time to create interactive visualizations that could fully utilize the concept. It took approximately 20 years for geographers and cartographers to truly begin to take advantage of the space-time cube concept. As an early example of its use, Figure 24.14 portrays a space-time cube that Mei-Po Kwan (2000) created illustrating the daily activity paths (the z-axis is the time of day in this case) of African Americans (in purple) and Asian Americans (in green) in Portland, Oregon.[10] Although somewhat difficult to see in this static image, the space-time paths of Asian Americans are more spatially scattered than those for African Americans, which appeared to concentrate in the eastern portion of the study area that Kwan examined—the difficulty of seeing this difference illustrates the need to interact with the space-time cube, which Kwan was able to do.[11]

As a more recent example of the space-time cube, consider Figure 24.15, in which Menno-Jan Kraak (2014) uses the cube to display the time component of Minard's classic map of Napolean's Russian Campaign of 1812 (we introduced Minard's map in Chapter 21). On the right is Kraak's adaptation of Minard's map, and on the left, the path of the campaign is shown in the space-time cube. Kraak argues that Minard's 2-D map is useful for showing *where* things occurred, but it fails to show key *temporal*

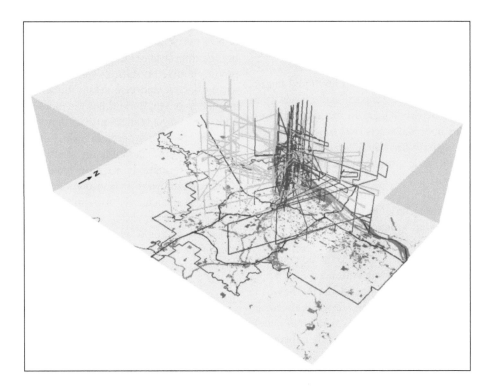

FIGURE 24.14 A space-time cube depicting the daily activity paths of African Americans (in purple) and Asian Americans (in green) in Portland, Oregon. (From Kwan, M.-P. (2009) "Space-time paths." In *Manual of Geographic Information Systems*, ed. M. Madden, pp. 427–442. Bethesda, MD: American Society for Photogrammetry & Remote Sensing; reprinted with permission from the American Society for Photogrammetry & Remote Sensing, Bethesda, Maryland, asprs.org.)

aspects. In this case, the *space-time path* in the cube clearly shows that there was a period in Moskva (Moscow) during which Napolean's army did not move (the halfway point along the line's length), as the orange path is vertical during this period, indicating that time changed, but space did not change.[12]

As our emphasis in this chapter is on animation, we do not have space to review numerous applications of the space-time cube. A review of early work with the space-time cube and a discussion of its potential can be found in Kraak (2008). For a more recent summary of work with the space-time cube, see Kveladze et al. (2015, 2019), who also describe experiments to (1) examine the usability of the cube in a geovisual analytics environment; and (2) test the effectiveness of various visual variables within the cube.[13] We also touch on the space-time cube in Section 21.8, where we consider its use in geovisual analytics and flow mapping.

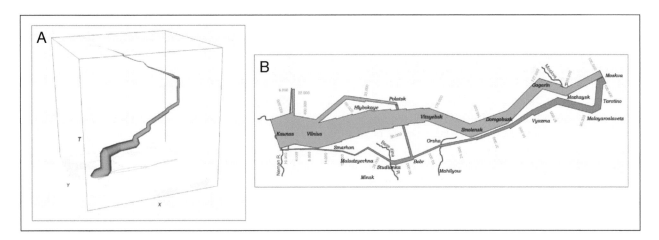

FIGURE 24.15 Napolean's Russian campaign shown in (A) a space-time cube and (B) an adaptation of Minard's classic flow map. (After p. 113 of Kraak, M.-J. (2014) *Mapping Time: Illustrated by Minard's Map of Napoleon's Russian Campaign of 1812*. Redlands, CA: ESRI Press; courtesy of Menno-Jan Kraak.)

24.11 SUMMARY

In this chapter, we examined several matters pertaining to map animation. We saw that the visual variables utilized for static maps can also be used on animated maps; for example, we might utilize proportional circles (the visual variable size) to depict changes in the magnitude of COVID-19 infections in census tracts over time. Animated maps, however, also utilize other visual variables, including duration, rate of change, order, display date, frequency, and synchronization.

We examined several *temporal* animations for a variety of thematic mapping approaches. One of the keys to developing an effective temporal animation is to make use of the congruence principle, which states that the content and format of the graphic should correspond to the content and format of the concepts to be conveyed. Animated flow maps are particularly illustrative of this principle as the flows of a phenomenon over time can be represented by movement over time (e.g., airplanes in Peterson and Wendel's animated atlas of air traffic). We saw, however, that other thematic mapping methods can also take advantage of this principle if they are carefully designed. For instance, increases and decreases in circle size can be seen as "apparent motion" and thus be seen as increases and decreases in the phenomenon being mapped.

Interestingly, we found that some design guidelines for static maps may not apply to animated maps. For instance, Harrower found that unclassed choropleth maps were more appropriate for animated maps, contradicting cartographers' traditional recommendation for using *classed* choropleth maps. As another example, specialized legends may be required for animations, as we cannot expect viewers of animations to easily view both the animation of a spatial pattern and an associated traditional legend simultaneously.

We examined several *nontemporal* animations. Particularly intuitive is the *Wind Map*, in which the speed and direction of the wind are represented by animated streamlets—this is another clear application of the congruence principle. Fly-overs are a common form of nontemporal animation. Although fly-overs attract the viewer's attention and are considered "cool," we saw that there are a number of potential problems with them including oblique perspective, their focus on realism and thus information overload, visual occlusion, and user disorientation.

We argued that the most effective animations are those which permit users to start and stop them and control their pace. Furthermore, for some applications, it may be desirable to provide users with a broader set of interactive tools so that they can explore the underlying data in greater depth. We illustrated the notion of interactivity with some early work by Mark Harrower and the recently developed CoronaViz application. We will delve into the use of interactivity in animation in greater depth in the next chapter, where we discuss the software MapTime.

In viewing animations for this chapter, we found many of them difficult to understand, raising the question of whether an animation is truly effective. One alternative to animation is the small multiple; unfortunately, studies comparing the effectiveness of animation and small multiple have produced mixed results, as sometimes animation is deemed more effective and vice versa. Rather than seeing whether animation or small multiples are more effective, we should understand *how* readers utilize these techniques and *why* they are effective for certain tasks.

Since we suspect that many readers will want to develop their own animations, we have provided several guidelines for developing your own animations. Key elements of these guidelines include: (1) utilizing the congruence principle, if possible; (2) providing sufficient interactivity (at a minimum, users should be able to start, stop, and change the pace of the animation); (3) carefully linking the legend with the animated spatial distribution (e.g., possibly using sound to assist in interpreting the legend); and (4) testing your animation with a small sample of viewers.

The bulk of animations that we discussed are displayed in 2-D space and thus do not make use of the third dimension. An alternative to animation is the space-time cube in which the *z*-axis of 3-D space is used to depict time. Although space-time cubes enable the detection of information that cannot be seen or is difficult to see in an animation, a downside is that symbols for different locations block one another, and thus user interaction generally is required to interpret the space-time cube.

24.12 STUDY QUESTIONS

1. Using "The Sex Ratio" animation (http://www.edstephan.org/Animation/sexratios.html), answer the following questions: (A) Which visual variable(s) are utilized? (B) Explain whether you consider the color scheme effective. (C) Is the number of classes appropriate for the animation?

2. Go to https://www.youtube.com/watch?v=mWW_A52Vo9M and run the UPS Flight Animation. (A) Which visual variable(s) are utilized? (B) Discuss the key time periods when UPS planes arrive in and leave Louisville, KY. (C) Discuss the nature of the temporal legend and indicate any problems that you had in interpreting the map when utilizing this legend. How might you improve upon this legend? (Note: The online version of the software does not allow you to change the speed of the animation.)

3. Using Max Galka's Two Centuries of U.S. Immigration (http://metrocosm.com/animated-immigration-map/): (A) Which visual variable(s) are utilized? (B) Discuss whether the animation seems to support the summary text provided below the animation. (C) Indicate any problems that you had in interpreting the animation. (D) Suggest a potential enhancement to the animation.

4. Using the proportional circle animation of job gains and losses (https://tipstrategies.com/geography-of-jobs/), identify major periods of job losses and gains

for the U.S. as a whole, and identify the historical event(s) associated with these losses and gains.

5. (A) Using the interactive video map at http://co2. digitalcartography.org/, enlarge and center the map on North America, and describe the character of CO_2 that you see above that continent over the course of the year. Generally, when does the level of CO_2 peak and when is it lowest? (B) Can you hypothesize any reasons for the timing and location of high and low values of CO_2 in this case? (C) Discuss any limitations that you see in this interactive video map.

6. Access the bubble plot at https://www.gapminder. org/tools/#$chart-type=bubbles&url=v1, and change "Life Expectancy" to "Child Mortality." (A) Run the animation and describe, in general, the situation for 1800, 1900, 1950, and 2018. (Remember that you can start and stop the animation whenever desired.) (B) Select just Japan and the United States and run the animation. Describe how the two countries change over time.

7. (A) Using the voter turnout animation (https://dsl. richmond.edu/voting/voterturnout.html), discuss why it is important to be able to interact with map animations. (B) What other forms of interactivity and data would you like to see available for the voter turnout application? Explain your choices.

8. Use the Web to find an animation that intrigues you, and then answer the following questions: (A) Which visual variable(s) are utilized? (B) Summarize what the animation tells you about the phenomenon being mapped? (C) Overall, how successful is the animation? Can you suggest any improvements to the animation? (Note: Some examples of animations can be found at https://infogram.com/blog/map-examples-from-the-web/.)

NOTES

1 These animations can also be viewed at https://www.youtube.com/user/theflightatlas.

2 For other related animations, see https://vimeo.com/33712288 and https://vimeo.com/66477874, both of which are referenced in Wood et al. (2014).

3 This is part of the Voting America website (https://dsl.richmond.edu/voting/).

4 In the study, regions were defined as either an entire state or a combination of counties within a metropolitan area.

5 The TIP website indicates red is used for losses, but on our monitors, this appeared more like an orange.

6 Personal communication, Mark Harrower, 2007.

7 We selected these visual variables based on statements made by Viégas and Wattenberg at https://vimeo.com/48625144.

8 Data for CoronaViz were acquired from the Johns Hopkins University COVID-19 Resource Center (https://coronavirus.jhu.edu).

9 For a broader overview of cognitive issues associated with animation, see Harrower (2007a).

10 Kwan termed this a *space-time aquarium*.

11 For a summary of Kwan's work with space-time cubes, see Kwan (2009).

12 The example of the space-time cube shown here is part of Kraak's (2014) larger work *Mapping Time: Illustrated by Minard's Map of Napoleon's Russian Campaign of 1812.* Here you will find a wide range of approaches for mapping time including animation, small multiples, and the space-time cube. For an animation illustrating the nature of the space-time cube involving Napolean's Russian Campaign, see https://www.youtube.com/watch?v=laXh2cgE2g0.

13 To view examples of the space-time cube in action, see https://www.youtube.com/watch?v=QyOO7DljKwI. One solution to the difficulty of interacting with the space-time cube in 3D space is to utilize an immersive virtual environment—for an example, see https://vimeo.com/370107834.

REFERENCES

Campbell, C. S., and Egbert, S. L. (1990) "Animated cartography: Thirty years of scratching the surface." *Cartographica 27*, no. 2:24–46.

Cutler, M. E. (1998) "The effects of prior knowledge on children's abilities to read static and animated maps." Unpublished MS thesis, University of South Carolina, Columbia, SC.

DiBiase, D., MacEachren, A. M., Krygier, J. B., et al. (1992) "Animation and the role of map design in scientific visualization." *Cartography and Geographic Information Systems 19*, no. 4:201–214, 265–266.

Dorling, D. (1992) "Stretching space and splicing time: From cartographic animation to interactive visualization." *Cartography and Geographic Information Systems 19*, no. 4:215–227, 267–270.

Fabrikant, S. I., Rebich-Hespanha, S., Andrienko, N., et al. (2008) "Novel method to measure inference affordance in static small-multiple map displays representing dynamic processes." *The Cartographic Journal 45*, no. 3:201–215.

Fish, C., Goldsberry, K. P., and Battersby, S. (2011) "Change blindness in animated choropleth maps: An empirical study." *Cartography and Geographic Information Science 38*, no. 4:350–362.

Gershon, N. (1992) "Visualization of fuzzy data using generalized animation." *Proceedings, Visualization '92*, Boston, MA, pp. 268–273.

Griffin, A. L., MacEachren, A. M., Hardisty, F., et al. (2006) "A comparison of animated maps with static small-multiple maps for visually identifying space-time clusters." *Annals of the Association of American Geographers 96*, no. 4: 740–753.

Hägerstrand, T. (1970) "What about people in regional science?" *Papers of the Regional Science Association 24*, no. 1:7–21.

Harris, R. Migliozzi, B., and Chokshi, N. (2020) "13,000 missing flights: The global consequences of the coronavirus."*The New York Times*, https://www.nytimes.com/interactive/2020/02/21/business/coronavirus-airline-travel.html?referringSource=articleShare.

Harrower, M. (2001) "Visualizing change: Using cartographic animation to explore remotely-sensed data." *Cartographic Perspectives*, no. 39:30–42.

Harrower, M. A. (2002) "*Visual benchmarks: Representing geographic change with map animation.*" Unpublished Ph.D. dissertation, The Pennsylvania State University, University Park, PA.

Harrower, M. (2007a) "The cognitive limits of animated maps." *Cartographica 42*, no. 4:349–357.

Harrower, M. (2007b) "Unclassed animated choropleth maps." *The Cartographic Journal 44*, no. 4:313–320.

Harrower, M., MacEachren, A., and Griffin, A. L. (2000) "Developing a geographic visualization tool to support earth science learning." *Cartography and Geographic Information Science 27*, no. 4:279–293.

Harrower, M., and Sheesley, B. (2005) "Moving beyond novelty: Creating effective 3-D fly-over maps." *Proceedings, 22nd International Cartographic Conference*, A Coruña, Spain, pp. 1–11.

Harrower, M., and Sheesley, B. (2007) "Utterly lost: Methods for reducing disorientation in 3-D fly-over maps." *Cartography and Geographic Information Science 34*, no. 1:19–29.

Ho, Q. V., Lundblad, P., Åström, T., et al. (2012) "A web-enabled visualization toolkit for geovisual analytics." *Information Visualization 11*, no. 1:22–42.

Jenny, B., Liem, J., Šavrič, B., et al. (2016) "Interactive video maps: A year in the life of Earth's CO_2." *Journal of Maps 12*, no. S1:36–42.

Johnson, H., and Nelson, E. S. (1998) "Using flow maps to visualize time-series data: Comparing the effectiveness of a paper map series, a computer map series, and animation." *Cartographic Perspectives* no. 30:47–64.

Kettunen, P., and Oksanen, J. (2019) "Motion of animated streamlets appears to surpass their graphical alterations in human visual detection of vector field maxima." *Cartography and Geographic Information Science 46*, no. 6:489–501.

Kostelnick, J. C., Land, J. D., and Juola, J. F. (2007) "Judgments of size change trends in static and animated graduated circle displays." *Cartographic Perspectives*, no. 57:41–55.

Koussoulakou, A., and Kraak, M. J. (1992) "Spatio-temporal maps and cartographic communication." *The Cartographic Journal 29*, no. 2:101–108.

Kraak, M.-J. (2008) "Geovisualization and time: New opportunities for the space-time cube." In *Geographic Visualization: Concepts, Tools and Applications*, ed. by M. Dodge, M. McDerby and M. Turner, pp. 293–306. Chichester, UK: Wiley.

Kraak, M.-J. (2014) *Mapping Time: Illustrated by Minard's map of Napoleon's Russian Campaign of 1812*. Redlands, CA: ESRI Press.

Kveladze, I., Kraak, M.-J., and van Elzakker, C. P. (2015) "The space-time cube as part of a GeoVisual analytics environment to support the understanding of movement data." *International Journal of Geographical Information Science 29*, no. 11:2001–2016.

Kveladze, I., Kraak, M.-J., and van Elzakker, C. P. J. M. (2019) "Cartographic design and the space-time cube." *The Cartographic Journal 56*, no. 1:73–90.

Kwan, M.-P. (2000) "Interactive geovisualization of activity-travel patterns using three-dimensional geographical information systems: A methodological exploration with a large data set." *Transportation Research Part C 8*:185–203.

Kwan, M.-P. (2009) "Space-time paths." In *Manual of Geographic Information Systems*, ed. by M. Madden, pp. 427–442. Bethesda, MD: American Society for Photogrammetry and Remote Sensing.

Lobben, A. (2008) "Influence of data properties on animated maps." *Annals of the Association of American Geographers 98*, no. 3:583–603.

MacEachren, A. M. (1995) *How Maps Work: Representation, Visualization, and Design*. New York: Guilford.

Midtbø, T. (2001) "Visualization of the temporal dimension in multimedia presentations of spatial phenomena." *Proceedings of the 8th Scandinavian Research Conference on Geographical Information Science*, Ås, Norway, pp. 213–224.

Midtbø, T., Clarke, K. C., and Fabrikant, S. I. (2007) "Human interaction with animated maps: The portrayal of the passage of time." *Proceedings of the 11th Scandinavian Research Conference on Geographical Information Science*, pp. 45–60.

Moellering, H. (1976) "The potential uses of a computer animated film in the analysis of geographical patterns of traffic crashes." *Accident Analysis & Prevention 8*:215–227.

Muehlenhaus, I. (2014) *Web Cartography: Map Design for Interactive and Mobile Devices*. Boca Raton, FL: CRC Press.

Patton, D. K., and Cammack, R. G. (1996) "An examination of the effects of task type and map complexity on sequenced and static choropleth maps." In *Cartographic Design: Theoretical and Practical Perspectives*, ed. by C. H. Wood and C. P. Keller, pp. 237–252. Chichester, England: Wiley.

Peterson, M. P. (1993) "Interactive cartographic animation." *Cartography and Geographic Information Systems 20*, no. 1:40–44.

Peterson, M. P. (1999) "Active legends for interactive cartographic animation." *International Journal of Geographical Information Science 13*, no. 4:375–383.

Samet, H., Han, Y., Kastner, J., et al. (2020) "Using animation to visualize spatio-temporal varying COVID-19 data." In *1st ACM SIGSPATIAL International Workshop on Modeling and Understanding the Spread of COVID-19*, Seattle, Washington, pp. 53–62.

Simons, D. J., and Chabris, C. F. (1999) "Gorillas in our midst: Sustained inattentional blindness for dynamic events." *Perception 28*:1059–1074.

Slocum, T. A., and Egbert, S. L. (1993) "Knowledge acquisition from choropleth maps." *Cartography and Geographic Information Systems 20*, no. 2:83–95.

Thrower, N. J. W. (1959) "Animated cartography." *The Professional Geographer 11*, no. 6:9–12.

Thrower, N. J. W. (1961) "Animated cartography in the United States." *International Yearbook of Cartography*, pp. 20–29.

Tobler, W. R. (1970) "A computer movie simulating urban growth in the Detroit region." *Economic Geography 46*, no. 2 (Supplement):234–240.

Traun, C., and Mayrhofer, C. (2018) "Complexity reduction in choropleth map animations by autocorrelation weighted generalization of time-series data." *Cartography and Geographic Information Science 45*, no. 3:221–237.

Traun, C., Schreyer, M. L., and Wallentin, G. (2021) "Empirical insights from a study on outlier preserving value generalization in animated choropleth maps." *ISPRS International Journal of Geo-Information 10*, no. 4:1–21.

Treves, R., and Skarlatidou, A. (2018) "What path and how fast? The effect of flight time and path on user spatial understanding in map tour animations." *Cartography and Geographic Information Science 45*, no. 2:128–139.

Tversky, B., Morrison, J. B., and Betrancourt, M. (2002) "Animation: Can it facilitate?" *International Journal of Human–Computer Studies 57*: 247–262.

Wood, J., Beecham, R., and Dykes, J. (2014) "Moving beyond sequential design: Reflections on a rich multi-channel approach to data visualization." *IEEE Transactions on Visualization and Computer Graphics 20*, no. 12:2171–2180.

25 Data Exploration

25.1 INTRODUCTION

In using the communication model approach (see Chapter 1), cartographers traditionally produced a single "best" map that presumably communicated a particular message to map users. With the development of interactive graphical approaches, cartographers began to realize that developing a single "best" map was not necessary nor even desirable. Rather it would be more useful for the cartographer and map users to be able to explore data using a variety of interactive tools and the visual information processing capability of the human brain, a process known as **data exploration**. As the size of data sets has expanded, however, we have begun to realize that there are limitations of the data exploration approach. This has led to the development of **geovisual analytics**, which combines the notion of data exploration with the computational power of computers to assist humans' cognitive abilities in making sense of large geospatial data sets. Data exploration is the focus of the present chapter, and geovisual analytics is covered in Chapter 26.

In this chapter, we summarize the basic goals of data exploration and introduce a broad range of methods that are used to explore geographic data (e.g., varying the general form of symbology used, linking maps with other forms of display, highlighting portions of a data set, and toggling individual themes on and off). We then introduce several examples of data exploration software/applications. For some examples, the software is no longer available, but we discuss these to provide a historical sense of the evolution of data exploration. Space does not permit us to discuss the full range of data exploration tools that have been developed. In addition to those discussed here, others that you may wish to examine include DataShine (https:/datashine.org.uk; O'Brien and Chesire (2016)); HazMatMapper (https://geography.wisc.edu/hazardouswaste/map/, Nost et al. (2017)); and Spatial Analysis using ArcGIS Engine and R (SAAR, https://hymokoo.weebly.com/applications.html, Koo et al. (2018)).

25.2 LEARNING OBJECTIVES

- Define data exploration and contrast it with geovisual analytics.
- Summarize four broad goals for exploring data and give an example of each.
- List at least six of the methods for exploring data and give an example of each method.
- Define sequencing and explain why it might be useful when exploring data.
- Explain why small multiples and change maps might be helpful in interpreting map animations.
- Explain how interactive data exploration software can ease the interpretation of unclassed choropleth maps.
- Define a boxmap data classification and explain how it can detect outliers in data.
- Define a linked micromap plot and discuss how it enables us to focus on local pattern perception.
- Interpret a parallel coordinate plot in a data exploration environment.

25.3 GOALS OF DATA EXPLORATION

In Chapter 1, we indicated that, like the larger notion of geovisualization, data exploration is most frequently a private activity in which unknowns are revealed in a highly interactive environment. What are these *unknowns*, or, more specifically, what are the goals of data exploration? The following are four broad goals for data exploration; in each case, we provide an example using attributes associated with census tracts in Los Angeles, California:

1. Identify the spatial pattern for a single attribute at one point in time. (Example: What was the spatial pattern of housing costs in 2020?)
2. Compare spatial patterns for two or more attributes at one point in time. (Example: How did the spatial pattern for housing costs compare with the spatial pattern for the percentage of Hispanics in the population in 2020?)
3. Identify how the spatial pattern for a single attribute changes over time. (Example: How has the spatial pattern of housing costs changed with each decennial census?)
4. Compare spatial patterns for two or more attributes to see how they covary over time. (Example: How has the relationship between housing costs and the percentage of Hispanics changed over time?)

It can be argued that these goals are no different from those for static paper maps, and this is true. One difference, however, is that the character of spatial patterns might not become clear until we manipulate the map in a highly interactive graphics environment. For instance, if, for the first goal, we make a single choropleth map of housing costs, our visual interpretation might be a function of the color scheme—we might visualize a quite different pattern using a red–blue diverging scheme than with a blue sequential scheme. In a data exploration environment, we can change

DOI: 10.1201/9781003150527-28

the color scheme instantaneously and thus learn something about the spatial pattern of housing costs that was "hidden" in a single-map solution.

Another difference between a paper map and a data exploration environment is that the latter provides linkages to other views of the data in tables, graphs, and numerical summaries. For instance, when viewing a bivariate map of housing costs and the proportion of Hispanics in the population, you might also utilize a scatterplot depicting the relationship between the attributes. Examining this scatterplot might help you better understand the spatial pattern appearing on the maps of these attributes. Alternatively, you might notice a group of points that cluster together on the scatterplot and wonder where they occur on the map. To determine the geographic location of the clustered points, interactive techniques allow you to select the clustered points in the scatterplot, which causes these same points to be highlighted on the map. Clearly, this function cannot be accomplished using a static map.

25.4 METHODS OF DATA EXPLORATION

In this section, we summarize a broad range of methods characteristic of data exploration software. Obviously, individual software packages will not include all of the methods, as each piece of the software normally has specific (and thus limited) purposes. As you read about (and, we hope, use) data exploration software, you might find it useful to consider which of these methods have been implemented and how they are utilized. It should be noted that there is no widely agreed-on set of methods for exploring spatial data. The methods listed here were culled from the works of Buja et al. (1996), Roberts (2008), Edsall et al. (2009), and Roth (2012).

25.4.1 Manipulating Data

Methods for manipulating data include standardizing data (e.g., adjusting raw totals for enumeration units to account for the area over which the totals are distributed), transforming attributes (e.g., taking logarithms of data), and various methods of data classification. The first two of these methods can be handled in spreadsheet programs, but it is arguably easier for users if they are included within data exploration software.

25.4.2 Varying the Symbolization

One approach to varying the symbolization is to change the general form of symbolization. For instance, in Figure 4.10, we showed how different views of wheat harvested in Kansas could be obtained using four different methods of symbolization: choropleth, isopleth, proportional symbol, and dot. In a data exploration environment, such views can often be generated quickly, although certain symbolization methods (e.g., the dot map) are not trivial to create.

It is also possible to vary the precise manner in which the symbolization is implemented. For instance, if a choropleth

map is deemed appropriate, then numerous color schemes can be employed. As we saw in Chapter 15, alternative color schemes can lead to dramatically different-looking maps. The power of data exploration software is that not only can the overall color scheme be changed, but the detailed relation between color and data can also be modified dynamically. For instance, with the CommonGIS software (see Section 25.5.5), we can drag a break point separating two colors in a diverging scheme and have the result displayed instantaneously.

25.4.3 Manipulating the User's Viewpoint

Manipulating the user's viewpoint refers to modifying the "view" that users have of the map, with the data and method of symbolization held constant. Zooming and panning are perhaps the most common viewpoint manipulations. Such functions are critical, given the relatively small size of standard display screens and the large datasets that we sometimes wish to display. Sophisticated software will automatically adjust the level of generalization as a function of the degree of zoom (e.g., the names of census tracts might not be visible at a small scale but would appear at a larger scale).

Closely associated with zooming is **lensing**, in which only a portion of the display is enlarged, thus enabling the viewer to see detail in an area of interest while the surrounding area is distorted and reduced in importance. This is often accomplished using a *fisheye* approach in which an area of interest (the *focus*) is shown in detail, and the surrounding area (the *context*) is greatly distorted.[1] Daisuke Yamamoto and colleagues (2009) modified the basic focus+context approach by creating a *glue* area between the focus and context areas, thus eliminating distortion in both areas. For an illustration of their approach, see https://tk-www.elcom.nitech.ac.jp/demo/fisheye.html (select the option "Demo movie for the Focus+Glue+Context map with multiple Focuses and tools").[2]

The ability of users to change the viewpoint is, of course, crucial with 3-D maps, as high points on the surface can obscure lower points. Also important with 3-D maps is the means by which a user navigates through the landscape (e.g., using a mouse vs. using a joystick and the associated on-screen controls that are available) and how the user is provided a sense of orientation. Because these notions are particularly relevant to virtual environments, we also consider them in Chapter 28.

25.4.4 Multiple Map Views

Multiple map views refers to displaying more than one map image on the computer screen at a time. One form of multiple map view is the **small multiple**, in which maps are displayed for individual time steps or attributes. Thus, if we have data on housing costs for decennial censuses, we could include a map for each decennial census. Alternatively, for a particular decennial census, we might create separate maps for percent Hispanic, median income, housing costs, and percentage of students attending public schools. Another form of multiple map view is to show the same data set using different symbolization methods, as was discussed in Section 25.4.2. In general, multiple map views take advantage of our ability to compare things visually.

25.4.5 Linking Maps with Other Forms of Display

The ability to link maps with tables, graphs, and numerical summaries (known as **coordinated multiple views**) is a critical aspect of data exploration software. Users should be able to highlight data in tables, graphs, or maps and have the highlighted specifications carried through to the other forms of display (we will see this in Section 25.5.9). Numerical summaries provide a useful complement to various forms of graphical display. For a single attribute, numerical summaries often include measures of central tendency and dispersion, such as the mean and standard deviation. In Chapter 3, we argued that measures of spatial autocorrelation also could be useful in determining whether a spatial pattern might have resulted by chance. For multiple attributes, more complex numerical summaries would be appropriate (e.g., for two numeric attributes, the user might want to know the correlation coefficient).

25.4.6 Highlighting Portions of a Data Set

Fundamental to data exploration is the ability to highlight portions of a data set. For example, you might want to highlight census tracts that are less than 25 percent Hispanic, or you might wish to display the percentage of Hispanics only for tracts that voted for a particular political party. Brushing and focusing are two terms that are often used in association with highlighting. According to Roth (2012, 391–392), **brushing** consists of two steps: a selection step and a manipulation step (where highlighting or some other manipulation takes place). For example, a user might select census tracts that are adjacent to a river and highlight those tracts with a particular color. Roth notes that **focusing** has been defined in several ways: (1) as providing more detail, (2) as synonymous with filtering, and (3) as the secondary action combined with

brushing. For our purposes, we will equate focusing with filtering. For example, our selection of census tracts having less than 25 percent Hispanic would be a form of focusing.

Traditionally, highlighting has been accomplished using a bright saturated color such as a bright saturated red. Amy Griffin and Anthony Robinson (2015) explored the use of leader lines as an alternative highlighting approach. To illustrate, Figure 25.1 compares the use of color and leader lines on a coordinated multiple view consisting of a map and a parallel coordinate plot. Using map displays similar to Figure 25.1, Griffin and Robinson found that leader lines were just as effective as color, thus permitting color to be reserved for other thematic information.

25.4.7 Probing the Display

An obvious advantage of interactive displays is the ability to probe the display: to hover over (or point to) a location and discover the value of an attribute (or attributes) associated with that location. **Probing** avoids the difficulty that we may have in determining a value associated with a location by using a legend. For example, on the unclassed choropleth map shown in Figure 1.8B, it may be unclear which data value we should associate with a particular state, but with the appropriate software, we can simply hover over the state and get the precise value. It should be noted, however, that this capability does not solve the question of whether unclassed maps are appropriate, as being able to determine precise data values for locations does not necessarily mean that the perception of the overall spatial pattern will be enhanced.

25.4.8 Toggling Individual Themes On and Off

In GIS software, a variety of miscellaneous resources are often available as layers that can be toggled on or off. This *toggling* is obviously useful for exploring the relationships between various attributes. When a large number of layers are present, it may be essential to toggle certain layers off in order to see key relationships between attributes.

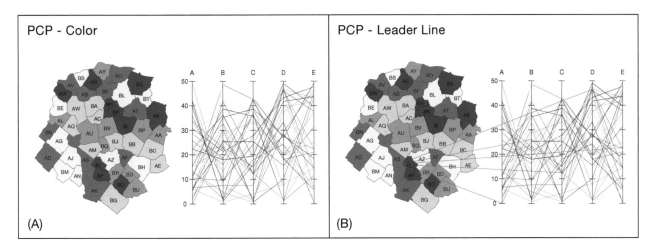

FIGURE 25.1 Comparison of two methods for highlighting: color (A) and leader lines (B). (Courtesy of Amy Griffin & Anthony Robinson.)

25.4.9 ANIMATION

Animation, which we discuss more fully in Chapter 24, is often an integral part of data exploration software. The key with data exploration software is that the user has considerably more control over an animation than is possible with canned animation "movies." With data exploration software, users might be able to stop an animation, redefine the data–symbol relationship, and rerun the animation. An example of this is the MapTime software discussed in Section 25.5.4.

25.4.10 ACCESS TO MISCELLANEOUS RESOURCES

Throughout this discussion, we have assumed that you have collected data that you wish to visualize. In practice, as you explore spatial data, particularly with either multiple attributes or spatiotemporal data, you will find that you need additional information. In some cases, this information will be in a format similar to what you have already acquired. For instance, in relating housing costs to percent of Hispanic population for census tracts in Los Angeles, you might want to examine public expenditures for education, another attribute for these census tracts. In other cases, you might wish to collect a different form of information. For instance, in studying housing costs, you might want to be able to click on a census tract and see what typical houses look like in that tract. Ideally, it would be desirable if you could readily import this sort of information directly into your data exploration software, possibly via the Web.

25.4.11 HOW SYMBOLS ARE ASSIGNED TO ATTRIBUTES

With multiple attributes (i.e., a multivariate map), an important decision often involves which attributes are assigned to which symbols. For instance, for the combined star and snowflake glyph symbol (Section 22.4.2.4), the appearance will depend on the assignment of attributes to the rays of the glyph. In Figure 25.2, the eight attributes on the left are arbitrarily assigned, whereas, on the right, the attributes have been ordered from smallest to largest beginning at the top and progressing clockwise. As another example, the order in which attributes are displayed on a parallel

coordinate plot will influence the look of the plot. Referring back to Figures 3.8 and 3.9, remember that for highly positively correlated attributes, line segments on the parallel coordinate plot tend to be parallel, whereas, for highly negatively correlated attributes, line segments tend to intersect one another in the middle of the graph.

25.4.12 AUTOMATIC MAP INTERPRETATION

In discussing data exploration thus far, we have focused on the ability of the user to interpret spatial patterns. In complex multiple-attribute situations or with spatiotemporal data, it might be useful to utilize methods that automatically interpret the data or at least assist the user in finding useful patterns in the data. This leads us to the notion of *data mining*, a relatively complex topic that we consider in the following chapter.

25.5   EXAMPLES OF DATA EXPLORATION

25.5.1 MOELLERING'S 3-D MAPPING SOFTWARE

Hal Moellering's (1980) 3-D mapping software was an early attempt to demonstrate the potential of data exploration. With Moellering's software, a user could explore a 3-D digital elevation model (DEM) in real-time, a process that Moellering termed "surface exploration." Because Moellering's software depended on expensive hardware, he developed a video (Moellering 1978) to demonstrate software capabilities. In addition to illustrating surface exploration, the video portrayed two animations of spatiotemporal data: the growth of the U.S. population from 1850 to 1970 and the diffusion of farm tractors in the United States from 1920 to 1970. These animations were created using attractive, colored 3-D prism maps.

Although Moellering's software suggested great potential for both data exploration and animation, for approximately ten years following his effort, there was little development in data exploration (at least by geographers), in part because many geographers lacked the necessary hardware and software. The lack of progress in animation, in particular, was lamented by Craig Campbell and Stephen Egbert (1990). Fortunately, considerable advancements have taken place since 1990, as this and other chapters in this book illustrate. Today it is possible to generate maps similar to Moellering's on a personal computer right on one's desktop.

25.5.2 EXPLOREMAP AND MAP SEQUENCING

Another early data exploration effort was ExploreMap, which focused on exploring univariate data normally displayed as a choropleth map (Egbert and Slocum 1992). An interesting feature of ExploreMap was a **sequencing** option, in which the map was built piece by piece rather than displaying the entire map at once (as on printed maps). In particular, the map elements were displayed in the following order: title, geographic base, legend title, and boxes, and then each class was displayed (from low

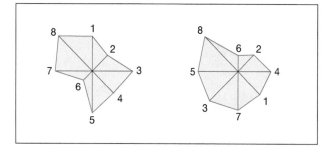

FIGURE 25.2 Different ways of assigning attributes to the combined star and snowflake glyph symbol. On the left, the attributes are arbitrarily assigned; on the right, they are assigned in order from the lowest to the highest attribute.

to high). Two arguments can be made for sequencing: (1) it enhances map understanding by providing users "chunks" of information (Taylor 1987), and (2) it emphasizes the quantitative nature of the data (i.e., that the data are ordered from low to high values). However, experimental tests of sequencing have produced mixed results. A study by Slocum et al. (1990) found that although more than 90 percent of readers favored sequencing over the traditional static (printed) approach, objective measures of map use tasks revealed no significant difference between sequenced and static approaches. As a result, Slocum et al. argued that sequenced maps might be preferred simply because of their novelty. A later study by Patton and Cammack (1996) revealed that sequencing is more effective when the time to study the map is limited.

25.5.3 PROJECT ARGUS

Project Argus, a collaborative venture of research laboratories at the University of Leicester and Birkbeck College in England, was created to illustrate a broad range of possible data exploration functions. Jason Dykes was responsible for developing most of the software and wrote papers (1996; 1997) summarizing its capability. Here we consider the so-called "enumerated software" that was intended for data collected for enumeration units. Although this software is no longer widely utilized, we consider it because it represents an important early attempt to develop sophisticated data exploration software.

The enumerated software could symbolize data in a variety of forms, including choropleth maps, proportional circle maps, and Dorling's cartograms, and display up to three attributes at a time. Interestingly, proportional circle maps were, by default, displayed using a redundant symbology of size and gray tones; the gray tones could, however, also depict a second attribute, thus producing a bivariate map.

To illustrate some of the enumerated software's capability, consider Figure 25.3, which portrays a view created with a sample data set for the U.S. state of Illinois that was included with the software. The three attributes mapped are percent black, percent of the population 25 years and older with any college education, and percent below the poverty level. Across the top of the screen, the attributes are shown as individual choropleth maps (from left to right).

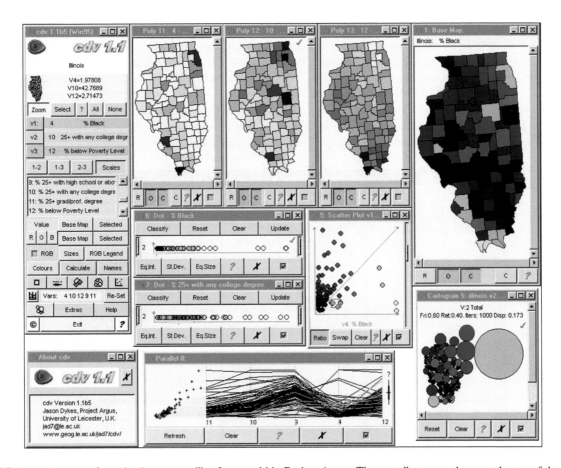

FIGURE 25.3 A screen from the "enumerated" software within Project Argus. Three attributes are shown at the top of the screen: percent black, percent of population 25 years and older with any college education, and percent below the poverty level. In the top right is a bivariate choropleth map of the first two attributes; high values on these two attributes are represented by increasing amounts of yellow and blue, respectively. In the central portion are point graphs and a scatterplot of the first two attributes. The lower and lower right views show a parallel coordinate plot and Dorling's cartogram method, respectively. (Courtesy of Jason Dykes, Department of Computer Science, City, University of London—@jsndyks.)

Choropleth maps are shown (as opposed to proportional circle maps) because the maps are based on standardized data. By default, unclassed choropleth maps are shown, although several classification methods were available within the software.[3]

In the middle portion of the screen, we see point graphs (1-D scatterplots) of two of the attributes: percent black and percent of the population 25 years and older with any college education. We can compare these plots with the previous maps and see that the overall light appearance of the percent black map is a function of a positive skew (i.e., outliers occur toward the positive end of the number line). To the right of the point graphs is a scatterplot illustrating the relationship between the percent black and education attributes (displayed along the x and y axes, respectively). We can see that when the percent black is low, a large range of education percentages is possible, but that the larger percent black values are associated with lower percent education values.

In the top right of the screen, we see a bivariate choropleth map (introduced in Section 22.3.2.1) of the percent black and education attributes. Higher values on these attributes are represented by increasing amounts of yellow and blue, respectively, whereas the lowest values are represented by the absence of these colors or black in the extreme case. Note that the two highest percent black values shown in the scatterplot (they have low scores on the education attribute) are depicted as a bright yellow in the extreme southern part of Illinois. Overall, the map is relatively dark because of the concentration of both attributes in the lower-left portion of the scatterplot. Also, note that the colors displayed on the map are shown within dots on the scatterplot.

It must be recognized that the static nature of a book prevents a full understanding of the dynamic capability of the enumerated software. For example, the software permitted designating an outlier on a point graph and highlighting that point on all other displays currently in view. Alternatively, it was possible to highlight a subrange of values on an attribute. To illustrate these concepts, in Figure 25.3, a small cluster of dots in the scatterplot (in green) has a small gray outlined box drawn around it, and these same observations are highlighted in green on other selected views. Also shown in Figure 25.3 is a parallel coordinate plot and Dorling's cartogram method (in this case, illustrating the prominence of Cook County, where the city of Chicago is located).

25.5.4 MapTime[4]

MapTime permits users to explore temporal data associated with fixed point locations using three major approaches: animation, small multiples, and change maps. Stephen Yoder (1996), the developer of MapTime, argued that animation is a logical solution for showing changes over time because "… the cartographic presentation [is not only] a scale model of space, but of time as well" (30). Animations of point data are easiest to interpret when data change relatively gradually over time. For example, an animation of the change in the population of U.S. cities from 1790 to 1990 (distributed with MapTime) is easy to follow because city populations gradually increase and decrease. In contrast, an animation of the change in water quality at point locations along streams and rivers exhibits much sharper increases and decreases and thus is harder to interpret.

In the context of temporal data, a small multiple consists of a set of maps, one for each time element for which data have been collected. As an example, Figure 25.4 portrays a

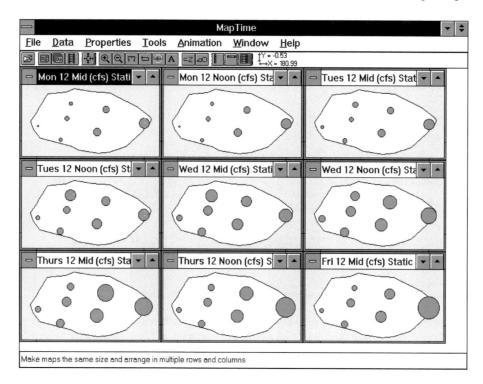

FIGURE 25.4 A small multiple created within MapTime. Each map represents one point in time for hypothetical stream discharge data at selected points. (Courtesy of Stephen C. Yoder.)

small multiple of stream discharge for seven collection stations within a hypothetical drainage basin at 12-hour intervals over a four-day period (the data are distributed with MapTime). It is presumed that the streams flow from west to east, that circle size is proportional to stream discharge (in cubic feet per second), and that heavy rainfall occurs between 6 PM Monday and 3 AM Tuesday throughout the region. The small multiple reveals an initial increase in discharge for upstream stations, followed by a later increase for downstream stations; essentially, a pulse of water moves through the system. The pulse of water seen in the small multiple is also detectable in an animation of the stream discharge data, but it is not easy to discern. Thus, small multiples are often a necessary complement to animations. Small multiples can also assist in contrasting two arbitrary points in time. For example, one might examine the beginning and ending multiples for stream discharge and note approximately how much change has taken place over the period at each location.

Change maps explicitly show the change that takes place between two points in time; in MapTime, change can be computed in terms of the raw data (magnitude change), in percent form, or as a rate of change. Population data for 196 U.S. cities from 1790 to 1990 provide a good illustration of the need for change maps. When these raw data are animated (see Animation_MapTime.mp4)[5], one perceives a major growth in city populations in the Northeast beginning in 1790, with an apparent drop in population for some of the largest northeastern cities from about 1950 to the present. However, it is not immediately clear which cities decrease in population when these decreases occur and

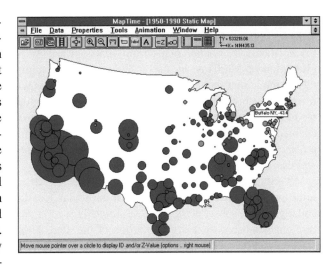

FIGURE 25.5 A change map created using MapTime. Red and blue circles represent population decreases and increases, respectively, between 1950 and 1990. Note the distinctive region of population losses in the Northeast; this pattern was not easily revealed in an animation of the data over this time period. (Courtesy of Stephen C. Yoder.)

what the magnitudes of these decreases are. In contrast, a map showing the percent of population change between 1950 and 1990 reveals a distinctive pattern of population decrease throughout most of the Northeast, as shown in Figure 25.5.

An interesting possibility is to combine the character of small multiples and change maps in a single presentation. This is illustrated in Figure 25.6, where see a small multiple

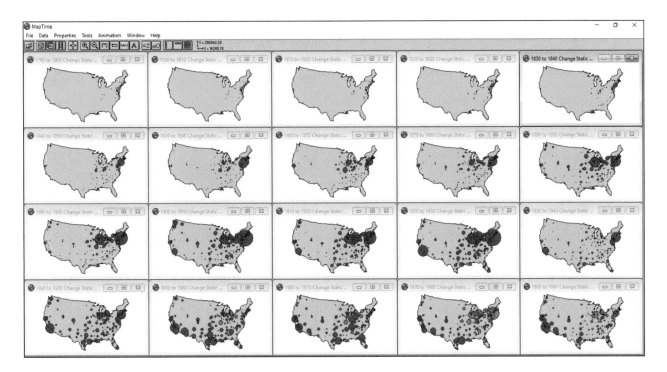

FIGURE 25.6 A small multiple of change maps created using MapTime. Population increases are shown by blue circles, whereas population decreases are shown by red circles. The population decreases in the northeastern United States from 1950 to the present were not easily recognized in either an animation or a raw small multiple of the data. (Courtesy of Stephen C. Yoder.)

of a series of change maps for the U.S. population data from 1790 to 1990; note the distinctive losses in population in the northeast for change maps from 1950 onward. In an evaluation of MapTime by users, Slocum and colleagues (2004) found that such a small multiple of change maps was favored over both animation and a simple raw small multiple for showing change over time. Animation, however, was deemed useful for examining general trends and providing a *sense* of change over time, whereas the raw small multiple was useful for comparing arbitrary time periods.

As with other exploration software, MapTime can determine precise values associated with a particular location and highlight a subset of the data. Additionally, a zoom feature enables users to enlarge a portion of the map. These features can be used to explore a single moment in time or be implemented throughout an animation. Those who wish to experiment with different circle-scaling methods (as described in Chapter 18) also will find that circle-scaling exponents can easily be changed in MapTime.

25.5.5 CommonGIS

Developed between 1998 and 2001, CommonGIS was created to provide a wide range of users with easy access to visualizations of attribute data associated with either enumeration units or point locations. Software development was funded by the CommonGIS Consortium, which was composed of seven public and private companies from five countries of the European Union. The roots of CommonGIS were largely the Descartes software (formerly IRIS), which was developed by Gennady Andrienko and Natalia Andrienko (1999).[6] Although CommonGIS is no longer commercially available, the Andrienkos make the software freely available as V-Analytics (http://geoanalytics.net/V-Analytics/), illustrating the transition that we mentioned from data exploration to geovisual analytics.

CommonGIS includes numerous interesting capabilities for exploring spatial data. Consider the unclassed choropleth map of birth rate for European countries shown in Figure 25.7A. To the right of the map, we see a vertical point graph (CommonGIS terms this a "dot plot") of the data displayed alongside the range of possible colors representing birth rates. This point graph displays the distribution of the data along the number line—it just happens to be a vertical plot rather than the horizontal plot that we introduce in Chapter 3. Note that the point graph reveals a positively skewed distribution, largely due to a single outlier, Albania, which has a birth rate of 21.71 (note that Albania is labeled on the map). Because of this outlier, most of the countries appear to be similar in color, thus making it difficult to visualize which countries have the next highest birth rates after Albania. We can improve the ability to discriminate countries if we remove the outlier and extend the color scheme over the remaining countries. This is accomplished by dragging the small dark blue triangle at the top of the point graph to just above the next highest observation. The result is shown in Figure 25.7B, in which the outlier is depicted by a small brown triangle (in Albania) and a second point graph is drawn (to the left of the initial one) representing the reduced data set. Now we see that Moldova, Iceland, and Macedonia clearly have the highest birth rates after Albania.

In Chapter 15, we introduced the notion of diverging color schemes, in which two colors diverge from a central light point. To illustrate their use in CommonGIS, we'll utilize the elderly population data shown for administrative divisions of Portugal in Figure 25.8. For this map, we specified a blue-red diverging scheme but initially mapped the data with just the red portion of the scheme, as shown in Figure 25.8A. We then selected the "Dynamic map update" option (shown in the lower right of the map) and dragged the "black line with arrows marker" along the legend. As we did so, we could see "how blue shades (corresponding to a low % greater than or equal to 65) spread

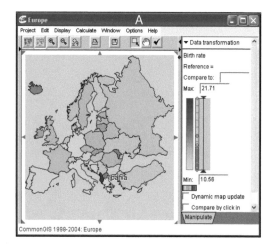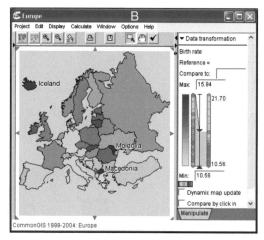

FIGURE 25.7 Data exploration in CommonGIS involving the removal of an outlier: (A) an unclassed choropleth map of all data for birth rates in Europe; (B) the Albanian outlier is removed. (Courtesy of Knowledge Discovery Department Team, Fraunhofer Institute IAIS, http://www.iais.fraunhofer.de.)

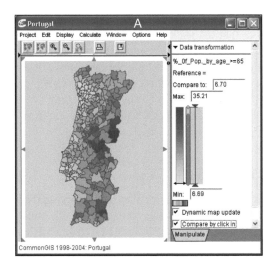

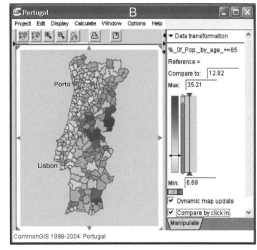

FIGURE 25.8 Data exploration in CommonGIS using a dynamic, diverging color scheme: (A) an unclassed choropleth map of the percent of the population greater than or equal to 65 in divisions of Portugal; (B) a diverging color scheme is applied to a portion of the data range, illustrating that the lowest values occur around the biggest cities of Portugal such as Lisbon and Porto. (Courtesy of Knowledge Discovery Department Team, Fraunhofer Institute IAIS, http://www.iais.fraunhofer.de.)

from the areas around the biggest cities of Portugal such as Lisbon and Porto, first along the coast and then into the inner parts of the country" (CommonGIS documentation 2002). Figure 25.8B shows one point in the dynamic update where we can see that the lowest values occur around the biggest cities in Portugal. An interesting question would be to determine how much of this interpretation is based on the dynamic update and how much is based on the diverging scheme. To assist you in answering this question, we provide a video demonstrating this approach: CommonGIS_Dynamic_Update.mp4.[7]

An important component of CommonGIS is its capability to display spatiotemporal data. In addition to providing visualizations similar to those we described for MapTime (e.g., animation, small multiples, and change maps), CommonGIS utilizes a **value flow diagram** to display change in an attribute over a time interval on a single map. Figure 25.9 displays value flow diagrams of the burglary rate from 1960 to 2000 for the 48 contiguous U.S. states; the x-axis of each diagram represents the year, and the y-axis represents the burglary rate. Andrienko and Andrienko (2004, 417) argue that value flow diagrams

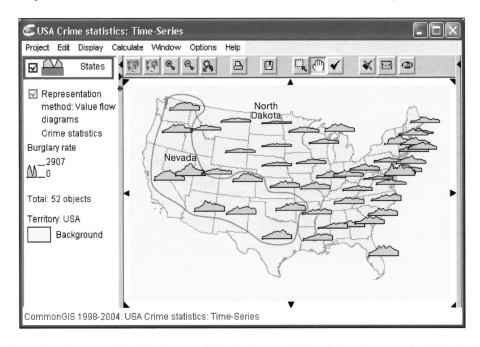

FIGURE 25.9 Value flow diagrams utilized in CommonGIS; the diagrams depict the burglary rates in the United States from 1960 to 2000. The solid line encloses diagrams having a similar temporal structure. Note that those enclosed by the solid line are all red in Figure 25.10 (i.e., they have burglary rates above the median). (Courtesy of Knowledge Discovery Department Team, Fraunhofer Institute IAIS, *http://www.iais.fraunhofer.de.*)

permit answering the following questions related to a time interval:

1. How do attribute values at a given place vary over time?

   For instance, in the case of Nevada, the burglary rate is relatively low at the beginning of the period, increases to a maximum near the middle of the period, and then decreases to a level similar to where it began.

2. Are the behaviors at locations l_1 and l_2 similar or different?

   If we contrast Nevada with North Dakota, we see that North Dakota's rate is much lower throughout the time period.

3. How are different behavior patterns distributed over the territory? Are there spatial clusters of similar behaviors?

   To illustrate the notion of spatial clusters, we have drawn a line around several value flow diagrams in the west and southwest that appear similar in structure (i.e., relatively high crime rates, with a peak near the middle of the period). This, of course, is a subjective process, given that our cluster differs slightly from the one that the Andrienkos created (see Andrienko and Andrienko 2006, 101).

To further highlight the capability of CommonGIS to explore the burglary rate data, Figure 25.10 displays value flow diagrams using the median for all years (for the country) as a comparison value: values above the median are shown in red, and values below the median are shown in blue. Here we can contrast Nevada and North Dakota again and note that Nevada is always above the median value,

whereas North Dakota is always below the median. We also note that the states we identified as similar in structure in Figure 25.9 all appear in red throughout the time period. It is important to stress that we have only touched on some of the capabilities of CommonGIS. Other options, such as linking the map of value flow diagrams with graphical displays of the data for each enumeration unit, can best be appreciated by actually using the software.

25.5.6 GEODA

GeoDa is intended for the exploratory analysis of spatial data, with an emphasis on examining the statistical significance of that data. The initial version was developed by Luc Anselin and his colleagues at the University of Illinois, Urbana-Champaign. The most recent version is now available through The Center for Spatial Data Science at the University of Chicago (https://spatial.uchicago.edu/geoda).

To illustrate the capability of GeoDa, we will utilize two attributes: the percentage of foreign-born residents for counties in Florida from 2006 to 2010 (the same data that we used in Chapter 5) and the percentage of Catholics in these same counties for 2010. The latter data are provided every ten years by the Religious Congregations & Membership Study (RCMS) (http://www.usreligioncensus.org/). Since religious data were unavailable for three of the counties (Gilchrist, Liberty, and Union), these counties are shown in the background color on maps in this section.

In addition to the traditional data classification techniques that we covered in Chapter 5, GeoDa includes two additional techniques: the percentile map and the boxmap. These techniques focus on extremes in the data, which is often important in a statistical analysis. The **percentile map** is a special

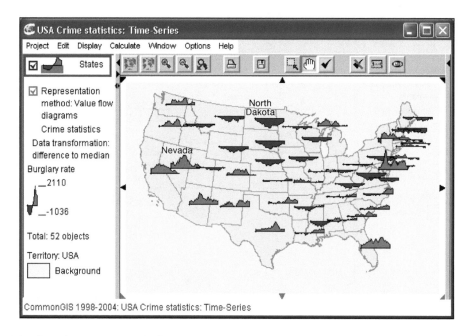

FIGURE 25.10 Value flow diagrams utilized in CommonGIS. The diagrams are based on the same burglary data as Figure 25.9, but values above the median are shown in red, and values below the median are shown in blue. Note that value flow diagrams having a similar structure in Figure 25.9 (i.e., those enclosed by the solid line) are red in this figure. (Courtesy of Knowledge Discovery Department Team, Fraunhofer Institute IAIS, http://www.iais.fraunhofer.de.)

case of a quantile map in which class limits are adjusted to accentuate extreme values. Specifically, the following six-class limits can be created based on percentiles: <1%, 1% to <10%, 10% to <50%, 50% to <90%, 90% to <99%, and 99% and above. In the case of the Catholic data (Figure 25.11A), we can see that there are no data values in the lowest class, but there is one data value in the highest class (in east central Florida in Indian River County). If you examine the histogram of percent catholic in Figure 25.11C, this should make sense, as we see that most of the data are concentrated in the lower classes, but we do see a slight gap between the bar for the highest data value and the bar for adjacent data values.

The **boxmap** is a special case of a four-class quantile map in which outliers (values distinctly different from the rest of the data) are placed in classes by themselves. The boxmap thus indicates whether the most extreme values are truly different from the rest of the data. Using an approach developed by John Tukey (see Section 3.7 for an introduction to Tukey's work), we can define outliers as follows:

$$\text{Values} > UQ + 3 \times (UQ - LQ)$$

$$\text{Values} < LQ - 3 \times (UQ - LQ)$$

where UQ and LQ are the upper and lower quartiles, respectively. GeoDa actually uses somewhat different formulas than these, but the principle is the same. Looking at the boxmap for the Catholic data in Figure 25.11B, we see

that no values for the Catholic data are considered outliers (no values fall in either the lowest or highest class); thus, although the percentile technique suggested that the highest value was extreme, the boxmap method indicates it is not. In contrast for the foreign-born data in Figure 25.11D, Miami-Dade County in the extreme south is identified as a distinctive outlier at the high end of the data (recall our dispersion graph in Figure 5.1).

GeoDa includes a range of approaches for visualizing and exploring bivariate and multivariate data. For example, imagine that you wish to analyze the relationship between the percent foreign born and the percent Catholic for counties in Florida. Figure 25.12 illustrates some of the maps and graphs that you might use in your analysis. Note first that in 25.12A and 25.12B, we have displayed quantile maps of each attribute. This is reasonable since we indicated in Section 22.3.1.1 that quantile maps are useful for map comparison. The overall spatial patterns on these maps suggest a strong correlation between these attributes. Surprisingly, however, note that the points in a scatterplot of the two attributes (Figure 25.12C) do not fit a regression line very well. The reason for this lack of fit is that the distribution of both attributes is positively skewed. When we transform both attributes (performing a square root transformation for percent Catholic and a log transformation for percent foreign born), the points in the scatterplot appear much more closely aligned with the regression line (Figure 25.12D).[8]

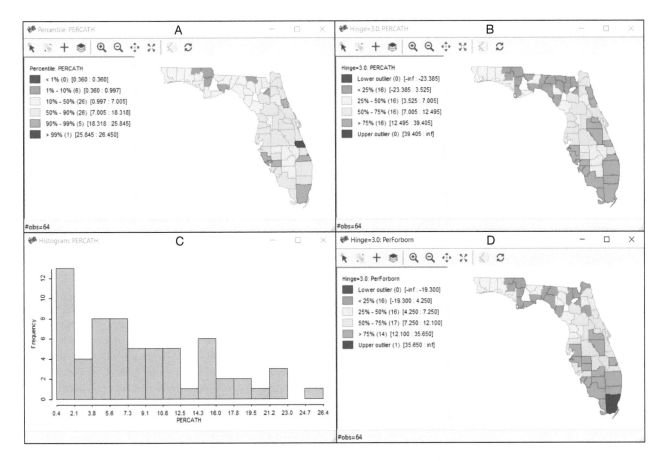

FIGURE 25.11 Visualizing extreme values in GeoDa using data for counties in Florida: (A) a percentile map of percent Catholic; (B) a boxmap of percent Catholic; (C) a histogram of percent Catholic; and (D) a boxmap of percent foreign born. (Courtesy of Luc Anselin.)

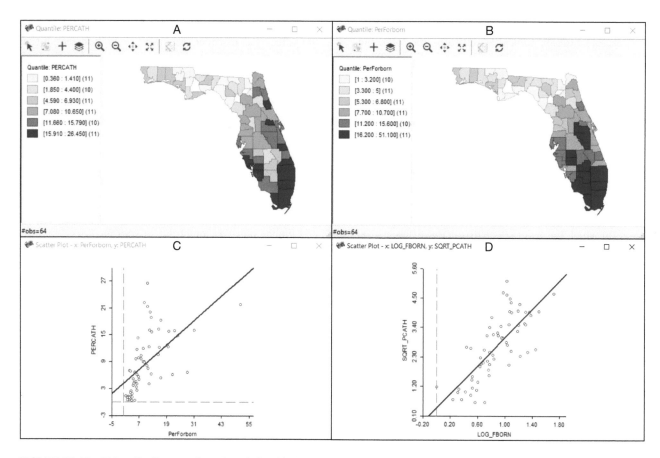

FIGURE 25.12   Using GeoDa to analyze the relationship between percent foreign born and percent Catholic for counties in Florida; quantiles maps of percent Catholic and percent foreign born are shown in A and B, and scatterplots of the raw and transformed data are shown in C and D. (Courtesy of Luc Anselin.)

In addition to being able to easily manipulate maps and graphs, a key aspect of GeoDa is the availability of dynamic linkages between maps, graphs, and tables. Thus, any object selected (or *brushed*) in one image is highlighted on all other images. For instance, we might highlight dots that do not fit a regression line (such as several of the dots well above the regression line in the raw scatterplot in Figure 25.12) and see where these observations lie on other maps and graphics. Alternatively, we might define a rectangle on a map; as we brush the rectangle over the map, enumeration units falling within this rectangle will be highlighted on the map and on all other maps, graphs, and tables. Remember that we argued in Section 25.4.5 that such linkages are a key component of data exploration software.

Another intriguing aspect of GeoDa is the Map Movie option, which provides a variant of the sequencing approach previously described for ExploreMap (Section 25.5.2). In the case of GeoDa, the process begins by displaying the entire mapped distribution semi-transparently, and then individual enumeration units (as opposed to classes of data) are sequenced (or *animated,* in GeoDa terms). Most useful is a Cumulative option, in which enumeration units remain in their class color once they are sequenced (for a video of the approach, see Sequencing_GeoDa.mp4).[9] In our opinion, sequencing individual enumeration units is more useful than sequencing an entire class (as was done in ExploreMap), as the process gives

the reader time to evaluate the location of individual enumeration units and associated spatial pattern(s) that are created.

25.5.7   Micromaps

Although the software that we have described provides a broad range of approaches for interacting with and exploring spatial data, most of the maps and associated graphics are not novel and have been used previously in non-interactive environments (paper and static computer displays). In contrast, in this section, we consider **micromaps**, which are a set of small maps used to analyze one or more attributes distributed spatially. The notion of micromaps is similar to the small multiple, but we will see that graphical and statistical information are also often linked with these maps. The idea of micromaps was developed by Daniel Carr and colleagues. If the notion of micromaps piques your interest, you may wish to examine the book *Visualizing Data Patterns with Micromaps* by Daniel Carr and Linda Pickle (2010). In this book, Carr and Pickle describe three kinds of micromaps: linked, conditioned, and comparative. We will focus on just the first two here.

25.5.7.1   Linked Micromaps Plot

A **linked micromaps plot** consists of three basic elements, as illustrated in Figure 25.13: a legend (the column labeled "States"), graphics associated with one or more attributes

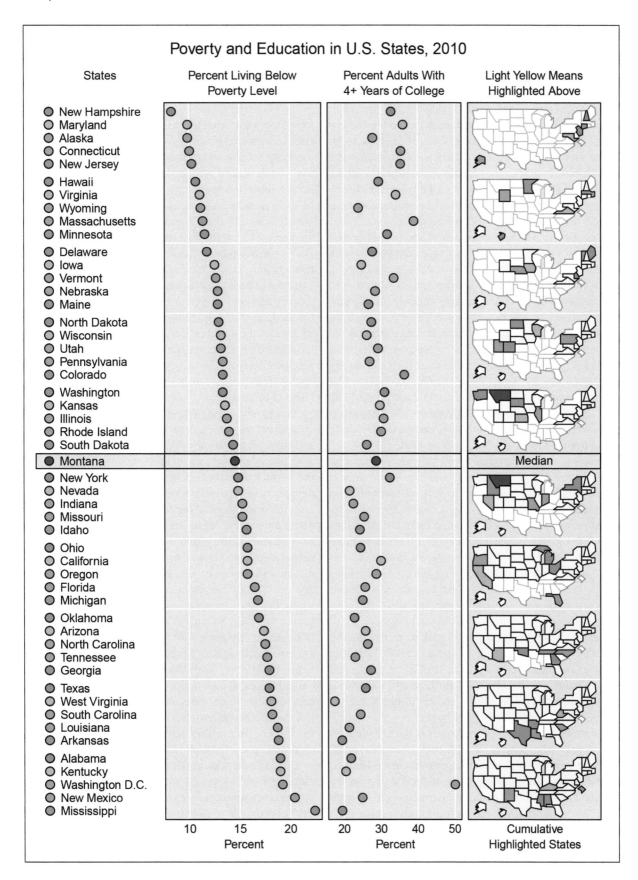

FIGURE 25.13 Example of a linked micromaps plot. The main attribute of interest is the percent living below poverty level, which is sorted from low to high values in the second column and depicted in a series of micromaps in the fourth column. (Image was created using software available at http://mason.gmu.edu/~dcarr/LinkedMicromaps/; data source: U.S. Bureau of the Census.)

(the second and third columns), and a series of small maps (the micromaps). The primary attribute of interest (in this case, "Percent Living Below Poverty Level") is plotted graphically from low to high values (from top to bottom) as colored dots in the second column and displayed spatially in the micromaps. The term *linked* is derived from the notion that information in one column can be related to information in another column because of the corresponding color. Thus, if we look at the first state, New Hampshire, we see that it is coded in red, it has the lowest poverty level, it has a relatively high value on the education variable (Percent Adults with 4+ Years of College), and it appears on the first micromap with several northeastern states plus the state of Alaska (in the lower left).

Rather than providing a depiction of the overall spatial pattern (as on a traditional choropleth map), the purpose of a linked micromaps plot is to focus on local pattern perception, which is achieved by ordering the micromaps from low to high data values based on the poverty attribute. In examining the micromaps, note how the lower poverty values tend to occur in northeastern and northcentral states, whereas the higher poverty values tend to occur in southern states. In a sense, the set of maps serves as a form of map sequencing (as was discussed for ExploreMap and GeoDa), except that we see all elements of the sequence in this static image. Note that the light yellow color is used to represent all enumeration units that have been depicted in a previous micropmap as we progress from the top to the bottom micromap.

In looking at the micromaps, you may also notice that individual states appear distorted and that some appear larger or smaller than we might expect on an equivalent (equal-area) map projection. Using a notion developed by Mark Monmonier (1992), enumeration units are intentionally distorted so that smaller enumeration units can be more easily identified (Washington, DC is shown just off the east coast of the United States). Even with such modifications, however, small enumeration units will be difficult to detect; thus, in static (printed) form, no more than about 100 enumeration units are possible in a linked micromaps plot.

Inclusion of the education attribute allows us to examine the relation between poverty level and education. Note that low values on the poverty attribute tend to be associated with high values on the education attribute and vice versa. The trend shown here for 2010 is similar to one that Carr and Pickle plotted for 2000 in their book.

Software for creating linked micromaps is available through the R programming system. The R package micromapST (https://cran.r-project.org/web/packages/micromapST/index. html; Pickle et al. 2015) enables the production of plots for U.S. state data, and the R package micromap (https://cran. r-project.org/web/packages/micromap/micromap.pdf; Payton et al. 2015) enables the production of plots for a broader set of data.

25.5.7.2 Conditioned Micromaps

Conditioned micromaps permits a visual examination of two attributes that might explain the spatial pattern of a third attribute (the dependent attribute). The explanatory attributes

are also considered the conditioning attributes, hence the term *conditioned* micromaps. To illustrate, consider Figure 25.14, which provides a set of conditioned micromaps intended to assist in explaining lung cancer mortality rates for white men ages 65 to 74 for health service areas (i.e., counties or aggregates of counties) in the United States. In the legend along the top of the color plate, you can see that lung cancer mortality rates are depicted in three colors: blue for low values, gray for moderate values, and red for high values. Values at the bottom of this legend are lung cancer mortality rates, whereas values above the legend are the percent of white men ages 65 to 74 in each class. In this example, the attributes considered (because of their potential relationship to lung cancer) are percent below the poverty level and annual precipitation. Although precipitation might seem like a strange attribute to relate to lung cancer, Carr and colleagues (2002) argued that high precipitation was historically advantageous for growing tobacco and could lead to greater time spent indoors and that the associated high humidity could increase chemical activity on the lung surface. Carr and colleagues (2005) provided an alternative argument for using precipitation—that low precipitation would be associated with high particulate levels in the atmosphere, which would be associated with higher levels of lung cancer.

The annual precipitation data are depicted in the three columns of maps (health service areas having low, medium, and high precipitation values are shown from left to right, with the values for each column indicated in the legend at the bottom of Figure 25.14). In an analogous fashion, the percent below poverty level data are shown in the three rows of maps (health service areas having low, medium, and high poverty levels are displayed from bottom to top, as indicated in the legend to the right). Thus, similar to a scatterplot, low values of precipitation and poverty are represented in the lower left, and high levels of precipitation and poverty are represented in the upper right; the actual symbols shown on each map, however, represent the lung cancer mortality rate data. We can see that a large number of health service areas with high lung cancer mortality rates are associated with high precipitation and a high percentage below the poverty level (the red region in the southeastern United States, shown in the upper right map). This association (or correlation) does not necessarily mean that there is a cause-and-effect relationship, but it does suggest that these and other related attributes should be examined in order to explain the high mortality rates in the Southeast.[10]

Conditioned micromaps are best appreciated by experimenting with associated software, which can be found at http://mason.gmu.edu/~dcarr/CCmaps/. There you will find that it is possible to dynamically change the class limits on any of the three attributes, thus developing different visualizations of the relationships among the attributes.

25.5.8 ViewExposed

ViewExposed is data exploration software that focuses on visualizing the vulnerability to natural hazards in counties

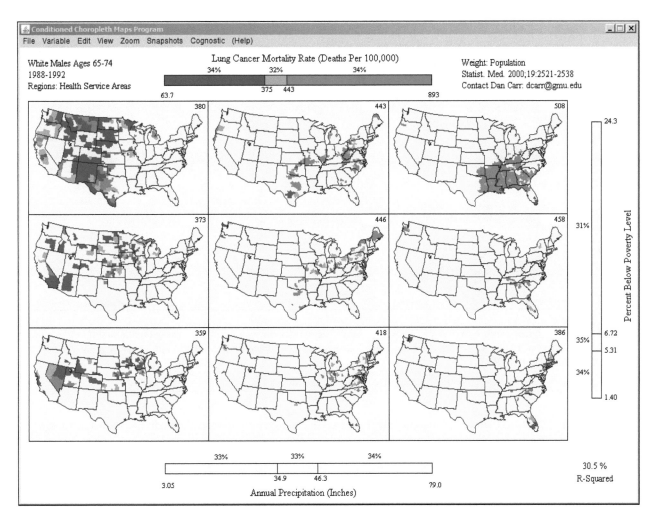

FIGURE 25.14 Example of conditioned maps. The attribute being studied is the lung cancer mortality rate for white men ages 65 to 75. Potential explanatory attributes—annual precipitation and percent below the poverty line—are depicted from left to right, and from bottom to top, respectively. (After "Conditioned choropleth maps and hypothesis generation," by D. B. Carr, D. White, and A. M. MacEachren, *Annals of the Association of American Geographers*, copyright © The Association of American Geographers, reprinted by permission of Taylor & Francis Ltd., http://www.tandfonline.com on behalf of The Association of American Geographers, www.aag.org.)

and municipalities of Norway. We chose this example because (1) it is readily available via the Web (http://folk.ntnu.no/opach/tools/viewexposed/); (2) it has a relatively easy-to-use interface that illustrates coordinated multiple views and the use of parallel coordinate plots; and (3) several papers are available that describe its development and an evaluation of its effectiveness (Opach and Rød (2013, 2014) and Golebiowska et al. (2017)).

Opach and Rød (2013) argue that visualizing vulnerability is especially important today, given the anticipated negative effects of climate change. To represent vulnerability, they developed an Integrated Vulnerability Index that combines an Exposure Index and a Social Vulnerability Index (the latter is based on various social factors such as age and social status). They split the Exposure Index into three major hazard indexes: exposure to floods, exposure to landslides, and exposure to storms. An administrative official (or a resident) might be interested in what their

municipality's overall vulnerability is, how that overall value is broken down into the components of vulnerability, and how their vulnerability compares with county and national vulnerability. They also might want to explore their location on the map and consider what geographic factors might be affecting their vulnerability.

With this brief introduction as a basis, we will now discuss the online version of ViewExposed. As we do so, we encourage you to work with the software. Figure 25.15 is a screenshot of ViewExposed in which we have clicked on one municipality, Aurland, within the county Sogn og Fjordane. Note that the boundary of Aurland is highlighted in red on the map, and a brown line on the parallel coordinate plot indicates its scores on the various indexes. Here we can see that the Integrated Vulnerability is moderate (about 55), but the Exposure Index is relatively high (about 74). Furthermore, there is a drastic difference in the components of the Exposure Index, with floods, landslides, and

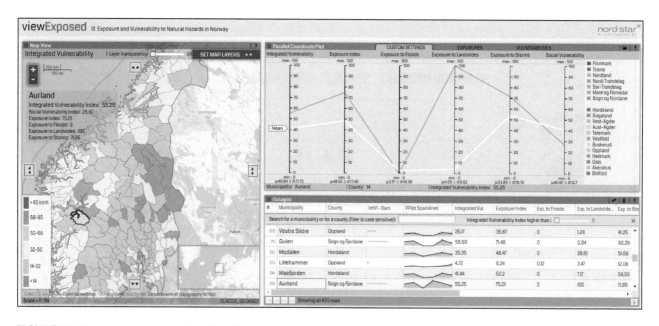

FIGURE 25.15 A screen capture from ViewExposed showing: the municipality of Aurland, Norway highlighted in red on the map; a parallel coordinate plot with Aurland and the state mean displayed; and a table with Aurland highlighted in yellow. (Courtesy of Tomasz Opach and Jan Ketil Rød.)

storms scoring 0, 100, and about 72, respectively. Finally, we can see that the Social Vulnerability is relatively low (about 28). We can also see how Aurland compares to the entire country of Norway, as mean values for all municipalities are shown in white. Here the largest difference is clearly the exposure to landslides, as the national mean is about 5 and Aurland is at 100.

In the lower right of Figure 25.15, we see a table of the data, where Aurland is highlighted in yellow because we have selected that municipality. Here we can get more precise values for the various indexes, confirming what we estimated in the parallel coordinate plot. Note that there are two graphic components to this table. One is the column "IntVI – Bars" in which a horizontal bar is drawn proportional to the magnitude of the Integrated Vulnerability. The second graphic component is in the column "PPlot Sparklines," where we see line graphs (**sparklines**) of values corresponding to the values found in the parallel coordinate plot.[11] The purpose of sparklines, in this case, is to ease the interpretation of the values shown in the parallel coordinate plot when lots of observations are shown. (You can get a better feel for the nature of sparklines if you hover over them.)

Figure 25.15 represents the simplest possible display in ViewExposed. A more complex example is shown in Figure 25.16, where we have clicked on the county of Sogn og Fjordane in the parallel coordinate plot. Now we see that all municipalities within the county have been displayed within the parallel coordinate plot and that the average of these municipalities is shown in yellow. Comparing the line for Aurland with the other parallel coordinate plot lines, we can see that several other municipalities in the county

have higher Integrated Vulnerability values, with two noteworthy in their exposure to floods. You can also click on the "Integrated otVul." Column in the table and thus order the municipalities in terms of Integrated Vulnerability. We encourage you to experiment further with this interface, as we will pose a study question later that asks you to interpret some associated data.

25.5.9 Using Tableau to Create Interactive Data Visualizations

Tableau (www.tableau.com) is a software company that produces data visualization and visual analytics applications for many of the world's leading businesses, research labs, and non-profits. The company was founded in 2003 by researchers who previously had worked at Stanford University's Department of Computer Science. The motivation to create Tableau stemmed from their shared interests in developing visualization tools for exploring and visualizing relational databases. As a result of these database interests, Tableau connects to the major database engines such as MySQL, PostgresSQL, and SQL Server. Along with numerous standard visualization tools including parallel coordinate plots, bubble charts, and treemaps, Tableau also provides the ability to map spatial data using choropleth and proportional circle symbology. The ability to quickly and easily build highly interactive dashboards that seamlessly link various visualization tools together is one of the main advantages of working with Tableau. Tableau provides a suite of licensed products such as Tableau Desktop, Tableau Server, and Tableau Mobile. These products allow the developer to save their

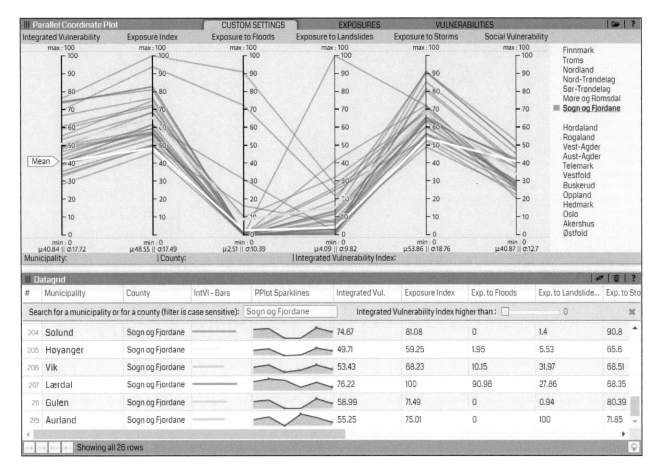

FIGURE 25.16 A screen capture from ViewExposed showing: all municipalities within the county of Sogn og Fjordane displayed in the parallel coordinate plot, along with the mean for the municipalities in yellow; and the associated table of those same municipalities. (Courtesy of Tomasz Opach and Jan Ketil Rød.)

applications locally. Aside from these licensed products, Tableau offers Tableau Public (https://public.tableau. com/en-us/s/), which is a free platform downloadable as a stand-alone desktop app. Tableau Public incorporates the same dashboard interface development tools found in their licensed products, thus allowing users to connect data with various maps and graphics to create interactive visualizations ("vizzes" in Tableau) that are sharable online. However, unlike Tableau Desktop, for example, any visualization created through Tableau Public is stored on Tableau's server and can only be published online.

We will consider a Tableau Public visualization of the spatial patterns of Bigfoot sightings in the United States and Canada between 1870 and 2017. The Bigfoot visualization can be explored at this site: https://public.tableau.com/app/profile/robert.zupko/viz/Zupko_Lab9/BigfootintheUnitedStates. This visualization was created by Robert Zupko, a student enrolled in GEOG 486 "Cartography and Visualization" in spring 2021 while pursuing a degree in Penn State's Master of Geographic Information Systems. The Bigfoot visualization (Figure 25.17) is divided into three dashboards, each focusing on a different Bigfoot topic. The dashboards are shown in the three grey boxes at the top-left of the window.

For example, the second dashboard, titled "Yearly breakdowns imply rarity," focuses on mapping the location of Bigfoot sightings between 1870 and 2017 and provides a measure of uncertainty for each reported citing. There are two components to this dashboard: a colored dot map at the top and a stacked histogram at the bottom. Dots appearing on the map are color coded according to the year of the sightings (early years are dark red while later years are light orange). Below the map is a histogram showing the distribution of the number of Bigfoot sightings by year, with each year classified according to the uncertainty in each sighting report. Here, uncertainty is divided into three classes and assigned three colors. Class A (blue) represents sighting reports that are the most credible (lowest uncertainty), class B (orange) indicates sighting reports that have a greater uncertainty, and class C (red) is associated with sightings derived from second or third-hand reports leading to the highest uncertainty.

Interactivity through pointing-clicking and linking-brushing highlight the exploration options available with Tableau Public dashboards. For example, mousing-over individual dots on the map produces pop-up windows that report the class (A, B, or C), year of the sighting, latitude,

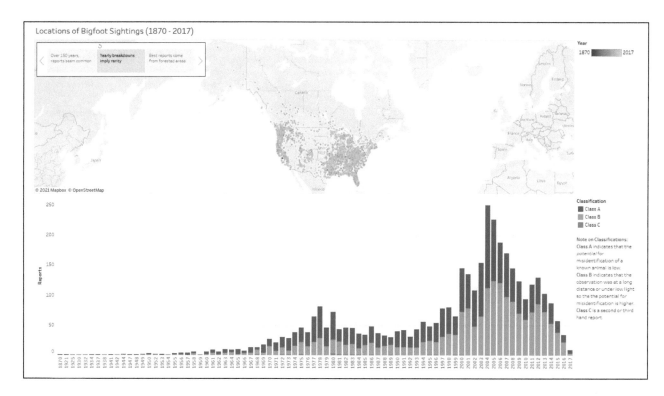

FIGURE 25.17 The Tableau Public Bigfoot "Yearly breakdowns imply rarity" dashboard. (Courtesy of Robert Zupko.)

and longitude of the sighting, and any associated text from the sighting report. Figure 25.18 shows the results of linking-brushing operations that are possible through the dashboard. Here we have used the mouse to draw a box around multiple dots in an area in southeast Ohio on the map; as a result, the bars on the histogram will simultaneously highlight corresponding to the years associated with the highlighted dots. Looking at Figure 25.18, we see that for southeast Ohio, Bigfoot sightings tended to occur after 1977, with a peak in 2004, and that the uncertainty is

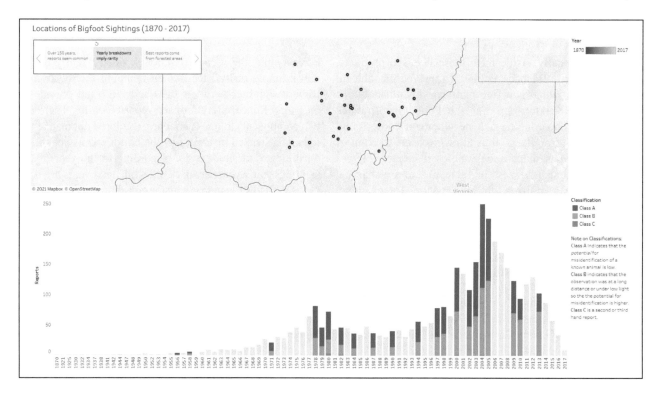

FIGURE 25.18 Highlighted dots on the map in southeast Ohio are linked to the uncertainty classes in the histogram. (Courtesy of Robert Zupko.)

split roughly equally between classes A and B, with very few sightings falling in class C. Users can also highlight data on the histogram and see the results on the map. For instance, the user can highlight a single year (1978) or multiple years (1978–1988) and see the spatial distribution of Bigfoot sightings appear on the map. Individual uncertainty classes for a given year can also be selected, and their patterns visualized on the map (e.g., for a given year, click on a specific bar color such as class B—orange). Since any visualization created by Tableau Public is publicly available, the dashboard also allows the user to download an image of the dashboard or share the dashboard through social media.

Since we have outlined only some of the capabilities of the Bigfoot visualization, we encourage you to examine it more fully at the site listed above. Other examples of Tableau Public applications can be found at https://public. tableau.com/app/discover/viz-of-the-day. A distinct advantage of Tableau Public applications is that you can develop them without any programming background. We encourage you to examine the learning resources available for Tableau (https://public.tableau.com/s/resources) and consider developing your own data visualizations.

25.6   SUMMARY

In the process of data exploration, either the cartographer or map users are able to explore spatial data using a variety of interactive tools and the visual information processing capability of the human brain. With very large datasets, the computational power of computers is also necessary—this leads us into the realm of geovisual analytics, which enables us to make sense out of very large datasets using the techniques of data mining, which we consider in Chapter 26. Some of the methods for exploring data include: manipulating data (e.g., classifying data), varying the symbolization (showing both choropleth and isopleth maps), manipulating the user's viewpoint (this is critical in 3-D mapping), multiple map views (e.g., the small multiple), linking maps with other forms of display (e.g., tables, graphs, and numerical summaries in coordinated multiple views), highlighting portions of a dataset (brushing and focusing are two common terms), probing the display (to acquire values of points and areas), and toggling themes on and off.

Sequencing (a form of animation for nontemporal data) involves displaying either map elements or individual data values in a particular sequence. For example, in GeoDa we saw how the data for enumeration units could be displayed in sequence from low to high values, emphasizing where the low and high data values occur. In examining MapTime, we saw how the animation of temporal data can be enhanced through the use of small multiples and change maps.

In examining CommonGIS, we saw how the interpretation of an unclassed map can be enhanced by removing an outlying value from the choropleth classification. We also saw how a change in the breakpoint of a diverging color scheme (achieved through a dynamic map update) could enhance the interpretation of spatial patterns. Finally, we saw how a static value flow diagram could be used to interpret temporal data associated with enumeration units (thus providing an alternative to animation, which is more frequently used to display temporal data).

Micromaps are a set of small maps used to analyze the spatial distribution of one or more attributes. The notion is similar to the small multiple, but the emphasis is on local pattern perception rather than comparing the overall pattern on multiple maps. Conditioned micromaps permit a visual examination of attributes (explanatory attributes) that might explain the spatial pattern of a third attribute (the dependent attribute). The explanatory attributes are also considered the conditioning attributes, hence the term conditioned micromaps. The data classifications for each conditioning attribute break the overall distribution into a set of micromaps that allow us to see how a dependent attribute varies as a function of the explanatory attributes.

The ViewExposed software provided an illustration of how an interactive data exploration system consisting of coordinated multiple views could be used to examine a real-world problem, the vulnerability to hazards in counties and municipalities of Norway. In addition to seeing how the parallel coordinate plot could be used to interpret a variety of indexes associated with hazards, we also introduced the use of sparklines, a small line graph technique that generally does not include axes or labels and thus can be imbedded in either a table or text.

Finally, we introduced a Bigfoot Tableau data visualization. We hope that an examination of this and other Tableau visualizations will encourage you to consider developing your own interactive visualizations using Tableau Public.

25.7   STUDY QUESTIONS

1. Imagine that you are working for a GIS firm, and your boss has asked you to make one county-level choropleth map of the United States illustrating the percentage of the adult population who have at least a bachelor's degree. She indicates that you can select whatever classification method and color scheme you deem appropriate but that you should make only one map. Explain why this single-map solution might be inappropriate.

2. Explain the concept of coordinated multiple views. How might such a concept enhance an examination of the relationship between the percentage of voting for a particular candidate and the percentage of Protestants within voting precincts of a major metropolitan area?

3. Throughout this textbook, we have emphasized that classed choropleth maps are much more common than unclassed choropleth maps. Explain how methods of data exploration could arguably obviate the need to create classed choropleth maps.

4. Imagine that you have created an animated proportional symbol map showing the magnitude of oil production from each of the 100 largest oil fields around the world over the last 50 years. Discuss how data exploration approaches might enhance an interpretation of such an animation.

5. Define the concept of map sequencing and discuss how it might enhance the interpretation of choropleth maps.

6. Discuss the use of the "dynamic map update" in CommonGIS. Provide your opinion as to whether this technique enhances your interpretation of spatial patterns.

7. Given that you have data on the % Black for decennial censuses from 1900 to 2000 for each of 75 census tracts within an urban area discuss how you could symbolize these data on a single static map.

8. Use the linked micromap plot in Figure 25.13 to contrast northeastern states (ME, NH, VT, MA, CT, RI, NY, PA, NJ) with southeastern states (NC, TN, SC, GA, AL, MS, and FL) in terms of the percent living below poverty level. Also, indicate how the percent living below poverty level relates to the percent of adults with 4+ years of college.

9. Use ViewExposed (http://folk.ntnu.no/opach/tools/viewexposed/) to contrast the vulnerability to natural hazards for municipalities in Nordlund and Hedmark counties.

NOTES

1 The fisheye concept is similar to the variable scale map projection approach discussed in Section 8.8.3, in which the mapped area of focus is set at a larger scale and the area surrounding is set at a smaller scale.

2 To access the demo, we recommend using Firefox. For a broad overview of lenses, see Tominski et al. (2017).

3 For a discussion of whether classed or unclassed maps should be used when comparing choropleth maps, see Section 22.3.1.1.

4 MapTime is available as the zipped file MapTime.zip, which can be downloaded from Chapter 25 at www.routledge.com/9780367712709.

5 The file Animation_MapTime.mp4 can be downloaded from Chapter 25 at www.routledge.com/9780367712709.

6 For a discussion of the roots of CommonGIS, see http://geoanalytics.net/and/KDworkshopPaper2004/KDworkshop.html.

7 The file CommonGIS_Dynamic_Update.mp4 can be downloaded from Chapter 25 at www.routledge.com/9780367712709.

8 Here we have used a basic regression that does not consider the spatial autocorrelation in the data; more sophisticated approaches in GeoDa enable you to also consider the autocorrelation in the data.

9 The file Sequencing_GeoDa.mp4 can be downloaded from Chapter 25 at www.routledge.com/9780367712709.

10 The numbers in the upper right of each map are the weighted averages of lung cancer mortality rates for health service areas appearing on that map.

11 For an introduction to sparklines, see Tufte (2006).

REFERENCES

Andrienko, G. L., and Andrienko, N. V. (1999) "Interactive maps for visual data exploration." *International Journal of Geographical Information Science 13*, no. 4:355–374.

Andrienko, N., and Andrienko, G. (2004) "Interactive visual tools to explore spatio-temporal variation." *Proceedings of the Working Conference on Advanced Visual Interfaces*, Gallipoli, Italy, pp. 417–420.

Andrienko, N., and Andrienko, G. (2006) *Exploratory Analysis of Spatial and Temporal Data: A Systematic Approach.* Berlin: Springer-Verlag.

Buja, A., Cook, D., and Swayne, D. F. (1996) "Interactive high-dimensional data visualization." *Journal of Computational and Graphical Statistics 5*, no. 1:78–99.

Campbell, C. S., and Egbert, S. L. (1990) "Animated cartography: Thirty years of scratching the surface." *Cartographica 27*, no. 2:24–46.

Carr, D. B., and Pickle, L. W. (2010) *Visualizing Data Patterns with Micromaps.* Boca Raton, FL: CRC Press.

Carr, D. B., White, D., and MacEachren, A. M. (2005) "Conditioned choropleth maps and hypothesis generation." *Annals of the Association of American Geographers 95*, no. 1:32–53.

Carr, D. B., Zhang, Y., and Li, Y. (2002) "Dynamically conditioned choropleth maps: Shareware for hypothesis generation and education." *Statistical Computing & Statistical Graphics Newsletter 13*, no. 2:2–7.

Dykes, J. (1996) "Dynamic maps for spatial science: A unified approach to cartographic visualization." In *Innovations in GIS 3*, ed. by D. Parker, pp. 177–187, Color Plates 4–5. Bristol, PA: Taylor & Francis.

Dykes, J. A. (1997) "Exploring spatial data representation with dynamic graphics." *Computers & Geosciences 23*, no. 4:345–370.

Edsall, R., Andrienko, G., Andrienko, N., et al. (2009) "Interactive maps for exploring spatial data." In *Manual of Geographic Information Systems*, ed. By M. Madden, pp. 837–858. Bethesda, MD: American Society for Photogrammetry and Remote Sensing.

Egbert, S. L., and Slocum, T. A. (1992) "ExploreMap: An exploration system for choropleth maps." *Annals of the Association of American Geographers 82*, no. 2:275–288.

Golebiowska, I., Opach, T., and Rød, J. K. (2017) "For your eyes only? Evaluating a coordinated and multiple views tool with a map, a parallel coordinated plot and a table using an eye-tracking approach." *International Journal of Geographical Information Science 31*, no. 2:237–252.

Griffin, A. L., and Robinson, A. C. (2015) "Comparing color and leader line highlighting strategies in coordinated view geovisualizations." *IEEE Transactions on Visualization and Computer Graphics 21*, no. 3:339–349.

Koo, H., Chun, Y., and Griffith, D. A. (2018) "Integrating spatial data analysis functionalities in a GIS environment: Spatial Analysis using ArcGIS Engine and R (SAAR)." *Transactions in GIS 22*, no. 3:721–736.

Moellering, H. (1978) "A demonstration of the real time display of three-dimensional cartographic objects." *Videotape.* Columbus: Department of Geography, Ohio State University.

Moellering, H. (1980) "The real-time animation of three-dimensional maps." *The American Cartographer 7*, no. 1:67–75.

Monmonier, M. (1992) "Authoring graphic scripts: Experiences and principles." *Cartography and Geographic Information Systems 19*, no. 4:247–260, 272.

Nost, E., Rosenfeld, H., Vincent, K., et al. (2017) "HazMatMapper: An online and interactive geographic visualization tool for exploring transnational flows of hazardous waste and environmental justice." *Journal of Maps 13*, no. 1:14–23.

O'Brien, O., and Chesire, J. (2016) "Interactive mapping for large, open demographic data sets using familiar geographical features." *Journal of Maps 12*, no. 4:676–683.

Opach, T., and Rød, J. K. (2013) "Cartographic visualization of vulnerability to natural hazards." *Cartographica 48*, no. 2:113–125.

Opach, T., and Rød, J. K. (2014) "Do choropleth maps linked with parallel coordinates facilitate an understanding of multivariate spatial characteristics?" *Cartography and Geographic Information Science 41*, no. 5:413–429.

Patton, D. K., and Cammack, R. G. (1996) "An examination of the effects of task type and map complexity on sequenced and static choropleth maps." In *Cartographic Design: Theoretical and Practical Perspectives*, ed. by C. H. Wood and C. P. Keller, pp. 237–252. Chichester, England: Wiley.

Payton, Q. C., McManus, M. G., Weber, M. H., et al. (2015) "micromap: A Package for Linked Micromaps." *Journal of Statistical Software 63*, no. 2:1–16.

Pickle, L. W., Pearson, J. B., and Carr, D. B. (2015) "micromapST: Exploring and communicating geospatial patterns in US state data." *Journal of Statistical Software 63*, no. 3:1–25.

Roberts, J. C. (2008) "Coordinated multiple views for exploratory geovisualization." In *Geographic Visualization: Concepts, Tools and Applications*, ed. by M. Dodge, M. McDerby and M. Turner, pp. 25–48. Chichester, UK: Wiley.

Roth, R. E. (2012) "Cartographic interaction primitives: Framework and synthesis." *The Cartographic Journal 49*, no. 4:376–395.

Slocum, T. A., Robeson, S. H., and Egbert, S. L. (1990) "Traditional versus sequenced choropleth maps: An experimental investigation." *Cartographica 27*, no. 1:67–88.

Slocum, T. A., Sluter, R. S., Kessler, F. C., et al. (2004) "A qualitative evaluation of MapTime, A program for exploring spatiotemporal point data." *Cartographica 39*, no. 3:43–68.

Taylor, D. R. F. (1987) "Cartographic communication on computer screens: The effect of sequential presentation of map information." *Proceedings of the 13th International Cartographic Conference*, Volume 1, Morelia, Mexico, pp. 591–611.

Tominski, C., Gladisch, S., Kister, U., et al. (2017) "Interactive lenses for visualization: An extended survey." *Computer Graphics Forum 36*, no. 6:173–200.

Tufte, E. (2006). *Beautiful Evidence*. Cheshire, CT: Graphics Press.

Yamamoto, D., Ozeki, S., and Takahashi, N. (2009) "Focus+Glue+Context: An improved fisheye approach for Web map services." *ACM GIS '09*, Seattle, Washington, pp. 101–110.

Yoder, S. C. (1996) "The development of software for exploring temporal data associated with point locations." Unpublished MA thesis, University of Kansas, Lawrence, KS.

26 Geovisual Analytics

26.1 INTRODUCTION

Those dealing with geospatial data are increasingly finding that they must deal with very large datasets, sometimes referred to as **Big Data**. Probably the most obvious example is in the realm of remote sensing; for instance, the NASA Earth Observing System Data and Information System (EOSDIS) manages more than 9 petabytes of data and adds 6.4 terabytes of data *every day* (Blumenfeld 2017).[1] But large datasets are by no means restricted to remote sensing—think about the cell phones and in-vehicle navigation systems that many of us use on a regular basis. These devices utilize location-aware technologies (LATs) such as GPS to provide location information that can be linked with other attributes, and when the results are combined for people across a city and for multiple points in time, millions of records of information are easily possible. Similarly, when providing volunteered geographic information (VGI) via the Web, any information that you provide is likely to be combined with the results of others; given the fact that there are potentially several billion users of the Web, it is easy to conceive of the monstrous datasets that can result.

Researchers are now realizing that they cannot use visualization alone to understand spatial patterns associated with these large datasets. Rather, it is necessary to combine our human visual processing capabilities with the analytical capabilities of computers. The result is the rapidly growing field of **visual analytics**, which is termed **geovisual analytics** when geospatial data are used. This chapter focuses on this rapidly burgeoning area and the associated area of **data science**.

We begin our discussion in Section 26.3 by reviewing the nature of Big Data in greater depth, stressing its potential but also its limitations. In Section 26.4, we delve into the origins of visual analytics and geovisual analytics and discuss their characteristics in greater depth. In Section 26.5, we introduce the notion of the **self-organizing map (SOM)**, a common computational approach that has been used in geovisual analytics. In the remainder of the chapter, we cover a range of examples that illustrate the use of geovisual analytics.

26.2 LEARNING OBJECTIVES

- Using at least five criteria, contrast traditional small data with Big Data.
- Explain why you might want to be cautious in letting Big Data "speak for itself."
- Define geovisual analytics and explain why geovisual analytics involves both computational methods and visualization.

- Explain the notion of a SOM and discuss the two roles that it can play in geovisual analytics: as a clustering approach and as a dimensionality reduction approach.
- Define a heat map and indicate how it can assist in the interpretation of data associated with point locations.
- Define a mosaic diagram and indicate how it can enhance the interpretation of spatiotemporal data.
- Discuss computational approaches for topic modeling and sentiment analysis associated with Twitter data, and indicate how the results can be displayed in mapped form.
- Explain the activities undertaken by those who contribute to OpenStreetMap (OSM), and describe how we might evaluate the nature of those contributions using geovisual analytics.

26.3 CHARACTERISTICS AND LIMITATIONS OF BIG DATA

Oftentimes, when defining Big Data, the emphasis is on the sheer magnitude of the data, but there are numerous other characteristics of Big Data that distinguish it from traditional small data. To illustrate, we'll discuss the seven criteria shown in Table 26.1, which is a modification of a table developed by Rob Kitchin and Gavin McArdle (2016).[2] *Volume* refers to the size of datasets, which of course, are larger for Big Data. In examining Table 26.1, we see that small data are generally less than 100 observations, whereas Big Data can be thousands to millions of observations. There is no magic dividing line between small data and Big Data—we thus can consider a number of observations between 100 and several thousand as a transition zone between small data and Big Data. *Velocity* refers to the speed with which data are generated. Small data may require a considerable amount of time to collect. For example, if you wish to get 30 people's opinions about a new mapping technique and you want to conduct face-to-face interviews, you may spend several days collecting your data. In contrast, Big Data can be collected in near real-time; for instance, you might post an app for a new mapping technique to the Web and start getting people's reaction to the technique (via recorded mouse clicks) in a few minutes.

Variety refers to the various kinds of data that are collected. The collection of small data tends to be tightly structured and purposeful in nature—you set up an interview or you design an experiment to collect a certain type of data. In contrast, Big Data are often collected in unstructured situations, and so the range of possible data types is large—you might analyze Twitter feeds, YouTube

DOI: 10.1201/9781003150527-29

TABLE 26.1

Comparison of Small Data and Big Data on Seven Criteria

Criterion	Definition	Small Data	Big Data
Volume	Size of dataset	Generally, less than 100 observations	Thousands to millions of observations
Velocity	Speed of generation	Days to weeks	Near real-time
Variety	Kinds of data	Tends to be structured	Greater variety and often unstructured
Exhaustivity	How is the population represented?	Samples are used	Entire population is used
Resolution	How detailed are the data?	Coarse	Fine
Relationality	Ease of relating data to other data	Generally difficult	Relations are easily established
Flexibility	Ease of expanding the database	Moderately difficult	Easy

Source: After Kitchin and McArdle (2016, 2).

videos, or clicks on various locations on a map that you have posted. *Exhaustivity* refers to the extent to which you examine the population of interest. With small data, the presumption is that you collect a sample of data because you don't have the time and money to examine the entire population (as we discussed in Chapter 3). With large data, we often have access to the entire population; for example, given the proper computer resources, we can examine all Twitter feeds about a topic for a particular time period.

Resolution refers to the fineness (or alternatively coarseness) of the data collected. Small data have a relatively coarse resolution, whereas Big Data have a fine resolution. For instance, census data traditionally are reported for enumeration units rather than individuals (to protect privacy), whereas Big Data can readily provide data about individuals (as in a Twitter feed), including the location of that individual. *Relationality* refers to the ability to easily relate one dataset to another via a common field (which is typically found in GIS databases); although relationality is commonplace with Big Data, it oftentimes is not implemented when producing small data. Finally, *flexibility* refers to the ability to easily add new fields of data to the database and expand the size of the database. Such features are common in the sophisticated computational environments characterized by Big Data.

Although this comparison of small and Big Data suggests numerous advantages for Big Data, Rob Kitchin (2013) noted several reasons why we should be cautious in embracing Big Data. One problem is that there has been a tendency for some to jump to an analysis of the data without first thinking about potential relationships within the data. Traditionally, when working with small data, researchers developed hypotheses about the data and then tested those hypotheses using the collected data. Instead, some working with Big Data have argued that hypotheses are unnecessary and that we should simply "let the data speak for itself." Although this approach may be useful for uncovering initial ideas or relationships that you hadn't thought of, a danger is that it does not make use of past hypotheses and theories that have been developed. For example, you might discover a high correlation between two attributes that is completely spurious when you consider the cause and effect relationships that have been established by prior research.

The second potential problem is that emphasizing Big Data may denigrate the utilization of small data studies, which can be quite beneficial. For instance, cartographers have often used *focus group studies* (e.g., Kessler 2000), a qualitative methods approach in which small groups of 4 to 6 potential map users are asked their thoughts about a particular mapping technique. Such small group studies can be extremely useful in setting the stage for studies that involve a much larger number of users and more focused research questions. The third potential problem is that the results of any Big Data analysis are a function of the particular algorithms that are used in the analysis. In this context, Mei-Po Kwan (2016) argues that it is not the Big Data per se that is the problem but rather a failure to consider the nature of the algorithms that are used to process the data. Do we fully understand the algorithms that are used, and how do we select between two algorithms that produce dramatically different results?

26.4   WHAT IS GEOVISUAL ANALYTICS?

The term *visual analytics* was developed by the U.S. National Visualization and Analytics Center (NVAC), which was formed in response to September 11, 2001, terrorist attacks on the United States. NVAC's research agenda and the notion of visual analytics initially were defined in the book *Illuminating the Path: The Research and Development Agenda for Visual Analytics* by Jim Thomas and Kristin Cook (2005). Thomas and Cook defined **visual analytics** as "the science of analytical reasoning facilitated by interactive visual interfaces" that should "detect the expected and discover the unexpected" (p. 4). Although visual analytics was initially developed for U.S. homeland security purposes, today, the term is used to describe a much broader range of applications, both in the United States and around the world.

When geospatial data are considered, the term geovisual analytics is often used. For our purposes, we will define **geovisual analytics** as computer-based systems that combine the visualization power of humans with the computational power of computers to make sense out of large geospatial datasets (Big Data). These geospatial datasets are often multivariate (they involve numerous attributes), and they often involve a temporal component (we refer to the datasets as spatiotemporal). "Making sense" out of these datasets can be equated to what Thomas and Cook termed "discover the unexpected"; in the context of geospatial data, this sense-making normally will involve finding previously unknown spatial patterns. Integrating human visualization with computational methods generally requires a highly interactive graphical interface so that the two approaches can easily be used in tandem.

The visualization portion of geovisual analytics is often described as following Ben Shneiderman's (1996) **information-seeking mantra**, which includes overview, zoom and filter, and details on demand. In the overview stage, the analyst visualizes a summary of the data. In the zoom and filter stage, the analyst uses interactive techniques that are likely to reveal interesting patterns or data subsets that might merit further investigation. Finally, in the details on demand stage, the analyst focuses on patterns identified in the zoom and filter stage and forms or validates hypotheses (Keim et al. 2004, 41). Since very large datasets often make it difficult to visualize an overview of the data, Daniel Keim identified a **visual analytics mantra**, which consists of the following five stages: "Analyze first, Show the Important, Zoom, filter and analyze further, and Details on demand" (Keim et al. 2008, 164).

In their book, *Visual Analytics for Data Scientists*, Natalia Andrienko and colleagues (2020) provide a comprehensive overview of visual analytics (and geovisual analytics) and their relation to data science. They indicate that the basic steps of **data science** involve selecting and processing data, exploring the data to identify patterns, verifying the patterns, developing models, and communicating results. Within data science, this traditionally has been accomplished by emphasizing computational approaches, whereas in visual analytics, the emphasis has been on visualization and human reasoning. Andrienko et al. argue that data science will be most effective if the computational methods are combined with visualization and reasoning approaches (22).

Visualization for geovisual analytics can be accomplished using the varied techniques that we introduce throughout this book, with an emphasis on the interactive approaches that we discussed in Section 25.4. Andrienko et al. introduce numerous additional visualization approaches, some of which we will illustrate in the examples in Section 26.6. Computational approaches for geovisual analytics can be divided into two broad categories: *spatialization* and *grouping* (Andrienko et al. 2020).[3] These can be most easily understood if we think of our basic input data as a data matrix, where the attributes are columns and the observations are rows. Spatialization essentially involves reducing a large number of attributes to two or three dimensions that can be visualized. A classic example is principal component analysis in which each component is a linear combination of the original attributes (Davis 2002). Grouping involves combining observations into groups, typically involving a

cluster analysis approach. We discussed the concept of hierarchical clustering in Section 22.5; in the next section, we introduce a nonhierarchical clustering method, the SOM.

26.5   THE SELF-ORGANIZING MAP (SOM)

In this section, we consider the notion of the **SOM**. We base our discussion largely on André Skupin and Pragya Agarwal's (2008) "What Is a Self-Organizing Map?," which is the first chapter of *Self-Organizing Maps: Applications in Geographic Information Science*, an entire book dedicated to SOMs.

The SOM is a type of *artificial neural network* (ANN). Inspired by the notion of biological neural networks, ANNs consist of a set of interconnected nodes (or neurons). An example of an ANN based on a remote sensing application is shown in Figure 26.1A. Here each of the circles

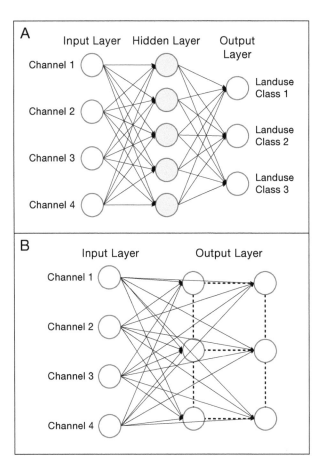

FIGURE 26.1   Structure of ANNs for supervised (A) and unsupervised (B) approaches using remote sensing channels (or bands) as input and land use classes as output. A self-organizing map (SOM) generally corresponds to the unsupervised approach. Note that there are six land use classes in the unsupervised approach, as indicated by the 3 × 2 matrix of output nodes. (From Skupin, A. and Agarwal, P. (2008) "Introduction: What is a self-organizing map?" In *Self-Organizing Maps: Applications in Geographic Information Science*, ed. by P. Agarwal and A. Skupin, pp. 1–20. Chichester, England: John Wiley. Copyright © 2008 John Wiley & Sons Ltd.)

represents a node, and we can see that the nodes have been divided into an *input layer*, a *hidden layer*, and an *output layer*. The nodes for the input layer consist of pixel information for each of four channels (bands) of a remotely sensed image. The nodes for the output layer consist of three potential land use classes for pixels in the image. Using the input node information as a basis, the key to ANNs is determining the weights to assign to each of the incoming links for the nodes in the output layer. The process of determining these weights is known as *training the network*. The details of this processing are often hidden from the user, and thus the term "hidden layer" makes some sense.

The ANN that we have just described is known as a *supervised* approach because we presumed that we had determined the potential land use classes before training the network. An alternative approach, again using our remote sensing example, would be an *unsupervised approach*, where we do not know the potential land use classes beforehand (Figure 26.1B). Note that here we have the same nodes in the input layer, but we have no hidden layers, and our output layer consists of a matrix of 3 × 2 nodes. Normally, you would experiment to see what number of nodes works best with your particular data set. The SOM is generally associated with the unsupervised approach.

Skupin and Agarwal stress that the SOM can be viewed as either a clustering approach or a dimensionality reduction (a spatialization) approach. For our purposes in this section, we consider the SOM as a clustering approach. In the examples section of the chapter (Section 26.6.5), we consider a paper by Skupin and Esperbé (2011), where the SOM is used as a dimensionality reduction approach.

In Chapter 22, we introduced the notion of cluster analysis and noted that traditionally there were two basic approaches to cluster analysis, the hierarchical and the non-hierarchical approaches. In the hierarchical approach, each observation begins as a separate cluster, and the clusters are progressively combined. An example of this is the unweighted pair-group method using arithmetic averages (UPGMA) that we covered in detail in Chapter 22. In contrast, non-hierarchical clustering methods do not presume a hierarchy. An example of non-hierarchical clustering is the

k-means method: in this method, we assume a desired number of clusters and then determine cluster membership by moving observations among clusters such that we minimize the sum of the Euclidean distances of each observation to the centroid of each cluster.

Like the *k*-means method, the SOM method is also non-hierarchical, and it makes use of the Euclidean distance metric, but a key difference is that the nature of the output layer provides an explicit expression of the topological relationship between clusters. To illustrate, we used the SOM to cluster the same data that we clustered using the UPGMA method in Chapter 22: the scores of 57 counties of New York State on the four attributes percent population change, percent unemployed, percent African American, and infant mortality rate. To accomplish this, we utilized the V-Analytics software (http://geoanalytics.net/V-Analytics/), which was originally CommonGIS (as discussed in Chapter 25). This software essentially duplicates the classic SOM_PAK approach developed by Teuvo Kohonen and colleagues (1996).

In order to express the topological relations between clusters, the output nodes normally are set up as a square matrix. To produce a result at least somewhat comparable to the seven-cluster solution we created using UPGMA, we used a 3 × 3 matrix of output nodes, which produced nine clusters (Figure 26.2). The mean z-scores for the four attributes on the nine clusters are shown in Table 26.2. As with the UPGMA method in Chapter 22, we named the clusters in Figure 26.2 as a function of whether the z-scores were close to ±1.0 or ±2.0 (values near ±1.0 represent moderately high or low scores, whereas values near ±2.0 represent distinctly high and low scores). An illustration of the topological relation between SOM clusters is that clusters near one another in the legend of Figure 26.2 are more similar to one another than those further away from one another. For example, note how clusters 1 and 2 are more similar to one another than are clusters 3 and 7 (e.g., cluster 1 has high infant mortality and cluster 2 has moderately high infant mortality, whereas cluster 3 has moderately high % African American and cluster 7 has low % African American). We have utilized a color scheme to depict the

TABLE 26.2

Mean z-Scores for Attributes on Each of the SOM Clusters

Cluster	% Population Change	% Unemployed	% African American	Infant Mortality Rate
1	−0.20	0.62	−0.16	1.85
2	−1.20	−0.78	−0.08	0.92
3	−0.81	−0.65	1.20	−0.05
4	−0.49	1.16	−0.59	0.31
5	−0.34	−0.38	−0.48	−0.05
6	0.50	−0.86	1.22	−0.29
7	−0.39	0.75	−0.80	−0.78
8	0.66	−0.45	−0.73	−0.64
9	1.54	−0.82	0.57	−0.47

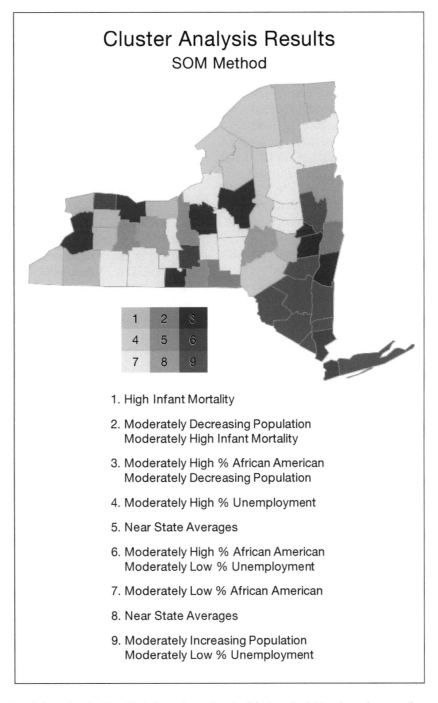

FIGURE 26.2   Results of clustering the New York State data using the SOM method. Note how clusters adjacent to one another in the 3×3 output matrix (the legend) are more similar to one another than those far apart.

clusters that reflects the topological similarity of the clusters in the legend (e.g., note that the colors for clusters 1 and 2 are similar to one another). For more advanced treatment of the use of color in SOMs and associated maps, see Andrienko et al. (2010).

26.6   EXAMPLES OF GEOVISUAL ANALYTICS

In this section, we cover several examples of geovisual analytics. Since geovisual analytics is a rapidly evolving discipline, space does not permit us to cover the full breadth of possibilities. Other examples that you may wish to examine include Slingsby et al. (2010) and Wood et al.'s (2011) work with *spatial treemaps*; Ferreira et al.'s (2011) visualization of bird populations (http://www.birdvis.org/); Lins et al.'s (2013) work with *nanocubes*; von Landesberger et al.'s (2016) MobilityGraphs approach for visualizing flows of people; Wang et al.'s (2017) Name Profiler Toolkit for linking people's names with demographic data; and Curry et al.'s (2019) use of crowd-sourced data to examine mass migration events. For still more examples, see the Further Reading section.

26.6.1 TAXIVIS: A SYSTEM FOR VISUALIZING TAXI TRIPS IN NYC

TaxiVis is a software tool that was developed by Nivan Ferreira and colleagues (2013) to explore and visualize taxi trips in New York City (NYC). At the time TaxiVis was developed, there were approximately 500,000 taxi trips each day in NYC. In total, Ferreira et al. were provided taxi data for three years encompassing over 540 million taxi trips, which is clearly indicative of the Big Data often dealt with in geovisual analytics. Data for each trip included the pickup and drop-off locations, along with a variety of other data including the traveled distance, fare amount, and tip amount. Information on the route trajectory was not available.

Ferreira et al. argued that although tools like R, MATLAB, and ArcGIS can analyze this sort of data, they cannot handle the entire dataset in a single setting. Rather, users must examine slices of the dataset; moreover, users must be familiar with the user interfaces and associated commands found in these tools. As a result, they developed TaxiVis. Although TaxiVis is intended explicitly for examining the taxi data, they argue that their approach could be applied to similar origin-destination data for other applications (2152).[4]

Ferreira et al. developed potential user queries for TaxiVis on the basis of Donna Peuquet's (1994) classic Triad Framework, in which information is stored relating to *where* (a location-based view), *what* (an object-based view), and *when* (a time-based view). We stress this framework because it could be utilized in other geovisual analytics applications. The framework permits answering three kinds of questions:

1. when + where → what: This permits determining the objects present (what) at particular times (when) and locations (where). For example, we could ask, "which taxi trips occurred between 10 and 11 AM on May 2, 2012, in Greenwich Village?"
2. when + what → where: This permits determining the locations (where) for a set of objects (what) at particular times (when). For example, we could ask, "what was the spatial distribution of taxi trips in which there was no tip after 5 PM on May 13, 2011?"
3. where + what → when: This permits determining the times (when) pertaining to a set of objects (what) found at particular locations (where). For example, we could ask, "what was the temporal distribution in 2012 for taxi trips exceeding 2 kilometers in NYC?"

An example of a basic visualization provided by TaxiVis is shown in Figure 26.3, where blue and red dots are used to show individual pickups and drop-offs, respectively, for a four-hour period for the area near the Port Authority Bus Terminal in NYC. Using individual dots works fine for a short time interval and when the area covered is small, as is the case here. In other situations, however, different approaches are required. As an example, Figure 26.4 displays a small multiple of heat maps of taxi trips in the Manhattan area of NYC around the time that Hurricane Sandy struck the east coast of the United States in 2012. In this instance, a **heat map** of different lightnesses of yellow

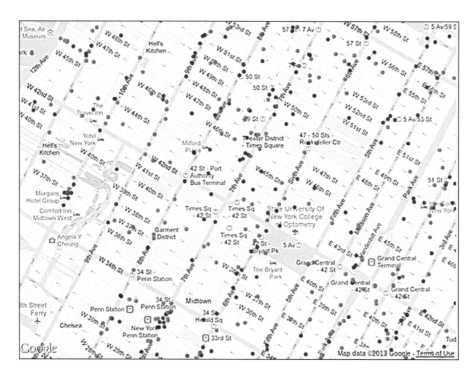

FIGURE 26.3 An image from TaxiVis showing taxi pickups and drop-offs (represented by blue and red dots, respectively) in the area near Port Authority Bus Terminal in NYC. (From http://vgc.poly.edu/projects/taxivis/#video; courtesy of Nivan Ferreira.)

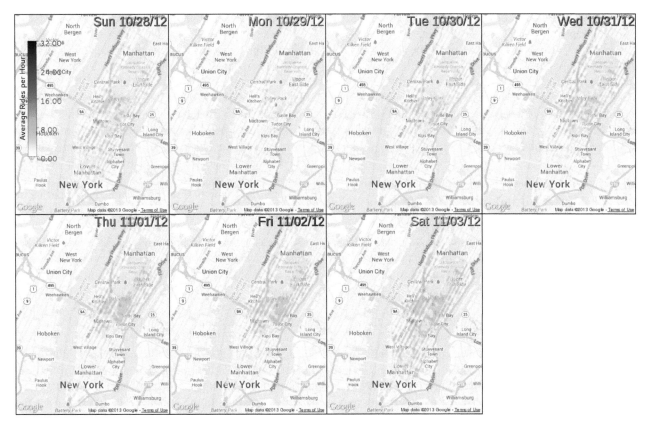

FIGURE 26.4 A small multiple of heat maps from TaxiVis showing the density of taxi trips in the Manhattan area of NYC around the time that Hurricane Sandy struck the east coast of the United States in 2012. (© 2013 IEEE. Reprinted, with permission, from Ferreira, N., Poco, J., Vo, H. T., et al. (2013) "Visual exploration of Big spatio-temporal urban Data: A study of New York City taxi trips." *IEEE Transactions on Visualization and Computer Graphics* 19, no. 12:2149–2158.)

and orange is used to depict the density of dots (representing taxi trips) in the region. If we attempted to show the individual dots, they would overlap one another, but if we overlay a grid on the map and compute the density of dots within a search radius of each grid cell, we can create a smooth representation (the resulting heat map) similar to the unclassed isarithmic map shown in Figure 17.15D.[5]

Note that the darker yellow and oranges in Figure 26.4 represent a higher density of taxi trips.

Another example of a visualization from TaxiVis is the portrayal of taxi trips in the neighborhoods shown in Figure 26.5. Here Harlem and East Harlem are outlined in blue, Midtown is outlined in orange and red, and East, West, and Greenwich village are outlined in green. The heat

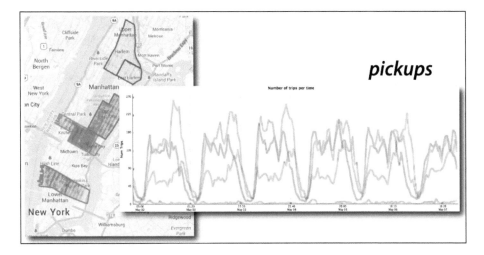

FIGURE 26.5 A heat map and graph from TaxiVis displaying taxi pickups for several neighborhoods in NYC. (© 2013 IEEE. Reprinted, with permission, from Ferreira, N., Poco, J., Vo, H. T., et al. (2013) "Visual exploration of Big spatio-temporal urban Data: A study of New York City taxi trips." *IEEE Transactions on Visualization and Computer Graphics* 19, no. 12:2149–2158.)

map shows the total pickups for the first week of May 2011, and the graph shows changes in pickups over the course of the week. Ferreira et al. note that the heat map reflects the social inequalities of taxi service, as "people in Harlem have long complained about the lack of taxi service in their neighborhood" (2156). The graph clearly supports this difference as we can see a blue line (for the Harlem area) near the bottom of the graph with lines for the other neighborhoods well above this line.[6]

26.6.2 Mosaic Diagrams: A Technique for Visualizing Spatiotemporal Data

When working with Big Data, we often find that we need to utilize unusual visualization approaches to handle a large amount of data that needs to be displayed. In this section, we consider one of these approaches, the mosaic diagram, that was utilized by Gennady Andrienko and Natalia Andrienko (2010) to examine the results of GPS-tracking of 17,241 cars in Milan, Italy, during a one-week period. Initially, this dataset consisted of more than two million records.

Andrienko and Andrienko indicated that such movement data can be viewed in two ways, as a situation-oriented view and as a trajectory-oriented view. In the *situation-oriented view*, we are interested in the spatial positions of all moving entities and the values of movement-related attributes such as speed and direction *for some time moment*. In the *trajectory-oriented view*, we are interested in the movement of individual entities and of the collective movement of multiple entities. For example, we might be interested in the trajectory of a single car over a 4-hour period and the general trend of trajectories for a group of ten cars over a 4-hour period. Due to space limitations, we focus here on just the situation-oriented view. In their paper, Andrienko and Andrienko present computational procedures for handling the trajectory-oriented view.

One approach for handling the situation-oriented view is to create an animation of the Milan GPS car data. To illustrate this concept, Andrienko and Andrienko displayed three frames of an animation as a small multiple (Figure 26.6). Here an equally spaced grid has been laid over Milan and the median speed in each grid cell has been computed over hourly intervals; higher speeds are depicted by larger circle sizes. All three frames of the animation suggest that speeds are lower in the central part of the city—it turns out that a system of belt roads around the periphery of the city enables higher speeds there. We can also see variation from frame to frame; for example, comparing the 17–18-h and 19–20-h frames, speeds seem generally higher on the belt roads in the latter frame.

A more complex and intriguing example of a situation-oriented view is to use the mosaic diagrams shown in Figure 26.7A. **Mosaic diagrams** provide a graphical representation of data that would normally be shown in contingency tables (Friendly 1994). Note that the mosaic diagram shown on the left-hand side of Figure 26.7A consists of 24 rows (for each hour in a day) and 7 columns (for each day of the week). As indicated in the legend just above the diagram, each grid cell in the diagram is used to indicate the median speed of cars for a particular hour and day at a particular location. In the middle of the figure, we see that mosaic diagrams have been laid on top of the city of Milan in a grid format: the value for a grid cell within a particular mosaic diagram is a function of all cars falling within the diagram on the map for the hour and day associated with that grid cell.

Representative examples of mosaic diagrams for the Milan data are shown in Figure 26.7B. Comparing these representative examples with those shown on the map reveals that the central portion of the city is characterized by relatively slow speeds regardless of the time of day or day of the week. In contrast, some of the areas where belt

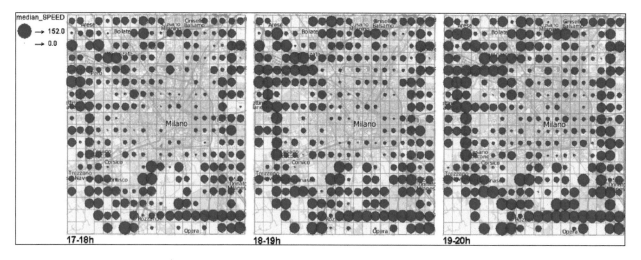

FIGURE 26.6 A situation-oriented view of the Milan GPS data consisting of three frames of an animation. An equally spaced grid has been laid over Milan and the median speed in each grid cell has been computed for each of the three hourly intervals; higher speeds are depicted by larger circle sizes. (From "A general framework for using aggregation in visual exploration of movement data," by G. Andrienko and N. Andrienko, *The Cartographic Journal*, copyright © The British Cartographic Society 2010, reprinted by permission of Taylor & Francis Ltd., http://www.tandfonline.com on behalf of The British Cartographic Society.)

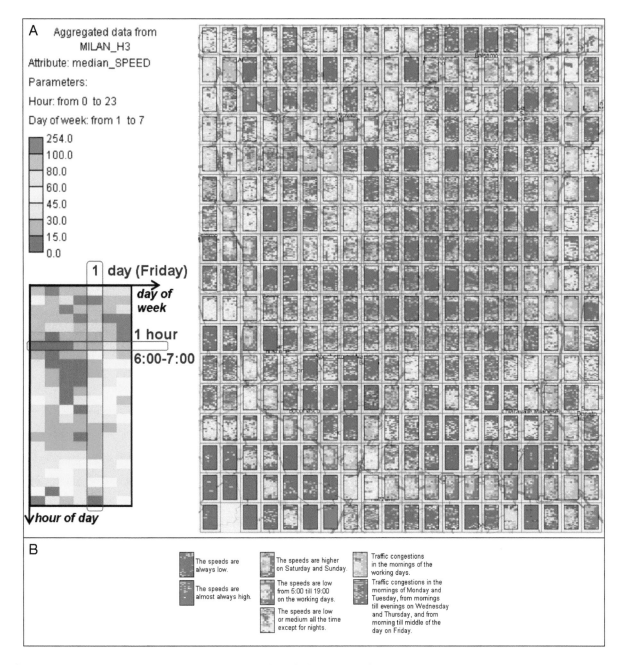

FIGURE 26.7 (A) A situation-oriented view of the Milan GPS data in which mosaic diagrams have been placed within equally sized grid cells overlaying Milan. The diagrams show the median speed for each hour of the day for each day of the week in a grid cell. (B) Representative examples of mosaic diagrams for the Milan data. (From "A general framework for using aggregation in visual exploration of movement data," by G. Andrienko and N. Andrienko, *The Cartographic Journal*, copyright © The British Cartographic Society 2010, reprinted by permission of Taylor & Francis Ltd., http://www.tandfonline.com on behalf of The British Cartographic Society.)

roads are found (on the periphery of the city) tend to have relatively fast speeds regardless of the time of day and day of the week; note in particular two of the cells in the northwest. Some readers may have difficulty seeing these characteristics due to the red–yellow–green diverging color scheme. Although this **traffic light color scheme** is logical because red is associated with slow speeds and green with fast speeds (the concept is similar to a traffic light that uses red for stop and green for go), those readers with a red–green color deficiency may not be able to see the difference between red and green. As discussed in Section 15.4.3, however, there are alternative diverging color schemes that could be utilized for the mosaic diagram.

26.6.3 CarSenToGram: An Approach for Visualizing Twitter Data

CarSenToGram is an approach developed by Caglar Koylu and colleagues (2019) to analyze and visualize tweets (from Twitter) related to President Donald Trump's suspension of the entry of people into the United States from seven predominately Muslim countries on January 27, 2017. Using the Twitter Streaming API, Koylu et al. collected tweets that contained keywords related to immigration, and Muslim refugees and immigrants (e.g., "immigrant," "Muslim," and "refugee"), for four weeks before and four weeks after Trump's travel ban.[7] There were more than 72,000,000 "immigration" tweets generated by more than 6,000,000 users worldwide, obviously reflecting the Big Data nature of this problem. Since the focus was on the reaction to the travel ban within the United States, only English tweets that were generated within the United States were considered. To further clean the data, retweets were deleted, and tweets

related to nudity and pornography were filtered. After these modifications, approximately 2,000,000 tweets remained for analysis.

The next step was to determine the topics mentioned within the tweets, which is known as *topic modeling.* This was accomplished using Latent Dirichlet Allocation (LDA), a **machine learning** approach that uses computational methods to determine an appropriate set of topics represented in the tweets.[8] Koylu et al. experimented with numerous models for each of the eight weeks in the time period of interest, ultimately deciding that there were 160 possible topics. They then clustered these 160 topics to create the six topics shown at the top of Figure 26.8 (the names

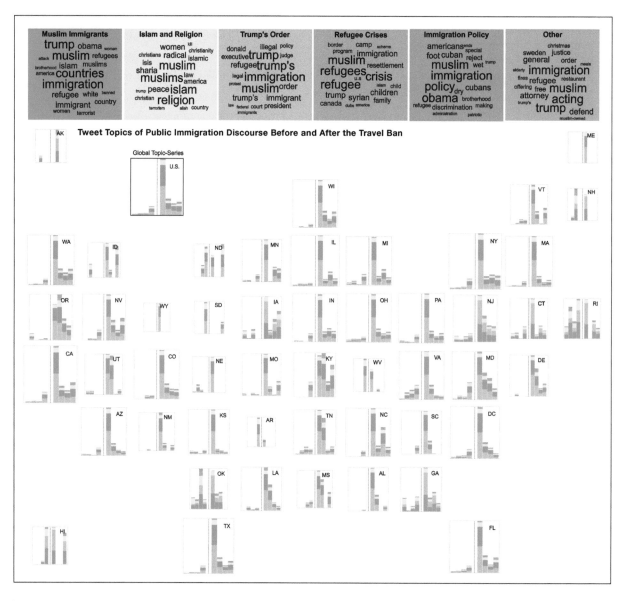

FIGURE 26.8 Topic View in CarSenToGram in which spatiotemporal data associated with tweets are depicted in a noncontiguous rectangular cartogram. By default, each state is scaled on the basis of the total number of tweets in that state. Within each state, bar graphs are shown for each of eight weeks depicting the relative distribution of the six topics shown above the map. The dashed line depicts the date of President Trump's travel ban. (After Koylu, C., Larson, R., Dietrich, B. J., et al. (2019). First published in *Cartography and Geographic Information Science* 46(1), p. 65. Reprinted with permission from the Cartography and Geographic Information Society. Courtesy of Caglar Koylu.)

of the topics are shown at the very top of each box).[9] Each tweet was then assigned to one of these six topics, as a function of which topic the tweet was most highly associated with (based on probability scores from the LDA).

Koylu et al. then conducted a *sentiment analysis*, which provides the sentiment (or opinion) associated with each of the tweets (Liu 2015). This was accomplished using TextBlob (https://textblob.readthedocs.io/en/dev/), which automatically assigns a score ranging from −1 (an extremely negative view) to +1 (an extremely positive view) to each tweet.

To portray the results of the topic modeling and sentiment analysis, Koylu et al. provide an interactive web application (https://ryan-p-larson.github.io/paper/). We encourage you to examine this application as we discuss its components. By default, the application displays a Topic View of tweets in which the spatiotemporal distribution of tweets for the six topics is portrayed using a noncontiguous rectangular cartogram (Figure 26.8). Rectangles are used for each state so that it is possible to easily display the bar graphs for topics shown within each state, and the noncontiguous character of the cartogram allows states to be placed near their correct geographic location. Each state is scaled based on the total number of tweets, although the scaling is adjusted so that states with a small total number of tweets are still legible. The Tile Sizing option in the upper right can be utilized to scale the rectangles based on population.[10] In Figure 26.8, a rectangle for the overall United States is shown above the rectangles for Idaho (ID) and North Dakota (ND) (in the interactive application, the rectangle for the United States appears *between* ID and ND).

Each rectangle for a state in Figure 26.8 displays bar graphs for each of the eight weeks showing the number of tweets for each week and the relative contribution of each of the six topics. A dashed line indicates the date of the travel ban. You should examine the distribution of tweet topics within individual states and also consider how you would summarize the overall spatial pattern of these tweet topics. You can get a better feel for the contribution of each topic by clicking on the word cloud associated with that topic (other topics will appear more transparent). You can also get more detail on individual states by simply clicking on them. For instance, Figure 26.9 shows details for the state

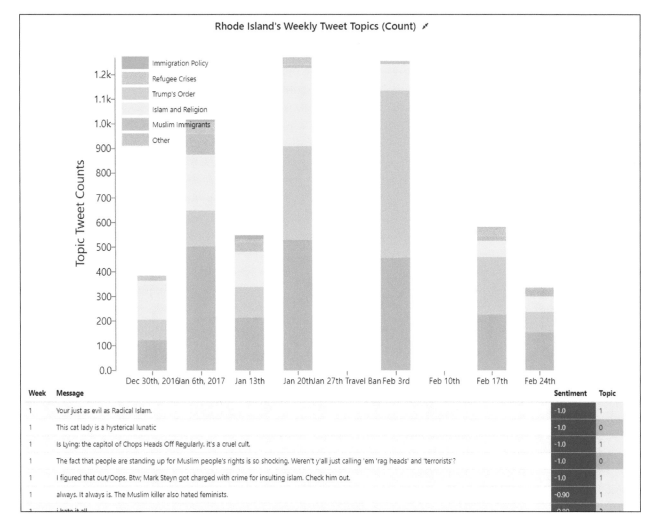

FIGURE 26.9 A detailed view of bar graphs in CarSenToGram depicting the tweets for Rhode Island for four weeks prior to and four weeks after the travel ban. Below the graphs is a list of representative tweets along with sentiment scores and the topic most strongly associated with each tweet. (Courtesy of Caglar Koylu.)

of Rhode Island. In addition to providing a clearer view of the relative contribution of the six topics, this view provides a list of tweets along with the sentiment rating and the topic which the tweet is most closely aligned with. Tweets between −0.33 and +0.33 are not included as those do not reflect strong views in either direction.

An alternative view of the data is the Distribution View shown in Figure 26.10. This view shows just the number of tweets, ignoring a breakdown into the six topics, with orange and blue bars used for tweets before and after the travel ban, respectively. You can see the difference between this view and the Topic View if you focus on one state and toggle between the two views. The detailed topic information is now missing, but the overall character of the number of tweets before and after the travel ban is much clearer.

A final intriguing display is the Sentiment View shown in Figure 26.11, where we see a summary of the number of extreme negative and positive sentiment scores (those less than or equal to −0.33 and those greater than or equal to +0.33) shown in orange and blue, respectively. In this case,

the total area of each rectangle is a function of the total number of extreme scores, ignoring those close to 0. You should examine each state and think about how the relative number of negative and positive sentiments compare with one another and how the values for one state compare with other states (i.e., what is the overall spatial pattern?).

26.6.4 CROWD LENS: A TOOL FOR VISUALIZING OPENSTREETMAP CONTRIBUTIONS

Crowd Lens for OpenStreetMap (OSM) is a web tool developed by Sterling Quinn and Alan MacEachren (2018) that enables professional users of OSM to evaluate the characteristics of those individuals who make contributions to the OSM database. Although intended for professional users of OSM, it is our experience that Crowd Lens can be used by nonprofessionals to develop an appreciation for the character of OSM contributors. As we discussed in Chapter 1, OSM is an example of a spatial database that has been developed using VGI: the volunteers (or the "crowd")

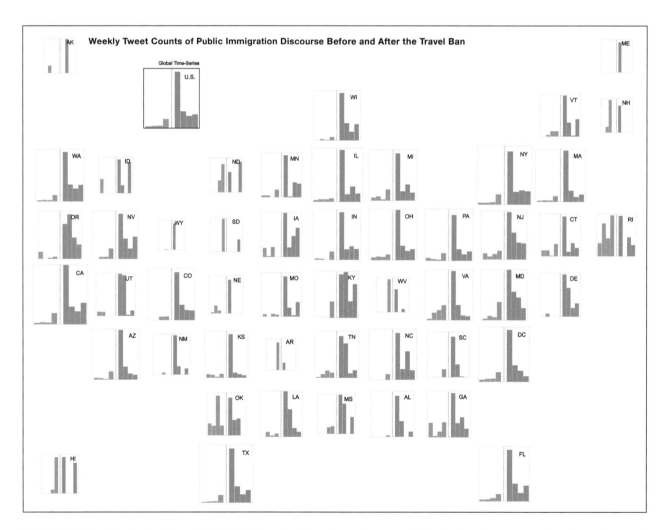

FIGURE 26.10 Distribution View in CarSenToGram depicting the spatiotemporal distribution of tweets, ignoring the topics associated with the tweets. Tweets before the travel ban are shown in orange, while tweets after the travel ban are shown in blue. (After Koylu, C., Larson, R., Dietrich, B. J., et al. (2019). First published in *Cartography and Geographic Information Science* 46(1), p. 64. Reprinted with permission from the Cartography and Geographic Information Society. Courtesy of Caglar Koylu.)

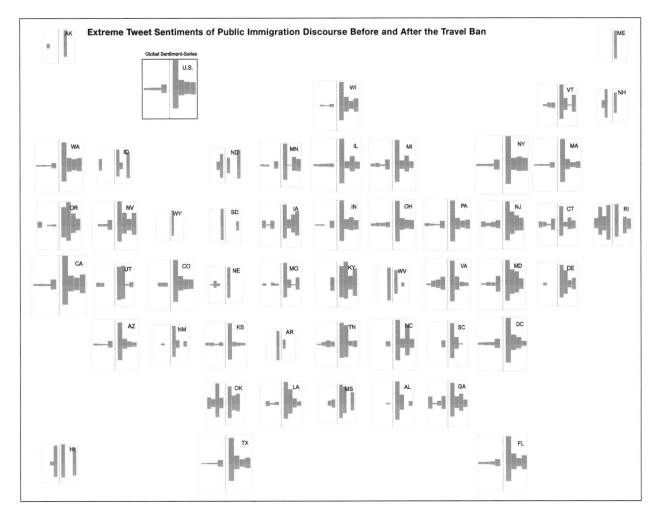

FIGURE 26.11 Sentiment View in CarSenToGram depicting a summary of extremely negative and positive sentiment scores (in orange and blue, respectively). (After Koylu, C., Larson, R., Dietrich, B. J., et al. (2019). First published in *Cartography and Geographic Information Science* 46(1), p. 68. Reprinted with permission from the Cartography and Geographic Information Society. Courtesy of Caglar Koylu.)

are those individuals who wish to make contributions to the database. OSM fits into the framework of Big Data in two ways: (1) the size of the database now rivals those developed by commercial and government entities, and (2) over 2 million people have contributed to the database (141).

Although the functions used within Crowd Lens could be used to analyze the entire OSM system, Crowd Lens is intended as a proof-of-concept, and so the focus is on a sample of six small cities located on six continents. For illustrative purposes, we will work with Hereford in the United Kingdom.[11] We encourage you to access Crowd Lens (http://sterlingquinn.net/apps/crowdlens/) and experiment with the software as you read through the following discussion. As illustrated in Figure 26.12, the interface for Crowd Lens is divided into four basic parts: (A) a list of contributors for the map being displayed; (B) a map highlighting the changes made to OSM by a contributor; (C) filters that can be applied to the list of contributors in A to focus on particular characteristics; and (D) a panel summarizing the characteristics of a particular contributor. We'll now discuss each of these.

Examining the list of contributors (Figure 26.12A), we can see that 265 people have contributed to the database for Hereford. By default, these are arranged as a small multiple of maps sorted from high to low on the basis of the number of days that they have edited the database for Hereford ("Days active here"). Note that a yellow line is drawn around the contributor with the largest number of days of editing, and this user's changes are highlighted on the map (Figure 26.12B). You can simply click on one map in the small multiple to see the results for a particular contributor displayed in the larger map. You should use the scroll bar to the right of the contributor list to visualize the full range of contributor contributions. One distinctive characteristic you should discover is that the majority of contributors make very few changes to OSM. This was something that Quinn and MacEachren found professionals frequently commenting on when they asked them to provide feedback on Crowd Lens. In the drop-down box at the top of the contributor list, you will see that you can order contributors on the basis of other attributes; for instance, if you select "Changesets here," contributors will be listed by

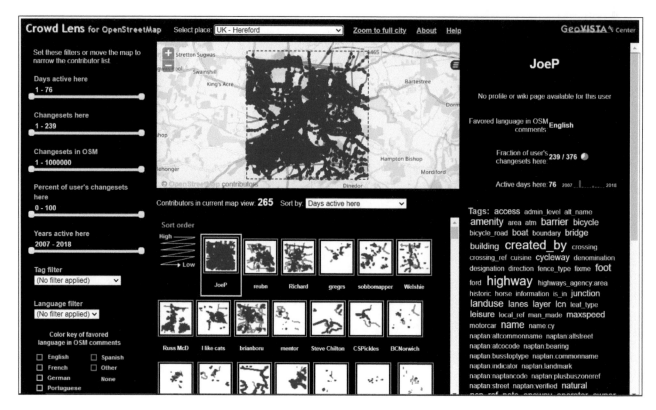

FIGURE 26.12   The interface for Crowd Lens for OpenStreetMap (OSM): (A) The list of contributors to the geographic region of interest (in this case, Hereford, UK); (B) a map from OSM highlighting the geographic region edited; (C) filters that can be applied to the list of contributors in A to focus on particular characteristics; and (D) a panel summarizing the characteristics of a particular contributor. (Courtesy of Sterling Quinn and Alan MacEachren.)

the number of changesets (in OSM, a changeset is a group of changes made over a short period of time; https://wiki. openstreetmap.org/wiki/Changeset).

To illustrate the use of the filters shown in Figure 26.12C, we can change the "Percent of user's changesets here" from "0–100" to "70–100," thus requiring that the bulk of the contributor's changesets in OSM should be in Hereford. We can argue that this is desirable because the resulting contributors should be familiar with Hereford, as that is where they have chosen to develop most of their changesets. Looking at the resulting small multiple of contributors in Figure 26.13, however, we see that the pool of contributors is small and many of them make few changes to the database. This is likely due to the fact that the small percentage of contributors who make the bulk of changes to OSM do so in a variety of spatial contexts (Hereford is just one of those contexts).

Finally, let's turn to the panel summarizing the characteristics of a particular user (Figure 26.12D). To illustrate the use of this parameter, we have used the same parameters as Figure 26.12, but we have chosen "brianboru" as a contributor from the contributor list. In Figure 26.14, we see that this contributor's favored language in OSM comments is English, the fraction of user's changesets is very small (88 out of 62,713), the active days in Hereford is 16, and a "tag cloud" illustrates the contributor's relative contribution to various tags within OSM. Even though brianboru's fraction

of changesets is very small, note that brianboru ranks ninth in days active in Hereford, and the map indicates that this contributor has made substantial changes to the database for Hereford. This again supports the notion that there are "power" contributors who make lots of changes in many distinctly different geographic locations. You can find out more about individual contributors by clicking on "User profile" at the top of the contributor characteristics panel. Using such information, Quinn and MacEachren (2018, 152) noted that 15 out of 34 (44 percent) of the contributors who modified Johnstown in 2015 identified themselves as mapping on behalf of Mapbox, the company that sells web services based largely on OSM data. Thus, it is not strictly volunteers who contribute to OSM.

26.6.5   USE OF A SOM FOR SENSE-OF-PLACE ANALYSIS

André Skupin and Aude Esperbé (2011) developed an innovative approach for using a SOM to uncover the structure in large multivariate datasets. Although they did not use the term geovisual analytics in their paper, their work obviously fits into this framework. Their intention was to combine a broad range of attributes to create a "sense-of-place" of a geographic region: they wished to consider attributes "… in terms of their potential relationship to the sensory experience of a place, i.e., its smells, sights, sounds, and broad physiological impact" (291). For this purpose, they

FIGURE 26.13   An image from Crowd Lens showing the results when the filter "Percent of user's changesets here" is changed from 0–100 to 70–100. Compare with Figure 26.12. (Courtesy of Sterling Quinn and Alan MacEachren.)

considered 69 attributes for the 200,000+ block groups in the United States, including attributes related to climate, topography, soil, geology, land use/land cover, and characteristics of the population, thus combining both physical and human attributes.

Rather than use the SOM as a *clustering technique* (as we did in Section 26.5), Skupin and Esperbé used the SOM to uncover the *structure* in this large multivariate dataset; they argued that this approach is an alternative to the more traditional approaches of multidimensional scaling (MDS) and principal component analysis (PCA). For this purpose, they created an extremely high-resolution SOM consisting of 250,000 neurons (500 in the x-direction and 500 in the y-direction); they argued that this would enable each of the 200,000+ block groups to have a chance of occupying its own neuron. As with other researchers, they found that with this very large number of neurons, it was most effective to train the neural network in two stages; in the first stage, the

network was trained for 60,000 runs, and then the results of this training were submitted to a second stage for 2 million runs. The total training time was 6.5 days using a dual-processor 2.33 GHz Xeon CPU! Clearly, we have a ways to go before this approach can be used in a timely fashion, but they viewed their work as a proof of concept and indicated that studies are underway to parallelize the SOM algorithm and thus improve its efficiency.

Skupin and Agarwal (2008) suggested three possible approaches for visualizing the results of a SOM analysis: (1) visualization of the SOM itself; (2) mapping data onto the SOM; and (3) linking the SOM with other visualizations. To visualize the SOM, Skupin and Esperbé displayed individual component planes of the SOM, one for each of the 69 attributes, where each plane represented the neuron weights of the respective attribute. For ease of visualization, in Figure 26.15, we display just the five planes that Skupin and Esperbé focused on in their discussion: Forest

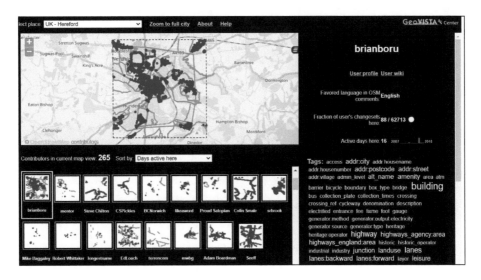

FIGURE 26.14   An image from Crowd Lens showing the characteristics of the user brianboru. (Courtesy of Sterling Quinn and Alan MacEachren.)

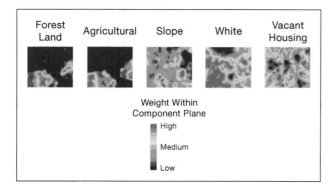

FIGURE 26.15   Five of the 69 component planes resulting from Skupin and Esperbé's (2011) analysis of 200,000+ block groups and 69 attributes. (Reprinted from *Journal of Visual Languages and Computing*, Vol. 22, no. 4, Skupin, A. and Esperbé, A., "An alternative map of the United States based on an n-dimensional model of geographic space," pp. 290–304, Copyright (2011), with permission from Elsevier.)

Land, Agricultural, Slope, White, and Vacant Housing. For these planes, they stressed focusing on the bottom left corner. First, they noted the relationship between Forest Land and Agricultural, indicating that high values on one plane tended to be associated with low values on the other plane. They discussed the notion of forested land being converted to agricultural land as depicted in this area of the SOM, noting that this did not happen for steep slopes (here, they contrasted the Slope plane with the previous two planes). They also suggested that given typical patterns of land use conversion, it could be concluded that this area of the SOM represents regions originally dominated by forest cover, which

were then converted to agricultural land. They argued that forested land that escaped conversion was a function of the steep slopes (contrast the Slope plane with the Forest Land plane). Furthermore, they noted that the non-white population was absent from this part of the SOM (contrast the White plane with the Forest Land plane). Finally, they noted that the forested, steeper portions of this region of the SOM have among the highest proportions of vacant housing, likely reflecting vacation/seasonal homes (contrast the Vacant Housing plane with the Forested Land and Slope planes).

To map data onto the SOM, Skupin and Esperbé determined which of the 250,000 neurons each of the original vectors (the 200,000 block groups with its associated attributes) matched most closely. The result is shown in Figure 26.16, where brown tones indicate areas with a high density of block groups and blue tones indicate areas with a low density of block groups. Skupin and Esperbé argued that areas of low density represent gaps between block groups. To make these gaps more explicit, they conducted a *k*-means clustering of the original input data (the 200,000+ block groups and 69 attributes) and overlaid the clusters; note how the boundaries for these clusters coincide with the low-density blue areas in Figure 26.16.

As an illustration of linking the SOM with other visualizations, consider Figure 26.17. Here the clusters depicted originally in Figure 26.16 have been applied to both the SOM and a map of the United States at the block group level. The patterns on the map look much more complex than those on the SOM because the process of self-organization in the SOM generates relatively smooth transitions between neighboring neurons—these neighboring areas on the SOM may

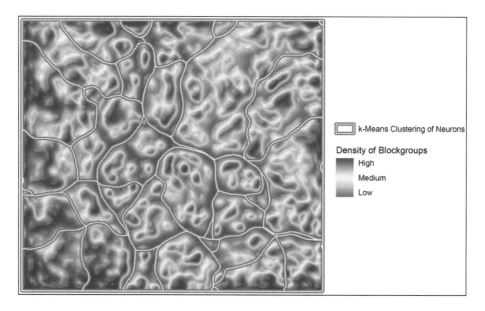

FIGURE 26.16   A visualization from Skupin and Esperbé that illustrates mapping the original data for the 200,000+ block groups onto the SOM. Also depicted are cluster boundaries resulting from a *k*-means clustering of the original data. Note how the cluster boundaries coincide with low-density areas. (Reprinted from *Journal of Visual Languages and Computing*, Vol. 22, no. 4, Skupin, A. and Esperbé, A., "An alternative map of the United States based on an n-dimensional model of geographic space," pp. 290–304, Copyright (2011), with permission from Elsevier.)

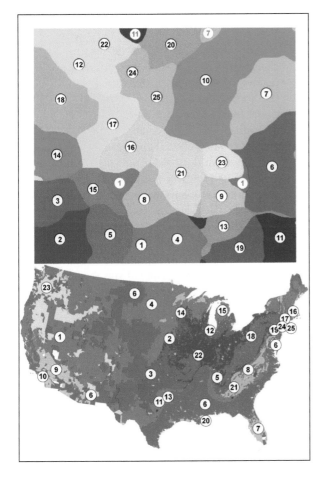

FIGURE 26.17 An illustration of linking a SOM with a traditional map of the data depicted in the SOM. In this case, the results of the *k*-means clusters from Figure 26.16 are shown in the SOM and on the map. Note how some of the colors shown in the SOM (the yellows and greens) are not readily apparent on the map. Skupin and Esperbé indicated that a zoom was necessary in order to visualize the nature of these clusters. (Reprinted from *Journal of Visual Languages and Computing*, Vol. 22, no. 4, Skupin, A. and Esperbé, A., "An alternative map of the United States based on an n-dimensional model of geographic space," pp. 290–304, Copyright (2011), with permission from Elsevier.)

not necessarily be adjacent on the map. In comparing the two images, you will also note that many of the colors in the SOM are not readily apparent on the map (for example, note the absence of greens and yellows). These colors tend to occur in urban areas, which can only be seen through a zoom operation. Skupin and Esperbé illustrate a couple of zoom operations that show how these initially hidden clusters make sense when the map is viewed at a more detailed level. For example, they illustrate and discuss the nature of clusters in New Orleans. We encourage you to examine Skupin and Esperbé's paper to develop a more nuanced view of their approach.

26.7 SUMMARY

Geovisual analytics are computer-based systems that combine the visualization power of humans with the computational power of computers to make sense out of large

geospatial datasets (Big Data). Both computational methods and visualization are essential because visualization alone cannot allow us to interpret extremely large geospatial datasets that are often multivariate and temporal in nature. As the name implies, Big Data are characterized by a large *volume*, but they also have higher *velocity*, greater *variety*, greater *exhaustivity*, higher *resolution*, greater *relationality*, and greater *flexibility* when compared to small data.

The SOM is an example of a computational technique used in geovisual analytics. The SOM can serve as either a grouping approach or a dimension reduction approach, which are two of the general computational approaches used in geovisual analytics. We illustrated the grouping approach by using the SOM to cluster data for New York State that we had previously clustered using a hierarchical UPGMA method, and we illustrated the dimension reduction approach by describing Skupin and Esperbé's study that created a sense-of-place for geographic locations in the United States by combining 69 physical and cultural attributes for more than 200,000 block groups.

In the process of introducing various examples of geovisual analytics, we introduced various visualization approaches, including the heat map and mosaic diagram. Rather than having individual dots overlap one another, a heat map is created by overlaying a grid on the map, computing the density of dots within a search radius of each grid cell, and using a color scheme to symbolize the resulting densities. Mosaic diagrams create a visual expression of data that normally would be displayed in tabular form. For example, rather than creating a table of cells displaying the number of cars passing by a particular point for each hour of the day and each day of the week, we can convert the numbers to a color scheme that can be easily visualized.

Interactivity is often a key element of geovisual analytics applications. We examined two applications that made use of interactivity. With CarSenToGram, we were able to explore several approaches for visualizing the nature of tweets associated with President Donald Trump's travel ban to the United States from seven predominately Muslim countries. Here we also saw an intriguing use of a noncontiguous rectangular cartogram. With Crowd Lens for OSM, we saw how providing users with a range of options for interacting with the data could enable users to learn interesting characteristics about contributors to OSM.

26.8 STUDY QUESTIONS

1. For each of the following, indicate whether the resulting data would be considered small data or Big Data, and briefly indicate why: (A) results from a focus group to determine people's opinion about a new map animation technique; (B) results of the Christmas Bird Count (https://tinyurl.com/Results-Christmas-Bird-Count); (C) recordings of the actions (e.g., options chosen, regions of the map clicked on, and time spent between each of these actions) from 472 people who used a

geovisual analytics application over the course of 10 minutes; and (D) the routes taken by all users of a 500-bicycle pay-per-ride system in New York City over the course of a month.

2. Contrast the following two datasets using the criteria volume, velocity, variety, and exhaustivity shown in Table 26.1: (A) An instructor wishes to determine whether students learn better when using computer-based physical models as compared to when they manipulate actual physical models in their hands. She has one lab of 20 students work with the computer-based models and another lab of 25 students work with the actual physical models. Over the two-hour lab, she watches them work with the different learning approaches and tests their knowledge of the various models at the end of the lab. (B) National Geographic has just released a novel visualization of the ocean floor, and the instructor wants to get the public's reaction to the new visualization, and so she examines all tweets that deal with "National Geographic visualizations" for a week following the introduction of the method.

3. Contrast the map of cluster analysis results for the SOM in Figure 26.2 with the map of cluster analysis results for UPGMA in Figure 22.23.

4. Discuss what the heat maps in Figure 26.4 tell you about taxi trips around the time that Hurricane Sandy struck New York City. What other heat maps would you wish to create in order to fully evaluate the impact of the hurricane on taxi trips? (Note that the hurricane reached New York City on Monday, October 29. For info on hurricane impacts, see https://en.wikipedia.org/wiki/Effects_of_Hurricane_Sandy_in_New_York.)

5. Answer either A or B: (A) Critique the color scheme that is utilized in Figure 26.7 and suggest an alternative color scheme. (B) Regarding the mosaic diagrams shown in Figure 26.7, answer the following questions: (1) In the text, we indicated that the central portion of Milan is characterized by relatively slow speeds regardless of the time of day; where else do you see this characteristic? How might you explain the slow speeds in these areas? (2) Where do you see the most distinctive traffic congestion in the mornings of working days?

6. Respond to the following questions regarding CarSenToGram: (A) Using Figure 26.10, describe the spatial pattern for U.S. states of weekly tweet counts before and after the travel ban. (B) Using Figure 26.11, describe the spatial pattern for U.S. states of extreme tweet sentiments before and after the travel ban.

7. Answer the following questions for Crowd Lens for OpenStreetMap (http://sterlingquinn.net/apps/crowdlens/) using the United States—Johnstown data: (A) How many people contributed to this OpenStreetMap database between 2007 and 2018? Which contributors were the top five in terms of "day active here"? What do the maps of the changes made by contributors suggest about the relative contributions of the various contributors? (B) Sorting the data by "changesets here," which contributors are now the top five? Using the filters, determine how many contributors have made 70 percent or more of their changesets here (i.e., with the Johnstown database). Does the work by these contributors represent a substantial portion of the changes to the Johnstown database? Explain your response. (C) Using the filters, determine which contributors have created 5,000 or more changesets in OSM and at the same time have made between 0 and 20 percent of their changesets with this database. What does the result tell you about the nature of the contributors to the Johnstown database?

NOTES

1 A petabyte is 1×10^{15} bytes, and a terabyte is 1,000,000 megabytes.

2 For additional criteria associated with Big Data, see Li et al. (2020).

3 Together, these can be considered techniques for *data mining*.

4 The source code for TaxiVis is available via github (https://github.com/ViDA-NYU/TaxiVis). A video illustrating the capability of TaxiVis is also available at http://vgc.poly.edu/projects/taxivis/#video.

5 For a discussion of algorithms for creating heat maps with Big Data, see Perrot et al. (2015).

6 Their paper includes a similar map and graph for drop-offs.

7 They excluded tweets for the day of the travel day because they felt that a large number of tweets for that day biased the results.

8 For an introduction to topic modeling and LDA, see https://tedunderwood.com/2012/04/07/topic-modeling-made-just-simple-enough/.

9 The names of topics were based on commonly occurring words in tweets and on the contents of sample tweets within each topic.

10 Note that under the Data View, a Relative option is available that provides a clearer picture of the relative contribution of each of the six topics for each of the eight weeks. It is important to realize, however, that this ignores the magnitude of the associated tweets.

11 The other cities are Hervey Bay, Australia; Johnstown (Pennsylvania), United States; Kadiri, India; Suhum, Ghana; and Tres Arroyos, Argentina.

REFERENCES

Andrienko, G., and Andrienko, N. (2010) "A general framework for using aggregation in visual exploration of movement data." *The Cartographic Journal 47*, no. 1:22–40.

Andrienko, G., Andrienko, N., Bremm, S., et al. (2010) "Space-in-time and time-in-space self-organizing maps for exploring spatiotemporal patterns." *Computer Graphics Forum 29*, no. 3:913–922.

Andrienko, N., Andrienko, G., Fuchs, G., et al. (2020) *Visual Analytics for Data Scientists.* Cham, Switzerland: Springer.

Blumenfeld, J. (2017) "Getting petabytes to people: How EOSDIS facilitates Earth observing data discovery and use," https://earthdata.nasa.gov/learn/articles/tools-and-technology-articles/getting-petabytes-to-people-how-the-eosdis-facilitates-earth-observing-data-discovery-and-use, accessed 2/1/2021.

Curry, T., Croitoru, A., Crooks, A., et al. (2019) "Exodus 2.0: Crowdsourcing geographical and social trails of mass migration." *Journal of Geographical Systems 21,* no. 1:161–187.

Davis, J. C. (2002) *Statistics and Data Analysis in Geology* (3rd ed.). New York: Wiley.

Ferreira, N., Lins, L., Fink, D., et al. (2011) "BirdVis: Visualizing and understanding bird populations." *IEEE Transactions on Visualization and Computer Graphics 17,* no. 12:2374–2383.

Ferreira, N., Poco, J., Vo, H. T., et al. (2013) "Visual exploration of Big spatio-temporal urban Data: A study of New York City taxi trips." *IEEE Transactions on Visualization and Computer Graphics 19,* no. 12:2149–2158.

Friendly, M. (1994) "Mosaic displays for multi-way contingency tables." *Journal of the American Statistical Association 89,* no. 425:190–200.

Keim, D., Andrienko, G., Fekete, J.-D., et al. (2008) "Visual analytics: Definition, process, and challenges." In *Information Visualization: Human-centered Issues and Perspectives,* ed. by A. Kerren, J. T. Stasko, J.-D. Fekete, et al., pp. 154–175. Berlin: Springer-Verlag.

Keim, D. A., Panse, C., Sips, M., et al. (2004) "Visual data mining in large geospatial point sets." *IEEE Computer Graphics and Applications 24,* no. 5:36–44.

Kessler, F. (2000) "Focus groups as a means of qualitatively assessing the U-boat narrative." *Cartographica 37*(4):33–60.

Kitchin, R. (2013) "Big data and human geography: Opportunities, challenges and risks." *Dialogues in Human Geography 3,* no. 3:262–267.

Kitchin, R., and McArdle, G. (2016) "What makes Big Data, Big Data? Exploring the ontological characteristics of 26 datasets." *Big Data & Society 3,* no.1:1–10.

Kohonen, T., Hynninen, J., Kangas, J., et al. (1996) *SOM_PAK: The Self-Organizing Map Program Package. Technical Report A31.* Otaniemi, Finland: Helsinki University of Technology.

Koylu, C., Larson, R., Dietrich, B. J., et al. (2019) "CarSenToGram: Geovisual text analytics for exploring spatiotemporal variation in public discourse on Twitter." *Cartography and Geographic Information Science 46,* no. 1:57–71.

Kwan, M.-P. (2016) "Algorithmic geographies: Big Data, algorithmic uncertainty, and the production of geographic knowledge." *Annals of the American Association of Geographers 106,* no. 2:274–282.

Li, Y., Yu, M., Xu, M., et al. (2020) "Big data and cloud computing." In *Manual of Digital Earth,* ed. H. Guo, M. F. Goodchild and A. Annoni, pp. 325–355. Berlin: Springer.

Lins, L., Klosowski, J. T., and Scheidegger, C. (2013) "Nanocubes for real-time exploration of spatiotemporal datasets." *IEEE Transactions on Visualization and Computer Graphics 19,* no. 12:2456–2465.

Liu, B. (2015) *Mining Opinions, Sentiments, and Emotions.* New York: Cambridge University Press.

Perrot, A., Bourqui, R., Hanusse, N., et al. (2015) "Large interactive visualization of density functions on big data infrastructure." *Proceedings of the IEEE 5th Symposium on Large Data Analysis and Visualization,* Chicago, Illinois, pp. 99–106.

Peuquet, D. J. (1994) "It's about time: A conceptual framework for the representation of temporal dynamics in geographic information systems." *Annals of the Association of American Geographers 84,* no. 3:441–461.

Quinn, S. D., and MacEachren, A. M. (2018) "A geovisual analytics exploration of the OpenStreetMap crowd." *Cartography and Geographic Information Science 45,* no. 2:140–155.

Shneiderman, B. (1996) "The eyes have it: A task by data type taxonomy for information visualizations." *Proceedings of the IEEE Symposium on Visual Languages,* Washington, DC, pp. 336–343.

Skupin, A., and Agarwal, P. (2008) "Introduction: What is a self-organizing map?" In *Self-Organizing Maps: Applications in Geographic Information Science,* ed. by P. Agarwal and A. Skupin, pp. 1–20. Chichester, England: John Wiley.

Skupin, A., and Esperbé, A. (2011) "An alternative map of the United States based on an n-dimensional model of geographic space." *Journal of Visual Languages and Computing 22,* no. 4:290–304.

Slingsby, A., Wood, J., and Dykes, J. (2010) "Treemap cartography for showing spatial and temporal traffic patterns." *Journal of Maps 6,* no. 1:135–146.

Thomas, J. J., and Cook, K. (eds.). (2005) *Illuminating the Path: The Research and Development Agenda for Visual Analytics.* Los Alamitos, CA: IEEE Computer Society.

von Landesberger, T., Brodkorb, F., Roskosch, P., et al. (2016) "MobilityGraphs: Visual analysis of mass mobility dynamics via spatio-temporal graphs and clustering." *IEEE Transactions on Visualization and Computer Graphics 22,* no. 1:11–20.

Wang, F., Hansen, B., Simmons, R., et al. (2017) "Name Profiler Toolkit." *IEEE Computer Graphics and Applications 37,* no. 5:61–71.

Wood, J., Badawood, D., Dykes, J., et al. (2011) "BallotMaps: Detecting name bias in alphabetically ordered ballot papers." *IEEE Transactions on Visualization and Computer Graphics 17,* no. 12:2384–2391.

27 Visualizing Uncertainty

27.1 INTRODUCTION

We often think of maps as truthful representations of reality. For example, imagine viewing a state-level choropleth map of the United States entitled "Percent of the Population That Exercises Regularly." If a shade in the legend has class limits of 21 to 25 percent, then we are apt to presume that states with that shade do, in fact, fall in the 21 to 25 percent range. In reality, this presumption probably would not be correct because the map likely would be based on a sample—thus, there would be a **margin of error (*MOE*)** around the sampled value. Such error in maps is a form of **uncertainty**; in this chapter, we consider approaches for visualizing this uncertainty.

In Section 27.3, we consider the basic elements of uncertainty. Although the term uncertainty is frequently used in the cartographic literature, the terms "reliability" and "quality" are also used (i.e., if data are uncertain, they are also unreliable and of poor quality). We will consider five categories for assessing data quality: lineage, positional accuracy, attribute accuracy, completeness, and logical consistency. We will see that the category of attribute accuracy has received the most attention in the literature pertaining to visualization.

In Section 27.4, we briefly describe the three general ways in which uncertainty can be depicted: (1) separate maps are created for an attribute and its associated uncertainty (i.e., maps are *adjacent*); (2) the attribute and its uncertainty are displayed on the same map (i.e., maps are *coincident*); and (3) dynamic maps are utilized (i.e., either the user is able to interact with maps or animation is utilized). Generally, the "coincident maps" approach has been most commonly utilized, although several software packages have been developed for interacting with data and exploring its associated uncertainty.

Section 27.5 considers various visual variables that can be utilized to depict uncertainty. Broadly speaking, we can group these into **intrinsic** and **extrinsic visual variables**, depending on whether the visual variable depicting uncertainty is visually separable from the visual variable depicting the actual attribute of interest (extrinsic variables are separable, whereas intrinsic variables are not). We will see that although some traditional visual variables (e.g., size and shape) can be utilized, specialized visual variables have been developed for depicting uncertainty, including **crispness**, **resolution**, and **transparency**.

In Section 27.6, we cover applications of visualizing uncertainty. By looking at numerous applications, we hope to give you a feel for the broad range of approaches that have been developed for visualizing uncertainty. We will consider approaches for handling the uncertainty in choropleth maps, visualizing climate change uncertainty, visualizing uncertainty in decision-making, and using interactivity and animation to evaluate uncertainty. In the last section of the chapter, we consider the possibility of using sound to represent uncertainty. Since the argument can be made that the visual channel can become overloaded by having to visualize both an attribute and its uncertainty, it may be beneficial to offload some of the visual processing to the aural channel.

27.2 LEARNING OBJECTIVES

- Describe the character of the three basic approaches for depicting uncertainty and give an example of each: maps are adjacent, maps are coincident, and dynamic maps.
- Define intrinsic and extrinsic visual variables and give examples of each.
- Define three visual variables that have been developed explicitly for visualizing uncertainty.
- Explain the concepts of *MOE* and coefficient of variation (*CV*) as they apply to visualizing uncertainty.
- Explain how confidence levels (*CLs*) can be used to deal with uncertainty in choropleth mapping.
- Contrast the maximum likelihood estimation (MLE) approach for choropleth data classification with the optimal data classification approach described in Chapter 5.
- Contrast pattern-based methods with color-based methods for depicting the uncertainty of climatic change and indicate which approach is likely to be more effective.
- Describe an application of visualizing uncertainty in decision-making.
- In general, indicate whether dynamic approaches (i.e., interaction and animation) are likely to be more effective than static maps for evaluating uncertainty, and discuss situations in which you would use either interactivity or animation.
- Indicate why sound might be useful for depicting uncertainty and describe the sound dimensions that are apt to be most useful.

27.3 BASIC ELEMENTS OF UNCERTAINTY

Uncertainty can arise from a variety of sources, including the raw data on which a map is based, the manner in which these data are processed, and the manner in which

DOI: 10.1201/9781003150527-30

a visualization is created (Pang 2008). As an example of *uncertainty in the raw data*, consider the attribute "percent foreign born," which we listed for counties in Florida in Table 5.1. If we consider the value of 2.6 for Dixie County, we are apt to think that this is the "correct" value for that county, but the values in this table are based on a sample of data (as part of the American Community Survey), and thus are estimates of population values.[1] We will examine the uncertainty in the foreign born data in depth in Section 27.6.1. As an example of *uncertainty in processing data*, consider the problem of interpolating values between known control points. In Chapter 17, we found that there are a variety of possible algorithms, each producing a potentially different set of interpolated values. Thus, for any particular location on a map, we can think of the set of interpolated values as constituting uncertainty in the data. Finally, an example of *uncertainty in the visualization* is the various approaches that are used to illuminate 3-D scenes (Pang 2008).

Although the term **uncertainty** is commonly used in the literature to describe the potential variation in values of an attribute at a spatial location, it should be recognized that uncertainty can arise in a wide variety of instances. Some of the earliest work on uncertainty was done in association with developing spatial data transfer standards (SDTS) (MacEachren et al. 2005), where the term *data quality* was used rather than uncertainty. In particular, the U.S. Federal Information Processing Standard 173 (FIPS 173, National Institute of Standards and Technology 1994) listed five categories for assessing data quality: lineage, positional accuracy, attribute accuracy, completeness, and logical consistency. **Lineage** refers to the historical development of digital data. For example, if the data were acquired from a paper map, you might want to know the scale and projection of the map, whether digitizing or scanning was used to generate the digital data, and when the data were converted

from analog to digital form. **Positional accuracy** refers to the locational accuracy of geographic features, both horizontally and vertically. On a USGS topographic sheet, you might wish to examine the accuracy of a stream's position or the spot height of a mountain. **Attribute accuracy** refers to the accuracy of features found at particular locations. For example, in remote sensing, you might be interested in the accuracy of a land use/land cover classification. **Logical consistency** describes "the fidelity of relationships encoded in the data structure of the digital spatial data" (National Institute of Standards and Technology, 23). For instance, we might ask whether all polygons close correctly (in GIS terminology, this would be termed topological correctness). Finally, **completeness** includes "information about selection criteria, definitions used and other relevant mapping rules. For example, geometric thresholds such as minimum area or minimum width must be reported" (National Institute of Standards and Technology, 24).[2]

Based on the work of Thomson et al. (2005), Alan MacEachren and colleagues (2005) developed Table 27.1, which provides a broad typology for uncertainty associated with geospatial information. Looking at the left-hand column, we can see that the typology utilizes some of those terms mentioned above (e.g., accuracy, completeness, consistency, lineage) but includes other terms (precision, currency, credibility, subjectivity, and interrelatedness). Looking at the columns in the table, we also see that the terms can be applied to attributes, locations, and time. The examples that we cover in this chapter will largely fall in the "Accuracy/Error" and "Attribute Examples" cell of the table,[3] but we provide this table to illustrate the broad range of uncertainty issues that you need to be aware of.

Our focus in this chapter is on methods for *visualizing* the uncertainty in spatial data. You should be aware that sometimes you might need to *purposely create* uncertainty in the data. For example, you may wish to map the locations

---

TABLE 27.1

A Typology for Uncertainty Associated with Geospatial Information

Category	Attribute Examples	Location Examples	Time Examples
Accuracy/error	Counts, magnitudes	Coordinates, buildings	+/− 1 day
Precision	Nearest 1,000	1 degree	Once per day
Completeness	75% of people reporting	20% of photos flown	2004 daily/12 missing
Consistency	Multiple classifiers	From/for a place	5 say Mon; 2 say Tues
Lineage	Transformations	#/quality of input sources	# of steps
Currency	Census data	Age of maps	$C = T_{present} - T_{info}$
Credibility	U.S. Analyst interpretation of financial records <...> informant report of financial transaction	Direct observation of training camp <...> email interception with reference to training camp	Time series air photos indicating event time <...> anonymous call predicting event time
Subjectivity	Fact <...> guess	Local <...> outsider	Expert <...> trainee
Interrelatedness	All info from same author	Source proximity	Time proximity

Source: From MacEachren et al. (2005) First Published in *Cartography and Geographic Information Science* 32, no. 3:147. Reprinted with Permission from the Cartography and Geographic Information Society.

of deaths due to certain diseases or crimes but find that there are laws protecting the confidentiality rights of the victims. Along these lines, researchers have developed guidelines for masking the locations of confidential point data and thus purposely introduced uncertainty into the point data (e.g., Leitner and Curtis 2004, 2006).

27.4 GENERAL METHODS FOR DEPICTING UNCERTAINTY

Alan MacEachren (1992) suggested three general methods for depicting uncertainty. First, individual maps can be shown for both an attribute and its associated uncertainty (maps are *adjacent*). For example, we might create one map showing the "Percent of the Population That Exercises Regularly" and a second map showing a confidence interval for each enumeration unit depicted on the first map. Second, the attribute and its uncertainty can be displayed on the same map—given the appropriate visual variables (e.g., gray tones for one and parallel lines for the other), we can conceive of overlaying one map on the other (maps are *coincident*). Note that these first two methods correspond to the "maps compared" and "maps combined" approaches presented for bivariate and multivariate mapping in Chapter 22. The third method is to create a dynamic display that allows users to either interact with or animate the attribute and its uncertainty. For example, the user might be permitted to toggle back and forth between the display of the attribute and its uncertainty or be permitted to view an animation reflecting uncertainty in the data.

27.5 VISUAL VARIABLES FOR DEPICTING UNCERTAINTY

If we assume the "maps combined" approach, visual variables for depicting uncertainty can be broken down into two broad groups: intrinsic and extrinsic (Gershon 1998). As the name implies, **intrinsic visual variables** are intrinsic to the display; for example, to depict uncertainty, we might vary the saturation of colored tones on a choropleth map. **Extrinsic visual variables** involve adding objects to the display, "such as dials, thermometers, arrows, bars, [and] objects of different shapes" (Gershon 1998, 44). Below we provide some basic examples of intrinsic and extrinsic visual variables. In Section 27.6, we will provide additional examples of their usage.

27.5.1 SOME EXAMPLES OF INTRINSIC VISUAL VARIABLES

One approach for depicting uncertainty is to utilize the basic visual variables introduced in Section 4.5. In some cases, these visual variables are suitable, whereas in other cases, they can be confusing. A suitable usage would be Tissot's indicatrix (distortion ellipse), in which the visual variables size and shape (of an ellipse) represent the ability of various map projections to maintain correct size and angular relationships at point locations (see Figure 27.1).

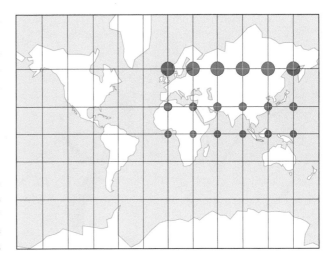

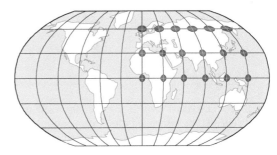

FIGURE 27.1 The visual variables size and shape represent the ability of various map projections to maintain correct size and angular relationships at point locations. Here the visual variables are easily understood. (Adapted from Abler, Ronald F. *Geography's Inner Worlds: Pervasive Themes in Contemporary American Geography*. New Brunswick: Rutgers University Press, 1992. Copyright © 1992 by Rutgers, the State University. Reprinted by permission of Rutgers University Press.)

Potentially confusing would be utilizing the visual variable size to depict the uncertainty of stream position (e.g., using a wider line to indicate greater uncertainty) because a wide line normally would be associated with greater discharge (see Figure 27.2).

Of those visual variables introduced in Section 4.5, MacEachren (1992) initially argued that saturation is particularly logical for depicting uncertainty, with "pure hues used for very certain information [and] unsaturated hues for uncertain information" (14). However, in a later experimental study, MacEachren and colleagues (2012) found that saturation was one of the least effective visual variables for depicting uncertainty.[4] In contrast, in a more recent study utilizing eye-tracking, Jan Brus and colleagues (2019) found that saturation was the most effective approach. Clearly, there is disagreement among researchers on the effectiveness of saturation for depicting uncertainty.[5]

In addition to the standard visual variables introduced in Section 4.5, MacEachren (1995, 275–276) proposed the visual variable **clarity**, which, he argued, could be subdivided into three additional visual variables: crispness, resolution, and transparency. **Crispness** (the polar opposite

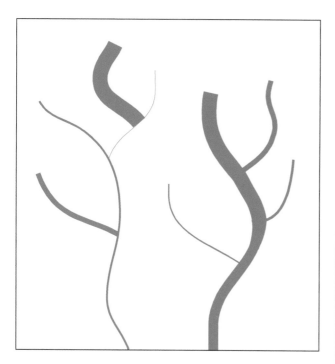

FIGURE 27.2 The visual variable size (e.g., the width of a line) depicts data uncertainty associated with stream position. Here the level of data uncertainty might be misinterpreted as stream discharge. (After McGranaghan 1993, 17.)

would be **fuzziness**) refers to the sharpness of boundaries (or area fills): A crisp boundary would represent more certain data, whereas a fuzzy boundary would represent uncertain data. For example, Figure 27.3 illustrates the uncertain nature of the boundary where grits is consumed in the United States (grits is a southern food made of cornmeal). **Resolution** is the level of detail in the spatial data underlying an attribute, where a lower (coarser) resolution equals greater uncertainty. For example, Figure 27.4 illustrates different resolutions for a raster database. **Transparency**

FIGURE 27.4 Example of differing resolutions for a raster database; the low-resolution image would have greater uncertainty. (Republished with permission of Guilford Publications Inc., from p. 277 of *How Maps Work: Representation, Visualization, and Design* by A.M. MacEachren, 2004; permission conveyed through Copyright Clearance Center, Inc.)

is the ease with which a theme can be seen through a "fog" placed over that theme; more certain data can be easily seen through the fog, whereas uncertain data cannot be easily seen (see Figure 27.5).

27.5.2 Some Examples of Extrinsic Visual Variables

Alex Pang and colleagues (1997) at the Santa Cruz Laboratory for Visualization & Graphics were early proponents of using extrinsic visual variables to depict uncertainty. Wittenbrink et al. (1996) described their approach as "verity visualization," as it "suggests the quality of the data that is exactly what it purports to be or is in complete accord with the facts." Examples of their work with extrinsic

FIGURE 27.3 The visual variable crispness depicts the uncertain nature of the boundary of grits in the United States.

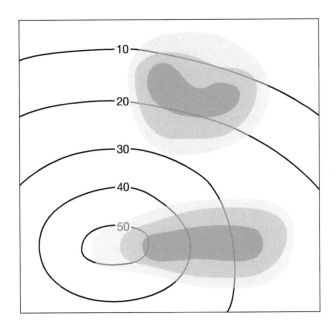

FIGURE 27.5 The visual variable transparency is used to depict the uncertainty of the data. A light "fog" indicates reliable data, whereas a dark fog indicates uncertain data. (After MacEachren 1992, 15.)

visual variables, and associated uncertainty *glyphs*, can be found at http://avis.soe.ucsc.edu/uglyph.html. Although the resulting glyphs might appear complex, Wittenbrink and colleagues showed that people could extract useful information from them.

Another example of uncertainty glyphs is Helen Mitasova and colleagues' (1993, 555) efforts to map the uncertainty of interpolation methods. They stated:

> Smoothly interpolated surfaces, while visually pleasing and valuable for analysis of data, sometimes imply that the data has greater spatial resolution than is the case ... The use of glyphs helps to highlight which sampling sites were most important in calculating the interpolated concentrations ....

27.6 APPLICATIONS OF VISUALIZING UNCERTAINTY

27.6.1 HANDLING THE UNCERTAINTY IN CHOROPLETH MAPS

Since choropleth maps are commonly utilized, numerous methods have been developed for depicting the uncertainty associated with these maps. To illustrate some of these methods, we will work with the percent foreign born data utilized in Chapter 5. Table 27.2 displays the **estimate** of the percent foreign born for Florida counties from 2006 to 2010 along with a measure of uncertainty (the **margin of error** or *MOE*) and the *CV* for the associated *MOE*. The *MOE* values provide us a 90 percent confidence interval for each of the percent foreign born values: for example, for Dixie County, we are 90 percent confident that the true

TABLE 27.2

Estimate, Margin of Error (*MOE*), and Coefficient of Variation (*CV*) for Percent Foreign Born Data for Florida Counties from 2006 to 2010

County	Estimate	MOE	CV
Baker	1.0	0.4	24.3
Liberty	1.0	0.6	36.5
Bradford	1.7	0.4	14.3
Wakulla	2.0	0.6	18.2
Taylor	2.1	0.7	20.3
Holmes	2.3	0.6	15.9
Dixie	2.6	1.9	44.4
Gilchrist	2.6	0.8	18.7
Nassau	2.7	0.5	11.3
Franklin	3.1	1.4	27.5
Jackson	3.1	0.7	13.7
Gulf	3.2	0.8	15.2
Columbia	3.3	0.7	12.9
Washington	3.3	1.3	23.9
Calhoun	3.6	1.2	20.3
Madison	3.8	1.1	17.6
Jefferson	3.9	2.6	40.5
Putnam	4.2	0.6	8.7
Lafayette	4.3	3.7	52.3
Levy	4.4	0.9	12.4
Santa Rosa	4.7	0.5	6.5
Walton	4.9	0.8	9.9
Hamilton	5.0	0.9	10.9
Citrus	5.3	0.5	5.7
Bay	5.4	0.5	5.6
Gadsden	5.6	0.8	8.7
Suwannee	5.8	1.0	10.5
Union	5.8	2.7	28.3
Clay	5.9	0.5	5.2
Escambia	5.9	0.3	3.1
Sumter	5.9	0.6	6.2
St. Johns	6.3	0.5	4.8
Leon	6.5	0.4	3.7
Hernando	6.6	0.4	3.7
Okaloosa	6.8	0.5	4.5
Volusia	7.7	0.4	3.2
Marion	7.8	0.3	2.3
Brevard	8.6	0.4	2.8
Lake	8.7	0.4	2.8
Duval	9.0	0.3	2.0
Charlotte	9.2	0.5	3.3
Pasco	9.2	0.4	2.6
Alachua	10.2	0.5	3.0
Martin	10.4	0.6	3.5
Indian River	10.5	0.7	4.1
Polk	10.7	0.4	2.3
Pinellas	11.2	0.3	1.6
Highlands	11.3	1.0	5.4
Seminole	11.5	0.4	2.1
Sarasota	11.6	0.4	2.1
Flagler	12.1	1.1	5.5
Okeechobee	12.1	1.2	6.0
Manatee	12.3	0.5	2.5
Hillsborough	15.1	0.3	1.2
Lee	15.3	0.5	2.0
St. Lucie	15.6	0.6	2.3
Monroe	16.2	1.2	4.5
Glades	16.3	2.9	10.8
Orange	19.1	0.4	1.3
DeSoto	19.6	2.4	7.4
Osceola	19.7	0.8	2.5
Hardee	21.3	2.1	6.0
Palm Beach	22.3	0.3	0.8
Collier	23.6	0.7	1.8
Hendry	27.8	2.6	5.7
Broward	30.9	0.4	0.8
Miami-Dade	51.1	0.3	0.4

Source: Estimate and *MOE* values from American Community Survey, https://www.census.gov/programs-surveys/acs/data.html.

value for percent foreign born is within 1.9 units of the estimate of 2.6 (or we are 90 percent confident that the true value falls in the interval 0.7 to 4.5). Since larger estimated percent foreign born values tend to have larger margins of error, a preferred measure of uncertainty is the *CV*, which is computed as follows (Sun and Wong 2010):

$$CV = \frac{\text{Standard error}}{\text{Estimate}}$$

Since the standard error is computed as *MOE*/1.645, we have:[6]

$$CV = \left[\frac{MOE}{1.645}\right] / \text{Estimate} \times 100$$

Here we have multiplied by 100 so that the *CV* is in percentage form. For Dixie County, we have:

$$CV = \left[\frac{1.9}{1.645}\right] / 2.6 \times 100 = 44.4\%$$

Note that the data in Table 27.2 have been sorted on the basis of the percent foreign born data. If you scan this table from top to bottom, you will see that low percent foreign born values are associated with relatively high *CV* values (the correlation between the two attributes is $r = -0.48$).[7]

Figure 27.6 provides a visualization of the percent foreign born estimates for a quantiles classification along with the associated uncertainty values, as represented by the *CV* values, again using a quantiles classification. Here we can see a spatial expression of the uncertainty, as the highest *CV* values are clearly in the northern part of the state where the lowest percentage foreign born values are found. This tells us that we should be cautious in interpreting the estimates in the northern part of the state.

Although Figure 27.6 provides us an idea about the uncertainty in the percent foreign born data, the map reader must visualize both the percent foreign born data and its associated uncertainty. As an alternative, researchers have explored approaches for incorporating uncertainty information directly into the data classification associated with the estimates. We will consider two approaches that have been developed for this purpose.

27.6.1.1 Using Confidence Levels (CLs) to Create Class Breaks

Min Sun and colleagues (2015) have developed a **class separability** approach for incorporating uncertainty into data classification by selecting class breaks that maximize the confidence that observations in two adjacent classes are significantly different from one another. A key formula in this process is the computation of the **confidence level** ($CL_{i,j}$) associated with assigning two enumeration units i and j to separate classes:

$$CL_{i,j} = \Phi\left(\frac{|\bar{x}_i - \bar{x}_j|}{\sqrt{SE_i^2 + SE_j^2}}\right)$$

where $\bar{x}_i$ and $\bar{x}_j$ are the estimates (means) for the two enumeration units, SE_i^2 and SE_j^2 are the standard errors of these estimates, and Φ is a function returning the probability of the expression in parentheses. If the estimates for two enumeration units are clearly significantly different from one another, then the resulting probability will be high (e.g., exceeding 90 percent). The separability, $S_{A,B}$, between any

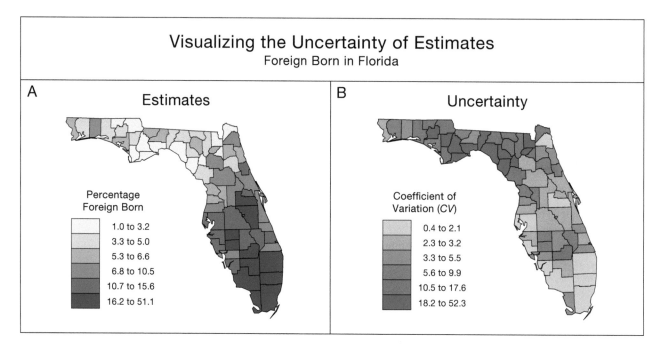

FIGURE 27.6 A visualization of the estimates and the uncertainty of those estimates for the foreign born data shown in Table 27.2: (A) The estimates of percent foreign born; (B) the uncertainty of the estimates.

two classes A and B is then defined as a minimum of the above *CL* values for all pairwise combinations of enumeration units in each class:

$$S_{A,B} = \min_{i \in A, j \in B} \left(CL_{i,j} \right), i \neq j$$

Thus, all observations in two classes are used to determine the *CL* associated with the break between two classes. Ultimately, the goal is to find a set of class breaks that maximize the magnitude of the *CL* values associated with the class breaks.

The software SEER*CMapper (http://geospatial.gmu.edu/seer-cmapper/index.htm) can be utilized to (1) determine an ideal set of *CL* values (those that maximize the *CL* values associated with class breaks); and (2) evaluate the *CL* values for traditional classification methods that do not consider the uncertainty in the data. In Figure 27.7A, we see the result of computing an ideal set of *CL* values for the Florida foreign born data, and in Figure 27.7B, we see the *CL* values associated with the optimal method of classification used in Figure 5.2D. In Figure 27.7A, we see that all of the *CL* values are above 94 percent, indicating that estimates in adjacent classes are distinctly different from one another. One limitation of the class separability approach is that in trying to maximize *CL* values, we sometimes find that one or more classes will have a limited number of observations: in this case, three classes contain only one observation. In Figure 27.7B, we see that four of the *CL* values are 85 percent or better, but one of them is distinctly deficient, with a value of only 14.0 percent. This result should make sense, however, as Figure 27.6B showed that there was considerable uncertainty in the northern part of the state, where the

two classes associated with the 14 percent value are found. It should be noted that SEER*CMapper permits you to experiment with moving class breaks to see their effect on *CL* values and the classification error.

27.6.1.2 Using Maximum Likelihood Estimation to Create Class Breaks

Wangshu Mu and Daoqin Tong (2019) have developed a more complex approach for incorporating uncertainty into data classification that utilizes *maximum likelihood estimation* (MLE).[8] Their technique finds representative values within classes that will maximize the homogeneity within each class while also considering the uncertainty associated with the data. In a sense, we can think of this as an optimal solution while also considering uncertainty. In their most recent implementation of the approach (Mu and Tong 2020), they also consider the robustness of the results. In order to utilize robustness, the user is asked to provide two parameters: $\hat{\alpha}$ and $\widehat{R_\alpha}$. $\hat{\alpha}$ is the minimum percentage of enumeration units that are correctly assigned to a class, and $\widehat{R_\alpha}$ is the minimum probability associated with $\hat{\alpha}$. For instance, if we set $\hat{\alpha} = 70$ and $\widehat{R_\alpha} = 90$, we are specifying that at least 70 percent of the enumeration units in each class must be correctly assigned and that the probability associated with these correct assignments must be 90 percent or greater. Software for implementing the maximum likelihood approach can be found at https://figshare.com/articles/code/MapRobust/10005647/1. Using this software, we experimented with various values of $\hat{\alpha}$ and $\widehat{R_\alpha}$. Figure 27.8 compares the result of using $\hat{\alpha} = 80$ and $\widehat{R_\alpha} = 80$ (A) and $\hat{\alpha} = 90$ and $\widehat{R_\alpha} = 90$ (B). The most distinctive difference between these two maps is in the northern part of

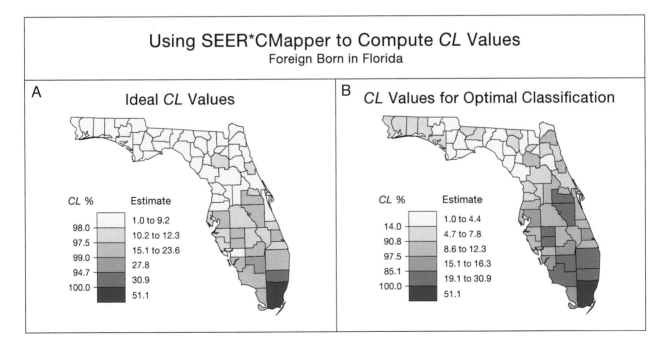

FIGURE 27.7 Using the software SEER*CMapper to compute confidence levels (*CLs*) for creating class breaks on choropleth maps: (A) A choropleth map resulting from utilizing an ideal set of *CLs*; (B) the *CLs* resulting from the optimal classification used for Figure 5.2D.

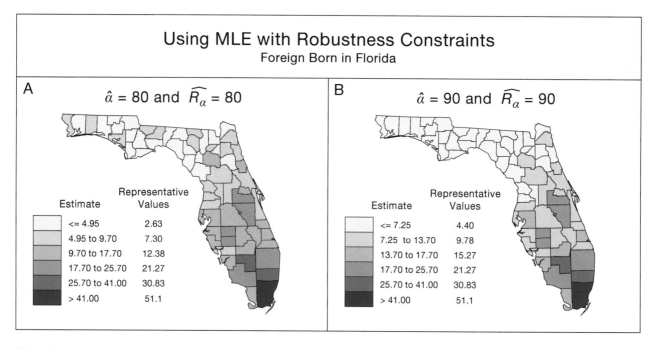

FIGURE 27.8 Using maximum likelihood estimation (MLE) with robustness constraints to develop an optimal solution that considers uncertainty in the data: (A) The result using $\hat{\alpha} = 80$ and $\widehat{R_\alpha} = 80$; (B) the result using $\hat{\alpha} = 90$ and $\widehat{R_\alpha} = 90$.

the state, where we see that the majority of counties have been placed in the lowest class in Figure 27.8B—this makes sense as this is the area where there was the greatest uncertainty in the data.

The class limits shown in Figure 27.8 are based on those provided by the above software. Here the lower limit of classes 2–6 matches the upper limit of the preceding class. Values equal to a lower class limit would be included in the preceding class. Mu and Tong (2020, 17–18) note "… that some cartographers might snap class breaks to the minimum and maximum values in the class to give readers … information about the class range [but] this practice could potentially affect the robustness metric."

Two additional characteristics of the maximum likelihood approach are that it can handle nonnormal distributions of uncertainty and will avoid overlapping classes that are sometimes found with some methods of handling uncertainty.[9] Although the software mentioned above was not developed to handle nonnormal distributions, Mu and Tong (2019) discuss an example in which they experimented with a nonnormal distribution.

27.6.1.3 Using the SAAR Software to Visualize Uncertainty

Hyeongmo Koo and colleagues (2018) developed the Spatial Analysis using ArcGIS Engine and R (SAAR) to enable general GIS users to explore spatial data and utilize spatial data analysis (SDA) tools in the ArcGIS environment and to enable advanced GIS users to customize SDA tools with minimal effort. Here we focus on just the SAAR capability to visualize uncertainty; those interested in data exploration (see Chapter 25) may wish to examine the broader SAAR capability.

SAAR provides three basic approaches for visualizing uncertainty: overlaid symbols on a choropleth map, variations of color within circles on proportional circle maps, and composite symbols.[10] In Figure 27.9, we consider three approaches for depicting uncertainty on choropleth maps with overlaid symbols: (A) changing the saturation of color within a fixed circle size (in this case, a more saturated red indicates a greater uncertainty); (B) changing the spacing of lines (more closely spaced lines indicates greater uncertainty); and (C) changing the size of proportional circles (larger circles indicate greater uncertainty). With the exception of the uncertainty for proportional circles (which are unclassed), we have depicted the Florida foreign born data estimates and their uncertainty (as expressed by the *CV*) with a quantiles classification.[11]

27.6.2 Visualizing Climate Change Uncertainty

The phenomenon of climate change has received a lot of press in recent years, and an important aspect of climate change is the uncertainty associated with it. To help visualize the uncertainty associated with climate change, we will examine a study by David Retchless and Cynthia Brewer (2016) that evaluated a wide variety of methods for representing the uncertainty of the global temperature change from 1980–2010 to 2071–2100. Figure 27.10 shows one of the maps that Retchess and Brewer evaluated in their study. In this case, the change in surface temperature from 1980–2010 to 2071–2100 is represented by a sequential yellow–orange–red color scheme selected from ColorBrewer (https://colorbrewer2.org/) and the uncertainty of that change is represented by varying sizes of a pattern of squares, with the least certain temperature change data

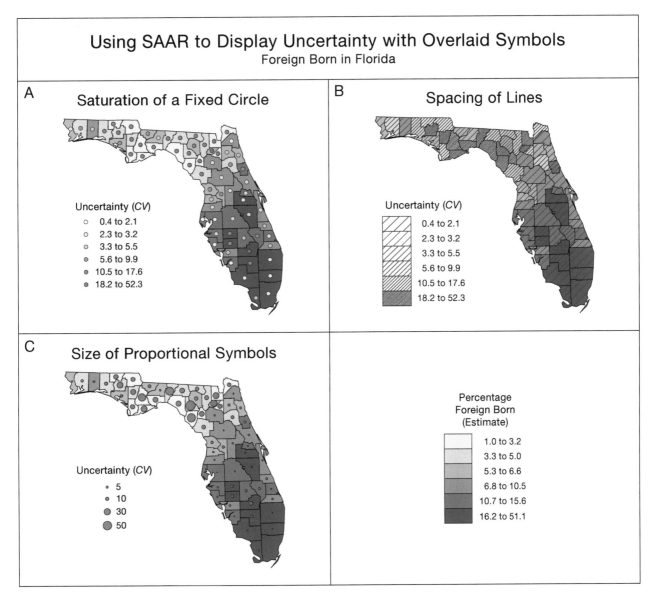

FIGURE 27.9 Three approaches for depicting uncertainty in SAAR by overlaying symbols on a choropleth map of the estimates: (A) Changing the saturation of color within a circle of fixed size; (B) changing the spacing of lines; and (C) changing the size of proportional circles.

represented by the smallest squares. This yellow-orange-red scheme (or some variation of it) was used on all of the maps in the study because of the public's familiarity with the approach. Note that the attributes temperature change ("Degrees warmer") and "Certainty" have been overlaid on top of one another (the attributes are coincident)—Retchless and Brewer argued that past research has shown that coincident maps lead to more accurate assessments than adjacent maps (1145). Note that the legend depicts "certainty" rather than "uncertainty" because the concept of certainty was deemed easier to understand (e.g., it avoids the double negative of "low uncertainty").

Figure 27.11 shows the broad range of visual variables that Retchless and Brewer evaluated for depicting uncertainty. Each of the images in the figure represents a strip (including Africa) that was taken from the full world map

that was tested. The first five images In Figure 27.11 use some variation of color to represent the certainty attribute. In Figure 27.11A, chroma (or saturation) is used to depict certainty, with less certain changes represented by a lower saturation. As we indicated in Section 27.5.1, saturation has been a common visual variable used to depict uncertainty, but its effectiveness has varied. Sometimes saturation has been combined with lightness (or value), as in Figure 27.11B. Here less certain changes are represented by a lower saturation and a lighter color. Parts C and D of Figure 27.11 utilize partially opaque lightnesses of gray for the certainty attribute, with the darkest gray used for the least certain change; the idea is to create a "fog" for the less certain areas by using the visual variable transparency. Parts C and D of Figure 27.11 contrast classed and unclassed approaches, which is an issue we discussed in

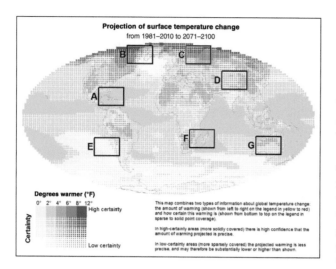

FIGURE 27.10 One of the maps used in the study by Retchless and Brewer (2016). The attribute of temperature change from 1981–2010 to 2071–2100 is represented by a yellow–orange–red color scheme, and the certainty of that change is represented by a pattern of squares of differing sizes. (Republished with permission of John Wiley & Sons—Books, from "Guidance for representing uncertainty on global temperature change maps," by D. P. Retchless and C. A. Brewer, *International Journal of Climatology* 36, no. 3, 2016; permission conveyed through Copyright Clearance Center, Inc.)

depth in Section 15.6. *White mixing* in Figure 27.11E is an approach developed by Kaye et al. (2012) in which different amounts of white are added to an RGB color to create different levels of saturation, with the greatest amount of white added to the lowest level of certainty. Note that white mixing uses a light blue for the smallest temperature change class and follows IPCC guidelines for completely "whiting-out" the least certain class (Retchless and Brewer 2016, 1152).

The last three images in Figure 27.11 use some element of pattern to represent the certainty attribute. Using an idea initially developed by MacEachren et al. (1998), the visual variable spacing is used in Figure 27.11F, with black and white closely spaced lines used for the lowest certainty, sparser gray and white lines used to represent higher certainty, and no lines drawn within the most certain areas. The *spot* technique in Figure 27.11G utilized a variation of an approach suggested by the Intergovernmental Panel on Climate Change (IPCC)—in Retchless and Brewer's variation, the least certain class is completely "whited-out," the second-lowest certainty class uses a disordered speckle of white dots, the most certain class uses a regular grid of black dots, and the second-highest certainty class uses no overlay. As previously indicated, Figure 27.11H varies the sizes of a pattern of squares, with the least certain data represented by the smallest squares.

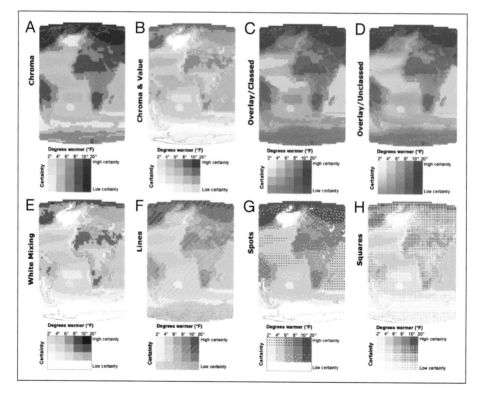

FIGURE 27.11 The eight techniques for depicting certainty in the study by Retchless and Brewer (2016). Note that the temperature change attribute is represented by a yellow–orange–red or some variation of that (e.g., A and B include purple and E includes a light blue). (Republished with permission of John Wiley & Sons—Books, from "Guidance for representing uncertainty on global temperature change maps," by D. P. Retchless and C. A. Brewer, *International Journal of Climatology* 36, no. 3, 2016; permission conveyed through Copyright Clearance Center, Inc.)

To evaluate the effectiveness of the above certainty approaches, 274 participants were randomly assigned to either one of the eight techniques for depicting uncertainty or to one of two control groups (one that showed adjacent temperature change and certainty maps and one that showed just the temperature change data).[12] The participants performed two map tasks. First, they ranked the seven regions shown in Figure 27.10 from the largest overall increase in temperature to the smallest overall increase in temperature. Second, they ranked the same seven regions from highest overall certainty to lowest overall certainty. Those participants who utilized one of the eight coincident maps also were shown a gallery of the eight coincident maps and asked:

> If you had to take this survey again, which of the following ways of representing temperature change and uncertainty do you think would be easiest to use? (1153)

The results revealed that the three pattern-based methods were more effective than the five color-based methods for depicting the certainty information, as the three pattern-based methods were not significantly different from the control approach with adjacent maps, and the five color-based methods were all significantly worse. The spot technique was the most effective of the three pattern-based methods. If participants had to take the survey again, the spot approach (Figure I27.11G) was clearly the most preferred (it was chosen by 32.2 percent of participants), but white mixing (Figure 27.11E) and overlay unclassed (Figure 27.11D) were also highly desired (they were chosen by 16.9 and 14.1 percent of participants). Retchless and Brewer noted that ultimately the choice of a method for depicting certainty would depend on a map maker's goals (1157). For example, if areas showing certainty information are relatively large (it is possible to see a pattern in the areas), then the spot technique may be the most effective. However, if some areas showing certainty information are small (pattern is not detectable in these areas), then a color-based approach (such as white mixing may be appropriate).

27.6.3 Visualizing Uncertainty in Decision-Making

A variety of studies have examined the utilization of uncertainty in decision-making. Here we look at two examples. For a review of studies involving uncertainty visualization in decision making, see Kinkeldey et al. (2017).

27.6.3.1 Visualizing the Uncertainty of Water Balance Models

Daniel Cliburn and colleagues (2002) utilized various visualization techniques to allow decision-makers to examine the uncertainty of water balance models for terrestrial regions of the world. The basic output of a water balance model is an indication of whether each geographic location has a water surplus or deficit, along with an associated magnitude. Uncertainty arises because modelers must determine

appropriate input parameters for the model (e.g., which historical temperature and precipitation data should be used) and which global circulation models (GCMs) should be used to predict the future water balance. Although GCMs generally predict global warming (and thus less water available), there is spatial variation within models, with cooling occurring at some locations (and thus more water available).

Following Gershon's notion of intrinsic and extrinsic visual variables, Cliburn et al. proposed a range of intrinsic and extrinsic methods for depicting uncertainty, with an emphasis on 3-D (technically, $2\frac{1}{2}$-D) surfaces. Figure 27.12 illustrates an intrinsic approach that was utilized. In Figure 27.12A, the height of the 3-D surface provides an indication of water availability based on historical temperature and precipitation data. Areas of the surface above 0 mm (note the scale in the lower-left) represent a water surplus, and areas below 0 mm represent a deficit. Obviously, it is not easy to see both surpluses and deficits in this static image; this problem was handled by permitting users to manipulate the image (e.g., rotate and zoom in) via a

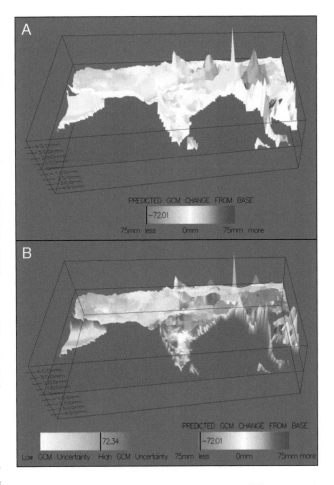

FIGURE 27.12 Visualizations developed by Cliburn et al. (2002): (A) Height of the surface shows water surpluses and deficits based on historical data, whereas the surface is colored to reflect changes anticipated based on average GCM model data; (B) transparency (an intrinsic approach) is used to depict areas that are uncertain. (Courtesy of Daniel C. Cliburn.)

wall-sized display (Section 28.4.2). Note that Figure 27.12A is colored using an orange–purple diverging scheme, indicating how the average of GCMs compares with the historical data; in this case, orange and purple indicate less and more water, respectively, in comparison to historical data. Recall from Section 15.4.2 that the orange–purple scheme is particularly effective from the standpoint of color naming, and it also avoids the problems of color vision impairment and simultaneous contrast (Table 15.2). Figure 27.12B depicts the uncertainty in the GCM predictions using transparency (here, the more transparent the image, the greater the variation among the GCMs, and hence the greater the uncertainty). Note how this affects our view of orange and purple peaks in the northeast part of Figure 27.12A—apparently, there is disagreement among the models regarding what will happen at these locations.

Figure 27.13 is an example of an extrinsic method utilized by Cliburn et al. In this case, orange and purple bars represent the range of GCM predictions at a particular location. For example, if only an orange bar appears, then at least one GCM predicted less water available (and none predicted more water available), whereas if both orange and purple bars appear, then at least one GCM predicted less water and at least another GCM predicted more water. Small pyramid-like symbols at the end of the bars denote which GCMs were associated with the extreme low or high point on a bar (note the legend in the lower left). Obviously, the complexity of this image requires that users be able to manipulate it, which was possible.

Interestingly, although all decision-makers who were asked to utilize this water balance application felt that it could be helpful in the decision-making process, they were skeptical about the usefulness of the uncertainty information. They noted that many decision-makers do not necessarily like to see uncertainty; rather than having to choose a side of an issue, the decision-makers would like to know what choice is best. On this basis, Cliburn et al. proposed the development of a visual abstract reasoning network that could assist decision-makers in choosing a course of action when having to deal with uncertainty (948).

27.6.3.2 Visualizing the Uncertainty of Forecasted Hurricane Paths

Figure 27.14 displays the approach used by the U.S. National Hurricane Center to display the uncertainty associated with the path of a hurricane. The white region shows the "potential track area" for Hurricane Laura beginning on August 25, 2020. At the top of the image, we see the explicit statement that "The cone contains the probable path of the storm center but does not show the size of the storm." The intention here is thus to show the uncertainty associated with the hurricane path, with the wider white region indicating a greater uncertainty in the hurricane path over time. Unfortunately, viewers often misinterpret maps using this approach when viewing news reports, assuming that the size of the cone represents either the size or the intensity of the hurricane.

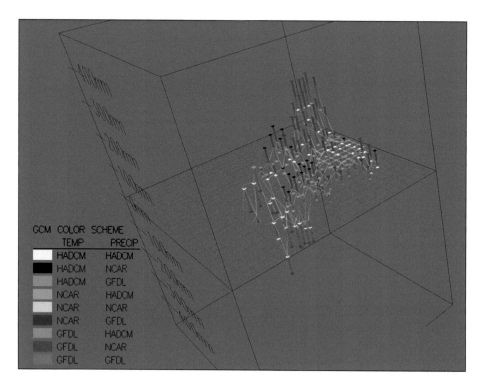

FIGURE 27.13 Extrinsic uncertainty visualization developed by Cliburn et al. (2002). Orange and purple bars represent the range of GCM predictions at a particular location, and small pyramid-like symbols at the end of bars denote which GCMs are associated with the extreme low or high point on a bar. (Courtesy of Daniel C. Cliburn.)

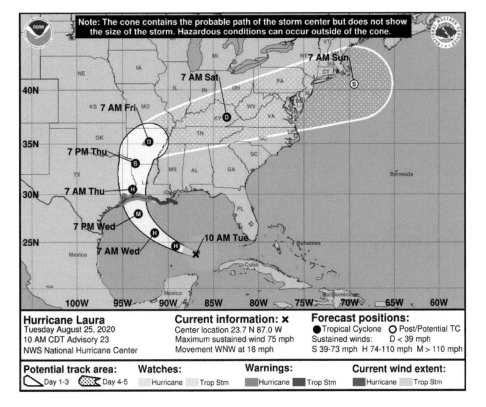

FIGURE 27.14 An example of a map used by the U.S. National Hurricane Center to display the uncertainty associated with the path of a hurricane. (From https://www.nhc.noaa.gov/aboutcone.shtml; accessed October 12, 2021.)

Although several studies have examined this issue, we'll consider one by Ian Ruginski and colleagues (2016) in which they compared a traditional *cone-centerline* approach (Figure 27.15A) with four other approaches: *centerline-only*, *ensemble*, *fuzzy-cone*, and *cone-only* (Figure 27.15B–E). Figure 27.14 does not show an explicit centerline, but such centerlines have been common on maps showing hurricane paths. Each line in the ensemble is intended to show numerous possible predicted paths for the hurricane. Previous research by Cox et al. (2013) suggested that the ensemble approach could be more effective than the cone-centerline approach, but Ruginski et al. wanted to conduct a detailed analysis of these approaches. The fuzzy-cone approach essentially makes use of the crispness visual variable introduced in Section 27.5.1—the idea is that the fuzziness would suggest the uncertainty of the hurricane track.

Ruginski et al. tested the effectiveness of the above five methods in a controlled experiment with 200 participants. Using one of the five methods shown in Figure 27.15, participants were asked to evaluate the damage (on a seven-point Likert scale) to oil rig locations at various distances from the centerline of a hurricane path and at two time intervals (24 hours and 48 hours). Figure 27.16 shows a summary of one of six hurricane paths that were tested along with the potential oil rig locations. After completing the Likert scale evaluation, participants also were asked to think aloud while completing 12 additional trials, thus providing insight into how participants made their damage

evaluations. Finally, participants also responded to a set of 9 true–false questions (for example, one question was "The display shows the hurricane getting large over time.").

The results revealed that the ensemble approach was clearly more effective than the other approaches. This was most obvious for the damage ratings to the oil rigs at the 48-hour time interval, as they were significantly lower for the ensemble approach at this interval, indicating that those viewing the ensemble did not consider the hurricane getting larger over time. This was also supported by two of the true–false questions, as a significantly *lower* percentage of ensemble participants found the "hurricane getting larger over time" and "the extent of the damage of the hurricane getting greater over time." Furthermore, the think aloud protocol suggested that those using the ensemble used different words to describe their processing; for instance, the ensemble group frequently mentioned "intensity" and "depth of color," whereas the cone-centerline group mentioned "the curve" and "the size of the hurricane."

27.6.4 Examples of Interactivity and Animation

In an extensive review of user studies that analyzed the visualization of uncertainty, Christoph Kinkeldey and colleagues (2014) noted that the bulk of the studies involved static maps rather than interactive or animated ones. This confirms our impression that interactive and animated approaches for visualizing uncertainty have not been as common as static ones. Although not as common, there

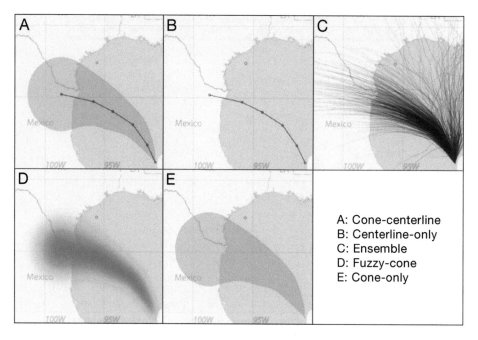

FIGURE 27.15 The five methods that Ruginski et al. (2016) examined in their study of approaches for depicting the uncertainty associated with the path of hurricanes: (A) Cone-centerline, (B) centerline-only, (C) ensemble, (D) fuzzy-cone, and (E) cone-only. The red dot is an example of an oil rig location for which the damage was estimated. (After Ruginski, I. T., Boone, A. P., Padilla, L. M., et al. (2016) "Non-expert interpretations of hurricane forecast uncertainty visualizations." *Spatial Cognition & Computation* 16, no. 2:154–172, Taylor & Francis Ltd., https://www.tandfonline.com/journals/hscc20.)

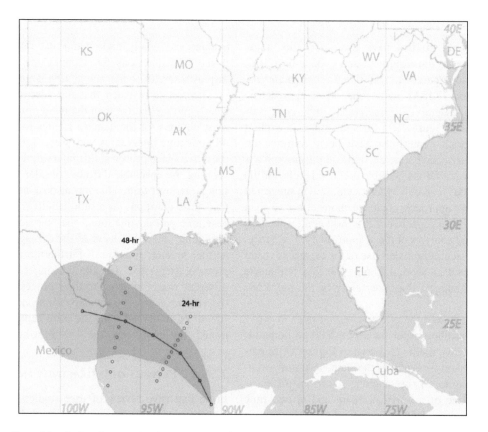

FIGURE 27.16 One of the six hurricane paths that were tested in the study by Ruginski et al. (2016). In this case, the path is depicted using the cone-centerline approach; also shown are the 24-hour and 48-hour time intervals along with the possible oil rig locations used for those time intervals. Only one oil rig would be shown at a time in the actual testing. (After Ruginski, I. T., Boone, A. P., Padilla, L. M., et al. (2016) "Non-expert interpretations of hurricane forecast uncertainty visualizations." *Spatial Cognition & Computation* 16, no. 2:154–172, Taylor & Francis Ltd., https://www.tandfonline.com/journals/hscc20.)

certainly are instances in which interactive and/or animated approaches can be useful. For instance, as we worked with the SAAR software described in the preceding section, it was apparent that interaction could enhance the interpretation of the coincident maps. Since the spacing of lines technique in Figure 27.9B makes the underlying estimates difficult to interpret when the lines are closely spaced, we found that toggling the line spacing layer off and on could assist in interpreting the coincident map. Alternatively, we could use the Identify tool in SAAR to determine the precise value for an enumeration unit for both coincident layers. Static maps generally cannot provide the latter capability unless a large amount of text is shown on the map.

One of the difficulties with illustrating interactive and animated approaches is that the resulting interactive software and animations frequently are either not distributed or become unusable as computer technology evolves. Thus, we will point you to several papers that describe the use of interactivity and/or animation. In the realm of interactivity, an important early development was Howard and MacEachren's (1996) Reliability Visualization (R-VIS) software that was intended to enable environmental scientists and policy analysts to examine levels of dissolved inorganic nitrogen (DIN) in Chesapeake Bay using isarithmic maps. A more recent interactive example is Kunz et al.'s (2011) software prototype for visualizing the uncertainty associated with natural hazards. If you are considering developing an interactive system for visualizing uncertainty, we feel that this paper provides a number of useful suggestions and approaches. Other interactive examples that deal with more complex multivariate data include Sanyal et al.'s (2010) development of Noodles, a software tool for modeling the uncertainty associated with *numerical weather model ensembles*; and Slingsby et al.'s (2011) development of a tool to explore the uncertainty associated with *geodemographic classifiers*, which use cluster analysis to combine numerous demographic and lifestyle attributes and assign geographic areas to the resulting clusters.

The following are some papers dealing with animation that Kinkeldey et al. pointed to in their review of user studies involving the visualization of uncertainty. Evans (1997) was one of the first to explore the effectiveness of displays of uncertainty from the standpoint of the map user. Using land use/land cover maps, Evans compared the effectiveness of a static coincident map in which both land use/ land cover and its reliability were shown with an animated map in which the reliability information was "flickered" on and off and an interactive version in which the reliability information was toggled on and off by the user. The results revealed little difference between the coincident and animated maps and that the toggling approach was less effective. In an examination of uncertainty in remotely sensed images, Blenkinsop et al. (2000) compared a static grayscale image of uncertainty with both serial and random animations of the grayscale image and found that the static image was preferred. In a similar fashion, a study by Aerts et al. (2003) involving an urban growth model found

that a static adjacent view was slightly favored over an animation in which a model and its uncertainty were toggled on and off automatically. Overall, Kinkeldey et al. summarized the results of evaluating interactive and animated approaches by stating:

> All in all, there is evidence that animated views have a potential to successfully represent uncertainty when static solutions are not feasible but [there is] little evidence that they perform better (or even as well) as more traditional static depictions. (381)

They noted, however, that animation can be useful for developing an initial impression of uncertainty.

One intriguing paper that was not covered by Kinkeldey et al. was Peter Fisher's (1996) work with dot maps, as it did not involve a user study. In Fisher's approach, a dot map was first created using approaches similar to those described in Section 19.3, in which dots are ultimately placed using some variant of a random approach. Since there is some uncertainty in this random approach, Fisher animated the resulting dot map by selecting one dot at random, deleting that dot, and locating a replacement dot within the enumeration unit containing the deleted dot. He thus intended the movement of dots on the resulting animated map to illustrate the uncertainty associated with dot placement (Figure 27.17 portrays two frames from one of Fisher's animations). Although Fisher's notion of animating dot maps is intriguing, we found the animations disconcerting. One problem was that relatively dense enumeration units appeared to be just as uncertain as less dense units, although intuitively, it seems that the position of dots within a dense area should appear more certain. The second problem was that a purely random placement of dots could lead to strange concentrations of dots (note the circular arrangement of dots in the western portion of Connecticut in the top map in Figure 27.17). These problems might be ameliorated by a revised dot placement algorithm that is not purely random such as those we described in Section 19.3.3.2.

27.7 USING SOUND TO REPRESENT DATA UNCERTAINTY

Throughout this chapter, we have presumed that *visual* variables are used to depict both an estimate and its uncertainty. This is potentially problematic because the visual channel can become overloaded, as both an estimate and its uncertainty must be dealt with in the visual channel (e.g., consider the challenge of interpreting the overlaid maps shown in Figure 27.9). An alternative is to offload uncertainty information into the *aural* channel. For example, we might show the choropleth map in Figure 27.9 and have a user point to locations and vary the pitch, with a higher pitch representing greater uncertainty in the data. If we take such an approach, we need to know whether the pitch is an appropriate **aural variable** to represent uncertainty and whether there might be more appropriate aural variables.

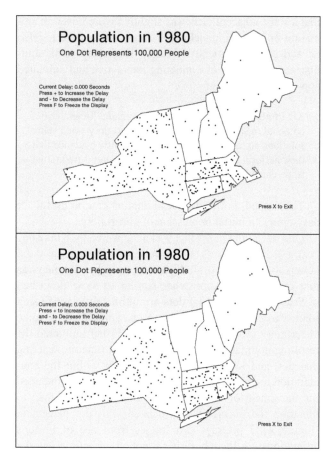

FIGURE 27.17 Two dot maps from an animation used to illustrate uncertainty. The number of dots within each data collection unit (state, in this case) is held constant, but the dots are placed at different random locations for each frame of the animation. (After Fisher, P. F. 1996. First published in *Cartography and Geographic Information Systems* 23, no. 4:200. Reprinted with permission from the Cartography and Geographic Information Society.)

An experiment by Andrea Ballatore and colleagues (2019) has dealt with precisely this issue. Ballatore et al. tested the aural variables shown in Table 27.3, which they termed **sound dimensions**. These sound dimensions were derived from the abstract sound variables developed by John Krygier (1994), which we introduced in Figure 4.13. Comparing Figure 4.13 and Table 27.3, we can see that Ballatore et al. have added a number of sound dimensions to the original abstract sound variables, including noise, clarity, and harmony. A description of each sound dimension is given in Table 27.3 along with two sample values.

In their experiment, Ballatore et al. created both synthetic (machine-produced) and natural sounds. The synthetic sounds were generated using the Minim library for Processing 2.1 (http://code.compartmental.net/minim) and/or the Audacity (https://www.audacityteam.org) audio editing software. Natural sounds were based on violin recordings. Where possible, both synthetic and natural sounds were used for each sound dimension, although only synthetic sounds were used for three variants of the noise dimension and two variants of the timbre dimension.

TABLE 27.3

Sound Dimensions Utilized by Ballatore et al.

Sound Dimension	Description	Sample Values
Loudness	A sound's perceived amplitude	Quiet, loud
Duration	Length of a sound	Short, long
Location	Where a sound appears in space	Left, right
Pitch	Perceived frequency of a sound	C, D
Register	Octave in which a note or melody is played	C4, C5
Attack	How a sound increases in volume before reaching its highest point (the sustain level)	Short, long
Decay	How a sound decreases in volume from the sustain level before reaching zero amplitude	Short, long
Rate of change	How rapidly the sound changes	Slow tempo, fast tempo
Noise	Unwanted signal or distortion; mathematically, the degree of randomness in a complex sound spectrum	Noisy, clear
Clarity	Amount of original sound heard, rather than distorted or hidden in noise	Distorted, undistorted
Timbre	Hard to define. Encompasses all aspects not captured by the other parameters. Brightness, tone quality	Flute, trumpet
Order	Order of execution of different sounds	Regular, irregular
Harmony	Consonance or dissonance between notes played at the same time	Consonant, dissonant

Source: After Ballatore et al. (2019) First Published in *Cartography and Geographic Information Science* 46, no. 5:389–390. Reprinted with Permission from the Cartography and Geographic Information Society.

For each of the 26 sound stimuli tested (15 synthetic and 11 natural), 33 participants were asked to indicate which of two sounds "was coming from a source with BETTER/ WORSE data quality" (393). Note that in a fashion similar to the Retchless and Brewer (2016) study examined in Section 27.6.2, Ballatore et al. did not pose this question with respect to "data uncertainty"; instead, they felt that participants had a better feel for the notion of "data quality." The results of the experiment revealed that the most effective sound dimensions were loudness (a louder sound indicated greater data quality, with synthetic sounds more effective), order (a regular order indicted greater data quality, with natural sounds more effective), and clarity (a non-distorted order indicated greater data quality, with natural sounds more effective). Interestingly, the results revealed that pitch, which we suggested might be used for depicting uncertainty, should not be used, as near random responses ensued with its use (399). The results of the Ballatore et al. study are limited by the fact that it did not involve spatial tasks, but the study represents an important first step in evaluating the potential of sound for depicting data uncertainty.

27.8 SUMMARY

In this chapter, we have considered the notion of uncertainty in geospatial data and various approaches for visualizing this uncertainty. We saw that uncertainty can arise from a variety of sources, including the raw data on which a map is based, the manner in which these data are processed, and the manner in which a visualization is created. The creation of an isarithmic map depicting precipitation recorded at weather stations provides a good illustration of various sources of uncertainty. Uncertainty in the raw data is illustrated by the possibility that incorrect values can be recorded at individual weather stations; uncertainty due to processing is reflected in the variety of interpolation methods that can be utilized to estimate values between weather stations (the control points); and uncertainty of visualization is a function of whether a classed or an unclassed isarithmic map is created.

There are three general methods for depicting uncertainty: separate maps can be created for an attribute of interest and its associated uncertainty (i.e., maps are adjacent); the attribute and its uncertainty can be displayed on the same map (i.e., maps are coincident); and dynamic maps can be utilized (i.e., interactive tools and animation are possible). Generally, the "coincident maps" approach has been most commonly used, although a number of interactive and animated tools have been developed (we experimented with both SEER*CMapper and SAAR in this chapter).

Visual variables for depicting uncertainty can be split into intrinsic and extrinsic categories. Intrinsic visual variables are intrinsic to the display; for example, uncertainty might be depicted by varying the transparency of the colors composing an isarithmic map. Extrinsic visual variables involve adding objects to the display, such as the glyphs that Wittenbrink et al. (1996) used for depicting the uncertainty of wind speed and direction. Specialized visual variables for depicting uncertainty include crispness, resolution, and transparency.

We covered a broad range of applications involving the visualization of uncertainty, including handling the uncertainty in choropleth maps, visualizing climate change uncertainty, visualizing uncertainty in decision-making,

and examples of interactivity and animation. In terms of handling uncertainty in choropleth maps, we stressed the importance of the *MOE* and the associated *CV*. We described two approaches for data classification for choropleth maps that incorporate uncertainty into the data classification, one based on *CL*s and one based on MLE. When visualizing climate change uncertainty, we found that pattern-based approaches tended to be more effective than color-based approaches for depicting certainty information (users had difficulty understanding the concept of "uncertainty" and so the term "certainty" was utilized). We considered two examples for visualizing the uncertainty in decision-making, one dealing with the uncertainty in water balance models and one dealing with the uncertainty of forecasted hurricane paths. In the latter case, we saw that a conventional cone-centerline approach can lead to incorrect decisions when estimating the damage associated with hurricanes and that an ensemble approach is likely to lead to better estimates of damage. Although interactive tools and animation can sometimes be useful for evaluating uncertainty, research thus far generally has not shown that animation is more effective than a static map, although it can be useful for developing an initial impression.

We concluded the chapter by considering the possibility of using sound to represent uncertainty. This seems logical as the visual channel may become overloaded by having to consider both an estimate of an attribute and its associated uncertainty. A study by Ballatore et al. revealed that the sound dimensions of loudness, order, and clarity appeared to be effective for representing uncertainty. Interestingly, participants in this study also found the term "uncertainty" awkward, and so when participants were asked to evaluate sounds, they were asked to evaluate them in terms of "data quality" rather than "uncertainty."

27.9 STUDY QUESTIONS

1. Indicate which visual variable is being used to depict uncertainty in each of the following cases, and briefly explain your choice: (A) on an isarithmic map depicting the likelihood for burn potential, some areas appear to be covered by "fog" whereas other areas appear "clear" and are thus easily interpreted; (B) some boundaries between countries on a world map are fuzzy, whereas others are crisp; and (C) on a remotely sensed image, some areas of the image are depicted with 30-meter cells, whereas other areas are depicted with 120-meter cells.

2. Define the terms intrinsic visual variables and extrinsic visual variables and give two examples of each.

3. (A) Specify a 90 percent confidence interval for Bradford County for the data shown in Table 27.2. (B) Presuming a 90 percent confidence interval, show how to compute a *CV* for Bradford County.

4. Explain the logic of how the *CL* values shown in Figure 27.7A were computed.

5. Contrast the maximum likelihood method (MLE) for considering uncertainty in a data classification with the optimal method described in Chapter 5.

6. Using the three approaches for depicting uncertainty shown in Figure 27.9: (A) explain which method is most effective in showing the distribution of uncertainty; (B) explain which method is most effective in visually correlating the map of the estimates with the map of uncertainty.

7. (A) Describe how "certainty" is represented using white mixing in Figure 27.11E and spots in Figure 27.11G, and describe how easy it is for you to interpret the resulting maps. (B) Indicate how effective these techniques were in the Retchless and Brewer study, and indicate a situation in which you would use each technique.

8. If a government official had to make a decision using the system developed by Cliburn et al. in Section 27.6.3.1, explain why they might find the system awkward to work with.

9. (A) Why is the approach for depicting the uncertainty in a hurricane path shown in Figure 27.14 considered undesirable? (B) What approach would you suggest as an alternative to the approach shown in Figure 27.14, and how would you justify your suggested approach?

10. Presume that a researcher develops two approaches for depicting the uncertainty associated with a new variant of COVID-19. In one approach, the uncertainty is overlaid on a static map of the estimates. In the second approach, the uncertainty is overlaid on the static map of the estimates, but the uncertainty is animated in a sequenced fashion by repeatedly going through the classes depicted. Which approach would you suspect would be most appropriate?

11. (A) In general, why might sound be useful for depicting the uncertainty on a map? (B) Define the difference between synthetic and natural sounds. (C) Which sound dimensions are likely to be most useful for depicting uncertainty (or data quality) and should synthetic or natural sounds be used?

NOTES

1 Details on the sampling procedure for the 2006–2010 data used for Table 5.1 can be found here: https://www.census.gov/library/publications/2010/acs/archive.html.

2 These five categories and two additional ones (semantic accuracy and temporal information) were described in detail in *Elements of Spatial Data Quality*, which was published by the International Cartographic Association (ICA) Commission on Spatial Data Quality (Guptill and Morrison 1995).

3 An exception is the work of Ślusarski and Jurkiewicz (2020), which examines the attribute, locational, and temporal uncertainty associated with objects found on Polish topographic maps.

4 More effective than saturation was fuzziness (i.e., crispness), location, value (lightness), and arrangement.

5 For a review of numerous studies that have examined the use of saturation to depict uncertainty, see Kinkeldey et al. (2014, 378).

6 The value 1.645 is associated with a 90 percent confidence interval, which is the standard used for American Community Survey data.

7 Since the relationship between percent foreign-born and *CV* is nonlinear, a nonparameteric correlation is arguably more appropriate. A common nonparametric correlation is Spearman's rho, which revealed an even stronger inverse relationship, with a correlation of −0.80.

8 For an introduction to MLE, see Eliason (1993).

9 For a discussion of the issue of overlapping classes, see Wei et al. (2017).

10 An example of a composite symbol is to vary circle size as a function of an estimate and to vary the thickness of the boundary of the circle as a function of uncertainty.

11 In addition to providing several ways of visualizing both estimates and their uncertainty, SAAR also includes the capability to incorporate uncertainty into the data classification procedure. Rather than using the MLE approach developed by Mu and Tong, SAAR utilizes an approach developed by Koo and colleagues (2017).

12 The certainty map was a smaller image matching the size of legends used in the other maps and was shown in lightnesses of gray.

REFERENCES

Aerts, J. C. J. H., Clarke, K. C., and Keuper, A. D. (2003) "Testing popular visualization techniques for representing model uncertainty." *Cartography and Geographic Information Science 30*, no. 3:249–261.

Ballatore, A., Gordon, D., and Boone, A. P. (2019) "Sonifying data uncertainty with sound dimensions." *Cartography and Geographic Information Science 46*, no. 5:385–400.

Blenkinsop, S., Fisher, P., Bastin, L., et al. (2000) "Evaluating the perception of uncertainty in alternative visualization strategies." *Cartographica 37*, no. 1:1–13.

Brus, J., Kucera, M., and Popelka, S. (2019) "Intuitiveness of geospatial uncertainty visualizations: A user study on point symbols." *Geografie 124*, no. 2:163–185.

Cliburn, D. C., Feddema, J. J., Miller, J. R., et al. (2002) "Design and evaluation of a decision support system in a water balance application." *Computers & Graphics 26*:931–949.

Cox, J., House, D., and Lindell, M. (2013) "Visualizing uncertainty in predicted hurricane tracks." *International Journal for Uncertainty Quantification 3*, no. 2:143–156.

Eliason, S. R. (1993) *Maximum Likelihood Estimation: Logic and Practice.* Thousand Oaks, CA: SAGE Publications.

Evans, B. J. (1997) "Dynamic display of spatial data-reliability: Does it benefit the map user?" *Computers & Geosciences 23*, no. 4:409–422.

Fisher, P. F. (1996) "Animation of reliability in computer-generated dots maps and elevation models." *Cartography and Geographic Information Systems 23*, no. 4:196–205.

Gershon, N. (1998) "Visualization of an imperfect world." *IEEE Computer Graphics and Applications 18*, no. 4:43–45.

Guptill, S. C., and Morrison, J. L. (1995) *Elements of Spatial Data Quality.* Kidlington, Oxford, England: Elsevier Science.

Howard, D., and MacEachren, A. M. (1996) "Interface design for geographic visualization: Tools for representing reliability." *Cartography and Geographic Information Systems 23*, no. 2:59–77.

Kaye, N. R., Harley, A., and Hemming, D. (2012) "Mapping the climate: guidance on appropriate techniques to map climate variables and their uncertainty." *Geoscientific Model Development 5*:245–256.

Kinkeldey, C., MacEachren, A. M., Riveiro, M., et al. (2017) "Evaluating the effect of visually represented geodata uncertainty on decision-making: Systematic review, lessons learned, and recommendations." *Cartography and Geographic Information Science 44*, no. 1:1–21.

Kinkeldey, C., MacEachren, A. M., and Schiewe, J. (2014) "How to assess visual communication of uncertainty? A systematic review of geospatial uncertainty visualisation user studies." *The Cartographic Journal 51*, no. 4:372–386.

Koo, H., Chun, Y., and Griffith, D. A. (2017) "Optimal map classification incorporating uncertainty information." *Annals of the American Association of Geographers 107*, no. 3: 575–590.

Koo, H., Chun, Y., and Griffith, D. A. (2018) "Integrating spatial data analysis functionalities in a GIS environment: Spatial Analysis using ArcGIS Engine and R (SAAR)." *Transactions in GIS 22*, no. 3:721–736.

Krygier, J. B. (1994) "Sound and geographic visualization." In *Visualization in Modern Cartography*, ed. by A. M. MacEachren and D. R. F. Taylor, pp. 149–166. Oxford, England: Pergamon.

Kunz, M., Grêt-Regamey, A., and Hurni, L. (2011) "Customized visualization of natural hazards assessment results and associated uncertainties through interactive functionality." *Cartography and Geographic Information Science 38*, no. 2:233–243.

Leitner, M., and Curtis, A. (2004) "Cartographic guidelines for geographically masking the locations of confidential point data." *Cartographic Perspectives* no. 49:22–39.

Leitner, M., and Curtis, A. (2006) "A first step towards a framework for presenting the location of confidential point data on maps—Results of an empirical perceptual study." *International Journal of Geographical Information Science 20*, no. 7:813–822.

MacEachren, A. M. (1992) "Visualizing uncertain information." *Cartographic Perspectives* no. 13:10–19.

MacEachren, A. M. (1995) *How Maps Work: Representation, Visualization, and Design*. New York: Guilford.

MacEachren, A. M., Brewer, C. A., and Pickle, L. W. (1998) "Visualizing georeferenced data: Representing reliability of health statistics." *Environment and Planning A 30*:1547–1561.

MacEachren, A. M., Robinson, A., Hopper, S., et al. (2005) "Visualizing geospatial information uncertainty: What we know and what we need to know." *Cartography and Geographic Information Science 32*, no. 3:139–160.

MacEachren, A. M., Roth, R. E., O'Brien, J., et al. (2012) "Visual semiotics & uncertainty visualization: An empirical study." *IEEE Transactions on Visualization and Computer Graphics 18*, no. 12:2496–2505.

McGranaghan, M. (1993) "A cartographic view of spatial data quality." *Cartographica 30*, no. 2&3:8–19.

Mitasova, H., Brown, W., Gerdes, D. P., et al. (1993) "Multidimensional interpolation, analysis and visualization for environmental modeling." *GIS/LIS Proceedings*, Volume 2, Minneapolis, MN, pp. 550–556.

Mu, W., and Tong, D. (2019) "Choropleth mapping with uncertainty: A maximum likelihood–based classification scheme." *Annals of the American Association of Geographers 109*, no. 5:1493–1510.

Mu, W., and Tong, D. (2020) "Mapping uncertain geographical attributes: Incorporating robustness into choropleth classification design." *International Journal of Geographical Information Science 34*, no. 11: 2204–2224.

National Institute of Standards and Technology. (1994) *Federal Information Processing Standard 173 (Spatial Data Transfer Standard Part 1, Version 1.1)*. Washington, DC: U.S. Department of Commerce.

Pang, A. (2008) "Visualizing uncertainty in natural hazards." In *Risk Assessment, Modeling and Decision*, ed. by A. Bostrom, S. French and S. Gottlieb, pp. 261–294. Berlin: Springer.

Pang, A., Wittenbrink, C., and Lodha, S. (1997) "Approaches to uncertainty visualization." *The Visual Computer 13*, no. 8:370–380.

Retchless, D. P., and Brewer, C. A. (2016) "Guidance for representing uncertainty on global temperature change maps." *International Journal of Climatology 36*, no. 3: 1143–1159.

Ruginski, I. T., Boone, A. P., Padilla, L. M., et al. (2016) "Non-expert interpretations of hurricane forecast uncertainty visualizations." *Spatial Cognition & Computation 16*, no. 2:154–172.

Sanyal, J., Zhang, S., Dyer, J., et al. (2010) "Noodles: A tool for visualization of numerical weather model ensemble uncertainty." *IEEE Transactions on Visualization and Computer Graphics 16*, no. 6:1421–1430.

Slingsby, A., Dykes, J., and Wood, J. (2011) "Exploring uncertainty in geodemographics with interactive graphics." *IEEE Transactions on Visualization and Computer Graphics 17*, no. 12:2545–2554.

Ślusarski, M., and Jurkiewicz, M. (2020) "Visualisation of spatial data uncertainty: A case study of a database of topographic objects." *ISPRS International Journal of Geo-Information 9*, no. 1:1–15.

Sun, M., and Wong, D. W. S. (2010) "Incorporating data quality information in mapping American Community Survey data." *Cartography and Geographic Information Science 37*, no. 4:285–300.

Sun, M., Wong, D. W., and Kronenfeld, B. J. (2015) "A classification method for choropleth maps incorporating data reliability information." *The Professional Geographer 67*, no. 1:72–83.

Thomson, J., Hetzler, B., MacEachren, A., et al. (2005) "A typology for visualizing uncertainty." *Conference on Visualization and Data Analysis 2005 (part of the IS&T/ SPIE Symposium on Electronic Imaging 2005)*, San Jose, CA, pp. 1–12.

Wei, R., Tong, D., and Phillips, J. M. (2017) "An integrated classification scheme for mapping estimates and errors of estimation from the American Community Survey." *Computers, Environment and Urban Systems 63*:95–103.

Wittenbrink, C. M., Pang, A. T., and Lodha, S. K. (1996) "Glyphs for visualizing uncertainty in vector fields." *IEEE Transactions on Visualization and Computer Graphics 2*, no. 3:266–279.

28 Virtual Environments and Augmented Reality

28.1 INTRODUCTION

From reading daily newspapers, watching television, or playing computer games, you are probably familiar with the term **virtual reality (VR)**—it likely brings forth the image of donning a head-mounted display (HMD) and experiencing a simulation of reality. Although this notion is correct, in this chapter, we will see that another terminology is often used and that a broader set of technologies are possible. In Section 28.3, we define terms that are closely associated with VR. A **virtual environment (VE)** is a 3-D computer-based simulation of a real or imagined environment that users are able to navigate through and interact with; for example, you might "fly through" the Grand Canyon while sitting at your computer. VEs of interest to geographers and other geoscientists can be termed **geospatial virtual environments (GeoVEs)**. A **mixed reality (MR)** combines a real-world experience with a virtual representation, such as overlaying the boundaries of "old-growth" forests on your view of a mountainous environment. Most commonly, MRs are referred to as **augmented reality (AR)** because the real world is dominant (i.e., virtual representations are used to augment reality).

Section 28.4 describes some of the technologies for creating VEs, including personalized, wall-sized, head-mounted, and room-format displays. Section 28.5 discusses four "I" factors that are relevant to designing GeoVEs: immersion, interactivity, information intensity, and intelligence of objects. In Section 28.6, we answer several questions that are commonly posed regarding VEs—our hope is that our responses to these questions will enhance your understanding of VEs. In Section 28.7, we provide several examples of the use of the relatively new area of AR. Although cartographers are excited about the potential offered by VEs and AR, we should recognize that there are health, safety, and social issues associated with their use—we briefly consider these issues in Section 28.8.

Since we cannot hope to thoroughly cover the rapidly evolving areas of VEs and AR in this single chapter, we encourage you to survey other books, articles, and research centers that focus on these topics. Examples of books include the *Handbook of Virtual Environments* (Hale and Stanney 2015) and Jason Jerald's (2016) *The VR Book*. A chapter by Arzu Çöltekin and colleagues (2020) in the book *Manual of Digital Earth* provides an extensive overview of VEs and AR, along with related developments in geovisualization. Research centers that focus on either geographic or cartographic applications of VEs or AR include the Center for Immersive Experiences (https://immersive.psu.edu/) at

Penn State University and the Immersive Analytics Lab (https://ialab.it.monash.edu/) at Monash University.[1]

28.2 LEARNING OBJECTIVES

- Define the terms VE and AR and discuss how they relate to one another on the reality-virtuality continuum.
- Discuss the difference between tangible and intangible VEs and provide examples of each.
- Define the terms immersion and presence and indicate the relationship of these terms to one another.
- Contrast the sense of immersion provided by personalized, wall-size, head-mounted, and room-format displays.
- Explain the concept of stereoscopic vision and indicate how it can be achieved using passive and active stereo technology.
- Discuss why enhancing the realism in VEs may not necessarily be desirable.
- Explain why specialized thematic symbols might be desirable in VEs and provide examples of such symbols.
- Describe the nature of VE technology that could enhance collaborative decision-making.
- Define the term Digital Globe and indicate intangible and tangible examples of such globes.
- Discuss how computer interaction techniques in VEs and AR differ from those traditionally used in a desktop computer environment.

28.3 DEFINING VEs AND AR

Early notions of VR promoted the idea of simulating the tangible real world—the everyday world that we see, hear, touch, smell, taste, and move through. A pioneer in this area was Ian Sutherland (1968), who developed the first HMD. He described the initial system as follows:

> Our objective in this project has been to surround the user with displayed three-dimensional information .... We can display objects beside the user or behind him which will become visible to him if he turns around .... The objects displayed appear to hang in the space all around the user. (757)

An indication of how far ahead of the times Sutherland was is that it took almost 25 years before geographers started to utilize VR. One of the first researchers to realize

DOI: 10.1201/9781003150527-31

the potential of VR for geography was Ian Bishop (1994). Although Bishop did not use the term VR and focused largely on vision, it was clear that this was what he had in mind:

> The non-scientific audience .... wants abstraction minimized, information content maximized .... with the whole package digestible and non-threatening. This suggests the use of a visual realism approach [that] shows information consumers what will or might happen under a variety of conditions and permits them to explore the alternative environments using their natural sensory perceptions. (Republished with permission of John Wiley & Sons—Books, from p. 61 of "The role of visual realism in communicating and understanding spatial change and process," by I. Bishop (1994), In *Visualization in Geographical Information Systems*, ed. by H. M. Hearnshaw and D. J. Unwin; permission conveyed through Copyright Clearance Center, Inc.)

Figure 28.1 is an example of some of Bishop's early work—a simulation of a proposed lake created as a result of filling an existing open-cut coal mine with water.

Although it makes sense to simulate the tangible real world, there are instances in which we might need to display something intangible. For instance, imagine depicting the concentration of ozone in the atmosphere both vertically and horizontally at some point in time. Because we cannot see ozone, we must create an abstract representation of it. To achieve this, we might immerse individuals in a computer-simulated environment in which they travel through a virtual atmosphere, seeing the concentration of ozone as they move. The concentration of ozone could be depicted by the density of small balls having a uniform size and color, with a high density of balls indicating a high ozone concentration.

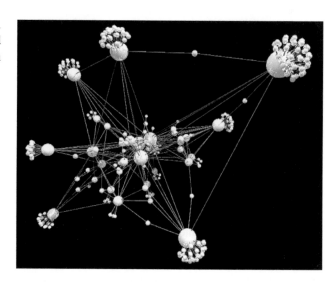

FIGURE 28.2   A depiction of the spatial structure of a portion of the Web. (Courtesy of Bob Hendley.)

Here the term *virtual reality* seems inappropriate, given that humans cannot see ozone; as a result, researchers have chosen to use the term *virtual environment* to encompass both tangible and intangible simulations.

As a more abstract example, consider Figure 28.2, which was developed by Andrew Wood and colleagues (1995) to portray the spatial structure of a portion of the Web. Here the structure is based not on geographic location but rather on the relationship between topics presented via the Web, with more similar topics shown together. Each sphere represents a Web page, with the links between spheres representing the links between Web pages and the size of a sphere representing a function of the number of linking Web pages. The idea is that we could utilize the resulting 3-D space to avoid becoming "lost in hyperspace" (Dodge and Kitchin 2001, 126). In this case, the ultimate reality is a set of zeros and ones in the computers used to depict this information, and thus the term VE is again appropriate.

For our purposes, we define a **virtual environment** as a 3-D computer-based simulation of a real or imagined environment that users are able to navigate through and interact with. Normally, VEs are experienced visually, although ideally, it would be possible to utilize the full range of senses, including sound, touch, smell, taste, and body movement.

Although early efforts focused on the potential of immersing oneself in a VE, more recent work has suggested that a combination of reality and VEs might also be useful. This notion is depicted in Figure 28.3, where we see a real-world-to-VE continuum. The "real world" end of the continuum represents what we perceive in the real world, whereas the VE end represents a simulation of an environment. In between, we see **mixed realities (MRs)**, which represent a combination of the real world and a virtual representation. The notion of MRs can be further subdivided into **augmented reality (AR)** and **augmented virtuality (AV)**, depending on whether the real world or the virtual representation is emphasized.

FIGURE 28.1   An example of Ian Bishop's early work on visual realism—a simulation of a proposed lake created as a result of filling an existing open-cut coal mine with water. (From Slocum, T. A., Armstrong, M. P., Bishop, I., et al. (1994) "Visualization software tools." In *Visualization in Modern Cartography*, ed. by A. M. MacEachren and D. R. F. Taylor, pp. 91–122. Oxford: Elsevier.)

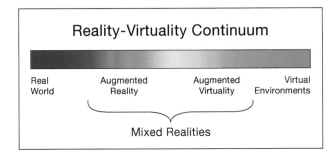

FIGURE 28.3 The real-world-to-virtual-environment continuum. (After Drascic and Milgram 1996.)

Although there is some logic in the use of the separate terms for AR and AV, AR is much more commonly used, even when AV might be appropriate. For instance, in early work in geography, Nick Hedley (2003) developed an approach for displaying a VE on a large card held in a user's hands. Users could rotate the card just as they would rotate an object in the real world, thus providing a potentially more intuitive approach for examining 3-D landscapes than traditional mouse-based interfaces. Although Hedley described this work as a form of AR, it can be argued that the approach placed greater emphasis on interpreting the VE and thus might be considered a form of AV.

28.4 TECHNOLOGIES FOR CREATING VEs

Technologies for creating VEs are important because they impact how we experience those environments. Nick Hedley (2003) argued that technology for creating VEs can be broken down into three parts: the display, the hardware controls, and the graphical user interface (GUI). For instance, with desktop computers, we might use an LCD display, a mouse and keyboard as controls, and Microsoft Windows as the GUI. Normally, in describing the different technologies for visualization purposes, emphasis is placed on the display, and so we do that here. You should bear in mind, however, that the affordances possible with a particular display might be a function of the controls and GUI.

VEs can be created using five basic forms of display technologies: personalized, wall-sized, head-mounted, room-format, and drafting-table format.

28.4.1 PERSONALIZED DISPLAYS

Personalized displays refer to displays for *individuals* such as those found on a desktop computer, laptop, tablet, or phone. Although the term *desktop display* normally has been used rather than personalized display, we argue that the increasing capability to create sophisticated visualizations on a variety of personal devices makes personalized displays a more appropriate term. Personalized displays traditionally have been the most common approach for depicting VEs. Software packages for creating and interacting with 3-D VEs on personalized devices can be divided into two groups: those that display a relatively small portion of

Earth's surface (the base map is essentially a 2-D environment, although the symbols extend in the third dimension), and those that display either the entire Earth or a substantial portion of it (known as **virtual globes** or **virtual worlds**). Examples of the former include ArcGIS's ArcScene, Visual Nature Studio, Blender, and SketchUp. Examples of virtual globes include ArcGIS's ArcGlobe,[2] Google Earth, and NASA WorldWind.

You may be thinking that personalized displays do not seem to produce the 3-D character that we would normally associate with a VE, but it turns out that there are numerous potential 3-D depth cues that can be utilized to suggest a VE, as indicated in Table 28.1. Here you can see that the depth cues have been split into monocular pictorial, monocular motion-produced, and binocular. On a 2-D display screen, all of the monocular cues can be used to create a 3-D environment.

Traditionally, most personalized displays did not provide enhancements to make the VE appear more realistic (i.e., to look truly 3-D). In recent years, however, it has been increasingly common to present separate images to the left and right eye, thus enabling a **stereoscopic view** or **true 3-D view** (Buchroithner et al. 2012), just as we have in the real world (thus providing the **stereoscopic depth cue** that is shown at the bottom of Table 28.1). Stereoscopy can be accomplished in three ways: passive, active, and autostereoscopic. In a **passive display**, images for the left and right eye are created simultaneously by the computer/display system, each with a different polarization of light. The user then wears lightweight specialized polarized glasses, which interpret the two images and produce a true 3-D image.[3] In an **active display**, images for the left and right eye are cycled back and forth on the display screen, and the user wears shutter glasses that are in sync with the display screen. In general, passive display systems are less fatiguing on the eye-brain system, but the associated computer hardware is more expensive. With **autostereoscopic displays**, no specialized glasses are required.[4] Although autostereoscopic displays sound ideal, they are more expensive and are typically characterized by a limited number of viewing angles known as "sweet spots." Presently, autostereoscopic displays are not as commonly used as passive and active ones, but this may change as the technology evolves. Manfred Buchroithner and Claudia Knust (2013) have promoted their use in cartography.

28.4.2 WALL-SIZE DISPLAYS

A **wall-sized display** covers a large portion of a wall by tiling images from either multiple projectors or multiple display panels. For instance, the display shown in Figure 28.4 measured 7.6 × 1.5 meters (25 × 6 feet), covered 120° of the visual field, provided a resolution of 5,760 × 1,200 pixels, and was created using three graphics subsystems (and associated projectors). Given their large size and the large rooms within which they are found, it makes sense that wall-size displays can be utilized effectively by groups of individuals.

TABLE 28.1

3-D Depth Cues Utilized in Virtual Environments and Augmented Reality

General Type	Cue	Definition
Monocular pictorial	Occlusion	Partially hidden object is seen as more distant
	Relative height	Objects higher in the field of view are seen as more distant
	Familiar size	Distance is based on our prior knowledge about the size of familiar objects
	Linear perspective	Lines or objects that appear to converge are seen as more distant
	Atmospheric perspective	The more air and particles that we look through, the more distant objects appear
	Texture gradient	Increasing the density of textural elements suggests a greater distance away
	Shadows	Shadows cast by objects suggest the distance and 3-D character of the objects
	Blurring	Objects out-of-focus (blurred) may be seen as more distant (or alternatively closer)
	Shading	Areas away from a light source are shaded, as in shaded relief
Monocular motion-produced	Motion parallax	Objects moving by rapidly appear closer than objects moving more slowly
	Kinetic depth	Moving 3-D objects are more likely seen as 3-D than static 3-D objects
Binocular	Convergence	Distance of an object determined by angle formed between the object and the eyes
	Stereoscopic depth	Distance of an object is a function of the difference in the images formed on the retinas of the eyes

Note: After Goldstein and Brockmole (2017) and Ware (2013).

The one shown in Figure 28.4 was utilized to assist groups of decision-makers in evaluating the uncertainty associated with global water balance models (Cliburn et al. 2002).

A limitation of projection-based systems like the one shown in Figure 28.4 is that the resulting resolution is relatively coarse given the large areal extent of the resulting image. A solution to this problem is to combine

FIGURE 28.4 An example of a wall-sized display for creating a VE. The image portrays decision-makers examining the uncertainty associated with a water balance model. (Courtesy of James R. Miller.)

numerous LCD panels to create a high-resolution image. For instance, researchers at the Electronic Visualization Laboratory (EVL) at the University of Illinois at Chicago developed LambdaVision (https://www.evl.uic.edu/cavern/lambdavision/display.html), which combined 55 LCD panels, each with a resolution of 1,600 × 1,200, for a total display of 100 million pixels). The areas between panels (called *bezels*) appear awkward to deal with (see Figure 28.5), but the above link describes several applications that have effectively used this approach.

The **GeoWall** is a specialized form of a wall-sized display that was developed by geologists and computer scientists to display stereoscopic images to large groups of people at a modest cost.[5] In its basic form, a passive GeoWall is comprised of four components (see Figure 28.6; Halfen et al. 2014, 525):[6]

- A computer with a dual-head graphics card capable of sending two different images at the same time
- Two DLP projectors equipped with polarizing filters
- A display screen capable of preserving light polarization
- Passive glasses for users

FIGURE 28.5 An early panel-based wall-size display. Note the bezels (the dark areas between each LCD panel). Subsequent panel-based systems have decreased the size of these bezels. (Courtesy of Electronic Visualization Laboratory, the University of Illinois at Chicago.)

GeoWalls have been primarily used to teach geology and physical geography, but the technology could certainly be used to display stereoscopic thematic maps in cartography classes. One disadvantage of the GeoWall is that in its basic form, the resolution is limited; for example, a typical resolution is $1,024 \times 768$ pixels. To address this issue, researchers at EVL developed the GeoWall-2, which like Lambda Vision, used an array of 15 LCD panels, with each panel powered by its own computer (https://www.evl.uic.edu/cavern/pg2/).

Those working with stereoscopic displays need to be aware that not all people can see in stereo, and this is true whether stereo images are created on a wall-size display or on an individual display screen. Colin Ware (2013), a specialist in working with stereoscopic displays, indicates that "as much as 20% of the population may be stereo blind" (259).

FIGURE 28.6 A typical GeoWall setup; key elements include a computer with a dual-head graphics card, two DLP projectors with polarizing filters, a display screen capable of preserving the polarization, and passive glasses for users. (Courtesy of Electronic Visualization Laboratory, the University of Illinois at Chicago.)

Similarly, in a study done with a GeoWall, Daniel Hirmas and colleagues (2014) found that in a test of stereopsis, almost 20 percent of students thought that they could not see a 3-D effect, but interestingly many of these same students were able to answer questions correctly regarding stereoscopic images. Another potential limitation of the GeoWall is that the angle of the student relative to a centerline from the front to the back of the room should not exceed 45 degrees. Hirmas et al. confirmed that this was a necessary guideline as students exceeding this angle performed more poorly.

28.4.3 HEAD-MOUNTED DISPLAYS

The term **head-mounted display (HMD)** derives from the means used to create such a display—placed on the user's head is a helmet-like device that shields the real-world view and provides images of the VE to each eye (Figure 28.7A). HMDs normally provide separate images to the left and the right eyes, thus enabling the user to have a stereoscopic view. The HMD includes a head-tracking device so that the view of the virtual representation is a function of the position of the user's head. Because the real world is not visible when using HMDs, specialized control devices such as a *data glove* are utilized. Traditionally, HMDs were criticized because they were a burden for users to wear (due to their weight), their resolution was low (a

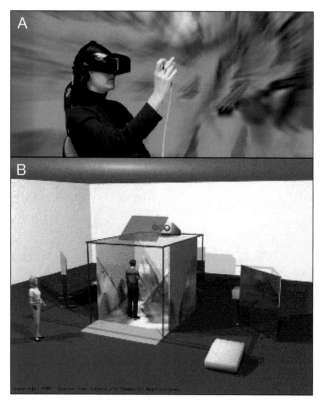

FIGURE 28.7 Examples of two displays for creating VEs: (A) Head-mounted and (B) room-format (CAVE). (Figure A courtesy of William Winn; Figure B courtesy of Electronic Visualization Laboratory, the University of Illinois at Chicago.)

typical high resolution was 640 × 480 addressable pixels), and the field of view was narrow (the view was rectangular like on a traditional computer screen). Fortunately, these limitations have gradually been improved, and the recent release of novel HMD systems by companies such as Oculus (https://www.oculus.com/) and Vive (https://www.vive.com/us/) has improved things dramatically. No doubt some of you are using such technology in the gaming community. It is also noteworthy that you can build your own HMD-like device using Google Cardboard (https://arvr.google.com/cardboard/get-cardboard/), which enables you to view VEs on your smartphone.

28.4.4 ROOM-FORMAT AND DRAFTING-TABLE FORMAT DISPLAYS

Room-format displays were developed in response to the limitations of HMDs. The most frequently cited example is the CAVE, which was initially developed by EVL. The CAVE provides a room-sized view of a VE by projecting images onto three walls and the floor (Figure 28.7B).[7] As users stand or move in the CAVE, a head tracker on one user governs the view that all users see through stereo glasses; the tracker and glasses are much less obtrusive than traditional HMDs. Interaction with the VE is achieved through a 3-D mouse, which consists of a joystick for navigation and programmable buttons for interaction. One advantage of the CAVE is that users can see each other—in that sense, the CAVE can be considered a MR, except that the emphasis is clearly on viewing the VE.

Drafting-table format displays, such as the ImmersaDesk (https://www.evl.uic.edu/research/1770), were developed as a less expensive substitute for the CAVE, which can cost several hundred thousand dollars. With the ImmersaDesk, the VE was projected onto a single screen that was tilted at a 45° angle (the logic is that looking down in the CAVE is important—the tilt helped simulate this). Again, a head tracker on one user governed the view that other users could see, and a specialized 3-D mouse was used to interact with the simulated environment.[8]

In addition to their high cost, CAVEs have several other limitations: users must stand for long periods of time, arm fatigue results from holding a wand, there is fan noise from the projectors, users must wear active stereo glasses, the resolution is relatively poor, and colors tend to be dark (Marai et al. 2019). As a result, researchers at EVL developed CAVE2, which combines the features of the original CAVE with the characteristics of high-resolution tiled displays. CAVE2 consists of 72 LCD panels arranged in a 320-degree circle (Figure 28.8). The panels are controlled by 18 computers that provide both 2-D and passive stereoscopic capability. Fourteen cameras permit tracking up to 20 objects (e.g., glasses and controllers associated with users). As you can see in Figure 28.8, users do not necessarily have to stand, and they can easily interact with one another. The resulting system is even more expensive than the CAVE and tiled wall-size displays (approximately 1 million dollars),

FIGURE 28.8 The CAVE2 environment: 72 LCD panels are arranged in a 320-degree circle, users can view images either in 2-D or in true 3-D stereo, and 14 cameras permit tracking up to 20 objects (e.g., glasses and controllers associated with users). (Courtesy of Electronic Visualization Laboratory, the University of Illinois at Chicago.)

but Marai et al. describe several applications that have benefited from this approach and also various new hardware technologies that have been implemented at the University of Illinois at Chicago as a result of developing CAVE2.[9]

28.5 THE FOUR "I" FACTORS OF VEs

Based on the work of Michael Heim (1998), Alan MacEachren and colleagues (1999) proposed four "I" factors important to creating GeoVEs: immersion, interactivity, information intensity, and intelligence of objects. Although initially introduced in association with VEs, it should be noted that these terms can also be applied to AR.

28.5.1 Immersion

Immersion is defined as ".... perceiving oneself to be enveloped by, included in, and interacting with an environment that provides a continuous stream of stimuli and experiences" (Witmer and Singer 1998, 225). The notion of immersion is closely tied to that of **presence,** which is defined "as the subjective experience of being in one place or environment when one is physically in another" (Witmer and Singer 1998, 225). Because a high level of immersion leads to a sense of presence, those discussing VEs often speak of the level of immersion, and we will do the same. Keep in mind, though, that the notion of presence is important—an indication of its importance is that at least two journals dealing with VEs focus on presence in their title (https://direct.mit.edu/pvar and https://dl.acm.org/journal/pres).

Each of the displays that we have discussed provides a different sense of immersion. Personalized displays in which stereoscopy is not used provide the lowest sense of immersion because both the real world outside the computer screen and the VE depicted on the screen are readily visible and because the images do not look truly 3-D (because they are not presented stereoscopically). For this reason, such

displays are sometimes termed non-immersive, but we prefer to think of them as semi-immersive since viewers often feel that they are part of the VE. For instance, in describing students using virtual field course software, Jason Dykes (2002, 71–72) argued that such displays provide a "strong sense of immersion, [in which] users frequently move with the surface, duck whilst navigating around obstacles, and express their desire to 'go' somewhere, or the fact that they are 'lost.'"

A wall-sized display provides a greater sense of immersion than personalized displays because a much greater portion of the visual field is covered, especially when one stands or sits close to the wall-size display. This still must be termed a semi-immersive system, though, as images do not look truly 3-D (unless multiple projectors are used to create stereoscopic images and specialized viewing glasses are used). In contrast, HMDs and CAVEs both provide a fuller sense of immersion because the real world can be hidden from view, images are presented stereoscopically, and senses other than vision (e.g., sound) are often used.

The display systems that we have described rely primarily on vision and, to a limited extent, sound. It is important to recognize that a sense of immersion is a function of more than vision and sound—it involves all of the senses, including touch, smell, taste, and body movement. Outside the field of geography, sophisticated systems have been developed that incorporate multiple senses. For instance, the National Advanced Driving Simulator (NADS) housed at the University of Iowa can be used to simulate the "feel" of driving (https://www.nads-sc.uiowa.edu/sim_nads1.php), and thus can provide a safe means of evaluating the effects of alcohol, drugs, and fatigue on driving.

Another example more closely related to geography is the Multisensory Apulia Touristic Experience (MATE), developed by Vito Manghisi and colleagues (2017) to provide potential tourists a sense of what it would be like to visit areas around Apulia, Italy. Manghisi et al. argued that a sense of presence could be enhanced if multiple senses

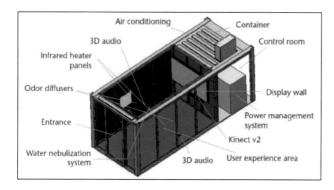

FIGURE 28.9 The Multisensory Apulia Touristic Experience (MATE) used to provide potential tourists a geographic sense of areas around Apulia, Italy. The larger room on the left is the "user experience area," and the small room on the right is the "control room." (© 2017 IEEE. Reprinted, with permission, from Manghisi, V. M., Fiorentino, M., Gattullo, M., et al. (2017) "Experiencing the sights, smells, sounds, and climate of southern Italy in VR." *IEEE Computer Graphics and Applications* 37, no. 6:19–25.)

are engaged and if there is agreement on what these senses are telling you. To achieve this enhanced presence, they designed a mobile container, which consists of a user experience area and a control room, as shown in Figure 28.9. Users can see images and videos projected on a large-format 85-inch monitor (essentially a wall-size display) and experience the environment of the projected region by rapidly changing the temperature, humidity, and ambient scents. This is achieved by using an oversized air conditioning system, a high-wattage infrared heating panel, a water nebulization system (for simulating rain and fog), and an aspirator (to rapidly clean the air). Interestingly, Manghisi chose 2-D images rather than 3-D ones to avoid cybersickness, which we consider in Section 28.8. You can read more about the MATE system here: https://tinyurl.com/MATE-Summary.

28.5.2 Interactivity

A basic issue in interacting with VEs is navigating through them and understanding one's current orientation and location; as a result, users often report feeling lost in VEs (Dykes 2002). This is an issue that has yet to be solved by those developing VEs, although some research has been done. For example, Sven Fuhrmann and Alan MacEachren (2001) observed geoscientists navigating a desktop GeoVE via a traditional VRML Web browser[10] and found that users were often disoriented and thus not satisfied with the navigation and orientation capabilities provided. As an alternative, they developed a GUI based on a flying saucer metaphor, which they argued would be appropriate because:

- Its movement characteristics are commonly understood, and movement in all directions is expected to be possible.
- In science fiction, flying saucers operate in geographic environments.

- Expertise in controlling flying saucers is not an issue because no user has training using the GUI.
- Because a flying saucer has no apparent moving parts, users think of it as sophisticated technology with simple-to-learn controls.
- Flying saucers are simple in shape, providing a clean user interface.

Fuhrmann and MacEachren tested the usability of their GUI and found that navigation was indeed easier than with the standard VRML browser and that users liked the interface. A shortcoming, however, was that many users still claimed to be "lost." The problem was that while users had a natural egocentric view of the VE, they were not able to orient themselves in the flying saucer GUI because of missing spatial information. To solve this problem, Fuhrmann and MacEachren added an overview map to the user interface showing the overall extent of the displayed landscape and a directional cone indicating the user's current position and orientation (Figure 28.10). In conducting a range of usability studies with this prototype, they found that users were less likely to be disoriented in a desktop GeoVE when provided with both the natural egocentric view and an enhanced overview map.

Another important issue concerning interactivity in VEs is determining appropriate tools for interacting with and modifying the VE. As a starting point, it seems reasonable to provide tools similar to those used outside VEs. For instance, based on Section 25.4, we presume that tools for brushing and focusing should be available. However, the 3-D character of the VE implies that other tools should also be available, such as those for picking up and rotating objects. Although a suitable set of tools for geography has yet to be fully developed, Joseph Gabbard (1997) summarized numerous interaction techniques for VEs that might serve as the basis for developing such tools.

A third issue concerned with interactivity in VEs is the nature of the control devices used to interact with the VE. Traditionally, users have interacted with computers using a windows-icons-menus-pointing device (WIMP) interface. Over the last 20 years or so, this has gradually changed as researchers have experimented with **multimodal interfaces** in which two or more novel modes of interaction are used, such as speech, lip movements, pen-based gestures, free-hand gestures, and head and body movements. Sharon Oviatt (2008) was a pioneer in championing these approaches; for a general discussion of multimodal interaction techniques in VEs, see Popescu et al. (2015). In Section 28.7.3, we consider the related problem of eliciting appropriate hand and foot gestures for interacting with maps in AR.

28.5.3 Information Intensity

Information intensity deals with the level of detail (LOD) presented in the VE. One issue relevant to LOD is how much detail should be shown at a particular scale. If the purpose

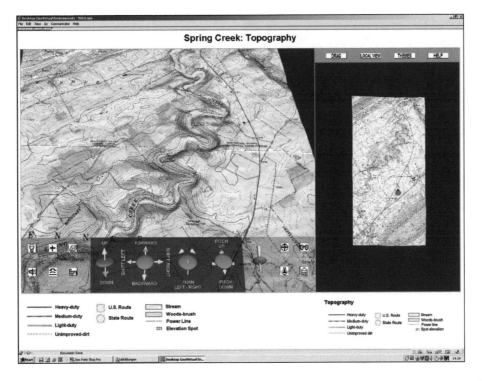

FIGURE 28.10   The user interface for Fuhrmann and MacEachren's software that utilized a flying saucer metaphor. On the left is the egocentric view; on the right is an overview map with a directional cone indicating the user's current position and orientation. (Courtesy of Sven Fuhrmann.)

is to simulate reality, then the inclination is to show considerable detail, but how much detail is appropriate? Alexey Voinov and colleagues (2018, 416) argue that rather than simply trying to show greater realism, we should consider the purpose of the VE and who the users are, which should make sense, as these are also the basic questions that we ask when designing a map (see Section 1.5). In the realm of thematic mapping, Mary Hegarty and colleagues (2009) found that although both undergraduates and post-graduate meteorologists preferred realistic maps, they actually performed better with weather maps that had less realism. Thus, there seems to be a general tendency to assume that greater realism will enhance interpretation when in fact, it may not. Hegarty et al. termed this result "naïve cartography."

28.5.4   Intelligence of Objects

Intelligence of objects refers to the notion that objects within a VE should exhibit some degree of behavior or intelligence, just as we would expect of real-world objects. For instance, if we attempt to move a small branch on a tree in a VE, we should expect the branch to move. Similarly, if we encounter a person (known as a **virtual assistant** or an **intelligent agent**) in a VE, we would like that person to exhibit intelligent behavior such as guiding us to a desired location. Given the role of virtual assistants in everyday life (e.g., Amazon's Alexa and Apple's Siri), it would seem natural that they also would be incorporated in VEs (Granell et al. 2021). Although the applications that we examined did not make use of virtual assistants, we did find them used in

the related disciplines of cultural heritage (Machidon et al. 2018) and urban design (Wang and Chen 2009). We would expect that as VEs and AR evolve, we will see greater use of virtual assistants.

28.6   SOME KEY QUESTIONS REGARDING VEs

28.6.1   Are Specialized Symbols Necessary for Thematic Maps Created in VEs?

Some symbols, such as 3-D cylinders, seem like a natural fit for VEs because their 3-D character matches the 3-D character of the VE, especially when viewing a virtual globe. For example, in Figure 28.11, 3-D cylinders are used to depict Internet usage for countries around the world— this image was created by Bjørn Sandvik using the Keyhole Markup Language (KML) and Google Earth (https://developers.google.com/kml/documentation/kml_tut); Sandvik was a pioneer in the use of thematic maps on virtual globes (see http://thematicmapping.org/engine/). But what about more traditional symbols, such as the proportional circles that we examined in Chapter 18? Should these symbols also be used on virtual globes, and how effective are they compared to other approaches? In this section, we consider two studies that have examined this more traditional symbol and also evaluated alternative symbols for virtual globes.[11]

Using the Google Earth virtual globe, Susanne Bleisch (2011) evaluated the effectiveness of the following six symbol approaches: (A) 2-D bars on billboards, (B) 2-D bars with reference frames on billboards, (C) 3-D bars, (D) 3-D

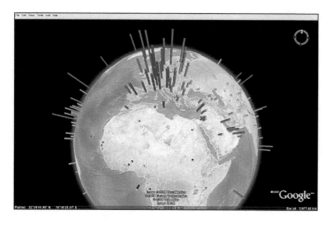

FIGURE 28.11 3-D cylinders created using KML in Google Earth to portray Internet users per 100 population. Note the clear digital divide, as some countries appear to have no cylinders whatsoever. (Courtesy of Bjørn Sandvik.)

bars with reference frames, (E) 2-D circles on billboards, and (F) 2-D circles with reference frames on billboards. A *billboard* is a 3-D graphic element that is placed at a fixed location in the virtual globe but that rotates so that its graphic face is always aligned vertically with the current viewpoint of the user. The billboard itself is transparent, but the graphic symbol (such as a circle) placed on it is legible. The *reference frame*, as the name suggests, provides a frame of reference for the symbol being estimated, with the size of the frame a function of the largest value for the data associated with that symbol. In an earlier study, Bleisch found that symbols with reference frames were interpreted

faster and with more confidence than symbols without reference frames.

Bleisch had 31 participants with either a geography or geomatics background perform two specific information tasks in a Google Earth environment in which the six symbol approaches were placed in a realistic-looking landscape setting. The results revealed that 2-D bars with reference frames produced the fastest and most accurate results, although 3-D bars with reference frames were almost as effective. 2-D circles (the traditional map symbol) clearly were the least effective. Although symbols with reference frames produced more accurate responses, they were responded to more slowly. One limitation of the study was that participants did not make frequent use of a navigation option even though they had that capability.

In a more recent study, Kadek Satriadi and colleagues (2021) evaluated the effectiveness of the five symbols shown in Figure 28.12 in a custom-built virtual globe for the Web: billboarded bars (bars placed on a billboard, in a fashion analogous to Bleisch), tangential bars (bars appear as though printed on the globe), normal bars (cylinders displayed perpendicular to the globe's surface), billboarded circles (circles placed on a billboard), and tangential circles (circles appear as though printed on the globe). A reference frame for symbols was not considered in this study. One important attribute that was examined was the angular distance between two symbols to be compared: values of 20°, 60°, and 120° were used; participants were permitted to rotate the globe but not zoom in. Fifty-two participants completed a specific

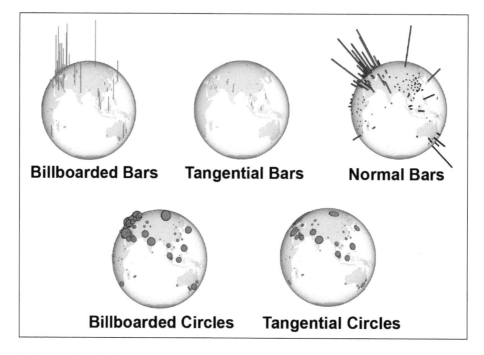

FIGURE 28.12 Symbols used by Satriadi et al. (2021) in a study examining the use of point symbols on virtual globes. (© by ACM, Inc. Figure 1 of Satriadi, K. A., Ens, B., Czauderna, T., et al. (2021) "Quantitative data visualisation on virtual globes." *CHI Conference on Human Factors in Computing Systems*, Yokohama, Japan, pp. 1–14.)

map task involving symbol comparisons. Some of the key findings were as follows:

- Circle-based approaches were significantly less accurate than bar-based approaches.
- There was no significant difference in accuracy for bar-based approaches.
- Circle-based approaches were responded to most quickly.
- Billboarded bars were responded to the fastest, and normal bars were responded to the slowest.
- Participants were most confident in their estimations with billboarded bars and normal bars.
- In terms of aesthetics, normal bars were most pleasing, while billboarded bars were least pleasing.
- Angular distance between symbols did not affect the results, as participants frequently rotated the globe to make estimations.

Overall, the results supported the use of normal bars (which the authors argued is a common solution), although one should recognize that estimates will be slower with this approach. One extension to both the Bleisch and Satriadi et al. studies would be to consider the ability of readers to detect spatial patterns using these symbologies. In total, both studies show that traditional proportional circle symbology is not as effective as billboarded bars and 3-D bars (termed normal bars in the Satriadi et al. study).

28.6.2 Are Stereoscopic Maps More Effective than Non-Stereoscopic Maps?

Since most personalized displays and the majority of wall-size displays do not have stereo capability, it is natural to ask whether stereoscopic maps are more effective than non-stereoscopic maps, given that stereo displays are costlier. More generally, we can ask whether stereo displays have proven more effective for a wide variety of applications. In this respect, John McIntire and colleagues (2014) conducted a review of 160 published articles that described over 180 experiments comparing the effectiveness of stereoscopic and non-stereoscopic displays. Overall, they found an improved performance for stereo in 60 percent of the experiments, only a marginal improvement for stereo (or mixed results) in 15 percent of the experiments, and no improvement in 25 percent of the experiments. Thus, although stereo can be effective, there have been many instances in which it has not been effective. Perhaps most important are situations in which they found stereo to be most effective:

- Depth-related tasks performed in the near-field (tasks in close proximity to the viewer, such as spatial manipulations of objects).
- Difficult/complex tasks and for tasks that are not well-learned.

- For spatial manipulation of real or virtual objects … and for finding, identifying, and classifying objects or imagery.
- For tasks involving judgments about object positions or distances, and for tasks involving spatial understanding, memory, or recall. (After McIntire et al. 2014, 23)

Let's attempt to apply these findings to a recent cartographic study by Vojtěch Juřík and colleagues (2020) that examined the effect of stereo and interactivity on the ability of people to extract information from a digital terrain model (DTM). The 91 participants in the study worked with four different map environments: (1) stereo/interactive, (2) stereo/static (not interactive), (3) non-stereo/static, and (4) non-stereo/interactive. There were three basic tasks performed by the participants: comparing the elevation of objects (cubes) in the DTM, comparing the elevation of water bodies in the DTM, and identifying the character of terrain profiles.[12] Overall, the results revealed that both stereo and interactivity tended to increase response times, with the longest response times for a combination of stereo and interactivity. Moreover, the static conditions generally produced a higher percentage of correct responses. Thus, stereo did not work as well as one might hope. The tasks do appear to deal with McIntire et al.'s "judgments about object positions" and "spatial understanding," but possibly they are not complex enough to take advantage of stereo. For the most complex task (the terrain profile), there appeared to be some advantage for stereo, as the highest percent of correct responses was obtained with a stereo/static condition.

An example of a study that showed a stronger stereo effect is the work of Dennis Edler and colleagues (2014), in which they utilized an autostereoscopic display. Edler et al. examined the ability of people to count point symbols on maps displayed in non-stereo and stereo environments. The maps consisted of a base layer (e.g., city locations and boundaries, and rivers) and point symbols representing ten different industries. For the non-stereo condition, all information was presented on a standard LCD non-stereo monitor. For the stereo condition, an autostereoscopic display was used to position the base information and point symbols at different distances from the reader. Approximately one-half of the point symbols were placed in a layer closest to the reader, and the remaining point symbols were placed in a second layer further away from the reader. The base information was placed in a third layer furthest away from the reader. The *projection layer*, the layer with the sharpest view and highest resolution, was placed between the two point-symbol layers (an animation illustrating the relative positions of the various layers can be found at https://dbs-lin.ruhr-uni-bochum.de/3D_Karten/video/). Given this use of multiple layers, we can think of the stereo approach as establishing a visual hierarchy of map information (although Edler et al. did not use this terminology).

To compare the non-stereo and stereo approaches, 83 participants performed a series of counting tasks involving the point symbols. The findings revealed no significant difference in the accuracy of counts for non-stereo and stereo, but the speed of counting was significantly faster for the stereo condition. Looking back at McIntire et al.'s situations in which stereo is deemed effective, the counts of a particular industry point symbol in the stereo condition seem to represent a "depth-related task" (a symbol layer is found at a particular depth) and involve "classifying objects" (determining a point symbol of a specific industry type), thus suggesting why the stereo approach was more effective.

28.6.3   What Are Some Examples of VEs That Make Use of Caves and Wall-Size Displays?

In this section, we consider two examples illustrating the effective use of a CAVE and a wall-size display, respectively. For other examples, see Li et al. (2010), Kuhlen and Hentschel (2014), and the Further Reading section.

28.6.3.1   Using a CAVE to Create Soils Maps

In this section, we consider Trevor Harris and Paddington Hodza's (2011) use of a CAVE to enable soil scientists to create soils maps in a collaborative digital environment. Traditionally, the creation of soils maps was largely a non-digital process that was cognitively demanding, time consuming, and non-collaborative (20). A major component of the traditional approach was the stereoscopic interpretation of aerial photograph pairs, again done individually and without links to a full range of digital data. As an alternative to this traditional approach, Harris and Hodza utilized a CAVE to develop an Experiential GIS (EGIS), which combined the immersive and collaborative character of the CAVE with GIS data and software.

To evaluate the effectiveness of the EGIS, four experienced soil scientists were asked to use EGIS to create a soils map of a previously mapped area that they had not been involved in mapping. Figure 28.13 provides a sense of what the CAVE environment looked like to the soil scientists. As you can see, the landscape was projected onto three walls and the floor; users wore active stereo glasses and a head-tracking device and navigated the scene using a hand-held wand. The soil scientists "reported a significantly improved ability to observe, interpret and analyze the patterns, variations and relationships in soil forming factors that previously had to be inferred from 2-D stereo pairs of analog aerial photos" (28). Moreover, the process was collaborative and was completed in about one-half of the time of traditional soils mapping. The soil scientists felt that the resulting soils map was at least comparable to the official soils map of the area. An analysis of the two maps by Harris and Hodza indicated that the official soils map was more detailed, but Harris and Hodza cautioned that this does not necessarily mean it was more accurate. Overall, both Harris and Hodza and the soil scientists expressed the considerable potential of this collaborative immersive approach for soils mapping.

FIGURE 28.13   The CAVE environment utilized in the EGIS system developed by Harris and Hodza (2011) to enable soil scientists to create soils maps collaboratively. (From Harris and Hodza 2011. First published in *Cartography and Geographic Information Science* 38(1), p. 25. Reprinted with permission from the Cartography and Geographic Information Society.)

28.6.3.2   Using a Wall-Size Display to Obtain Public Input on Climate Change Scenarios

We have read about climate change and its potential impacts, but how can we visualize those impacts at a local level and allow a broad range of stakeholders to have input on potential scenarios for land-use planning? To explore this question, Chen Wang and colleagues (2016) utilized a wall-size display (the Virtual Landscape Theatre (VLT)—https://www.hutton.ac.uk/learning/exhibits/vlt; Figure 28.14) to obtain input from a variety of stakeholders on potential land-use scenarios in Tarland in northeast Scotland. Four basic land-use scenarios were developed:

- World Markets: globalization and market forces dominate
- National Enterprise: protectionism at the national and European levels, particularly concerning food security
- Global Sustainability: global cooperation and binding treaties for a high degree of environmental protection
- Local Stewardship: decision-making dominated by local issues, particularly sustainable solutions to local food and energy needs (591–592)

Wang et al. created quantitative expressions of these various scenarios and then produced realistic-looking expressions of the results, such as those shown in Figure 28.15.

The resulting 3-D interactive images were displayed in the VLT, which had a screen measuring 5.5 × 2.25 meters (18.1 × 7.4 feet), with a curvature of 160° (thus providing

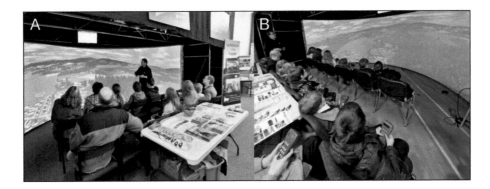

FIGURE 28.14 Views of the Virtual Landscape Theatre (VLT) utilized by Wang et al. (2016) to explore landuse scenarios in northeast Scotland. (A) a non-local audience (from Edinburgh); (B) a local audience from Tarland in northeast Scotland. (From Wang, C., Miller, D., Brown, I., et al. (2016) "Visualisation techniques to support public interpretation of future climate change and land-use choices: A case study from N-E Scotland." *International Journal of Digital Earth* 9, no. 6:586–605, Taylor & Francis Ltd., https://www.tandfonline.com/journals/tjde20.)

some sense of immersion, although stereoscopy was not possible). Approximately 20–25 people could comfortably view the screen at one time, and hand-held units permitted participants to vote on scenarios and to manipulate the image when allowed to do so by those directing a session. Formal sessions were created for six groups of stakeholders: land managers, natural heritage managers, planners, school and youth groups, university students, and the general public. This is noteworthy as it is our impression that such evaluations often do not include such a wide range of stakeholders. Those directing a session introduced the basic scenarios, recorded verbal comments made by stakeholders, and had stakeholders vote on various scenarios using the hand-held units. Following the introduction, individual

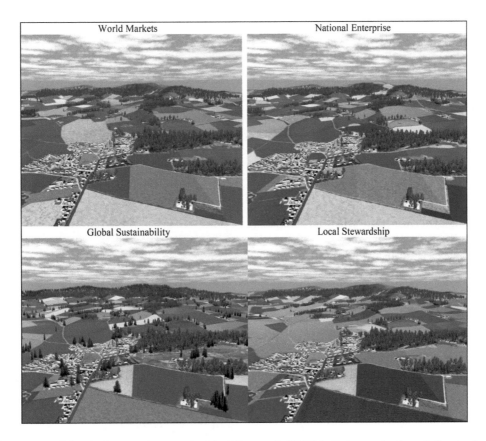

FIGURE 28.15 Images of four basic scenarios examined by Wang et al. (2016) in their examination of climate change impacts on land-use change. The visualizations for individual scenarios could be manipulated by either those managing a formal session or by participants in the session. (From Wang, C., Miller, D., Brown, I., et al. (2016) "Visualisation techniques to support public interpretation of future climate change and land-use choices: A case study from N-E Scotland." *International Journal of Digital Earth* 9, no. 6:586–605, Taylor & Francis Ltd., https://www.tandfonline.com/journals/tjde20.)

participants were also allowed to manipulate the various scenarios (e.g., to select desired locations for either wind turbines or a conservation area).

The Local Stewardship scenario was the most preferred general scenario, as approximately one-half of the participants preferred it. More than 80 percent of the stakeholders felt that the VLT was "… effective for capturing views on priorities for future land uses including the role of climate change in modifying existing options" (Wang et al. 2016, 600). One concern was the lack of sufficient texture on the ground vegetation, but Wang et al. argued that such further detail may distract from public engagement and "may lead to an assumption that future change is more predictable than it actually is in practice …" (600).

28.6.3.3 HMDs as a Potential Cost-Effective Solution for Collaborative Efforts

One limitation of CAVEs and wall-size displays is their high cost. As a solution to this limitation, researchers are starting to evaluate the potential of using HMDs for some collaborative efforts that might normally be done with CAVEs and wall-size displays. For instance, Maxime Cordeil and colleagues (2017) compared the effectiveness of HMDs with a CAVE-style environment for the collaborative analysis of connectivity in network data, which are utilized in graph analysis. They found no difference in the accuracy of results and in the ability to communicate in the two environments; moreover, responses were faster in the HMD condition and the CAVE-style condition "introduced an asymmetry in movement between collaborators" (441). Overall, these findings suggest considerable potential for collaborative tasks involving HMDs.

28.6.4 What Progress Has Been Made Toward Developing a Digital Earth?

In 1998, former U.S. Vice President Al Gore argued that the "desktop metaphor" employed by the Macintosh and Windows operating systems is insufficient for interacting with the vast amount of geographic data now available (e.g., 1-meter satellite imagery). Instead, Gore called for a new metaphor—a multiresolution, 3-D **Digital Earth**:

> Imagine …. a young child going to a Digital Earth exhibit at a local museum. After donning a head-mounted display, she sees Earth as it appears from space. Using a data glove, she zooms in, using higher and higher levels of resolution, to see continents, then regions, countries, cities, and finally individual houses, trees, and other natural and man-made objects. Having found an area of the planet she is interested in exploring, she takes the equivalent of a "magic carpet ride" through a 3-D visualization of the terrain. Of course, terrain is only one of the many kinds of data with which she can interact. Using the systems' voice recognition capabilities, she is able to request information on land cover, distribution of plant and animal species, real-time weather, roads, political boundaries, and population. She can also visualize the environmental information that she

and other students all over the world have collected …. She is not limited to moving through space, but can also travel through time. After taking a virtual field-trip to Paris to visit the Louvre, she moves backward in time to learn about French history, perusing digitized maps overlaid on the surface of the Digital Earth, newsreel footage, oral history, newspapers and other primary sources. (From Gore, 1999, 528; reprinted with permission from the American Society for Photogrammetry & Remote Sensing, Bethesda, Maryland, asprs.org/.)

As a result of Gore's vision, the first International Symposium on Digital Earth was held in Beijing, China, in 1999, and the Beijing Declaration on Digital Earth (https://tinyurl.com/Digital-Earth-Declaration) was produced. This led to the formation of the International Society for Digital Earth (http://www.digitalearth-isde.org/list-1-1.html), which promotes the *International Journal of Digital Earth* and other publications. In 2020, the society produced the *Manual of Digital Earth* (Guo et al. 2020), which provides an extensive overview of worldwide progress in developing a Digital Earth. The paper by Arzu Çöltekin and colleagues (2020) that we mentioned in the introduction to this chapter comes from Chapter 7 of this book.

Closely associated with the notion of a Digital Earth are **Digital Globes**, which show digital spatiotemporal information about Earth or other celestial bodies as an undistorted 3-D globe (Riedl 2011). Most Digital Globes are virtual (such as the default view in the ArcGlobe software), but particularly intriguing are Digital Globes that can be touched and also permit the display of digital images. For example, the National Oceanic and Atmospheric Administration (NOAA) has developed Science On a Sphere (SOS; https://sos.noaa.gov/What_is_SOS/), which provides a room-sized global display (Figure 28.16) that permits the animation of Earth processes such as storms, ocean temperature, and climate changes.[13] An interesting research question is whether animations shown on Digital Globes are more effective than those shown on 2-D map projections.[14]

FIGURE 28.16 A classroom viewing NOAA's Science On a Sphere, an example of a Digital Globe that can be used to animate Earth processes such as storms, ocean temperature, and climate change. (NOAA Photo by Will von Dauster.)

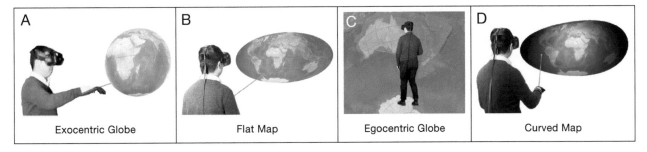

A	B	C	D
Exocentric Globe	Flat Map	Egocentric Globe	Curved Map

FIGURE 28.17 Various approaches for viewing worldwide maps in a VE. (From Figure 1 of Yang, Y., Jenny, B., Dwyer, T., et al. (2018) "Maps and globes in virtual reality." *Computer Graphics Forum* 37, no. 3:427–438; © 2018 The Author(s) *Computer Graphics Forum* © 2018 the Eurographics Association and John Wiley & Sons Ltd. Published by John Wiley & Sons Ltd.)

We conclude this section by considering a study by Yalong Yang and colleagues (2018) that considered various ways in which worldwide maps could be viewed in VEs. Consider Figure 28.17, which illustrates the four approaches that Yang et al. tested. The first approach (Figure 28.17A) is our traditional way of viewing virtual globes, in which we see the entire globe from an external viewpoint (it is an *exocentric globe*). The second approach is our traditional way of viewing maps outside VEs, as the entire world is viewed as a projection on a *flat map*. The third approach is a novel one in which our viewpoint is from within the virtual globe, and we see Earth's surface projected onto the surrounding 360° sphere (the result is an *egocentric globe*). A possible advantage of this approach is that if our viewpoint is near the center of the globe, then the inside surface of the globe is a constant distance away, thus reducing distortion. The fourth approach is also a novel one in which a map is projected onto a section of the sphere, and the viewer faces the concave side of the sphere (this is the *curved map*). This approach may also reduce distortion as the distance from the viewpoint to the map is relatively constant. Thirty-two participants in this study performed three basic tasks: distance comparison, area comparison, and direction estimation. Overall, the traditional exocentric approach was the most effective. Yang et al. indicated that this was surprising given that less of Earth's surface is visible in this approach (half of it is hidden), and it has the greatest perceptual distortion.

28.7 SOME RECENT EXAMPLES OF THE UTILIZATION OF AR

In this section, we consider several recent examples of the use of AR. We want to stress that we expect the use of AR to expand considerably as the technology continues to evolve. For instance, in a study of the potential use of AR virtual graphics in an urban environment, Quang Quach and Bernhard Jenny (2020, 471) noted that they wished to run an experiment in the outdoors, but that current AR technology (such as Microsoft HoloLens) was not designed for outdoor use and thus could not be used due to "limited contrast between virtual objects projected onto the glasses and the real world." As a result, they did their study indoors with a VE setup. In the future, we expect that such studies will move into the AR realm.

28.7.1 THE AUGMENTED REALITY SANDBOX

Perhaps the most attention-grabbing example of AR has been the augmented reality (AR) sandbox developed at the University of California, Davis (https://eos.org/science-updates/augmented-reality-turns-a-sandbox-into-a-geoscience-lesson). As the name suggests, AR Sandbox is indeed a sandbox, but it is nothing like what we played in when we were children. As you and your collaborators move sand around in the sandbox, a Microsoft Kinect 3-D camera mounted above the sandbox detects the distance to all points on the sand and a 3-D model is created of the sand landscape in real-time. A computer projector then utilizes the 3-D model to display contours and hypsometric tints onto the sand surface. Areas below sea level can be made to appear as water (with waves), and with appropriate hand gestures, it is possible to make it appear to rain, and see the resulting water flow over the land surface. Figure 28.18 shows an image taken while a sandbox was being used, but to truly appreciate the capability of the AR Sandbox, you should examine a video of it in use (https://www.youtube.com/watch?v=bA4uvkAStPc), or better yet find one to experiment with near you.

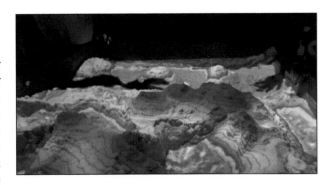

FIGURE 28.18 Image from an Augmented Reality Sandbox described at https://www.youtube.com/watch?v=bA4uvkAStPc. (Courtesy of Beals Science, LLC, https://www.bealsscience.com/.)

Several studies have evaluated the effectiveness of the AR Sandbox on student learning—here, we consider two examples. Karen McNeal and colleagues (2019) evaluated the potential of AR Sandbox for teaching students how to read topographic maps. They compared three AR approaches (unstructured play, a semi-structured lesson, and a structured lesson) with a control condition in which the AR approach was not used. Student performance on a Topographic Maps Assessment (TMA) test following completion of these approaches revealed no significant difference in the topographic map understanding among the four conditions. However, it was discovered that those using the AR approach who scored either low or high on a mental rotation test performed best on the TMA test.

In another study, Antoni Moore and colleagues (2020) compared the effectiveness of the AR Sandbox with conventional GIS software (they used ArcGIS) to teach terrain and hydrologic analysis. The results revealed that students performed tasks significantly faster and had significantly greater learning satisfaction with the AR approach. In terms of accuracy of the tasks performed, in general, there was no significant difference between the AR and GIS software approaches, although 3-D GIS tasks tended to be more accurate. Overall, Moore et al. felt that the AR approach could provide an engaging introduction to terrain and hydrologic concepts prior to having students conduct a more accurate analysis in GIS.

28.7.2 Using AR to Enhance an Understanding of Topographic Maps

Traditionally, the interpretation of topographic maps has been complicated by the fact that the 3-D structure of the real world is depicted on a flat 2-D map. Although various symbolization approaches (such as relief shading) can suggest a 3-D effect, a traditional paper map (or computer screen display without stereo) is still a 2-D representation. To attempt to address this deficiency, Carlos Carrera and Luis Asensio (2017) created 3-D representations of DTMs in Google SketchUp and then had students view the resulting 3-D maps in an AR environment on a tablet computer (they utilized iPads) using an AR-Media app. Students pointed the iPad camera at a marker on a piece of paper, and a corresponding 3-D image appeared above the marker.[15] Students then used hand gestures (while holding the iPad) to manipulate the realistic-looking 3-D image of the terrain.

Carrera and Asensio initially tested the effectiveness of their AR approach using a workshop of 63 students. A variety of symbolization approaches were utilized on the 3-D maps including orthophotos, contour lines, hypsometric tints, hypsometric tints with contour lines, and oblique shading with hypsometric tints and contour lines. Students initially worked with 2-D maps in the workshop and answered a series of questions about the 2-D maps. Then the students worked with the 3-D representations in the AR environment and checked the answers that they had created for the 2-D representations. To determine the extent of

learning, a Topographic Map Assessment (TMA) test was given prior to and after completion of the workshop. The test clearly revealed that there was a significant improvement in the ability to interpret topographic maps. The results of the workshop then were compared with a later experiment in which only 2-D maps were used (there was no AR component). There was still a significant improvement in the ability to interpret topographic maps, but the results were less pronounced than when AR was included. In a later paper, Carrera et al. (2018) noted that students in the initial workshop found the AR approach quite usable and motivating. Overall, these studies suggest great potential for AR in the interpretation of topographic maps, especially given the fact that students had not previously worked with AR technology.

28.7.3 Developing Novel Methods for Interacting with AR Environments

In Section 28.5.2, we introduced interactivity as one of the four I factors of VEs. We now consider how we might apply this I factor in AR by examining a study by Christopher Austin and colleagues (2020) that attempted to elicit appropriate hand and foot gestures for interacting with maps in AR. Austin et al. focused on immersive horizontal maps placed on the floor because they felt this approach offered the following advantages:

- Multiple users can collaborate and discuss in a face-to-face setting
- A room-size map can show fine details throughout a large geospatial area
- Users can gather and physically walk around an area of interest
- Users can interact in both egocentric and exocentric ways
- The third dimension of terrain and other map features are viewed with a natural perspective to give an engaging and immersive experience (After Austin et al. 2020. First published in *Cartography and Geographic Information Science* 47(3), p. 214. Reprinted with permission from the Cartography and Geographic Information Society.)

Users viewed a 3-D map of buildings In New York City (Figure 28.19) using an HTC Vive headset with a video camera attached that provided a video see-through AR experience (218).

The purpose of the study was not to determine whether this particular AR environment was effective but rather how users should interact with the environment. Thus, rather than users being *told* how to interact with the environment, they were asked to suggest appropriate hand or foot gestures following a particular map operation. For instance, users were shown the map panning in a particular direction and then were asked to provide a hand or foot gesture that would be appropriate to generate that movement.

FIGURE 28.19 The AR map of New York City used in Austin et al.'s (2020) study. (From Austin et al. 2020. First published in *Cartography and Geographic Information Science* 47(3), p. 218. Reprinted with permission from the Cartography and Geographic Information Society.)

Foot gestures may seem strange to those of us who have interacted with computers screens through point-and-click interfaces (using our hands), but remember that the AR map was displayed on the floor. Some of the key findings were as follows: For hand gestures, users preferred single-handed gestures for panning, creating point markers, and selecting point markers. In contrast, two-handed gestures were preferred for zooming and changing the height of the map. Users reported difficulty in designing foot gestures, and agreement rates were lower for foot gestures than for hand gestures. Common foot gestures for panning, rotating, and zooming involve having one foot support body weight while the other foot performs a sliding, pivoting, or dipping movement (224). This discussion of gestures may seem to move us away from cartography, but it illustrates the novel ways in which we may interact with maps in the future as we move away from traditional 2-D maps (whether printed or on-screen) and view these maps in a VE or AR.

28.7.4 HOLOGRAMS

We conclude this section by considering holograms, which provide still another approach for creating an AR. The basic concept behind **holograms** involves splitting a laser beam of light into two parts: one part that is reflected off a mirror and then strikes a recording plate, and one part that is reflected off the object of interest and reaches the same recording plate. These two beams of light make an interference pattern on the plate that provides detailed information about the object of interest. If the resulting plate is viewed under an appropriate light source, then the original 3-D object will appear in 3-D space, and the person viewing the object will not need to wear any specialized viewing glasses (Kasper and Feller 2001). Modern computer-based systems are more involved than this, but it is useful to begin with this conceptual basis. You may have heard about holograms because they are commonly utilized on driver's licenses and credit cards.

Interestingly, a hologram movie (an animation) depicting the changes in population by county in the U.S. from 1790 to 1970 was created by Geoffrey Dutton way back in 1978. Nick Chrisman (2006) described this rather tedious process in which "Each frame of the movie was ... exposed onto successive vertical strips of the holographic film, projected through one beam of laser, and merged with a reference beam so that the interference pattern is recorded on film." Perhaps it is not surprising that it was approximately 30 years before Sven Fuhrmann and colleagues started to experiment again with holograms. In 2009, Furhmann and colleagues were using fully computer-based systems to create cartographic representations, but their results were monochrome and static (Furhmann et al. 2009). By 2015, Furhmann and colleagues were demonstrating color and interactive holograms and discussing the possibility of real-time updates to holograms (Furhmann et al. 2015). As technology continues to evolve, it will be interesting to see what role holograms ultimately play in cartography.

28.8 HEALTH, SAFETY, AND SOCIAL ISSUES

Although we feel that VEs and AR have considerable potential for geovisualization, you should be aware that there are health, safety, and social issues that are often glossed over. Health and safety issues can range from something as simple as tripping over a power cord while immersed in a VE to discomfort associated with wearing an HMD to **cybersickness**, a form of motion sickness that can result from exposure to VEs. In reviewing various applications in VEs and AR for this chapter, the most common health issues that we saw were eyestrain associated with wearing active shutter glasses when viewing stereoscopic displays and cybersickness associated with using HMDs in VEs. You need to be aware of these sorts of problems, both as a potential user of VEs and AR and because you might be involved in evaluating the effectiveness of these technologies (and thus might need to warn users of potential health and safety issues). In this context, the edited volume by Hale and Stanney (2015) and the textbook by Jerald (2016) mentioned in the introduction contain several chapters dealing with health issues associated with VEs and AR.

One natural question to ask is whether health and safety issues differ for VEs and AR. Since we can see the real world in AR, it seems reasonable to hypothesize that there would be less cybersickness in AR. A recent paper by Claire Hughes and colleagues (2020) supports this hypothesis as they indicate that AR is unlikely to produce a full range of cybersickness symptoms but can produce visual discomfort/fatigue, difficulty focusing, and headaches.

Our review of the VE and AR applications revealed little discussion of social issues associated with geovisualization applications. We suspect that this is a function of the relative novelty of this technology in the geovisualization community. As the technology evolves, we would expect social issues of VEs and AR to become more common, just as they are in the gaming community. For instance, in reviewing social issues associated with VEs, Sandra Calvert (2015, 700) asks the following sorts of questions:

- How will the time spent in such VEs impact children's social skills?
- Will something innately be lost if people choose to interact with virtual rather than with real characters?...
- Will people feel lonely and isolated when they interact with embodied virtual agents or with peers that are not in their immediate environments or will life on the screen become part of everyday experiences? (Republished with permission of Taylor & Francis Group LLC—Books, from "Social impact of virtual environments," by S. L. Calvert (2015) In *Handbook of Virtual Environments: Design, Implementation, and Applications,* ed. K. S. Hale and K. M. Stanney; permission conveyed through Copyright Clearance Center, Inc.)

28.9 SUMMARY

In this chapter, we have explored the use of VEs and AR in geography and cartography. A VE can be defined as a 3-D computer-based simulation of a real or imagined environment that users are able to navigate through and interact with. An example would be if you donned a HMD and toured a simulation of how the 3-D landscape changed during major glaciations of North America. We saw that there were a variety of display technologies for creating VEs, including not only the HMD but also personalized displays, wall-size displays, and room-format displays. A distinct advantage of HMDs and CAVES is that they allow the user to become more immersed in the VE—the user might actually feel a sense of presence, of being in the VE, even though it is artificial. Today, most VEs are experienced visually, although it is possible to utilize the full range of senses, including sound, touch, smell, taste, and body movement.

In examining Table 28.1, we saw that there are a variety of depth cues useful in creating a 3-D VE. Most of these are monocular, meaning that they can be seen using only one eye, but the "stereoscopic depth" cue is often focused on because it allows us to see in "true 3-D." Although stereoscopy can provide attractive and realistic images, we should not assume that stereo images will necessarily be easier to interpret and analyze. For instance, the study by McIntire et al. revealed that stereoscopy improved performance in only 60 percent of the experiments examined. Thus, rather than assuming that stereoscopy will improve performance, you should carefully consider the tasks that users are performing.

We discussed four "I" factors that are relevant to designing geospatial VEs: immersion, interactivity, information intensity, and intelligence of objects. Although there is a tendency to assume that higher immersion is desirable (as in donning an HMD that shields you from the real world), some researchers have found that users work effectively with standard desktop systems, which have a low degree of immersion. Key issues in interacting with the geospatial VE are navigating through it and understanding one's current orientation and location. Information intensity refers to the LOD present in the GeoVE. Although there is an inclination to show considerable detail (a high level of realism), we should be aware that increased realism does not necessarily correspond to a higher level of understanding of the VE. Intelligence of objects refers to the notion that objects in a VE can exhibit intelligence, just as they do in the real world. Given the role of virtual assistants in everyday life (e.g., Amazon's Alexa), we expect that virtual assistants will become more common in VEs.

We saw that the 3-D character of virtual globes may require the use of specialized 3-D symbols that are not common in traditional 2-D maps. For instance, the study by Satriadi et al. (2021) revealed that normal bars (cylinders displayed perpendicular to the globe's surface) were most effective. The increasing use of these virtual globes also raises the question of how worldwide maps should be viewed in a VE. In this respect, the study by Yang et al. (2018) revealed that an *exocentric globe* approach was most effective (when we see the entire globe from an external viewpoint).

AR involves a mix of the real world and the VE, with the real world often receiving greater emphasis. Although AR constituted a relatively small portion of this chapter, you should be aware that this technology is evolving rapidly, and researchers are only just beginning to explore its potential. Examples of AR, such as the AR Sandbox suggest that this technology has tremendous potential for geography and cartography.

Finally, we touched on health, safety, and social issues associated with VEs and ARs. Health and safety issues can range from something as simple as tripping over a power cord to cybersickness, a form of motion sickness. Social issues deal with how the use of VEs and ARs will affect our views of the real world and our interaction with others. For instance, will there be a tendency for some to escape into VEs, and will it be difficult to interact with someone who utilizes an AR when you do not have access to that same environment?

28.10 STUDY QUESTIONS

1. Define the terms VE and AR and describe a geographic (or cartographic) example of each (use examples not covered in the textbook).
2. Define immersion and describe the sense of immersion provided by each of the following: a smartphone, a laptop computer, a wall-size display, and a HMD. (Assume that stereo is not available for the smartphone, laptop, and wall-size display.)
3. Define and provide examples of the following monocular depth cues: occlusion, linear perspective, atmospheric perspective, and shadows.

4. (A) Explain the concept of stereoscopic vision and indicate how stereoscopy can be accomplished using passive and active computer technology. (B) Assuming that you wish to present stereoscopic images to students in a classroom using a Geowall, discuss two key problems that you are likely to encounter.

5. Define the term Digital Globe and describe examples of intangible and tangible Digital Globes.

6. Describe how we have typically interacted with desktop and laptop computers and how this likely differs from how we interact with more immersive CAVEs.

7. Contrast the effectiveness of the following point symbols on a virtual globe: traditional proportional circles that appear to be printed on the globe, proportional circles that appear on see-through billboards, cylinders that are perpendicular to the globe's surface, and spheres that appear to touch the globe at each point location.

8. Imagine that a researcher wishes to illustrate how forests in Wisconsin have changed between 1850 and today (most old-growth trees were felled in the late nineteenth and early twentieth century). Make arguments for and against creating a highly realistic VE to illustrate this change over time.

9. You attend your local town hall meeting where the mayor is using a traditional computer projection system with PowerPoint to demonstrate her vision for land use in the local region over the next 20 years. At the end of the meeting, the mayor indicates that a benefactor has donated $100,000 to the city, and the board needs to consider how that money should best be spent. Make an argument for using this $100,000 to create a wall-size display system to conduct collaborative decision-making.

10. Discuss in general how you could use both the Augmented Reality Sandbox and GIS software to teach basic concepts of topographic maps, including the interpretation of contours, to high school students.

NOTES

1 Physical models can also be utilized to create representations of reality; we review those in Chapter 23.

2 ArcScene and ArcGlobe are associated with ArcMap; in ArcGIS Pro, these two pieces of software are implemented as scenes, with local and global options.

3 For an example of a passive system, see https://www.3d-pluraview.com/en/highlights-en.

4 For a simplified discussion of the basic approaches used in autostereoscopic displays, see https://www.3d-forums.com/threads/autostereoscopic-displays.1/.

5 The cost will depend on the computer knowledge of the individual setting up the system; it is possible to set up such a system for under $10,000.

6 Active systems are also possible; in that case, a single projector would suffice, but active glasses would be essential.

7 The software permits projection onto six surfaces, but normally only four are used.

8 For an example of a more recent ImmersaDesk, see https://www.evl.uic.edu/research/1913.

9 Papadopoulos et al. (2015) describe another modern alternative to CAVE2 that is known as the Reality Deck (https://www.youtube.com/watch?v=YCvzrvXwjS4).

10 At the time, VRML was the standard approach for creating interactive 3-D graphics; this has been superseded by X3D (https://www.web3d.org/x3d/what-x3d).

11 You may also wish to consider a study by White (2012) that compared traditional area symbolization (choropleth and prism maps) for world-based maps displayed in both ArcScene and ArcGlobe.

12 Technically, each participant worked with only one of the three tasks.

13 Another example of this technology is Pufferfish displays (https://pufferfishdisplays.com/).

14 Although normally viewed in nonstereo, a stereo version of SOS also has been created (Middleton et al. 2005).

15 For an example of AR that does not require the use of markers, see Dave McLaughlin's work (https://marmots.davemaps.com/).

REFERENCES

Austin, C. R., Ens, B., Satriadi, K. A., et al. (2020) "Elicitation study investigating hand and foot gesture interaction for immersive maps in augmented reality." *Cartography and Geographic Information Science 47*, no. 3:214–228.

Bishop, I. (1994) "The role of visual realism in communicating and understanding spatial change and process." In *Visualization in Geographical Information Systems*, ed. by H. M. Hearnshaw and D. J. Unwin, pp. 60–64. Chichester, England: Wiley.

Bleisch, S. (2011) "Toward appropriate representations of quantitative data in virtual environments." *Cartographica 46*, no. 4:252–261.

Buchroithner, M. F., Boulos, M. N. K., and Robinson, L. R. (2012) "Stereoscopic 3-D solutions for online maps and virtual globes." In *True-3D in Cartography: Autostereoscopic and Solid Visualization of Geodata*, ed. by M. Buchroithner, pp. 391–412. Berlin: Springer.

Buchroithner, M. F., and Knust, C. (2013) "True-3D in cartography—Current hard- and softcopy developments." In *Geospatial Visualisation*, ed. by A. Moore and I. Drecki, pp. 41–65. New York: Springer.

Calvert, S. L. (2015) "Social impact of virtual environments." In *Handbook of Virtual Environments: Design, Implementation, and Applications*, ed. K. S. Hale and K. M. Stanney, pp. 699–718. Boca Raton, FL: CRC Press.

Carrera, C.C., and Asensio, L. A. B. (2017) "Augmented reality as a digital teaching environment to develop spatial thinking." *Cartography and Geographic Information Science 44*, no. 3:259–270.

Carrera, C. C., Perez, J. L. S., and Cantero, J. (2018) "Teaching with AR as a tool for relief visualization: Usability and motivation study." *International Research in Geographical and Environmental Education 27*, no. 1:69–84.

Chrisman, N. (2006) *Charting the Unknown: How Computer Mapping at Harvard Became GIS*. Redlands, CA: ESRI Press.

Cliburn, D. C., Feddema, J. J., Miller, J. R., et al. (2002) "Design and evaluation of a decision support system in a water balance application." *Computers & Graphics 26*:931–949.

Çöltekin, A., Griffin, A. L., Slingsby, A., et al. (2020) "Geospatial information visualization and extended reality displays." In *Manual of Digital Earth*, ed. by H. Guo, M. F. Goodchild and A. Annoni, pp. 229–277. Berlin: Springer.

Cordeil, M., Dwyer, T., Klein, K., et al. (2017) "Immersive collaborative analysis of network connectivity: CAVE-style or head-mounted display?" *IEEE Transactions on Visualization and Computer Graphics 23*, no. 1:441–450.

Dodge, M., and Kitchin, R. (2001) *Mapping Cyberspace*. London: Routledge.

Drascic, D., and Milgram, P. (1996) "Perceptual issues in augmented reality." *Proceedings, SPIE*, Volume *2653*, San Jose, CA, pp. 123–134.

Dykes, J. (2002) "Creating information-rich virtual environments with geo-referenced digital panoramic imagery." In *Virtual Reality in Geography*, ed. by P. Fisher and D. Unwin, pp. 68–92. London: Taylor & Francis.

Edler, D., Huber, O., Knust, C., et al. (2014) "Spreading map information over different depth layers—An improvement for map-reading efficiency?" *Cartographica 49*, no. 3:153–163.

Fuhrmann, S., Holzbach, M. E., and Black, R. (2015) "Developing interactive geospatial holograms for spatial decision-making." *Cartography and Geographic Information Science 42*, no. sup1:27–33.

Fuhrmann, S., Komogortsev, O., and Tamir, D. (2009) "Investigating hologram-based route planning." *Transactions in GIS 13*, no. s1:177–196.

Fuhrmann, S., and MacEachren, A. M. (2001) "Navigation in desktop geovirtual environments: Usability assessment." *Proceedings, 20th International Cartographic Conference*, International Cartographic Association, Beijing, China, https://icaci.org/files/documents/ICC_proceedings/ICC2001/icc2001/file/f15040.pdf.

Gabbard, J. L. (1997) *"A taxonomy of usability characteristics in virtual environments."* Unpublished MS thesis, Virginia Polytechnic Institute and State University, Blacksburg, VA.

Goldstein, E. B., and Brockmole, J. (2017) *Sensation and Perception* (10th ed.). Boston, MA: Cengage Learning.

Gore, A. (1999) "The Digital Earth: Understanding our planet in the 21st century." *Photogrammetric Engineering & Remote Sensing 65*:528–530.

Granell, C., Pesántez-Cabrera, P. G., Vilches-Blázquez, L. M., et al. (2021) "A scoping review on the use, processing and fusion of geographic data in virtual assistants." *Transactions in GIS*, https://onlinelibrary.wiley.com/doi/abs/10.1111/tgis.12720.

Guo, H., Goodchild, M. F., and Annoni, A. (2020) *Manual of Digital Earth*. Berlin: Springer.

Hale, K. S., and Stanney, K. M. (eds.) (2015) *Handbook of Virtual Environments: Design, Implementation, and Applications* (2nd ed.). Boca Raton, FL: CRC Press.

Halfen, A. F., White, T., Slocum, T., et al. (2014) "A new stereoscopic (3D) media database and teaching strategy for use in large-lecture introductory geoscience courses." *Journal of Geoscience Education 62*, no. 3:515–531.

Harris, T. M., and Hodza, P. (2011) "Geocollaborative soil boundary mapping in an experiential GIS environment." *Cartography and Geographic Information Science 38*, no. 1:20–35.

Hedley, N. R. (2003) *"3D geographic visualization and spatial mental models."* Unpublished Ph.D. dissertation, University of Washington, Seattle, WA.

Hegarty, M., Smallman, H. S., Stull, A. T., et al. (2009) "Naive cartography: How intuitions about display configuration can hurt performance." *Cartographica 44*, no. 3:171–186.

Heim, M. (1998) *Virtual Realism*. New York: Oxford University Press.

Hirmas, D. R., Slocum, T., Halfen, A. F., et al. (2014) "Effects of seating location and stereoscopic display on learning outcomes in an introductory physical geography class." *Journal of Geoscience Education 62*, no. 1:126–137.

Hughes, C. L., Fidopiastis, C., and Stanney, K. M. (2020) "The psychometrics of cybersickness in augmented reality." *Frontiers in Virtual Reality 1*, Article 602954:1–12.

Jerald, J. (2016) *The VR Book: Human-Centered Design for Virtual Reality*. New York: ACM Books.

Juřík, V., Herman, L., Snopková, D., et al. (2020) "The 3D hype: Evaluating the potential of real 3D visualization in geo-related applications." *PLoS ONE 15*, no. 5: e0233353(1–18).

Kasper, J. E., and Feller, S. A. (2001) *The Complete Book of Holograms: How They Work and How to Make Them*. Mineola, NY: Dover.

Kuhlen, T. W., and Hentschel, B. (2014) "Quo vadis CAVE: Does immersive visualization still matter?" *IEEE Computer Graphics and Applications 34*, no. 5:14–21.

Li, J., Jarvis, C. H., and Brunsdon, C. (2010) "The use of immersive real-time 3D computer graphics for visualization of dilution of precision in virtual environments." *International Journal of Geographical Information Science 24*, no. 4: 591–605.

MacEachren, A.M., Edsall, R., Haug, D., et al. 1999. Virtual Environments for Geographic Visualization: Potential and Challenges. *NPIVM '99: Proceedings of the 1999 Workshop on New Paradigms in Information Visualization and Manipulation*. Kansas City, MO, pp. 35–40.

Machidon, O. M., Duguleana, M., and Carrozzino, M. (2018) "Virtual humans in cultural heritage ICT applications: A review." *Journal of Cultural Heritage 33*:249–260.

Manghisi, V. M., Fiorentino, M., Gattullo, M., et al. (2017) "Experiencing the sights, smells, sounds, and climate of southern Italy in VR." *IEEE Computer Graphics and Applications 37*, no. 6:19–25.

Marai, G. E., Leigh, J., and Johnson, A. (2019) "Immersive analytics lessons from the Electronic Visualization Laboratory: A 25-year perspective." *IEEE Computer Graphics and Applications 39*, no. 3:54–66.

McIntire, J. P., Havig, P. R., and Geiselman, E. E. (2014) "Stereoscopic 3D displays and human performance: A comprehensive review." *Displays 35*, no. 1:18–26.

McNeal, K. S., Ryker, K., Whitmeyer, S., et al. (2019) "A multi-institutional study of inquiry-based lab activities using the Augmented Reality Sandbox: Impacts on undergraduate student learning." *Journal of Geography in Higher Education 44*, no. 1:85–107.

Middleton, D., Scheitlin, T., and Wilhelmson, B. (2005) "Visualization in weather and climate research." In *The Visualization Handbook*, ed. by C. D. Hansen and C. R. Johnson, pp. 845–871. Amsterdam: Elsevier.

Moore, A., Daniel, B., Leonard, G., et al. (2020) "Comparative usability of an augmented reality sandtable and 3D GIS for education." *International Journal of Geographical Information Science 34*, no. 2:229–250.

Oviatt, S. (2008) "Multimodal interfaces." In *The Human-Computer Interaction Handbook: Fundamentals, Evolving Technologies, and Emerging Applications* (2nd ed.), ed. by A. Sears and J. A. Jacko, pp. 413–432. New York: Lawrence Erlbaum Associates.

Papadopoulos, C., Petkov, K., Kaufman, A. E., et al. (2015) "The Reality Deck—An immersive gigapixel display." *IEEE Computer Graphics and Applications 35*, no. 1:33–45.

Popescu, G. V., Trefftz, H., and Burdea, G. (2015) "Multimodal interaction modeling." In *Handbook of Virtual Environments: Design, Implementation, and Applications*, ed. by K. S. Hale and K. M. Stanney, pp. 411–434. Boca Raton, FL: CRC Press.

Quach, Q., and Jenny, B. (2020) "Immersive visualization with bar graphics." *Cartography and Geographic Information Science 47*, no. 6:471–480.

Riedl, A. (2011) "Digital globes: Their historical development." *Globe Studies*, no. 57/58:149–161.

Satriadi, K. A., Ens, B., Czauderna, T., et al. (2021) "Quantitative data visualisation on virtual globes." *CHI Conference on Human Factors in Computing Systems*, Yokohama, Japan, pp. 1–14.

Sutherland, I. E. (1968) "A head-mounted three-dimensional display." *Proceedings of AFIPS '68 (Fall, part 1)*, pp. 757–764.

Voinov, A., Çöltekin, A., Chen, M., et al. (2018) "Virtual geographic environments in socio-environmental modeling: A fancy distraction or a key to communication?" *International Journal of Digital Earth 11*, no. 4:408–419.

Wang, X., and Chen, R. (2009) "An experimental study on collaborative effectiveness of augmented reality potentials in urban design." *CoDesign 5*, no. 4:229–244.

Wang, C., Miller, D., Brown, I., et al. (2016) "Visualisation techniques to support public interpretation of future climate change and land-use choices: A case study from N-E Scotland." *International Journal of Digital Earth 9*, no. 6:586–605.

Ware, C. (2013) *Information Visualization: Perception for Design* (3rd ed.). Waltham, MA: Morgan Kaufman.

White, T. M. (2012) "Evaluating the effectiveness of thematic mapping on virtual globes." Unpublished MA thesis, University of Kansas, Lawrence, Kansas.

Witmer, B. G., and Singer, M. J. (1998) "Measuring presence in virtual environments: A presence questionnaire." *Presence 7*, no. 3:225–240.

Wood, A., Drew, N., Beale, R., et al. (1995) "HyperSpace: Web browsing with visualization." *Proceedings of the Third International World Wide Web Conference*, Darmstadt, Germany, pp. 21–25.

Yang, Y., Jenny, B., Dwyer, T., et al. (2018) "Maps and globes in virtual reality." *Computer Graphics Forum 37*, no. 3:427–438.

Glossary

abrupt phenomena: phenomena that change abruptly over geographic space, such as sales tax rates for each state in the United States.

abstract sound variables: analogous to visual variables, these are variables used to depict abstract sounds (e.g., the location and loudness of sound).

abstract sounds: sounds that have no obvious meaning and thus require a legend for interpretation (e.g., a mouse click on a census tract produces a loudness value as a function of income).

accommodation: the process in which the eye automatically changes the shape of the lens to focus on an image.

active display: an approach for creating a stereoscopic view in which images for the left and right eye are cycled back and forth on the display screen, and users wear shutter glasses that are in sync with the display screen.

additive primary colors: colors that are visually added (or combined) to produce other colors; for example, red, green, and blue are additive colors used in computer displays.

aggregation: a generalization operation that involves the joining together of multiple point features, such as a cluster of buildings.

aliasing: a staircase appearance of straight lines caused by displaying lines on a coarse raster graphics display.

amalgamation: a generalization operation that combines nearby polygons (e.g., a series of small islands in close proximity having a size and amount of detail that cannot be depicted at a small scale).

anaglyphs: two images are created, one in green or blue and the other in red; when viewed through special anaglyphic glasses, a 3-D view is produced.

analytical cartography: a theoretical branch of cartography developed by Waldo Tobler that deals with analytical topics such as cartographic data models, coordinate transformations and map projections, interpolation, and generalization.

ancillary information: information used to more accurately map data associated with enumeration units (e.g., when making a dot map of wheat based on county-level data, we avoid placing dots in bodies of water).

animated maps: maps characterized by continuous change, such as the daily weather maps shown on television newscasts that depict changes in cloud cover.

animated streamlets: small "comet-like trails" are used to show the direction and speed of a flowing phenomenon such as the wind; greater speed is shown by a wider, brighter, and more rapidly drawn streamlet, and direction is provided by drawing the streamlet in the direction of the flow.

animation: automatic dynamic change on a map that allows us to see either temporal changes in a spatial distribution (e.g., changes in ocean currents over the course of the year) or the character of a spatial distribution (e.g., a fly-over of the Grand Canyon).

anomalous trichromats: a less serious form of color vision impairment in which one of the three types of cone cells (red, green, and blue) is deficient; contrast with dichromats.

application software: software programs designed for specific purposes, such as graphic design, GIS, and remote sensing.

area cartogram: a map in which the areal relationships of enumeration units are distorted on the basis of an attribute (e.g., the sizes of states are made proportional to the number of deaths in each state due to AIDS).

areal phenomena: geographic phenomena, such as a forested region, that are two-dimensional in spatial extent, having both length and width; data associated with enumeration units can also be considered areal phenomena because each unit is an enclosed area.

areal weighting: a method of interpolation in which each source zone contributes to the target zone in direct proportion to the percentage of the source zone's area that is found in the target zone (thus, no ancillary information is utilized).

arrangement: a visual variable in which the marks making up symbols are arranged in various ways, such as breaking up lines into dots and dashes to create various forms of political boundaries.

aspect: (1) placement of a projection's center with respect to Earth's surface; common aspects include equatorial, polar, and oblique; (2) the direction in which a topographic slope faces (e.g., a northwest aspect would face the northwest).

attribute: a theme or variable that you might wish to display on a map (e.g., the predicted high temperature for each major U.S. city).

attribute accuracy: one means of assessing data quality (or uncertainty); refers to the accuracy of features found at particular locations (e.g., the accuracy of a land use/land cover classification).

augmented reality: a combination of a real-world experience with a virtual representation in which the real world is dominant, such as overlaying the boundaries of "old-growth" forests on your view of a mountainous environment; a subset of a mixed environment.

augmented virtuality: a combination of a real-world experience with a virtual representation in which the virtual representation is dominant; a subset of a mixed environment.

aural variable: a dimension of sound that can be used to represent either data or its uncertainty; for example, a louder sound could represent greater data certainty.

autostereoscopic display: a method of creating a stereoscopic display in which specialized glasses are not necessary.

available space: empty areas on a map in which map elements can be placed.

avatar: a representation of a person in a virtual environment.

axis of rotation: an imaginary line running through the North and South Poles about which Earth rotates.

azimuth: refers to specifying direction on a map and is customarily measured in a clockwise fashion, starting with geographic north as the origin and passing through 360°.

azimuthal projection: a projection in which all directions or azimuths along straight lines radiating from the map projection's center represent great circles on Earth; directions are preserved from the center of the map to any other point on the map.

balance: the organization of map elements and empty space that results in visual harmony and equilibrium.

balanced cartogram: a cartogram that avoids the extreme distortion found on conventional cartograms by setting a lower threshold on how small an enumeration unit can become.

balanced data: two phenomena coexisting in a complementary fashion, such as the percentage of English and French spoken in Canadian provinces.

bar scale: the map element that is a graphical expression of scale, resembling a small ruler.

base information: graphical symbols and types that provide a geographic frame of reference for thematic symbols.

Big Data: refers to very large data sets that are often used in geovisual analytics, such as those utilized in the field of remote sensing.

binary method: a method of dasymetric mapping in which there are only two possible outcomes for ancillary categories (e.g., when mapping population with land use/land cover, ancillary classes are either habitable or uninhabitable).

bipolar and ganglion cells: cells that merge the input arriving from rods and cones in the eyes.

bipolar data: data characterized by either a natural or a meaningful dividing point, such as the percentage of population change.

bird's-eye view: see panorama.

bivariate cartogram: describes a cartogram in which two attributes are displayed; one attribute (e.g., population) is represented by sizing enumeration units as a function of that attribute, and a second attribute is symbolized within each enumeration unit (e.g., choropleth symbolization is used to display the percent voting for a particular candidate).

bivariate choropleth map: the overlay of two univariate choropleth maps (e.g., one map could utilize lightnesses of cyan and the other lightnesses of red).

bivariate correlation: a numerical method for summarizing the relationship between two interval or ratio-level attributes.

bivariate mapping: the cartographic display of two attributes, such as median income and murder rate, for census tracts within a city.

bivariate-normal: when y values associated with a given x value are normal, and the x values associated with a given y value are also normal.

bivariate point symbol: a point symbol used to portray two attributes simultaneously (e.g., representing two attributes by the width and height of ellipses).

bivariate ray-glyph: used to map two attributes by extending straight line segments to either the right or the left of a small central circle.

bivariate regression: the process of fitting a line to two interval or ratio-level attributes; the dependent attribute is plotted on the y axis, and the independent attribute is plotted on the x axis.

blind spot: the small, visually insensitive region of the retina associated with where the optic nerve carries information to the brain.

block diagram: a technique developed by A. K. Lobeck for depicting the three-dimensional structure of geomorphologic and geologic features.

blue noise pattern: a distribution of points on a dot map in which there are large mutual distances between points and no regularity artifacts.

boundary error: if classed data are conceived as a prism map, boundary error describes how close the resulting cliffs come to matching the cliffs on an unclassed prism map of the data.

bounding rectangle: a rectangle that just touches but completely encloses an arbitrary shape.

box plot: a method of exploratory data analysis that illustrates the position of various numerical summaries along the number line (e.g., minimum, maximum, median, and lower and upper quartiles).

boxmap: a variant of quartile data classification in which outliers are placed in classes by themselves; thus, six rather than four classes are used.

brushing: when a user is able to select and highlight an arbitrary set of spatial entities while exploring data; for instance, a user might select census tracts that are adjacent to a river and highlight those tracts with a particular color.

bubble plot (or bubble chart): an animated graphic in which spatial entities (such as countries) are plotted on a Cartesian coordinate display as a function of their values on two attributes (providing their x and y locations), with each entity plotted as a circle based on its population; as time changes, the circles move to new locations.

callout: a method for labeling a spatial feature; the effect of overprinting is minimized by using a combination of a mask and a leader line.

Cartesian coordinates: a coordinate system in which the locations of points in a plane are referenced with respect to the x and y distance from two perpendicular intersecting axes that form an origin.

cartogram: a map that purposely distorts geographic space based on values of a theme (e.g., making the size of countries proportional to their population).

cartographic design: a partly mental, partly physical process in which maps are conceived and created.

cartographic scale: based on the representative fraction; a large cartographic scale portrays a small portion of Earth (contrast with geographic scale).

case: describes how a developable surface touches a reference globe's surface; the two common cases are tangent and secant.

cell type: the shape of each individual mark of a halftone pattern.

central meridian: the line of longitude at the center of a map about which the projection is symmetrical.

centrifugal force: the tendency for an object that is positioned on and in motion along a curved path to move away from the center of that rotating body.

centroid: see geographic center.

change blindness: the failure to see changes in one portion of an animated display when attending to other areas of the display.

change map: a map representing the difference between two points in time.

Chernoff face: distinct facial features that are associated with individual attributes (e.g., a broader smile depicts cities with a higher per capita expenditure on public schools).

chiaroscuro: see shaded relief.

chorodot map: a combination of choropleth and dot maps; the "dot" portion derives from using small squares within enumeration units, whereas the "choropleth" portion derives from the shading assigned to each square.

choropleth map: a map in which enumeration units (or data-collection units) are shaded with an intensity proportional to the data values associated with those units (e.g., census tracts shaded with gray tones that have an intensity proportional to population density).

chroma: see saturation.

CIE: an international standard for color that allows cartographers to precisely specify a color so that others can duplicate that color.

CIELAB: a widely used perceptually uniform color model for the CIE color model.

CIELUV: a perceptually uniform color model for the CIE color model that was originally recommended for graphic displays; it has been superseded by CIELAB.

clarity: a term used to summarize the following visual variables for depicting data uncertainty: crispness, resolution, and transparency.

class: the type of developable surfaces used to create a projection, including cylindrical, conic, and planar.

class interval: in creating an equal-interval map, the width that each class occupies along the number line.

classed map: a map in which data are grouped into classes of similar value, and the same symbol is assigned to all members of each class (e.g., a data set with 100 different data values might be depicted using only five lightnesses of gray).

clip art: pictures that are available in a digital format; such pictures can serve as the basis for creating pictographic symbols.

cluster analysis: a mathematical method for grouping observations (say, counties) based on their scores on a set of attributes.

CMYK: a system for specifying colors in which cyan (C), magenta (M), yellow (Y), and black (K) are used; this is common in offset printing.

cognition: the mental activities associated with map reading and interpretation, including the initial perception of the map, thought processes, prior experiences, and memory.

cokriging: a multivariate extension of kriging that allows other attributes to influence the interpolation of an attribute; for instance, we might use elevation to enhance the interpolation of precipitation data.

collaborative geovisualization: geovisualization activities in which several individuals are involved in the visualization process; for instance, individuals located at different universities might manipulate and visualize a climate-change model (ideally, the results of one individual's manipulations can be seen by other individuals immediately).

collapse: a generalization operation involving a conversion of geometry; for instance, a complex urban area might be collapsed to a point due to scale change and symbolized with a circle.

color composite: a digital color proof created on a color laser, ink-jet, or other printer.

color convention: a method of applying color that is commonly used and which is typically successful, when symbolizing certain types of map features in particular situations.

color lookup tables: used in association with a frame buffer to make rapid color changes in a graphics display.

color management system: a software application that identifies differences in gamuts among devices and corrects the variations in color introduced by each device.

color map: see color scheme.

color models: refers to various approaches for specifying color (e.g., RGB vs. CMYK).

color profile: an electronic file that describes the manner in which a particular device introduces color variations.

color ramping: a method for specifying colors on a graphics display in which the user selects two endpoints, and the computer automatically interpolates intermediate colors.

color scheme: a system for assigning colors to map symbols representing data; for example, on a choropleth map, we might assign lightnesses of blue to five classes representing percentage voting democratic.

color stereoscopic effect: occurs when colors from the long-wavelength portion of the electromagnetic spectrum appear nearer to a map reader than colors from the short-wavelength portion; this effect is enhanced with special viewing glasses.

color temperature scale: a scale used to categorize colors as they are emitted from an object (a black body) at different temperatures.

color vision impairment: the condition in which some individuals do not see the range of colors that most people do; males are most commonly affected, with red and green colors being most frequently confused.

color weaving: a method for multivariate choropleth mapping in which each attribute is represented by a different color, and the colors are combined using a sampling of independent pixels defined by a random noise function.

color wheel: a graphical device that allows us to see relationships between colors.

colorimeter: a physical device for measuring color; typically, color specifications are given in the CIE system.

command-line interface: a computer interface in which the user types commands specifying processes to be performed.

compaction index (CI): a measure of shape involving the ratio of the area of the shape to the area of a circumscribing circle.

comparative micromaps: a set of micromaps, typically indexed by time, along with a set of maps showing the changes between each pair of maps.

compare patterns: when two or more maps are analyzed visually, readers are expected to compare the patterns found on the maps.

complementary colors: opposite colors such as cyan and red that combine to produce a shade of gray.

completeness: one means of assessing data quality (or uncertainty); includes information about selection criteria and definitions used (e.g., the minimum area needed to define a water body as a lake).

compromise projection: a map projection that does not possess any specific property but, rather, strikes a balance among various projection properties.

conceptual point data: data that are collected over an area (or volume) but conceived as being located at one point for the purpose of symbolization (e.g., the number of microbreweries in each state).

conditioned micromaps: a spatial pattern for a single attribute is split into a series of micromaps in which the rows and columns of the maps correspond to two attributes that might explain the attribute that is displayed in the map series.

confidence level: the following formula is used to determine the confidence level (or probability) that two enumeration units i and j are in separate classes:

$$CL_{i,j} = \Phi\left(\frac{|\bar{x}_i - \bar{x}_j|}{\sqrt{SE_i^2 + SE_j^2}} \right)$$

where $\bar{x}_i$ and $\bar{x}_j$ are the estimates (means) for the two enumeration units, SE_i^2 and SE_j^2 are the standard errors of these estimates, and Φ is a function returning the probability of the expression in parentheses.

cone: one of three developable surfaces; produces the conic class of projections.

cones: a type of nerve cell within the retina that functions in relatively bright light and is responsible for color vision.

conformal projection: a map projection in which angular relationships around any point correspond to the same angular relationship on Earth's surface.

congruence principle: the principle that the content and format of the graphic should correspond to the content and format of the concepts to be conveyed (e.g., on an animated map, the movement of airplanes between two locations can be represented by the movement of small moving airplane icons).

conic projection: a map projection in which the graticule is conceptually projected onto a cone that is either tangent or secant to the developable surface; in most instances, conic projections represent meridians as straight lines and parallels as smooth curves about the North Pole.

constrained extended local processing routines: simplification algorithms that search beyond the immediate neighboring points and evaluate larger sections of lines.

constrained ratio: a level of measurement for data sets that are constrained to a fixed set of numbers, such as probability (the range is 0 to 1) or percentages (the range is 0 to 100).

contiguous cartogram: a cartogram in which topological relations between enumeration units are retained (i.e., enumeration units adjacent to one another on a conventional map will also be adjacent on the cartogram), and there is an attempt to preserve shape.

continuous flow maps: depict the movement of a continuous phenomenon such as the wind or ocean currents.

continuous phenomena: geographic phenomena that occur everywhere, such as the yearly distribution of snowfall in Wisconsin or sales tax rates for each state in the United States.

continuous-tone map: an unclassed isarithmic map.

contour lines: these represent the intersection of horizontal planes with the three-dimensional surface of a smooth, continuous phenomenon.

contour map: see isarithmic map.

contrast: the visual differences between map features that allow for differentiation of features and the implication of relative importance.

control points: points that are used as a basis for interpolation on an isarithmic map, such as weather station locations.

cool colors: colors associated with cool things, such as blue and green.

coordinated multiple views: the ability to link maps with tables, graphs, and numerical summaries in interactive graphical displays.

cophenetic correlation coefficient: when using cluster analysis, measures the correlation between raw resemblance coefficients (Euclidean distance values in the case of UPGMA) and resemblance coefficients derived from the dendrogram (normally referred to as the cophenetic coefficients).

cornea: the protective outer covering of the eye.

correlation coefficient: a numerical expression of the relationship between two interval-level or ratio-level attributes.

counts: a level of measurement in which individual objects, such as people, are counted.

crispness: a visual variable for depicting data uncertainty; for example, a sharp (or crisp) boundary would represent certain data, whereas a fuzzy boundary would represent uncertain data.

critical cartography: a subdiscipline of cartography that considers the social and political implications of mapping—that maps can have multiple meanings and hidden agendas and can be instruments of power; it also presumes that mapmakers can create maps that appear to counter traditional mapping practices (e.g., the maps might be *radical*, *disruptive*, *protesting*, or *subversive*).

critical GIS: a consideration of the social implications of using GIS.

critical points: those points that are significant for defining the structure of a linear or areal feature.

cross-blended hypsometric tints: a technique that combines traditional colors for hypsometric tints with colors that we might expect to find in the environment.

cross-hatched shading: a technique for bivariate choropleth mapping in which horizontal and vertical lines of varied spacing are used to represent two attributes.

cross validation: methods for evaluating the accuracy of interpolation for isarithmic mapping that involve comparing observed data values with predicted (or estimated) data values.

crustal velocities: describes the fact that Earth's landmasses are composed of rigid plates that are in constant motion driven by tectonic forces. This motion is estimated to be about 1.2 cm (0.6 inches) a year and is an important consideration when establishing accurate coordinates and heights.

cybercartography: a term used to describe the character of modern cartography; some of its characteristics include the capability to interpret spatial patterns using senses other than vision (e.g., hearing and touch); the use of multimedia formats and new communication technologies; and new research partnerships among academia, government, and the private sector.

cybersickness: a form of motion sickness that can result from exposure to virtual environments.

cyclical: a level of measurement appropriate for phenomena that have a cyclical character, such as angular measurements (the cycle is 360°).

cylinder: one of three developable surfaces; produces the cylindrical class of projections.

cylindrical projection: a map projection in which the graticule is conceptually projected onto a cylinder that is either tangent or secant to the developable surface; in most instances, cylindrical projections represent meridians and parallels as straight lines intersecting each other at 90°.

dasymetric map: like choropleth maps, area symbols are used to represent zones of uniformity, but the bounds of zones need not match enumeration unit boundaries.

data exploration: utilizing interactive graphical approaches and our visual information processing capability to explore spatial data; for example, we might view a choropleth map in both its unclassed and its classed forms.

data jack: triangular spikes are drawn from a square central area and made proportional to the magnitude of each attribute being mapped.

data quality: the notion that data for thematic maps are often subject to some form of error; see uncertainty.

data resolution: indicates the granularity of the data that are used in mapping (e.g., 1-meter remote sensing data are considered high resolution).

data science: refers to the interdisciplinary field that extracts knowledge and insight from Big data; it involves selecting and processing data, exploring the data to identify patterns, verifying the patterns, developing models, and communicating results. Traditionally, it has been accomplished by emphasizing computational methods, but data science will be most effective if it is combined with visualization and human reasoning.

data source: the map element that describes the origin of the data represented on a map.

data splitting: a method of cross validation for isarithmic mapping; control points are split into two groups, one to create the contour map and one to evaluate its accuracy.

data standardization: see standardized data.

deconstruction: the process of analyzing a map (or text) to uncover its hidden agendas or meanings.

degrees: a basic unit of measure utilized in specifying longitude and latitude; Earth encompasses 360° of longitude and 180° of latitude.

Delaunay triangles: the type of triangles commonly used in triangulation, a method of interpolation for isarithmic maps; the longest side of any triangle is minimized, and thus the distance over which interpolation takes place is minimized.

Demers cartogram: a cartogram in which squares are used to represent enumeration units, with the size of the square a function of an attribute associated with the units.

democratization of cartography: the notion that modern technology (e.g., the Web) provides virtually anyone the capability to design maps.

dendrogram: a treelike structure illustrating the resemblance coefficient values at which clusters in a cluster analysis combine.

descriptive statistics: used to describe the character of a sample or population, such as computing the mean square footage of a sample of 50 stone houses.

developable surface: a surface that can be flattened to form a plane without compressing or tearing any part of it; three commonly used developable surfaces for map projections are the cylinder, cone, and plane.

dichromats: a more serious form of color vision impairment in which one of the three types of cone cells (red, green, and blue) is absent; contrast with anomalous trichromats.

Digital Earth: a virtual globe promoted by U.S. Vice President Al Gore that would provide a wide range of users the ability to easily access and interact with the vast storehouse of the world's geospatial and aspatial information and knowledge.

digital elevation model (DEM): topographic data commonly available as an equally spaced gridded network.

Digital Globe: digital spatiotemporal information about Earth or another celestial body is displayed in an undistorted three-dimensional globe as either a virtual or a tactile display.

digital proof: a proof created on a digital printer rather than an offset press.

discrete phenomena: geographic phenomena that occur at isolated locations, such as water towers in a city.

dispersion graph: a graph in which data are grouped into classes, and the number of values falling in each class is represented by stacked dots along the number line.

displacement: a generalization operation that involves moving features apart to prevent coalescence (e.g., when a highway and a railroad follow a coastline in close proximity).

display date: one of the visual variables for animated maps; refers to the time that some display change in an animation is initiated (e.g., in an animation of population for U.S. cities, a circle for San Francisco appears in 1850).

distance cartogram: a map in which real-world distances are distorted to reflect some attribute (e.g., distances on a subway map might be distorted to reflect travel time).

distortion: involves altering the size of Earth's landmasses and the arrangement of Earth's graticule when they are projected to the two-dimensional flat map.

dithering: a process in which colors are created by presuming that the reader will perceptually merge different colors displayed in adjacent pixels (on a graphics display) or dots (on a printer).

diverging color map: see diverging color scheme.

diverging color scheme: a sequence of colors in which two hues diverge from a common light hue or a neutral gray, as in a dark red–light red–gray–light blue–dark blue scheme.

Dorling cartogram: a cartogram in which uniformly shaped symbols (typically circles) are used to represent enumeration units, with the size of the symbol a function of an attribute (typically population) associated with the units.

dot-density shading: a technique for mapping smooth, continuous phenomena (e.g., precipitation) in which closely spaced dots depict high values of the phenomenon, whereas widely separated dots depict low values.

dot gain: a phenomenon in which ink is forced to spread out due to pressures exerted between the cylinders of an offset lithographic printing press.

dot map: a map in which small symbols of uniform size (typically, solid circles) are used to emphasize the spatial pattern of a phenomenon (e.g., one dot might represent 1,000 head of cattle).

dot plot: as in the dispersion graph, dots are stacked, but the placement of a stack is determined by the raw data values rather than by classes resulting from the grouped frequency table.

dot size: how large dots are on a dot map.

drafting table–format display: a virtual environment is achieved by projecting images onto a large table-like screen that is tilted at a 45° angle; a head tracker on one user governs the view that the other users see, and a specialized 3-D mouse can be used to interact with the simulated environment.

draped image: a remotely sensed image (or other information; e.g., land use and land cover) is draped on a 3-D map of elevation.

duration: one of the visual variables for animated maps; refers to the length of time that a frame of an animation is displayed.

Ebbinghaus illusion: a phenomenon in which a circle surrounded by large circles will appear smaller than the same-sized circle surrounded by small circles.

edge bundling: a technique in which clutter on flow maps is handled by bundling nearby flows together.

electromagnetic energy: a waveform having both electrical and magnetic properties; refers to the way that light travels through space.

electromagnetic spectrum: the complete range of wavelengths possible for electromagnetic energy.

electronic atlas: a collection of maps (and databases) available in a digital environment; sophisticated electronic atlases enable users to take advantage of the digital environment through Internet access, data exploration, map animation, and multimedia.

ellipsoid: see oblate spheroid.

Encapsulated PostScript (EPS): a subset of the PostScript page description language that allows digital maps and other documents to be transported between software applications, and between different types of computers.

enhancement: a generalization operation that involves a symbolization change to emphasize the importance of a particular object (e.g., the delineation of a bridge over an existing road is often portrayed as a series of cased lines).

enumeration unit: a data-collection unit for thematic mapping such as a county or state.

equal-areas: a method of data classification in which an equal portion of the map area is assigned to each class.

equal-intervals: a method of data classification in which each class occupies an equal portion of the number line.

Equator: the origin for the system of latitude (0° latitude); serves as the dividing line between the Northern and Southern Hemispheres.

equatorial aspect: positioning the developable surface over the reference globe such that the Equator becomes the central latitude.

equidistant projection: a map projection for which distances from one point (usually the projection's center) to all other points are preserved compared to those distances on Earth's surface.

equiprobability ellipse: if data are bivariate normal, an ellipse can be drawn that encloses a specified percentage of the data.

equivalent (equal-area) projection: a map projection for which areas are preserved compared to the same areas on Earth's surface.

estimate: an estimated value based on a sample of data; for example, the American Community Survey might survey a sample of people in order to estimate the percent foreign-born in a county.

exact interpolator: see honoring the control point data.

exaggeration: a generalization operation in which the character of an object is amplified (e.g., exaggerating the mouth of a bay that would close under scale reduction).

expert system: a software application that incorporates rules derived from experts to make decisions and solve problems.

exploratory data analysis (EDA): a method for analyzing statistical data in which the data are examined graphically in a variety of ways, much as a detective investigates a crime; this is in contrast to fitting data to standard forms, such as the normal distribution.

exploratory spatial data analysis (ESDA): data exploration techniques that accompany a statistical analysis of spatial data.

extrinsic visual variables: when mapping uncertainty, these are visual variables that involve adding objects to the display, such as arrows, dials, and thermometers.

figure-ground: methods of accentuating one object more than another, based on the perception that one object stands in front of another, and appears to be closer to the map user.

Fisher–Jenks algorithm: a method for classifying data in which an optimal classification is guaranteed by essentially considering all possible classifications of the data.

fishnet map: a map in which a fishnet-like structure provides a three-dimensional symbolization of a smooth, continuous phenomenon.

Flannery correction: a method of adjusting the sizes of proportional circles to account for the perceived underestimation of larger circle sizes; a symbol scaling exponent of 0.57 is used.

flattening constant (f): expresses the degree to which an ellipse deviates from a circle; computed as $(a - b)/a$, where a and b are the semimajor and semiminor axes of the ellipse, respectively.

flow map: a map used to display the movement of some phenomenon between geographic locations; generally, this is done using "flow lines" of varying thickness.

fluorescent inks: specialized printing inks that produce brilliant, intense color.

fly-over (or fly-through): a form of animation in which the viewer is given the feeling of flying over or through a landscape.

focusing: when exploring data, a user filters the data; for example, a user might focus on census tracts that are less than 25 percent Hispanic.

font: a set of all alphanumeric and special characters of a particular type family, style, and size.

four-color process printing: the combination of the process colors (CMYK) with screening techniques, allowing for a very wide range of colors in print reproduction.

fovea: the portion of the retina where visual acuity is the greatest.

frame line: the map element—normally, a rectangle—that encloses all other map elements.

framed-rectangle symbol: a point symbol that consists of a "frame" of constant size, within which a solid "rectangle" is placed; the greater the data value, the greater the proportion of the frame that is filled by the rectangle.

frequency: one of the visual variables for animated maps; refers to the number of identifiable states per unit time, such as color cycling used to portray the jet stream.

fuzzy categories: a level of measurement in which category memberships are fuzzy, such as some remote sensing classification procedures.

gamut: a range of colors produced by a computer or a printing or display device.

general information: the type of information that one acquires when examining spatial patterns on thematic maps.

general-reference map: a type of map used to emphasize the location of spatial phenomena (e.g., a U.S. Geological Survey (USGS) topographic map).

generalization: the process of reducing the information content of maps due to scale change, map purpose, intended audience, or technical constraints.

geocentric latitude: latitude as computed on a reference ellipsoid is measured by an angle that results when a line at Earth's surface is drawn intersecting the plane of the Equator at Earth's center.

geocollaboration: see collaborative geovisualization.

geodesy: the field of study investigating Earth's size and shape.

geodetic datum: a combination of a specific reference ellipsoid and a geoid (also commonly referred to as a datum).

geodetic latitude: latitude as computed on a reference ellipsoid is measured by an angle that results when a perpendicular line at the reference ellipsoid's surface is drawn toward Earth's center.

geographic brush: as areas of a map are selected (e.g., using a mouse), corresponding dots are highlighted within a scatterplot matrix.

geographic center: the "balancing point" for a geographic region.

geographic information systems (GIS): automated systems for the capture, storage, retrieval, analysis, and display of spatial data.

geographic scale: the extent of a study area, such as a neighborhood, city, region, or state; here, "large scale" indicates a large area (contrast with cartographic scale).

geographic visualization: a private activity in which previously unknown spatial information is revealed in a highly interactive computer graphics environment.

geo-icon: a method for multivariate mapping in which adjacent pixels are combined in various ways to depict multiple attributes; for instance, the shape of each grouping of pixels could signify a particular attribute (e.g., a vegetation attribute could be depicted by a treelike pixel grouping).

geoid: describes undulating surface that would result on Earth if the oceans were allowed to flow freely over the continents, creating a single, undisturbed water body; the surface would have undulations reflecting the influence of gravity.

geometric datum: a datum that collects and defines information about latitude, longitude and crustal velocities.

geometric symbols: symbols that do not necessarily look like the phenomena being mapped, such as using squares to depict barns of historical interest; contrast with pictographic symbols.

geopotential datum: a datum that collects and defines information related to gravity variations resulting from the geoid.

geoslavery: the use of GPS and GIS technologies to monitor and control the movement of people.

geospatial virtual environments (GeoVEs): virtual environments of interest to geographers and other geoscientists.

geostatistical methods: numerical summaries for which spatial location is an integral part (e.g., centroid and spatial autocorrelation measures).

geovisual analytics: computer-based systems that combine the visualization power of humans with the computational power of computers to make sense out of large geospatial datasets (Big Data).

geovisualization: see geographic visualization.

GeoWall: a specialized form of wall-sized display that includes four components: a computer with a dual-head graphics card capable of sending two different images at the same time, two DLP projectors equipped with polarizing filters, a display screen capable of preserving light polarization, and passive glasses for users.

gerrymandering: the purposeful distortion of legislative or congressional districts for partisan benefit.

Gestalt principles: descriptions of the manner in which humans see the individual components of a graphical image, and organize the components into an integrated whole.

global algorithms: simplification algorithms that process an entire line at once (e.g., the Douglas–Peucker method).

glyphs: a term applied to multivariate point symbols when the attributes being mapped are in dissimilar units (i.e., not part of a larger whole).

goodness of absolute deviation fit (GADF): a measure of the accuracy of a classed choropleth map when the median is used as a measure of central tendency.

goodness of variance fit (GVF): a measure of the accuracy of a classed choropleth map when the mean is used as a measure of central tendency.

graduated dot map: a dot map in which dots of varying sizes are used.

graduated symbol map: see range-graded map.

graduated-symbol scaling: see range-graded scaling.

graphical user interface (GUI): a type of computer interface in which the user specifies tasks to be performed by pointing to the desired task, typically by using a mouse.

graticule: lines that represent latitude and longitude taken in combination that appear on a map.

gray curves (or grayscales): a graph (or equation) expressing the relation between printed area inked and perceived blackness.

great circle: represents the line along which a plane intersects Earth's surface when the plane passes through Earth's center.

gridded cartogram: a cartogram in which a raster grid of data values is overlaid on a desired country or territory, and the resulting grid cells are then sized as a function of the values occurring within each grid cell using the Gastner/Newman algorithm.

grouped-frequency table: constructed by dividing the data range into equal intervals and tallying the number of observations falling within each interval.

hachures: a method for depicting topography in which a series of parallel lines is drawn perpendicular to the direction of contours (and parallel to the slope).

halftone screening: the application of ink or toner in a pattern of equally spaced cells of variable size to create tints from a base color.

halo: a method for reducing the effect of overprinting by creating outlines of letters in type.

haptic variable syntax: refers to the variables involved in sensing spatial information with touch (haptically); for example, the vibration variable refers to the vibration we might feel when moving a tactile mouse over a map.

haptics: the use of touch to acquire spatial information (e.g., a user feels different vibrations when a force-feedback mouse is placed over map locations that have differing crime rates).

harmonious colors: similar colors such as blue and cyan.

head-mounted display (HMD): helmet-like device that shields a person from the real-world view and provides images of a virtual environment to each eye.

head/tail breaks: a method of classification appropriate for heavy-tailed (strongly positively skewed) distributions; classification is accomplished by recursively dividing the data above the mean as long as the data above the mean are heavy-tailed.

heat map: a map that has the appearance of a continuous-tone isarithmic map, but it is not based on interpolating values at point locations; rather, a grid is overlaid on a map of plotted points and the density of points is computed within a search radius of each grid cell, thus creating a smooth representation.

hexagon bin plot: a method of data display in which a set of hexagons is placed over a conventional scatterplot, and the hexagons are filled as a function of the number of dots falling within them.

high-fidelity process colors: colors based on the traditional process colors (CMYK) but including two or three additional colors that are mixed on the page via a printing device.

hill shading: see shaded relief.

histogram: a type of graph in which data are grouped into classes, and bars are used to depict the number of values falling in each class.

hologram: the 3-D character of an object (or scene) is recreated by splitting a laser beam of light into two parts: one part that is reflected off a mirror and then strikes a recording plate, and one part that is reflected off the object of interest and reaches the same recording plate. These two beams of light make an interference pattern on the plate that provides detailed information about the object of interest.

honoring the control point data: a term applied if, after an interpolation is performed, there is no difference between the original value of a control point and the value of that same point on the interpolated map.

horizontal datum: a specific reference ellipsoid to which accurate latitude and longitude values are referenced.

HSV: a system for specifying color in which hue (H), saturation (S), and value (V) are used; although the system utilizes common color terminology, colors are not equally spaced in the visual sense.

hue: along with lightness and saturation, one of three components of color; it is the dominant wavelength of light, such as red vs. green.

HVC color model: a system developed by Tektronix for specifying color; the terms hue (H), value (V), and chroma (C) are used, and colors are equally spaced from one another in the visual sense.

hypertext markup language (HTML): a relatively simple, text-based programming language that is used to define the content and appearance of Web pages.

hypsometric tints: the colored areas sometimes used between contour lines on a topographic map to represent different elevation ranges (e.g., progressing from green in lowland areas to yellow to brown to purple to white in highland areas).

iconic memory: refers to the initial perception of an object (in our case, a map or portion thereof) by the retina of the eye; contrast with short-term visual store and long-term visual memory.

illuminated choropleth map: a color scheme for a traditional choropleth map is combined with a planimetrically correct 3-D prism view.

illuminated contours: an approach for depicting topography in which both the color and width of contour lines are varied as a function of their angular relationship to a presumed light source.

immersion: one of the four "I" factors of geospatial virtual environments (GeoVEs); a psychological state in which a user is enveloped by, included in, and able to interact with an environment that provides a continuous stream of stimuli and experiences.

independent point routines: simplification algorithms that select coordinates based on their position along the line, nothing more (e.g., an nth point routine might select every third point).

inferential statistics: used to make an inference (or guess) about a population based on a sample.

information acquisition: the process of acquiring spatial information while a map is being used; contrast with memory for mapped information.

information intensity: one of the four "I" factors of geospatial virtual environments (GeoVEs); deals with the level of detail (LOD) presented in the VE.

information-seeking mantra: the visualization portion of geovisual analytics that includes overview, zoom and filter, and details on demand.

information visualization: involves the visualization and analysis of nonnumerical abstract information such as the topics that are discussed on the front page of a newspaper over a monthlong period.

inset map: the map element consisting of a smaller map used in the context of a larger map (e.g., a map that is used to show a congested area in greater detail).

intellectual hierarchy: a ranking of symbols and map elements according to their relative importance.

intelligence of objects: one of the four "I" factors of geospatial virtual environments (GeoVEs); refers to the idea that objects within a VE often exhibit some degree of behavior or intelligence.

intelligent agent: see virtual assistant.

intelligent dasymetric mapping (IDM): a dasymetric mapping approach that utilizes both the cartographer's domain knowledge and the relationship between enumeration units and ancillary information.

interactivity: one of the four "I" factors of geospatial virtual environments (GeoVEs); an important element of interactivity is developing suitable approaches for navigating through the VE.

intermediate frames: in an animation, the frames (or maps) associated with interpolated data; contrast with key frames.

International Color Consortium (ICC): an entity that has developed a standard for a vendor-neutral, cross-platform color profile, and architectures for color management.

International Dateline: the line coinciding approximately with the 180th meridian; when crossing it, a day is lost or gained depending on the direction of travel.

Internet: a global network of computers that permits sharing of data and resources.

interpolation: in isarithmic mapping, the estimation of unknown values between known values of irregularly spaced control points.

interquartile range: the absolute difference between the 75th and 25th percentiles of the data.

interrupted projection: a projection having several central meridians, where each central meridian (usually located over a landmass) creates a lobe.

interval: a level of measurement in which numerical values can be assigned to data, but there is an arbitrary zero point, such as for SAT scores.

intrinsic visual variables: when mapping uncertainty, these are visual variables that are intrinsic to the display and thus involve altering the existing symbology; for example, we might vary the saturation of colored areal symbols on a choropleth map.

inverse-distance weighting: a method of interpolating data for isarithmic maps; a grid is overlaid on control points, and estimates of values at grid points are an inverse function of the distance to control points.

isarithmic map: a map in which a set of isolines (lines of equal value) are interpolated between points of known value, as on a map depicting the amount of snowfall in a region.

isometric map: a form of isarithmic map in which control points are true point locations (e.g., snowfall measured at weather stations).

isopleth map: a form of isarithmic map in which control points are associated with enumeration units (e.g., assigning population density values for census tracts to the centers of those tracts and then interpolating between the centers).

isosurface: a surface of equal value within a true 3-D phenomenon (e.g., a surface representing a wind speed of 100 kilometers per hour in the jet stream).

Jenks–Caspall algorithm: an optimal data classification that is achieved by moving observations between classes through trial-and-error processes known as reiterative and forced cycling.

kerning: the variation of space between two adjacent letters.

key frames: in an animation, the frames (or maps) associated with collected data; contrast with intermediate frames.

kriging: a method of interpolating data for isarithmic maps that considers the spatial autocorrelation in the data.

labeling software: software developed for the automated positioning of type.

large-scale map: a map depicting a relatively small portion of Earth's surface.

latitude: an imaginary line that crosses Earth's surface in an east-to-west fashion; latitude is measured in degrees, minutes, and seconds north or south of the Equator.

latitude of origin: the latitude that is set as the origin for the x and y plotting coordinates.

leading: the vertical space between lines of type; equivalent to line spacing.

leave-one-out (LOO): a method of cross validation in which we (1) remove a control point from the data to be interpolated, (2) create an interpolated map using all other control points, and (3) compute the difference between the known and estimated value at that control point. These steps are repeated for each control point in turn.

legend: the map element that defines graphical symbols.

legend heading: the map element that is used to explain a map's legend and theme.

lens: the focusing mechanism within the eye.

lensing: a portion of a mapped display is enlarged, enabling the viewer to see detail in an area of interest, while the surrounding area is distorted and reduced in importance.

letter spacing: the space between each letter in a word.

level of measurement: refers to the different ways that we measure attributes; we commonly consider nominal, ordinal, interval, and ratio levels.

lightness: along with hue and saturation, one of three components of color; refers to how dark or light a color is, such as light blue vs. dark blue.

limiting variable method: a method of dasymetric mapping in which the density value for an ancillary class within a source zone is a function of the density value assigned to another ancillary class within that source zone.

lineage: one means of assessing data quality (or uncertainty); refers to the historical development of the data.

linear color map: see sequential color scheme.

linear-legend arrangement: a form of legend design for proportional symbol maps; symbols are placed adjacent to each other in either a horizontal or a vertical orientation.

linear phenomena: geographic phenomena that are one-dimensional in spatial extent, having length but essentially no width, such as a road on a small-scale map.

linked micromaps plot: a series of micromaps are linked with one or more attributes and a legend; the linking is established by the color for each enumeration unit that is identified in the legend, and the emphasis is on local pattern perception.

lobe: A portion of an interrupted projection described by a central meridian usually displaying a specific landmass, such as South America.

local processing routines: simplification algorithms that utilize immediate neighboring points in assessing whether a point should be deleted.

location: a visual variable that refers to the possibility of varying the position of symbols, such as dots on a dot map.

location-based services (LBS): the delivery of data and information services as a function of a user's location or projected location; for example, you might use a smartphone to determine the location of the nearest Thai restaurant.

logical consistency: one means of assessing data quality (or uncertainty); describes the fidelity of relationships encoded in the structure of spatial data (e.g., are polygons topologically correct?).

long-term visual memory: an area of memory that does not require constant attention for retainment; contrast with iconic memory and short-term visual store.

longitude: an imaginary line originating at the poles that crosses Earth's surface in a north-to-south fashion; longitude is measured in degrees, minutes, and seconds east or west of the Prime Meridian.

loxodrome: a line that intersects all meridians at the same angle.

luminance: brightness of a color as measured by a physical device.

machine learning: a branch of artificial intelligence in which computer algorithms are used to learn and interpret data; for example, a set of tweets might be analyzed to determine the underlying topics found in those tweets.

magnitude estimation: (1) a method for constructing gray-scales in which a user estimates the lightness or darkness of one shade relative to another; (2) a method for determining the perceived size of proportional symbols in which a value is assigned to one symbol on the basis of the value assigned to another symbol.

manual interpolation: in isarithmic mapping, the interpolation of unknown values by "eye."

map communication model: a diagram or set of steps summarizing how a cartographer imparts spatial information to a group of map users.

map complexity: a term used to indicate whether the set of elements composing a map pattern appears simple or intricate (i.e., complex); for example, does the pattern of gray tones on a choropleth map appear simple or intricate?

map design research: research focusing on which mapping techniques are most effective and why.

map editing: the critical evaluation and correction of every aspect of a map.

map elements: graphical and textual units that maps are composed of, including the title, legend, and scale.

map projection: a systematic transformation of Earth's graticule and landmass from the curved two-dimensional surface to a planimetric surface.

map reproduction: the printing of a map on paper or similar medium (print reproduction) or the electronic duplication of a map in digital form (nonprint reproduction).

map tour: see fly-over.

map user: the person or persons for whom a particular map has been designed and produced.

mapped area: the map element representing the geographic region of interest.

maps and society: a paradigm in cartography that assesses the relationship between modern mapping and potential societal impacts; technological developments in mapping, and the availability of spatial databases, has raised concerns over access, "power," privacy, and the democracy of mapping.

margin of error: an expression of the error associated with using a sample to represent a population; for example, we might estimate the percent foreign-born in a county to be 5.1 percent $\pm$ 2.2 percent. The value 2.2 percent is the margin of error. There is a confidence level associated with the margin of error (90 percent is a typical confidence level).

mask: a method for reducing the effect of overprinting by placing a polygon (e.g., a white rectangle) underneath the type but above the underlying graphics.

mathematical scaling: sizing proportional symbols in direct proportion to the data; thus, if a data value is 40 times another, the area (or volume) of the symbol will be 40 times as large as another symbol.

maximum-contrast colors: on a choropleth map, colors are selected so that they have the maximum possible contrast with one another.

mean: the average of a data set; computed by adding all values and dividing by the number of values.

mean absolute error (MAE): a measure of error for cross validation in which we compute the mean of the sum of the absolute value of residuals (the observed minus the predicted values).

mean bias error (MBE): a measure of error for cross validation in which we compute the mean of the sum of all residuals (the observed minus the predicted values).

mean center: a measure of central tendency for point data; computed by independently averaging the x and y coordinate values of all points.

mean–standard deviation: a method of data classification in which the mean and standard deviation of the data are used to define classes.

measures of central tendency: used to specify a value around which data are concentrated (e.g., the mean and median).

measures of dispersion: used to specify the dispersion or variability in a data set (e.g., the interquartile range and standard deviation).

median: the middle value in an ordered set of data.

memory for map information: remembering spatial information that was previously seen in map form; contrast with information acquisition.

merging: a generalization operation involving the fusing of groups of line features, such as parallel railway lines or edges of a river or stream.

meridian: see longitude.

metonic cycle: a time period of approximately 19 years that describes the period between when the new and full moon returns to the same day of the year.

micromaps: a set of small maps is used to analyze one or more attributes distributed spatially; graphical and statistical information are also often linked with these maps. Examples include linked micromaps and conditioned micromaps.

minimum-error projection: a projection having the least amount of total distortion of any projection in that same class according to a specified mathematical criterion.

minutes: a unit of measurement of the sexagesimal degree system; each degree contains 60 minutes.

mixed reality (MR): combines a real-world experience with a virtual representation; most commonly, the real world is dominant—this is termed augmented reality.

mobile mapping: users are able to access maps and related geospatial information as they move through Earth's environment.

mode: the most frequently occurring value in a data set.

modifiable areal unit problem: the notion that the magnitude of statistical measures, such as the correlation coefficient (r), can be affected by the level at which data have been aggregated.

moiré pattern: an unwanted print artifact resulting from the interplay of dots in overlying halftones.

monochromatic composite: a digital proof created by printing a map in black, white, and gray tones, typically on a laser printer.

mosaic cartogram: a cartogram in which a mosaic of small uniformly shaped geometric symbols (typically squares or hexagons) is displayed as a function of data that either consists of (or can be cast into) small integer units (e.g., electoral votes by states in the United States).

mosaic diagram: a graphical representation of data that would normally be shown in contingency tables; if these diagrams are associated with areas on a map, they can provide an intriguing view of a spatial distribution. An example is the display of car speeds for each hour of the day and day of the week, in rows and columns, respectively, of the diagram, with each diagram associated with a subarea of a map.

multimedia: the combined use of maps, graphics, text, pictures, video, and sound.

multimodal interfaces: interfaces that involve two or more novel approaches for interacting with the computer; for instance, you might point to a location without touching the screen (a free-hand gesture) and say, "Show me all homes within 100 meters of this location."

multiple regression: a statistical method for summarizing the relationship between a dependent attribute and a series of independent attributes.

multivariate dot map: a dot map in which distinct symbols or colors are used for each attribute to be mapped (e.g., corn, blueberries, and apricots could be represented by yellow, blue, and orange dots, respectively).

multivariate mapping: the cartographic display of two or more attributes; for example, we might simultaneously map ocean temperature, salinity, and current speed.

multivariate point symbol: the depiction of three or more attributes using a point symbol.

multivariate ray-glyph: a term used when rays are extended from a small interior circle, with the lengths of the rays made proportional to values associated with each attribute being mapped.

Munsell: a system for specifying color in which colors are identified using hue, value, and chroma; colors are equally spaced from one another in the visual sense.

Munsell curve: a curve developed by Munsell that expresses the relation between physical blackness (area inked) and perceived blackness; it is intended for smooth untextured lightnesses of gray and can also be used for a chromatic scheme that varies solely in lightness.

National Geodetic Vertical Datum 1929 (NGVD29): the datum that established heights for North America using 26 tidal gauge stations where mean sea level was measured over an approximate 19-year period. The NGVD29 is not a geopotential surface and was replaced by the North American Vertical Datum of 1988 (NAVD88).

National Spatial Reference System: is the official and consistent coordinate system that defines latitude, longitude, height, scale, gravity, orientation, and shoreline throughout the United States. The NSRS serves as the official geodetic control for all geospatial activities engaged in by federal agencies of the United States. As of this writing, the North American Datum of 1983 (NAD83) and the North American Vertical Datum of 1988 (NAVD88) comprise the NSRS.

natural breaks: a method of data classification in which a graphical plot of the data (e.g., a histogram) is examined to determine natural groupings of data.

Natural Earth: a visual representation that combines satellite-derived land cover data (represented by different colors) with shaded relief.

natural neighbors interpolation: presuming that Thiessen polygons have already been computed for a set of control points if you wish to interpolate a value for an arbitrary point, a new Thiessen polygon is computed for that point, and the proportion of overlap of that new polygon with the original polygons is used to determine the weight assigned to each of the control points associated with the original polygons.

neat line: the map element—normally a rectangle—that is used to crop the mapped area.

necklace map: eliminates the problem of symbol overlap on proportional symbol maps by displaying proportional symbols on a one-dimensional curve (the necklace) that surrounds the mapped region of interest.

nested-legend arrangement: a form of legend design for proportional symbol maps in which smaller symbols are drawn within larger symbols.

nested means: a method of data classification in which the mean is used to repeatedly divide the data set; only an even number of classes is possible.

nominal: a level of measurement in which data are placed into unordered categories, such as classes on a land use/land cover map.

nomograph: a graphical device that can assist in selecting dot size and unit value on a dot map.

non-photorealistic rendering (NPR): a computer-based approach in terrain rendering that attempts to imitate the look of images created with certain production techniques (e.g., you might want to imitate the appearance of Raisz's or Lobeck's work).

nontemporal animations: map animations that do not show change over time (e.g., a fly-over, in which a user has the sense of flying over a landscape).

normal distribution: a term used to describe the bell-shaped curve formed when data are distributed along the number line.

North American Datum of 1927 (NAD27): a horizontal datum providing geodetic control for North America defined by the Clarke 1866 reference ellipsoid and a geoid specification having Meades Ranch, KS, as the point where the geoid and the reference ellipsoid contact each other and assumed to be 0.

North American Datum of 1983 (NAD83): as of this writing, a horizontal datum providing geodetic control for North America defined by the Geodetic Reference System (GRS80) reference ellipsoid which is located at Earth's center of mass. NAD83 replaced NAD27. NAD83 will be replaced by the North American Terrestrial Reference Frame 2022 (NATRF2022).

North American-Pacific Geopotential Datum (NAPGD2022): The geopotential or "vertical" datum that will replace NAVD88 as a part of NSRS modernization efforts by NGS.

North American Terrestrial Reference Frame 2022 (NATRF2022): a geometric datum replacing NAD83 as a part of NSRS modernization efforts by NGS. Note that NATRF2022 will be accompanied by three additional reference frames (PATRF2022, CATRF2022, MATRF2022) where each derives their namesake from the specific tectonic plate (Pacific, Caribbean, Mariana).

North American Vertical Datum 1988 (NAVD88): as of this writing, the current vertical datum that determines the height for North America and serves as a geopotential surface. This datum identified a single fixed height tidal benchmark at Father Point/Rimouski, Quebec, Canada, located along the Gulf of Saint Lawrence. This datum replaced the National Geodetic Vertical Datum 1929 and is scheduled to be replaced by the North American-Pacific Geopotential Datum 2022 (NAPGD2022).

north arrow: the map element that points in the direction of north on a map.

North Pole: the geographic location in the Northern Hemisphere with a latitude of 90° North that serves as a point around which Earth's axis rotates.

North Star: the star (called Polaris) that happens to be located directly above the North Pole.

Northern Hemisphere: the portion of Earth extending from the Equator to the North Pole.

numerical data: a term used to describe the data associated with the interval and ratio levels of measurement, as numerical values are assigned to the data.

oblate spheroid: refers to the notion that Earth bulges at the Equator and compresses at the poles due to the centrifugal force caused by rotation.

oblique aspect: a map projection with a center that is aligned somewhere between the Equator and a pole.

offset lithographic printing press: a mechanical printing device that provides excellent print quality, high printing speed, and a significant decrease in the cost per unit as the number of copies increases.

offset lithography: a form of lithography in which ink is transferred to an intermediate printing surface before being transferred to the print medium.

one-dimensional scatterplot: see point graph.

opponent-process theory: the theory that color perception is based on a lightness–darkness channel and two opponent-color channels: red–green and blue–yellow.

optic nerve: the nerve that carries information from the retina to the brain.

optimal: a method of classification in which like values are placed in the same class by minimizing an objective measure of classification error (e.g., by minimizing the sum of absolute deviations about class medians).

optimal interpolation: a term sometimes applied to kriging, which attempts to minimize the difference between the estimated value at a grid point and the true (or actual) value at that grid point.

order: one of the visual variables for animated maps; refers to the sequence in which frames or scenes of an animation are presented.

ordinal: a level of measurement in which data are ordered or ranked but no numerical values are assigned, such as ranking states in terms of where you would like to live.

ordinary kriging: a form of kriging in which the mean of the data is assumed to be constant throughout geographic space (i.e., there is no trend or drift in the data).

orientation: (1) the indication of direction on a map, normally taking the form of an arrow or a graticule; (2) a visual variable in which the directions of symbols (or marks making up symbols) are varied, such as changing the orientation of short line segments to indicate the direction from which the wind blows.

origin-destination flow maps: used to portray flows between geographic locations when the actual route of flow is unimportant; an example would be a map showing the migration flows from Brazil to each country of the world.

outliers: unusual or atypical observations.

overprinting: a phenomenon that occurs when type is placed on top of another graphical object.

page description data: digital data consisting of a set of printing instructions that describe every graphical and textual component of a map.

page description language: a particular data structure used to describe page description data.

panorama: provides a view of the landscape that we might expect from a painting (which is how traditional panoramas were created), but careful attention is given to geography so that we can clearly recognize known features.

parallel: see latitude.

parallel coordinate plot (PCP): a method of graphical display in which attributes are depicted by a set of parallel axes and observations are depicted as a series of connected line segments passing through the axes.

partitioning: a method for constructing a Munsell curve in which a user places a set of grays of differing lightnesses between white and black such that the resulting grays will be visually equally spaced.

passive display: an approach for creating a stereoscopic view in which images for the left and right eye are created simultaneously by the computer/display system, each with a different polarization of light; users wear lightweight specialized polarized glasses to interpret the images.

percentile map: a method of data classification in which class limits are adjusted to accentuate extreme values; six classes are used and the limits are <1 percent, 1 percent to <10 percent, 10 percent to <50 percent, 50 percent to <90 percent, 90 percent to <99 percent, and >99 percent.

perceptual accuracy: refers to how easy and accurately people can make estimates of the size relations among enumeration units on a cartogram.

perceptual scaling: a term used when proportional symbols are sized to account for the perceived underestimation of large symbols; thus, large symbols are drawn larger than is suggested by the actual data.

perceptually uniform color model: refers to a color model in which colors are equally spaced in the visual sense.

personalized displays: graphical displays intended for individuals, such as those used with desktop computers, laptops, tablets, and phones.

perspective height: a visual variable involving a perspective 3-D view, such as using raised sticks (or lollipops) to represent oil production at well locations.

pexels: a multivariate point symbol consisting of small 3-D bars that can be varied in height, spacing, and color.

physical model: a truly 3-D model of Earth's landscape (i.e., we can hold and manipulate small models in our hands).

physical visualization: see physical model.

physiographic method: a method developed by Raisz for depicting topography in which Earth's geomorphological features are represented by standard, easily recognized symbols.

pictographic symbols: symbols that look like the phenomena being mapped, such as using diagrams of barns to depict the locations of barns of historical interest; contrast with geometric symbols.

pie chart: a modification of the proportional circle in which a portion of the circle is filled in (e.g., to represent the percentage of land area from which wheat is harvested).

pixels: the individual picture elements composing a raster image on a graphics display device.

plan oblique relief: a technique for shaded relief maps that causes relief to "stand up," thus providing a more easily seen 3-D view, and yet preserves the geometric accuracy of conventional planimetric shaded relief maps.

planar projection: a map projection in which the graticule is conceptually projected onto a plane that is tangent or secant to Earth's surface.

plane: one of the three developable surfaces; produces the planar class of projections.

platesetter: a device that produces a latent image on a printing plate.

plotting coordinates: a coordinate system used to draw latitude and longitude on a planimetric surface.

point graph: a type of graph in which each data value is represented by a small point symbol plotted along the number line.

point-in-polygon test: a test that is intended to determine whether a point falls inside or outside a polygon.

point of projection: where an imaginary light source is located when creating a projection.

point phenomena: geographic phenomena that have no spatial extent and can thus be termed "zero-dimensional," such as the location of oil well heads.

polar aspect: a map projection with a center that coincides with one of the poles.

polyconic projection: a class of projections that conceptually involves the placement of many cones over the reference globe, each with its own standard line producing curved meridians (except for a straight central meridian) and parallels represented as circular arcs.

polygonal glyph: a symbol formed by connecting the rays of a multivariate ray-glyph.

population: the total set of elements or things that one could study, such as all stone houses within a city.

Portable Document Format (PDF): a file format that allows digital maps and other documents to be transported between software applications and between different types of computers, and is capable of embedding features such as hyperlinks, movies, and keywords.

positional accuracy: one means of assessing data quality (or uncertainty); refers to the correctness of location of geographic features.

PostScript: the de jure standard page description language, produced by Adobe Systems, Inc.

power function exponent: the exponent used for the stimulus in the power function equation relating perceived size and actual size; this exponent is used to summarize the results of experiments involving perceived size of proportional symbols.

prepress: the phase of high-volume map reproduction consisting of various technologies and procedures that make offset lithographic printing possible.

presence: the subjective experience of being in one place or environment when one is physically in another.

press check: an event at which press proofs are made on an offset lithographic press shortly before a press run begins.

press proof: a printed map that is produced at a press check.

Prime Meridian: the internationally agreed-upon meridian with a value of $0°$ that runs through the Royal Observatory located in Greenwich, England.

principal components analysis: a method for succinctly summarizing the interrelationships occurring within a set of attributes.

Printer Control Language (PCL): a standard page description language produced by Hewlett Packard, Inc.

printer driver: software that converts digital map data from the native format of the application software into page description data.

printing plate: a sheet of aluminum (or polyester) containing a latent image, which is mounted on a roller in an offset lithographic printing press.

printing service bureau: a business that specializes in the creation of products such as printing plates and proofs and provides print-related services.

printing unit: an organization of cylinders in an offset lithographic printing press that transfers one base color of a map to the print medium.

prism map: a map in which enumeration units are raised to a height proportional to the data associated with those units.

proactive graphics: software that enables users to initiate queries using icons (or symbolic representations) as opposed to words.

probing: to hover over (or point to) a location and discover the value of an attribute (or attributes) associated with that location.

process colors: the subtractive primary colors (CMY) plus black (K); these colors are mixed on the page via a printing device.

Projected Augmented Relief Model (PARM): a system that permits the projection of images onto a physical model, thus permitting the themes for the model to vary and to show change over time.

projection selection guideline: a set of expert-developed rules and suggestions that are intended to help novices select an appropriate map projection.

prolate spheroid: refers to the historical notion that Earth bulges at the poles and is compressed at the Equator; today, it is known that an oblate spheroid (bulging at the Equator) is the actual shape.

proof: a preliminary representation of what a final reproduced map will look like.

proportional symbol map: a map in which point symbols are scaled in proportion to the magnitude of data occurring at point locations, such as circles of varying sizes representing city populations.

pseudocylindrical projection: a projection that represents lines of latitude as straight, parallel lines (similar to cylindrical projections), but the meridians are curved lines converging to the poles, which are represented as either lines or points.

pupil: the dark area in the center of the eye.

pycnophylactic interpolation: a method of interpolating data for isopleth maps; the data associated with enumeration units are conceived as prisms, and the volume within those prisms is retained in the interpolation process.

quantiles: a method of data classification in which an equal number of observations is placed in each class.

quasi-continuous tone method: a method for unclassed mapping in which only a portion of the data is shown using a smooth gradation of tones; very low and very high data values are represented by the lightest and darkest tones, respectively.

R: the stated radius of the reference globe.

radial flow map: a flow map that has a distinctive radial pattern, such as a map centered on Kansas showing migration from Kansas to other states as a result of the Dust Bowl in the 1930s.

rainbow color map: see rainbow color scheme.

rainbow color scheme: see spectral color scheme.

range: the maximum minus the minimum of the data.

range-graded map: a term used to describe a classed proportional symbol map.

range-graded scaling: a method of sizing proportional symbols in which data are grouped into classes, and each class is represented with a different-sized symbol.

raster: an image composed of pixels created by scanning from left to right and from top to bottom.

raster data model: a general approach for representing geographic features in digital form that uses rows and columns of square grid cells or pixels.

raster image processing: the conversion of a digital map into a raster image that can be processed directly by a raster-based printing device.

raster image processor (RIP): software, or a combination of software and hardware, that interprets page description data and converts them into a raster image that can be processed directly by a printing device.

rate of change: one of the visual variables for animated maps; defined as m/d, where m is the magnitude of change between frames or scenes, and d is the duration of each frame or scene.

ratio: a level of measurement in which numerical values are assigned to data, and there is a nonarbitrary zero point, such as with the percentage of forested land.

ratio estimation: a method for determining the perceived size of proportional symbols; two symbols are compared, and the viewer indicates how much larger or smaller one is than the other (e.g., one symbol appears five times larger than the other symbol).

raw table: a form of tabular display in which the actual data are listed from lowest to the highest value.

raw totals: raw numbers that have not been standardized to account for the area over which the data were collected (e.g., we might map the raw number of acres of tobacco harvested in each county, disregarding the size of the counties).

realistic sounds: sounds that have meaning based on our past experience, such as the sound of the wind; contrast with abstract sounds.

receptive field: a single ganglion cell corresponding to a group of rods or cones; receptive fields are circular and overlap one another.

rectangular cartogram: a cartogram created by sizing rectangles representing enumeration units as a function of an attribute being mapped.

rectangular point symbol: the width and height of a rectangle are made proportional to each of two attributes being mapped.

rectilinear cartogram: a generalization of the rectangular cartogram in which enumeration units are represented by rectilinear polygons rather than only rectangles.

reduced major-axis approach: the process of fitting a line to two attributes when we do not wish to specify a dependent attribute.

redundant symbols: used to portray a single attribute with two or more visual variables, such as representing population for cities by both size of circle and gray tone within the circle.

re-expression: an animation created by modifying the original data in some manner, such as by choosing subsets of a time series or by reordering a time series.

reference ellipsoid: a smooth, mathematically defined figure with a semimajor axis that is longer than the semiminor axis and is designed to approximate Earth's true shape to a better degree than would be a spherical model.

reference globe: a conceptual spherical model of Earth reduced to the same scale as the final map where the scale of the reference globe is defined by its radius, R.

refinement: a generalization operation that involves reducing a multiple set of features such as roads, buildings, and other types of urban structures to a simplified representation or "typification of the objects.

registration: the alignment of base colors in multicolor printing.

relief inversion effect: see terrain inversion effect.

remote sensing: the acquisition of information about Earth (or other planetary body) from high above its surface via either reflected or emitted radiation; this is most commonly done via satellites.

representative fraction (RF): expresses the relationship between map and Earth distances and is expressed as a ratio of map units/earth units (e.g., an RF of 1:25,000 indicates that 1 unit on the map is equivalent to 25,000 units on the surface of Earth).

resemblance coefficients: in cluster analysis, the values that express the similarity of each pair of observations (e.g., the Euclidean distance between two observations).

resolution: (1) for graphic displays, the number of addressable pixels; for printers, the number of dots per inch; (2) a visual variable for depicting data uncertainty; for example, coarser raster cells indicate greater uncertainty in the data.

retina: the portion of the eye on which images appear after they pass through the lens.

RGB: a system for specifying color in which red (R), green (G), and blue (B) components are used, as when a pixel on an LCD monitor is lit with different intensities of red, green, and blue.

RGB/CMY color wheel: a color wheel in which red, green, and blue are separated from each other by cyan, magenta, and yellow.

rhumb line: see loxodrome.

ring map: one attribute is shown with a traditional mapping approach (e.g., the choropleth map) and surrounding concentric rings depict other attributes.

rods: a type of nerve cell within the retina that functions in dim light and plays no role in color vision.

room-format display: provides a room-sized view of a virtual environment by projecting images onto walls and the floor (e.g., the CAVE); a head tracker on one user governs the view that all users see through stereo glasses.

root mean square error (RMSE): a measure of error for cross validation in which we compute the square root of the mean of the sum of squared residuals (the observed minus the predicted values).

rosette: a pattern of halftone cells (consisting of the process colors) that results when moiré patterns are avoided through the appropriate specification of halftone screen angles.

rough draft: a preliminary printout of a map that is used to evaluate an evolving design.

RYB color wheel: a color wheel containing red, yellow, and blue—colors that have traditionally been considered primary colors for use in painting.

sample: the portion of the population that is actually examined, such as sampling only 50 stone houses out of the 5,000 existing within a city.

satellite geodesy: the use of Earth-orbiting satellites to derive very accurate information regarding Earth's size and shape.

saturation: along with hue and lightness, one of three components of color; can be thought of as a mixture of gray and a pure hue: very saturated colors have little gray, whereas desaturated colors have a lot of gray.

scale: provides an indication of the amount of reduction that has taken place on a map.

scale factor (SF): the ratio of the scale found at a particular location on a map compared to the stated scale of the map; departure from a scale factor of 1.0 indicates distortion.

scatterplot: a diagram in which dots are used to plot the scores of two attributes on a set of x and y axes.

scatterplot brushing: subsets of data within a scatterplot matrix are focused on by moving a rectangular "brush" around the matrix; dots falling within the brush are highlighted in all scatterplots.

scatterplot matrix: a matrix of scatterplots is used to examine the relationships among multiple attributes.

scientific visualization: the use of visual methods to explore a broad range of problems that scientists deal with; examples include medical imaging and the visualization of molecular structure and fluid flows.

screen angle: the angle at which lines of halftone cells are oriented.

screen frequency: the spacing of halftone cells within a given area, expressed in lines per inch (lpi).

screening: the lightening of graphics to reduce visual weight by reducing the amount of ink or toner applied to the print medium.

secant case: describes a developable surface that intersects the reference globe along two separate lines, usually two parallels of latitude.

seconds: a unit of measurement in the sexagesimal degree system; each minute contains 60 seconds.

self-organizing map (SOM): a type of *artificial neural network* (ANN) that can be used for either clustering or dimensionality reduction.

semivariance: a measure of the variability of spatial data in various geographic directions (e.g., we might want to determine whether temperature values are more similar to one another along a north–south axis than along an east–west one).

semivariogram: a graphical expression of semivariance; the distance between points is shown on the x axis, and the semivariance is shown on the y axis.

sentence case: type composed of lowercase letters, with the first letter of each sentence set in uppercase.

sequence of preferred locations: a method for reducing the effect of overprinting by placing type associated with point symbols in specific, ranked locations.

sequencing: displaying a map piece by piece (e.g., displaying the title, geographic base, legend title, and then each class in the legend).

sequential color scheme: a sequence of colors that is characterized by a gradual change in lightness, as in varying lightnesses of a blue hue.

serifs: short extensions at the ends of major letter strokes.

sexagesimal system: the system for specifying locations on Earth's surface in degrees, minutes, and seconds (a base 60 system).

shade: a darker version of a base color achieved by adding small amounts of black.

shaded relief: a method for depicting topography; areas facing away from a light source are shaded, whereas areas directly illuminated are not; slope might also be a factor, with steeper slopes shaded darker.

shadowed contours: a method for depicting topography in which only the width of the contour lines varies, with the thinnest lines facing the source of illumination and the thickest lines on the side opposite the source of illumination (where shadows are more likely).

shape: a visual variable in which the form of symbols (or marks making up symbols) is varied, such as using squares and triangles to represent religious and public schools.

shape accuracy: refers to how well the shapes of enumeration units are maintained on a cartogram relative to an untransformed map.

short-term visual store: an area of memory that requires constant attention (or activation) to be retained; contrast with iconic memory and long-term visual memory.

simplification: involves the elimination or "weeding" of unnecessary coordinate data when generalizing a linear or areal feature.

simultaneous contrast: when the perceived color of an area is affected by the color of the surrounding area.

size: (1) a visual variable in which the magnitudes of symbols (or marks making up symbols) are varied (e.g., using circles of different magnitudes to depict city populations); (2) how large a typeface is; normally expressed in points, where a point is 1/72 of an inch.

sketch map: a rough, generalized hand drawing that represents a developing idea of what a final map will look like.

skewed distribution: when data are placed along the number line, the distribution appears asymmetrical; most common is a positive skew, in which the bulk of data are concentrated toward the left, and there may be outliers to the right.

slope: the steepness of a topographic surface.

small circle: represents the line along which a plane intersects Earth's surface when the plane does not pass through Earth's center.

small multiple: many small maps are displayed to show the change in an attribute over time or to compare many attributes for the same time period.

small-scale map: a map depicting a relatively large portion of Earth's surface.

smooth phenomena: phenomena that change gradually over geographic space, such as the yearly distribution of snowfall in Wisconsin.

smoothing: shifts the position of points making up a line to improve the appearance of the line.

snowflake: see polygonal glyph.

soft proof: a representation of a digital map on a computer screen.

sonification: using abstract sounds to represent spatial data.

sound dimension: see aural variable.

soundscape: the perceived acoustic environment at a landscape scale, which could include sounds from human activity, biological organisms, and geophysical processes.

South Pole: the geographic location in the Southern Hemisphere with a latitude of 90° South that serves as the point around which Earth's axis rotates.

Southern Hemisphere: the portion of Earth extending from the Equator to the South Pole.

space-time cube: rather than animating temporal changes on a 2-D map, the third dimension (the *z*-axis) is used to show changes over time.

spacing: a visual variable in which the distance between marks making up a symbol is varied, such as shading counties on a choropleth map with horizontal lines of varied spacing.

sparklines: a very small line chart that is typically drawn without axes or coordinates; as such, it can be imbedded within either text or cells of a spreadsheet.

spatial autocorrelation: the tendency for like things to occur near one another in geographic space (technically, this is termed positive spatial autocorrelation).

spatial dimension: a term that describes whether a phenomenon can be conceived of as points, lines, areas, or volumes.

spatial treemap: a treemap is a visualization method for displaying hierarchical data using nested rectangles; the spatial treemap is a geographic extension of this in which the mapped region is depicted as a contiguous rectangular cartogram, and each of the rectangles is then subdivided to represent the attribute(s) of interest.

spatialization: the process of converting abstract information to a spatial framework in which visualization is possible (e.g., converting the nature of topics that are discussed on the front page of a newspaper over a monthlong period to a spatial framework).

specific information: the information associated with particular locations on a thematic map.

spectral color scheme: a sequence of colors for a choropleth or isarithmic map in which colors span the visual portion of the electromagnetic spectrum (sometimes referred to as a ROYGBIV scheme).

spot colors: colors that are mixed before they reach the printing device.

standard deviation: the square root of the average of the squared deviations of each data value about the mean.

standard distance (SD): a measure of dispersion for point data; essentially, a measure of spread about the mean center of the distribution.

standard line: a line on a map projection with a scale factor (SF) of 1.0; thus, distortion is zero.

standard point: a point on a map projection with a scale factor (SF) of 1.0; thus, distortion is zero.

standardized data: when raw totals are adjusted to account for the area over which the data are collected; for example, we might divide the acres of tobacco harvested in each county by the areas of those counties.

star: see multivariate ray-glyph.

static maps: maps whose information content is unchanging, such as a traditional printed map; many online maps are also static, as their information content does not change while the user looks at the map.

statistical accuracy: refers to whether the relations among attribute values are properly represented by the relations among the sizes of enumeration units representing those attribute values in a cartogram.

statistical map: see thematic map.

stem-and-leaf plot: a method for exploratory data analysis in which individual data values are broken into "stem" and "leaf" portions; the result resembles a histogram but provides greater detail for individual data values.

stereo pairs: a term describing when two maps of an area are viewed with a stereoscope, which enables the reader to see a three-dimensional stereoscopic view.

stereoscopic depth cue: distance of an object is a function of the difference in the images formed on the retinas of the eyes; ultimately, differences in the horizontal position of objects on the retinas are processed in the brain to produce depth perception.

stereoscopic view: separate images are presented to the left and right eye, providing the viewer with a truly 3-D view of the landscape or 3-D map (this takes advantage of the stereoscopic depth cue).

Stevens curve: a grayscale that is sometimes used for smooth (untextured) gray tones; some cartographers have recommended that it be used for unclassed maps.

stochastic screening: the application of ink or toner in a pattern of very small, pseudorandomly spaced dots of uniform size to create tints from a base color.

subtitle: the map element that is used to further explain the title.

subtractive primary colors: the colors (cyan, magenta, and yellow) associated with printing inks or pigments that can be mixed to create a wide range of colors.

sun elevation: the angle between the horizon and the height of the sun in the sky.

synchronization: one of the visual variables for animated maps; refers to the temporal correspondence of two or more time series.

tabular error: if classed data are conceived as a prism map, tabular error is the difference in height between prisms on this map and those on an unclassed prism map of the data.

Tanaka's method: a method for illuminated contours in which the width of a contour line at a point is defined as the cosine of the angle formed between the aspect of that point and the azimuthal angle of illumination.

tangent case: describes a developable surface that intersects the reference globe along one line, usually a parallel of latitude or point.

telecommunications flow map: depicts flows of telecommunications technology, including the Internet and its associated information spaces.

temporal animations: map animations that show changes over time (e.g., a map animation that displays the path of a forest fire over the course of a week).

terrain inversion effect: a visual effect associated with topography wherein concave structures (such as valleys) appear as convex structures (such as ridges), and vice versa.

thematic map: a map used to emphasize the spatial distribution (or pattern) of one or more geographic attributes (e.g., a map of predicted snowfall amounts for the coming winter in the United States).

thematic mapping software: software that is expressly intended for the creation of thematic maps; MapViewer is a traditional example.

thematic symbols: graphical symbols that directly represent a map's theme or topic.

Thiessen polygons: polygons enclosing a set of control points such that all arbitrary points within a polygon are closer to the control point associated with that polygon than to any other control point.

thin-plate splines: a method of interpolation for isarithmic maps in which a mathematical surface is fit to the data such that the roughness of the surface is minimized.

three-class method: a method of dasymetric mapping in which the phenomenon of interest is assigned to three possible classes (e.g., when mapping population with three land cover categories, we might assign 15 percent of the population to forest, 30 percent to grassland, and 65 percent to urban).

three-dimensional bars: the height of 3-D-looking bars are made proportional to each of the attributes being mapped.

three-dimensional scatterplot: a diagram that uses small point symbols in which the scores for observations on three attributes are plotted on a set of x, y, and z axes; viewing all points in the plot requires interactive graphics.

3-D printers: a machine that enables the creation of a physical object from a three-dimensional digital model (e.g., such a machine could create a 3-D model of a landscape that you could hold and manipulate in your hands).

tint: a lighter version of a base color.

tint percentage: the degree to which the appearance of an ink or toner is lightened when employing screening techniques.

Tissot's indicatrix: a graphical device plotted on a map projection that illustrates the kind and extent of distortion at infinitesimally small points.

title: the map element that provides a succinct description of a map's theme.

title case: type composed of lowercase letters with the first letter of each word set in uppercase.

tone: a color that has been "muted" through the addition of equal parts black and white, or gray; a desaturated base color.

topological accuracy: refers to whether the spatial adjacencies of enumeration units normally found on an untransformed map are maintained on a cartogram.

traffic light color scheme: a color scheme based on the colors of a traffic light (e.g., red, yellow (or orange), and green).

trajectory-based flow maps: used to depict flows when the actual route of flow is deemed important; an example would be a map showing the routes of members of a wolf pack over the course of several months.

transparency: a visual variable for depicting data uncertainty; refers to the ease with which a theme can be seen through a "fog" placed over that theme.

trapping: a series of techniques used to minimize the effects of misregistration in multicolor printing.

triangulation: (1) a method of surveying where two points having known locations are established and the distance between them is measured; angles from this baseline are then measured to a distant point, creating a triangle, which can then serve as a basis for other triangles; (2) a method of interpolating data for isarithmic maps in which original control points are connected by a set of triangles.

trichromatic theory: the theory that color perception is a function of the relative stimulation of blue, green, and red cones.

trivariate choropleth map: the overlay of three univariate choropleth maps; this approach generally should be used only when attributes add up to 100 percent, such as percent voting Republican, Democratic, and Independent.

true point data: data that can actually be measured at a point location, such as the number of calls made at a telephone booth over the course of a day.

true 3-D phenomena: a form of volumetric phenomena in which each longitude and latitude position has multiple attribute values depending on the height above or below a zero point (e.g., the level of ozone in Earth's atmosphere varies as a function of elevation above sea level).

true 3-D view: see stereoscopic view.

2½-D phenomena: a form of volumetric phenomena in which each point on the surface is defined by longitude, latitude, and a value above (or below) a zero point (e.g., Earth's topography).

type: words that appear on a map.

type family: a group of type designs that reflect common characteristics and a common base name.

type size: the size of type, expressed in points.

type style: variations of type design within a given type family.

typeface: type of a particular type family and style.

typography: the art or process of specifying, arranging, and designing type.

ubiquitous cartography: the ability of users to create and use maps in any place and at any time to resolve geospatial problems.

uncertainty: the notion that data for thematic maps are often subject to some form of error; uncertain data are also considered to be unreliable and of poor quality.

unclassed map: a map in which data are not grouped into classes of similar value, and thus each data value can theoretically be represented by a different symbol (e.g., a data set with 100 different data values might be depicted using 100 different lightnesses of gray on a choropleth map).

unconstrained extended local processing routines: simplification algorithms that search large sections of a line, with the search terminated by the geomorphological complexity of the line, not by an algorithmic criterion.

uniform color model: see perceptually uniform color model.

uniform sky illumination model: a method for creating shaded relief in which the light source is presumed to be evenly distributed throughout the sky.

unipolar data: data that have no obvious dividing point and do not involve two complementary phenomena, such as per capita income in African countries; contrast with bipolar and balanced data.

unit value: the count represented by each dot on a dot map.

unit-vector density map: a map in which the density and orientation of short fixed-length segments (unit vectors) are used to display the magnitude and direction, respectively, of a continuous vector-based flow, such as wind speed and direction.

univariate cartogram: used to describe a cartogram in which only a single attribute is displayed; enumeration units are sized as a function of that attribute (e.g., population).

universal kriging: a form of kriging that accounts for a trend (or drift) in the data over geographic space.

unordered hues: colors that are based upon distinctly different hues and which are not in the order found in the color spectrum.

unweighted pair–group method using arithmetic averages (UPGMA): a method for computing resemblance coefficients when clusters in a cluster analysis are combined; involves averaging the coefficients between observations in the newly formed cluster with coefficients for observations in existing clusters.

uppercase: type composed of all uppercase letters.

value: see lightness.

value-by-alpha map: an alternative to the cartogram that utilizes the visual variable transparency to portray an *equalizing attribute* (e.g., population) that is normally used to size enumeration units in a conventional cartogram.

value-by-area map: see cartogram.

value flow diagram: a graph that displays the change in an attribute over a time interval; multiple value flow diagrams permit the analysis of spatiotemporal data on a single static map.

variable: see attribute.

variable scale projection: a map projection with a scale that is manipulated in specific ways to emphasize certain land areas at different scales.

vector data model: a general approach for representing geographic features in digital form that uses discrete points, lines, and polygons.

verbal scale: the map element that is a textual expression of scale, such as "one inch to the mile."

vertical datum: a reference surface (often mean sea level) for elevations on Earth's surface.

virtual assistant: within a virtual environment, a computer-produced assistant that can assist in interpreting the virtual environment—ideally, this assistant should have human-like qualities (also called an intelligent agent).

virtual environment (VE): a three-dimensional computer-based simulation of a real or imagined environment that users are able to navigate through and interact with.

virtual globe: a digital display of the entire Earth or a substantial portion of it (e.g., ArcGIS's ArcGlobe, Google Earth, and NASA World Wind).

virtual world: see virtual globe.

virtual reality (VR): a popular term used to describe three-dimensional computer-based simulations of a real or imagined environment that users are able to navigate through and interact with.

visible light: the portion of the electromagnetic spectrum to which the human eye is sensitive.

visual analytics (VA): a term developed by the U.S. National Visualization and Analytics Center (NVAC); defined as "the science of analytical reasoning facilitated by interactive visual interfaces" that should "detect the expected and discover the unexpected."

visual analytics mantra: an alternative approach to the classic "information-seeking mantra" used in geovisual analytics; the visual analytics mantra consists of the following five stages: Analyze first, Show the Important, Zoom, filter and analyze further, and Details on demand.

visual benchmarks: reference points with which other frames of an animation can be compared (e.g., for a proportional-circle map animation of AIDS by country, yellow and red circles could represent the minimum and the maximum number of new AIDS cases for each country over the time period of the animation).

visual cortex: the first place in the brain where information from both eyes is handled; technically, the visual cortex is divided into various subareas, and the first of these subareas is the primary visual cortex.

visual hierarchy: the graphical representation of an intellectual hierarchy.

visual pigments: the light-sensitive chemicals found within rods and cones.

visual variables: the perceived differences in map symbols that are used to represent spatial phenomena.

visual weight: the relative amount of attention that a map feature attracts.

volunteered geographic information: geotagged information provided by volunteers that is collected from the Web (e.g., OpenStreetMap).

wall-sized display: a graphic display covering a large portion of a wall; this is accomplished by combining either multiple projected images or multiple display panels.

warm colors: colors associated with warm things, such as red and yellow.

wavelength of light: the distance between two wave crests associated with electromagnetic energy.

Web 2.0: a variant of the Web that was common in the early decades of the twenty-first century; key components included the ability to harness the collective intelligence of Web users; lightweight programming models that enable the creation of map mashups; and the ability to work with social media sites such as Facebook and Twitter.

weighted isolines: the width of contour lines is made proportional to the associated data; thus, labeling of contour lines is not required.

Williams curve: a gray scale that is recommended for coarse (textured) areal shades.

word spacing: the space between words.

World Wide Web (or Web): a system of Internet file servers that support the hypertext transfer protocol.

Zip-a-Tone: in traditional manual cartography, areal symbols could be cut from preprinted sheets of Zip-a-Tone and stuck to a base map.

zoom function: in an interactive graphics environment, an area to be focused on is enlarged; one approach is to enclose a desired area with a rectangular box and click the mouse.

Index

Note: Page numbers in **bold** denote tables, page numbers in *italics* denote figures. Page numbers of the form XnY denote footnote Y on page X.